D1265488

PHYSICS

Foundations and Frontiers

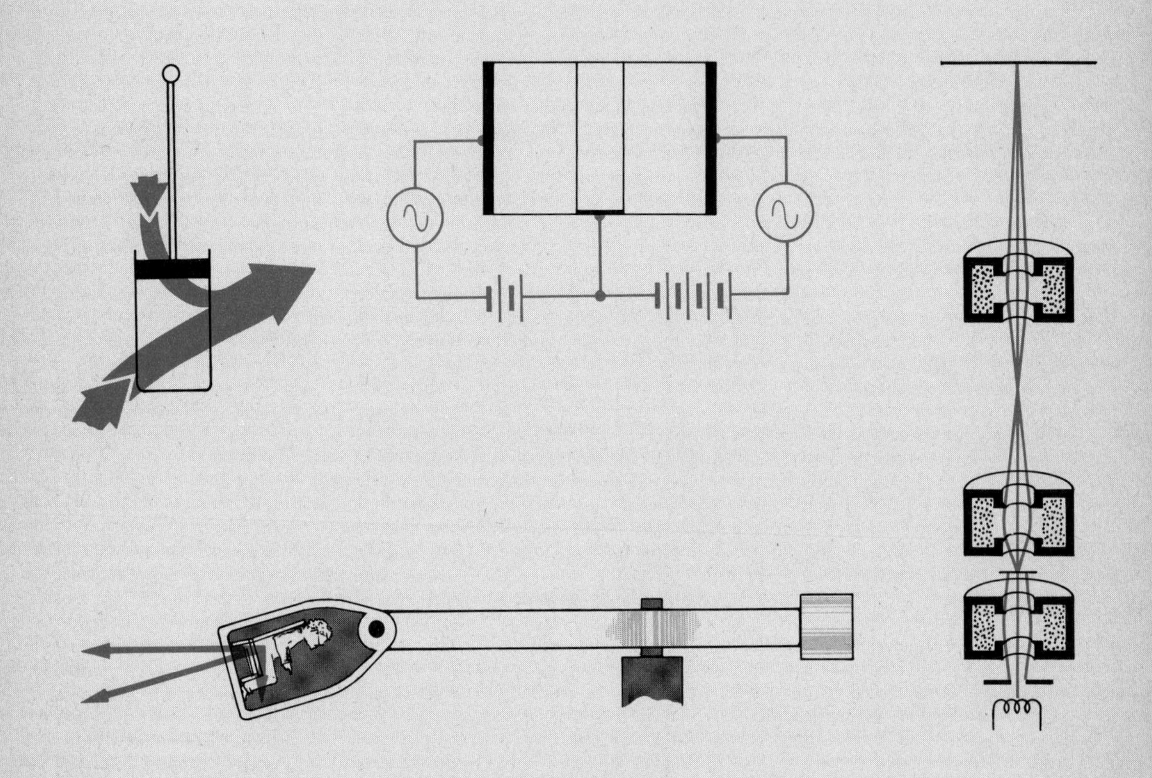

PHYSICS

Foundations and Frontiers

THIRD EDITION

GEORGE GAMOW

JOHN M. CLEVELAND
Department of Physics
University of Colorado

PRENTICE-HALL, INC., Englewood Cliffs, New Jersey

Library of Congress Cataloging in Publication Data

Gamow, George

Physics : foundations & frontiers.

Includes index

1. Physics. I. Cleveland, John M., joint author.
II. Title.
QC23.G195 1976 530 75-35752
ISBN 0-13-672535-X

10 9 8 7 6 5 4 3 2 1

Printed in the United States of America

Prentice-Hall International, Inc., *London*
Prentice-Hall of Australia Pty. Limited, *Sydney*
Prentice-Hall of Canada, Ltd., *Toronto*
Prentice-Hall of India Private Limited, *New Delhi*
Prentice-Hall of Japan, Inc., *Tokyo*
Prentice-Hall of South-East Asia Private Limited, *Singapore*

Contents

3 Bodies in Motion 44

4 Energy and Momentum 66

5 Rotational Mechanics 88

6 Gravitation and Orbits 104

18 The Electrical Nature of Matter 325

19 Light Rays 346

20 Light Waves 371

24 The Structure of Atoms 443

25 The Wave Nature of Particles 461

26 Radioactivity and the Nucleus 475

27 Artificial Nuclear Transformations 493

28 The Structure of the Nucleus 507

29 Large-Scale Nuclear Reactions 517

30 Particles, Particles! 535

31 Biophysics 548

Preface

The third edition of this text is designed to serve the same purpose as the first two editions: a one-year course in introductory physics, requiring only a prior exposure to high school algebra.

A substantial and growing fraction of the students taking such a course are biologically oriented. To better serve the needs and interests of this group, many examples throughout the text have been given a biological slant. Along with the other final chapters on Geophysics and Astrophysics, the chapter on Biophysics has also been retained and extensively revised. Although these revisions may make the flavor more palatable to most students, the basic diet of physics has remained unchanged.

In my own teaching from the second edition, I found it advisable to rearrange the sequence of some of the chapters. This rearrangement has been incorporated in the third edition, and I am certain that those who use the book will find it to be a helpful improvement.

In the extensive rewriting, and in the many decisions on deletions, additions and rearrangements, the contributions of the late George Gamow's creative imagination have been sadly missed. I hope he would approve of what has been done.

JOHN M. CLEVELAND

1

Our Place in the Universe

1-1 The Large and the Small

In our everyday life we encounter objects of widely differing sizes. Some of them are as large as a mountain and others as small as a grain of dust. When we go much beyond these limits, either in the direction of much larger objects or in the direction of much smaller ones, it becomes increasingly difficult to grasp their actual sizes.

Objects that are much larger than mountains, such as our earth itself, the moon, the sun, the stars, and stellar systems, constitute what is known as the *macrocosm* (Greek for "large world"). Very small objects such as bacteria, atoms, and electrons belong to the *microcosm* (Greek for "small world"). If we use one of the standard scientific units, the *meter* (abbreviated "m": 1 m = 39.37 inches) or the *centimeter* (abbreviated "cm": 1 cm = 0.01 m; 2.54 cm = 1 inch) for measuring sizes, objects belonging to the macrocosm will be described by very large numbers, and those describing the microcosm by very small ones. Thus the diameter of the sun is 1,390,000,000 m, and the diameter of a hydrogen atom is 0.000000000106 m.

Scientists, however, would ordinarily express such very large and very small numbers in a different and more useful way. The diameter of the sun in meters is given by the number 139 followed by (and here we must stop and count them) seven zeros. Each zero means a multiplication by 10, so we can write this number as 139×10^7. It is customary, but by no means necessary, to express such numbers with the numerical part (the *coefficient*) made to be between 1 and 10. This can be done by dividing the 139 by 100, to give 1.39. However, if we do this, we must avoid changing the whole quantity, by multiplying the exponential part by 100, to give 1.39 followed by *nine* zeros, or 1.39×10^9 m.

We can similarly express very small numbers in exponential notation if we recall that 10^{-3}, for example, means $1/01^3$, or 0.001. For the diameter of the hydrogen atom, given as 0.000000000106 m, we can move the decimal point 10 places to the right—the same thing as multiplying by 10^{10}—to get 1.06. To balance this multiplication we must then

FIG. 1-1 Space and time scales of the universe.

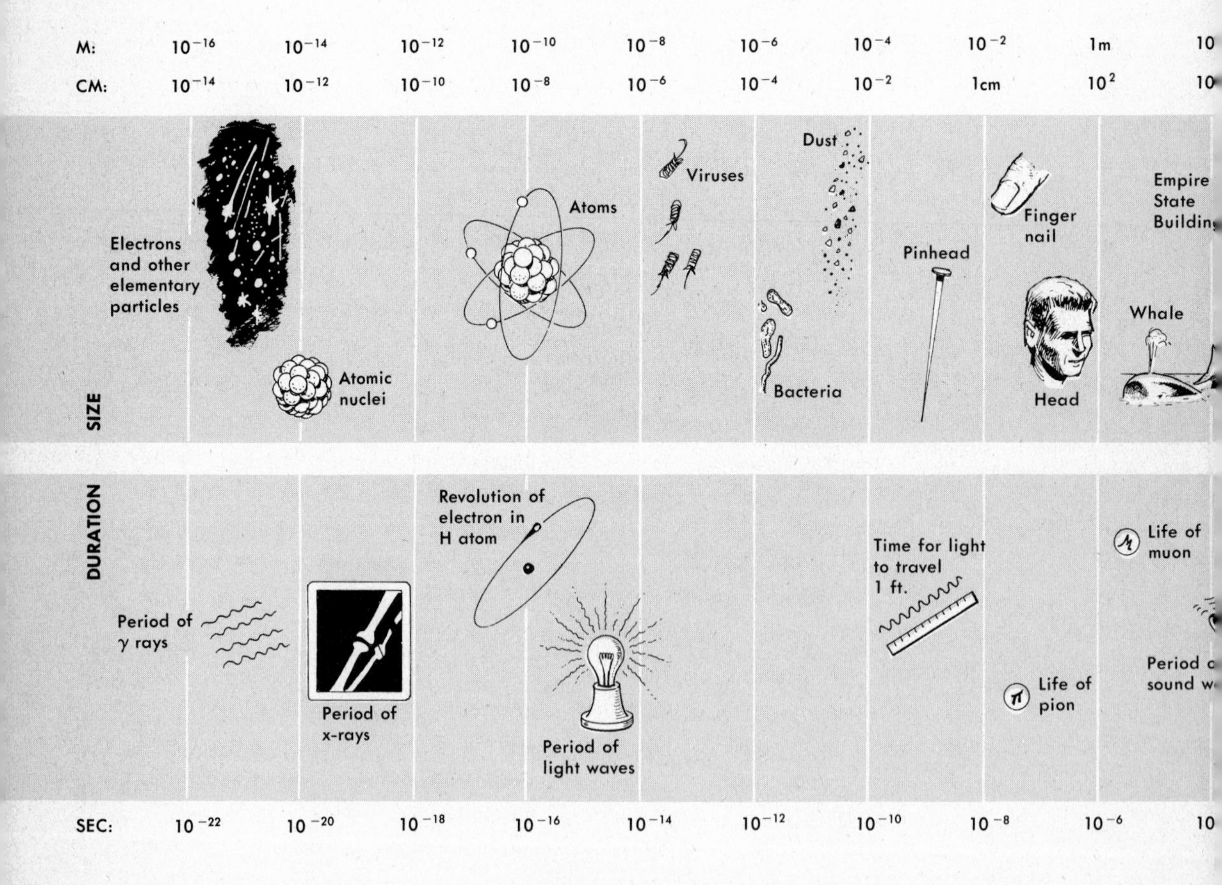

divide the number by 10^{10}, which is the same as multiplying by 10^{-10}. Combining these gives the diameter of the hydrogen atom as 1.06×10^{-10} m.

The top strip of Fig. 1-1 gives a rough idea of the range of sizes encountered in the universe around us. In most graphs we are accustomed to have each scale division mean the *addition* of some number. In the representation of Fig. 1-1, however, each equal distance means an equal *multiplication*. The distance from each vertical white line to the next means multiplication by 100, or 10^2. On a graph of this sort we can represent the range of sizes from the diameter of the so-called "elementary particles" (about 10^{-13} cm, or 10^{-15} m), to the diameter of clusters of giant stellar galaxies (about 10^{26} cm, or 10^{24} m). On this sort of a scale the human head is about midway in size between an atom and the sun.

The bottom strip of Fig. 1-1 shows an equally wide variation in time

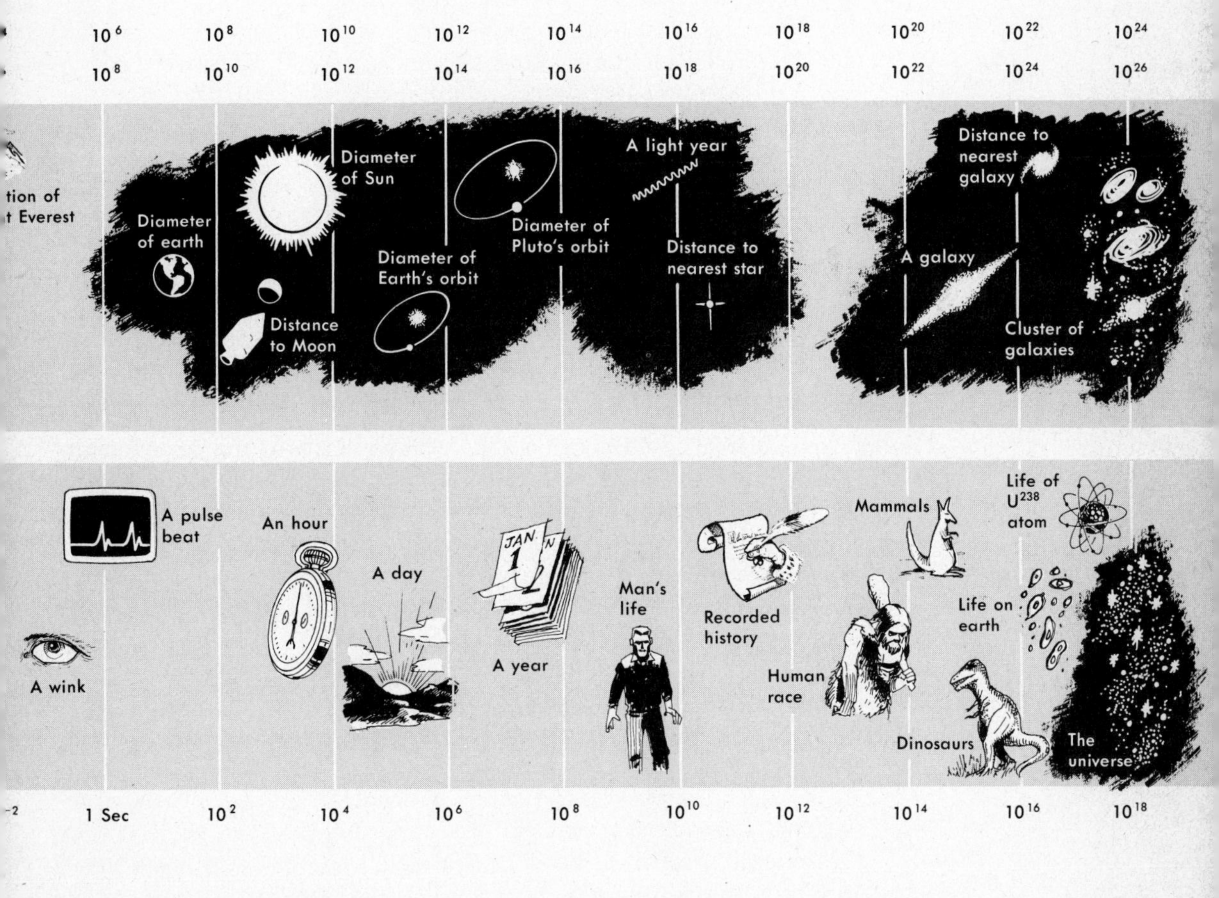

intervals. The range, on the same scale as the distance graph, goes from the period of gamma rays (i.e., the time between one gamma-ray wave and the next) to the age of the universe according to some cosmological theories. By comparing distances on this graph, we see that there are more heartbeats in a man's lifetime than there are lifetimes in the age of the universe.

This sort of graphical notation is very effective in giving us a visual representation of various sizes and times, but we also need to be able to deal with quantities of all sorts in a more analytical way.

1-2 The Arithmetic of Exponents

To see how this "exponential notation" works, let us quickly review some of the rules for working with exponents. Suppose we want to multiply 10^2 by 10^5. Since $10^2 = 10 \times 10 = 100$, and $10^5 = 10 \times 10 \times 10 \times 10 \times 10 = 100{,}000$, these two numbers multiplied together give us 10,000,000, which is 10^7. It is easier, however, just to write $10^2 \times 10^5 = 10^7$; thus to multiply exponential numbers we simply add exponents. From this explanation you might guess, and correctly, that to divide, you subtract exponents. You can easily check this by dividing 10^5 by 10^2, which gives 1000, or 10^3. But suppose we had wanted to divide 10^2 by 10^5? Following the rule for division, we would subtract 5 from 2 and get the answer 10^{-3}, which represents the number 1/1000, or $1/10^3$.

As an example, let us work out how many times larger than a hydrogen atom the sun is. We could, of course, simply divide 1,390,000,000 by 0.000000000106, but in doing this it would be very hard to keep the decimal point straight. If, instead, we use 1.39×10^9 and 1.06×10^{-10}, the calculation becomes quite simple, as the numerical multipliers and the powers of 10 can be handled separately:

$$\frac{1.39 \times 10^9 \cancel{\text{m}}}{1.06 \times 10^{-10} \cancel{\text{m}}} = 1.31 \times 10^{9-(-10)} = 1.31 \times 10^{19}.$$

Notice that in the fraction above, both numerator and denominator are given in the same units—meters. These cancel out to give the answer 1.31×10^{19} without any units, which is just as it should be. The sun's diameter *is* 1.31×10^{19} times as large as the diameter of a hydrogen atom no matter what units are used to measure them.

With numbers expressed in this exponential notation, raising to various powers is fairly obvious. For example, $(1.20 \times 10^3)^2$ is the same as $(1.20)^2 \times (10^3)^2 = 1.44 \times 10^6$. To square 10^3 we must multiply the exponent by 2. Similarly, $(10^5)^3 = 10^{15}$.

In taking roots of numbers, we must be a little more careful. Since raising to a power means *multiplying* the exponent by the power, it is reasonable to expect that taking the cube root, for example, would mean we must *divide* the exponent by 3. The cube root of $10^{12} = 10^4$. But how do we deal with the cube root of 10^8? Following the rule, it must equal

$10^{8/3} = 10^{2.67}$. This is true, but without using logarithms it is hard to know that $10^{2.67} = 464$. In this case, it is easy to *make* the exponent evenly divisible by 3: $10^8 = 100 \times 10^6$. We can now take the cube root in two parts:

$$\sqrt[3]{10^8} = \sqrt[3]{100} \times \sqrt[3]{10^6}.$$

From a slide rule or a table, we can find that the cube root of 100 is 4.64, and the cube root of $10^6 = 10^{6/3} = 10^2$. For the answer we have 4.64×10^2, or 464.

1-3 Units Used in the Physical Sciences

Until the beginning of the nineteenth century, the situation in the field of weights and measures was not much better than the linguistic situation at the Tower of Babel. The units of length varied from country to country, from town to town, from one profession (such as tailoring) to another (such as carpentry) and were mostly defined, rather loosely, by reference to various parts of the human body. Thus an "inch" was defined as a thumb-width, a "hand" or "palm" (still used for measuring the height of horses) as the breadth of a hand, a "foot" as the length of a British king's foot, a "cubit" as the distance from the elbow to the tip of the middle finger, a "fathom" (used in measuring ocean depths) as the distance between the tips of the middle fingers of the two hands when the arms are outstretched in a straight line, etc. In the year 1791, the French Academy of Sciences recommended the adoption of an international standard of length and suggested that the unit of length be based on the size of the earth. This unit, called the *meter*, was intended to be one ten-millionth of the distance from the pole to the equator. The Academy prepared a "standard meter"—the distance between two fine lines engraved on a platinum–iridium bar. Later measurements of the earth, more accurate than those available to the French Academy, showed that the earth is somewhat larger than they had supposed. The exact relation between the meter and the polar circumference of the earth is of no practical importance, however, and the original length of the bar itself has been retained as the standard of length for scientific measurements everywhere. The original meter is kept at the Bureau des Poids et Mésures in Sèvres (not far from Paris), and accurate copies have been distributed to most of the countries of the world.

In recent years, it has become possible to measure lengths much more accurately by optical means than by locating two engraved lines on a metal bar. Accordingly, in 1960 the International Bureau of Weights and Measures decided to define the meter as exactly 1,650,763.73 wavelengths of the orange-red light emitted by the gas krypton-86. (Later in the book we shall see what this terminology means.) Now, if all the standard meter bars in the world were melted down into paperweights, we would still be able by this optical definition to reestablish

FIG. 1-2 This balance, designed by the United States National Bureau of Standards, is used to compare government and industrial copies with the primary United States kilogram.

the meter, and more accurately than we ever could by measuring and comparing metal bars.

Although commerce and engineering in English-speaking countries employ measurements made in feet, inches, and yards (units inherited from England along with our language), scientific measurements of length and distance are nearly always in metric units.

Along with the standard unit of length, the metric system also introduced a new unit for amount of matter, or mass. Disposing of pounds and ounces, it uses the *gram* (g), which was intended to be equal to the mass of a cubic centimeter of water at a temperature a few degrees above the freezing point, at which it has its greatest density. (Later more accurate measurement showed the maximum density of water to be only 0.999973 g/cm^3; but for our purposes, we can use an even 1 g/cm^3.) A standard *kilogram* (1000 grams, abbreviated kg) was made of platinum and iridium alloy; the original is kept together with the original meter in Sèvres and copies are distributed all over the world. While the gram and the kilogram are the standard units used in physical measurements, we also use *milligrams* (thousandths of a gram, mg) and *micrograms* (millionths of a gram μg) to express the mass of very small amounts of matter.

Science has not yet devised a more accurate way of measuring the mass of a body, i.e., the amount of matter it contains, than by comparing it with accurate standards on a very sensitive balance. With the balance shown in Fig. 1-2, masses can be compared with an accuracy of 7 parts in 10^9. Thus all of our most accurate measurements of mass are still given in terms of the actual standard kilogram.

The master clock, part of which is shown in Fig. 1-3, symbolizes a third physical unit: the unit of time. A day is divided into 24 hours, each hour is subdivided into 60 minutes, and each minute further divided into 60 seconds, abbreviated s. This system of time measurement is based

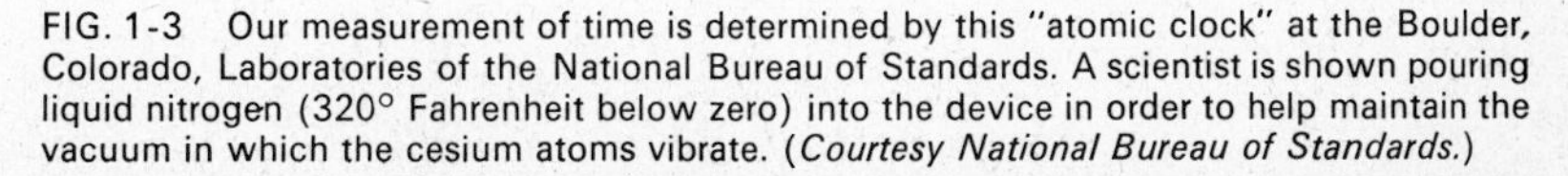

FIG. 1-3 Our measurement of time is determined by this "atomic clock" at the Boulder, Colorado, Laboratories of the National Bureau of Standards. A scientist is shown pouring liquid nitrogen (320° Fahrenheit below zero) into the device in order to help maintain the vacuum in which the cesium atoms vibrate. (*Courtesy National Bureau of Standards.*)

upon that used in ancient Babylon and Egypt, and not even the French Revolution was able to convert it into a decimal system. In the measurement of time intervals much shorter than a second, however, the decimal system is used, and we speak about *milliseconds* (thousandths of a second, ms) and *microseconds* (millionths of a second, μs).

Yet the definition of a second as 1/86,400 of the average length of a day throughout the year, though serving well enough to get us to breakfast on time, is not entirely satisfactory for modern scientific purposes. Astronomers have discovered that the rate at which the earth spins on its axis is not exactly constant, so that the actual length of the day varies slightly from year to year and from century to century. In order to keep the length of the second constant for scientific purposes, our National Bureau of Standards now defines the second as the time it takes an atom of cesium to make 9,192,631,770 internal vibrations.

The clock in Fig. 1-3 counts and keeps track of the vibrations of cesium atoms with an accuracy of 1 part in 2×10^{11}; it would have to run for 6000 years before gaining or losing a single second. The rotation of the earth is not nearly this steady. For this reason one year may be slightly longer or shorter than the next. In order to keep our clocks and our calendars in step with each other, "leap seconds" may be added to (or subtracted from) a year. The year 1972, for example, was 2 seconds longer than normal. One leap second was added just before midnight of June 30 and again on December 31 to give this year two 61-second minutes.

Having defined the units for length, mass, and time, we can express through them the units for other physical quantities. Thus one unit of velocity could be a *centimeter per second* (cm/s), the unit of density a *gram per cubic centimeter* (g/cm^3), etc. The above have been expressed in the system of units known as the CGS system (for centimeter–gram–second). The MKS (meter–kilogram–second) system is also in common use. These two decimal systems, related to each other by simple powers of 10, are accepted by scientists all over the world and represent a definite advantage over the Anglo–American system of units in which speed, for example, may be expressed at will in "feet per second," "miles per hour," or even in "furlongs per fortnight."

In common usage, we have a foot composed of 12 inches, a yard composed of 3 feet, and a mile that is 1760 yards or 5280 feet long (unless it is a nautical mile of 6076 feet). We use the ounce, equal to 437.5 grains (unless it is a troy ounce, used in pharmacy and in weighing precious metals, which is 480 grains), the pound of 16 ounces (unless it is a troy pound of 12 troy ounces), the ton of 2000 pounds (unless it is a long ton of 2240 pounds), etc. Calculations are much easier in the metric system, whose various subdivisions are all related by factors of 10 and whose relationships are indicated by the following standard prefixes, the most common of which are in boldface type:

giga	(G)	$= 10^9$	1 gigavolt	(GV)	$= 10^9$ volts
mega	(M)	$= 10^6$	1 megabuck	(M$)	$= 10^6$ dollars
kilo	(k)	$= 10^3$	1 kilometer	(km)	$= 10^3$ meters
hecto		$= 10^2$	1 hectogram		$= 10^2$ grams
deca		$= 10$	1 decaliter		$= 10$ liters
deci	(d)	$= 10^{-1}$	1 decibel	(dB)	$= 10^{-1}$ bel
centi	(c)	$= 10^{-2}$	1 centimeter	(cm)	$= 10^{-2}$ meter
milli	(m)	$= 10^{-3}$	1 millivolt	(mV)	$= 10^{-3}$ volt
micro*	(μ)	$= 10^{-6}$	1 microwatt	(μW)	$= 10^{-6}$ watt
nano	(n)	$= 10^{-9}$	1 nanosecond	(ns)	$= 10^{-9}$ second
pico	(p)	$= 10^{-12}$	1 picofarad	(pF)	$= 10^{-12}$ farad

* The word "micron," as well as "micrometer," is used to mean 10^{-6} m. "Micrometer" is also a measuring device, but the context should always prevent any confusion in meaning.

1-4 Mixtures of Units

In comparing the sizes of the sun and the hydrogen atom, we were fortunate in having both expressed in the same unit of measurement. We may not always be so lucky. How does a 100-meter dash compare, for example, with a 100-yard dash? In order to make the comparison we must either convert the 100 meters to yards or the 100 yards to meters. Let us convert the meters to yards, using the known fact that 2.54 cm = 1 inch. Then,

$$100\,\cancel{\text{m}} \times \frac{100\,\cancel{\text{cm}}}{1\,\cancel{\text{m}}} \times \frac{1\,\cancel{\text{in}}}{2.54\,\cancel{\text{cm}}} \times \frac{1\,\cancel{\text{ft}}}{12\,\cancel{\text{in}}} \times \frac{1\text{ yd}}{3\,\cancel{\text{ft}}} = 109.4\text{ yd}$$

$$\frac{100\text{ m}}{100\text{ yd}} = \frac{109.4\text{ yd}}{100\text{ yd}} = 1.094.$$

Or we could have converted the 100 yards into meters:

$$100\,\cancel{\text{yd}} \times \frac{3\,\cancel{\text{ft}}}{1\,\cancel{\text{yd}}} \times \frac{12\,\cancel{\text{in}}}{1\,\cancel{\text{ft}}} \times \frac{2.54\,\cancel{\text{cm}}}{1\,\cancel{\text{in}}} \times \frac{1\text{ m}}{100\,\cancel{\text{cm}}} = 91.4\text{ m}$$

$$\frac{100\text{ m}}{100\text{ yd}} = \frac{100\text{ m}}{91.4\text{ m}} = 1.094 \text{ again.}$$

In the first example above, we have started with the 100 m as a numerator. Each *conversion factor* is a fraction equal to 1, because the numerator and the denominator each represent exactly the same length. A dimension in the numerator can be canceled with the same dimension in the denominator, just as numbers can. When we do this in the example, we are left with only the dimension of yards for the answer.

As another example, we can convert the speed of light, 3.00×10^{10} cm/s, into mi/hr. This is a double conversion: We must convert the distance measurement from centimeters to miles and also the time measurement from seconds to hours.

$$\frac{3 \times 10^{10}\ \cancel{\text{cm}}}{\cancel{\text{s}}} \times \frac{1\ \cancel{\text{in}}}{2.54\ \cancel{\text{cm}}} \times \frac{1\ \cancel{\text{ft}}}{12\ \cancel{\text{in}}} \times \frac{1\ \text{mi}}{5.28 \times 10^3\ \cancel{\text{ft}}} \times \frac{3.6 \times 10^3\ \cancel{\text{s}}}{1\ \text{hr}}$$

$$= 0.0671 \times 10^{10}$$

$$= 6.71 \times 10^8\ \text{mi/hr}.$$

The speed of light is the basis of a distance measurement used by astronomers: the *light-year*, which is the distance a ray of light travels through space in a year.

$$\frac{3 \times 10^{10}\ \text{cm}}{\cancel{\text{s}}} \times \frac{60\ \cancel{\text{s}}}{1\ \cancel{\text{min}}} \times \frac{60\ \cancel{\text{min}}}{\cancel{\text{hr}}} \times \frac{24\ \cancel{\text{hr}}}{\cancel{\text{day}}} \times \frac{365\ \cancel{\text{days}}}{\text{yr}}$$

$$= 9.46 \times 10^{17}\ \text{cm/yr}.$$

The light-year is thus 9.46×10^{17} cm. Or, in English units,

$$\frac{6.71 \times 10^8\ \text{mi}}{\cancel{\text{hr}}} \times \frac{24\ \cancel{\text{hr}}}{\cancel{\text{day}}} \times \frac{365\ \cancel{\text{days}}}{\text{yr}} = 5.88 \times 10^{12}\ \text{mi/yr}$$

to give the light-year as 5.88×10^{12} mi.

The following table may be helpful in converting measurements to different units. A more complete table is on p.

1 angstrom (A) = 10^{-8} cm = 10^{-10} m
1 foot (ft) = 30.48 cm = 0.3048 m
1 meter (m) = 39.37 in = 3.281 ft
1 mile (mi) = 1.609 km
1 kilometer (km) = 0.6215 mi
1 pound (lb) = weight* of 453.6 g
Weight* of 1 kilogram (kg) = 2.205 lb

* Since the weight of a gram or kilogram is slightly different at different locations on the earth, we must specify that this weight is to be measured at sea level, at 45° latitude.

1-5 A Better Return on Your Investment

Like other courses, a physics course is more than merely so many hours of credits. It represents an investment of time and effort from which you expect a return that can be measured in terms of accomplishment, satisfaction, and grades. As with other investments, the more time and effort you expend, the greater the returns you can expect.

Some students, however, find that the rate of return on this investment turns out to be less than it should be. Study time may not always seem to pay off in commensurate accomplishment and satisfaction. Perhaps the reason is that the time and effort are not being used effectively. Study methods that work for other courses may not be the best for physics. Of course people differ; minds and attitudes differ—there can be no one hard-and-fast set of rules for everyone. Nevertheless, give the following suggestions a try:

1. Read the assigned text material *once*—quickly but thoughtfully. If you don't understand it all, don't worry. This is merely an orientation to the general ideas, and a preliminary exposure to new terms and new relationships. With practice, though, this first step will become more and more effective, so that less and less remains to struggle with later.

2. This is a *don't*—what *not* to do. Many students start out with the idea that a really good job of studying the physics text consists of repeating Step 1 four or five or six times. This does *not* do it. The ideas you did not understand after the first reading, you probably still don't understand after the sixth. You have largely wasted your time and effort.

3. Go back and *study* the text material *once*. The text serves as a medium to transfer ideas, pictures, relationships, and (to a lesser degree) facts, from the author's mind to your own. At first, you may need to check yourself after every sentence. "Do I really *understand* what this one sentence meant?" If not, read it again and *think* about it. With practice, this sentence-by-sentence study can be expanded to paragraphs and sections. The idea is *not* to memorize, but to *understand*. Memorizing physics—statements, equations, the *words* of definitions and new ideas—will contribute little to your progress. If you *understand* and *visualize*, it is remarkable how little you will need to deliberately memorize.

4. The many worked-out examples and derivations in the text are not to be just read through with a head-nodding in vague agreement. *Work* through them yourself *with pencil and paper*. (Many newspapers sell pads of newsprint trimmings quite cheaply. Many physics departments have stacks of obsolete computer printouts they are glad to get rid of. The backs of these are good enough.) Think your way through from one step to the next. *Don't merely copy*. A good typist copies with a smooth flow into her eyes and out of her fingers without the material she copies leaving any impression on the way.

5. *Use* the illustrations. Most of them are visual representations of text material. Study them as you study the related text; each will reinforce your understanding of the other.

6. When you believe you understand the assigned material, turn to the Study Supplement. Work through the questions and problems in order. A most important factor here is to *write your answer or solution on paper* for each question before you go on to the next. Hopefully, you can work through your entire assignment without trouble. You can then check your own *written* answers and solutions with those in the Study Supplement. If you come to a problem you cannot answer, do not (repeat—**not**!) immediately turn to the solutions for help. Give yourself a chance to think about it a minute or so. Try the next question—it may give a clue to a new line of thought. If not, then turn to the solution. If you run into too many of these snags, it may be a hint that you need to study that part of the text again.

7. Give yourself a break. Study something else, or (if your schedule permits it) wait until the next day. A little time for your subconscious to digest things is very valuable. Then try problems from the text.

8. Keep up with your assignments. It is easy to argue yourself into going to class or lecture unprepared: What you learn there will make your studying easier. There is some truth in this argument, but the balance is heavily weighted in the other direction. When you go to class or lecture fortified with a good foundation of study, demonstrations make sense; different approaches to the material are reasonable; the myriad byways and side issues not covered in the text are understandable. For some courses it is perfectly feasible to set aside one large block of time to do an entire week's assignments. In physics, this does not work so well. There are too many new ideas, too many new relationships, and too much logical structure that must be thoughtfully knit in place. Frequent small bites are much more digestible than an occasional big meal.

Questions

(1-1)

1. Express these numbers in exponential notation: (a) 563, (b) 0.0012, (c) 43,900,000, (d) 0.0000007190.

2. Use exponential notation to write the following: (a) 93,000,000, (b) 0.00006893, (c) 5280, (d) 0.0400.

3. Express the following in an exponential form that is *not* a fraction: (a) $1/10^4$, (b) $1/10^{-4}$, (c) $2/10^6$, (d) $0.0043/10^{-4}$.

4. Write these numbers in an exponential form that is *not* a fraction: (a) $1/10^{-3}$, (b) $1/10^5$, (c) $396/10^6$, (d) $0.00012/10^{-7}$.

5. From Fig. 1-1: (a) About how many times larger than its nucleus is an atom? (b) About how many revolutions does a hydrogen-atom electron make while light is traveling 1 ft? (c) What is about 10,000,000,000 times as large as a pinhead?

6. From Fig. 1-1: (a) About how many atoms side-by-side would reach from sea level to the top of Mt. Everest? (b) What has a lifetime about 10^{15} times that of a muon? (c) About how many times more distant than the moon is the nearest star?

(1-2)

7. What are the values of the following fractions?

(a) $\dfrac{8 \times 10^{-7} \times 6 \times 10^4}{3 \times 10^{-10} \times 4 \times 10^5}$ (b) $\dfrac{3.14 \times 10^8 \times 0.912}{6420 \times 1.13 \times 10^{-5}}$

8. What are the values of the following fractions?

(a) $\dfrac{6 \times 10^9 \times 4 \times 10^{-3}}{2 \times 10^{-4} \times 3 \times 10^{15}}$ (b) $\dfrac{7.19 \times 10^{-5} \times 3420}{18.0 \times 10^6 \times 2.61 \times 10^{-8}}$

9. The mass (which roughly means the amount of matter) of a proton is 1.67×10^{-27} kg. How many protons would be required to make 1 kg?

10. The mass (which roughly means the amount of matter) of an electron is 9.11×10^{-28} g. How many electrons would be required to make 1 g?

11. What is the value of (a) $(6 \times 10^2)^3$? (b) $(1.2 \times 10^{-5})^3$? (c) $(150{,}000)^2$?

12. What is the value of (a) $(3 \times 10^{-6})^3$? (b) $(4.6 \times 10^2)^2$? (c) $(46{,}000)^3$?

13. What is the square root of (a) 10^8? (b) 9×10^8? (c) 10×10^8? (d) 10^9? (e) 6.4×10^9? (f) 463×10^{11}?

14. What is the square root of (a) 10^6? (b) 4×10^6? (c) 10×10^6? (d) 10^7? (e) 4.9×10^7? (f) 316×10^9?

15. What is the cube root of (a) 10^9? (b) 27×10^9? (c) 100×10^9? (d) 10^{11}? (e) 6.4×10^{10}? (f) $\pi \times 10^{10}$?

16. What is the cube root of (a) 10^6? (b) 8×10^6? (c) 10×10^6? (d) 10^7? (e) 2.16×10^8? (f) $\pi \times 10^8$?

(1-3)

17. (a) How many millimeters (mm) are there in a kilometer (km)? (b) How much larger than a nanosecond (ns) is a millisecond (ms)? (c) A microwatt (μW) is what fraction of a kilowatt (kW)?

18. (a) A gigaparsec (Gpsc) is how many kiloparsecs (kpsc)? (b) How many picofarads (pF) are there in a microfarad (μF)? (c) What fraction of a kilovolt (kV) is a millivolt (mV)?

(1-4)

19. A drill is $\frac{1}{8}$ in. in diameter. Express this in centimeters.

20. A man is 6 ft 1 in. tall. What is his height in meters?

21. The average diameter of the earth is 7927 miles. What is this in centimeters? in meters? in kilometers?

22. The diameter of a common size of European bullet is 7 mm. Express this diameter both in inches and in feet.

23. Light of a certain color is composed of a train of waves, each 0.437 μ long. How many of these waves are there in a foot?

24. Light of a certain color is composed of a train of waves, each 5890 Å long. How many of these waves are there in an inch?

25. A 31-story building is 132 m tall. What is the height of each story in feet?

26. If 325 sheets of paper make a stack 1 in. high, what is the thickness of a sheet of paper in centimeters?

27. How far (in feet) does light travel in 1 ns?

28. How long (in seconds) does it take light to travel 1000 ft?

2

Bodies at Rest

2-1 Equilibrium and Forces

Since we are going to study bodies in motion and what makes them move, the easiest place to start is with bodies that are not moving at all. Such objects are in a state of *static equilibrium*. As we shall see in the next chapter, any body moving in a straight line at constant speed is also in equilibrium; the stationary objects we shall consider now are merely special cases in which the constant speed happens to be zero. Let us examine a few of these special cases.

A book is probably lying on your desk—it is motionless and in static equilibrium. What do we know about the forces acting on the book? Unless there is a high wind, there are no forces tending to move it horizontally across the desk. We know, however, that there is a vertical force acting on it, trying to move it downward toward the earth's center. This gravitational attraction toward the earth is the book's *weight*; you can feel this if you hold the book in your hand. We know, too, that in order to keep the book from falling, you must push upward with a force equal to its weight. If you withdraw your hand, the downward

FIG. 2-1 Two equal and opposite forces, not acting along the same line, exert a torque on the log.

force of gravity acts unopposed to bring the book crashing to the floor. We can then be confident that the book lying on the table is also being acted on by two forces: the downward pull of gravity and the equal and opposite upward push of the tabletop.

If we want to call an upward force "+" and a downward force "−", a little more elegant way of saying the same thing is to state that all the forces acting on a body in equilibrium add up to zero.

This rule of course applies to other directions as well as to up and down. If you push the book north with a force of 5 lb, the book will move unless your push is opposed by some southward 5-lb force.

2-2 Equilibrium and Torques

Even if we were sure that all the forces acting on a body added up to zero, this alone would not guarantee that the body would be in equilibrium. In Fig. 2-1 the log on the ground is being acted on by two equal and opposite forces. Tractor A pulls one end of the log east with a force of 2000 lb; tractor B pulls the other end west with an equal 2000-lb force. These two forces add up to zero, but we know the log will *not* lie motionless; instead, it will begin to rotate, as shown. The rotating effect of a force depends, not only on how big the force is, but also on where it is applied. In Fig. 2-2A, for example, it will be very hard to turn the rusted nut. The turning effect, or *torque*, is the force F multiplied by the distance d_1, called the *lever arm*.

$$\textit{Torque} = \textit{force} \times \textit{lever arm}.$$

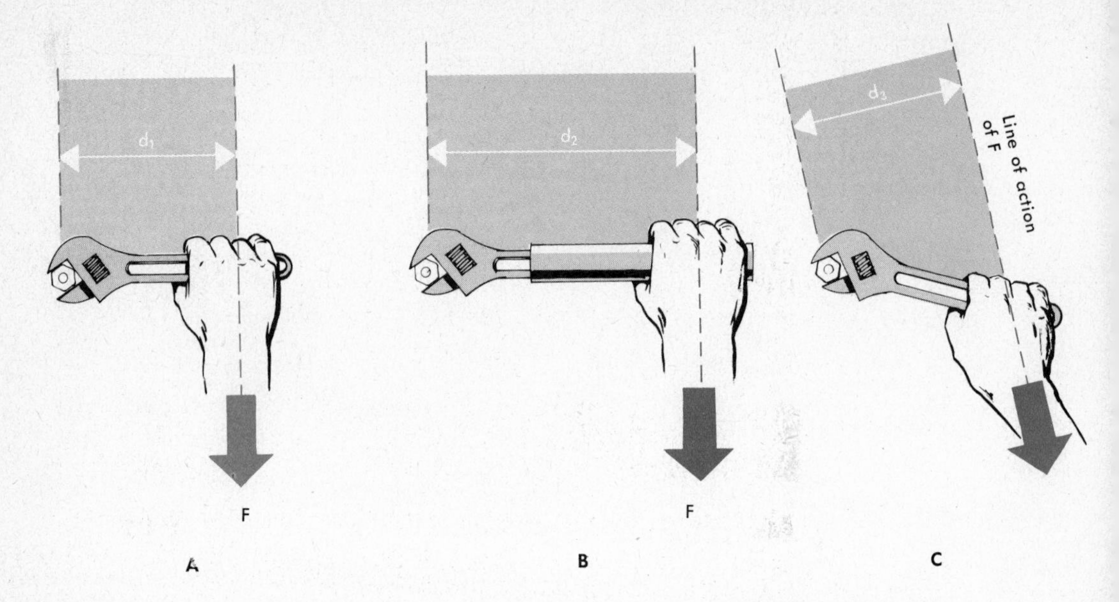

FIG. 2-2 The torque produced by a force depends on the force and on its lever arm.

By slipping a piece of pipe over the wrench handle as in Fig. 2-2B, the distance, or lever arm, is increased to d_2, and the torque Fd_2 is made much greater without any increase in the force.

We must note here that the lever arm is always measured ***from the center of rotation perpendicular to the line of action of the force F.*** This is seen to be the case for d_1 and d_2 in Fig. 2-2A and B, in which the pull *is* perpendicular to the handle of the wrench. In Fig. 2-2C, however, the pull is at an angle and the situation is different. The distance d_3 (which is the lever arm) must be measured from the center of the nut in the direction shown, perpendicular to the dashed *line of action* of the force *F*, i.e., the prolongation of the arrow representing the pull of the hand.

Figure 2-3 shows a light bar with weights suspended from its ends, balanced on a narrow support at *B*. If we were to balance such a bar with weights as shown, we would find that the distances *BC* and *AB* would have to be in the same proportion as the weights hung from *A* and *C*, respectively; in this particular case, in the proportion, or ratio, of 3 to 1. If we imagine the bar trying to turn about *B*, force *A* will have a torque of $3 \times 2 = 6$ in.-lb counterclockwise and will be just balanced by the clockwise torque of force *C*, which equals $1 \times 6 = 6$ in.-lb. (Since we obtain torque by multiplying a distance times a force, the units, or dimensions, of torque are always expressed as distance–force units—in this case, inches × pounds, or in.-lb.)

At point *B*, the support does of course exert an upward force on

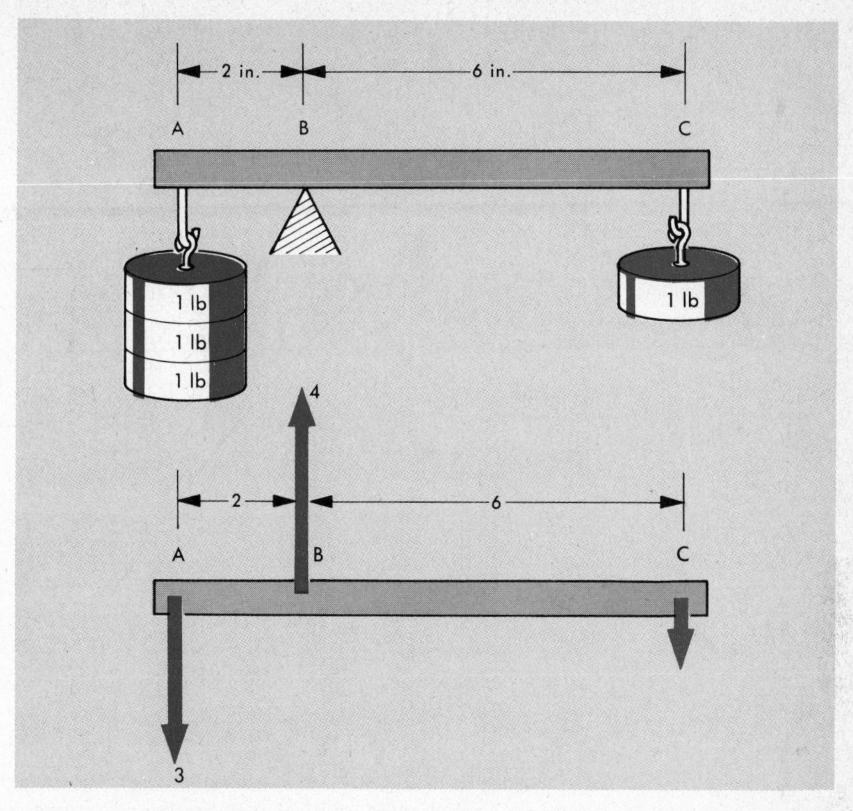

FIG. 2-3 For a body in equilibrium, the forces in one direction equal the forces in the opposite direction, and the clockwise torques equal the counterclockwise torques.

the balanced bar, but since the force passes *through B*, its lever arm (measured from *B*) is zero and so is its torque.

In this example we did not need to know the upward force at *B*. It would, however, have been easy to find, since in equilibrium the forces up must equal the forces down: $F_B = 3 + 1 = 4$ lb. With this knowledge that $F_B = 4$ lb upward, we can check the torques on the bar around *A* or *C* or any other point and get the same results, that the clockwise and counterclockwise torques are equal. Around *A*, for example, the force due to the 3-lb weight passes *through A* and hence its lever arm and its torque around *A* are zero. The upward force at *B* tries to rotate the bar around point *A* in a counterclockwise direction; the weight pulling down at *C* tries to rotate it clockwise around *A*. So we can write for torque clockwise = torque counterclockwise that

$$1 \times 8 = 4 \times 2.$$

Using $+F$ and $-F$ to indicate upward and downward forces and $+\tau$ (the Greek letter "tau") and $-\tau$ for clockwise and counterclockwise torques, we can put the conditions for equilibrium into a very simple form:

$$\sum F = 0 \quad \text{and} \quad \sum \tau = 0.$$

The $\sum$ is the Greek capital letter "sigma," used to mean "the sum of." These relationships can be checked out as follows for the lever of Fig. 2-3:

$$\sum F = 4 - 3 - 1 = 0.$$

$$\sum \tau = (1 \times 8) - (4 \times 2) = 0 \qquad \text{(around } A\text{)}$$

$$\sum \tau = (1 \times 6) - (3 \times 2) = 0 \qquad \text{(around } B\text{)}$$

$$\sum \tau = (4 \times 6) - (3 \times 8) = 0 \qquad \text{(around } C\text{).}$$

2-3 Center of Gravity

In objects that posses a definite shape, we can single out an important point known as the *center of gravity* (CG). The force of gravity acts, of course, on all parts of any given object, but the center of gravity can be defined by saying that objects subjected to gravity behave as if there were only a single force applied at that point. If a body is supported at its center of gravity, it will be in balance and have no tendency to move or rotate in any direction.

For any symmetrical body of uniform composition, the center of gravity is the same as its geometrical center. In the case of an object of more complicated shape, the center of gravity can always be found by suspending it on a string attached first at one point and then at another point on its surface. An object always comes to rest with its

FIG. 2-4 A simple experimental method for locating the center of gravity (CG) of an irregularly-shaped figure.

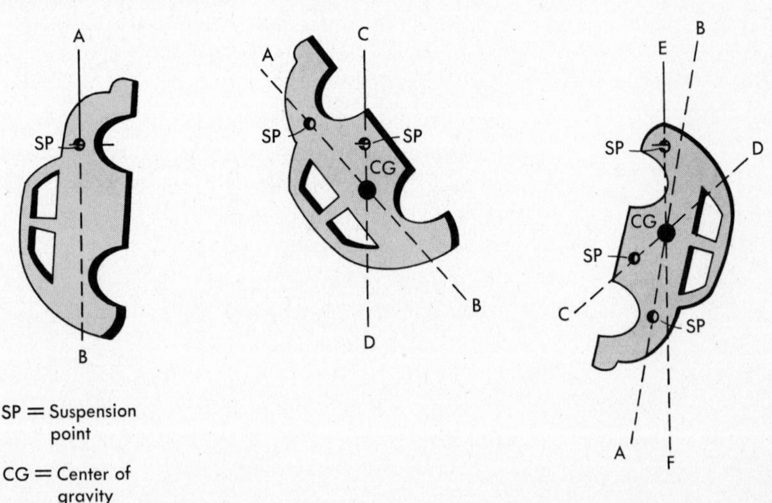

center of gravity directly under the point of suspension. Suppose we cut an object out of plywood with the shape shown in Fig. 2-4. If we suspend it at point *A*, it will hang in the way shown in Fig. 2-4 (left), and its center of gravity must be located somewhere on the line *AB*. If we suspend it at another point *C*, the object will hang as shown in Fig. 2-4 (center) and the center of gravity must be somewhere on the line *CD*. Thus the exact location of the center of gravity is determined by the intersection of the lines *AB* and *CD*. It is unnecessary to suspend it from a third point *E* (right) because the vertical line *EF* must also pass through the intersection of the previous two lines.

In our investigation in the preceding section, concerning the light bar with the weights at its ends, the center of gravity of the whole arrangement was at the balance point *B*. In this problem, we took advantage of the word "light" to ignore the effect of gravity on the bar itself. Most real bars and sticks, however, are heavy enough so that their own weights must also be taken into account if we are to expect accurate answers. Let us go back to this example, as illustrated in Fig. 2-3, and assume that the bar itself weighs 1 lb. Now, where will the balance point be with the weights still attached as they were?

Figure 2-5 shows the same bar, with the addition of the force marked *W*, representing the weight of the bar itself. Although gravity pulls

FIG. 2-5 Including its own weight in balancing a bar similar to the one considered weightless in FIG. 2-3.

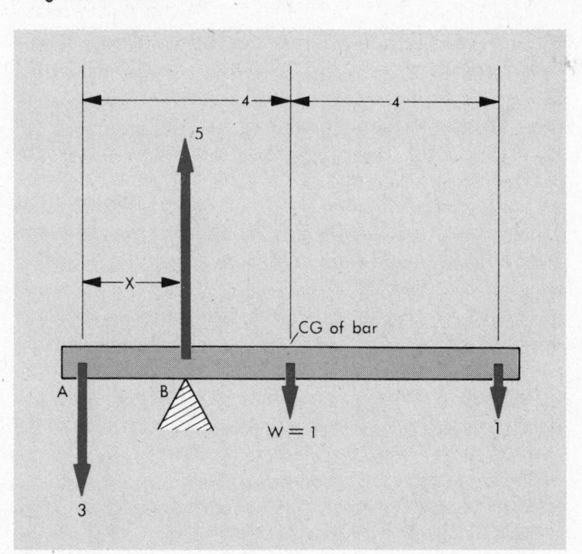

equally along the entire length of the bar, we can consider the entire pull of 1 lb to be exerted at the center of gravity of the bar. We can assume that the bar is uniform, so that its center of gravity will be at its center, 4 in. from each end. Although the upward supporting force B can be calculated to be 5 lb by using $\sum F = 0$, we do not now know where this point of balance will be located. Assume it to be x inches from the left end. Then, with the bar balanced and in equilibrium, the total torques of all the forces acting on the bar will add up to zero, no matter what point we choose to pick as a center of rotation. Point A is as good as any. So, using A as the center and putting torques clockwise = torques counterclockwise, we get

$$(3 \times 0) + (1 \times 8) = 5 \times x$$

$$5x = 8$$

$$x = \tfrac{8}{5} = 1.6 \text{ in.}$$

2-4 Vectors

Some physical quantities, such as age, amounts of money, and temperature, have magnitude only—that is, they can be described by a single number that tells the amount of the property we are interested in: 30 years, \$27.19, 75°, etc. Such quantities are called *scalars*. There are other quantities, however, that have direction as well as magnitude. We cannot describe a force, for example, merely by saying that it is equal to 30 lb. The direction of the force must also be given if we are to have a complete picture of it. Such quantities are called *vectors*. Vector quantities can be very conveniently represented by arrows. The length of the arrow, drawn to some convenient scale, represents the magnitude of the quantity, and the arrow is pointed in a direction parallel to the direction of the vector quantity.

FIG. 2-6 The graphical addition of vectors.

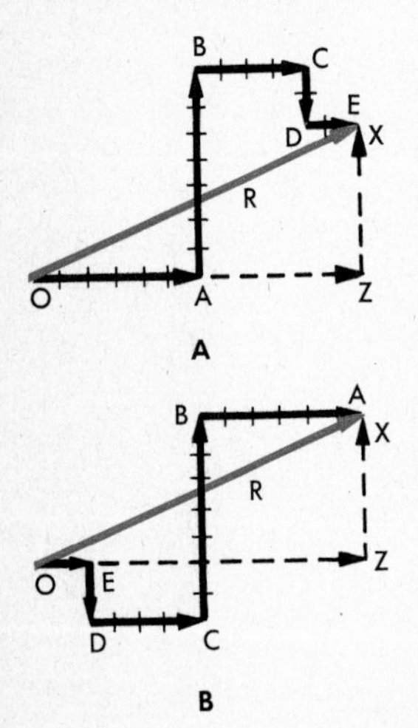

Probably the simplest example of the application of vectors is in their use to represent distances or displacements. Figure 2-6A shows a walking tour in which a man starting from O goes 6 km east, 7 km north, 4 km east, 2 km south, and 2 km east to at last arrive at X. We have represented each segment of this journey by one of the separate vectors A, B, C, D, and E; note that the tail of each one starts from the point of the preceding one.

This is a graphical method of vector addition; the *vector sum*, or *resultant*, of the five vectors A, B, C, D, and E is the single vector R that goes from O to X. We can find the magnitude, or length, of R by sketching in the east–west line OZ and the north–south line from Z to X. The line OZ is $6 + 4 + 2 = 12$ km long; ZX is $7 - 2 = 5$ km. The Pythagorean theorem tells us that

$$R = \sqrt{(12)^2 + (5)^2} = 13.$$

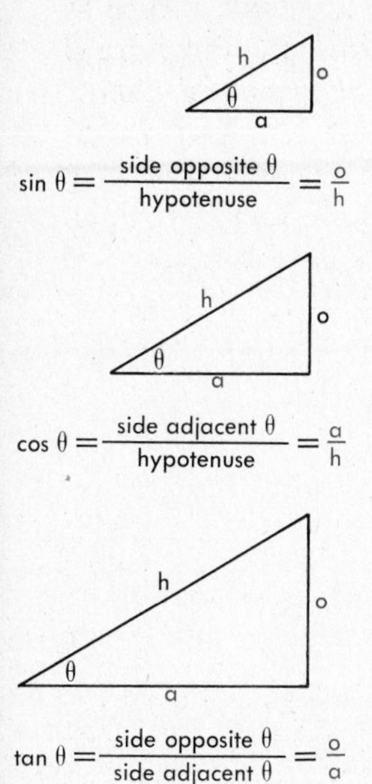

FIG. 2-7 Trigonometric functions of an angle.

(Note that this is something quite different from the total 21 km the man has walked.)

It is a good thing to know that it makes no difference in what order vectors are added. Figure 2-6B shows them added in the order *E, D, C, B, A*. A quick check will show that *OZ* (the east–west *component* of *R*) is again 12, that *ZX* (the north–south *component* of *R*) is 5, and that *R* is identical with the *R* of Fig. 2-6A.

We now have half the information we need to describe a vector. It would be convenient to be able to say also that "*R* points in a direction so many degrees north of east." Of course, we could lay a protractor on the figure (if it is drawn accurately to scale) and thereby find the angle *ZOX* to be about 23°. But it is generally desirable to use some more accurate and more convenient analytical method for finding out about angles. Fortunately for us, the patient labors of the men who computed the tables of trigonometric functions have made this easy; a condensed table is given on p.—.

Figure 2-7 shows the right-triangle relationships that define the three principal trigonometric functions—the *sine* (abbreviation: sin), the *cosine* (cos), and the *tangent* (tan). The angle θ (the Greek letter *theta*) is the same in all the various-sized triangles shown, which means that all the triangles have the same shape. Therefore, the ratio o/h or a/h or o/a will be the same, regardless of the size of the triangle; in other words, these ratios depend *only* on the angle θ, which determines the shape of the triangle.

We can now return to Fig. 2-6 and see that the tangent of angle $ZOX = \frac{5}{12} = 0.417$; $\sin ZOX = \frac{5}{13} = 0.385$ and $\cos ZOX = \frac{12}{13} = 0.923$. From any one of these, we can find from the table (or from a slide rule) that angle $ZOX = 22.6°$ and we can thus give the direction of *R* as 22.6° north of east.

FIG. 2-8 Tail-to-point and parallelogram methods for the graphical addition of vectors.

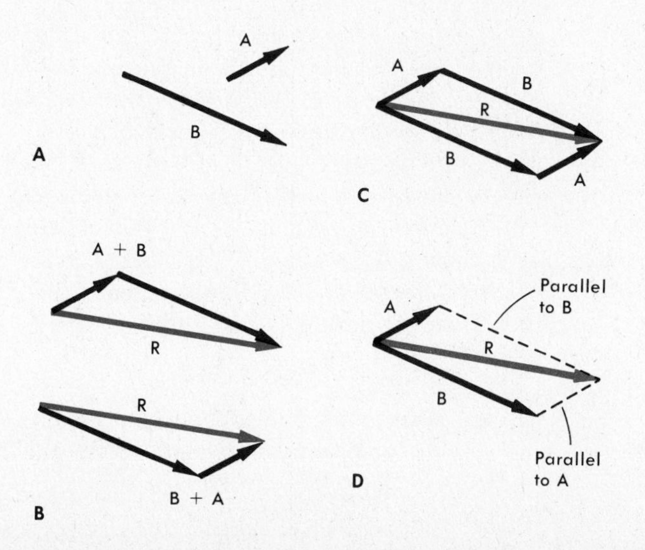

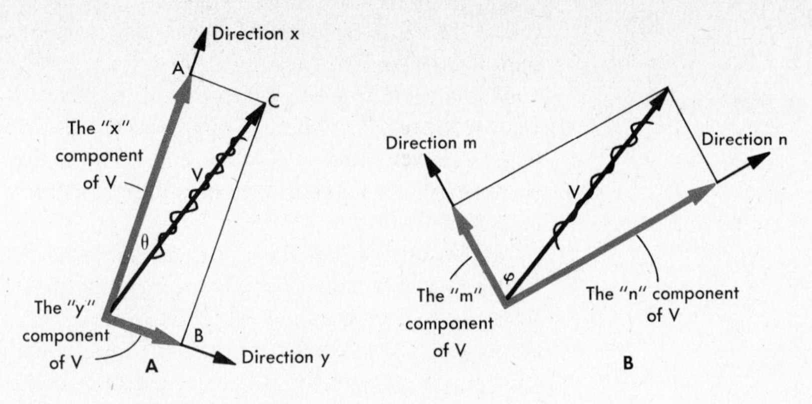

FIG. 2-9 Resolution of a vector into components.

In this discussion we have followed the logical procedure of adding vectors by starting the tail of one on the point of another. (If we want to *subtract* a vector, all we need do is reverse its direction by putting the arrow on the opposite end and then *add* it.)

There is another graphical method for adding two vectors, called the *parallelogram* method. Figure 2-8A shows the vectors A and B, which we are to add together. As in Fig. 2-8B, we can either add $A + B$ by the tail-to-point scheme or do the same for $B + A$. In either case we find the same resultant—that is, the R's are the same length and in the same direction. So in Fig. 2-8C the two parts have been put together on the same resultant. This justifies Fig. 2-8D, which illustrates the parallelogram method. A and B have been put together tail-to-tail. The dashed lines show the completion of a parallelogram; the resultant R is the diagonal of the parallelogram, drawn from the junction of the tails of A and B.

For many purposes it is useful to be able to replace a vector by two or more *components*—that is, by two or more vectors which will add together to give the original vector. Almost always we will want these components to be at right angles to one another, and Fig. 2-9 shows how this can be done in a process called *resolving* a vector into components. We see that the x component of vector V is $V \cos \theta$, and the y component (which is equal to AC) is $V \sin \theta$. Similarly, if we had for some reason chosen different directions for our components, we would have V's m component equal to $V \cos \phi$ (Greek letter *phi*), and its n component equal to $V \sin \phi$. The original vector V is shown marked out because it has been replaced by its components, which add up vectorially to produce the same effect as V itself.

2-5 Oblique Forces

The use of vectors and their components makes it possible for us to put the idea of equilibrium on a more secure and definite footing. In the examples we have used so far, the applied forces have always been parallel so that it was easy to see that the "ups" just nullified the "downs" and that the "lefts" exactly counterbalanced the "rights." For torques, also, we heretofore have considered only forces that were perpendicular to given lever arms so the torques were easy to calculate. Most actual examples, though, include forces and torques that are not so considerately arranged.

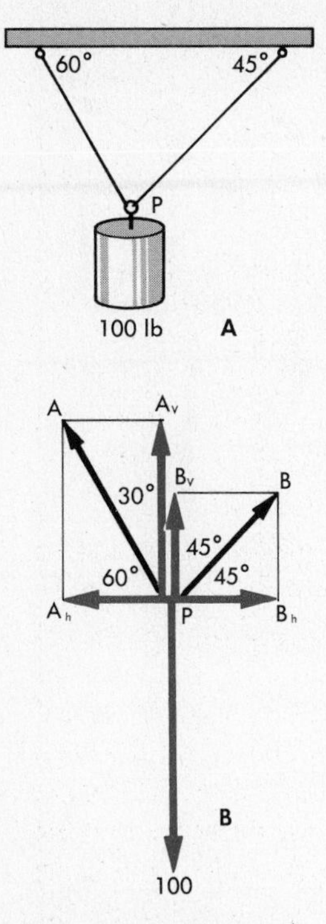

FIG. 2-10 Resolution of oblique vectors into vertical and horizontal components.

For example, look at a weight suspended from ropes as shown in Fig. 2-10A. The pulls of the ropes on the knot P will be along the directions of the ropes, as indicated by vectors A and B in Fig. 2-10B. Also acting on the knot is the pull of gravity on the suspended weight—100 lb straight down. By resolving A and B into vertical and horizontal components, we can as simply as before deal with "ups" and "downs" and "lefts" and "rights." A_v (the vertical component of A) is $A \cos 30° = 0.866A$, and A_h (the horizontal component of A) is $A \cos 60°$ (or, if you prefer, $A \sin 30°) = 0.500A$. Similarly, $B_v = 0.707B$ and $B_h = 0.707B$. Putting "ups" equal to "downs" and "rights" equal to "lefts," we have

$$0.866A + 0.707B = 100$$

and

$$0.500A = 0.707B.$$

The second equation gives us $A = 1.414B$. If we substitute this value for A into the first equation, we get

$$0.866 \times 1.414B + 0.707B = 100$$

$$1.932B = 100$$

$$B = 51.76 \text{ lb}.$$

Substituting this value for B back into one of our first equations, we find that

$$A = 73.19 \text{ lb}.$$

The example we have just considered was one in which all the forces acted together at a single point so that it was not necessary to deal with torques. In many cases, however, it will be necessary to compute the torque of an oblique force. Figure 2-11A shows a hinged wall shelf used as a desk. The shelf is supported on each side by a chain fastened as indicated on the drawing. The shelf itself weighs 20 lb and a 40-lb typewriter is placed on the shelf midway between the chains, with its center of gravity 10 inches back from the front edge. With this symmetrical arrangement, each chain and each hinge carries half of the total load. In order to find the tension in each chain and the reaction at each hinge, we can start by *isolating* one end of the shelf, shown in Fig. 2-11B, and showing all the forces acting on it.

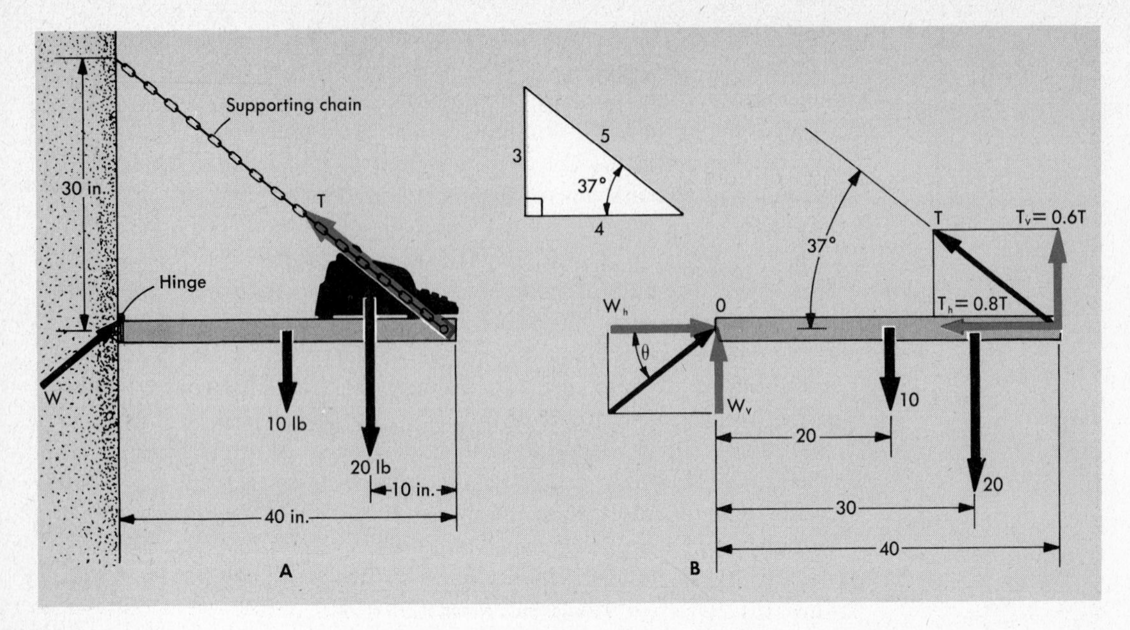

FIG. 2-11 Resolution of force vectors T and W into components at right angles to each other.

Forces W and T can be dealt with quite easily if we break them up into rectangular components as shown. From the given dimensions, we can see that the pull of the chain is along the hypotenuse of a 3–4–5 triangle, which is a right triangle whose smallest angle is about 37°. As seen from the drawing, the sine of this angle is $\frac{3}{5}$ and the cosine, $\frac{4}{5}$. Figure 2-11B shows us (forces up = forces down and forces left = forces right) that $W_v + 0.6T = 10 + 20$ and that $W_h = 0.8T$, but this is not enough information to let us solve for any of the three unknowns. For further information we must turn to torques. Because the shelf is stable and stationary, it is not turning about any axis whatever; this means that around any axis we choose, the sum of the clockwise torques must equal the sum of those acting to turn it in a counterclockwise direction. Since we are thus free to choose any axis we like, let us exercise this choice intelligently to save as much work as we can. A look at Fig. 2-11B tells us that it will be to our advantage to choose point O, the hinge, since three of our four force components pass through O and thus have zero torque about it. Therefore, setting the torques tending to turn the shelf clockwise about O equal to those acting in a counterclockwise direction, we have

$$10 \times 20 + 20 \times 30 = T_v \times 40$$

so that

$$T_v = 20 \text{ lb}$$

and

$$T = \frac{20}{0.6} = 33.3 \text{ lb.}$$

Substituting this value back into the force equations, we get

$$W_v = 10 \text{ lb}$$

and

$$W_h = 26.7 \text{ lb.}$$

If we also want the direction of the hinge reaction, we see that $\tan \theta = W_v/W_h = 0.375$, and so $\theta = 20.55°$.

2-6 Friction

So far, we have ignored one of the most important classes of forces in the world—the forces of friction. We all realize that if it were not for the friction between our shoes and the floor, we could not walk; cars move and stop only through the forces of friction between tires and road. (We are reminded of this in winter when roads are icy and these frictional forces very small.) Experiment shows that the frictional force between any two sliding surfaces is directly proportional to the force with which the surfaces are pressed together. The proportionality constant is called the *coefficient of friction* between the surfaces, and in most textbooks is assigned the symbol μ (the Greek letter *mu*), so that

$$F_{fr} = \mu \times F.$$

The value of μ depends on the materials that are sliding and is only very slightly affected by other factors, such as speed or contact area.

Let us take an iron block 2 in. by 3 in. by 6 in. and slide it along a concrete walk. The weight of the iron block is about 10 pounds, and the coefficient of friction is close to 0.3. From this we can compute that the force needed to drag the iron block will be $10 \times 0.3 = 3$ pounds. We will find that it makes almost no difference whether the iron block stands on end or lies flat on the walk; and whether we move it along rather slowly or at a higher speed, it will still take a 3-pound pull (see Fig. 2-12).

A simple experiment that demonstrates the action of friction can be carried out with a ruler or a golf club. Take a ruler and support each end by your index fingers. Now move you hands closer and closer together and note that the stick will slide over your right finger, then over the left one, then over the right, then over the left again, etc. Finally, when your two fingers come together, the ruler will still be in equilibrium, and its middle point (which in this case, of course, is the center of gravity) will be located between the two fingers. If you repeat the

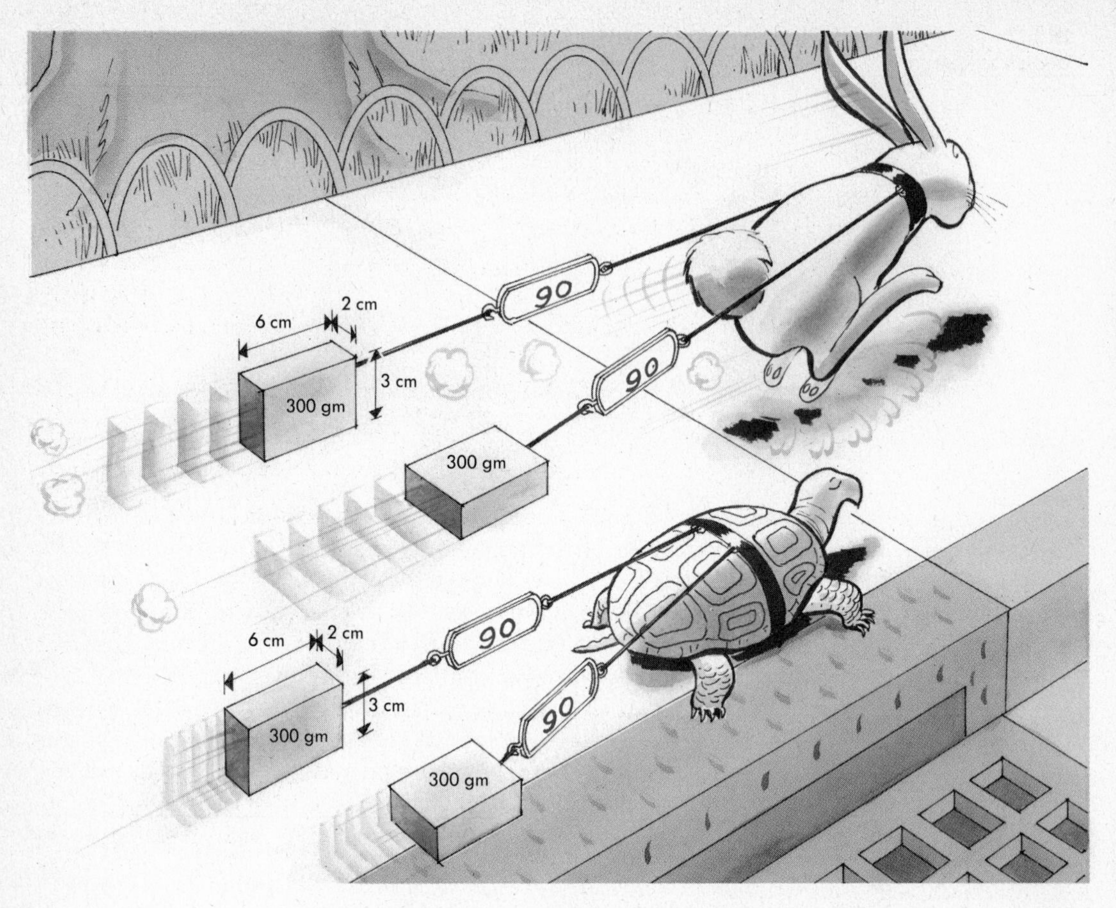

FIG. 2-12 The position of the block has no effect on the force of friction, which depends only on the nature of the surfaces and the force pressing them together (here, the weight of the block).

same experiment with a golf club, which has a heavy piece of iron fastened to one end, you will find that the center of gravity is located closer to the heavy end of the club. The alternating sliding of the ruler or the golf club over your fingers is caused by the fact that the friction force between any two objects sliding on one another is larger the more strongly they are pressed together, and by the fact that the finger located closer to the center of gravity is supporting the larger fraction of the total weight of the ruler or the club. The object will always slide on the finger supporting the least weight, which is the one farthest from the center of gravity. As the golf club slides first on one finger and then on the other, its center of gravity will always remain between them.

As another example, consider a man climbing a 40-lb, 13-ft ladder with its base 5 ft out from a smooth vertical wall, as shown in Fig. 2-13A. What must be the coefficient of friction between the foot of the ladder and the ground, if the man is to be able to climb to the top without the ladder slipping? Figure 2-13B isolates the ladder, and shows all of the forces acting on it. Because the man is to be able to climb to the top, with the ladder then still just barely in equilibrium, the man's weight of 180 lb is shown at the top. The wall is smooth ($\mu = 0$), so there can be no frictional force parallel to the wall; the force W must therefore be perpendicular to the wall. If we assume the ladder to be uniform, its 40-lb weight acts at its center. The push of the ground G is unknown in both direction and magnitude, and it will be convenient to break it into its components G_v and G_h. From Fig. 2-13B we can immediately write

$$G_v = 40 + 180 = 220 \quad \text{and} \quad G_h = W.$$

The best choice for the axis around which to calculate torques is the base of the ladder. Around this axis, the two unknown forces G_v and G_h both have zero torque and are thus eliminated from the equation:

$$40 \times 2.5 + 180 \times 5 = W \times 12$$

or

$$W = 83.3 \text{ lb.}$$

FIG. 2-13 Forces and torques acting on an inclined ladder.

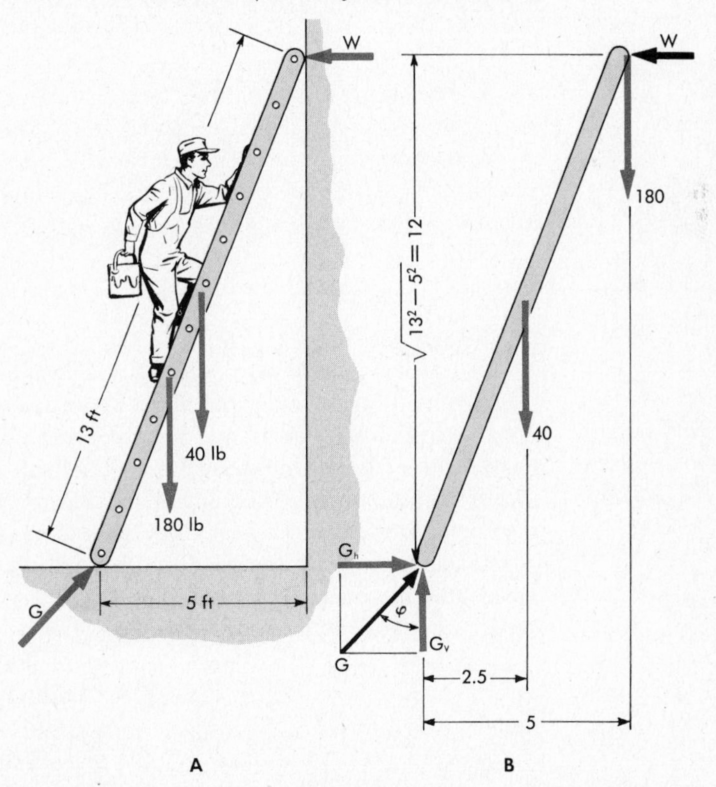

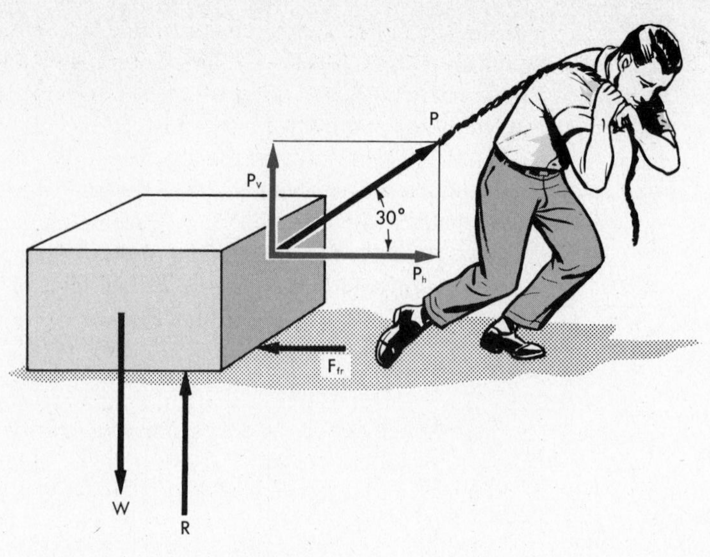

FIG. 2-14 The oblique force P both pulls the box along and reduces the force with which the box presses against the floor.

This value, substituted back into the force equations, gives us

$$G_h = 83.3 \text{ lb.}$$

The coefficient of friction at the base of the ladder cannot be less than G_h/G_v, or 0.38, if the man is to climb safely to the top. The tangent of ϕ, the angle at which the push of the ground is inclined from the vertical, is also 0.38, which makes $\phi = 20.8°$. The quantity G can be determined as

$$\sqrt{G_v^2 + G_h^2} \quad \text{or} \quad \frac{G_v}{\cos\phi} \quad \text{or} \quad \frac{G_h}{\sin\phi},$$

all of which give $G = 235$ lb.

A little more complicated example is shown in Fig. 2-14, which shows a box being dragged along the floor by a rope that makes an angle of 30° with the floor. Let us say that the box weighs 500 lb and that its coefficient of friction with the floor is 0.25. How hard will the man have to pull? Since the pull of gravity W is vertical and the reaction between the floor and the box can be considered in terms of its vertical and horizontal components R and F_{fr} (friction) and the motion of the box is horizontal, it would seem a good idea to break the oblique pull P into vertical and horizontal components also. This gives us

$$P_v = P \sin 30° = 0.5P$$

and

$$P_h = P \cos 30° = 0.866P.$$

In order to keep the box sliding along the floor, P_h need be only equal to the retarding force of friction F_{fr}. This frictional force is $0.25R$. (NOT $0.25W$! P_v helps lift the box upward so that $R = W - P_v$.) We can now write quite simply

$$\begin{aligned} P_h &= F_{fr} \\ &= 0.25R \\ &= 0.25(W - P_v). \end{aligned}$$

From this, we get

$$\begin{aligned} 0.866P &= 0.25(500 - 0.5P) \\ &= 125 - 0.125P \end{aligned}$$

or

$$0.991P = 125$$

and

$$P = 126.1 \text{ lb}.$$

2-7 Fluid Pressure

Liquids do not have a definite shape of their own and readily assume the shape of any vessel they are poured into. But in doing so, a liquid retains its total volume and a gallon of water will remain a gallon whether it is poured into a flat dish or into a tall, narrow container.

Consider water in a cylindrical glass. Because of its weight, the water exerts a force on the bottom of the glass, a force the same as that which would be produced by a cylindrical piece of ice if the water were frozen and the glass walls removed. This force is distributed over the entire bottom of the glass so that each square centimeter of the area of the bottom carries its own equal share of the load. This *force per unit area* is called the *pressure* P, equal to the total force divided by the area over which it is exerted:

$$F = PA \quad \text{and} \quad P = \frac{F}{A}.$$

All equations must be equations in every sense of the word—not only must the numerical values be the same on both sides, but so must the dimensions, or units in which the various quantities are measured. If, for example, we have a pressure of 10 lb/in^2 exerted on a surface whose area is 3 ft^2, we would get the following for F:

$$F = PA = \frac{10 \text{ lb}}{\text{in}^2} \times 3 \text{ ft}^2 = \frac{30 \text{ lb-ft}^2}{\text{in}^2} = ???$$

These units tell us nothing at all, and we see that in order to get a sensible answer we must either convert the in^2 into ft^2 or vice versa.

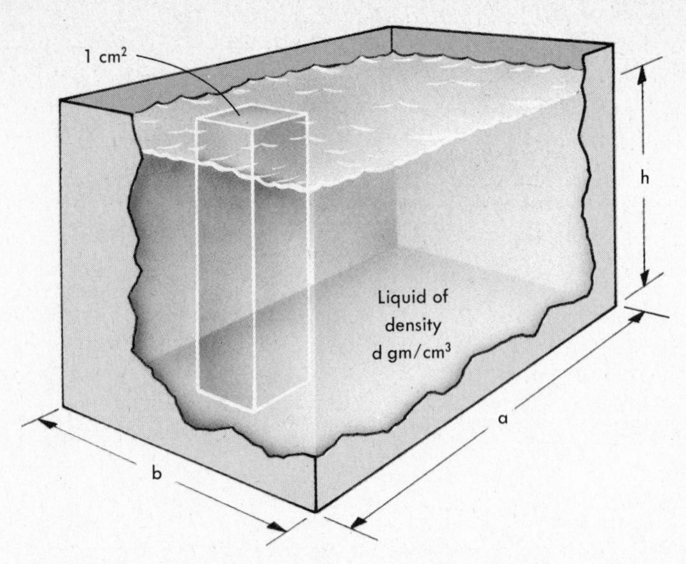

FIG. 2-15 Each square foot of the bottom of the tank supports a column of liquid that weighs *hd* pounds. This pressure (*hd*) times the area of the bottom (*ab*) gives the total force on the bottom (*abhd*).

If we do this, we can then cancel the area units in the numerator with the similar ones in the denominator and come out with a reasonable answer of so many pounds. Let us choose to convert the area into square inches: Since there are $12^2 = 144 \text{ in}^2$ per ft^2, the area is $3 \times 144 = 432 \text{ in}^2$, and we have

$$F = \frac{10 \text{ lb}}{\cancel{\text{in}^2}} \times 432 \cancel{\text{in}^2} = 4320 \text{ lb}.$$

Or, if we prefer, we can convert the pressure: Since $P = 10 \text{ lb/in}^2$ means that there is a force of 10 pounds on every square inch, the force on each square foot will be 144 times as great, and the pressure can be expressed as $10 \times 144 = 1440 \text{ lb/ft}^2$. Then

$$F = \frac{1440 \text{ lb}}{\cancel{\text{ft}^2}} \times 3 \cancel{\text{ft}^2} = 4320 \text{ lb} \quad \text{as before.}$$

Figure 2-15 represents a tank, and shows an imaginary column of liquid 1 ft square and extending h ft from the bottom to the top surface of the liquid. The volume of this column is $h \text{ ft}^3$, and its weight will be h times the weight of a single cubic foot. This latter quantity, the weight of 1 ft^3 (or 1 cm^3 or 1 m^3) of any substance, is the *density* of the substance, and we can let it be represented by the letter d. The weight of the column is thus hd pounds, and this weight is supported by 1 ft^2 of the bottom.

Since force per unit area is what we mean by *pressure*, we can use this general formula for the pressure at any depth in a liquid:

$$P = hd.$$

If the tank is, say, 5 ft × 8 ft in size, and is 4 ft deep, filled with oil whose density is 55 lb/ft³, we have for the pressure

$$P = hd = 4 \text{ ft} \times 55 \text{ lb/ft}^3 = 220 \text{ lb/ft}^2.$$

For the total force on the bottom of the tank, we have

$$F = PA = 220 \text{ lb/ft}^2 \times 5 \text{ ft} \times 8 \text{ ft} = 8800 \text{ lb}.$$

We used the bottom of the tank in making our calculation, but the same arguments will hold for the pressure at any depth. Here too, as in any quantitative work, we must keep our eyes on the units. As an example, we can calculate the pressure due to the overlying water at the bottom of the ocean at a depth of exactly 2 mi. The density of sea water is 64.0 lb/ft³, and if we merely substitute in the equation, we have

$$P = hd = 2 \text{ mi} \times \frac{64 \text{ lb}}{\text{ft}^3} = \frac{128 \text{ lb-mi}}{\text{ft}^3},$$

FIG. 2-16 The pressure in vessels of different shape is the same at equal depths of the fluid in them.

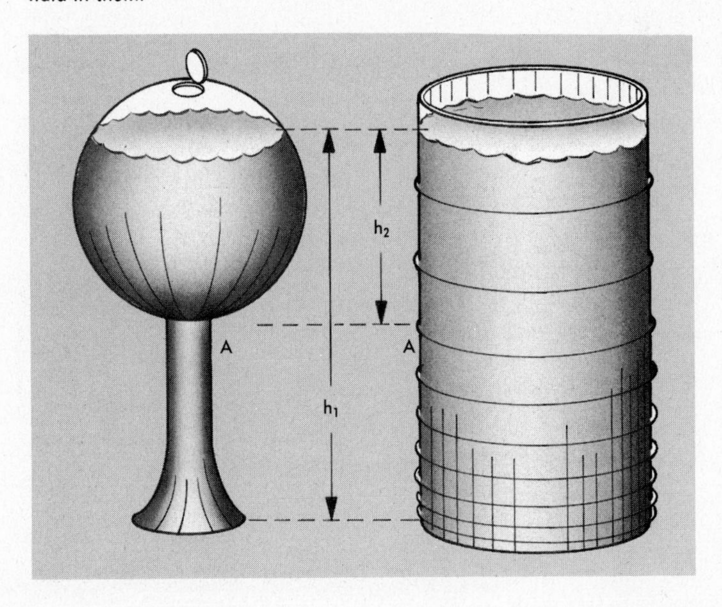

which makes as little sense as force expressed in lb-ft^2/in.^{2}. But if we convert the 2 mi into $2 \times 5280 = 10{,}560$ ft, the answer will be perfectly reasonable:

$$P = hd = 10{,}560 \text{ ft} \times \frac{64.0 \text{ lb}}{\text{ft}^3} = 676{,}000 \text{ lb/ft}^2.$$

Water, or any other fluid, since it does not have a rigid shape, cannot resist a pressure exerted on it in only one direction, the way, for example, a block of steel can when it resists being squeezed in a vise. Instead, liquids squash out in all directions and so exert equal pressure in all directions. Thus our formula $P = hd$ is equally useful in calculating the pressure against the wall of a container at any depth, no matter at what angle the wall happens to be. For this reason, the shape of the container makes no difference. In Fig. 2-16, if both vessels are filled with a liquid of density d, the pressure at the bottom is $h_1 d$ for both, and at the points marked A the pressure in both is $h_2 d$.

2-8 Pascal's Law

Since the pressure on any small piece of a fluid is the same in all directions, ***any increase of pressure on a fluid in a closed container will be transmitted equally to every part of the fluid.*** This basic law was discovered by the French physicist Blaise Pascal (1623–1662) and carries his name. Imagine a closed vessel with two vertical cylinders of different diameters protruding from its upper part (Fig. 2-17). These cylinders are fitted with pistons that can be loaded with various numbers of heavy weights. If we place one weight on the piston in the narrower

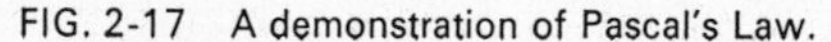

FIG. 2-17 A demonstration of Pascal's Law.

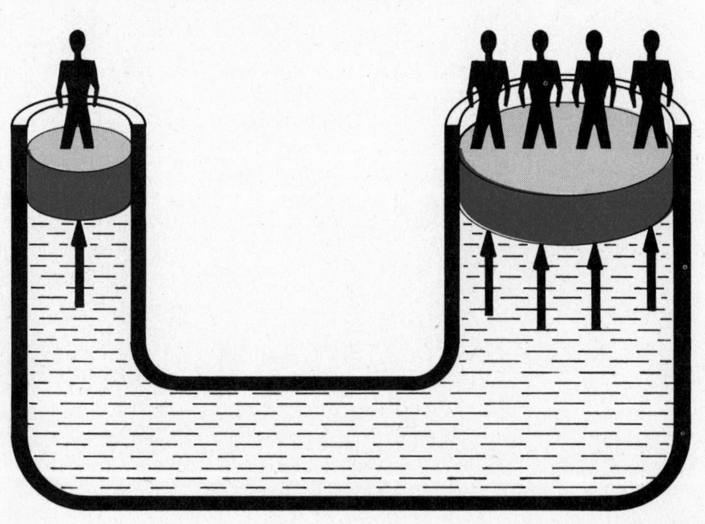

cylinder, it will produce a pressure within the liquid, and this same pressure will be transmitted to all parts of the vessel, including the surface of the larger piston. Since, however, the area of that piston is larger, the total force acting on it will be larger, too. In the example shown in Fig. 2-17, the cylinder on the right is twice as large in diameter, so the areas of the two pistons stand in the ratio of 4 to 1. Therefore, since the total *force* of hydrostatic pressure acting on the right piston will also be four times larger, we must place four weights on it to maintain the equilibrium. The principle described above forms the basis of the hydraulic press in which the pressure created within a liquid by a comparatively small force acting on a small piston exerts a much stronger force on another piston of considerably larger diameter.

2-9 Archimedes' Law

We turn now to the important subject of solids immersed in fluids. Everybody knows that a piece of wood will float in water because its density is smaller than that of water (i.e., it has less weight per unit volume) and that a piece of metal will sink because its density is greater. Although a solid metal object will not be entirely supported by water and so will sink to the bottom, the fact that it is submerged in a fluid will diminish its apparent weight.

Figure 2-18 shows a rectangular block of metal suspended from scales on a string, first in air and then submerged in a container of some liquid whose density is d g/cm^3. We should not be surprised to notice that the scale reading when the block is submerged is less than the reading when

FIG. 2-18 A block weighed in air (A), and weighed when submerged in a tank filled with liquid of density d (B).

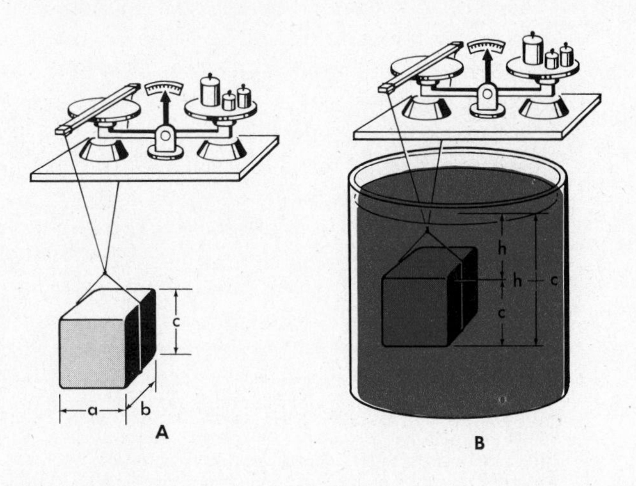

it hangs freely in the air. Apparently the liquid is exerting an upward buoyant force so that the string has to hold less than the actual weight. Let us figure out what we might expect.

In Fig. 2-18B, there is a downward pressure equal to hd on the top of the block. So the total downward force on the top is $F = PA = hd \times ab = hdab$. But there is also an *upward* pressure against the bottom of the block; the bottom is of course submerged deeper than the top and is $h + c$ cm below the water surface. The upward pressure is thus $(h + c)d$, and the upward force is $(h + c)d \times ab = hdab + cdab$. The buoyancy is the difference between the larger upward force and the smaller downward force:

$$\begin{aligned} \text{Buoyancy} &= F_{\text{up}} - F_{\text{down}} \\ &= hdab + cdab - hdab \\ &= cdab = d \times abc \\ &= \text{density of fluid} \times \text{volume of fluid displaced} \\ &= \text{weight of fluid displaced.} \end{aligned}$$

Although we have worked this out for a rectangular block, the result applies to an object of any shape and leads us to

Archimedes' law: Any body, either wholly or partly submerged in a fluid, is buoyed up with a force equal to the weight of the fluid displaced.

This famous law was discovered by Archimedes, who (as the story goes) thought of it while sitting in a bathtub and then, in his excitement, rushed wet and naked through the streets of Alexandria, shouting "*Eureka!*" ("I have found it!") The populace of the city was not impressed by the great discovery, however; undoubtedly they thought he had merely found a missing cake of soap in the tub. Whether the above account is true or not, a more credible story is that Archimedes used this law while checking the authenticity of a golden crown that was suspected of being made of a gold-plated cheap metal rather than pure gold. He had to determine its worth without breaking or scratching its surface so he weighed it when it was suspended on a string in a basin of water and then compared the result with its normal weight. Let us imagine that the crown weighed 2700 g* normally and that when it

* The student should be warned at this point that giving the *weight* of an object in grams is not technically correct. Weight is a *force*—roughly, the pull of gravity on the object; *mass* (as given in Chapter 1) is a measure of the quantity of matter in the object. Different units are used for force and for mass. As we will see in Chapter 3, in many relationships, mass and weight must (repeat: *must*) each be measured in the strictly proper unit. In this example, however, this distinction is not necessary. It will be helpful if, in this section, you take the word "gram" to mean "the weight of a gram."

was submerged in water, it appeared to weigh only 2560 g. The crown's loss of weight was 2700 − 2560 = 140 g and was due to the buoyant effect of the water, which, by Archimedes' law, was just the weight of the water the crown displaced. The density of water is exactly 1 g/cm^3, so the volume of water displaced by the crown equaled 140 cm^3, which was, of course, also the volume of the crown. Since density equals weight divided by volume, we can figure the density of the crown as 2700/140 = 19.3 g/cm^3, which is just the density of gold, and so Archimedes declared the crown-maker to be an honest man.

Archimedes' law applies also, of course, to floating objects only partially submerged in water. In this case, the weight of a floating object, such as a ship, is the same as the weight of the water it displaces.

We can try another example, a bit more complicated. Suppose a 200-g stone weighs 140 g when submerged in water and 150 g when submerged in oil. Because the stone loses 60 g in weight when it is in water, it must displace 60 g of water and its volume is therefore 60 cm^3. This gives a density of 200/60 = 3.33 g/cm^3 for the stone. In the oil, the 60-cm^3 stone loses only 50 g of weight so 60 cm^3 of oil must weigh 50 g and the density of the oil is 50/60 = 0.83 g/cm^3.

Questions

(2-2) **1.** A steel rod weighing 47 lb is supported at its midpoint by a clamp. What upward force does the clamp exert on the rod?

2. A board weighing 35 lb is balanced at its midpoint on a pivot. What upward force does the pivot exert on the board?

3. A mechanic hangs two buckets on the rod of Question 1. One bucket, weighing 80 lb, is hung 3.5 ft from the clamp; the other bucket is placed 5.0 ft on the other side of the clamp, and weighs 56 lb. (a) Is the rod still in balance on the clamp? (i.e., do the two buckets exert equal and opposite torques about the clamp?) (b) What total upward force does the clamp exert on the rod?

4. Two children use the board of Question 2 as a seesaw. A girl weighing 60 lb sits 6 ft from the pivot; a 72-lb boy sits 5 ft from the pivot. (a) Is the seesaw still in balance? (i.e., do the two children exert equal and opposite torques about the pivot?) (b) What total upward force does the pivot exert on the board?

5. Calculate the torque on the rod of Question 3 about the point where the 80-lb bucket is hung. Does $\Sigma \tau = 0$ about this point?

6. Calculate the torque on the plank of Question 4 about the point at which the boy is sitting. Does $\Sigma \tau = 0$ about this point?

(2-3) **7.** An irregular steel bar 18 ft long weighs 200 lb. Its center of gravity is 8 ft from one end. A support is placed under each end. What weight is held by each of the supports?

8. A tapered pole 12 ft long weighs 48 lb, and balances at a point 4 ft from one end. Two men (one at each end) now lift the pole. How much force does each of the men have to exert?

9. At a highway checkpoint, the front wheels of a truck are driven onto platform scales, which read 3600 lb. The truck moves forward so its rear wheels (only) are on the scales, which now read 6200 lb. Where is the center of gravity of the truck? (Its front and rear axles are 12 ft apart.)

10. Each end of a tapered pole 8 ft long is hung from a spring balance. The balances read 6 lb and 15 lb, respectively. Where is the center of gravity of the pole?

11. A tapering bamboo pole is 6 ft long and weighs 10 lb. It is found to balance at a point 2.5 ft from the large end. How much weight must be hung from the small end of the pole to make it balance at its midpoint?

12. A man has caught a large fish that he wishes to weigh. He has two spring scales, each able to weigh up to 10 lb, but the fish weighs more than 10 lb. So he takes a light stick 3 ft long, suspends each end of the stick from one of the scales, and hangs the fish on the stick. The two scales now read 6 lb and 8 lb, respectively. (a) What does the fish weigh? (b) At what point on the stick did he hang the fish? (c) If he had hung the fish from a point on the stick just 1 ft from one of the scales, what would each of the scales have read?

13. Figure 2-Q13 sketches the forces around the foot of a 200-lb man standing on the toes of one foot. (a) What is force W? (b) Assume the downward thrust of the tibia and the upward pull of the gastrocnemius muscle are both vertical. (The fibula contributes little here and is omitted from the sketch.) Calculate T and G.

14. Figure 2-Q14 shows a normal 165-lb man and an obese 270-lb man. The circled sketch to the left is a very schematic diagram of the forces involved in the lower back. W is the part of the body weight supported by the lower back, assumed to be two-thirds of the total, acting at the CG as indicated. V is the compressive force acting on one particular vertebra, and M is the pull of the back muscles, keeping the back from bending under the weight it supports. The details of the attachments of W and M are not shown. Assume M, V, and W are all vertical. Calculate M and V for each man.

(2-4) **15.** Look up the following: cos 19°, sin 58°, tan 27°, sin 37°, tan 11°, cos 71°.

16. Look up the following: sin 23°, tan 87°, cos 60°, tan 30°, sin 60°, cos 43°.

FIG. 2-Q13

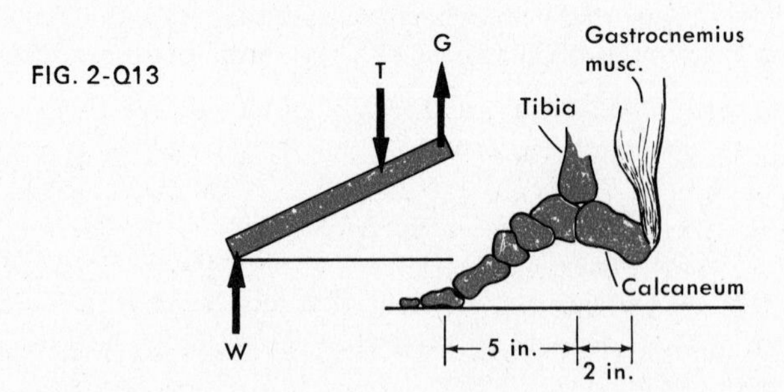

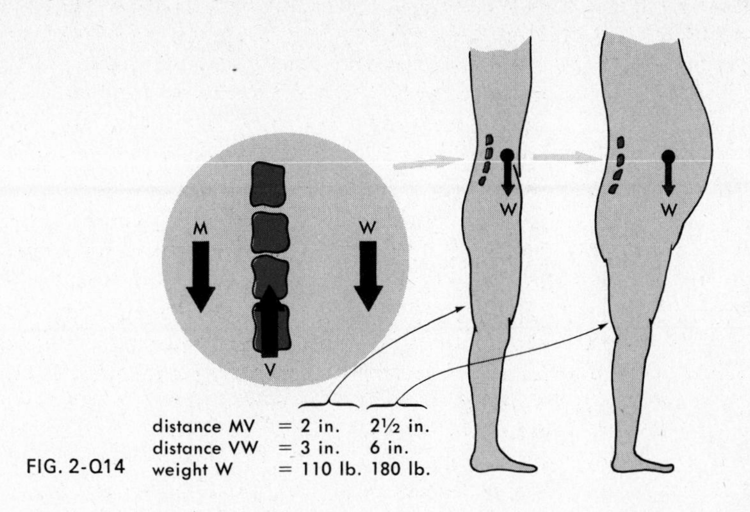

FIG. 2-Q14

17. A vector 30 units long points 30° north of east. Calculate (a) its northerly component and (b) its easterly component.

18. As the crow flies, Smithville is 26° south of west from Jonesville and is 35 mi distant. Smithville is (a) how far south of Jonesville? (b) how far west?

19. Given three vectors: α = 10 units 15° south of east, β = 25 units 37° east of north, γ = 20 units 18° west of south. (a) Add the vectors graphically in at least two different ways, such as $R = \alpha + \beta + \gamma$ and $R = \beta + \gamma + \alpha$. (b) Break α, β, and γ into their N–S and E–W components, and add these components to find the N–S and E–W components of the resultant.

20. Given the following three vectors: A = 5 units north, B = 10 units 30° east of south, and C = 10 units 45° east of north. (a) Choose some convenient scale, and add the three vectors graphically in at least two different ways, such as $R = A + B + C$ and $R = C + A + B$. (b) Break A, B, and C into their N–S and E–W components, and add these components to find the N–S and E–W components of the resultant.

(2-5) **21.** What is the magnitude and direction of R in Question 19?

22. What is the magnitude and direction of R in Question 20?

23. Two tractors are attached to a stump. Because of the terrain, one tractor must pull in a direction 30° north of east, and the other 37° east of south. Assuming each tractor can exert a pull of 5000 lb, in what direction will the stump move, and what total force will be exerted on it?

24. A charged particle moves in a region where it is influenced by two electric fields: Field A exerts on the particle a force of 2.80×10^{-16} newton in a direction 29° east of south; field B, a force of 1.40×10^{-15} newton directed 34° north of east. What is the total force on the particle, and in what direction does it act?

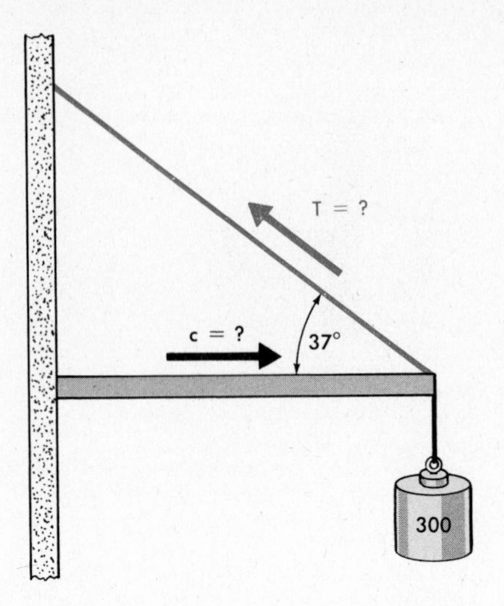

FIG. 2-Q25

25. A horizontal strut of negligible weight is supported by a wire brace as shown in Fig. 2-Q25. The suspended weight is 300 lb, and the wire brace makes an angle of 37° with the strut. What is (a) the tension in the wire? (b) the compressional force on the strut?

26. A 250-lb weight is suspended from a horizontal strut supported by a wire brace, similar to the arrangement of Fig. 2-Q25. The angle between wire brace and strut is 30°, and the weight of the strut is negligible. Calculate (a) the tension in the wire and (b) the compressional force on the strut.

27. A uniform horizontal 10-ft strut (whose center of gravity is therefore at its midpoint) weighs 80 lb. It carries a 120-lb load and is supported by a guy wire, as in Fig. 2-Q27. (a) What is the tension in the guy wire? (b) What are the horizontal and vertical components of the push of the wall against the strut?

28. A uniform 8-ft strut supported by a guy wire supports a 120-lb load in an arrangement similar to that of Fig. 2-Q27. The strut weighs 80 lb, and the load is suspended from a point 5 ft from the wall. The angle between strut and wire is 26°. (a) What is the tension in the wire? (b) What are the horizontal and vertical components of the force exerted by the wall on the end of the strut?

29. Figure 2-Q29 shows a 20-lb weight held in a hand on a horizontal outstretched arm. The arm is supported largely by the pull of the deltoid muscle. Using the dimensions on the sketch, calculate (a) the resulting tension in this muscle and (b) the total reaction S at the shoulder. (The 18-lb load shown represents the weight of the arm itself, acting at the CG of the arm.)

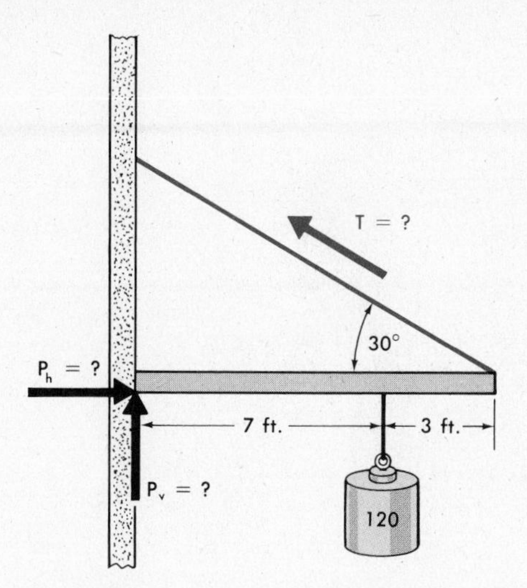

FIG. 2-Q27

30. Figure 2-Q30 shows a man bending over to pick up a 50-lb weight. On the diagram, the additional 40 lb at *S* represents the weight of the head and arms; at *T*, the 80 lb is the weight of the trunk of the body acting at its CG. In this *very* simplified arrangement, calculate (a) the tension in the erector spinae muscles and (b) the total resultant force of the pelvis against the base of the spine. (This problem may suggest why physicians recommend that such loads be picked up by the leg muscles, with the spine erect.)

FIG. 2-Q29

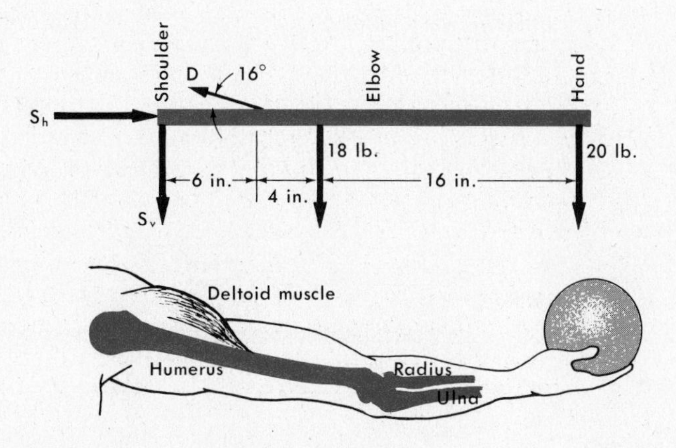

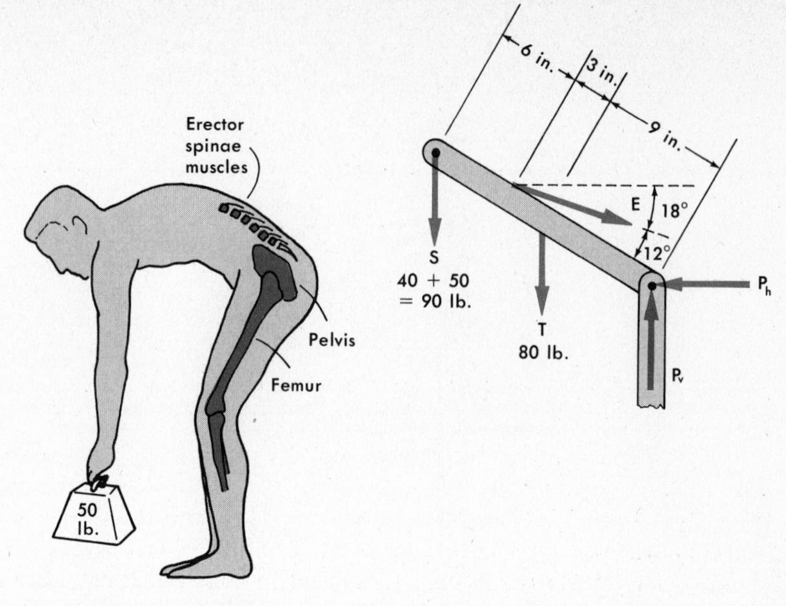

FIG. 2-Q30

31. A body is acted on by three forces, as shown in Fig. 2-Q31. The body is not in equilibrium, but it is possible to add one force that will hold it in equilibrium. What is the magnitude and direction of this force, and where on the body must it be applied?

32. A body is acted on by three forces, as shown in Fig. 2-Q32. The body is not in equilibrium, but it is possible to add one force that will hold it in equilibrium. What is the magnitude and direction of this force, and where on the body must it be applied?

(2-6) 33. A heavy machine is being shoved into position by a jack that applies a horizontal force to it. The machine weighs 7000 lb, and the coefficient of friction between its base and the floor is 0.35. How much force must the jack exert?

FIG. 2-Q31

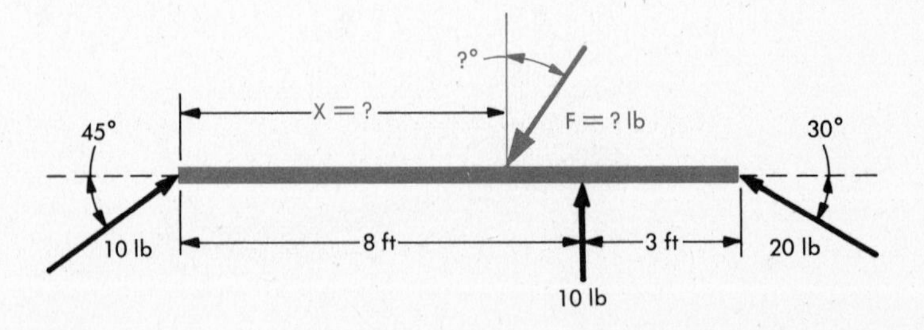

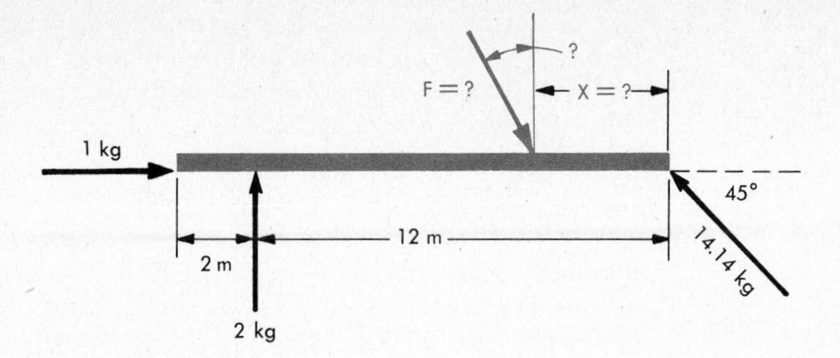

FIG. 2-Q32

34. A box weighing 120 lb requires a horizontal force of 25 lb to drag it across a level floor. What is the coefficient of friction between the box and the floor?

35. Would the experiment with the golf club on the fingers have the same outcome if the shaft were covered with fine sandpaper from the center of gravity to the end of the handle and the other end were smoothly polished? Explain.

36. In the experiment bracketing the center of gravity of the golf club between fingers, would the outcome be the same if one wore a smooth silk glove on the left hand, and a rough leather glove on the right? Explain.

37. A man stands at the midpoint of a 20-ft ladder that leans against a smooth vertical wall. The coefficient of friction between the bottom of the ladder and the floor is 0.20. How far out from the base of the wall may the foot of the ladder be placed, if it is not to slip? (Neglect weight of ladder.)

38. In order to frisk a suspect, the police have him lean forward with his arms horizontal and his hands against a smooth wall. His hands are greasy, so there is no friction between hands and wall. The coefficient of friction between his shoes and the floor is 0.40, and his center of gravity is midway between his shoulders and his feet. What is the minimum angle θ between the suspect's straight body and the floor that can be allowed without the suspect falling?

39. A man slides a 200-lb box along a floor ($\mu = 0.30$) at constant speed, with a push directed 37° down from the horizontal. How hard must he push?

40. A reluctant 120-lb dog ($\mu = 0.50$) is pulled along a path by a rope making an angle of 40° with the horizontal. How strong a pull is required?

(2-7) **41.** As given in the text, the density of water is very nearly 1 g/cm^3. What is its density in lb/ft^3?

42. Knowing the density of water in g/cm^3 and lb/ft^3 from Question 41, calculate the density in lb/ft^3 of the oil and mercury mentioned in Question 46.

43. A piece of iron 6.0 cm × 12.0 cm × 15.0 cm has a mass of 8.42 kg. What is the density of the iron in g/cm^3?

44. The density of copper is 8.90 g/cm^3. What is the mass of a block of copper 4.0 cm × 5.0 cm × 10.0 cm?

45. (a) A bottle weighs 213 g, and has an internal volume of just 800 cm^3. The bottle is filled with oil, and is then found to weigh 861 g. What is the density of the oil? (b) If you did not know the volume of the bottle, what would be an easy and accurate way of finding it?

46. A volumetric flask holds exactly 200.00 cm^3 and weighs 230 g empty. What would it weigh when filled with (a) water? (b) oil, of density 0.82 g/cm^3? (c) mercury, of density 13.6 g/cm^3?

47. A swimming pool 15 ft × 30 ft is 10 ft deep and filled with seawater (density = 1.03 g/cm^3). (a) What is the density of seawater in lb/ft^3? (See Question 41.) (b) What is the pressure on the bottom of the pool? (c) What is the total force on the bottom of the pool?

48. A cylindrical storage tank 4 ft in diameter is filled to a depth of 8 ft with oil whose density is 0.90 g/cm^3. (a) What is the density of the oil in lb/ft^3? (See Question 41.) (b) What is the pressure on the bottom of the tank? (c) What is the total force on the bottom of the tank?

49. The bottom 15 cm of a vertical glass tube 125 cm long is filled with mercury; the tube is then filled to the top with carbon tetrachloride (density 1.59 g/cm^3). What is the pressure at the bottom of the tube?

50. A tank 5 m high is half filled with water and then is filled to the top with oil of density 0.85 g/cm^3. What is the pressure at the bottom of the tank due to these liquids in it?

(2-8) **51.** In a hydraulic jack, the piston that raises the load is 3 inches in diameter; the operating force is applied to a piston 0.5 inch in diameter. How many pounds of force must be applied to raise a load of 6 tons?

52. A hydraulic jack has a piston 0.5 cm in diameter on which a force is applied, and a piston 6 cm in diameter that raises a load. How much force must be applied to lift a load of 2000 kg?

(2-9) **53.** A stone weighing 750 g appears to weigh only 500 g when submerged in water. (a) What is its loss in weight? (b) What is the upward buoyant force on the stone? (c) What weight of water does the stone displace? (d) What is the volume of this amount of water? (e) What is the volume of the stone? (f) What is the density of the stone?

54. A stone weighing 250 g appears to weigh only 150 g when submerged in water. (a) What is the volume of the stone? (b) What is the density of the stone?

55. A 1520-g metal bolt appears to weigh 1202 g when submerged in carbon tetrachloride. (See Question 49.) (a) What is the volume of the bolt? (b) What is its density?

56. A metal block weighing 3000 g appears to weigh only 2600 g when submerged in oil of density 0.80 g/cm^3. (a) What is the volume of the block? (b) What is its density?

57. In order to find the density of some oil, a piece of metal is weighed in air (300 g), then in water (200 g), and then in the oil (220 g). (a) What is the volume of the piece of metal? (b) What is the density of the metal? (c) What is the density of the oil?

58. In order to find the density of some acid, we first weigh a glass stopper in air (250 g), then in water (150 g), and then in the acid (125 g). (a) What is the volume of the stopper? (b) What is the density of the stopper? (c) What is the density of the acid?

59. A block of wood floats on oil (density 0.80 g/cm^3) with 60 percent of its volume submerged. What is the density of the wood?

60. A block of wood (density 0.60 g/cm^3) floats on water. What fraction of the volume of the block is submerged?

61. A boat loaded with logs of light wood (density less than 1 g/cm^3) floats in a swimming pool. A man in the boat throws the logs overboard. What happens to the water level of the pool? Explain.

62. If a boat loaded with rocks floats in the middle of a swimming pool and a man in the boat throws the rocks overboard, what will happen to the water level of the pool? Explain. (At a scientific meeting, this question was put to Dr. Gamow, the physicist J. Robert Oppenheimer, and Nobel prize-winner Felix Bloch. All three of them, not thinking too carefully, gave the wrong answer!)

63. Air is a fluid (as all gases are), and Archimedes' principle applies to bodies in air as well as to those submerged in liquids. A man needs a very precise value for the weight of a plastic cube exactly 10 cm on a side. On an accurate balance, he balances the cube on one side against 2980.2 g of brass weights (density 8.60 g/cm^3) on the other side. What is the weight of the plastic? (Take the density of air as 1.30×10^{-3} g/cm^3.)

64. A balloon with a volume of 3000 m^3 is filled with helium, which has a density of 1.80×10^{-4} g/cm^3. Assume the density of air to be 1.30×10^{-3} g/cm^3. (a) Calculate the pull on the mooring rope if the empty and collapsed balloon plus all its equipment weighs 1600 kg. (b) What would the pull be if the balloon were filled with hydrogen instead of helium? (The density of hydrogen is about 9×10^{-5} g/cm^3.)

3

Bodies in Motion

3-1 The Measurement of Motion

In order to investigate the behavior of moving bodies, we must first analyze the general idea of motion itself. What properties or characteristics of motion can we measure or calculate to help us describe the behavior of a moving body? The *distance* traveled and the *time* required to travel the distance are two obvious factors, and the idea of *average speed* is one that our own traveling has made familiar to all of us: If it takes 5 hours to drive to a city 200 miles away, our average speed is 200 miles/5 hours = 40 mi/hr. Without thinking of it, we have made use of the relationship

$$v_{av} = \frac{d}{t},$$

where the letter v stands for speed. Multiplying both sides of this equation by t, we get

$$d = v_{av}t.$$

If we average 30 mi/hr, in 4 hr we travel a distance of 30 mi/hr × 4 hr = 120 mi.

It is obvious that v_{av} alone tells us nothing about the speed at any instant during the time interval we have considered. A car slows, stops, and speeds up again many times during a four-hour trip and we must have some way of taking into account the way in which the speed of a body changes if we want to describe its motion properly. In order to do this, we make use of the idea of *acceleration*, which is defined as *the rate of change of velocity*.

So far we have been talking about the *speed* of a moving body—why do we instead use the word *velocity* when we define acceleration? In ordinary conversation, we generally use "speed" and "velocity" interchangeably, but the physicist gives somewhat different meanings to the two words. *Speed refers only to the rate at which distance is covered, without regard for direction; velocity includes both speed and direction.* A car going north at 40 mi/hr has the same *speed* as a car going east at 40 mi/hr, but their *velocities* are different because the directions are different. When we need to make this distinction, *speed is a scalar*, and *velocity is a vector*.

Later on, when we discuss rotating bodies and bodies moving in curved paths, we will see that a change in direction represents as real an acceleration as a change in speed; for this reason, acceleration is defined as the rate of change of *velocity*, which includes both. For the present, however, we will not consider changes in direction, so that our velocity changes will be only changes in speed.

If we step on the gas pedal of a car, the speed of the car will change—for example, from 5 mi/hr to 40 mi/hr in 10 s. Put into mathematical form, our definition of acceleration is

$$a = \frac{\Delta v}{t}.$$

(The Greek capital letter Δ—*delta*—is the mathematician's abbreviation for "the change in.")

Substituting our figures into this equation, we get

$$a = \frac{40 \text{ mi/hr} - 5 \text{ mi/hr}}{10 \text{ s}} = \frac{35 \text{ mi/hr}}{10 \text{ s}} = 3.5 \text{ (mi/hr)/s}$$

which is to say that in each second the speed has increased by 3.5 mi/hr. These are somewhat clumsy (although perfectly proper) units and acceleration is more often given in ft/s/s, abbreviated ft/s^2 (or cm/s^2, or m/s^2).

If a body starts from rest, moving with a constant acceleration of a ft/s^2, at the end of 1 s it will have a velocity of a ft/s; at the end of

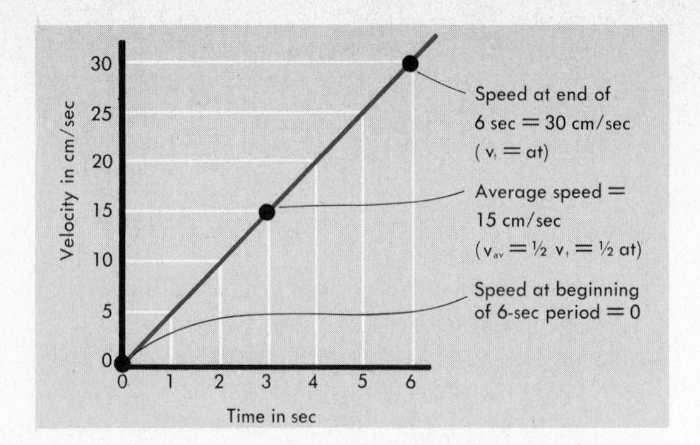

FIG. 3-1 Time vs. velocity for uniformly accelerated motion.

2 s its velocity is $2a$ ft/s, and so on. We can write this in a general form:

$$v_t = at,$$

where v_t is the velocity at the end of t seconds.

Figure 3-1 is a graph of this relationship for a numerical example. Velocity (or speed) has been plotted against time for a uniformly accelerated motion, starting from rest—that is, $v = 0$ at $t = 0$. Since the increase in velocity is the same for each second, the graph is a straight line. At $t = 6$ s, v is 30 cm/s, and we have, from $v_t = at$,

$$30 \text{ cm/s} = a \times 6 \text{ s}$$

$$a = \frac{30 \text{ cm/s}}{6 \text{ s}} = 5 \text{ cm/s}^2.$$

In a straight-line relationship such as this, the *average* velocity v_{av} during any time interval is equal to the velocity at the middle of the interval. This average velocity (starting from rest) is just half the final velocity. On the graph, the v_{av} is 15 cm/s, just half of the final velocity of 30 cm/s, which we know to be $v_t = at$. For this case, then, $v_{av} = \frac{1}{2}at$.

To get the distance d that the body will have moved, we need only multiply this *average* speed by the time. (A car averaging 40 mi/hr will go 200 mi in 5 hr.) We can now write

$$d = v_{av} \times \text{time} = \tfrac{1}{2}at \times t = \tfrac{1}{2}at^2.$$

By eliminating the time from the two equations we already have, we can get a third relationship. We know $v_t = at$, and solving for t we get

$t = v_t/a$ so that $t^2 = v_t^2/a^2$. Now we can substitute this value for t^2 in the equation $d = \frac{1}{2}at^2$:

$$d = \frac{1}{2} \times a \times \frac{v_t^2}{a^2} = \frac{v_t^2}{2a}$$

and

$$v_t^2 = 2ad.$$

So far, we have developed equations of motion that can be used *only when the accelerated motion starts from rest*. These will be much more useful if we can extend them to include bodies that are originally moving with a velocity we can call v_o.

The equation $v_t = at$ tells us how much the speed has increased because of the acceleration. This amount needs only to be added to the original speed, and we have

$$v_t = v_o + at$$

for this more general case.

The next equation, $d = \frac{1}{2}at^2$, gives the distance covered because of the acceleration; if the body were originally moving at the speed v_o, in t seconds it would have gone a distance $v_o t$ without any acceleration. By adding these together, we find

$$d = v_o t + \tfrac{1}{2}at^2.$$

In a similar manner, our last equation becomes

$$v_t^2 = v_o^2 + 2ad.$$

Let us tabulate these equations, with a comment on each:

1. $v_t = v_o + at$ (does not include the distance)
2. $d = v_o t + \frac{1}{2}at^2$ (does not include the final velocity)
3. $v_t^2 = v_o^2 + 2ad$ (does not include the time).

Which one (or ones) we want to use will depend on what data we have given in a problem and what answer we are trying to find. For example, what is the acceleration of a car that goes from a speed of 20 ft/s to 80 ft/s in 15 s? We know v_o, v_t, and t; we want to find a. The first equation includes these items so we write

$$80 = 20 + 15a$$

$$15a = 80 - 20 = 60$$

$$a = \frac{60}{15} = 4 \text{ ft/s}^2.$$

We might now want to find out how far the car moves during this period of acceleration. Since the acceleration is now known, we can use the third equation:

$$(80)^2 = (20)^2 + 2 \times 4 \times d$$

$$8d = 6400 - 400 = 6000$$

$$d = \frac{6000}{8} = 750 \text{ ft.}$$

(Can you get this same answer in another way?)

In this example, the acceleration, the velocity, and the distance have all been in the same direction. Often, however, this is not the case, and we must carefully watch + and − signs. Consider a car traveling north at 30 m/s. The driver applies the brake and reduces his speed by 5 m/s each second. This rate of reducing speed is *deceleration*, which subtracts from his speed rather than adds to it. It is actually acceleration toward the south. If we choose + to indicate vectors pointing north, the acceleration will be negative: $a = -5 \text{ m/s}^2$. To find how long it will take him to stop, we may use our original equation

$$v_t = v_o + at.$$

Since v_t, the final speed, is zero, we find

$$0 = +30 - 5t$$

or

$$t = 6 \text{ s.}$$

If we had chosen south to be our positive direction, we would have written

$$0 = -30 + 5t$$

and

$$t = 6 \text{ s.}$$

3-2 The Cause of Motion

Even a baby soon learns that there is some sort of a direct connection between force and motion—a push or a pull will often cause things to move. Exactly *how* force and motion are related, however, is not nearly so obvious. In fact, the basic ideas concerning motion itself were not really investigated until the beginning of the seventeenth century. Then the Italian physicist Galileo analyzed the relationships among distance, speed, and time, somewhat as we have done in the preceding section; he was probably the first man to understand the concept of acceleration and to appreciate its importance in the study of motion.

The year Galileo died in Italy (1642), the great physicist and mathematician Isaac Newton was born in England. Newton built on the foundations so well laid by Galileo, and his further studies led him to three fundamental laws of motion.

Newton's first law of motion: A stationary body will remain motionless, and a moving body will continue to move in the same direction with unchanging speed unless it is acted on by some unbalanced force.

This law is an expression of the property of matter called *inertia*, which describes its resistance to having its velocity changed in any way. The law also recognizes that the velocity of a body *can* be changed, but only by the application of a net force. Let us return to the book on the desk, and give it a slight sidewise push. If the push is a gentle one, the book does not move. We know why it does not; our push is opposed, or balanced, by the equal and opposite force of friction. Only if our push exceeds the opposing force of friction does the book begin to move.

The situation is less obvious when we push the book across the desk with constant speed in a straight line—in other words, with an unchanging velocity. In this case also, the force of friction exactly balances, or opposes, the force of the push, so that the net force is zero. If we push harder, we exceed the force of friction and the book speeds up; if we stop pushing, the unopposed force of friction quickly brings the book to a halt.

All this is well and good, but it does not bring us to grips with the main problem: Exactly what happens when the forces on a body do *not* add up vectorially to zero? For many centuries, scientists and philosophers had missed the point of this question, because they tried in some uncertain way to relate the force and the *velocity* that it presumably caused. Galileo almost grasped the point, and Newton grasped it clearly; the direct relationship is not between force and velocity, but between force and *acceleration*.

Newton's second law of motion: The acceleration given to a body by a force applied to it is directly proportional to the force, and is in the same direction as the force. If the same force is applied to bodies of different masses, the accelerations produced will be inversely proportional to their masses.

3-3 Mass and Weight

Before we put Newton's second law into the form of an equation, we should pause and give some thought to the meaning of *mass*, which is such an important part of the law. Matter of any kind has two universal characteristics: It is pulled by the force of gravity—that is, it has *weight*—and it resists being accelerated—that is, it has *inertia*. We can use either of these characteristics to measure the mass of a body.

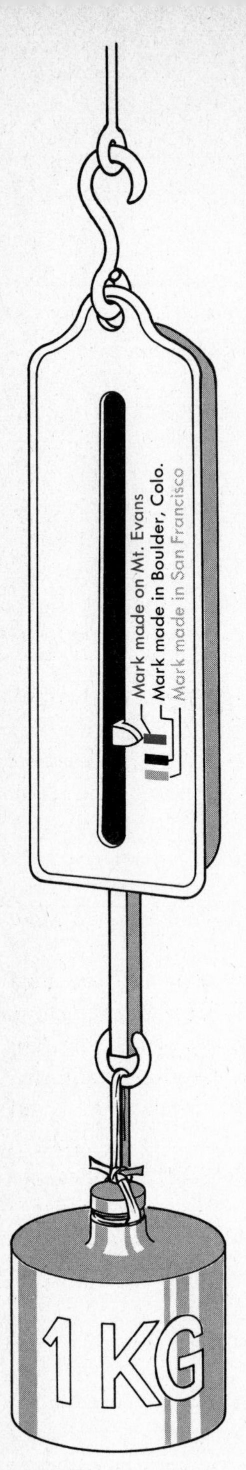

FIG. 3-2 A kilogram weighs different amounts in different locations.

Weight is something that can be measured directly by means of a spring balance. For example, let us take a standard kilogram (a piece of metal that the manufacturer has made to be an accurate duplicate of the standard kilogram in the vault at Sèvres) and hang it from a spring balance. Before we mark the position of the indicator, however, we should pause and consider. Suppose we make a mark when we are in Boulder, Colorado; then we go to San Francisco, California—here (because we are at a lower altitude and are hence closer to the center of the earth) the pull of gravity is stronger, the kilogram mass actually weighs more, and the mark will be in a slightly different place. Conversely, if we take the spring balance and standard kilogram to the high peak of Mount Evans in the Colorado Rockies, the weight is less, and we shall have to make a third mark (Fig. 3-2).

Which of these three marks should we choose to be permanently engraved? What does the manufacturer of spring balances do? Since most of the population lives not far above sea level, he probably marks it a little above the San Francisco mark and hopes for the best. The ordinary spring balance is rather crude and does not pretend to any great accuracy. If it is a little off for the customers in Death Valley or in Boulder, no harm is done, as the user does not expect precision anyway.

The point of this discussion is that the kilogram and the gram are units of *mass* and that their weights vary from place to place. If we use another, and more common, sort of balance—a platform or an analytical balance (Fig. 3-3)—we find that this confusion disappears. Select a stone that will exactly balance the kilogram in Boulder. It will also exactly balance the kilogram in San Francisco or on Mount Evans, because, although the pull of gravity varies from place to place, at any one location the pulls on the kilogram and on the stone will always remain equal. What we can say from "weighing" of this sort is that the mass of the stone is the same as the mass of the kilogram—or, more briefly, that the *mass* of the stone is one kilogram. What the actual *weight* of the stone is we do not know and cannot compute until we have some way of taking into account exactly how strong the pull of gravity is at the particular location we are interested in.

In modern commerce and engineering, the pound (unlike the gram and the kilogram) is *not* used as a unit of mass. Instead, its weight at 45° latitude and at sea level is used as a unit of *force*. With this definition in mind, we can successfully calibrate our spring balance. Take a standard pound to Calais, Maine, or Portland, Oregon (both of these cities nearly fulfill the requirements for latitude and altitude), hang it on the balance and you can confidently mark the indicator position as "1 lb." Now if we take our apparatus to Boulder or Mount Evans, we need not be surprised to find that a standard pound does not weigh a pound. Actually, the pull of gravity is smaller, and its weight *is* less than a pound.

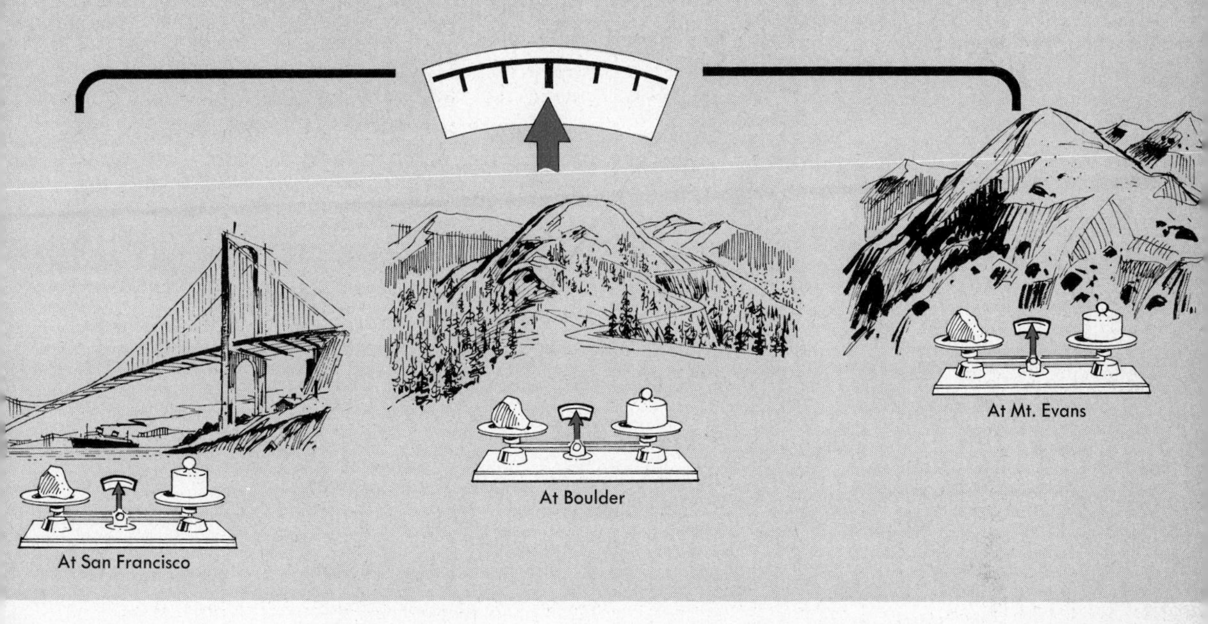

FIG. 3-3 The mass of an object, determined by comparison with a standard mass, is the same in all locations.

Suppose we were set down in some unknown location and were handed a stone from the roadside. With a platform balance and a set of calibrated masses (incorrectly called "weights," a misnomer we must learn to live with), we could accurately determine the mass of the stone in grams or kilograms but would be unable to measure its weight. On the other hand, an accurately calibrated spring balance would tell us that the weight of the stone was exactly so many pounds but would leave us ignorant about its mass. We need some procedure for relating mass and weight, and for this we must turn to the other property all material objects have—inertia.

Newton's second law states that (for equal masses) acceleration is proportional to force, and (for equal forces) acceleration is inversely proportional to mass. We can write this algebraically as

$$a = k \times \frac{F}{m}.$$

We have defined length and time units with which to measure a, and mass units (the gram and the kilogram) with which to measure m in the metric systems. We have a force unit (the pound) in the British system. So far we have no force unit in the metric systems, and no mass unit in the British system. It would seem to be very convenient if these missing units were defined in such a way as to make the proportionality constant k equal to 1 in Newton's second law. It also seemed so, to the nineteenth-century scientists who then defined them. When we use these units, we may write $a = F/m$, or more usually

$$F = ma.$$

These specially defined units are indicated in boldface type:

System	*Force*	*Mass*	*Acceleration*
CGS	**dyne**	gram	cm/s^2
MKS	**newton**	kilogram	m/s^2
British	pound	**slug**	ft/s^2

One dyne is the force necessary to give a mass of one gram an acceleration of one centimeter/second^2.

One newton (N) is the force necessary to give a mass of one kilogram an acceleration of one meter/second^2.

One slug is the mass to which a force of one pound will give an acceleration of one foot/second^2.

As an example of the use of these units, let us find what net force is needed to bring to a stop, in 2 min, a 200-metric ton* train traveling at 24 m/s. If we arbitrarily select the forward direction to be positive, the necessary Δv is $0 - 24 = -24$ m/s. The required acceleration is $\Delta v/t$ so $a = -24/120 = -0.20$ m/s^2. The mass (in the MKS system we are using) must be in kilograms; $m = 200{,}000 = 2 \times 10^5$ kg. Then

$$\begin{aligned} F &= ma \\ &= 2 \times 10^5 \times (-0.20) = -4 \times 10^4 \text{ N}. \end{aligned}$$

The negative sign on the force in the answer shows it to be toward the rear—opposite to the direction we selected to be positive.

Now consider a car whose mass is 100 slugs. If it can exert a braking force of 1600 lb, how many feet will it travel in coming to a stop from a speed of 60 ft/s? We can easily calculate the distance if we know the acceleration; since we know both the mass and the force, the acceleration follows simply from Newton's second law:

$$F = ma; \qquad a = \frac{F}{m} = \frac{1600}{100} = 16 \text{ ft/s}^2.$$

* A metric ton is 1000 kg, about 10 percent more than our English ton.

Then

$$v^2 = v_0^2 + 2ad$$

$$0 = (60)^2 + 2 \times 16 \times d; \qquad d = \frac{-3600}{32} - 113 \text{ ft.}$$

(We have implicitly made the acceleration of braking positive when we made the force that caused it positive in the equation above. This put the positive direction toward the rear of the car, and the -113 ft thus indicates that the car will travel *forward* 113 ft before it stops.)

Again, we might deal with an electric field that exerts a force of 1.8×10^{-10} dyne on an electron, whose mass is 9×10^{-28} g. Under these conditions, what is the acceleration of the electron?

$$F = ma; \qquad a = \frac{F}{m} = \frac{1.8 \times 10^{-10}}{9 \times 10^{-28}} = 2 \times 10^{17} \text{ cm/s}^2.$$

3-4 Weight and Falling Bodies

The Leaning Tower of Pisa, besides being one of the architectural wonders of the world, is also inseparably connected with the history of physics because of the part it played in an experiment that was alleged to have been performed more than three centuries ago by the famous Italian scientist Galileo. From the upper platform of the Tower, the story has it, Galileo simultaneously released two spheres, a heavy one made of iron and a lighter one made of wood (Fig. 3-4). In spite of a great difference in weight, both spheres dropped side by side and hit the ground at almost the same moment.

Is this experimental result of Galileo's what we should expect? Assume that the two balls have masses of 10 kg and 1 kg, respectively. Then the weight of the iron ball is also just 10 times as great as the weight of the wooden ball. The weights of the balls are the forces that cause them to accelerate downward, so we come to the conclusion that although the iron ball is 10 times more massive, the accelerating force is also 10 times larger, so that the two balls would have the same acceleration and would therefore travel downward side by side.

Let us follow the idea of falling bodies a little further and measure the downward acceleration of a kilogram mass dropped in, say, Eagle City, Alaska. Although there are other more precise ways of doing it, we could in principle determine this acceleration by accurately measuring the time needed for any massive body to fall any given distance. Then, from $d = \frac{1}{2}at^2$, we could calculate the acceleration. This downward acceleration of a freely falling body (universally referred to by the letter g) has been measured in Alaska and found to be 982.18 cm/s^2, or 9.8218 m/s^2. Assume we have dropped a mass of exactly 1 kg—a force of 1 N would give it an acceleration of 1 m/s^2; to give it an acceleration of 9.8218 m/s^2, a force of 9.8218 N would be required.

FIG. 3-4 Galileo's experiment with falling bodies.

We therefore see that in Eagle City the kilogram must weigh just 9.8218 N. Repeating the same experiment in Panama City, Canal Zone, we would find that at this location $g = 9.7824\ \text{m/s}^2$ and could conclude that here the kilogram weighs 9.7824 N. In general, we can say, for any body anywhere,

$$\text{weight} = \text{mass} \times \text{acceleration due to gravity} = m \times g.$$

(We must remember, of course, to keep the systems of units straight: newtons, kilograms, and meters/second^2; dynes, grams, and centimeters/second^2; pounds, slugs, and feet/second^2.)

The study of falling objects is an excellent way to learn about the basic laws of motion. Since ordinary free fall is much too fast for convenient observation, it is helpful to slow the fall with an ingenious device known as Atwood's machine (Fig. 3-5). This consists essentially of a long vertical support with a light, nearly frictionless pulley on top and a collection of calibrated metal masses that can be attached in any desired quantities to the ends of the string that runs over the pulley. With equipment as simple or as fancy as our resources allow (an ordinary two-meter stick and a stopwatch will give reasonably good results), we can measure d and t, and so investigate the relationship among force, mass, and acceleration.

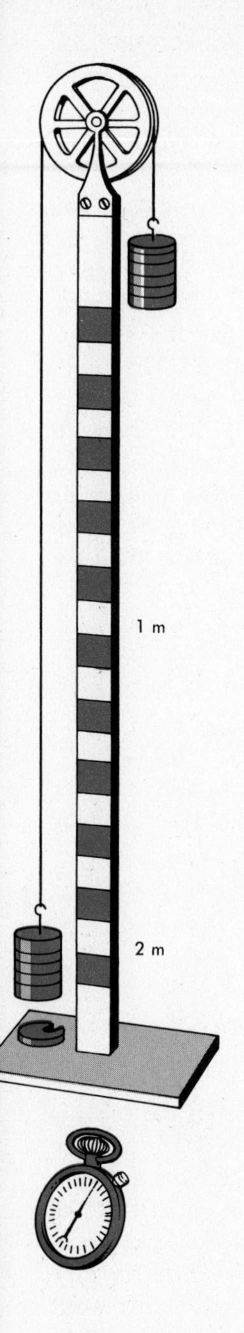

FIG. 3-5 An Atwood machine for "diluting" gravity.

Consider an Atwood machine on which we have placed exactly 900 g on one side, and 800 g on the other. Obviously the 900-g side will start to descend as soon as we release it. What will be its downward acceleration? If 800 g were on both sides, the weights would be balanced, and there would be no acceleration; the accelerating force is the unbalanced weight of the extra 100 g on the heavy side.

Our trouble now is that we do not know just what 100 g weighs. If this book were to be used only in Eagle City, Alaska, we could say at once that 100 g has a weight of mg dynes $= 100 \times 982.18 = 98{,}218$ dynes. But in the Canal Zone, 100 g weighs only 97,824 dynes; and so on. Let us reconcile ourselves to an error of a few tenths of a percent and assume that $g = 980\ \text{cm/s}^2$ or $9.8\ \text{m/s}^2$ or $32\ \text{ft/s}^2$ everywhere on the surface of the earth.

So for the Atwood machine the accelerating force is $100 \times 980 = 98{,}000$ dynes, or 0.98 N. This force serves to accelerate *both* masses—a total of $900 + 800 = 1700$ g, or 1.700 kg, and

$$a = \frac{F}{m} = \frac{98{,}000}{1700} = 57.6\ \text{cm/s}^2$$

or

$$\frac{0.980}{1.700} = 0.576\ \text{m/s}^2.$$

With this knowledge of a, we can go on to determine distances, times, speeds, etc., in any way that is required.

Let us ask another question about the Atwood machine as it was described above. What is the tension in the string? If we prevent any motion by holding the 900-g mass, the string will have to support only the weight of the 800-g mass, which will cause a tension of $800 \times 980 = 784{,}000$ dynes. However, if we hold the 800-g mass, the tension in the string will be the weight of the 900-g mass, or 882,000 dynes. Is it safe to assume that when the masses on the machine are accelerating freely, the tension will be somewhere between these two values? We can solve this problem by concentrating on one of the masses—the 800-g one, say—and putting the rest of the machine entirely out of mind. (This is called *isolating* one of the factors of a complex problem. See Fig. 3-6.)

We see only a mass of 800 g, which has an upward acceleration of 57.6 cm/s^2. The upward acceleration must be caused by a net upward force of $F = ma = 800 \times 57.6 = 46{,}080$ dynes. This net force is the resultant of the two forces acting on the mass: (1) the tension in the string, which is T dynes upward and (2) the weight of the mass, which is 784,000 dynes downward. Since T must obviously be greater than the weight, we can write

$$T - mg = 46{,}080$$
$$T = 46{,}080 + mg$$
$$T = 46{,}080 + 784{,}000 = 830{,}000 \text{ dynes.*}$$

This answer can be checked by isolating the 900-g mass in the same way. Since it is accelerating downward, its weight is greater than the tension T, and we have

$$mg - T = ma$$
$$900 \times 980 - T = 900 \times 57.6$$
$$T = 900 \times 980 - 900 \times 57.6$$
$$T = 882{,}000 - 51{,}800 = 830{,}000 \text{ dynes.}$$

3-5 Inclined Planes

In Fig. 3-7, a boy is shown coasting down a 20° slope. If the snow is reasonably dirty, we may assume the coefficient of friction to be 0.25. In this example, the motion must be along the slope: the boy's acceleration and the force that causes it must also be along, or parallel to, the slope. But the pull of gravity W is stubbornly vertical. This can easily be taken care of by resolving W into components parallel to the slope (F_{par}) and perpendicular to it (F_{perp}). Component F_{par} will act to accel-

* Since our value of g (980 cm/s^2) is somewhat uncertain in the third figure, we are certainly not justified in retaining more than three significant figures in our answer. It has accordingly been rounded off to 830,000.

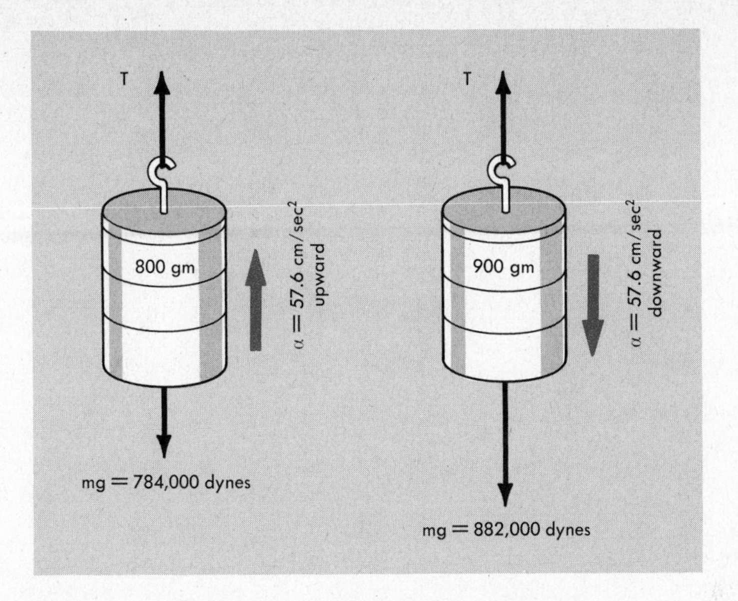

FIG. 3-6 Isolating each of the "weights" on the Atwood machine.

erate the sled downhill; F_{perp}, being perpendicular to the direction of motion, can itself have no effect on the motion. It is, however, the force normal to the surface which presses sled and snow together; it is exactly what we need in order to figure the retarding force of friction (F_{fr}). So let us calculate the net force accelerating the sled downhill:

$$\begin{aligned} F_{\text{net}} &= F_{\text{par}} - F_{\text{fr}} = F_{\text{par}} - \mu F_{\text{perp}} \\ &= W \sin 20° - 0.25 \times W \cos 20° \\ &= W \times 0.342 - 0.25 \times W \times 0.940 = 0.107\ W. \end{aligned}$$

FIG. 3-7 Resolution of the pull of gravity into components perpendicular and parallel to the direction of motion.

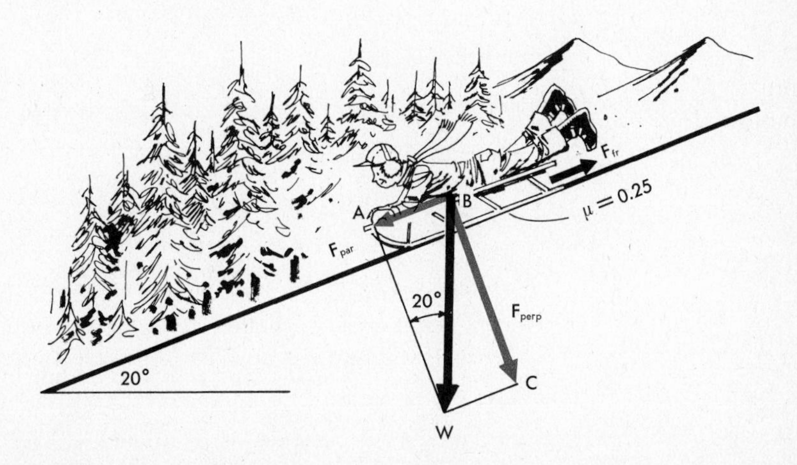

The mass of the sled is not given; it can be assumed to be m in any system we choose. The weight W will then be mg in this same system, and we have

$$F = ma$$

$$0.107\ mg = ma.$$

The mass m cancels out, showing that the results would be the same for a sled of any mass, and then

$$a = 0.107\ g.$$

The same result could be reached by reasoning that since a force equal to the weight of a body gives it an acceleration g, then a force of 0.107 times its weight would give it an acceleration of 0.107 g in any system of units.

3-6 Projectile Motion

Because all unsupported bodies have a constant downward acceleration produced by the pull of gravity, the equations of uniformly accelerated motion can be applied to falling bodies, to rocks thrown into the air, or to golf balls and bullets. As a simple case, consider a boy who stands on a bridge 10 m above the water and throws a stone nearly vertically upward with a speed of 20 m/s. How high will the stone rise? We need to pause a minute to think of a piece of information not explicitly stated in the problem—at the height of its rise the stone will momentarily be motionless. The question can now be rephrased: At what distance d will the final velocity v_t be zero?

We are dealing with up-and-down motion; let us distinguish these directions by calling upward "+" and downward "−". This gives us that $v_o = +20$ m/s, because v_o is upward. The acceleration $a = -9.8\ \text{m/s}^2$, since the acceleration of gravity is downward. Then, from $v_t^2 = v_o^2 + 2ad$, we have

$$(0)^2 = (20)^2 + 2(-9.8)d$$

from which

$$d = \frac{-400}{-19.6} = +20.4\ \text{m}.$$

The positive value tells us that this distance is *above* the point from which the stone was thrown, because we have chosen the upward direction to be +.

We might further ask how long it will be before the stone hits the water. When it does, it will be 10 m below the bridge, so that d will then be −10. (The distance d does *not* represent the total distance traveled, but is the distance of the stone from the point of beginning; as it passes the boy on the way down, d is at that instant zero.) For this calculation, we may use the second equation and find

$$-10 = 20t - \tfrac{1}{2} \times 9.8t^2$$

or

$$4.9t^2 - 20t - 10 = 0$$

from which

$$t = -0.45 \text{ s} \quad \text{or} \quad +4.53 \text{ s}.$$

The latter answer, happening *after* the stone was thrown, is obviously the one we want.

The choice of + and − was perfectly arbitrary. We would have come out with the same answers if the choice had been reversed. It would be a good idea to rework this same example, using upward as −, and downward as +.

Most projectiles are not considerate enough to move up and down in a strictly vertical direction. What would happen if the boy on the bridge had thrown the stone at 20 m/s exactly horizontally, instead of vertically? If we neglect the frictional resistance of the air, the only force acting on the stone is the pull of gravity, and so its only acceleration is 9.8 m/s^2 downward. This downward acceleration, since it has no horizontal component, cannot have any effect on the horizontal motion of the stone, which will continue at 20 m/s until it strikes the water. Similarly, this horizontal component of the motion will have no effect on the gravitationally accelerated up-and-down component.

Figure 3-8 shows a ball being thrown exactly horizontally at the same instant that another ball is dropped. Since the initial velocity of the thrown ball has no vertical component, its downward motion is the same as that of the dropped ball. The horizontal motion proceeds unchanged by the downward acceleration, so that the actual trajectory is the combination of the constant horizontal component and the accelerated downward component. Projectile problems can all be handled by considering the horizontal and vertical components of the motion separately and independently. As an example, Fig. 3-9 shows a golfer who has given a ball an initial velocity of 36 m/s, 30° above the horizontal, from a tee that is 15 m below the surrounding terrain. How far will his drive carry? (This is the horizontal distance x on the drawing.) We can start by breaking v into its horizontal and vertical components: $v_h = 36 \cos 30° = 31.2$ m/s, and $v_v = 36 \sin 30° = 18.0$ m/s. The vertical part of the ball's actual flight, $OBAC$, can be duplicated by the path $OB'A'C'$ of an imaginary ball thrown upward with a speed of 18 m/s. How long will it take this imaginary ball to reach C'? When it

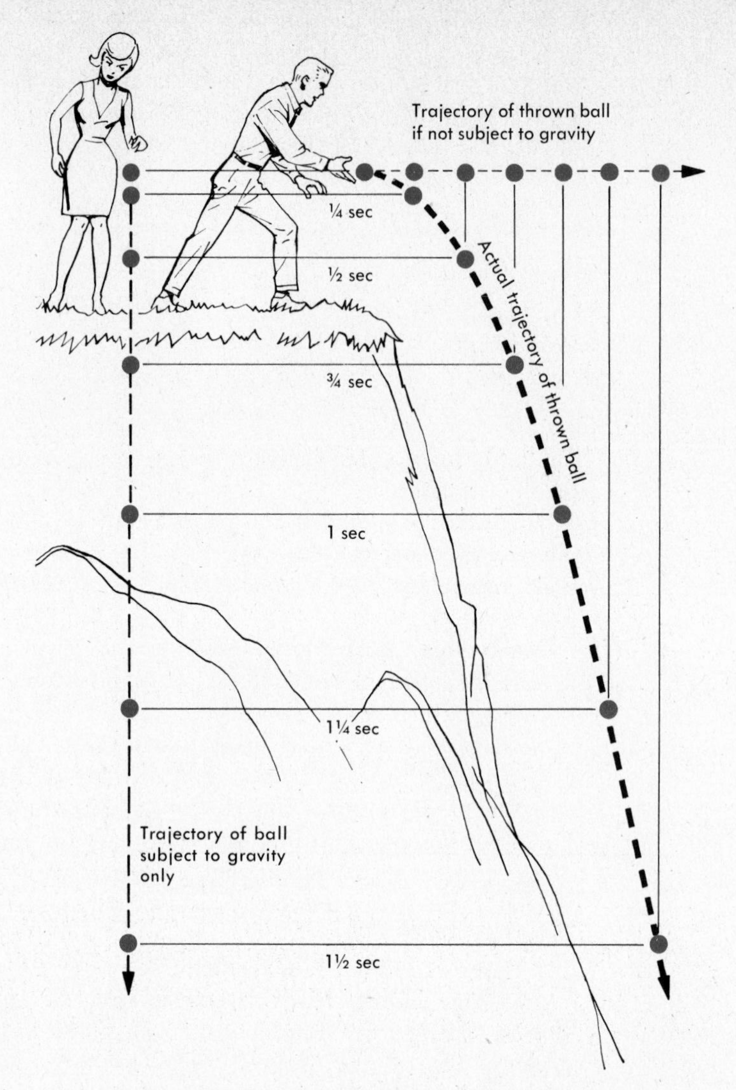

FIG. 3-8 Trajectories of ball when dropped, and when thrown horizontally.

does, the actual ball will be at C and will strike the ground. Dealing now with only the vertical component of motion we can write

$$d = v_0 t + \tfrac{1}{2}at^2$$

$$15 = 18t - 4.9t^2$$

$$4.9t^2 - 18t + 15 = 0,$$

from which

$$t = \frac{18 \pm \sqrt{(18)^2 - 4 \times 4.9 \times 15}}{2 \times 4.9}$$

$$= 1.28 \text{ s} \quad \text{or} \quad 2.40 \text{ s}.$$

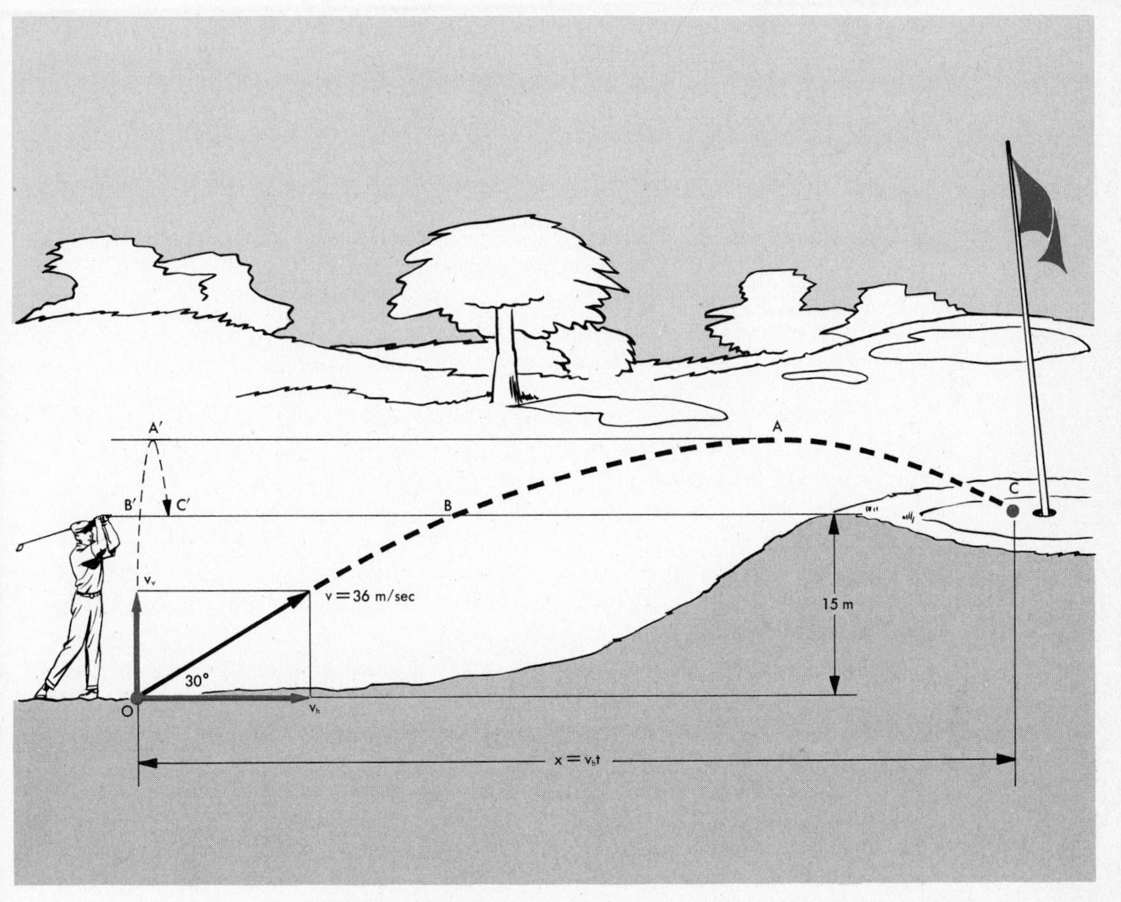

FIG. 3-9 Trajectory of a projectile launched with a velocity having both horizontal and vertical components.

The answer 1.28 s gives a time at which the ball is at the required level, 15 m above the tee, but it refers to point *B*, on the way up; the answer we need is 2.40 s, the time at which the ball reaches *C*. During all this time, v_h will have been continuing on at 31.2 m/s, and the ball will have traveled the horizontal distance

$$x = 2.40 \times 31.2 = 74.9 \text{ m}.$$

Questions

(3-1) **1.** A cyclist can accelerate from 5 mi/hr to 30 mi/hr in 40 s. (a) What is his average acceleration? (b) How far does he travel during his 40 s of acceleration?

2. A car is capable of accelerating from rest to a speed of 60 mi/hr in 10 s. (a) What is the average acceleration during these 10 s? (b) How far does it travel?

3. An electron traveling 8×10^6 cm/s enters an electric field that stops it in a distance of 4 cm. What is the acceleration of the electron?

4. A car accelerates from rest to 90 ft/s in a distance of 400 ft. What is its acceleration?

5. An electron traveling 4×10^6 cm/s enters an electric field that gives it an acceleration of 10^{12} cm/s^2, in the same direction as its initial velocity. (a) How long a time will it take for the electron to double its initial velocity? (b) Through what distance will the electron travel in this time?

6. A proton in an experimental chamber is moving east at 3×10^6 cm/s. An electric field is suddenly switched on, which gives the proton an acceleration of 10^{12} cm/s^2 in a direction *opposite* its initial velocity. (a) How long a time will it take to give the proton a speed of 6×10^6 cm/s? (b) In what direction will it be traveling when it attains this speed? (c) When it attains this speed, how far will it be, and in what direction, from the point where it was when the field was switched on?

7. A beam of ions has a velocity of 2.0×10^6 cm/s when it enters an accelerative electric field. It is necessary that the ions strike a cathode 30 cm away in just 8 μs. (a) What constant acceleration must the ions be given in order to accomplish this? (b) How fast will they be traveling when they strike the cathode?

8. A train traveling 20 m/s must reach a checkpoint 10 km distant in 6 min if it is to maintain its schedule. (a) What constant acceleration will bring the train to the checkpoint on time? (b) How fast will the train be traveling when it passes the checkpoint?

9. An electron, started from rest with uniform acceleration by an electric field, travels a distance of 2 cm in 2.5×10^{-8} s before it strikes the anode. (a) What is its acceleration? (b) With what speed does it strike the anode?

10. A ball rolls down an inclined track 2.0 m long in 4 s. (a) What is its acceleration? (b) What is its speed at the bottom of the track?

(3-3) **11.** If one were out to make money by buying pinto beans at one altitude and selling them for the same price at another altitude, should he buy or sell at the higher altitude location? (Weighing is done on a spring balance.)

12. If silver were selling everywhere at $2.30 per troy ounce, and presuming both buyers and sellers are willing to use your spring balance, should you buy in Eagle City, Alaska, and sell it in the Canal Zone, or vice versa? How much would you have to buy and sell to make a profit of $1000?

13. What net force is needed to give a mass of 250 kg an acceleration of 30 cm/s^2? (b) 30 m/s^2?

14. What net force is needed to give a mass of 450 g an acceleration of 12 cm/s^2? (b) 12 m/s^2?

15. A net force of 3.0 lb gives an acceleration of 15.0 ft/s^2 to a body on a frictionless horizontal track. What is the mass of the body?

16. On a horizontal frictionless track, a force of 0.25 lb gives a body an acceleration of 4.0 ft/s^2. What is the mass of the body?

17. A constant force F pushes an originally motionless mass of 400 g a distance of 3 m in 4 s, across a horizontal frictionless table. (a) What is its acceleration? (b) What is the force?

18. A mass of 3 slugs is motionless on a horizontal frictionless table. A constant force F pushes it a distance of 4 ft in 2 s. (a) What is its acceleration? (b) What is the force?

19. Work Question 17(b) if there is a coefficient of friction equal to 0.20 between the mass and the table.

20. Work Question 18(b) if there is a coefficient of friction equal to 0.25 between the mass and the table.

21. A locomotive with a mass of 30 metric tons is pulling a 500-metric ton train at 72 km/hr. (a) What is the weight of the locomotive, in newtons? (b) The engine transmits force for accelerating the train to only the locomotive wheels. If the coefficient of friction between locomotive wheels and track is 0.15, what is the greatest possible accelerating force that can be applied to the train and locomotive? (c) What is the maximum acceleration the train can have? (d) Assuming that the wheels of the cars of the train, as well as of the locomotive, have brakes, what distance will be required to bring the train to a stop?

22. A locomotive weighing 24 tons draws a 256-ton train at 45 mi/hr. See Question 21 for details of acceleration and braking. (a) What is the mass of the train and locomotive? (b) If the coefficient of friction between wheels and track is 0.18, what is the maximum acceleration of the train? (c) What distance is required to bring the train to a stop?

(3-4) **23.** A coffee cup is dropped from a height of 4 ft. How long does it take to hit the carpet?

24. How long will it take for a dropped stone to hit the bottom of a well 120 ft deep?

25. A flowerpot dropped from the top of a building strikes the ground 3 s later. How tall is the building?

26. A stone dropped from the top of a cliff hits the ground below 4 s later. How high is the cliff?

27. On Planet X, a dropped stone falls 48 m in 4 s. (a) What is g on Planet X? (b) What is the weight of a 10-kg mass on Planet X?

28. On Planet U, a dropped ball falls 198 ft in 3 s. (a) What is g on Planet U? (b) What is the mass of an object that weighs 154 lb on Planet U?

29. An Atwood's machine has weights of 20 lb and 18 lb on a cord passing over a frictionless pulley. (a) What will be the downward acceleration of the 20-lb weight? (b) When released from rest, how long will it take the heavier weight to descend 4 ft?

30. An Atwood's machine is composed of masses of 1000 g and 1020 g suspended from a frictionless pulley. (a) What will be the downward acceleration of the heavier mass? (b) When released from rest, how long will it take the heavier weight to descend 200 cm?

31. What is the tension in the cord in Question 29?

32. What is the tension in the cord in Question 30?

33. Consider a horizontal frictionless table with a light frictionless pulley mounted at its edge. From a 2.50-kg weight on the table, a horizontal string passes over the pulley; the end of the string hangs downward, and attached to this end is a mass of 0.60 kg. What is the acceleration of the weight on the table?

34. A frictionless horizontal table has a light frictionless pulley mounted at its edge. From a 2000-g weight on the table, a horizontal string passes over the pulley; the other end of the string hangs downward from the pulley, and attached to it is a 250-g weight. What is the downward acceleration of the hanging weight?

35. If there was a coefficient of friction of 0.10 between weight and table in Question 33, how far would the hanging weight descend in 2 s?

36. If there was a coefficient of friction equal to 0.09 between the weight and the table in Question 34, how fast would the weight be moving at the end of 1.5 s?

37. What is the string tension in (a) Question 33? (b) Question 35?

38. What is the string tension in (a) Question 34? (b) Question 36?

39. A 50-kg man stands on a spring scale in an upward-moving elevator. The elevator decelerates 2 m/s^2 as the elevator stops. What is the scale reading during the deceleration? (This scale is calibrated in newtons; scales calibrated in these units are to be found only in physics texts and physics laboratories.)

40. A 180-lb man stands on a spring scale in an elevator moving downward. If the elevator decelerates 4 ft/s^2, what will be the reading on the scale?

41. A man stands on a spring scale in a moving elevator. As the elevator stops, the scale reading goes from 160 lb to 190 lb, and then back to 160 lb when the elevator stops moving. (a) Was the elevator going upward or downward? (b) What was its acceleration while stopping?

42. A man stands on a spring scale in a moving elevator in Paris. As the elevator comes to a stop, the scale reading changes from 60 kg to 45 kg and then back to 60 kg when the elevator stops moving. (a) Was the elevator going upward or downward? (b) What was its acceleration while stopping?

(3-5)

43. A frictionless plane is inclined at an angle of 25° with the horizontal. What is the acceleration of an object sliding down this plane?

44. An object slides down a frictionless inclined plane making an angle of 20° with the horizontal. What is the acceleration of the object along the plane?

45. In Question 43, include a coefficient of friction equal to 0.15.

46. In Question 44, include a coefficient of friction equal to 0.20.

47. A 200-lb package marked "FRAGILE" slides down a 20° loading ramp; the coefficient of friction is 0.20. With what force, parallel to the ramp, must the loaders restrain the package so it will slide down at a constant speed?

48. A 1000-lb boat is slid down a ramp making an angle of 15° with the horizontal; the coefficient of friction is 0.30. Will the boat slide into the water of its own accord? If not, what push (parallel to the ramp) will be required?

49. A book will slide at constant speed down an incline that makes an angle θ with the horizontal. Show that the coefficient of friction between book and cline iins equal to tan θ.

50. A ramp is inclined at 26° to the horizontal. A timber slides down this ramp at constant speed. What is the coefficient of friction between timber and ramp? (See Question 49.)

51. A bale of rubber latex will not slide down a ramp inclined 15° with the horizontal, because the coefficient of friction is 0.40. What velocity must the bale be given at the top of the ramp to cause it to slide down and come to rest exactly at the bottom? The ramp is 20 ft long.

52. The upper end of a ramp 13 ft long is 5 ft higher than its low end. A man flicks a package of cigarettes up the ramp (μ = 0.15) with such a speed that it comes to rest exactly at the top of the ramp. What was the initial speed of the package?

(3-6) **53.** A man atop a building 50 m high shoots a stone upward with a slingshot at a speed of 30 m/s. How long before the stone lands on the ground at the base of the building?

54. A man stands by a well 120 ft deep and throws a stone upward at 20 ft/s. The stone comes down into the well. How long a time after it is thrown will the stone strike the bottom of the well?

55. On Planet Y, a mass of 2 kg is thrown upward with a speed of 10 m/s, and is caught as it comes down 8 s later. What is the weight of the 2-kg mass?

56. On Planet X, a man throws a 500-g mass upward with a speed of 20 m/s, and catches it as it comes down 20 s later. What does the 500-g mass weigh?

57. A golf ball is driven with a speed of 40 m/s in a direction 30° above the horizontal. (a) To what height will the ball rise? (b) What horizontal distance will the ball travel?

58. A baseball is batted with a speed of 120 ft/s in a direction 37° above the horizontal. (a) To what height will the ball rise? (b) How far (measured along the horizontal) will the ball travel?

59. A soccer punt starts at an angle of 45° above the horizontal, and travels a horizontal distance of 30 yards. (a) To what maximum height does the ball rise? (b) With what speed did it leave the player's toe?

60. A football is punted at an angle of 60° above the horizontal. (a) With what speed did it leave the punter's toe if it travels 50 yards, measured horizontally on the field? (b) To what maximum height does the ball rise?

61. A plane is climbing at an angle of 30° above the horizontal at a speed of 300 m/s. When it is at an altitude of 1200 m, it releases a bomb. (a) How many seconds pass before the bomb strikes the ground? (b) How far beyond the point of release does the bomb strike?

62. A plane, diving at an angle of 30° below the horizontal with a speed of 150 m/s, releases a bomb when it is 1000 m above the ground. (a) How many seconds pass before the bomb strikes the ground? (b) How far beyond the point of release will the bomb strike the ground?

4

Energy and Momentum

4-1 Work and Potential Energy

Let us return now to two cases discussed earlier: (1) two forces applied to the ends of a solid bar supported at a certain point in between, and (2) two forces applied to the pistons of different diameters in two connected cylinders. We assume here that the bar and the pistons have a negligibly small weight and that friction can be disregarded. These two cases of lever arms or areas with ratios of 1:2 are shown in Fig. 4-1A and B.

From our knowledge of levers and Pascal's law, we know that a downward force on the left side of either of these machines will be able to raise a weight on the right side that is twice as great as the force. This "magnification" of a force—the ratio of the force the machine can exert to the force that must be exerted on the machine—is called the *mechanical advantage* of the machine; in each of the above cases the mechanical advantage is 2.

Along with this advantage, however, is a disadvantage. For a downward push, of, say, 1 ft on the left side, the weight on the right is raised

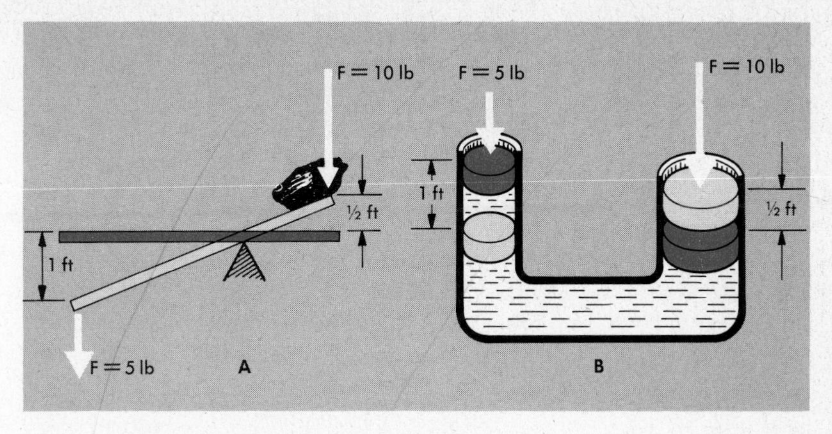

FIG. 4-1 Equilibrium conditions for the forces applied to the ends of a lever, to the pistons of a hydraulic press, or to any other machine of equal mechanical advantage.

in both cases only $\frac{1}{2}$ ft, so what we have gained in force we have lost in distance. In other words, the product of force and distance is the same on both sides. This product of force times the distance moved *in the direction of the force* is the physicist's definition of *work*.

$$\text{Work} = \text{Force} \times \text{distance}^*$$

$$W = Fd$$

It must, of course, be expressed in units that have the dimensions of a force times a distance, such as a foot-pound, or a dyne-centimeter, or an ounce-inch. In the simple examples of Fig. 4-1, we might consider a force of 5 lb on the left side moving through a distance of 1 ft, and thus doing 5 ft-lb of work on the machines. The machines in turn exert a force of 10 lb to lift a weight on the right side through a distance of $\frac{1}{2}$ ft, thereby doing $10 \times \frac{1}{2} = 5$ ft-lb of work on the stone or other restraining force.

This may also be taken as an example of the very important principle of *conservation of energy*. This principle says that work, or *energy* (which is simply the ability to do work), never appears from nowhere and never vanishes. The 5 ft-lb of work done on the machine merely transformed into the equal amount of work the machine did in raising the stone. This cannot be the end of the story, though, for what has happened to the work done on the stone?

The stone was raised, against the pull of gravity, $\frac{1}{2}$ ft higher than it was before, and it is now capable of exerting a force of 10 lb (its weight) through a distance of $\frac{1}{2}$ ft, in returning to its original position. As long

* Torque is also the product of force × distance, and is measured in what appears to be the same units. But bear in mind that for torque, *d* is the component of distance measured *perpendicular* to the force; in work, *d* is the component of distance *parallel* to the force.

as the stone is elevated, the 5 ft-lb of work done on it is stored, ready to be released whenever the stone is lowered. It has, in other words, the potentiality of doing this amount of work, which is called its *gravitational potential energy*.

The work an elevated object can do is of course (like all work in the physics sense) measured by force times distance; in the case of gravitational potential energy, the force is the weight of the object, and the distance is the object's height. But this immediately raises a question. From where should we measure the height? The answer is a very generous one. Measure it from any place you find convenient! We shall be interested only in *changes* in energy, and these changes will be the same no matter what we choose.

In the simple example of the stone and the lever, it might have been easiest to measure the height h from the original position of the stone. We would then say it had zero potential energy (PE) in this location, and 5 ft-lb of PE after it had been raised, thus giving it a 5-ft-lb *increase* in PE. However, a student from the seashore solving such a problem in a Colorado laboratory might prefer to refer his zero to sea level. If, for example, the elevation of the stone was originally 5612.3 ft above sea level, the student would be entitled to say that the stone had a PE of $5612.3 \times 10 = 56{,}123$ ft-lb. The work of the lever would raise it to an elevation of 5612.8 ft and a PE of 56,128 ft-lb, again giving the same answer—that its PE had increased by 5 ft-lb.

Another student might prefer to use the ceiling of the laboratory as his zero. He would run into only a slight complication, because if the PE were zero at the ceiling, it would be negative on the laboratory bench. If the stone were originally 9 ft below the ceiling, its PE would be $-9 \times 10 = -90$ ft-lb; when raised, it would be -85 ft-lb. Since -85 is greater than -90, he also would conclude that the gravitational potential energy had been increased by 5 ft-lb.

The notion of potential energy is not necessarily associated only with the force of terrestrial gravity. A tightly wound spring or a gas compressed in a metal cylinder is also able to produce mechanical work that can be measured in the same units. Potential energy in chemical, rather than mechanical, form is stored in an automobile fuel tank filled with gasoline or in a charge of high explosive in an artillery shell. Potential energy in still another form lies in the nuclear energy of the fissionable fuel rods in a nuclear reactor.

4-2 Kinetic Energy

If we do work on any body, the principle of conservation of energy tells us that the work done must all be accounted for in some way. Let us apply this to the situation illustrated in Fig. 4-2. A cart of mass m is stationary on a perfectly smooth, frictionless, horizontal path. Now apply to it a force F, until the cart has been moved through a distance d. The work done on the cart, Fd, has not been dissipated by friction, and

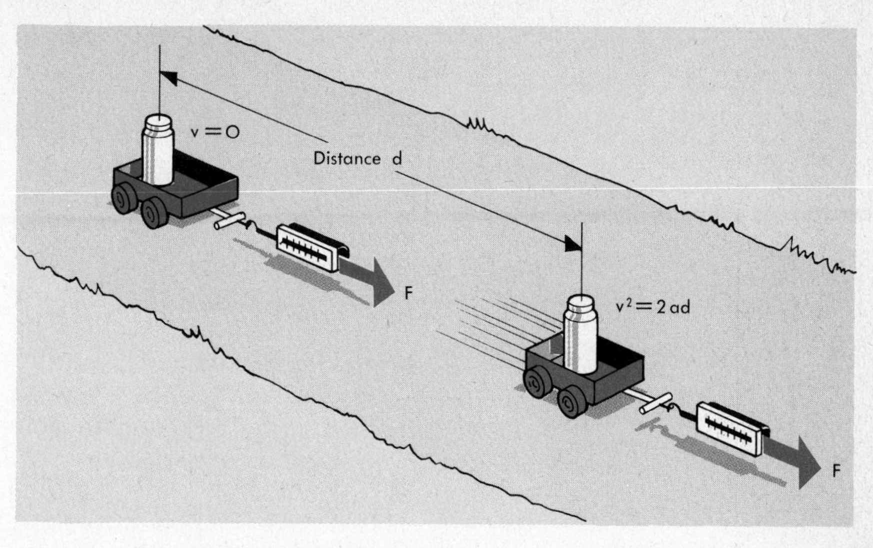

FIG. 4-2 Conversion of work into kinetic energy.

there has been no change in potential energy. The applied work must therefore all be represented by the energy the cart has because of its velocity v; that is, the kinetic energy (KE) of the mass is equal to the work done on it:

$$\mathrm{KE} = Fd.$$

For bodies starting from rest, where $v_0 = 0$, we know from our equations of uniformly accelerated motion that $v^2 = 2ad$, which gives $d = v^2/2a$. We also know that $F = ma$, and substituting these values for F and d, we get

$$\mathrm{KE} = Fd = ma \times \frac{v^2}{2a} = \frac{1}{2}mv^2.$$

Kinetic energy, like potential energy, is equivalent to work, and is measured in the same units of force × distance. Here is a tabulation that may help keep the units in mind:

System	*Mass Unit*	*Velocity Unit*	*Energy Unit*
CGS	g	cm/s	dyne-cm, or **erg**
MKS	kg	m/s	N-m, or **joule** (J)
British	slug	ft/s	ft-lb.

The two new names, *ergs* and *joules*, are very convenient abbreviations, because they are the units most commonly used to express energy in scientific work all over the world. Since 1 newton = 10^5 dynes, and 1 meter = 10^2 centimeters, it follows that 1 joule = 10^7 ergs.

An example will help tie together some of these ideas. Imagine an inclined ramp 50 ft long, the far end of which is 6 ft higher than its beginning. A man shoves a 240-lb box up the ramp by exerting a steady push of 80 lb. Let us assume that the force of friction between the box and the ramp is 40 lb. How fast will the box be moving when it reaches the top of the ramp? The total amount of work the man does can be determined by multiplying the force of his push by the distance through which he pushes the box:

$$Fd = 80 \times 50 = 4000 \text{ ft-lb.}$$

We must now consider what becomes of this 4000 ft-lb of work. Since the force of friction is given as 40 lb, the work that goes into overcoming friction and that is thus converted into heat is $40 \times 50 = 2000$ ft-lb. When it reaches the end of the ramp, the box will have been raised through a vertical distance of 6 ft, and consequently it will have gained $240 \times 6 = 1440$ ft-lb of potential energy.

This will account for $2000 + 1440 = 3440$ ft-lb of the 4000 ft-lb of work the man has done. Left over, we have $4000 - 3440 = 560$ ft-lb of work that can only have gone into increasing the speed and thus the kinetic energy of the box. We need now only determine how fast the box must be moving to give it 560 ft-lb of kinetic energy. To do this, we can compute directly from $\text{KE} = \frac{1}{2}mv^2$ (remembering that since the 560 is in foot-pounds, the mass must be expressed in slugs):

$$560 = \frac{1}{2} \times \frac{240}{32} \times v^2$$

$$v^2 = \frac{560 \times 2 \times 32}{240} = 149.3$$

$$v = 12.2 \text{ ft/s.}$$

Problems concerning work may often involve a force that acts in a direction different from the direction of the motion of the body it is acting on. In Fig. 4-3A is a cart being pulled up a hill by a force F, which is not parallel to the direction of motion. Figure 4-3B shows F resolved into two components: F_{perp}, perpendicular to the motion, and F_{par}, parallel to the motion. The work done is F_{par} times the distance d. Since F_{par} is $F \cos \theta$, we see that the general expression for the work done by a force is $Fd \cos \theta$. Because F_{perp} has no component in the direction of d, it does not do any work at all.

4-3 Energy Interchanges

In many examples of mechanical motion, there is an interchange between kinetic and potential energies. Thus, if we hold a Ping Pong ball in our hand some distance above the floor, it has potential energy but no kinetic energy. If we release it, it falls faster and faster toward the floor, and

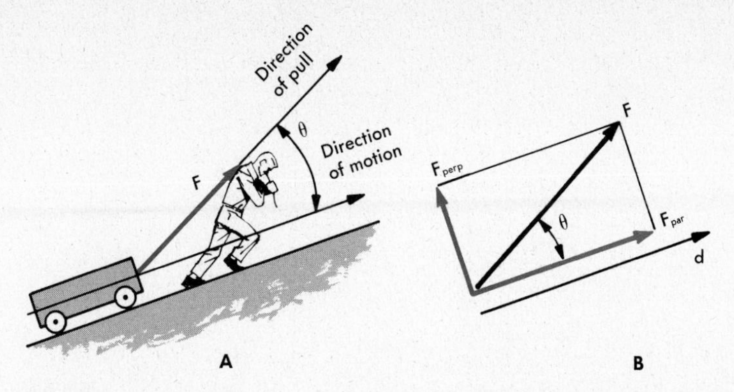

FIG. 4-3 The work done by a force acting in a direction different from the direction of motion.

when it reaches the floor, it will have no potential energy left; all of it will have been turned into kinetic energy. At the moment of impact with the floor, the ball will stop for a split second, and all its kinetic energy will be turned into the potential energy of the elastic deformations in its body. (In the case of an inelastic lead ball, the kinetic energy will be transformed into heat—by a process that we shall explain later—and it will not bounce.) The elastic energy is then changed back into kinetic energy, and this sends the Ping Pong ball up into the air, with the result that the kinetic energy is turned into gravitational potential energy. The ball will rise to approximately its original height. This process is repeated again and again until the friction forces gradually rob the system of its initial energy and the ball comes to a standstill on the floor.

A pendulum is another example of interchange of energy between PE and KE. Consider the pendulum shown in Fig. 4-4. At the ends of its swing (A and C), the pendulum bob is momentarily motionless and therefore has no kinetic energy. However, at these two points it is at its maximum elevation, a distance h above its elevation at B, in the center of its swing. If we take B as our zero, then at A and C its gravitational potential energy is mgh, while its kinetic energy is zero. At B the situation is reversed; its PE is zero, and its speed and KE are maximum. We can easily determine the speed of the pendulum bob at B by calling on the principle of the conservation of energy. If we neglect the small loss of energy due to air friction and the bending of the supporting string, we can conclude that the gain in KE from A to B must exactly equal the loss in PE:

$$\tfrac{1}{2}mv^2 = mgh$$

or

$$v = \sqrt{2gh}.$$

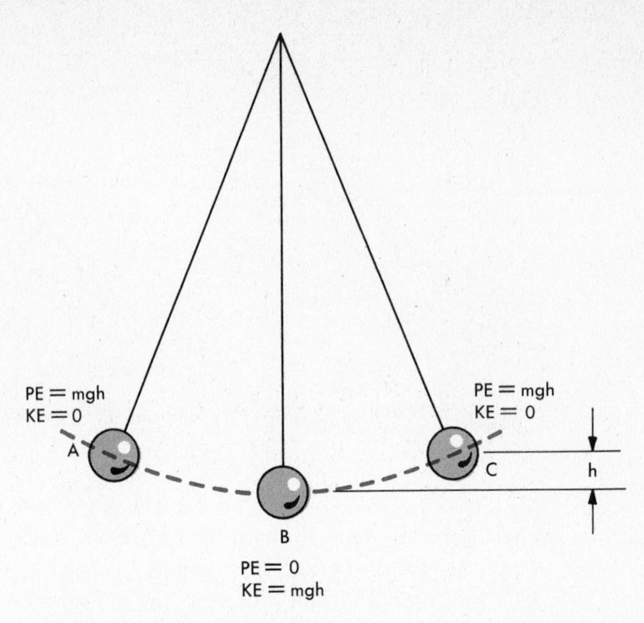

FIG. 4-4 Conversion of potential energy into kinetic energy, and kinetic energy into potential energy in a swinging pendulum.

4-4 Bernoulli's Principle

Consider a liquid flowing through a pipe of varying size (Fig. 4-5). Since liquids are almost incompressible, the volume per second passing through the large section at (1) must be the same as that passing through the small section at (2). In order to do this, the liquid must speed up as it enters the small section; that is, between (1) and (2) the liquid is accelerated to the right. In order to provide the force to cause this acceleration, the pressure at (1) must be greater than the pressure at (2). Similarly, the liquid must be decelerated when it leaves (2) and reenters the large section of pipe to the right; this means that there must be a retarding force toward the left, or that, again, the pressure is larger in the larger pipe. This fact can be easily demonstrated by attaching narrow vertical pipes to the three parts of our horizontal pipe, as shown in Fig. 4-5. The water in the middle pipe will stand lower and thus indicate a lower pressure. The statement that ***in the regions where the velocity of fluid is smaller, the pressure is higher, and vice versa,*** is known as the *principle of Bernoulli*, after a Swiss physicist, Daniel Bernoulli (1700–1782), who discovered it.

The principle of conservation of energy enables us to deal quantitatively with Bernoulli's principle. Since the same amount of liquid per second must flow through every part of the pipe, we can see that

$$v_1 A_1 = v_2 A_2.$$

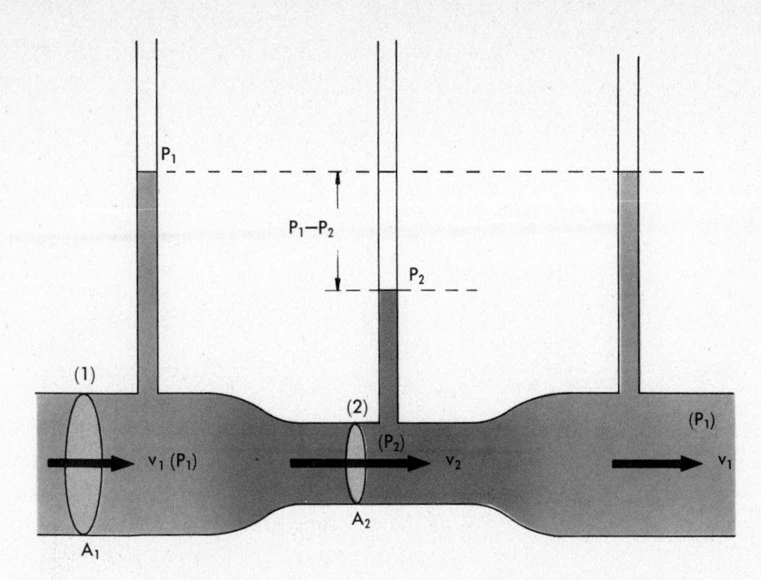

FIG. 4-5 The pressure is lower in the narrow part of the tube, where the fluid moves faster.

Imagine at location (1) a unit volume of the liquid (1 cm^3, or 1 ft^3, or 1 m^3). Its mass is numerically equal to its density d, and it has a kinetic energy of $\frac{1}{2}dv^2$. Our pipe is horizontal, so the gravitational potential energy of the liquid will not change as it moves along the pipe. However, there is another form of energy, due to pressure, that we must take into account. Imagine a hole 1 cm square cut in the side of the pipe. Now into this hole stuff a 1-cm cube of the liquid. (This is *very* difficult to do and you will be much better off just imagining it.) The force against the 1-cm^2 face of the cube is numerically equal to the pressure, and you must push it in a distance of 1 cm, which makes the work you do also numerically equal to the pressure p. This amount of work p can obviously be reclaimed by allowing the cube to emerge into the open again; thus p represents the energy the 1 cm^3 of fluid has because of its pressure. The total energy (ignoring gravitational PE) of the unit volume is then

$$\tfrac{1}{2}dv_1^2 + p_1.$$

By the same arguments, the energy of a unit volume of liquid at point (2) will be $\frac{1}{2}dv_2^2 + p_2$. If we ignore the loss of energy by friction, then by the principle of conservation of energy we get

$$\tfrac{1}{2}dv_1^2 + p_1 = \tfrac{1}{2}dv_2^2 + p_2.$$

As an example, take a pipe 10 cm in diameter, narrowing to a diameter of 5 cm. It contains oil of density 0.8 g/cm^3 flowing 100 cm/s in the large section of tube. What pressure difference could we expect between the large section and the small section of pipe? We have that

$$p_1 - p_2 = \tfrac{1}{2}d(v_2^2 - v_1^2).$$

In the narrowed section, since the diameter is halved, the area will be only one-fourth as great; v_2 must therefore be $4 \times v_1$, or 400 cm/s. Numerically,

$$\begin{aligned} p_1 - p_2 &= 0.5 \times 0.80(400^2 - 100^2) \\ &= 6 \times 10^4 \text{ dynes/cm}^2. \end{aligned}$$

This scheme is extensively used to measure the flow of fluids. A gauge reads the pressure difference between the normal pipe and a special narrowed section; if we know the ratio of pipe diameters, we need merely a reversal of the example given above to find the flow velocity.

The aspirators found in chemistry labs use Bernoulli's principle—water flows through an attachment with a narrowed section at high speed. An outlet from this narrowed part [as at (2) in Fig. 4-5] provides a pressure less than atmospheric, for suction filtration, etc. In automobile carburetors the air drawn into the cylinders likewise passes through a narrow section at high speed; the reduced pressure here draws in a spray of gasoline to mix with the air, ready for burning.

4-5 Power

Webster's New Collegiate Dictionary gives ten separate meanings for the word "power." Although the nine other meanings are perfectly legitimate in ordinary conversation, in physics we must confine ourselves to only one. Power is *the* ***rate*** *of doing work or the* ***rate*** *at which energy is converted from one form into another.*

As an example, we can figure the power needed to raise a 5000-lb elevator a distance of 120 ft in 30 s. The total work that must be done is 5000 lb × 120 ft = 600,000 ft-lb. Even a small motor could do this amount of work if it were allowed a long enough time. Our problem, however, says that the job must be done in 30 s, so the hoist must work at the rate of

$$\frac{600{,}000}{30} = 20{,}000 \text{ ft-lb/s}.$$

Ordinarily, for such things as hoists and engines, a larger unit of power is used: 1 *horsepower (hp) equals 550 foot-pounds per second (ft-lb/s)*. So, if the hoist mechanism were 100 percent efficient—that is, if none of its work were lost in overcoming friction—the elevator would require an engine of

$$\frac{20{,}000}{550} = 36.4 \text{ hp}.$$

No perfectly frictionless machine exists, however, and some allowance must be made for the work lost by being converted into heat by friction. In any machine, the work it does (its *output*) will be less than the energy supplied to it, or the work done on it (its *input*), because of these frictional losses. The term *efficiency*, when applied to a machine of any kind, means merely the ratio of its output to its input:

$$\text{Efficiency} = \frac{\text{output}}{\text{input}}.$$

Efficiency must obviously always be less than 1, and is commonly given as a percentage.

If the hoist mechanism is 75 percent efficient, this means that only 75 percent, or 0.75, of the engine input is usable in raising the elevator; 0.25 is wasted as frictional heat. To find the actual power of the required engine, we must put down that only 0.75 of the input to the hoist must equal 36.4 hp, or

$$0.75x = 36.4$$

$$x = \frac{36.4}{0.75} = 48.5 \text{ hp}.$$

In many applications, *all* the power supplied to a machine is used up in overcoming friction. This is true of an automobile moving at uniform speed on a level road, or of an airplane in horizontal flight. Neither potential nor kinetic energy is being increased; the entire engine output is consumed in the friction of moving parts, of the tires against the road, and, most important at high speed, in the friction of the air. The primary definition of work is $F \times d$; since power is the rate of doing work, we can say that

$$\text{power} = \frac{\text{work}}{\text{time}} = \frac{Fd}{t} = F \times \frac{d}{t} = Fv.$$

Therefore, if an automobile engine develops 200 hp when the car is moving at 75 mi/hr, the total force of friction can be easily figured. To have everything expressed in a consistent set of units, we write

$$200 \text{ hp} = 200 \times 550 = 110{,}000 \text{ ft-lb/s}$$

and

$$75 \text{ mi/hr} = \frac{75 \times 88}{60} = 110 \text{ ft/s}.$$

Substitution of these figures into $P = Fv$ gives us

$$110{,}000 \text{ ft-lb/s} = F \text{ lb} \times 110 \text{ ft/s}$$

or

$$F = 1000 \text{ lb force of friction}.$$

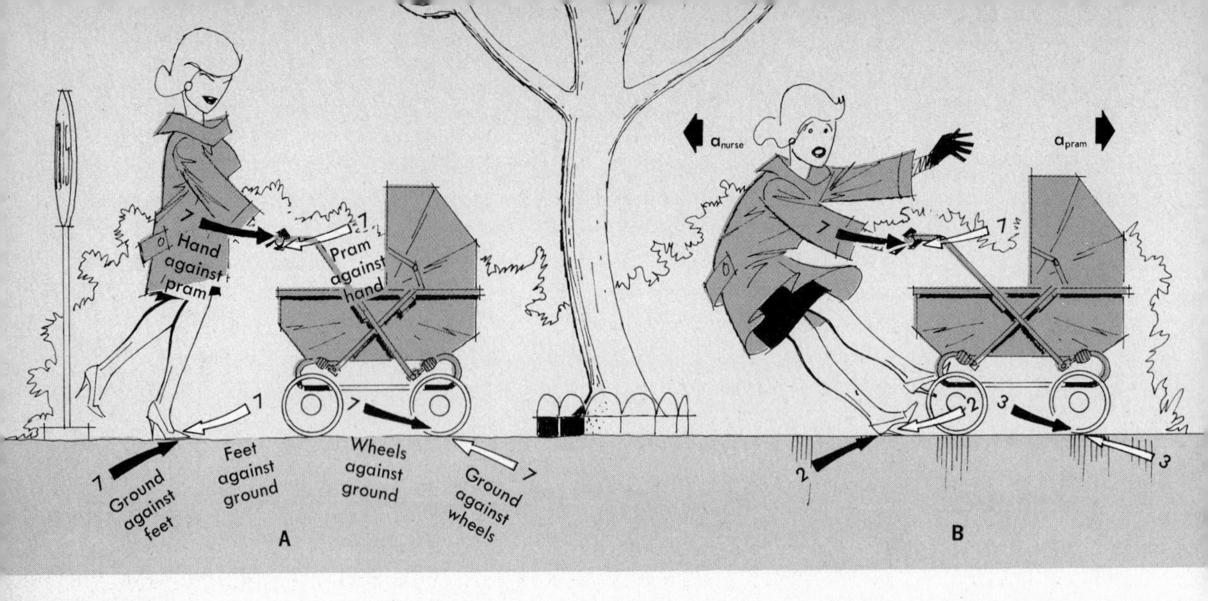

FIG. 4-6 Newton's third law: action and reaction are equal and opposite.

In the widely used MKS system of units, power is, of course, expressed in joules per second. This unit of power has been given the name *watt* (W) in honor of James Watt, the eighteenth-century British inventor of the steam engine. *One watt equals a power of one joule per second.*

$$1 \text{ W} = 1 \text{ J/s}.$$

Although the watt is frequently used to describe electrical power, its definition is fundamentally mechanical and, like any other power unit, can be used to measure the rate of conversion of any kind of energy. We might, for example, determine the power expended by a 70-kg man who takes 12 s to run up a flight of stairs 6 m high. His weight is $70 \times 9.8 = 686$ N. The total work done is $686 \text{ N} \times 6 \text{ m} = 4120$ N-m, or 4120 J. The power exerted, then, is $4120 \text{ J}/12 \text{ s} = 342$ J/s, or 342 W.

4-6 Action and Reaction

Newton's third law of mechanics—at least as important as his second law—says that ***if one body (A) exerts a force on another body (B), then B must exert an equal and opposite force on A.***

Figure 4-6A shows a nurse pushing a perambulator along a rough walk at a constant speed. All goes well, and three pairs of forces illustrating Newton's third law can be seen on the drawing. Let us say there is a 7-lb frictional force between the perambulator and the ground; this means the wheels must push *forward* against the ground with a force of 7 lb, while the ground pushes *backward* against the wheels with an equal 7-lb force. Similarly, the push of the nurse on the perambulator and the push of the perambulator back on her hands are another equal and opposite pair, each amounting to 7 lb, and exactly opposite in direction. There is another pair of 7-lb forces between the nurse's shoes and the ground. In fact, as far as we know, there is no such thing as a single force; forces *always* occur in equal and opposite pairs, each member of the pair being exerted on a *different* body. In Fig. 4-6A, we can see two equal and opposite forces acting on the nurse; one exerted by the perambulator, and one exerted by the ground. These are *not* a third law pair—they are both acting on the same body; each one is a member of a different pair. In this example, they happen to be equal and opposite only because the nurse happens to be pushing the perambulator at a constant speed on a level path.

In Fig. 4-6B, the situation is different. The rough gravel has given way to a patch of ice. The nurse, not noticing, has continued her 7-lb push; but on the slick surface the maximum forces of friction have been reduced to 3 lb for the perambulator, and to 2 lb for the smooth soles of the nurse's shoes. Again the frictional forces and the hand–handle forces occur in equal and opposite pairs, but the separate forces on the nurse and on the perambulator are no longer balanced. The nurse now experiences a net force of $7 - 2 = 5$ lb to accelerate her backward; the $7 - 3 = 4$-lb net force on the perambulator sends it shooting forward.

Since the forces of action and reaction between two objects are equal to one another and since the acceleration communicated to any object by a given force is inversely proportional to the mass of that object, the acceleration of the more massive of the two interacting objects will be smaller than that of the lighter one. When we shoot a rifle, for instance, the powder gases in the barrel force the bullet out toward the target, but they also push equally strongly toward the rear of the barrel and produce what is known as *recoil*. The recoil in an ordinary rifle is relatively small because the rifle is much more massive than the bullet, but even then, it is quite appreciable.

4-7 Momentum

We have already looked at one very useful quantity that depends on mass and motion—kinetic energy. Another quantity that is at least equally important is *momentum*, which is merely the product of the mass of a body times its velocity:

$$\text{Momentum} = \text{mass} \times \text{velocity}$$

The letter "p" is customarily used to stand for "momentum," so we can write this as

$$p = mv.$$

As we shall see when we look at the concepts of Einstein's Special Theory of Relativity, the mass of a body increases as its velocity increases, but this change is negligibly small except at really enormous speeds. For the present, let us confine ourselves to modest speeds of no more than a few thousand miles per second, so that the mass of a body can be considered to be constant. Under these conditions, a change in the momentum of a body must mean a change in its velocity, and vice versa.

Now go back to Newton's second law, although its connection with momentum may not be apparent at first glance:

$$F = ma.$$

Acceleration is defined as the rate of change of velocity, so instead of a we can write $(v_t - v_o)/t$. When this is substituted for a in Newton's second law, it gives

$$F = m \times \frac{(v_t - v_o)}{t} = \frac{mv_t - mv_o}{t} = \frac{p_t - p_o}{t} = \frac{\Delta p}{t}.$$

Put into words, this statement that the net force applied to a body is equal to the *rate* at which the body's momentum changes is the way Newton himself stated his second law. Rearranging the equation by multiplying it by t gives

$$Ft = \Delta p.$$

On the left side of the equation, the quantity Ft—the product of a force and the time during which it acts—is called *impulse*; on the right side of the equation is the change in the body's momentum. The fact that the impulse given a body equals its change of momentum leads us directly to the very important concept of the *conservation of momentum.*

Consider two billiard balls colliding on a smooth table. During the fraction of a second that the balls are in actual contact, each is slightly compressed, and in springing back to spherical shape each exerts a force on the other that will change its speed and direction. From Newton's third law, we know that these forces are at all times exactly equal and exactly opposite in direction, and the time during which these collision forces act is obviously identical for each ball. Thus each ball, during the collision, receives an impulse equal to the impulse received by the other ball, but opposite in direction. (Impulse and momentum are both vector quantities, since each is made up of a vector multiplied by a scalar.) It follows, then, that each ball undergoes a change in momentum equal and opposite to the change of momentum of the other ball. So if these changes are added (vectorially, remember!), they add up to zero. The fact that the total momentum change is always zero allows us to make the statement that ***in any collision or other interaction between bodies, the total momentum of the interacting bodies (considered vectorially) is the same afterward as it was before,*** which is one way of stating the law of *conservation of momentum.*

Before

X=?
20 ft/sec

After

FIG. 4-7 Conservation of momentum in a collision.

As an example, consider a 2000-lb car traveling at 80 ft/s that crashes into the rear of a 16,000-lb truck moving in the same direction at 10 ft/s. The small car bounces backward from the collision with a speed of 20 ft/s. What is the speed of the truck after the impact? (See Fig. 4-7.) In this example, the motions are in one dimension along a straight line, and we can take care of vector directions by assigning a + sign for motion to the right and a − sign for motion to the left. Before the collision, the total momentum of the two vehicles is

$$80 \times 2000 + 10 \times 16{,}000 = 320{,}000 \text{ lb-ft/s}.$$

Afterward, it is $-20 \times 2000 + 16{,}000x$. If we set the momentum before the collision equal to the momentum after, we get

$$320{,}000 = 16{,}000x - 40{,}000$$

or

$$x = 22.5 \text{ ft/s}.$$

Strictly speaking, we have used incorrect units for the momentum in the example above; in the English system mv is in slug-ft/s. These units could have been introduced by dividing the 2000 lb and the 16,000 lb weights by g, to give $320{,}000/g$ slug-ft/s on the right. But we would then immediately simplify the equation by multiplying both sides by g, to give the same equation we started with. *So long as the units are the same on both sides of the equal sign, we have a valid equation*, whether the cars are measured in pounds, slugs, grains, kilograms, ounces, or Martian pjotniks.

If you were to check the total kinetic energy of the two cars before and after the collision, you would find that more than 38 percent of the original KE had been lost; this lost KE was used up (and converted into heat) in the work devoted to crumpling fenders, etc. When some mechanical energy is lost in this way, the collision is called *inelastic*. There is no such thing as a perfectly elastic collision between ordinary bodies—that is, a collision in which no energy is lost. Glass or hard steel balls may come fairly close to it, however. A golf ball dropped on the pavement will bounce for a number of times before coming to rest, but each bounce will be to a lesser height than the previous one, thus giving us an indication that the ball has lost energy in each collision with the earth. At each impact with the pavement, the ball is slightly flattened. This causes the fibers in the ball to rub against each other and thus to use up energy by converting it to frictional heat that cannot be recovered. (We should add, however, that collisions between individual atoms and molecules are generally perfectly elastic.)

Although mechanical energy is never completely conserved in any collisions between real bodies, *momentum always is*. Two equal lumps of putty traveling at equal speeds in opposite directions will stick together and come to rest when they collide, thus using up all their KE in the work of rubbing putty particles against one another. Their total momentum, though, has not changed. Before collision, the two equal vectors representing their momentums point in opposite directions and hence add up (vectorially) to zero. After collision, the total momentum is obviously still zero.

4-8 Rocket Propulsion

As an approach to considering the great rockets that launch astronauts with tons of equipment up into their orbits, let us first consider a man firing a gun. Figure 4-8A shows the gun and bullet before firing. Both are stationary, and the total momentum is undeniably zero. A fraction of a second after firing (Fig. 4-8B), the 30-g bullet has left the barrel with a muzzle velocity of 250 m/s. It therefore has a momentum of 7.5×10^5 g-cm/s, or 7.5 kg-m/s, toward the right. In order that the total momentum may remain zero, the 5-kg gun must have gained a momentum $mv = 7.5$ kg-m/s, toward the left. Its recoil velocity is then $7.5/5 = 1.5$ m/s, toward the left. In this case, the gun is quickly stopped by the marksman's shoulder.

Suppose, though, that the gun were mounted on a frictionless carriage, and could fire one bullet per second for a minute. At the end of the minute, after 60 shots had each given it an additional velocity of 1.5 m/s, the gun would be moving at $1.5 \times 60 = 90$ m/s. We would get exactly

FIG. 4-8 Conservation of momentum in firing a gun.

the same result if the gun (actually, a small model rocket) were able to squirt 30 g/s of water or hot gases or anything else out of its muzzle with a relative velocity of 250 m/s. In the previous section, we saw that the force resulting from such a procedure is equal to the rate of change of momentum. In this example, 30 g/s have been given a velocity change of 250 m/s, to give a rate of change of momentum of 250 m/s × 0.030 kg/s = 7.5 kg-m/s^2. The resulting force on the rocket is thus 7.5 N. (Note that the dimensions check: 1 N = 1 kg-m/s^2.) Acting on the 5-kg rocket, this force produces an acceleration $a = F/m = 7.5/5 = 1.50$ m/s^2. After 60 seconds, $v = at = 1.5 \times 60 = 90$ m/s, which confirms our first result calculated from the bullet firing.

This same procedure can be used to calculate the thrust given one of the great modern satellite-launching rockets. The Titan III-C, together with its load, has a weight of 1.40×10^6 lb. Its first stage, responsible for starting the rocket on its way and for penetrating most of the atmosphere, burns solid fuel at a maximum rate of 9400 lb/s. At sea level, the exhaust gases resulting from this burning have a velocity of 7600 ft/s. Here, in order to calculate the thrust in *pounds*, we must have the rate of change of momentum in *slug*-feet/second^2. So we must convert the fuel-burning rate from 9400 lb/s to $9400/g = 9400/32.2 = 292$ slugs/s. Since each slug of fuel has its velocity change from 0 to 7600 ft/s, the total rate of change of momentum (which equals the force, or thrust) becomes $292 \times 7600 = 2.22 \times 10^6$ lb.

The *net* force on the rocket is the difference between this thrust and the rocket's weight, or $2.22 \times 10^6 - 1.40 \times 10^6 = 8.2 \times 10^5$ lb. This net

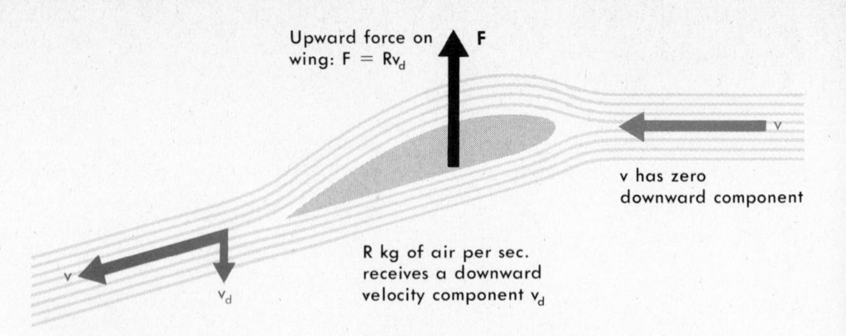

FIG. 4-9 Downward thrust of air passing around airplane wing equals the upward thrust on wing.

upward force would give the rocket an upward acceleration of $a = F/m = 8.2 \times 10^5/(1.40 \times 10^6/32.2) = 18.9$ ft/s^2, or 0.59 g.

An airplane wing acts on the same principle. In flight, it must give downward momentum to air at a rate sufficient to provide the needed upward force to keep the plane aloft. Figure 4-9 sketches a wing moving through the air from left to right. It is drawn as though the wing were stationary, with the air flowing by it from right to left (as though in a wind tunnel). As the air passes over the tilted wing, R kg/s is deflected so that it is given a downward velocity component v_d. Then, for the upward force on the wing, F,

$$F = v_d \text{ m/s} \times R \text{ kg/s} = v_d R \text{ kg-m/s}^2$$

$$= v_d R \text{ N.}$$

Golf and baseball players, as well as aviators, are also vitally concerned with conservation of momentum. Figure 4-10A shows a baseball speeding through the air without any spin. The onrushing air neatly parts and goes around each side of the ball. When the air on the sides comes back together on the back side of the ball, considerable turbulence forms, with resultant heat from internal friction in the air, which is the principal source for the loss of energy for any projectile. But in this case there is no change in the direction of the airflow or the path of the ball. In Fig. 4-10B the same ball is spinning about an axis perpendicular to its velocity; the passing air is dragged along with the spin, so that the recombination of the airstream is *not* opposite the point of separation. The air is thus (in the drawing) given a Δp upward; the ball *must* receive an equal Δp downward, which changes the direction of its velocity—thereby a curve ball or a slice!

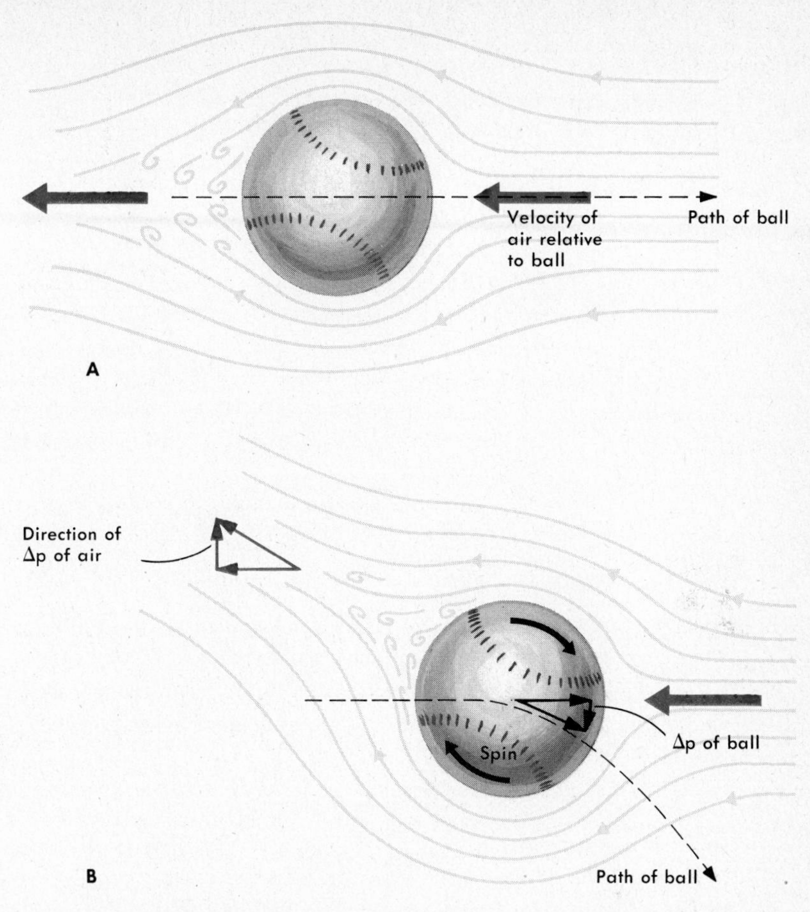

FIG. 4-10 A spinning baseball or golf ball moves in a curved path as a result of conservation of momentum.

Questions

(4-1) 1. A horse exerts a force of 250 lb in pulling a loaded wagon. How much work does he do per mile of travel?

2. It requires a force of 600 N to push a car at constant speed along a level road. How much work is done in pushing it $\frac{1}{2}$ km?

3. A storage tank holds 10 metric tons of water, and is 25 m above the pond from which it is filled. (a) How much work is done in filling the tank? (b) How much more potential energy does the water in the tank have than 10 metric tons of water in the pond?

4. A 180-lb man climbs to a roof 40 ft above the ground. (a) How much work did he do against the pull of gravity? (b) By how much did he increase his potential energy?

5. A 1500-lb magnet is on a stand 3 ft above a laboratory floor. What is its PE (a) with reference to the floor? (b) with reference to the ceiling, 12 ft above the floor? The magnet is now raised by 8 inches. Now, what is its PE with reference to (c) the floor? (d) the ceiling?

6. A 150-kg engine is suspended from a chain hoist 2.5 m below the garage ceiling, which is 4.0 m above the floor. What is its PE with reference to (a) the ceiling? (b) the floor? The engine is lowered 70 cm. Now, what is its PE with reference to (c) the ceiling? (d) the floor?

(4-2)

7. A nickel coin has a mass of about 5 g. How many ergs of work are needed to pick up a nickel from the floor and place it on a table 80 cm high?

8. How much work in joules must be done to hoist a 750-kg safe to the top of a building 35 m high?

9. The electric field in an X-ray tube exerts a force of 3.2×10^{-16} dyne on an electron, through a distance of 20 cm. How much kinetic energy is imparted to the electron?

10. An alpha particle, initially moving slowly, enters an electric field that exerts on it a force of 4×10^{-9} dyne. What is the kinetic energy of the alpha particle after having traveled 5 cm in the electric field (in ergs, and also in joules)?

11. From what height must a car be dropped in order for it to crash with as much KE as it has when traveling at 60 mi/hr?

12. By what factor is a car's kinetic energy increased if its speed is tripled?

13. How fast must a proton move to have as much KE as an electron traveling 8×10^6 m/s? (A proton is about 1800 times more massive than an electron.)

14. How fast does a 1500-kg car have to move in order to have as much kinetic energy as a 150-g bullet traveling 700 m/s?

15. It requires a horizontal force of 3×10^4 dynes to keep a block moving at constant speed across a level table. What is the KE of the block after a horizontal push of 7×10^4 dynes has been applied to it for a distance of 800 cm?

16. A push of 800 N is applied to the car of Question 2 when it is stationary. Through what distance must this force be exerted to give the car a speed of 3 m/s? (The mass of the car is 1400 kg.)

17. A 48-lb boy slides down a slide whose bottom is 6 ft below its top. He reaches the bottom with a speed of 16 ft/s. How much energy was used up in overcoming friction?

18. A 6080-lb truck, starting from rest, coasts down a hill whose foot is 25 ft lower than its top. When the truck reaches the bottom of the hill it has a speed of 30 ft/s. How much energy was used up in overcoming friction on the way down?

19. A 4800-lb truck coasts 200 ft down a 12° hill; the road then changes to an uphill 8°, and the truck coasts 150 ft along this uphill slope before stopping. What was the average force of friction?

20. A bicyclist (total weight 160 lb) coasts 300 ft down a hill sloping at 10°; the road then climbs up the opposite side of the valley at a 10° slope, and he coasts 100 ft up this road before he comes to a halt. What was the average force of friction?

(4-3) 21. A pendulum is pulled to one side until its center of gravity is 3 inches higher than when hanging vertical, and then is released. What is its speed as it passes through the midpoint of its swing?

22. The pendulum of Question 21 is pulled aside and released. When it swings through its midpoint, its speed is 6 ft/s. At the point of release, how much higher is its center of gravity than when hanging vertical?

(4-4) 23. Oil of density 0.86 g/cm^3 flows through a horizontal pipe at a speed of 50 cm/s. The pipe (cross-sectional area, 30 cm^2) is smoothly narrowed to a cross-sectional area of 10 cm^2, and pressure gauges are placed in the large and the small sections of pipe. What is the difference in their pressure readings?

24. A horizontal pipe of 10 cm^2 cross-sectional area is smoothly reduced to a cross-sectional area of 5 cm^2. The pipe carries water, which flows through the larger section with a speed of 30 cm/s. What is the difference in pressure between the large and small sections?

25. A horizontal pipe of 10 cm^2 cross-sectional area is smoothly reduced in size to a cross-sectional area of 5 cm^2. A gauge shows that the difference in pressure between the large and small sections of pipe is 1200 dynes/cm^2. How many cm^3/s of water are flowing through the pipe?

26. Oil of density 0.90 g/cm^3 flows through a pipe whose cross-sectional area is 30 cm^2. This pipe is smoothly reduced to a cross-sectional area of 10 cm^2. Gauges in the large and small pipe sections show a pressure difference of 3000 dynes/cm^2. How many cm^3 of oil flow through the pipe per second?

(4-5) 27. The tank of Question 3 is to be filled in 40 min. What is the power the pump must deliver?

28. The man in Question 4 makes the climb in 50 s. What power is he exerting (in ft-lb/s, and in horsepower)?

29. A ski lift must raise a total load of 2500 lb each minute, up a slope whose difference in elevation is 1500 ft. If the mechanism is 70 percent efficient, what is the horsepower of the engine?

30. A motor and construction hoist mechanism are 60 percent efficient. What is the minimum horsepower of the motor needed to raise a load of $2\frac{1}{2}$ tons to a height of 120 ft in 2 min?

31. A load of 800 kg is to be hoisted to an elevation of 35 m in 3 min. What is the power in watts of the driving motor, assuming it to be 100 percent efficient? 70 percent efficient?

32. A tank on the roof of a building 20 m high holds 10 m^3 of water, and it must be filled from a pond on the ground in 20 min. What is the power in watts of the pump that will be required, assuming it to be 100 percent efficient? 60 percent efficient?

33. What is the retarding force of friction on a car that must exert 150 hp to drive it at 80 mi/hr on a level road?

34. A 200-hp engine can give an airplane a speed of 120 mi/hr in level flight. What is the force of friction at this speed?

35. A 250-W heater burns for 6 hr. How many joules of energy are converted from electrical energy into heat?

36. How many joules of electrical energy are converted into heat and light when a 50-W lamp burns for 8 hr?

(4-7) 37. How fast must a proton move in order to have the same momentum as an electron traveling 8×10^6 m/s? (Refer to Question 13, and compare answers.)

38. How fast does a 1500-kg car have to move in order to have as much momentum as a 150-g bullet traveling 700 m/s? (Refer to Question 14, and compare answers.)

39. An 8-lb block at rest on a level, frictionless table is given a constant 8-oz push until it is moving 2 ft/s. For how long a time was the push applied?

40. For how long a time must a push of 10,000 N be applied to a 50,000-kg rocket ship in order to increase its speed from 25,000 to 30,000 m/s?

41. A 16-ton rocket ship traveling freely in space at 6000 mi/hr is given a thrust of 5000 lb in the direction of its motion for 5 min. What is its velocity at the end of this thrust?

42. An alpha particle (mass = 6.6×10^{-24} g) with a velocity of 3×10^7 cm/s is subjected to a force of 10^{-9} dyne in the direction of its motion for 10^{-6} s. What is the resulting velocity of the particle?

43. A 400-g glider on an air track is moving 10 cm/s when it is overtaken by a 600-g glider moving at 15 cm/s. The two gliders lock together. What is their speed after this collision?

44. A 2000-lb car traveling 60 mi/hr collides with a 6000-lb truck going in the same direction at 30 mi/hr. The bumpers lock together. What is the speed of the locked-together cars immediately after the collision?

45. An alpha particle moving east with a velocity of 10^5 cm/s scores a direct hit on a stationary proton. (The mass of an alpha particle is four times the mass of a proton.) The collision is perfectly elastic. What are the velocities (speed and direction) of the two particles after collision?

46. A proton moving east with a velocity of 10^5 cm/s scores a direct hit on a stationary alpha particle. (The mass of an alpha particle is four times the mass of a proton.) The collision is perfectly elastic. What are the velocities (both speed and direction) of the two particles after collision?

47. A 160-lb man sits on a level frictionless surface. A friend throws him a 1-lb ball at a speed of 90 ft/s, which he catches and holds. (a) What is the speed of the seated man after catching the ball? (b) The seated man fumbles the catch; after escaping his hands, the ball proceeds in its original direction at 30 ft/s. Now, what is the speed of the seated man?

48. A bullet weighs 10 g and has a speed of 300 m/s. It is fired into a stationary 5-kg block of wood on level frictionless ice, and remains imbedded in the wood. (a) What is the velocity of the block after contact? (b) Suppose the bullet passes through the block and emerges with a speed of 100 m/s and unchanged direction. Now, what is the speed of the block?

(4-8) 49. A 20-kg model rocket has an exhaust velocity of 2000 m/s. At what rate must it burn fuel to have an acceleration of 1 g?

50. Suppose the rocket in Question 40 has an exhaust velocity of 2500 m/s. At what rate must it burn fuel to develop its thrust of 10^4 N?

51. A hose has a diameter of 8 cm, and delivers a jet of water through a nozzle 2 cm in diameter. Water flows through the hose at a speed of 1 m/s. What is the backward force of the nozzle?

52. A fire hose is 3 inches in diameter and its nozzle is 1 inch in diameter. If the speed of the water through the hose is 10 ft/s, with what force must the nozzle be held to prevent it from flying backward?

53. If the rocket ship of Question 40 were in open space away from perceptible gravitational forces, what would be the apparent "weight" of an astronaut aboard it during the described acceleration? The astronaut weighs 175 lb on the earth.

54. With what force would a 200-lb astronaut press against his couch in taking off in a Titan III-C rocket launch?

5

Rotational Mechanics

5-1 Equations of Rotational Motion

There are two ways in which a body can move. One type of movement, which we have already investigated, is called *linear*, or *translational*, and is a mere change in position. The other type is called *angular*, or *rotational*, movement, in which the body rotates about a line called an *axis*. Many motions, of course, are a combination of the two; a spiral punt spins as it soars through the air, and the wheels of a car rotate as they travel along the road.

In translational motion, our basic measure is the distance traveled; in rotation, we must start with the angle through which the body turns. It might seem good enough to measure this angle in degrees, or in terms of complete revolutions, but in practice the relationships and equations are much simpler and easier to comprehend if we use another unit called the *radian* (rad).

Figure 5-1A shows what a radian is. Take a circle of any size and lay off along its circumference an arc equal in length to the radius of the circle. This arc measures an angle of 1 radian. Since the circumference is $2\pi r$, it is apparent that there are 2π radians in a full circle, or revolution.

Imagine a wheel of radius r, with a point C marked on its rim (Fig. 5-1B). If we turn the wheel on its axis through an angle of θ rad, point C moves in an arc through a distance d to a new position, C_1. If θ is 1 rad, the distance C–C_1 will equal r; if θ is 2 rad, C–C_1 will be $2r$. In general, $d = \theta r$; that is, the distance moved by any point on a rotating body is the angle (in *radians*!) through which the body turns, multiplied by the distance from the point to the axis of rotation.

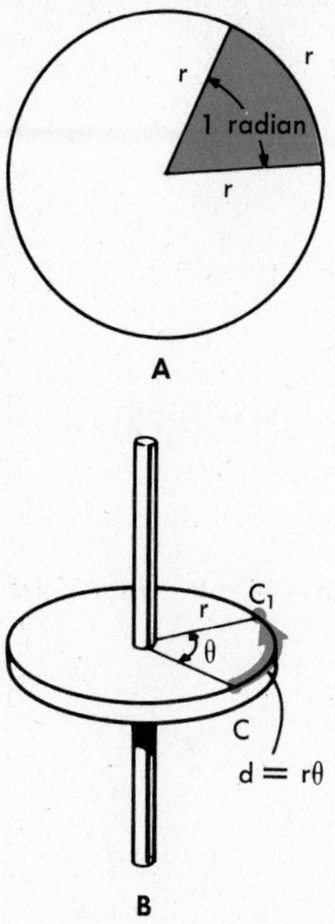

Linear velocity and angular velocity are similarly related. If our wheel rotates at a uniform rate through θ radians in t seconds, a point at distance r from the axis will have moved through a distance θr in the same time t. The angular speed ω is θ/t rad/s, and the linear velocity of point C is $v = \theta r/t$. From this we see that $v = \omega r$. Further analysis would show that for a point at a distance r from the axis of a wheel whose angular speed is changing, the linear acceleration of the point along its curved path is r times the angular acceleration of the wheel, measured in radians/second^2. Angular acceleration α is of course defined as the rate at which ω changes, or $\Delta\omega/t$. Stated as an equation, $a = \alpha r$. Here are these three simple relationships tabulated:

$$d = r\theta \qquad v = r\omega \qquad a = r\alpha.$$

FIG. 5-1 The radian (A), and a rotating disk (B).

By starting from basic definitions of angular velocity and angular acceleration, we can easily derive equations of rotational motion exactly as they were derived for translational motion:

$$\omega_t = \omega_o + \alpha t$$

$$\theta = \omega_o t + \tfrac{1}{2}\alpha t^2$$

$$\omega_t^2 = \omega_o^2 + 2\alpha\theta.$$

5-2 Moment of Inertia

An applied force causes a mass to accelerate and move from one location to another. In the rotational analogy, an applied torque causes a body to rotate, and it seems apparent that there must be some properties of a rotating body we can use to find a relationship corresponding to $F = ma$. Let us investigate this by considering the simplest possible rotating body, as shown in Fig. 5-2. It consists of a rod (imagined to be massless) extending from an axis of rotation O to a small mass m, at a distance r from O. Now, to the mass m apply a force F, as shown. The torque tending to rotate the body around O is Fr. From Newton's second law, a force F acting on a mass m will give it a linear acceleration of $a = F/m$. We know also that linear acceleration is related to angular acceleration

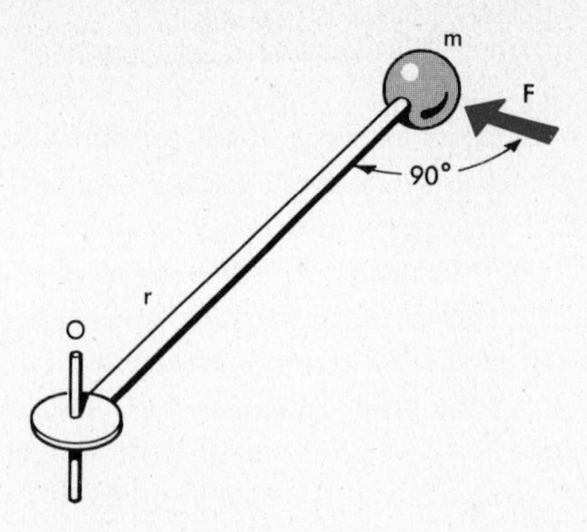

FIG. 5-2 A force applied to a small mass at the end of a massless rod causes it to rotate about the axis at O.

by the expression $a = r\alpha$. This gives two expressions for a, which we can set equal to one another:

$$\frac{F}{m} = r\alpha$$

from which

$$F = mr\alpha.$$

For rotational motion we need something including the torque τ instead of force. We can get this simply by multiplying the equation by r:

$$Fr = mr^2\alpha$$

or

$$\tau = mr^2\alpha.$$

The result is an equation exactly analogous to $F = ma$, except that instead of m we have mr^2. This expression, the mass of a particle times the square of its distance from the axis of rotation, is called the particle's *moment of inertia* and is usually designated by the letter I. Our rotational form of Newton's second law thus becomes

$$\tau = I\alpha.$$

We have worked this expression out for a single small particle, but what of the rods and wheels and cylinders we are most often concerned with when we need to consider rotation? Any large body can be thought of as being made up of a large number of small particles. Each particle contributes its own small mr^2, so the moment of inertia of the large body is the sum of the moments of inertia of the particles that make it up. For regularly shaped bodies of many sorts, the branch of mathematics

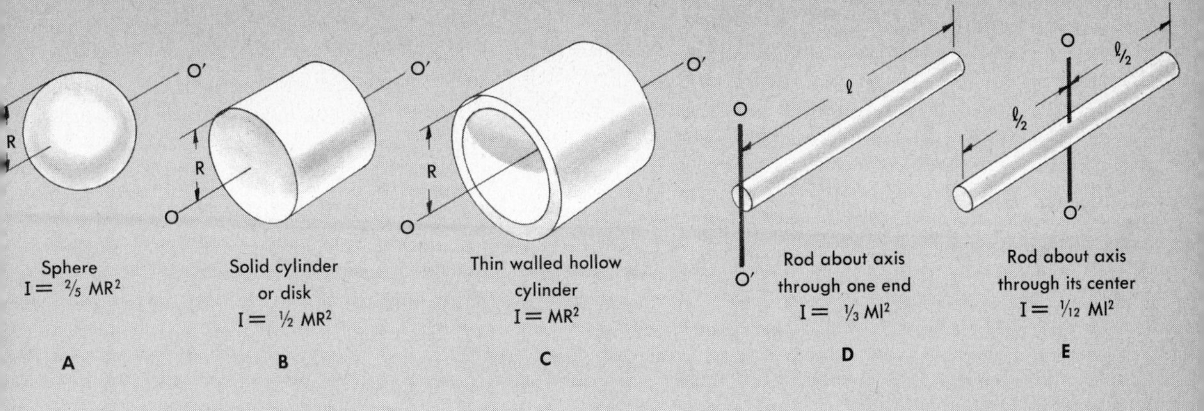

FIG. 5-3 Moments inertia of bodies of various shapes.

known as *calculus* gives these sums, some of which are shown in Fig. 5-3.

Without calculus, the only one of these we can satisfactorily confirm is the thin-walled hollow cylinder. If we consider the cylinder to be made of a multitude of small pieces of mass m, each small piece is the same distance R from the axis of rotation, and each has the same moment of inertia mR^2. If all these small moments of inertia are added together, the sum is clearly R^2 times the sum of the small masses, or MR^2, where M is the total mass of the cylinder.

5-3 Torque and Rotation

As an example of how to use moment of inertia in the rotational form of Newton's second law, consider a solid cylinder on a frictionless axle, similar to that shown in Fig. 5-3B. The weight of the cylinder is 128 lb, and it is 6 inches in *diameter*. A steady pull of 24 lb is exerted on a cord wrapped around the cylinder (Fig. 5-4). Starting from rest, how long a time will it take the cylinder to make 12 revolutions? We can first determine its angular acceleration from $\tau = I\alpha$, keeping a careful eye on units. If we use torque in lb-ft (*not* lb-in., because the distance unit must be the same as the distance unit in g, which we will take as 32 ft/s²), then the mass must be taken in slugs. This gives us

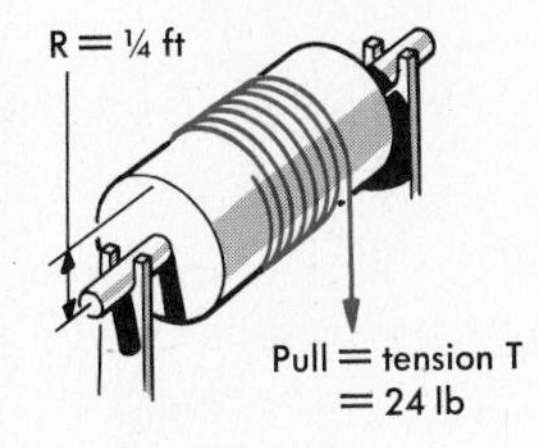

FIG. 5-4 Rotating cylinder given an angular acceleration by a 24-pound pull on a cord wrapped around it.

$$\tau = I\alpha = \tfrac{1}{2}MR^2\alpha$$

$$24 \times \tfrac{1}{4} = \tfrac{1}{2} \times \tfrac{128}{32} \times \tfrac{1}{16} \times \alpha$$

$$\alpha = 48 \text{ rad/sec}^2.$$

Once we have α, the rest is easy, for 12 revolutions is $12 \times 2\pi = 24\pi$ rad; and from $\theta = \tfrac{1}{2}\alpha t^2$,

$$24\pi = \tfrac{1}{2} \times 48t^2$$

$$t^2 = \pi$$

$$t = 1.77 \text{ s}.$$

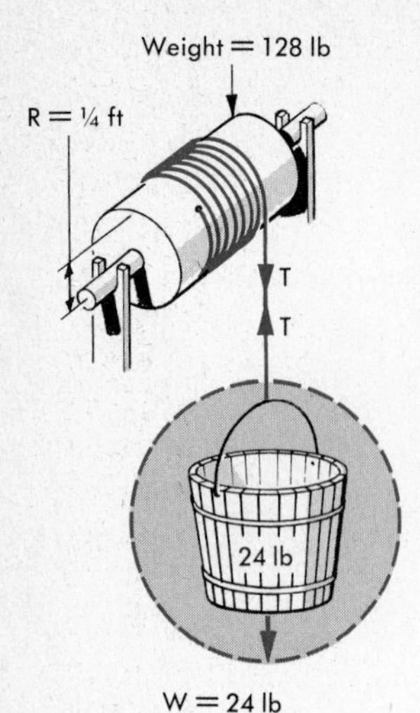

FIG. 5-5 Rotating cylinder given an angular acceleration by a 24-pound weight hung from a cord wrapped around it.

At first glance, it might appear that in the last example we could get the steady 24-lb pull we want by merely hanging a 24-lb weight on the end of the cord. A closer look, however, will convince us that this is not true. Figure 5-5 sketches the situation. If the axle of the cylinder is clamped so that it cannot turn, the weight will be motionless and the cord tension T will be 24 lb. But if the cylinder is free to turn on frictionless bearings, it will begin an angular acceleration, and the weight will have a linear acceleration downward. Let us look at just the descending weight, which is isolated in the dashed circle. It has a downward acceleration a, caused by the net force $24 - T$, so the cord tension must be less than 24 lb. $F = ma$ gives us

$$24 - T = \frac{24}{32}a = \frac{3a}{4}.$$

The torque of force T is what causes the cylinder to have an angular acceleration, and we can write another equation:

$$\tau = I\alpha = \tfrac{1}{2}MR^2\alpha$$

$$\tfrac{1}{4}T = \tfrac{1}{2} \times \tfrac{128}{32} \times (\tfrac{1}{4})^2\alpha = \tfrac{1}{8}\alpha$$

from which

$$\alpha = 2T.$$

We cannot solve two equations for three unknown quantities, but fortunately we know that $a = r\alpha$, which gives another relation between the unknowns:

$$a = \tfrac{1}{4}\alpha$$

or

$$\alpha = 4a.$$

Substitution into the equation $\alpha = 2T$ gives us

$$4a = 2T$$

$$T = 2a.$$

Another substitution into the first equation gives

$$24 - 2a = \tfrac{3}{4}a$$

from which

$$a = 8.73 \text{ ft/s}^2$$

and

$$T = 17.5 \text{ lb}.$$

5-4 Centripetal Force

Velocity, we have noted before, is a vector quantity. So far, we have considered only accelerations caused by a change in speed, but a vector can also be changed by changing its direction.

Figure 5-6 shows such a case. It represents a mass m at the end of a massless rod of length r, which rotates at constant speed about the axis OO' (Fig. 5-6A). A view looking down from the top is shown in Fig. 5-6B. Let us say that in a short time t, m moves from A to B through an angle θ. Its velocity will have changed from v_1 to v_2 during this time, a change in *direction* only; the speed is constant, so the length of the vector remains the same. Figure 5-6C shows a graphical determination of the change in velocity that has occurred. The velocity Δv must be added to v_1 to produce v_2; thus Δv represents the change in velocity during time t.

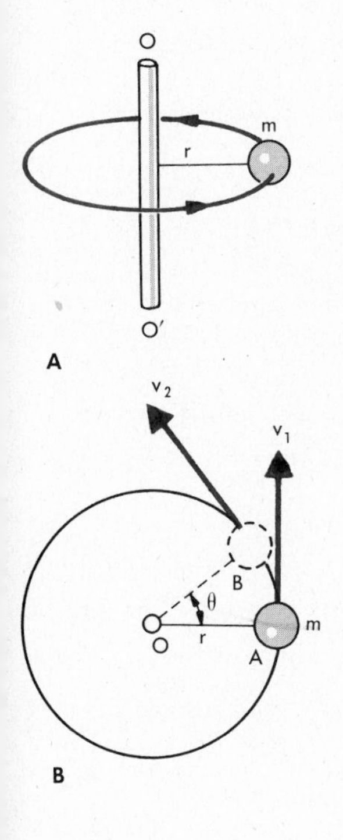

Notice that the triangle OAB is similar to the triangle formed by the vectors v_1, v_2, and Δv, because the angle θ is the same in both. In similar triangles (i.e., triangles of the same shape), corresponding sides are in the same proportions, so we can write

$$\frac{\Delta v}{v} = \frac{AB}{r} = \frac{vt}{r}$$

$$\Delta v = \frac{v^2 t}{r}.$$

(The subscript has been dropped from the v because we are now dealing with magnitudes only; since v_1 and v_2 are the same length, we do not need the subscripts to distinguish them.)

Acceleration a we have defined as the rate of change of velocity. The change Δv has occurred in the time t, so its *rate* of change is

$$a = \frac{\Delta v}{t} = \frac{v^2}{r}.$$

In some problems, it may be more convenient to use the angular speed ω rather than the linear speed v. We can make this change easily by substituting ωr for v, giving us

$$a = \omega^2 r.$$

Since a is a vector, we must also determine its direction. If, in Fig. 5-6C, we make the time interval, and hence also the angle θ, smaller and smaller, v_1 and v_2 become practically parallel, and Δv becomes perpendicular to either velocity (or to both). Thus Δv, and accordingly a also, point directly inward toward the center of the circle.

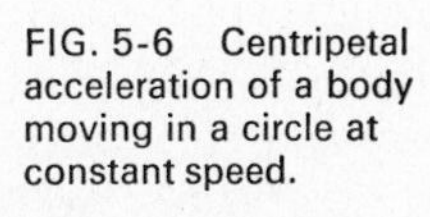

FIG. 5-6 Centripetal acceleration of a body moving in a circle at constant speed.

There can be no acceleration without a force to cause it, and we see at once that the center-directed acceleration of our mass is caused by the constant inward pull of the rod on which it is mounted. The force that the rod exerts on the mass is called the *centripetal* force; there is, of course, an equal and opposite *centrifugal* force that the mass exerts on the rod. The magnitude of these forces comes from our old friend:

$$F = ma; \qquad F = \frac{mv^2}{r} \quad \text{or} \quad F = mr\omega^2.$$

As an example, consider an ultracentrifuge 10 cm in diameter, spinning 1000 rev/s. In this device, $r = 5$ cm and $\omega = 2\pi \times 1000 = 6.28 \times 10^3$ rad/s. For the centripetal acceleration of its contents we have

$$a = r\omega^2 = 5 \times (6.28 \times 10^3)^2 = 1.97 \times 10^8 \text{ cm/s}^2.$$

This is $1.97 \times 10^8/980 = 2.01 \times 10^5$ times the normal acceleration due to gravity, and it would be described as $2.01 \times 10^5\ g$.

A more familiar example is that of a car rounding a curve. Let us take a 1500-kg car making a turn of $r = 40$ m on a level, unbanked road. The coefficient of friction between tires and road is 0.50. How fast can the driver try to make this curve? Figure 5-7A illustrates the situation. The force F necessary to give the car its centripetal acceleration comes from the force of friction between the tires and the road.

$$F = \mu mg = 0.50 \times 1500 \times 9.8 = 7350 \text{ N}.$$

We can find his maximum permissible speed by setting this $F = mv^2/r$:

$$7350 = \frac{1500v^2}{40}$$

$$v^2 = 7350 \times \frac{40}{1500} = 196$$

$$v = 14.0 \text{ m/s.} \quad \text{(about 31.3 mi/hr.)}$$

At a speed higher than this, the $F = \mu mg$ will not be able to accelerate the car inward enough for it to follow the curving roadway. The driver then has a choice—he can steer the car in a circle of greater r than the road, and thereby drive off the road to the right; or allow friction to do its inadequate best, and skid off the road to the right.

To permit higher speeds on turns, roads are often banked, or tilted. Figure 5-7B shows a section through a banked turn. The force F exerted by the roadway on the car must do two things: Its vertical component F_v must support the vertical weight of the car, mg; its horizontal component F_h must provide the necessary centripetal force mv^2/r. If the total F is perpendicular to the road, it will have no component parallel to the road,

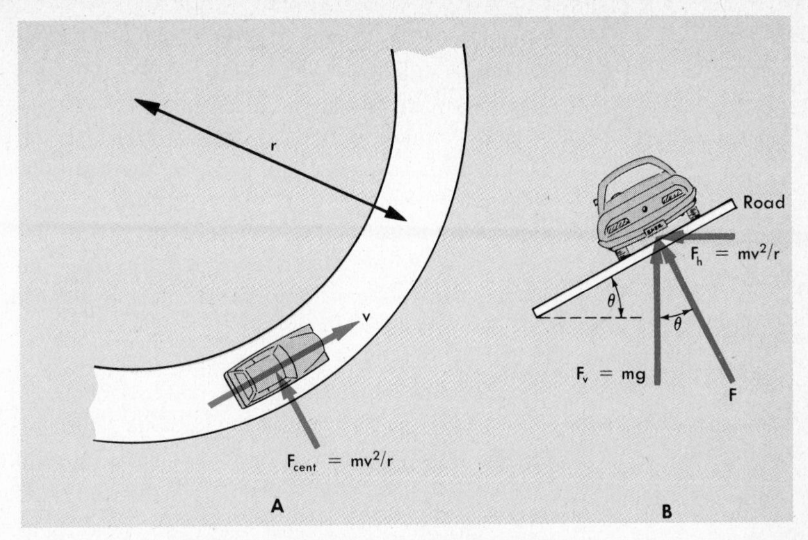

FIG. 5-7 On a properly banked curve, at the speed it was designed for, the total force of road against tires is perpendicular to the banked roadway.

and therefore no sideways friction between tires and road is needed. For this to be true, the road is banked at an angle θ, so that

$$\tan\theta = \frac{F_h}{F_v} = \frac{mv^2}{r \times mg} = \frac{v^2}{rg}.$$

5-5 Centrifugal Force

In the preceding section, only the center-directed centripetal force was mentioned—in fact, this was all that was needed to explain the behavior of bodies moving in curved paths. What of the "*centrifugal* force" we so often speak of? The "-fugal" is from the same root as "fugitive," and centrifugal force means a force directed *away* from the center of rotation. Figure 5-7B shows the force F exerted on the car by the road. From Newton's third law, we know there is an equal and opposite force exerted on the road by the car. This force does have a centrifugal horizontal component, but it is exerted on the road, not on the car.

Consider a man driving a car, with a smooth package on the seat beside him. As he makes a sharp left turn, the package slides across the seat to the right. From our point of view, standing by the roadside, the reason for this behavior is obvious: The friction against the seat is not enough to provide the package with a centripetal force that will cause it to follow a path with the same radius as the car. Instead, it follows a curve with larger r, and accordingly moves across the seat to the right.

The man in the car, however, sees the package slide to his right whenever he makes a left turn—this is a motion *away* from the center of his curved path. He would like to be able to apply Newton's $F = ma$ in his car as well as in the laboratory. The only way he can do this is to say there is a centrifugal force acting on the package. If he bothers to do a little experimenting, he will find that this hypothetical, or fictitious, force is exactly $F = mv^2/r$. The need for this fictitious centrifugal force arises because he wants to refer the motion to the car, which is itself accelerating. There is nothing unreal or illegal about this—if you want to use an accelerated frame of reference, it will merely be necessary to introduce such fictitious forces* in order to make things follow the laws of physics.

5-6 Rotational Work and Energy

Because work is required to make a heavy wheel spin and because the spinning wheel can be made to do work, it is apparent that kinetic energy is stored in a rotating body in much the same way that it is stored in a body having translational motion. To calculate the KE of a rotating wheel, we need only add the KE's of all the small particles that we consider the wheel to be made up of. Look back to Fig. 5-2 and imagine the mass m to be one small particle of a wheel revolving about axis OO'. If this small particle is moving at a linear speed v, its KE is $\frac{1}{2}mv^2$.

It would be very difficult, however, to add up the KE's of all the particles of the wheel if they were expressed in terms of v, because the speed of each particle depends on its distance from the wheel's axis of rotation.

We can get around this difficulty by converting our expression for KE into one in terms of ω, because the *angular* speed is the same for all the particles of a rotating body:

$$v = r\omega, \qquad \text{so} \qquad v^2 = r^2\omega^2.$$

Then

$$\text{KE} = \tfrac{1}{2}mv^2$$

becomes

$$\text{KE} = \tfrac{1}{2}mr^2\omega^2 = \tfrac{1}{2}I\omega^2.$$

Here the I is the moment of inertia of the whole rotating body, as given in Fig. 5-3, and we have an exact analog of the $\text{KE} = \frac{1}{2}mv^2$ used in translational motion.

In investigating the mechanics of translation, we saw that a body could be given energy—either potential or kinetic—by doing work on it. The same ideas apply to rotation. Of course, a body cannot be given gravitational potential energy merely by rotating it, but there are other forms of potential energy. You do work when you wind your watch

---

* In addition to centrifugal force, there is another fictitious force arising in rotational motion. This is the Coriolis force, which we will look at briefly in the chapter on geophysics.

and this is stored as PE in the wound-up spring. Work done in rotation is measured by the analog of force × distance—that is, torque × angle:

$$W = \tau\theta \quad (\theta \text{ in radians, naturally!}).$$

To give a moving body a certain amount of kinetic energy, we must do an equal amount of work on the body: $Fd = \frac{1}{2}mv^2$. Similarly, to give a rotating wheel, for example, a certain amount of rotational kinetic energy, we must also do an equal amount of work: $\tau\theta = \frac{1}{2}I\omega^2$.

Let us use the idea of rotational kinetic energy to compare the speeds with which a hoop and a solid disk reach the bottom of a slope. Figure 5-8 shows the hoop and disk ready to start their trip down, a journey that will consist of both rotation and translation. If no energy is lost in friction, the KE at the bottom of the slope must equal the PE at the top, or

$$\tfrac{1}{2}Mv^2 + \tfrac{1}{2}I\omega^2 = Mgh.$$

For the hoop,

$$I = MR^2$$

so

$$\tfrac{1}{2}Mv^2 + \tfrac{1}{2}MR^2\omega^2 = Mgh.$$

The mass of the hoop cancels out, which tells us that whether it is a wooden hoop or a lead hoop of the same size makes no difference, so that

$$\tfrac{1}{2}v^2 + \tfrac{1}{2}R^2\omega^2 = gh.$$

We are not through yet, since we know that $v = \omega R$. Thus $R^2\omega^2 = v^2$, and the equation becomes

$$v_{\mathrm{H}}^2 = gh \quad \text{or} \quad v_{\mathrm{H}} = \sqrt{gh}.$$

FIG. 5-8 Hoop and solid cylinder ready to race down an inclined plane.

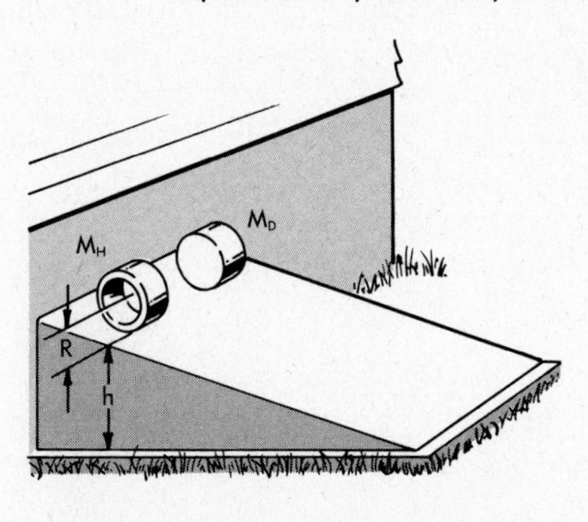

We see now that R has vanished, as the mass did, so the size of the hoop has no effect on the speed. A hoop made of paper, a hoop made of iron, a small hoop, and a large hoop will all roll down the slope side by side.

If you do the same computation for the solid disk, whose moment of inertia is $\frac{1}{2}M_D R^2$, you will find that

$$v_D = \sqrt{\tfrac{4}{3}gh}.$$

Therefore, the solid disk, which travels faster than the hoop, will reach the bottom first. How will a rectangular block do in this race, if it slides down the slope without friction?

5-7 Angular Momentum

Just as linear momentum was defined as mv, so *angular momentum*, which is the momentum a body has because of its rotation, is defined as $I\omega$. Newton's third law of action and reaction applies to torques as well as to forces, and the law of conservation of angular momentum can be proved in the same way as conservation of linear momentum.

The conservation of angular momentum can be very easily demonstrated by a student standing (or sitting) on a nearly frictionless rotating platform (Fig. 5-9A). Someone sets her spinning at a moderate speed that will presumably stay constant until it is gradually diminished by the unavoidable torque of friction. But if the passenger extends her arms (Fig. 5-9B), she at once slows down; when she returns her arms to their original position, she regains her former speed (Fig. 5-9C). The principle of conservation of angular momentum tells us that (neglecting the effect of the torque of bearing friction) the product $I\omega$ remains unchanged. So, when the rotating student increases her moment of inertia by extending her arms, her angular speed diminishes in the same proportion; when she reduces I, ω must immediately increase to keep the product $I\omega$ constant.

Spaceships will occasionally need to reverse the direction in which they point in flight—when, for instance, they want to come in for a landing on Mars tailfirst with their rockets blasting in order to brake the ship to a gentle stop. It has been suggested that this could be done by mounting a heavy wheel on an axis perpendicular to the axis of the ship. If the wheel is set rotating, the ship must rotate in the opposite direction so that the total angular momentum will continue to add up to zero. (The ship will turn much more slowly than the wheel, of course; since its I is enormously larger than that of the wheel, its ω must be correspondingly smaller.) When the ship has rotated far enough, one need only stop the wheel and the ship must also stop rotating.

5-8 The Gyroscope

A gyroscope is merely a massive wheel that can be set spinning at high speed. It is generally mounted in low-friction bearings on a frame so that it can be moved and handled while spinning. Its behavior may seem strange—ads for toy gyroscopes say "Fantastic!" and "Defies Gravity!" A closer look, however, will show that it is merely following the rules, like everything else we encounter.

Let us first consider a translational analog. Figure 5-10A shows a force F that is applied for a short time t, representing an impulse Ft. This impulse equals the change in momentum, Δp, given to any body to which it is applied. Figure 5-10B shows this impulse applied to a stationary body of mass m; it gives it a momentum $mv = Ft$ in the direction of F. Now (Fig. 5-10C), apply it to the same body moving with velocity v_1. The Δp vector, added to the initial momentum mv_1, gives a changed momentum mv_2. We have changed its direction by the small angle θ, which we can measure as $\Delta p/mv$ radians.

Vectors can be similarly used to represent characteristics of rotational motion. To do this, we use the *right-hand screw rule*. Imagine turning an ordinary screw in the direction a wheel is rotating; the vector representing ω points in the direction the screw would move, and is parallel to the axis of rotation. Alternatively, you can curl the fingers of your right hand to follow the rotation; your extended thumb will give the direction of the vector ω. Figure 5-11A shows a spinning wheel with its axle supported at one end in a bearing that is free to rotate about the stand on a vertical axis, and that can also pivot freely about the horizontal axis HH'. The right hand above the drawing has its fingers curled in the direction the wheel is spinning—the thumb shows the direction of ω. The screw, turned in the direction of spin, would be driven forward, to again give the same direction for the vector ω. The weight of the wheel, mg, produces a torque τ rotating clockwise around HH', as viewed from the lower right. The right-hand, or screw, rule applied to this torque about HH' gives the direction of the vector τ as indicated. Since I and t are both scalars, an impulse τt is in the same direction as τ, and the angular momentum of the wheel, $I\omega$, is in the same direction as ω.

Now let us set the wheel to spinning and see what will happen. $I\omega_1$ (Fig. 5-11B) is its initial angular momentum. Upon release of the axle, the torque $\tau = mgd$ is immediately applied. We can find what the effect of τ will be after some short time t: It will have produced an angular impulse τt (analogous to a linear impulse Ft). This impulse is equivalent to a change in angular momentum Δp, which must be added to the initial angular momentum $I\omega_1$. Figure 5-11C shows how this leads to a new angular momentum $I\omega_2$. Since the angular momentum of the spinning wheel lies along its axle, this means that the axle must have rotated in a horizontal plane through θ radians in the time t. The rotation of the axle in this way is called *precession*, and Ω, the angular speed of precession, is thus $\Omega = \theta/t$. From Fig. 5-11C, the angle θ in radians is seen to be $\tau t/I\omega$. This gives

$$\Omega = \frac{\tau t}{I\omega \times t} = \frac{\tau}{I\omega}.$$

A: Small I: Large ω:

B: Large I: Small ω:

C: Small I: Large ω:

FIG. 5-9 Demonstration that the angular momentum of a rotating body remains constant when no external torque is applied.

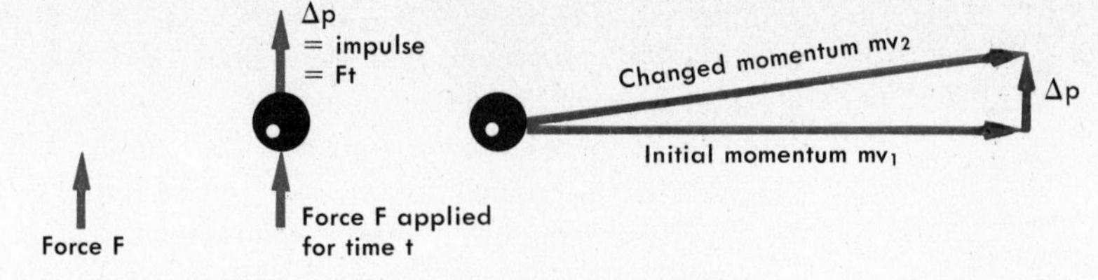

FIG. 5-10 An impulse ($Ft = \Delta p$) changes the direction of the ball's momentum.

The torque does *not* cause the spinning wheel to drop, but to swing around (in our example) in a counterclockwise direction, viewed from the top.

If we assume the wheel, which has a heavy rim, to be almost a hoop, we can figure its moment of inertia as MR^2. For a 500-g wheel 5 cm in radius, I would be 12,500 g-cm^2. If the distance d is 4 cm, τ is $4 \times 500 \times 980 = 1{,}960{,}000$ dyne-cm. A reasonable rotational speed would be 1800 rev/min, or $1800 \times 2\pi/60 = 188$ rad/s. Under these conditions, we could confidently predict that the wheel would precess at the rate of $1.96 \times 10^6/(1.25 \times 10^4 \times 1.88 \times 10^2) = 0.835$ rad/s.

FIG. 5-11 An angular impulse ($\tau t = \Delta I\omega$) changes the direction of angular momentum, and causes gyroscopic precession.

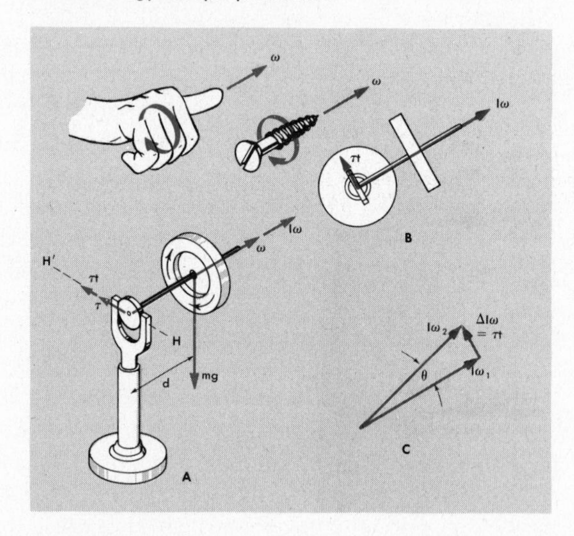

Questions

(5-1) **1.** Through how many radians does the earth rotate between noon at Greenwich (0° longitude), and noon at Denver (105° west longitude)?

2. A wheel has 12 equally spaced spokes. What is the angle between adjacent spokes, in radians?

3. (a) What is the angular speed of the earth's revolution around the sun, in rad/hr? (b) With this figure, determine the earth's orbital speed, taking the radius of its orbit to be 93×10^6 mi.

4. (a) What is the angular speed of the earth's rotation about its axis, in rad/hr? (b) With this figure determine the linear speed of a point on the earth's equator, using 8000 mi as the earth's *diameter*.

5. A centrifuge 10 cm in diameter rotates at 30,000 rev/min. (a) Convert this angular speed to rad/s. (b) What is the linear speed of the rim of the centrifuge?

6. A grinding wheel 8 inches in diameter rotates with an angular speed of 2400 rev/min. (a) Convert this angular speed to rad/s. (b) What is the linear speed of the rim of the wheel, in ft/s?

7. The centrifuge in Question 5 takes 4 min to attain its final speed. What is its average angular acceleration, in rad/s^2?

8. In Question 6, the wheel requires 4 s to obtain its final speed. What is its average angular acceleration, in rad/s^2?

9. A dental drill accelerates from rest to 9000 rev/min in 2 s. (a) What is its average angular acceleration? (b) How many revolutions does it make in coming up to full speed?

10. A wheel accelerates uniformly at 60 rad/s^2. Through how many revolutions does it turn to attain a speed of 3000 rev/min, starting from rest?

(5-2) **11.** What is the moment of inertia of a 160-lb solid cylinder 2 ft in diameter?

12. What is the moment of inertia of a 10-kg sphere 30 cm in diameter?

13. A slender rod 2 m long has a mass of 8 kg. What is its moment of inertia about an axis perpendicular to the rod at one end?

14. What is the moment of inertia of the rod in Question 13 about an axis perpendicular to the rod through its center?

(5-3) **15.** A circular saw (which can be considered to be a very short solid cylinder) is 8 inches in diameter, and weighs 4 lb. What torque is required to give it an angular acceleration of 80 rad/s^2?

16. A cylindrical grinding wheel has a mass of 3 kg and is 20 cm in diameter. (a) What is its moment of inertia? (b) What torque will be required to give it an angular acceleration of 120 rad/s^2?

17. An irregular 5-kg wheel is given an acceleration of 10 rad/s^2 by an applied torque of 2×10^7 dyne-cm. What is the moment of inertia of the wheel?

18. It is found that a torque of 8 oz-in. gives an irregularly shaped wheel an acceleration of 2 rad/s^2. What is the moment of inertia of the wheel?

19. What weight would have to be suspended from the cord of Fig. 5-4 in order to produce a 40-lb tension in it?

20. Refer to Fig. 5-4. What weight would have to be suspended from the cord in order to produce the 24-lb tension shown?

(5-4) **21.** A 2560-lb car rounds a level 300-ft radius curve at 40 mi/hr. With what frictional force must the paving press the tires toward the inside of the curve?

22. A 50-ton locomotive goes at 45 mi/hr around a curve whose radius is 800 ft. With what force do the rails press against the flanges of the wheels?

23. The mass of the moon is about 7.4×10^{25} g and the radius of its orbit around the earth is about 3.8×10^5 km. It makes one revolution around the earth in 27.3 days. (a) What is the force of gravitational attraction between earth and moon? (b) How much is this, in metric tons?

24. The mass of the earth is very nearly 6×10^{27} g and the radius of its orbit around the sun is about 1.50×10^{13} cm. (a) What is the force of gravitational attraction between earth and sun? (b) How much is this, in metric tons?

25. A small mass hangs from a thread 30 cm long. The mass is struck and given an initial velocity v that will carry the mass around in a vertical circle just fast enough so that the string will not become limp at the top of the circle. What is the required value of v? Let us lead up to this with a few intermediate questions: (a) What will be the necessary speed v_T at the top of the circle? (b) What is the KE of the mass at the top? (c) What is its PE? (d) What, then, must be the KE + PE at the bottom of the circle? (e) What is the PE at the bottom? (f) What KE must it have at the bottom? (g) What speed must it have at the bottom?

26. On Planet Z, where $g = 8$ m/s^2, a small mass hangs from a thread 20 cm long. The mass is struck and given an initial velocity v that will carry the mass around in a vertical circle just fast enough so that the string will not become limp at the top of the circle. What is the required value of v? (See Question 25, if necessary.)

27. Patrons of a certain amusement park concession stand with their backs against the wall of a cylindrical room 12 ft in diameter. The room rotates about its central axis with increasing speed until a certain angular velocity is reached. The operator then lowers the floor, and the patrons remain in position against the wall, held there by friction. What angular velocity must the room have, if $\mu = 0.30$?

28. A small carousel is 4 m in diameter and rotates once in 10 s. If we place a wood block on the floor at the outer edge of the carousel, will it stay in place or be flung off, if its coefficient of friction is 0.10?

(5-6) **29.** A sphere 10 cm in diameter has a mass of 3 kg, and rotates at 5 rev/s. What is its rotational KE?

30. A 10-kg cylinder 12 cm in diameter rotates about its axis at an angular speed of 2 rev/s. What is its rotational KE?

31. What will be the velocity of a rolling sphere, in terms of the height of a slope down which it rolls?

32. What will be the velocity of a sliding frictionless block, in terms of the height of a slope down which it slides?

33. A solid cylinder rolls down an inclined plane without slipping. What fraction of its KE is translational? rotational?

34. A hoop rolls down an inclined plane without slipping. What fraction of its KE is translational? rotational?

(5-7) **35.** A student with arms extended horizontally stands on a rotating platform, and is given an angular speed of 1 rev in 1.5 s. In this position, his I is 3 slug-ft^2. He brings his arms straight down at his sides, and speeds up to 1.5 rev/s. What is his I in his latter position?

36. A student standing on a ball-bearing platform with his arms at his sides has an $I = 2$ slug-ft^2. He is given an angular velocity of 1 rev/s; he then extends his arms out horizontally, increasing his I to 4 slug-ft^2. What is his angular velocity with arms extended?

37. In Question 35, does the student have the same KE in both positions? If not, explain.

38. Compare the student's rotational KE in his two positions of Question 36. If there is a change in KE, how do you account for it?

39. A spaceship has a moment of inertia about its center of 8×10^8 kg-m^2. The direction in which it points is controlled by a wheel with a heavy rim having a mass of 100 kg and a mean radius of 50 cm. How long will the wheel have to be rotated at 600 rev/min in order to rotate the ship by 90°? (Note that this can have no effect on the direction in which the ship travels.)

40. Two wheels are mounted side by side on the same axle. Wheel A, whose moment of inertia is 5×10^5 g-cm^2, is set spinning at 600 rev/min. Wheel B, with a moment of inertia of 2×10^6 g-cm^2, is stationary. A clutch now acts to join A and B so that they must spin together. (a) At what speed will they rotate? (b) Suppose the clutch acts gradually. Will the end result be the same as though they were joined suddenly? (Assume the bearings are frictionless.) (c) How does the rotational KE before joining compare with the KE afterward? (d) What torque will the clutch have had to transmit if A makes 10 revolutions *relative to B* during the operation of the clutch?

(5-8) **41.** Some planes (now obsolete) had massive engines that rotated counterclockwise as viewed by the pilot sitting behind them. When the plane nosed down to go into a dive, would it swing to the left or to the right as a result of gyroscopic action?

42. If the crankshaft and flywheel of a car rotate clockwise when viewed from the driver's seat, will the car tend to nose down or nose up when the car makes a right turn?

43. In the gyroscope of Fig. 5-11, as described in the last paragraph of Section 5-8, slide the wheel on its axle so the distance d is 3 cm. How fast (in rev/min) will the wheel have to spin in order to precess one revolution in 10 sec?

44. In the gyroscope of Fig. 5-11, as described in the last paragraph of Section 5-8, we slide the wheel on its axle until the rate of precession is just 1 rev in 10 s. What value does the distance d need to have?

6

Gravitation and Orbits

6-1 Newton's Law of Gravitation

It is interesting to apply the laws of centripetal force to the moon as it swings around the earth in a nearly circular orbit with a radius averaging 384,000 km. The moon takes 27.3 days to make one revolution about the earth, which in radians/second is

$$\omega = \frac{1 \text{ rev}}{27.3 \text{ day}} \times \frac{2\pi \text{ rad}}{1 \text{ rev}} \times \frac{1 \text{ day}}{24 \text{ hr}} \times \frac{1 \text{ hr}}{3600 \text{ s}}$$

$$= 2.67 \times 10^{-6} \text{ rad/s}.$$

We also know that

$$r = 384{,}000 \text{ km}$$

$$= 3.84 \times 10^{8} \text{ m}.$$

This gives us for the moon's centripetal acceleration toward the earth

$$\begin{aligned} a &= r\omega^2 \\ &= 3.84 \times 10^8 \times (2.67 \times 10^{-6})^2 \\ &= 2.74 \times 10^{-3}\ \text{m/s}^2. \end{aligned}$$

In the latter part of the seventeenth century, Newton did calculations similar to the ones above and made the proposal (very radical and controversial at the time) that the moon's earthward centripetal acceleration is caused by the same force that causes an apple to accelerate earthward when falling from a tree. The acceleration of the falling apple—g, which is about 9.80 m/s^2 or 980 cm/s^2—is much greater than the moon's earthward acceleration that we have just figured, but it seemed reasonable to Newton that the earth's force of gravitational attraction should become weaker with increasing distance. In fact, the figures can be quite satisfactorily reconciled by assuming that the gravitational pull decreases in proportion to the square of the distance from the earth's center; i.e., at twice the distance, the force is one-fourth as great; at three times the distance, it is one-ninth as great, etc.

The falling apple is 6380 km from the center of the earth, and the moon, at an average distance of 384,000 km, is farther by factor of 60.2. We should expect that the acceleration of the moon would be smaller than that of the apple by a factor of 60.2^2, or 3624. The fraction, 9.80/3624, equals 2.70×10^{-3} m/s^2, a figure that checks very closely with the 2.74×10^{-3} m/s^2 we obtained from purely geometrical considerations of the moon's orbit. (The agreement would have been exact if we had not avoided mathematical difficulties by neglecting a minor complication. In figuring the centripetal acceleration of the moon, we tacitly assumed the earth to be stationary while the moon circled around it; actually, both the moon and the earth revolve around their center of gravity, located about 3000 mi from the earth's center.)

Newton also tested his ideas on the orbits of the planets around the sun and finally proposed his famous *law of universal gravitation*: ***Every particle of matter in the universe attracts every other particle with a force proportional to the product of the masses of the particles, and inversely proportional to the square of the distance between them.***

In shorter algebraic form,

$$F = \frac{Gm_1m_2}{d^2}.$$

In Newton's day, it was not possible to evaluate the proportionality constant G, because the force of attraction between any masses that can be handled in the laboratory is very small indeed. A century after Newton, Henry Cavendish used a special type of balance to determine the value of G by measuring the force of attraction between known masses. Later more accurate work has given us the values

$$G = 6.67 \times 10^{-11} \text{ N-m}^2/\text{kg}^2 \text{ in MKS units,}$$

or

$$G = 6.67 \times 10^{-8} \text{ dyne-cm}^2/\text{g}^2 \text{ in CGS units.}$$

This means that two 1-kg masses 1 m apart will attract each other with a force of 6.67×10^{-11} N; or that two 1-g masses 1 cm apart will attract each other with a force of 6.67×10^{-8} dyne.

Once G is known, the mass of the earth can be readily determined. The force of attraction between the earth and a 1-kg mass is known to be the *weight* of the kilogram, which is about 9.80 N. The 1-kg mass is separated from the earth's center by the earth's radius, 6.38×10^6 m. Substituting these values in the equation, we get:

$$9.80 = \frac{6.67 \times 10^{-11} \times 1 \times m_e}{(6.38 \times 10^6)^2}$$

$$m_e = 5.98 \times 10^{24} \text{ kg} = 5.98 \times 10^{27} \text{ g.}$$

6-2 Weight on Other Planets

A handbook informs us that the planet Mars has a mass 0.1065 times the mass of the earth; its radius is 3430 km, or 3.43×10^6 m. Consider an astronaut who, with all his equipment, weighs 250 lb on the earth. What will he weigh when he sets foot on Mars? One perfectly straightforward method is to calculate the attraction between man and Mars. For data we have: $m_{\text{man}} = 250 \text{ lb} \times 0.454 \text{ kg/lb} = 1.135 \times 10^2$ kg; $m_{\text{Mars}} = 5.98 \times 10^{24} \times 0.1065 = 6.37 \times 10^{23}$ kg. The distance between them is the radius of Mars, $r = 3.43 \times 10^6$ m. This gives us

$$\text{Weight} = F = \frac{6.67 \times 10^{-11} \times 1.135 \times 10^2 \times 6.37 \times 10^{23}}{(3.43 \times 10^6)^2}$$

$$= 410 \text{ N.}$$

In order to make a comparison with the astronaut's earthly 250 lb, newtons must be converted to pounds:

$$4.10 \times 10^2 \text{ N} \times \frac{1 \text{ kg}}{9.80 \text{ N}} \times \frac{1 \text{ lb}}{0.454 \text{ kg}} = 92.1 \text{ lb.}$$

By using a little more reasoning and a little less arithmetic, we can arrive at the same answer by a different route. Let us consider separately the two factors responsible for the change in weight: Mars' different mass, and its different radius. If Mars were the same size as the earth, its smaller mass would make the man weigh only 0.1065×250 lb. And if Mars and the earth differed only in size, the astronaut would be closer to the center when on Mars, which would make him weigh more. This distance factor would be $(6.38 \times 10^6/3.43 \times 10^6)^2$. (Since we know the decreased distance will increase his weight, we need only to see that the larger radius is in the numerator, to make the fraction greater than 1.) Combining the two gives

$$\text{Weight} = 250 \times \frac{0.1065}{1} \times \left(\frac{6.38 \times 10^6}{3.43 \times 10^6}\right)^2 = 92.1 \text{ lb.}$$

(Note, however, that although his *weight* is less on Mars, his *mass* is unchanged.)

6-3 The Speed of Satellites

Consider a satellite circling the earth in an orbit of $r = 10^4$ km. At what speed is it traveling? Since it is moving in a curved circular path, it must have a centripetal acceleration directed toward the center of its orbit. It seems apparent that the centripetal force causing this acceleration is the gravitational attraction between satellite and earth. This can easily be stated algebraically:

$$\text{Centripetal force} = \text{gravitational attraction}$$

$$\frac{mv^2}{r} = \frac{GMm}{r^2}$$

where m is the mass of the satellite, M is the mass of the earth, and r is the radius of the orbit. A little canceling of m's and r's gives

$$v^2 = \frac{GM}{r}$$

$$v = \sqrt{\frac{GM}{r}}.$$

We can now answer the original question, which gave r to be 10^4 km, or 10^7 m.

$$v = \sqrt{\frac{6.67 \times 10^{-11} \times 5.98 \times 10^{24}}{10^7}}$$

$$= \sqrt{39.9 \times 10^6} = 6.32 \times 10^3 \text{ m/s}$$

$$= 6.32 \text{ km/s.}$$

It is worth noticing that m, the mass of the satellite, cancels out, and thus has no effect on the velocity needed for this particular orbit. The speed v is the same whether the satellite has a mass of a gram or a million tons. G and M are of course constants, so, for any satellite circling the earth, r is the only variable factor determining v. (Or we could put it the other way—that v determines r.)

If we look at two satellites (of the same central body), we see that for larger r, the fraction under the radical sign is smaller, so v is also smaller. The larger the orbit, the more slowly the satellite moves.

The *period* of a satellite is the time it takes to make one complete revolution around the central body. Figure 6-1 shows the orbits of the planets Earth and Mars. The orbit of Mars is larger than that of the earth, so we know that v_M is less than v_E. It is also obvious that Mars has a longer distance to travel in its period, T_M, than the earth does in its period, $T_E = 1$ yr. Thus Mars, moving more slowly than Earth, and also having a longer distance to go, must necessarily have a period greater than an Earth-year.

In the seventeenth century, a remarkably simple relationship between r and T was discovered by Johannes Kepler. We will look at this relationship in more detail later in the chapter.

6-4 Satellite Energy

It is easy to visualize the kinetic energy of the moon or of a satellite hurtling at high speed about the earth. We must not overlook, though, the fact that these bodies also have potential energy as well as kinetic energy. In our previous dealings with gravitational potential energy, it was necessary only to multiply the weight of a body by the height it was lifted above some chosen reference plane: The distances involved

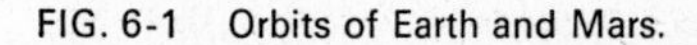
FIG. 6-1 Orbits of Earth and Mars.

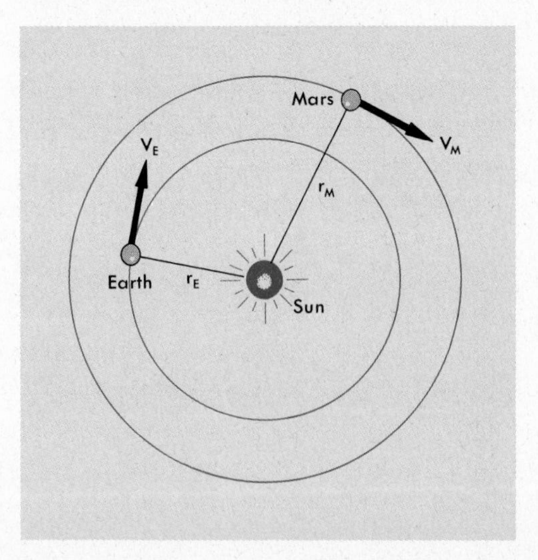

have been small enough for us to consider the pull of gravity to be constant. When we deal with distances of thousands of miles above the earth's surface, however, the pull of gravity is materially reduced, and this reduction must be taken into account.

Let us take the general case of two masses, M and m, and derive an expression for the work we must do against their gravitational attraction if we want to separate them. Figure 6-2 shows M and m; we shall compute the work needed to move m from point a to point f by dividing the whole distance into small steps and figuring the work for each small step separately.

The gravitational attractive force when m is at a will be GMm/r_a^2, where G is Cavendish's gravitational constant. At b, the force is GMm/r_b^2. If we take the *average* of the force at a and the force at b and multiply the average by the distance ab, we should get the work required to move m from a to b. There are several ways to take an average, and a calculus analysis of this problem shows that the proper one to use is the *geometrical average*. To find this average we multiply the forces together, and take the square root of their product.

By this method, we find that the average force between points a and b, which we can call F_{ab}, is

$$F_{ab} = \sqrt{\frac{GMm}{r_a^2} \times \frac{GMm}{r_b^2}} = \frac{GMm}{r_a r_b}.$$

The work done in moving m from a to b will be the force times the distance, which is

$$\begin{aligned} W_{ab} &= F_{ab}(r_b - r_a) \\ &= \frac{GMm}{r_a r_b}(r_b - r_a) \\ &= GMm\left(\frac{1}{r_a} - \frac{1}{r_b}\right). \end{aligned}$$

FIG. 6-2 The work done in moving apart masses that gravitationally attract each other.

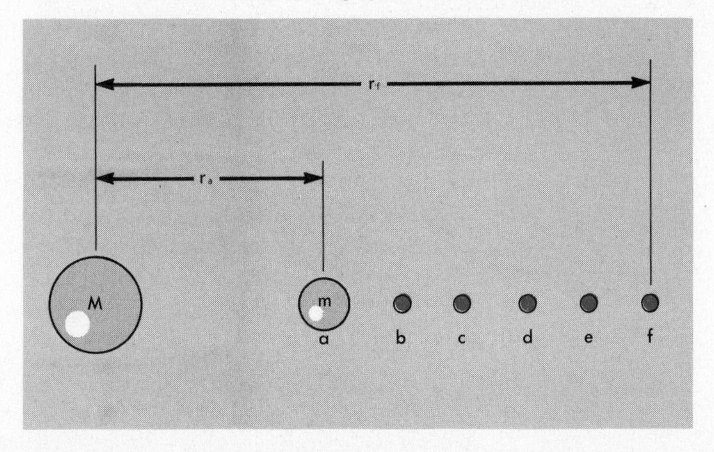

Similarly, the work done in moving *m* from *b* to *c* is

$$W_{bc} = GMm\left(\frac{1}{r_b} - \frac{1}{r_c}\right)$$

and so on until we reach point *f*.

We can now find the total work by adding all these pieces together. (Since each expression is multiplied by the factor *GMm*, let us lay *GMm* aside, and try not to forget to put it back when we are finished.) This gives

$$\begin{aligned}
& \frac{1}{r_a} - \frac{1}{r_b} \\
+ \quad & \frac{1}{r_b} - \frac{1}{r_c} \\
+ \quad & \frac{1}{r_c} - \frac{1}{r_d} \\
+ \quad & \frac{1}{r_d} - \frac{1}{r_e} \\
+ \quad & \frac{1}{r_e} - \frac{1}{r_f} \\
\hline
= & \frac{1}{r_a} + 0 + 0 + 0 + 0 - \frac{1}{r_f} = \frac{1}{r_a} - \frac{1}{r_f}.
\end{aligned}$$

So, putting our laid-aside factor back,

$$W_{af} = GMm\left(\frac{1}{r_a} - \frac{1}{r_f}\right).$$

With the expression above, we are able to compute the work needed to move *m* from any distance r_a from *M*, to any other distance r_f. The expression gives a somewhat surprising answer to the question, "How much work would be needed to move *m* to an infinite distance away?" The unthinking snap-judgment answer—"an infinite amount of work, of course"—sounds reasonable but is actually very wrong. Because of the inverse-square law of gravitation, as we go out farther and farther from any attracting body, the force gets small more rapidly than the distance gets large. As a result, the question has a definite, finite answer. As *m* is moved farther and farther away, r_f becomes increasingly large, and $1/r_f$ becomes smaller and smaller. At $r_f = \infty$ (∞ is the mathematician's symbol for infinity), $1/r_f$ becomes $1/\infty$, or zero. The amount of work needed to move *m* from point *a* to an infinite distance away thus becomes simply

$$W_{a\infty} = \frac{GMm}{r_a}.$$

In order to find a value for the potential energy, we must select a zero from which to measure it. The universal choice—because it makes the arithmetic so simple—is to call the PE zero when M and m are separated by an infinite distance. Since we must do work equal to GMm/r to increase their separation from r to ∞, the PE at distance r must be negative:

$$\text{PE} = -\frac{GMm}{r}.$$

The kinetic energy of the orbiting satellite is of course $\frac{1}{2}mv^2$. In Section 3 it was shown that $v^2 = GM/r$; substituting this for v^2 gives

$$\text{KE} = \frac{1}{2}m \times \frac{GM}{r} = \frac{1}{2}\frac{GMm}{r}.$$

The total energy is merely PE + KE:

$$\begin{aligned} E_{\text{total}} &= \text{PE} + \text{KE} \\ &= -\frac{GMm}{r} + \frac{1}{2}\frac{GMm}{r} = -\frac{1}{2}\frac{GMm}{r}. \end{aligned}$$

The choice made for PE = 0 at infinity leads to simple expressions for the energies of a satellite.

$$\begin{aligned} E_{\text{total}}\ (\text{negative}) &= \text{KE (positive)} \\ &= \tfrac{1}{2}\text{PE (negative)} \\ \text{PE (negative)} &= 2\text{KE (positive)} \\ &= 2E_{\text{total}}\ (\text{negative}) \\ \text{KE (positive)} &= E_{\text{total}}\ (\text{negative}) \\ &= \tfrac{1}{2}\text{PE (negative)}. \end{aligned}$$

6-5 Escape Velocity

A great deal has been written in popular articles about *escape velocity.* It is an important concept now that man has embarked on an exploration of the solar system. Escape velocity from the earth is given as about 7 mi/s, and is commonly defined as the speed a projectile must have in order to "break the bonds of gravity" or to "escape the earth's gravitational pull." Since the earth's gravitational pull extends to infinity (however weak it may be at great distances), escape velocity is apparently the speed a projectile must have at the earth's surface in order for it to be projected an infinite distance into space.

The work needed to move a projectile from the earth's surface (r distant from the earth's center) to infinity is GMm/r. Or we could say that we must increase its PE from $-GMm/r$ to zero, which is the same thing. This needed energy must come from the kinetic energy of the projectile, so we have

$$\frac{GMm}{r} = \frac{1}{2}mv^2$$

$$v^2 = \frac{2GM}{r}.$$

The letter M, of course, is the mass of the earth, which is 5.98×10^{24} kg. The fact that m cancels out of the formula shows that escape velocity is the same for bodies of all masses; whether they are molecules of the upper atmosphere or spaceships makes no difference. The radius of the earth, 6.38×10^6 m, is represented by r_e, and Cavendish's gravitational constant G is 6.67×10^{-11} in the MKS units we have used for M and r_e. By substituting these values in our equation we find

$$v^2 = \frac{2 \times 6.67 \times 10^{-11} \times 5.98 \times 10^{24}}{6.38 \times 10^6} = 12.5 \times 10^7$$

and

$$v = 1.12 \times 10^4 \text{ m/s}$$

$$= 11.2 \text{ km/s}$$

or

$$= 11.2 \times 0.6214 = 6.96 \text{ mi/s}.$$

The velocity needed to escape from another planet will in general be different from the 7 mi/s that we have figured for escape from the surface of the earth. It must be computed separately for each planet and will depend on the planet's radius and mass.

6-6 Weight and Weightlessness

"Weight" is a difficult thing to define. So far we have considered it to be entirely the result of the gravitational attraction between two bodies, one of them generally the earth. Let us say that you weigh 160.00 lb at sea level near the North Pole. (On a sensitive spring balance, of course—we are interested in *weight*, not *mass*.) Here you are 6.357×10^6 m from the earth's center. As you travel toward a warmer location at sea level near the equator, you calculate what you should expect to weigh at this new location, 6.378×10^6 m from the earth's center:

$$\text{Weight} = 160.00 \times \left(\frac{6.357}{6.378}\right)^2 = 158.94 \text{ lb}.$$

Disembarking at, say, Belem, Brazil, you step on the scale and find your weight is only 158.39 lb, rather than the expected 158.94. Auxiliary "weighings" on a balance indicate your mass is unchanged from that at the pole. What has happened to the missing 0.55 lb of weight? The answer is fairly obvious. At the pole there is no centrifugal force; on the equator you are moving 1 rev/day in a circle of the earth's radius. A little calculation shows the equatorial centrifugal force is equal to just 0.55 lb. Figure 6-3 shows the situation from two points of view. The point of view referred to the rotating (and therefore accelerated) earth is given in Fig. 6-3A. Here the man adds the "fictitious" centrifugal force and considers himself in equilibrium. From the viewpoint of an unaccelerated observer stationary in space, Fig. 6-3B, the man is *not* in equilibrium. He is subject to a net force of 0.55 lb toward the earth. This gives him the centripetal acceleration he must have to keep in his circular path.

In either case, however, the man's shoes press against the scales with a force of 158.39 lb, and this is what is generally accepted to be his weight. The values of g, the acceleration of free-fall, reflect the net effect of both gravitational pull and centrifugal force, so that weight equals mg is still exactly true.

The effects of a high-g environment on living organisms are fairly well-known and can be studied relatively easily. Pilots pulling out of steep dives can become temporarily blind or unconscious at centripetal accelerations of 5 or 6 g. Such accelerations can be easily reproduced and their effects studied in large centrifuges capable of holding a human

FIG. 6-3 Man standing on spring scales at the Equator, from two different points of view.

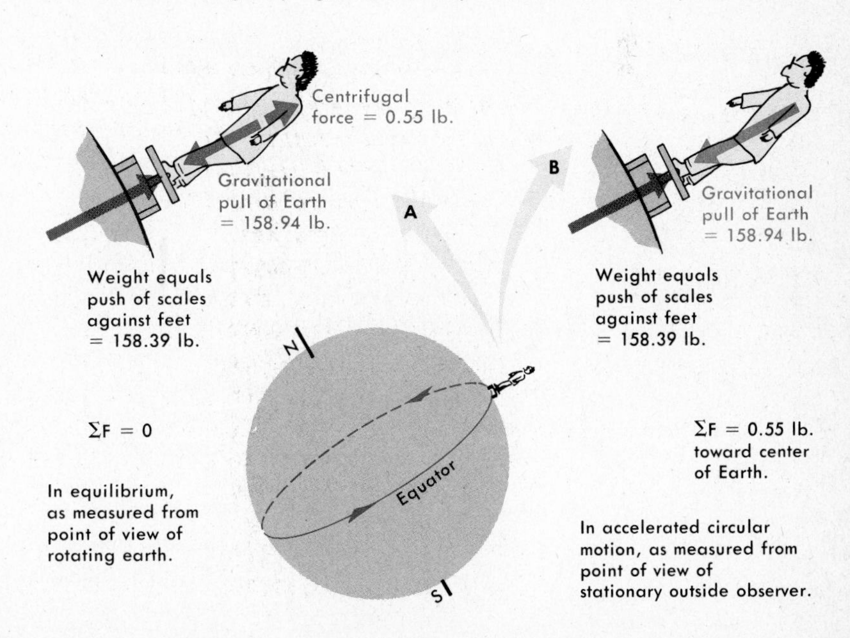

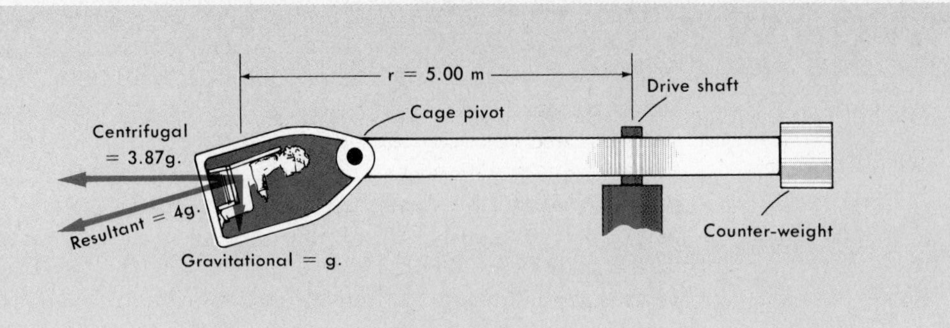

FIG. 6.4 A large centrifuge places a man in an artificially high-g environment.

being. Figure 6-4 sketches such a centrifuge, which is to be spun to subject the man to a total of 4 g. From the Pythagorean theorem, we see that his centripetal (or centrifugal) acceleration must be $\sqrt{15} \times g$, or 3.87 g, in order to give a total resultant of 4 g. From this,

$$r\omega^2 = 3.87g; \qquad \omega^2 = 3.87\frac{g}{r}$$

$$= 3.87 \times \frac{9.80}{5.00} = 7.59$$

$$\omega = \sqrt{7.59} = 2.75 \text{ rad/s}$$

$$= 2.75 \times \frac{60}{2\pi} = 26.3 \text{ rev/min.}$$

At this angular speed, the man will weigh four times what he does in his normal $1g$ environment. His blood will now weigh 4 g/cm^3 instead of its normal 1 g/cm^3. This "heavy" blood migrates to his legs and lower body, where it makes more room for itself by expanding the elastic blood-vessels there. This drains the blood supply away from retina and brain, causing the temporary blindness and unconsciousness that may occur.

Weightlessness is a more difficult condition to produce and study for any length of time. We define the weight of a body to be the force it exerts on whatever is supporting it. In free-fall there is no support, and, accordingly, no weight. It is easy to be weightless for a fraction of a second, merely by jumping off a chair or table; during the fall all parts of the body accelerate downward together. For this brief period, the shoulder does not support any weight from the arm, the back is not compressed by any weight of the torso—i.e., you are weightless. Skillful piloting can make a plane follow exactly along the path that a projectile would follow in a vacuum—any path, that is, with v_h constant, and a_v = 9.80 m/s^2 downward. In such a flight the passengers accelerate freely toward the earth without any contact with or support from the cabin walls. Unfortunately, however, these flights cannot last longer than a few seconds, which gives little opportunity for study of the effects of weightlessness.

In manned satellites circling either Earth or Moon, and during the free-fall trip between the two, the situation is the same as that in the plane guided along a projectile path in the atmosphere. Both cabin and passengers are constantly subjected to identical accelerations; both follow along identical orbits without any supporting forces being needed, or even possible. American and Russian astronauts have experienced weightlessness for weeks without any apparent lasting ill effect.

6-7 Kepler's Laws

So far, our calculating has been done as though the orbits we considered were perfectly circular. For many purposes this is a good enough approximation, although scientists have known for more than 300 years that the orbits of planets and other satellites are ellipses. Figure 6-5 sketches one way to draw an ellipse. A loose loop of string is slipped around a pair of pins stuck in a drawing board; a pencil keeping the string taut as it moves around the pins will draw an ellipse. The pins mark two geometrical points known as the *foci* (*foci*: plural of *focus*) of the ellipse.

The discovery that the orbits of planets and satellites are ellipses was made by the German mathematician and astronomer Johannes Kepler early in the seventeenth century. His discoveries can be summarized by three laws:

I. Each planet revolves around the sun in an elliptical orbit, with the sun at one focus of the ellipse.

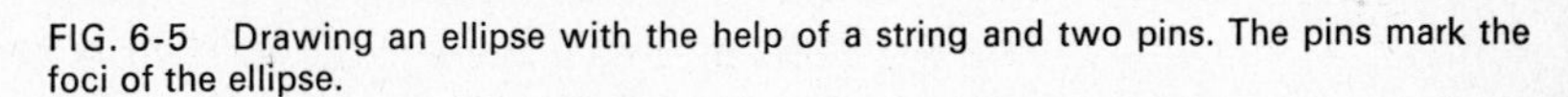
FIG. 6-5 Drawing an ellipse with the help of a string and two pins. The pins mark the foci of the ellipse.

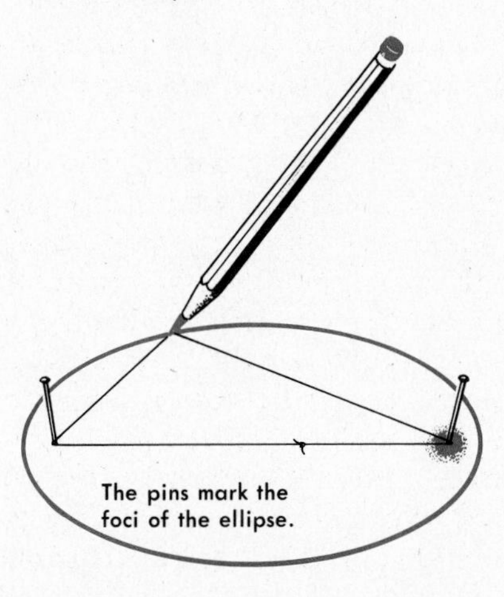

Kepler's first law can be shown to follow as a necessity from the inverse-square law of gravitation. The mathematical proof is too difficult and involved to be included here, but Kepler's second law may at least make it seem plausible.

II. The speed of a planet in its orbit varies in such a way that the radius connecting the planet and the sun sweeps over equal areas in equal times.

Figure 6-6 shows a very elliptical orbit typical of a comet—the orbits of the planets are more nearly circular. Astronomers observe this satellite of the sun to be at point 1. After some definite time t—say, 30 days later—it has moved to 2, and the radius drawn between satellite and sun has swept over the shaded area S12. Later, the body is observed to move from 3 to 4 in an equal 30-day time, with the radius sweeping over the area S34. Similarly, from 5 to 6, the radius sweeps over S56 in 30 days. Kepler, accurately plotting data from observations made by the Danish astronomer Tycho Brahe, found the areas S12, S34, etc., were all equal, as stated in his second law. From this, it is obvious that the speed of the satellite varies along its orbit, since the distances 1–2, 3–4, 5–6, covered in equal times, are proportional to the speed. The body moves fastest when closest to the sun, and more slowly as r increases.

Figure 6-6B indicates this is what we should expect. The vector F represents the gravitational attraction of S, and we can break this pull into components parallel to v and perpendicular to v, as shown. On the upper part of the drawing, the satellite is receding from the sun, and $F_{\parallel}$ is opposed to v, thereby slowing it down. Later, when the satellite is approaching S, $F_{\parallel}$ is in the same direction as v, thereby increasing the speed as r decreases. Throughout the orbit, $F_{\perp}$ provides a centripetal force that curves the orbit without affecting the speed.

III. The squares of the periods of any two planets are in the same ratio as the cubes of their average distances from the sun.

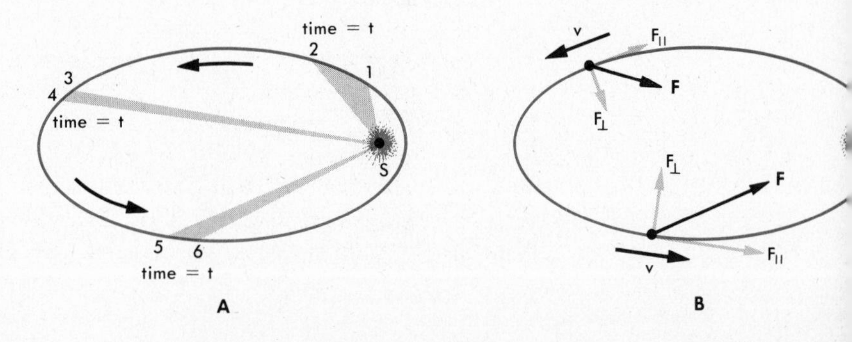

FIG. 6-6 Kepler's second law describes a planet's change in speed in different parts of its orbit.

The third law follows directly from the laws of mechanics. From Section 5 we know that

$$v_1^2 = \frac{GM}{r_1}.$$

where the subscript 1 refers to a particular satellite orbiting around a central body of mass M. The average speed also equals the circumference of the orbit divided by its period T_1:

$$v_1 = \frac{2\pi r_1}{T_1}; \qquad v_1^2 = \frac{4\pi^2 r_1^2}{T_1^2}.$$

We can set these two values for v_1^2 equal to each other and get

$$\frac{GM}{r_1} = \frac{4\pi^2 r_1^2}{T_1^2}$$

$$\frac{T_1^2}{r_1^3} = \frac{4\pi^2}{GM}.$$

For any other satellite of the same central body,

$$\frac{T_2^2}{r_2^3} = \frac{4\pi^2}{GM}.$$

Since they both equal $4\pi^2/GM$, it follows that

$$\frac{T_1^2}{r_1^3} = \frac{T_2^2}{r_2^3}.$$

which is one way of writing the statement of Kepler's third law. Although Kepler, with no theoretical backing for his laws, proposed them to apply only to planets circling the sun, it is apparent that they can be used for any satellites orbiting the same central body.

By the use of Kepler's third law, together with what we know about the moon, we can readily determine the period of any other satellite of the earth. For example, consider an artificial satellite whose perigee (closest point to the earth) is 3000 mi above the earth's surface and whose apogee (farthest point from the earth) is 4200 mi above the surface. Its *average* distance is 3600 mi, which is 3600 + 3960 = 7560 mi from the earth's center. We know that the moon is about 239,000 mi from the center of the earth and has a period of 27.3 days. Using Kepler's third law, we find

$$\frac{T_{\text{sat}}^2}{(7560)^3} = \frac{(27.3)^2}{(239{,}000)^3}$$

$$T_{\text{sat}}^2 = \frac{(27.3)^2 \times (7560)^3}{(239{,}000)^3}$$

$$= \frac{(2.73)^2 \times 10^2 \times (7.56)^3 \times 10^9}{(2.39)^3 \times 10^{15}}$$

$$= 236 \times 10^{-4}$$

$$T_{\text{sat}} = 15.4 \times 10^{-2} = 0.154 \text{ day}$$

$$= 3.70 \text{ hr.}$$

Questions

(6-1) 1. A ship of 2×10^4 metric tons is moored 50 m (center to center) from a ship of 4×10^4 metric tons. With what force do they gravitationally attract each other?

2. A young man (mass 80 kg) stands with his center of gravity 0.5 m from that of a girl whose mass is 50 kg. What is the *gravitational* attraction between them?

3. At an elevation of 6370 km above the earth's surface (this is equal to the radius of the earth), with what acceleration would a body fall toward the earth?

4. If a body were released from a spaceship at a distance of 1.92×10^5 km from the earth's center (this is half the distance to the moon), with what acceleration would it fall toward the earth?

5. What are the dimensions of G (the constant in the gravitational formula) in MKS units?

6. What are the dimensions of the gravitational constant G in CGS units?

(6-2) 7. An astronaut weighs 300 lb on Earth, including his life-support equipment. What would he weigh on the surface of Mercury? ($m_M = 0.055 m_E$, $r_M = 0.39 r_E$.)

8. How much would a 150-kg astronaut (including equipment) weigh on the surface of Jupiter? ($m_J = 318 m_E$, $r_J = 7.0 \times 10^4$ km.)

9. An adult green-type Martian weighs 87 pjotniks on his native planet. What would he weigh on the surface of Mercury? (See Question 7.)

10. An automated exploratory vehicle constructed by the Martians includes a weight suspended from a filament whose tensile strength is 153 pjotniks. How heavy a weight can be installed on Mars if the vehicle is to be sent to Jupiter? (See Question 8.)

(6-3) 11. An orbiting laboratory circles the earth 620 km above its surface. (a) What is the radius of its orbit? (b) What is its speed?

12. What is the speed of a satellite circling Jupiter 10,000 km above its surface? (See Question 8.)

(6-4) **13.** (a) How much work would be required to raise a mass of 1000 kg from the earth's surface to an altitude above the surface equal to the earth's diameter? (b) How does this compare with the erroneous answer we would find if we figured the earth's gravitational pull to remain constant at its surface value?

14. (a) How much work would be required to lift a load of 10^4 kg from the earth's surface to an altitude above the surface equal to the earth's radius? (b) How does this compare with the erroneous answer we would find if we figured the earth's gravitational pull to remain constant at the same value it has at the surface?

15. (a) How much work would be needed to remove a mass of 2 kg from the earth's surface to infinity? (b) If we decide to call the PE zero when at infinity, what is the PE of the 2-kg mass when on the earth's surface?

16. (a) How much work would be needed to remove a gram from the earth's surface to infinity? (b) If we decide to call the PE zero when at infinity, what is the PE of the gram when on the earth's surface?

17. The laboratory of Question 11 has a mass of 100 metric tons. (a) What is its KE, calculated as $\frac{1}{2}mv^2$? (b) Calculate its KE as done in Section 4. (c) What is its PE? (d) What is its total energy?

18. The satellite of Question 12 has a mass of 5000 kg. (a) What is its KE, calculated as $\frac{1}{2}mv^2$? (b) Calculate its KE as in Section 4. (c) What is its PE? (d) What is its total energy?

(6-5) **19.** Escape velocity from the earth's surface is 6.96 mi/s. (a) What would the escape velocity be if the earth's mass were multiplied by 9, its size being unchanged? (b) What would the escape velocity be if the earth's radius were multiplied by 4, its mass being unchanged?

20. Escape velocity from the earth's surface is 11.2 km/s. (a) What would the escape velocity be if the earth's mass were doubled, its size being unchanged? (b) What would the escape velocity be if the earth's radius were doubled, its mass being unchanged?

21. In terms of G, M, and r, write the expressions for escape velocity (v_e) and the velocity of a satellite just skimming the surface of an airless planet (v_0). (a) How are v_0^2 and v_e^2 related? (b) By what factor must you multiply v_0 to find v_e? (c) If the earth were airless, what would be the speed of a satellite orbiting just above the surface?

22. [See Question 21(a) and (b).] (a) The planet Mercury has a radius of 2.57×10^6 m, and a mass of 3.28×10^{23} kg. What is the speed of a satellite in a circular orbit 430 km above the planet's surface? (b) To what speed would this satellite have to be accelerated in order to escape from Mercury and return home?

(6-6) **23.** A medical research centrifuge has a radius of 20.0 ft, and carries a dog in a pivoted cage, as in Fig. 6-4. How many rev/min must this centrifuge rotate, to subject the dog to 3 g?

24. A small medical research centrifuge similar to that of Fig. 6-4 has a radius of 50 cm. A mouse is placed in the cage, and the centrifuge is spun at 50 rev/min. To how many g's is the mouse subjected?

(6-7) **25.** A comet has an extremely elongated elliptical orbit. At perihelion (closest to the sun) its distance from the sun is 9.0×10^{10} m, and its perihelion speed is about 5.4×10^{4} m/s. At aphelion (farthest from the sun) its distance is 5.7×10^{12} m. What is this comet's aphelion speed?

26. At perihelion (closest to the sun) the planet Mercury is 28.6×10^{6} mi from the sun. Its aphelion distance (farthest from the sun) is 43.4×10^{6} mi, and it has an aphelion speed of 23 mi/s. What is its speed at perihelion?

27. If the earth had an artificial satellite with an orbit of 119,500-mi radius (half the radius of the moon's orbit), what would be its period of revolution around the earth?

28. Mars has two moons: Deimos, the larger, orbits at a mean distance of 6.9 Martian radii from the center of Mars, and its period is about 30 hr. Phobos, the smaller moon, has a period of approximately 7.6 hr. How far (in Martian radii) is Phobos from the center of Mars?

29. The earth now has communications satellites for relaying radio and TV signals from one continent to another. Viewed from the earth these satellites appear to hang nearly motionless in the sky directly over some point on the equator. (a) What is the period in which they revolve about the earth's center? (b) What is the approximate radius of their orbits? (c) How far are they above the earth's surface?

30. What would be the radius of the orbit of a Martian satellite with a period of 20 hr? (Give answer in units of Martian radii. See Question 28.)

7

Elastic Vibrations

7-1 Young's Modulus

Solid bodies tend to maintain their shape and when deformed (within limits) by an external force they return to their original shape as soon as the force is removed. This property of solids is known as *elasticity* and is of great importance in many practical devices, such as, for example, those based on various types of springs. A fundamental law in this field, *Hooke's law of elasticity*, named after its discoverer Robert Hooke (1635–1703), states that ***the deformation of a solid body is proportional to the force acting on it,*** provided the force does not exceed a certain limit.

It will be easier to deal with Hooke's law quantitatively if we pause to define a pair of technical terms: "stress" and "strain." *Stress refers to the internal forces created within a material as a result of forces applied to it.* It is F/A, the applied force divided by the cross-sectional area of material that resists the force.

Imagine a weight of F dynes hung from the end of a wire that is 1 m long and that has a cross-sectional area of 2 mm^2. Since F is supported by 2 mm^2 of metal all along the wire, the stress in the wire is $F/2$

dynes/mm^2. If the wire were thicker, with a cross-sectional area of 4 mm^2, there would be more material to support F, and the stress would be reduced to $F/4$ dynes/mm^2.

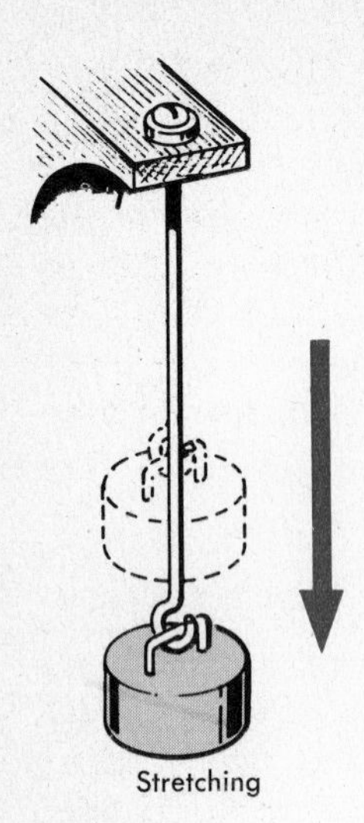

Strain is a measure of how much a body is deformed by a stress. The principal types of deformation are stretching (or compressing), bending, and twisting, as shown in Fig. 7-1. Let us look first at the simplest of these—the lengthening of a stressed wire. The strain of the wire is the fractional increase in length caused by the stress—that is, it is the elongation divided by the length. Mathematically, this is $\Delta l/l$, and since it is a length divided by a length, strain is a pure number without dimensions.

If we hang a weight on a piece of wire 1 m long, the wire will stretch a certain amount. The same weight hung on a piece of the same wire 2 m long will cause it to stretch twice as much, since each meter will elongate exactly the same as it did before. In both cases the strain $\Delta l/l$ is the same, and we can see that strain has the useful property of being independent of the length of the wire involved.

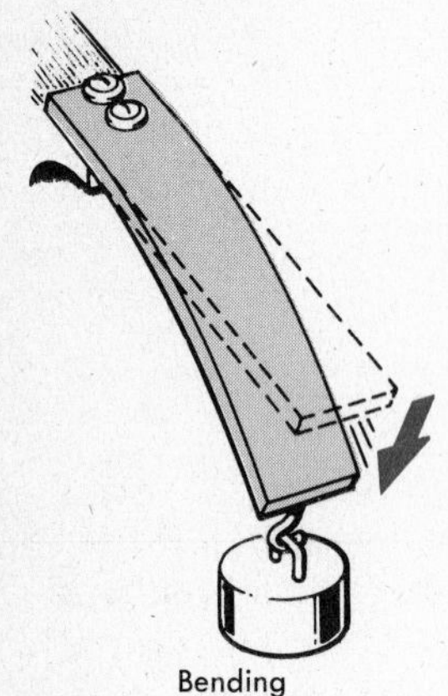

We can restate Hooke's law in a more useful way by saying, *stress is proportional to strain*, or

$$\frac{\text{stress}}{\text{strain}} = \text{a constant.}$$

This is true within certain limits. If the wire is stressed too much, it will no longer return to its original length when the stress is removed, and we say that we have exceeded the *elastic limit* of the material. Within the elastic limit, however, we can expand the equation above into

$$\frac{F/A}{\Delta l/l} = Y.$$

The proportionality constant Y is called the *Young's modulus* of the material, after a great English engineer, scientist, and philosopher of the early nineteenth century. We shall see his name again later in connection with optics. Here is Young's modulus for a few common materials:

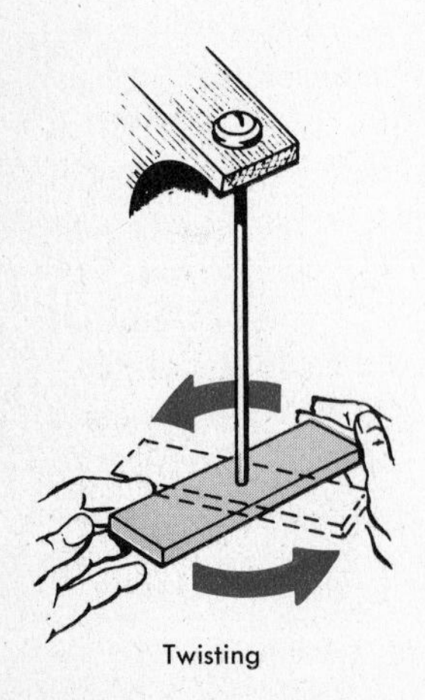

FIG. 7-1 Three types of elastic deformation.

Steel	2×10^{12} dynes/cm^2	or	3×10^7 lb/in^2
Copper	1×10^{12}		1.5×10^7
Aluminum	7×10^{11}		1×10^7
Bone	2×10^{11}		3×10^6
Wood	1×10^{11}		1.5×10^6

Since strain is dimensionless, it is apparent that Young's modulus must have the same dimensions as the stress—that is, force per unit area.

Let us see how this applies to a copper wire 0.4 mm in diameter and 3.0 m long. How much will this wire elongate if we use it to suspend a weight of 5 kg? First, we note that Young's modulus for copper is given in dynes/cm^2, so we must figure the stress on the wire in the same units. Since a gram weighs 980 dynes, the force is 5 kg $\times$ 1000 $\times$ 980 = 4.90 $\times$ 10^6 dynes. The circular cross-sectional area of the wire that resists this pull is $\pi r^2 = 3.14 \times (0.02)^2 = 1.26 \times 10^{-3}$ cm^2. From these values of force and area we can find the stress, which is $4.90 \times 10^6/1.26 \times 10^{-3} = 3.89 \times 10^9$ dynes/cm^2.

Since Young's modulus was defined as Y = stress/strain, it is apparent that strain = stress/Y, and we have

$$\text{strain} = \frac{3.89 \times 10^9}{1 \times 10^{12}} = 3.89 \times 10^{-3}.$$

This figure can be interpreted to mean that any piece of copper under a stress of 3.89×10^9 dynes/cm^2 will elongate 3.89×10^{-3} of its length (or, if it is compressed, will shorten by the same fraction). Thus the elongation of the wire will be

$$\begin{aligned} \Delta l &= \text{strain} \times l \\ &= 3.89 \times 10^{-3} \times 3.0 \\ &= 1.17 \times 10^{-2}\ \text{m} \\ &= 1.17\ \text{cm}, \quad \text{or} \quad 11.7\ \text{mm}. \end{aligned}$$

The elongation or the shortening of a coiled spring cannot be so easily figured from a knowledge of just its material and dimensions. As the spring is stretched, the principal stress on the wire of which it is made is twisting, and to this twisting is added some bending, as well as a small amount of straight tensile pull. It is too complicated a problem for us to go into, but we should note that within its elastic limit, a spring will nevertheless obey Hooke's law.

7-2 Simple Harmonic Motion

If we suspend a chunk of iron or lead from a rubber band or steel spring attached to the ceiling and by a gentle push set it bobbing up and down, we will find it moves with a very definite *period.* The period T of a regularly repeating motion such as this is the time needed to make a complete up-and-down cycle and is obviously the reciprocal of the frequency f, which is the number of cycles or complete vibrations the body makes in a unit of time:

$$T\ (\text{seconds/vibration}) = \frac{1}{f\ (\text{vibrations/second})}.$$

If the spring or rubber band obeys Hooke's law, the motion of the chunk of iron will be an example of *simple harmonic motion* (SHM). So

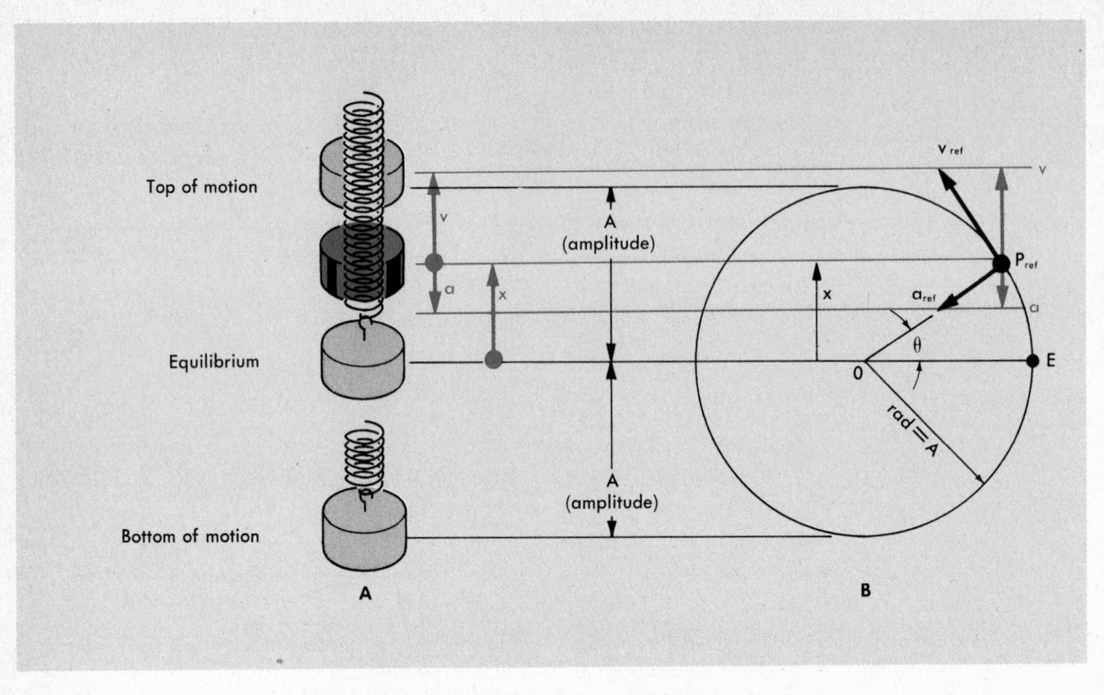

FIG. 7-2 Simple harmonic motion and the reference circle.

also will be the oscillation of a diving board, the vibration of a plucked violin string, or the swinging of a pendulum. Besides describing such back-and-forth movements, SHM lies behind all wave motion, and it will be worth our while to look at this special kind of motion more closely.

Figure 7-2A shows a weight suspended by a spring; with the weight hanging motionless, it is in the position marked "equilibrium." Here the upward pull of the spring just equals the pull of gravity on the weight. Now pull the weight down to "bottom of motion" and let it go; it will oscillate between "bottom of motion" and "top of motion" in SHM. The total range of travel of the weight will be $2A$. (A, the *amplitude*, is the distance from the equilibrium point to either extreme of motion.)

Let us temporarily forget the bobbing weight and turn to the geometrical fiction of Fig. 7-2B. On a line MM' parallel to the motion of the weight, mark the point O, level with the equilibrium position. With O as its center, draw a circle of radius equal to A, the amplitude of the motion. Around this circle (the *reference circle*), imagine that a *reference particle* P_{ref} is revolving with a constant angular velocity of ω rad/s. This angular velocity ω must be adjusted so that P_{ref} makes one revolution of 2π radians while the bobbing weight goes through one cycle. If f is the frequency in cycles per second, then $\omega = 2\pi f$ rad/s.

Now, just as the weight passes through equilibrium, click your imaginary stopwatch and start P_{ref} revolving from point E. The weight and P_{ref} are now synchronized, and will exactly follow each other up and down. What makes the idea of the reference point so useful, is that we can now easily calculate x, v, and a for P_{ref}—and for the bobbing weight at any instant, its x is the same as that of P_{ref}, and its v and a are merely the components of v_{ref} and a_{ref} in the direction of its motion, in this case the vertical components.

The angle shown as θ is ωt, so we have

$$x = A \sin \theta = A \sin \omega t$$

$$v = v_{ref} \cos \theta = \omega A \cos \omega t$$

$$a = a_{ref} \sin \theta = -\omega^2 A \sin \omega t = -\omega^2 x.$$

(The minus sign in the last equation is necessary because the direction of the acceleration is always opposite to the displacement x, measured from the equilibrium point.)

From Fig. 7-2B we can see that the maximum possible x is the amplitude A. The maximum v is v_{ref}, or ωA, and it occurs at equilibrium. The maximum acceleration is a_{ref}, and occurs at the top or bottom of the motion.

Since the frequency of any actual body moving in SHM is nearly always expressed in *cycles* (per second or per minute), rather than in the radian measure useful for P_{ref}, it will be helpful to use the relationship $\omega = 2\pi f$ to revise our last equations:

$$x = A \sin 2\pi ft$$

$$v = 2\pi fA \cos 2\pi ft$$

$$a = -\ 4\pi^2 f^2 x.$$

The force on the bobbing mass and the acceleration it causes are constantly changing, but Newton's $F = ma$ is always applicable. We can use this relationship to connect the motions we have been discussing with the physical characteristics of the spring and weight:

$$F = ma = m \times 4\pi^2 f^2 x$$

$$\frac{F}{x} = 4\pi^2 f^2 m.$$

The term F/x is the force per unit stretch of the spring, which tends to bring the mass m back to its equilibrium position when it is moved away from this position in either direction. If the spring obeys Hooke's law (and in general it will if it is not stretched too far), this ratio F/x will be the same for all amounts of stretch and is often called the *force constant* of the spring, or more simply the *spring constant*.

Since the period $T = 1/f$, this can be rewritten as

$$\frac{F}{x} = \frac{4\pi^2 m}{T^2}$$

$$T^2 = \frac{4\pi^2 m}{F/x}; \qquad T = 2\pi\sqrt{\frac{m}{F/x}}.$$

This last equation is remarkable for the things it does *not* contain. The amplitude A is missing, and so is g, the acceleration of gravity. It thus tells us that the period or frequency of the bobbing weight depends only on the mass of the weight and on the force constant of the spring. The period and frequency are the same for motion of large or small amplitude, and would be the same on the moon as on the earth.

As an example of SHM, let us take a bird watcher who sees a large bird light on the end of a slender tree limb, which is thus started oscillating. The bird makes 6 complete up-and-down bobs in 4 seconds. When the bird leaves, the watcher hangs a 1-kg weight on the limb and measures that the weight deflects the limb 12 cm. What was the mass of the bird? Since a 1000-g weight deflects the limb 12 cm, the force constant of the limb is $1000 \times 980/12$, or 8.17×10^4 dynes/cm. The frequency f is 6 vibr/4 s = 1.5 vibr/s. Whipping out his notebook, the ornithologist writes

$$8.17 \times 10^4 = 4\pi^2(1.5)^2 m$$

from which

$$m = 920 \text{ g}.$$

7-3 The Simple Pendulum

A simple pendulum theoretically consists of a massive particle of zero size, swinging at the end of a massless rod or string. A small, compact piece of heavy metal suspended from a thread comes close enough to these requirements for most practical purposes. Figure 7-3 is a diagram of a simple pendulum, made by suspending a small ball of mass m from the end of a thread of length l. The diagram shows it pulled aside from the vertical by an angle θ. We can resolve the weight of the ball mg into two components: $mg \cos \theta$, which pulls directly against the thread and has no tendency to make the ball move; and a component $mg \sin \theta$ at right angles to the thread, which urges the ball toward equilibrium. The horizontal displacement of the ball from its equilibrium position is $l \sin \theta$. We can see that F/x for the ball is $mg \sin \theta / l \sin \theta = mg/l$, which is a constant, because m, g, and l remain unchanged in value. Thus the restoring force is proportional to the displacement of the ball, and it moves in SHM. (We have calculated the displacement x as $l \sin \theta$, which is horizontal; the restoring force F, however, is perpendicular to the string and is therefore not quite horizontal and not quite parallel

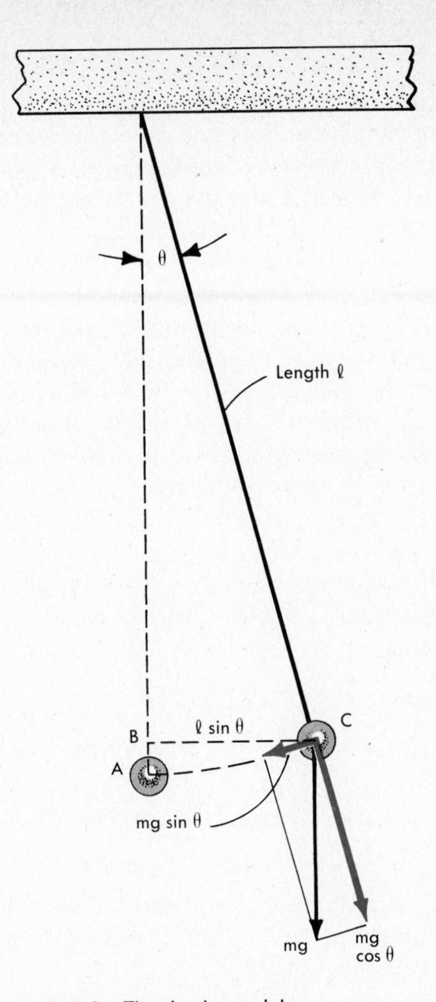

FIG. 7-3 The simple pendulum.

to x. Hence the motion of the pendulum is not exactly SHM; but if θ is small, the difference is negligible.)

Let us use this to derive an equation for the period T of the pendulum. Since $f = 1/T$, we can write

$$\frac{F}{x} = \frac{4\pi^2 m}{T^2}$$

from which

$$\frac{mg}{l} = \frac{4\pi^2 m}{T^2}$$

and

$$T = 2\pi \sqrt{\frac{l}{g}},$$

There are several things to observe in this formula, and to compare with the corresponding formula for the period of a mass bobbing on a spring. The amplitude does not appear in either equation. In the pendulum formula, the mass of the pendulum does not appear; if the mass were greater, its inertia and the pull of gravity would both increase by the same amount, and the motion would be unchanged. The gravitational acceleration g, missing for the bobbing mass, is present in the pendulum formula, as we should expect. The restoring force is *provided* by the pull of gravity; any change in g would have an effect on the pendulum equivalent to using a spring with a different force constant to suspend the bobbing mass. On the moon, a pendulum would have a longer period and a lower frequency, because the lunar g is smaller. In fact, special pendulums are used to measure small variations in g at different points on the earth's surface.

7-4 Rotational SHM

The idea of harmonic motion can easily be extended to include rotational motion. By analogy with translational motion, which we have already established,

$$\frac{F}{x} = 4\pi^2 f^2 m$$

becomes

$$\frac{\tau}{\theta} = 4\pi^2 f^2 I \quad \text{or} \quad \frac{4\pi^2 I}{T^2},$$

in which τ is the applied torque; θ is the twist that τ causes, measured in radians; and I is the moment of inertia about the axis of the twist. We now have

$$T = 2\pi\sqrt{\frac{I\theta}{\tau}} \quad \text{or} \quad 2\pi\sqrt{\frac{I}{\tau/\theta}}.$$

Figure 7-4 shows such a device (called a *torsion pendulum*), which consists of a very light crossbar suspended at its center by a thin wire or quartz fiber. A ball of mass m is placed on each end of the crossbar. If the crossbar is rotated away from its equilibrium position, the twist of the fiber will provide a restoring torque proportional to the angular displacement; hence the bar will rotate back and forth in rotational SHM. Almost all watches and some clocks are made to operate with this sort of pendulum.

The torsion pendulum is also used in determining G, the gravitational constant. Figure 7-4 shows two large masses M that are placed close to the balls of the pendulum and in the same horizontal plane, so that the gravitational attractions between M and m will twist the pendulum through a small angle from its equilibrium position. This angle can be accurately measured, and when the stiffness of the supporting fiber

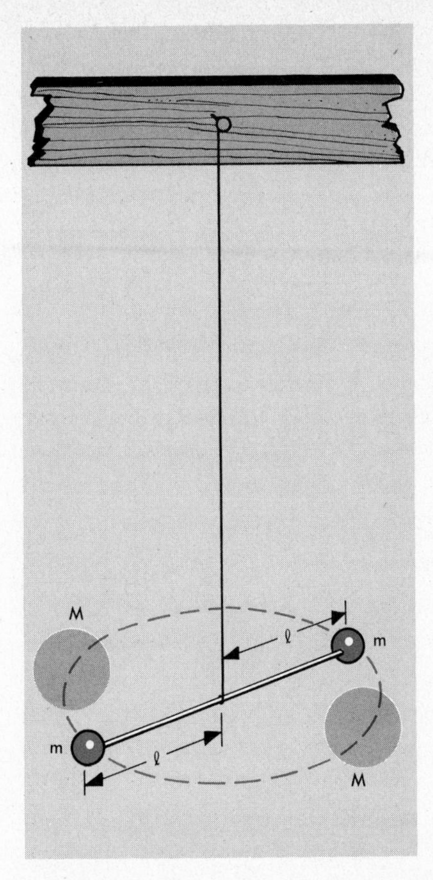

FIG. 7-4 A torsion pendulum, such as used to determine the gravitational constant G.

(τ/θ) is known, the deflecting torque and hence the gravitational pull between M and m can be computed. The stiffness of the fiber is determined by measuring the period of oscillation of the torsion pendulum when it is set to swinging by a gentle twist.

If, for example, the pendulum is made of two 10-g balls on a rod 20 cm long and has a period of 20 min, we can compute the stiffness of the fiber in this way: The moment of inertia of the pendulum (ignoring the light rod) will be $2ml^2$, or $2 \times 10 \times 10^2 = 2000$ g-cm^2. The period T is $20 \times 60 = 1200$ s. A simple substitution gives

$$\frac{\tau}{\theta} = \frac{4\pi^2 \times 2000}{(1200)^2} = 0.0548 \text{ dyne-cm/rad.}$$

In other words, a torque of 0.0548 dyne-cm would cause the fiber to twist through 1 radian, or about 57°.

7-5 Resonance

A very important notion in the study of all kinds of vibrations is that of *resonance*, which is *the specific response of a system which oscillates or vibrates with a certain period, to an external force that varies with the same, or nearly the same, period.* Consider a child on a swing. The swing is of course nothing but a simple pendulum, and its period is determined by the length of the ropes. In order to put the swing in motion and make it move with a larger and larger amplitude, the child must pull periodically on the ropes and stretch out his legs at the same time. But to be successful, these muscular efforts must be made with the same period as the natural oscillation period of the swing.

The situation can be demonstrated by the simple experiment shown in Fig. 7-5. Bar A is inserted through the holes in the supporting frame B, and a number of balls C_1, C_2, C_3, etc. are suspended from the bar on strings of different lengths. Another ball D is also suspended from the protruding end of bar A in such a way that the length of its string can be changed. If we make this length equal, say, to the length of the string of C_2, and set D to swinging, some of its energy will be transmitted

FIG. 7-5 The principle of resonance, as illustrated by a set of pendulums.

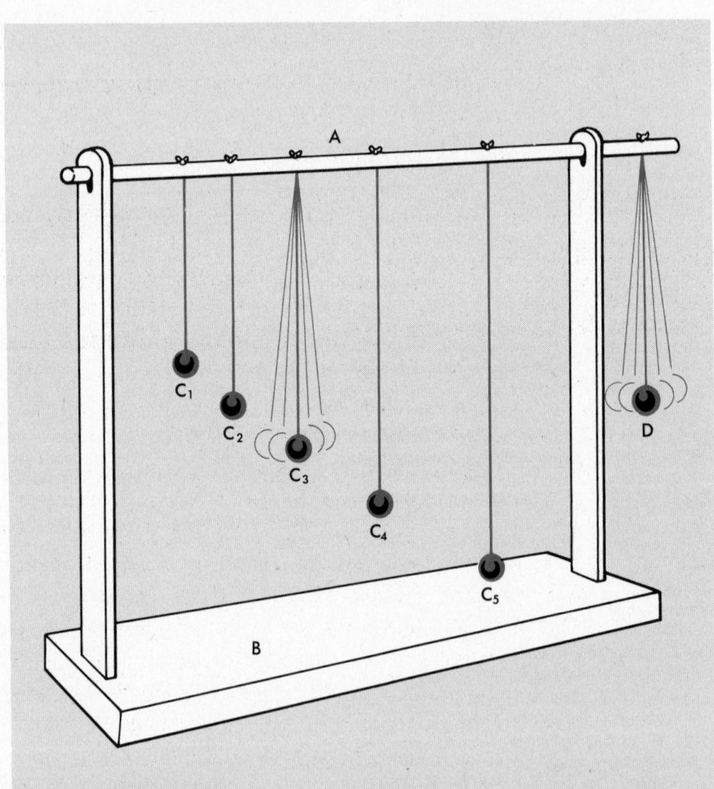

through small movements of the bar to the balls hanging inside the frame. Although the others will show only a very slight tendency to take up this motion, ball C_2 will begin to swing more and more until its amplitude becomes even larger than that of D. By changing the suspension length of D, we can in turn set in motion any of the other balls hanging from the bar.

The principle of resonance is used in the construction of one type of an instrument called a *tachometer*, which is used for measuring the speed of rotation of various motors. It consists of a number of steel strips of different lengths mounted on a common support. When put in contact with a running motor, the tachometer receives slight vibrations caused by the rotation of the motor's axis. The period of the motor's rotation will nearly coincide with the vibration period of one of the strips, and this particular strip will vibrate with an appreciable amplitude while all the other strips remain almost at rest. By reading the figures printed on a scale that runs along the row of strips, we can quickly find the number of revolutions per minute the motor is making.

Questions

(7-1) **1.** A load of 2000 kg is supported by a cylindrical column 10 cm in diameter. What is the stress in the column (a) in kg/cm^2? (b) in N/m^2?

2. A weight of 8000 lb is supported by a column 4 in × 4 in. What is the stress in the column?

3. A load of 20 kg is suspended from a metal strip 1 cm × 0.02 cm. What is the stress in the strip, in dynes/cm^2?

4. A wire 0.01 cm in diameter supports a load of 0.85 kg. What is the stress in the wire, in dynes/cm^2?

5. A steel tape 100 m long elongates 2 mm when a tensile force is applied. What is the strain in the tape?

6. A column 10 ft long supporting the floor of a shop shortens by 0.02 in. when a heavy machine is installed. What is the strain in the column?

7. The column in Question 1 is 5 m long, and is compressed by 0.06 mm. What is the Young's modulus of the material of which the column is made?

8. A piece of wire elongates by 10^{-3} of its length when a tensile stress of 2×10^9 dynes/cm^2 is applied to it. What is the Young's modulus of the material the wire is made from?

9. How much load could be suspended from an aluminum wire 2 mm in diameter and 5 m long, if it is to stretch no more than 1 mm?

10. How much load can be suspended from a steel wire 0.02 in. in diameter and 10 ft long, if it is to stretch no more than 0.1 inch?

11. What fraction of the load of Question 9 could be supported by (a) a copper wire of the same size? (b) a steel wire of the same diameter, but 15 m long? (The maximum permissible stretch is still 1 mm.)

12. What fraction of the load of Question 10 could be suspended from (a) an aluminum wire of the same size? (b) a copper wire of the same diameter, but 20 ft long? (The maximum permissible stretch is still 0.1 in.)

(7-2) **13.** An electron in an electromagnetic wave vibrates with a frequency of 6.25×10^{14} cycles (or vibrations)/second. What is the period of its oscillation?

14. What is the period of a vibratory motion whose frequency is 1600 cycles (or vibrations)/second?

15. The mass in SHM in Fig. 7-2 makes 6 cycles in 30 seconds. What is (a) its frequency? (b) its period? (c) the angular speed of P_{ref} in revolutions (or cycles) per second? (d) in radians per second?

16. The mass in SHM in Fig. 7-2 makes 10 cycles of motion in 4 s. (a) What is its frequency? (b) What is its period? (c) What is the angular speed of the corresponding reference point (P_{ref}) in revolutions (or cycles) per second? (d) What is the angular speed in radians per second?

17. A weight suspended from a spring bobs up and down. At what point or points on its path is (a) its velocity greatest? (b) its velocity zero?

18. A weight suspended from a spring bobs up and down. At what point or points on its path is (a) its acceleration greatest? (b) its acceleration zero?

19. A bottle floating on the sea moves up and down in approximate SHM as the waves pass by; its motion has an amplitude of 8 inches and a frequency of 40 cycles/min. What is (a) its maximum velocity, in ft/s? (b) its maximum acceleration?

20. A reciprocating part on a machine moves in SHM with an amplitude of 5 cm and a frequency of 300 cycles/min. What is (a) its maximum velocity? (b) its maximum acceleration?

21. A mass weighing 4.8 lb is hung from the lower end of a coiled spring, and causes the spring to elongate by 3 in. If the mass is now set into vertical oscillation, what will its frequency be, in cycles/min?

22. A 500-g mass is hung from the lower end of a coiled spring and causes the spring to elongate by 5 cm. If the mass is now set bobbing up and down, how many complete vibrations will it make in 1 min?

23. Rework Question 21, with the same spring, but using a 16-lb mass, and performing the experiment on the moon. (See Question 30.)

24. Rework Question 22, with the same spring, but using a 1000-g mass, and performing the experiment on the moon. (See Question 30.)

25. A flat horizontal platform moves up and down in SHM with an amplitude of 1 cm. A small object is placed on the platform. What is the maximum frequency the platform can have if the object is not to separate from it at any part of its motion?

26. A weight is suspended by a string whose upper end is fastened to a machine part that moves up and down in SHM with an amplitude of 2 in. What is the maximum frequency the machine can have if the string is not to become slack in any part of the cycle?

(7-3) **27.** A once-popular pendulum clock was the "80-beat" Seth Thomas clock, which made 40 complete oscillations/min. What is the length of the pendulum of this clock?

28. What is the length of a simple pendulum that will "beat seconds," i.e., that will have a period of 2 s?

29. How long would it take the clock of Question 27 to record an hour on Planet Z, where g is 1470 cm/s^2?

30. How long would it take the clock of Question 28 to record an hour if it were on the moon, where $g = 163$ cm/s^2?

(7-4) **31.** If, in Fig. 7-4 and the accompanying discussion in the text, the large balls each had a mass of 5 kg and were placed 10 cm (center-to-center distance) from the small balls, through what angle would the torsion pendulum twist?

32. The torsion pendulum that regulates the speed of a certain type of clock has a moment of inertia of 2000 g-cm^2 and a period of 30 s. How much torque would be needed to twist the suspending wire through an angle of 45°?

(7-5) **33.** A certain recording instrument whose mass is 12 kg is mounted on three springs to protect the instrument from vibration and shock. When the instrument is installed, each spring shortens by 15.5 mm. Would it be advisable to mount this recorder on a machine running at 240 rev/min? Explain.

34. A piece of machinery weighing 3200 lb is mounted on four springs, each bearing one-fourth of the weight. The force constant of each of the springs is 2050 lb/in. Would it be advisable to operate this machine at a speed of 300 rev/min? Explain.

8

Waves

8-1 Wave Pulses

Probably the simplest wave we can imagine is a single disturbance, or pulse, started by a sideways jerk of the hand, which then travels down a long rope tied to a wall at its far end. It will be easy to experiment with this device, and by using a stopwatch to measure the time it takes for the wave to travel the length of the rope, you can easily compute the velocity of the wave. You will find that the tighter the rope is pulled, the faster the wave will travel. You will also find that if you substitute a rope of greater *linear density* (m_ℓ)—that is, one that has a greater mass per unit length—the wave will travel more slowly than it did along the lighter rope pulled to the same tension. A little further investigation will show that the velocity of the wave is not affected by whether you jerk your hand quickly or slowly, or by whether you move it through a large or a small distance in starting the pulse, although the motion, of course, does influence the shape and size of the wave itself.

Study of Fig. 8-1 will suggest why wave velocity depends on the rope tension F and on the linear density m_ℓ. Segment 1 has no choice;

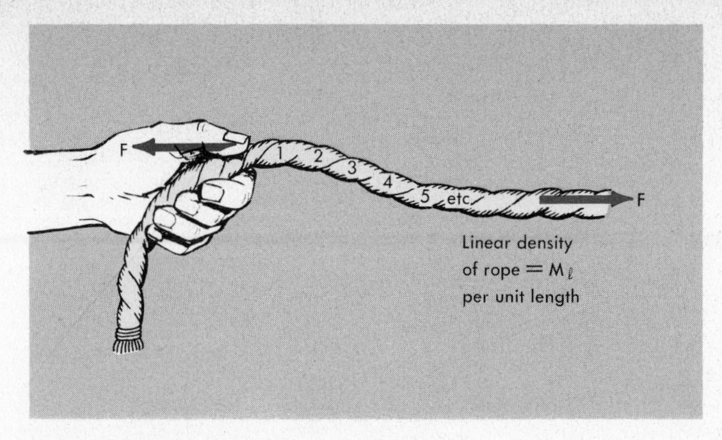

FIG. 8-1 Starting a transverse wave in a rope.

the hand forces it to move quickly upward and then back. In the drawing, segment 1 is at the top of the jerk and is about to start back. Segment 2 must follow along, and the greater the tension F, the more closely it will follow. Although segment 1 is about to start down, the kinetic energy of segment 2 will cause it to rise to the same distance as segment 1's maximum before it reverses its direction. Segment 3 follows segment 2, and so on, each segment duplicating the motion of the previous segment with a time delay that is less for a greater tension. Thus a greater tension results in a faster propagation of the wave. It is apparent, too, that if the mass of segment 2 is increased, the drag of segment 1 must operate for a longer time to bring segment 2 up to speed. This greater time delay in passing the motion along from segment to segment of the rope means that the wave travels more slowly in a rope of greater linear density. A mathematical derivation shows that velocity, tension, and linear density are related in the following way:

$$v = \sqrt{\frac{F}{m_\ell}}\,.$$

This formula applies to sideways waves of the sort we have been describing, traveling along a wire, a rubber tube, a long helical spring, a chain, or any other essentially one-dimensional wave carrier that is relatively flexible. It will *not* apply to stiff rods, planks, or similar objects in which the restoring force is due to the stiffness of the medium, rather than to a pure tensile stress such as we have considered in the example of the rope.

8-2 Wave Reflections

Up to now we have neglected to think of what becomes of a wave, or pulse; we have merely started it and forgotten about it. In general, two things may happen to it. If the rope is long enough, the energy of the pulse may be gradually reduced by friction with the air and by the

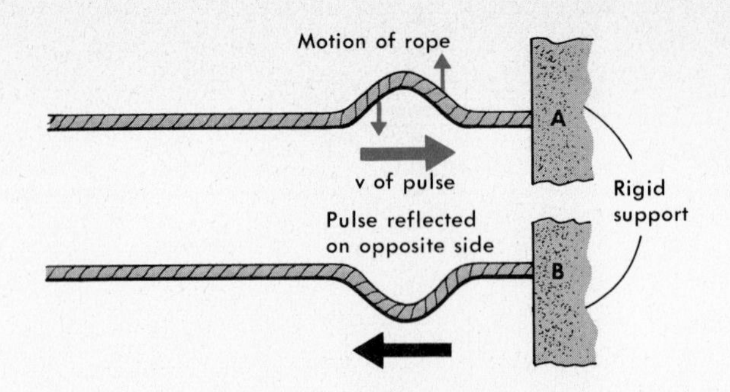

FIG. 8-2 Reflection of a wave pulse from a rigid support.

internal friction of rope fibers rubbing against one another until the amplitude has been reduced to zero and the disturbance vanishes. More often, however, the pulse will be reflected, either entirely or in part, from a discontinuity in the rope.

In the absence of any friction, the pulse started by the hand in Fig. 8-1 would continue down the rope undiminished. The moving disturbance of the pulse represents a moving region of energy; the displaced rope particles in the wave have both KE and PE. As the wave pulse moves from left to right in Fig. 8-2A, *all* the energy of each particle is passed on to its neighbor on the right, so that the rope is again straight and motionless after the pulse has passed.

When the wave strikes the rigid support at the end of the rope, it cannot pass any of its energy on to the support, because it can do no work on it. It can exert a force, but since the support does not move, the "d" in $W = Fd$ is zero, and so also is W. The energy must stay with the rope in the form of a reflected wave, as in Fig. 8-2B. Notice that the pulse is reflected on the *opposite* side of the rope.

Figure 8-3A shows a rope or wire with a discontinuity at the splice S, where there is an abrupt change from a higher linear density to a lower, as the wave passes from left to right. The last rope particle to the left of S is unable to give all its energy to the first particle to the right

FIG. 8-3 Partial reflection of a pulse at a point where the linear density changes.

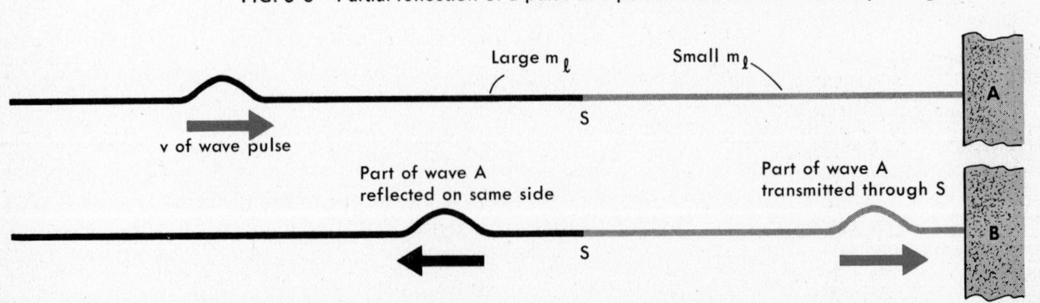

of S. In this case, for $W = Fd$, there is no difficulty with the d. But because of the smaller inertia of the lighter rope, the force F is reduced, so that only a part of the energy can be transmitted through S. The remainder causes a reflection, in this case on the *same* side, as shown in Fig. 8-3B.

8-3 Kinds of Waves

So far, we have considered only one special sort of wave—a pulse started by a sideways impulse, traveling down some one-dimensional medium. This is the simplest example of a *transverse* wave. As the disturbance passes along the rope or spring, the particles of the rope move at right angles—or transversely—to the direction of propagation of the wave. Figure 8-4A shows a metal bar being struck in a direction perpendicular to its length. The impact of the hammer will cause a bending-type deformation that will propagate along the bar just as we have previously considered other transverse waves moving along a rope. If we continue to hit the bar periodically, we shall produce a steady train of waves along it, as shown.

If, instead of striking the bar sideways, we hit it on the end as shown in Fig. 8-4B, we will produce a different kind of wave. The material

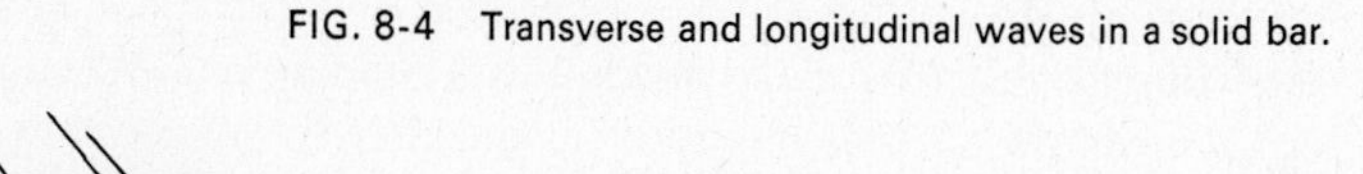

FIG. 8-4 Transverse and longitudinal waves in a solid bar.

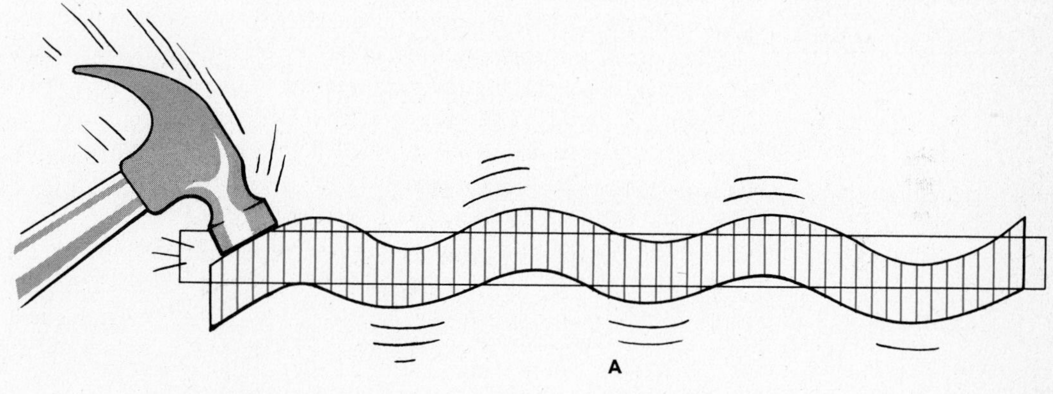

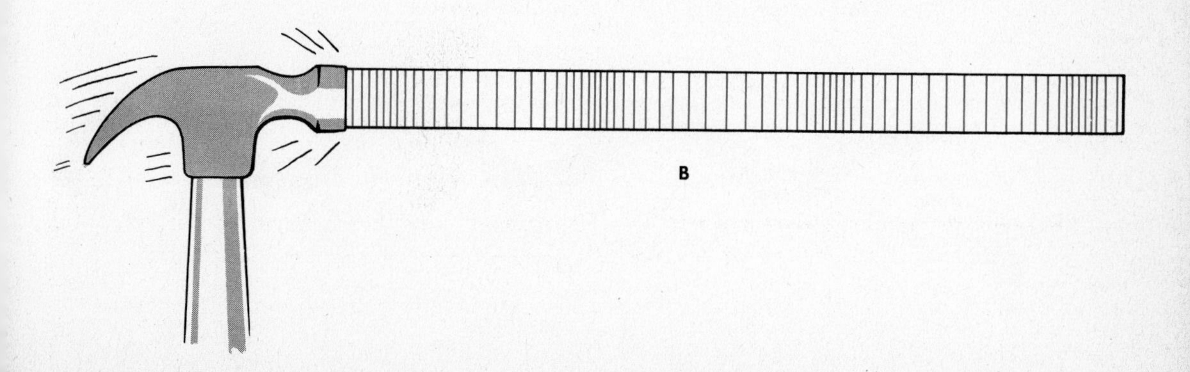

of the bar will be compressed by the impact, and this compression (followed by an expansion) will likewise travel along the bar. This is a *longitudinal* wave, in which the particles of the medium move back and forth along the direction of propagation of the wave. Here again, if we strike the bar periodically, we will produce a steady train of longitudinal, or compression, waves along the bar.

The two kinds of wave—transverse and longitudinal—will generally travel at different speeds because they depend on different characteristics of the medium: resistance to bending for the transverse wave and resistance to compression for the longitudinal wave.

Transverse waves can propagate only through solids, because their transmission depends on the rigidity of the medium. Liquids and gases therefore cannot transmit transverse waves. Longitudinal waves, however, depend only on resistance to compression, and are propagated through any material medium, because solids, liquids, and gases all resist a change in volume. Sound, most often transmitted through air, is the most common example of longitudinal, or compression, waves.

8-4 Periodic Wave Trains

Most wave phenomena of practical interest are concerned with long trains of equally spaced waves, rather than with the single pulses we have been observing. If the hand holding the rope in Fig. 8-5 moves up and down in simple harmonic motion, a snakelike series of waves will be sent traveling down the rope. A train of waves of this particular shape is called a *sine wave*, because the displacement of the rope is proportional to the sine of the SHM generating angle θ (see Fig. 7-2).

Such periodic trains of waves, whether they are sine waves or are of any other shape, have characteristics not possessed by simple pulses. If, say, the hand in Fig. 8-5 moves up and down f times per second, f complete waves per second will be generated. Wherever along the rope we choose to stand and count, we shall find that f waves per second will

FIG. 8-5 Generating a train of sine waves on a rope or wire.

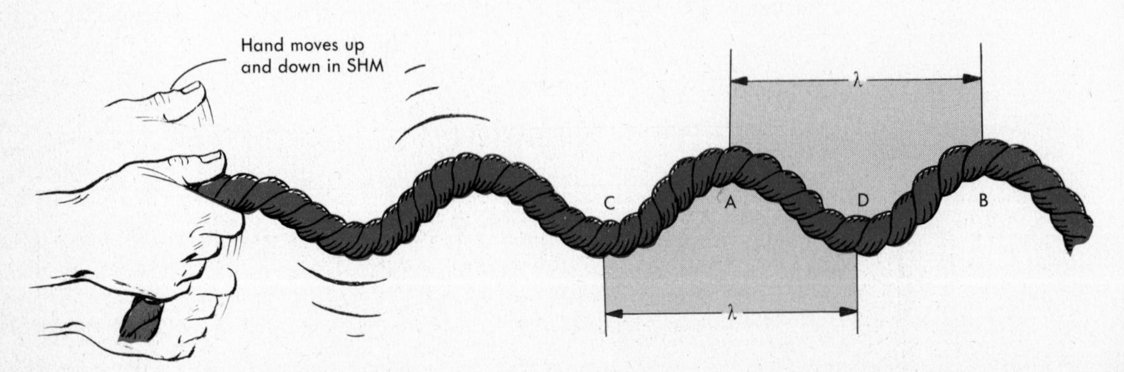

pass by; this—*the number of complete waves that pass by any fixed point in a unit of time—is the frequency* of the waves. Alternatively, we could measure the time required for a single complete wave to pass. *This time interval, measured between the passage of one crest and the next crest—or between the passage of consecutive troughs—is the period* of the waves. It should be apparent that the frequency (f, in waves per second) and the period (P, in seconds per wave) are each the reciprocal of the other:

$$f = \frac{1}{P} \quad \text{and} \quad P = \frac{1}{f}.$$

Another characteristic of a regular periodic wave train is its *wavelength*, generally indicated by λ, the Greek letter lambda. As the name suggests, *this is merely the length of a complete wave, measured from crest to crest* (A to B in Fig. 8-5) *or from trough to trough* (C to D).

There is a simple and very fundamental relationship among f, λ, and the speed of propagation of the waves v. Imagine yourself waiting in your car at a railroad crossing while a freight train passes by. Having nothing better to do, you find that 25 cars pass by per minute. And, being an old railroad man, you happen to know that each car is 42 ft long. So, if 25 cars, each 42 ft long, pass in a minute, $25 \times 42 = 1050$ ft of train pass per minute—and this is the train's speed: 1050 ft/min. So, transferring this analogy back to waves (or to any other moving periodic phenomenon), we discover that

$$v = f\lambda.$$

As an example, the note A above middle C on a piano should have a frequency of 440 vibr/s. If the speed of sound (which depends to some extent on the temperature) is 1100 ft/s, the wavelength of this sound is $\lambda = v/f = 1100/440 = 2.50$ ft.

8-5 Standing Waves

Suppose now that the far end of the rope in Fig. 8-5 is fastened to a wall, so that the train of waves is reflected backward from it. The rope will now be transmitting two wave trains simultaneously: the incoming hand-generated waves moving from left to right and the similar reflected wave train moving from right to left. To discover what will happen, let us first consider the situation in which a pair of single pulses traveling in opposite directions meet at some point along the rope.

An upward flick of the wrist will send a pulse along the top side of the rope, which will be reflected back as a pulse on the lower side. As this reflection leaves the wall, another upward flick will send another pulse head-on toward the returning reflection. Figure 8-6A shows this situation, with A rushing to meet the reflection B.

A surprising property that waves of any sort generally have is their

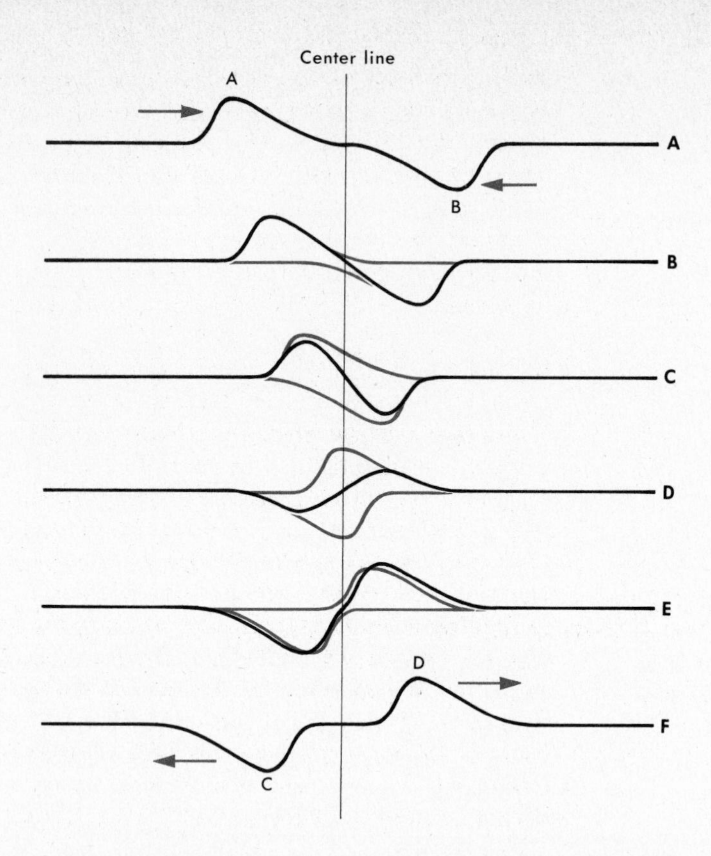

FIG. 8-6 Interference of two wave pulses passing each other on a rope.

ability to pass through each other in the same medium without being changed or distorted in any way. This is shown in Fig. 8-6B to F, which sketches a series of instantaneous pictures of the collision, representing the shape of the rope at intervals only a small fraction of a second apart.

In Fig. 8-6A the two wave pulses are almost touching, but have not yet interfered with one another. In Fig. 8-6B the two pulses overlap, and the colored lines show each pulse as it would have been *if it had proceeded undisturbed.* The black line, representing the rope itself, is the vector sum of the two interfering colored-line waves. Figures 8-6C, D, and E show the imaginary colored waves progressing along their separate ways, with their sum indicating the configuration of the rope, given as the black line. In Fig. 8-6F there is no longer any overlap, and pulses *C* and *D* continue undisturbed.

We should look back now over the six sketches and note that the center point of the rope on the center line, although two pulses have apparently been transmitted through it, has never stirred a hair's breadth from its original undisturbed position. How is it possible to transmit a wave along a rope if there is a piece of the rope in its path that does not move?

We can get ourselves off the uncomfortably sharp horns of this dilemma by saying that the waves were *not* transmitted through the center point at all! Pulse *D* in Fig. 8-6F is not pulse *A* at all; instead, it is pulse *B*, which has been reflected as though the center line were a rigid wall. Similarly, *C* is not *B* passing through, but the reflection of *A*. This seems a reasonable explanation, since *A*'s attempts to pull the center particle upward are exactly countered by *B*'s simultaneous downward pulls; the result is that, for each wave, the center particle is as rigid and unyielding as though it were a solid wall.

The perfect reflections of Fig. 8-6 are possible only when the two waves are identical in shape and size. If *A* and *B* were not the same shape, the center point would move to some extent, so that *C* would be composed of a partial reflection of *A*, plus a portion of *B* that is able to sneak through. A complicated mathematical analysis would show that *C* must nevertheless have exactly the same shape as *B*, so that although *C*'s origin is something much more complex, it *appears* as though *B* had traveled through undisturbed.

A similar analysis will show the motion of a rope when it is carrying simultaneously two similar trains of waves moving in opposite directions. Figure 8-7A shows two sine waves: The light solid line represents the incoming wave moving from left to right, and the dashed line represents the reflected wave moving away from the wall from right to left. Note that neither of these lines represents the configuration of the rope; the displacement of the rope at every point is the *sum* of the displacements of the two waves. In Fig. 8-7A, these everywhere add up to zero, so the heavy line, representing the rope itself, is perfectly straight. (Some portions, however, have an upward velocity, and some a downward.) In Fig. 8-7B, the incoming wave has moved slightly to the right and the reflected wave slightly to the left. If we go along from point to point, again adding the displacements of the two waves, we find as their sum the wavy heavy line, which represents the rope a fraction of a second later than in Fig. 8-7A. The diagrams proceed at intervals of a fraction of a second, each one showing the incoming and reflected waves to have moved slightly from their previous positions, and the heavy line in each diagram representing the actual shape of the rope at that instant.

There are a number of points along the rope where the rope has no motion. These points are called *nodes*, marked *N*. From the drawing it is evident that the nodes are spaced a half-wavelength apart. Between the nodes are points where the amplitude of the motion of the rope is twice the amplitude of the generating waves. These points are called *antinodes* (*A*) and are, of course, also spaced $\lambda/2$ apart. Looking at a rope vibrating as shown in Fig. 8-7, we would see no motion along the rope, as we would with a single traveling wave; it would appear to be merely oscillating up and down between the nodes. For this reason, such a wave pattern is called a *standing wave*.

Since the rope, or wire, cannot move where it is fastened to the wall,

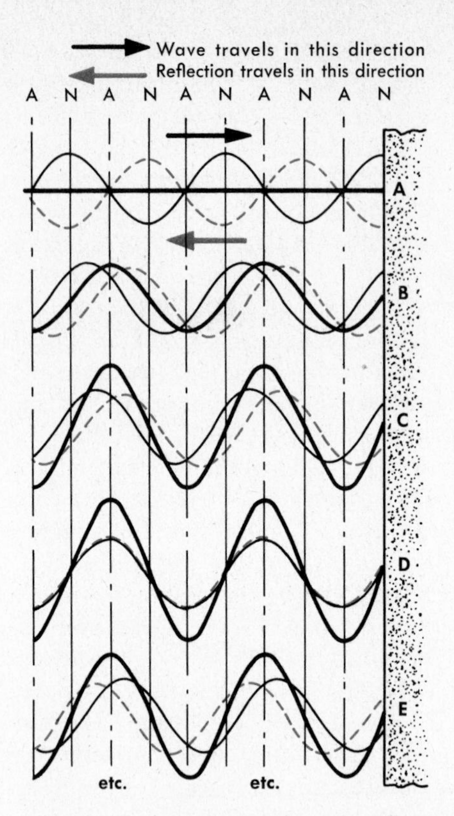

FIG. 8-7 Formation of standing waves in a rope or wire.

the wall must always be a node. And if, in Fig. 8-7, we clamp the rope rigidly at *any* node, the same standing wave will continue to vibrate between the clamp and the wall. This is the principle on which all stringed musical instruments operate.

Figure 8-8 shows a tightly stretched wire vibrating between two fixed points. When the wire is plucked, a hodgepodge of waves of all possible wavelengths are started along the wire. Most of these, since their reflections from the end points will not be in phase with one another, will fade out from interference with their own reflections. Their energy will not be lost, however, but will be transferred to those select vibrations whose wavelengths are just right to form nodes at the supports. The vibration of the longest wavelength (or lowest frequency, since $f = v/\lambda$ and v is the same for all the waves), shown in Fig. 8-8A, is called the *fundamental*; other modes of vibration, such as those shown in Fig. 8-8B and C, are called the *overtones* of this fundamental.

A plucked string normally will vibrate simultaneously in several modes. Figure 8-8D shows a wire vibrating in its fundamental, and first and second overtones.

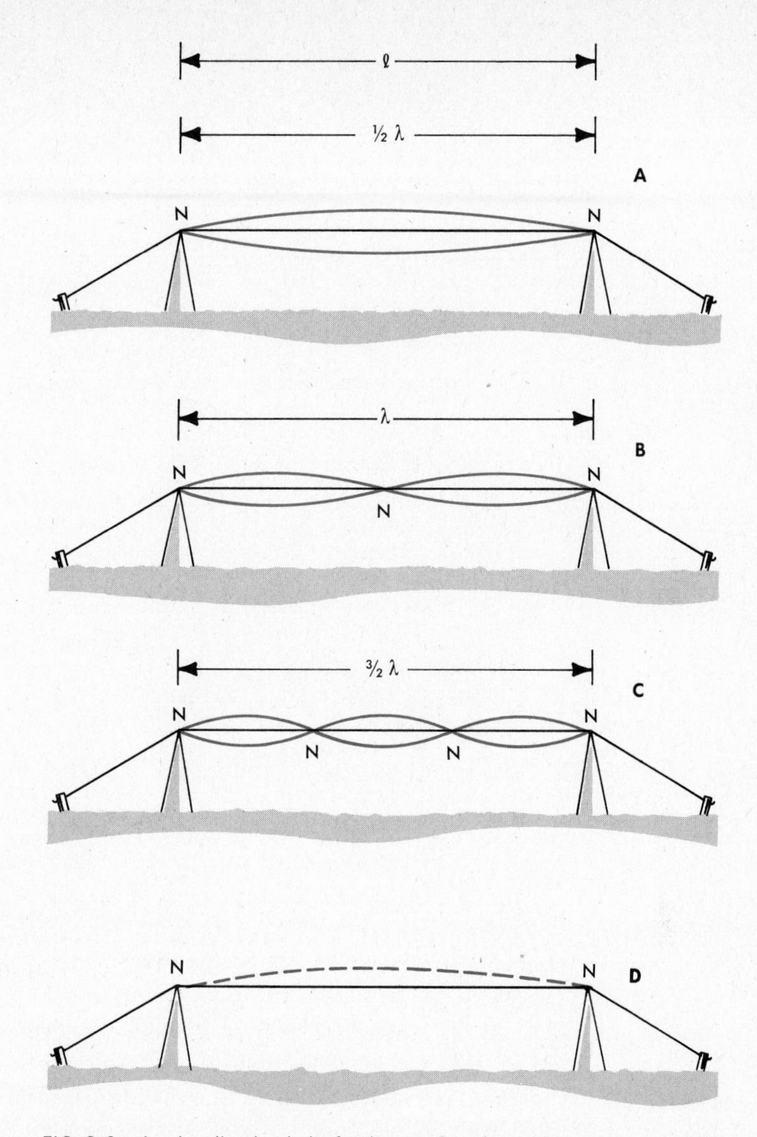

FIG. 8-8 A wire vibrating in its fundamental mode and in several overtones.

8-6 Surface Waves

We have so far considered only waves in a one-dimensional medium. In some ways it is easier to observe waves spreading out in two dimensions, such as the ripples on the surface of calm water. When a pebble or other object is dropped on the water surface, these ripples spread out from the disturbance with equal speeds in all directions, to form a series of concentric circles, as shown in Fig. 8-9. When these encounter an obstacle such as the straight wall of the pond in the drawing, the circular waves are reflected as circles. The reflected waves behave exactly as

FIG. 8-9 The waves caused by an object thrown into a pond will be reflected from the side of the pond. The reflected waves look as though they originated from an object tossed into an imaginary pond at an exactly opposite and equidistant point on the other side.

though they had been caused by an imaginary object dropped simultaneously on the surface of an imaginary pond on the other side of the wall, at the point marked A'. The points A and A' are exactly opposite each other, and are equidistant from the straight reflecting surface.

For a more detailed study of the propagation of surface waves, it is convenient to move from the pond into the laboratory and to generate waves by a more controllable means than tossing pebbles. Figure 8-10 shows a dish filled with water or mercury and an electrically driven vibrating strip that operates on much the same principle as an ordinary electric bell or buzzer. To the vibrating end of the strip we can attach a single needle, or two needles, or a long straight strip—all adjusted to barely touch the surface of the liquid and to serve as a source of regular wave trains as they oscillate up and down.

Figure 8-10 shows a series of concentric waves spreading out from a single vibrating point. As the point moves up and down, alternate crests and troughs move out from it with equal speeds in all directions, to form this simple pattern.

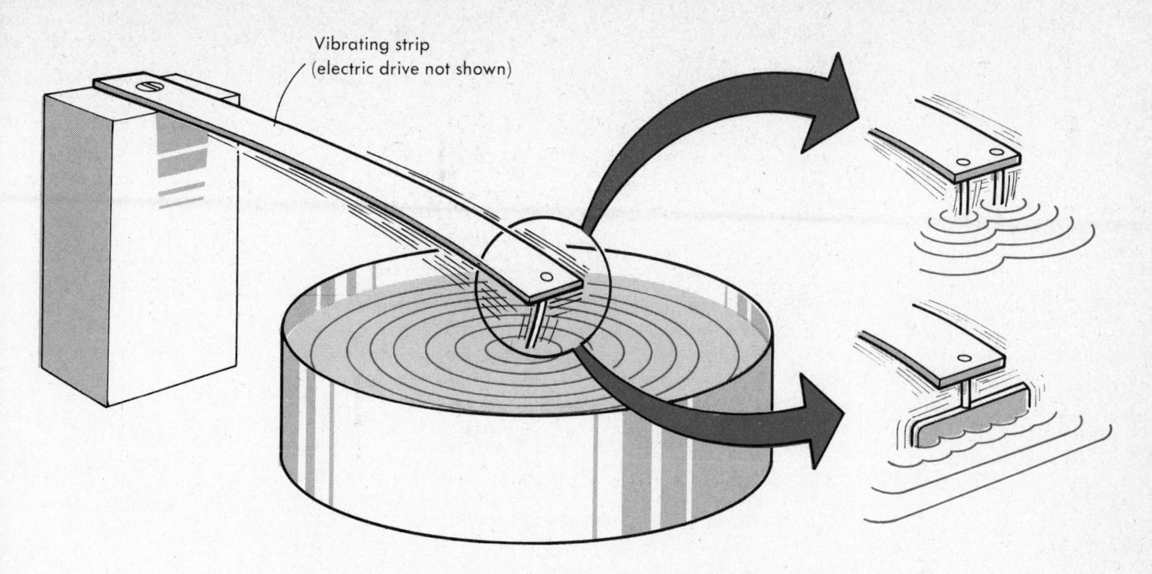

FIG. 8-10 A vibrating strip to generate waves on the surface of a liquid.

If we now replace the single needle with two needles, we should expect that each needle would send out an identical concentric pattern. This is true, but the fact that the two sets of circles overlap each other causes what is known as *interference* between them, with results as shown in Fig. 8-11. The surface of the liquid breaks up into a number of strips marked C and D on the photograph. Along the strips marked C (for *constructive* interference), the crests and troughs are deeper than they would be from either source alone; along the strips marked D (for *destructive* interference), the surface is relatively smooth and undisturbed.

An explanation of this behavior is shown in Fig. 8-12, which represents the two sources O_1 and O_2 of Fig. 8-11, and the line AB, drawn anywhere we wish. The point C_0 is equidistant from the two sources, and the waves arrive at C_0 *in phase*. That is to say, the crests arrive together and the troughs arrive together, so that their effects add up at this point, and the amplitude of motion of the surface is twice what it would be from either source alone.

If we go along AB a short distance to either side of C_0, we soon come to a point marked $D_{1/2}$, which is just half a wavelength farther from O_1 than it is from O_2. Here a crest from O_1 will arrive just in time to meet a trough from O_2, and vice versa, so that the two impulses will always add up to zero. This is destructive interference, and the surface at $D_{1/2}$ will remain undisturbed.

Farther along AB in either direction we shall come to points C_1 and C_2, at which the distances from the two sources will differ by λ, or by 2λ, etc. Wherever the difference is an integral number of wavelengths, the waves will arrive in phase to produce a point of maximum disturbance

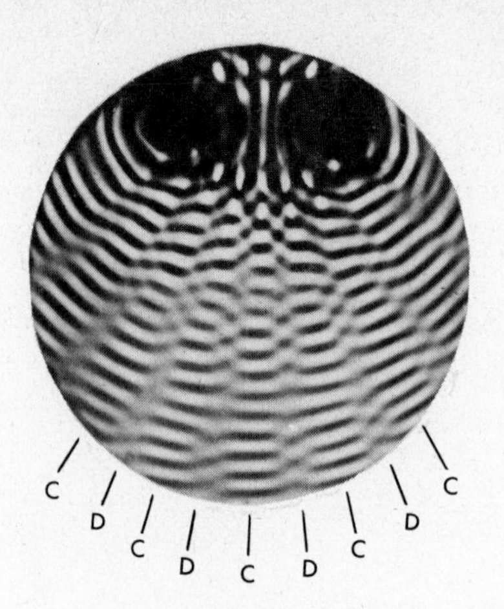

FIG. 8-11 Interference of two sets of concentric waves propagating from two points. *Courtesy The Ealing Corporation.*

of the surface. At $D_{3/2}$, $D_{5/2}$, etc., they will arrive exactly out of phase, so that each wave annuls the effect of the other.

If, in Fig. 8-11, we were to draw a number of lines like AB, we could plot along each of them similar points of maximum distrubance (antinodes) and points of zerö or minimum disturbance (nodes). Then, by connecting corresponding points, the nodal lines and antinodal lines could be traced out to follow their locations exactly, as shown in the photograph. (Students with some analytical geometry will recognize that these lines are hyperbolas.)

FIG. 8-12 The interference of two identical trains of waves coming from the points O_1 and O_2.

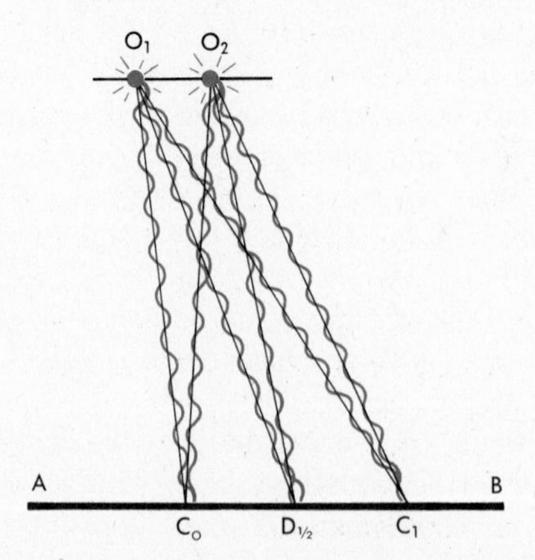

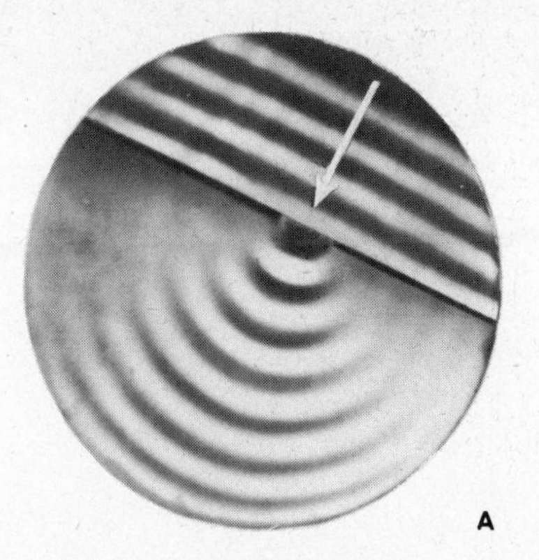

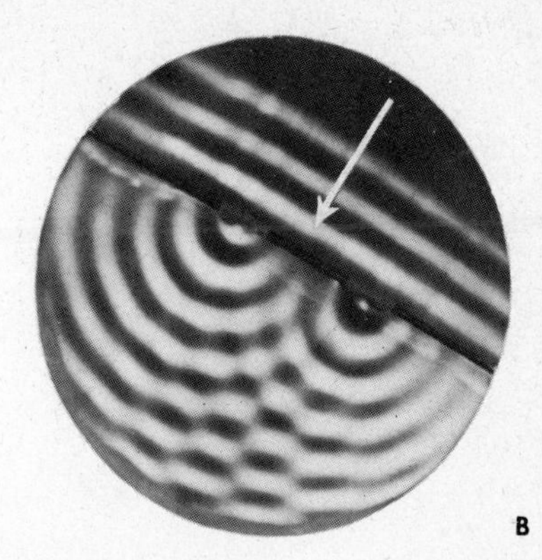

FIG. 8-13 The diffraction of a plane wave passing through a small opening in a breakwater (A); and the interference between two such wave patterns formed by two openings (B). *Courtesy The Ealing Corporation.*

8-7 Huygens' Principle

We could replace the vibrating needles on the wave generator by a long straight strip (Fig. 8-10), which would generate a series of straight-line waves propagating across the tank. (We could also generate waves that were practically straight lines, by placing a single vibrating needle at a great distance away. But since the waves become weaker with increasing distance, this would introduce too many practical difficulties.)

Now, if in the way of these straight waves we place a barrier with a single relatively narrow opening in it, the waves will pass through the opening and produce a pattern, beyond the barrier, similar to the pattern produced by a single oscillating needle (Fig. 8-13A). If the barrier, or "breakwater," has two openings, the pattern beyond it is similar to that produced by a pair of oscillating points (Fig. 8-13B).

In the paragraph above, we mentioned the wave "passing through" the opening in the breakwater. In a way, this statement is true, of course, but we shall do better to consider the matter from the point of view of Huygens. Looking up from the lower side of Fig. 8-13A, we could see the water in the slot of the breakwater oscillate up and down as the plane waves on the other side impinge against it. The disturbance caused by this oscillating water spreads out from the slot in concentric ripples, just as it does from the disturbance under an oscillating needle. In a similar way, the water rising and falling in the two slots of Fig. 8-13B produces two independent point sources of interfering waves.

These examples are illustrations of *Huygens' principle*, which says that ***every point along a wavefront serves as a source from which new waves spread out.*** This principle is true, even though we may not always have a convenient breakwater to shut off all its effects except those from one or two isolated points along a wavefront. The spreading of concentric ripples on an unobstructed surface, or the advance of plane waves,

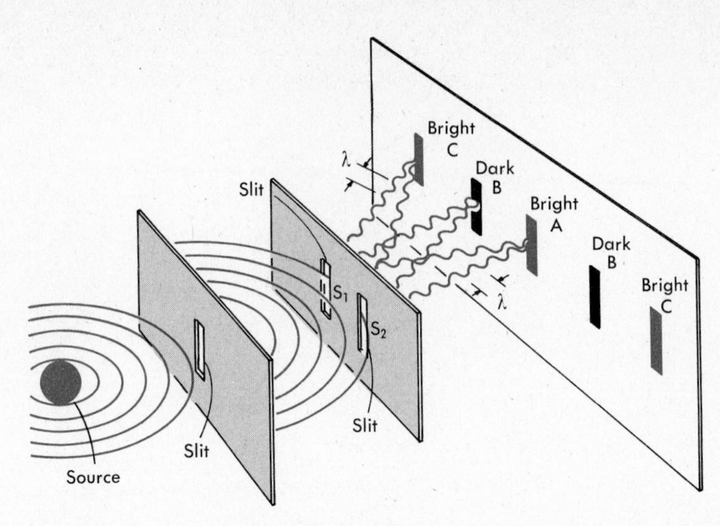

FIG. 8-14 Light waves from two slits arriving at a screen in phase and out of phase, to produce alternating light and dark bands on the screen.

results from the mutual interference, both backward and forward, among the wavelets sent out simultaneously from an infinite number of points along every wave.

8-8 Interference of Light Waves

We will not look into the nature and behavior of light in any detail for several chapters yet. At this point it will be enough to say that light is made up of an enormous number of trains of transverse waves called *photons*. Figure 8-14 shows a light source emitting photons, some of which fall on a narrow slit; the slit then serves, by Huygens' principle, as a very precisely localized source (Fig. 8-13A) sending photon waves on to a pair of closely-spaced slits, as in Fig. 8-13B. Light waves from these two slits then produce alternating bands of dark and light on the screen or photographic film. The bright band at A is equidistant from the slits S_1 and S_2, so that the waves arrive in phase to reinforce each other. Points B on the drawing are just a half-wavelength farther from one of the slits than from the other. Thus waves starting together from S_1 and S_2 arrive at B exactly out of phase and annul each other so that B remains dark. The bands at C are just one wavelength closer to one of the slits than to the other. The waves therefore arrive at C exactly one wavelength out of phase, which is equivalent to arriving in phase. At C we therefore have bright bands.

We can continue this reasoning to extend the series of alternating bright and dark bands on the screen. Whether a point will be dark or light depends on the length of the light-paths from the two light sources. If the path-lengths differ by any odd number of half-wavelengths, the point is dark; if the difference is any integral number of wavelengths, it will be bright. This result is no different from what would occur if we put a screen across the lower left part of Fig. 8-13B.

Questions

(8-1) **1.** Analyze the equation $v = \sqrt{F/m_\ell}$ to show what the units of m_ℓ must be, if v is in ft/s and F is in pounds.

2. Analyze the equation $v = \sqrt{F/m_\ell}$ to find what the units of F must be, if v is in m/s and m_ℓ is in kg/m.

3. A 100-m length of cord has a mass of 400 g. A piece of this cord is pulled with a tension of 5×10^6 dynes. With what speed would a transverse pulse travel along the cord?

4. A 1000-ft coil of wire weighs 12 lb. What is the speed of a transverse pulse along a length of this wire, when it is pulled taut by a force of 30 lb?

5. Why are strings of different m_ℓ used in different positions on a guitar? If the strings were all the same m_ℓ, what would it be necessary to do to compensate when tuning the instrument?

6. The bass strings of a piano are wrapped with a close helical winding of other wire. Why should this be done?

7. A 100-m length of wire has a mass of 800 g. This wire is spliced to a 200-m length of another wire, the same diameter as the first, but made of a material whose density is 3 times as great. The resulting 300 m of wire is pulled by a force equal to the weight of 30 kg. How long will it take for a transverse pulse to travel the length of the wire?

8. A cord 50 m in length has a mass of 1.2 kg. It is spliced to 100 m of another cord made of the same material, but of only half as great a diameter. The resulting 150 m of spliced cord is pulled taut by a force equal to the weight of 40 kg. How long will it take for a transverse pulse to travel the length of the cord?

(8-2) **9.** In Question 7, if a transverse pulse is started at the end of the lighter wire, two pulses should be expected to return. Explain, and calculate when the two pulses should be expected.

10. In Question 8, if a transverse pulse is started at one end, two pulses should return. Explain why, and calculate when the two pulses should be expected.

11. The pulse in Question 7 was started at the end of the 100-m length of wire, by striking down on it to start the pulse on the lower side. On which side of the wire—lower or upper—would you expect to see (a) the first reflection? (b) the second reflection?

12. The pulse in Question 8 was started by an upward blow at the end of the 50-m length of cord, to produce a pulse on the upper side. On which side of the cord—upper or lower—would you expect to find (a) the first reflection? (b) the second reflection?

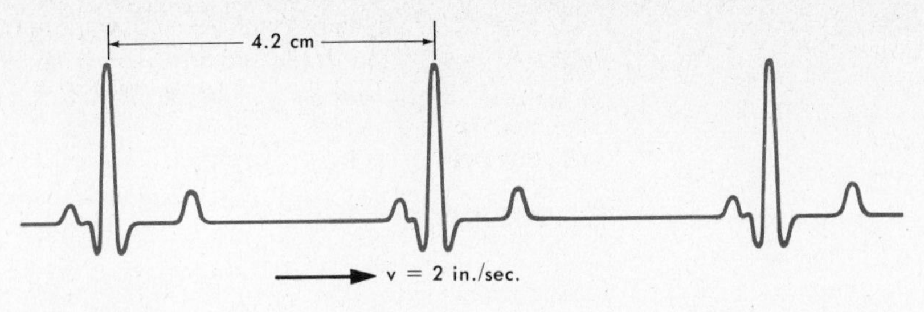

FIG. 8-Q16

(8-4) **13.** Light is a wave phenomenon that travels at a speed of 3.00×10^{10} cm/s. The wavelength of the red light emitted by a certain type of laser is 6.33×10^{-5} cm. What is the frequency of these light waves?

14. Radio waves are the same kind of electromagnetic waves as light, and travel at the same speed of 3.00×10^8 m/s. What is the wavelength radiated by station KLIR, at a frequency of 100.3 megacycles/second?

15. Cans of beans are placed at equal separations on a conveyor belt moving 75 ft/min. How far apart must the cans be placed if the belt is to deliver 125 cans per minute?

16. The illustrated normal electrocardiogram was recorded on tape moving at 2 inches/second. The patterns on the tape are about 4.2 cm apart. What is the patient's pulse rate, in beats per minute?

(8-5) **17.** A long wire is observed to be vibrating in the wind, and antinodes (where the wire is blurred by its motion) are seen to be 10 ft apart. At the same time, the wire emits a hum whose frequency is 50 cycles/second. With what velocity are transverse waves being transmitted along the wire?

18. A standing pattern of sound waves can be established in a glass tube, by reflection of the waves from pistons at each end. Cork dust in the tube will settle at the nodes, where the air is relatively undisturbed. If these nodes are measured to be 4.25 in. apart, (a) what is the wavelength of the sound? (b) what is its frequency? (Take the speed of sound to be 1120 ft/s.)

19. A 12-ft chain weighing 0.96 lb/ft supports a weight of 5000 lb. When struck at its midpoint it vibrates in its fundamental mode. What is the frequency of the sound it emits?

20. A tight guy wire 3.2 m long, with a linear mass of 4 g/cm, is twanged at its midpoint and sounds a tone whose fundamental frequency is 20 vibr/second. What is the tension on the guy wire?

21. In Fig. 8-8, assume l is 18 cm, and the fundamental frequency is 800 vibr/second. (a) What is the wavelength of the second overtone (Fig. 8-8C)? (b) At what speed do transverse waves move along the wire?

22. In Fig. 8-8, assume the wire to be 24 cm long, and the fundamental frequency to be 750 vibr/second. (a) At what speed do transverse waves travel along the wire? (b) What is the frequency of the *third* overtone?

(8-6) **23.** A liquid surface is disturbed by two vibrating points 8 cm apart, which oscillate together at a frequency of 12 cycles/s. The surface transmits these waves at a speed of 30 cm/s. (a) What is the wavelength λ of the surface waves?

(b) Consider the midpoint of the line connecting the oscillating points. Will the disturbances from the two points arrive in phase, or out of phase? Is this point a node, or an antinode? (c) How far along this line must we move to encounter the next antinode? (d) How many antinodes are there between the oscillating points? (e) How many nodes?

24. Two needles 9 cm apart vibrate in unison with a frequency of 15 cycles/s and generate waves on a liquid surface. The velocity of the waves is 25 cm/s. (a) What is the wavelength of the generated waves? (b) How many antinodal points (of maximum disturbance) will there be on the line joining the two vibrating points? (c) How many nodal points (of zero or minimum disturbance)?

25. Are any of the nodal points located in Question 23 actually points of zero disturbance? (Bear in mind that the amplitude of the waves diminishes with distance.)

26. On a line like AB in Fig. 8-12, is the disturbance actually zero at any of the nodal points? (Remember that the amplitude of the waves diminishes with distance.) Is the resultant disturbance less at $D_{3/2}$ or $D_{5/2}$?

9

Sound

9-1 Sound Transmission

In Chapter 8 we looked at some of the properties of waves traveling along one-dimensional carriers such as wires and ropes, and the two-dimensional waves on a liquid surface. When we deal with sound, we consider waves spreading out in three dimensions from some vibrating source, transmitted as longitudinal waves through the air or some other fluid.

If we watch road-builders blasting on a distant hillside, we see the puff of dust and rock somewhat sooner than we hear the sound of the explosion. And we know that the clap of thunder follows the lightning with a delay that depends on how far away the storm is. Since (by everyday standards) light propagates almost instantaneously, we can find the velocity of sound by timing the delay between seeing the road-builders' explosion and hearing its sound. By such experiments we would find the velocity of sound in air at ordinary temperature to be about 1100 ft/s, or about 330 m/s.

As an example, imagine that we see the dust and flying debris of an explosion on a hillside some distance away. We hear the bang of the explosion 14.8 seconds later. A survey (or more easily, careful measurement on an accurate map) shows the location of the explosion to be 3.083 miles away. For the speed of the sound, then, we have

$$v = \frac{3.083 \times 5280}{14.8} = 1100 \text{ ft/s}.$$

The velocity of sound in air is not appreciably affected by changes in pressure. We can qualitatively justify this experimental finding by comparing the transmission of the longitudinal waves in air with the way transverse waves are transmitted in a rope. If we increase the pressure on air (or any other gas), we may think of the resilience or "springiness" of the gas as being increased; this, we reason, should increase the speed with which a compression or rarefaction will be transmitted from one location in the gas to another location immediately adjacent, in the same way that increased tension speeds up the transmission of a transverse wave from one point on the rope to the next. There is a great difference however, between the effect of tension in the rope and pressure in the gas. An increased tension does not change the linear density of the rope enough to bother with. Air and other gases, however, are very easily compressible; as we shall see later, increasing the pressure on a gas reduces its volume and therefore increases its density in direct proportion to the increase in pressure. Increasing pressure thus tends to speed sound velocity by increasing the springiness of the gas; on the other hand, by making the gas more dense, it increases the inertia of the gas enough to completely nullify the gain. For this reason, a change in pressure has almost no effect on the speed of sound. An increase in temperature, though, tends to make a gas expand. Therefore, if the gas is heated, it must either expand and become less dense without changing the pressure, or, if the gas is confined so that it cannot expand, it will increase the pressure while the density remains the same. In either case, we see that sound must travel faster in warm gas than in cool.

A light gas, such as hydrogen, is much less dense than air is at the same pressure; therefore sound travels much more rapidly in hydrogen than in air. Similarly, in a dense gas like carbon dioxide, the velocity of sound is slower than it is in air.

For us to be able to speak of the frequency of a sound, the sound must consist of a train of a considerable number of waves. One way to produce a train of waves of definite frequency is to force some solid surface to vibrate back and forth by mechanical or electrical means, thus causing alternate compressions and rarefactions that spread out through the surrounding air. The high note of a soprano that reaches your ear from the radio or TV is caused by the vibration of the loudspeaker diaphragm, which is forced to move back and forth thousands of times per second by the rapidly fluctuating pull of a magnet. A card or light stick held against the teeth of a rotating gear wheel will move back and forth as each tooth strikes it, and so set up in the air sound waves of the same frequency as its own motion.

The sound of nearly all musical instruments (and of our own voices as well) depends on setting up standing waves in strings (piano, guitar, violin, etc.) or in air columns (organ, flute, trumpet, etc.).

Figure 8-8 showed a wire vibrating in several different *modes*; vibrations of this sort are the basis for all stringed instruments. The fingering of a violin or guitar changes the length of the string included between the end nodes and thus changes the frequency of the fundamental vibration. Take, for example, a string 20 in. long whose tension is adjusted so that its fundamental vibration has a frequency of 440 vibr/s. If a musician presses the string against a fingerboard, in effect reducing its length to 18 in., the fundamental wavelength has been reduced from 40 in. to 36 in. From our basic wave equation $v = f\lambda$, we see that the frequency must be increased in the same ratio as the decrease in λ. The shortened string will therefore vibrate with a fundamental frequency $f = 440 \times 40/36 = 489$ vibr/s.

A vibrating string is so small that by itself it would create very weak waves in the surrounding air. For this reason, all stringed instruments are provided with some form of sounding board, a large flat surface on which the strings are supported and which vibrates in unison with the strings. Sounding boards respond more readily to some frequencies than to others; hence some overtones are emphasized and others are suppressed. It is this difference in the relative strengths of the various overtones that makes a guitar, for example, sound different from a banjo, and a zither different from a harpsichord.

Most wind instruments are in effect tubes that are closed at one end and open at the other; the effective length is controlled by valves that expose openings along the tube. Organ pipes are of two kinds—open at both ends, or open at one end and closed at the other. (A pipe or tube closed at both ends would not be very useful as a musical instrument—at least one end must be open, so that the waves of alternating compression and rarefaction can escape into the surrounding atmosphere.)

Figure 9-1 shows tubes of these two types, with standing air waves sketched inside them. The layer of air next to a closed end cannot move, so a closed end must be a node; at the end open to the atmosphere, the vibrating air can move as freely as it is impelled to, so the open end must be an antinode. Like a plucked string or wire, the air column will vibrate at its fundamental frequency as well as in many overtones simultaneously. Figure 9-1 indicates only a few of the many possible overtones.

As an example, we can calculate the length of a closed-end organ pipe that is to produce a note whose fundamental frequency is 550 vibr/s. (A closed-end pipe has one end closed, and the other open.) The sound will be traveling in air at a speed of about 1100 ft/s. So from the equation

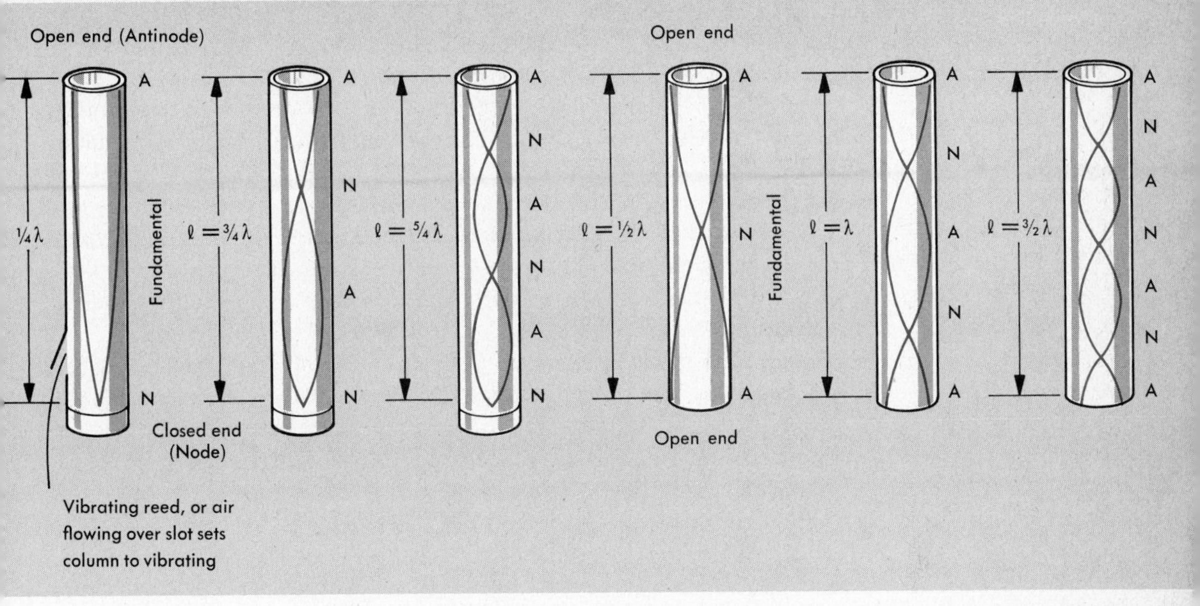

FIG. 9-1 The fundamental tone and some overtones in air columns vibrating in open-end and in closed-end tubes. (One end must always be open to allow the sound waves to escape.)

$\lambda = v/f$ we see that $\lambda = 2$ ft. Since the length of the tube is $\lambda/4$ (see Fig. 9-1), this gives us a required length of $\frac{1}{2}$ ft, or 6 in. (Actually this is not quite right; the antinode is not exactly at the open end but a slight distance beyond it, the distance depending on the wavelength and the tube diameter. For our purposes, we can neglect this refinement.)

9-2 Ultrasonics

The human ear cannot hear sound vibrations with a frequency of much less than 20 per second or more than about 20,000 per second. (As we get older, the high-frequency upper limit is reduced; many middle-aged people cannot detect frequencies higher than 10,000 per second.) A dog's ear, however, can hear sounds of considerably higher frequency, and this canine ability is often used in police work and by dog trainers. Dogs can receive "silent" orders from an *ultrasonic* whistle, the frequency of which is too high to be detected by a human ear.

Ultrasonic waves are not audible to human ears because their frequency is too high; they nevertheless carry much more energy than waves of equal amplitude but lower frequency. Compare a periodic wave A (in a rope, to simplify the matter) with another periodic wave B, which has a frequency twice that of A but is of the same amplitude. Each particle of wave B must travel the same distance back and forth in half the time needed by a particle in *A*. It must therefore have twice the average speed of an *A* particle and four times as much kinetic energy. The energy transmitted by a wave is thus seen to be proportional to the *square* of the frequency, and very-high-frequency ultrasonics carry enough energy to do many things that audible sound waves cannot. The most widespread use of ultrasonics is in the somewhat mundane job of cleaning, especially irregularly-shaped metal pieces, from dental bridges to ball-bearings and engine parts. The ultrasonic waves are generated in a tank of oil or other liquid; the energetic waves penetrate even deep crevices to shake loose the dirt adhering to any objects immersed in the ultrasonic bath.

Ultrasonics have application in several kinds of *echo sounding*. Probably the best-publicized sort of echo sounding is the sonar used by submarines, although this application generally uses sound in the audible range. A brief burst of sound is emitted in the water—if these spreading waves meet an obstacle in the water, some of the sound is reflected back. Measurement of the elapsed time between the transmission and the return of the reflected echo, and the direction from which the echo comes, gives an accurate location of the obstacle. Blind people unconsciously develop a similar method by which obstacles are sensed by the reflected echoes of their footsteps or the tapping of a cane.

Ultrasonic sonar is highly developed in bats. Audible (to human ears) sound would be of little use to a bat, because effective reflection takes place only from objects at least as large as the wavelength of the sound. Although it would enable them to avoid trees and the walls of the dark caves in which they live, it would be of no use in locating the night-flying insects on which most bats live, or in avoiding leaves and small branches. The bat's very short wavelength echo-ranging waves have a frequency of from 10^5 to 5×10^4 cycles/second; these are emitted in the form of short spurts of about 100 waves, repeated as often as 50 times a second. The silent intervals are presumably occupied by listening for reflected echoes.

Medical uses for ultrasonic vibrations are still very limited. One promising field of application is another aspect of echo ranging. X rays have certain deficiencies; as we will see later, X ray photographs are shadow pictures. Atoms of high atomic number (such as the phosphorus in bone) cast denser shadows. Many of the soft tissues are of nearly the same chemical composition in spite of their different structures, so X rays give little information about them. Short bat-like bursts of ultrasonic waves (of short wavelength to show finer detail) are partially reflected from any surface where the speed of the waves changes—as with discontinuities in the m_1 of a rope. Experienced operators can build up an internal picture through the electronic measurement of the time delay of the echoes reflected from the surface of an anatomic structure. This technique, still in its early infancy, has also been applied to examination of the brain, for which X rays are of limited value because of the shielding of the skull.

9-3 Supersonics and Shock Waves

Supersonics, which must not be confused with ultrasonics, is the study of the effects of objects that travel through a medium at a faster speed than the waves they generate. Nothing can move very fast through a solid, and even the most imaginative designers cannot yet dream of a submarine that travels faster than the speed of sound in water. So the practical problems of supersonics are essentially limited to planes and missiles that fly through the air at a speed greater than the speed of sound in air. In this situation, the moving object piles up disturbances—or waves—faster than they can get out of the way; this causes a single region of severe disturbance known as a *shock wave*.

The idea behind shock waves can be studied in a much more common and much more gentle form by considering the bow wave of a ship traveling at a speed greater than the speed of the surface waves it generates. We see such a ship in Fig. 9-2. The ship, moving with speed v_s, has traveled from A to C during the time in which the surface wave it generated at A has moved from A to A' at a speed of v_w. Similarly, the wave started at B has at this same instant reached B', and so on for the infinite

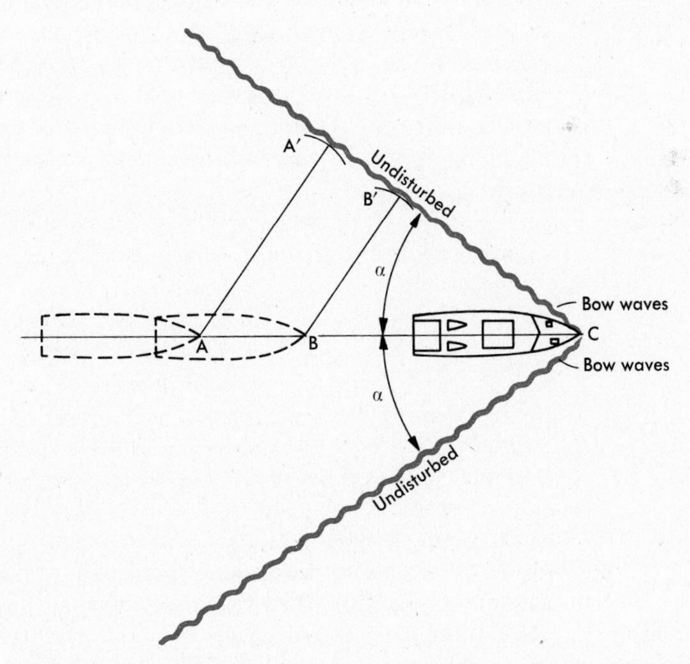

FIG. 9-2 The bow waves of a ship traveling faster than the speed of propagation of surface waves on the water.

number of points that could have been drawn between A and C. The angle α between a bow wave and the direction of motion of the ship depends on the relative velocities of the waves and the ship:

$$\sin \alpha = \frac{A'A}{AC} = \frac{v_w t}{v_s t} = \frac{v_w}{v_s}.$$

As an example, consider an ocean liner, cruising at 20 knots (1 knot = 1 nautical mile per hour = 1.151 miles per hour) and generating waves whose speed of propagation is 8 knots. We should expect the angle α between the path of the ship and the line of the "shock" bow wave to be given by

$$\sin \alpha = \frac{8}{20} = 0.400$$

$$\alpha = 23.6°.$$

A plane traveling faster than the speed of sound creates a similar disturbance in the air. Here, instead of the shock wavefront forming a V on the water surface, it forms a giant cone in the air, since the waves generated by the passing plane spread out in all directions (Fig. 9-3). On the surface of the cone, where the waves pile up, there is a sharp pressure difference. When this cone (known as the *Mach cone*, after the physicist Ernst Mach) strikes a house, it sounds like a loud thunderclap, and may even be strong enough to shatter windows. These sonic booms are becoming familiar in many parts of the country where supersonic planes pass overhead.

The half-angle of the cone, α, is of course given by the same relationship that applied to the bow wave of a ship:

$$\sin \alpha = \frac{\text{speed of sound}}{\text{speed of plane}}.$$

Supersonic speeds are often given in terms of their *Mach number*, which is simply the ratio of the speed of the plane to the speed of sound. At Mach 1.5, for example, the plane is traveling 1.5 times as fast as sound. A Mach speed cannot be directly converted into miles per hour without further information, because the speed of sound itself varies, primarily with the temperature. On a pleasant summer day near the earth's surface, Mach 1 (the speed of sound) may be about 750 mi/hr; 30,000 ft overhead, the temperature may be −70°F, and at this cold temperature Mach 1 would be about 640 mi/hr. If the speed of a plane or missile is given in Mach units, it follows that $\sin \alpha = 1/\text{Mach number}$.

FIG. 9-3 A shock wave (Mach) cone generated by the nose of an airplane flying faster than the speed of sound.

9-4 The Doppler Effect

Everyone has experienced (although many may not have observed) the Doppler effect. If a car, which is sounding its horn, rapidly approaches you, passes, and then continues on, a very noticeable drop in the pitch (or frequency) of the horn will occur at the moment of passing. A similar drop in pitch results when a moving observer passes a stationary bell or horn. This change in pitch caused by the motion of either the source of sound or the listener is called the *Doppler effect*, after the nineteenth-century Austrian physicist Christian Doppler.

Let us look first at a moving source. The top of Fig. 9-4A, at $t = 0$, shows a source of sound S in the process of emitting a wave. The source is moving toward the observer with a speed v_s. At a time T seconds (1 period) later, the previously emitted wave 3 has traveled toward the

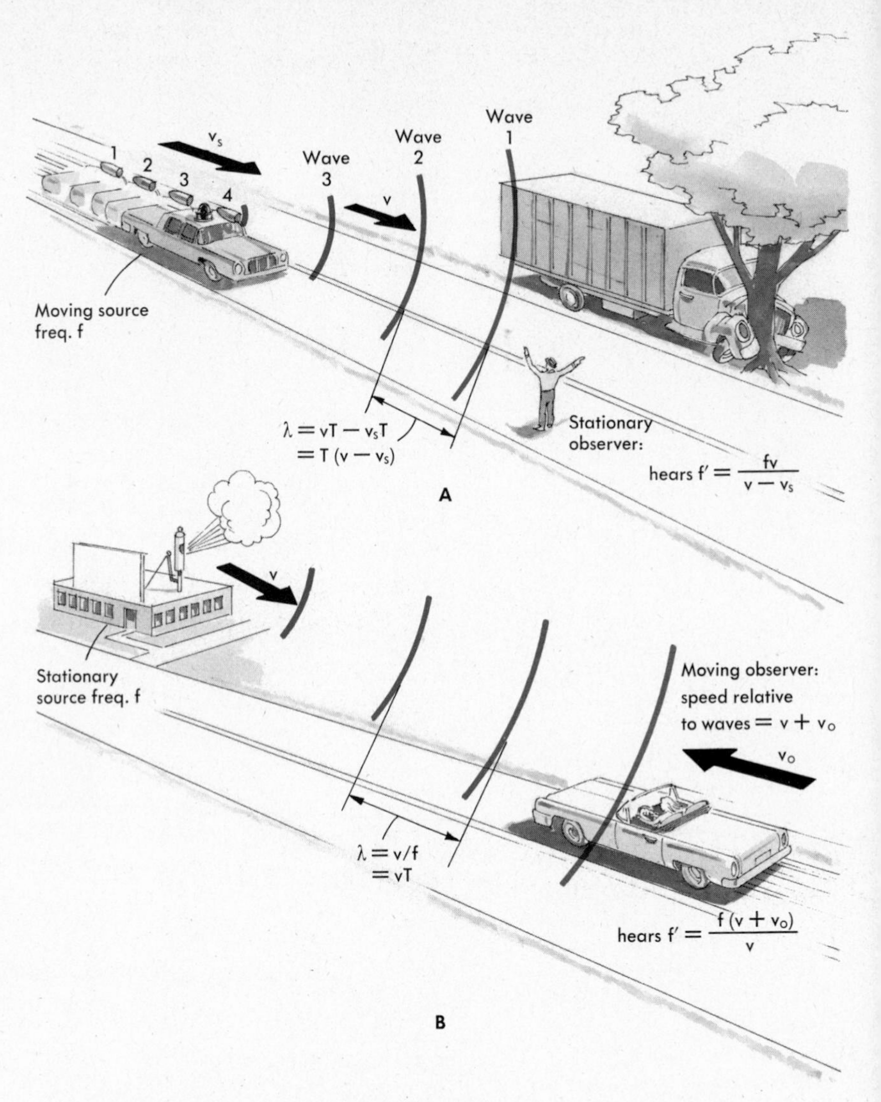

FIG. 9-4 The Doppler effect, as observed in sound.

observer a distance vT, while the source has moved a distance $v_s T$ in the same direction and is now emitting wave 4. The distance between the waves is, then, $vT - v_s T$, or $\lambda' = T(v - v_s)$. The frequency with which the waves strike the ear of the observer will thus be

$$f' = \frac{v}{\lambda'} = \frac{v}{T(v - v_s)}.$$

The actual frequency with which the source emits waves is $f = 1/T$. Thus the observed frequency, in terms of the actual emitted frequency, is

$$f' = f \times \frac{v}{v - v_s}.$$

If the source were receding from the observer, it is apparent that the equation would read

$$f' = f \times \frac{v}{v + v_s}.$$

Figure 9-4B shows the observer moving with a speed v_o toward a stationary source of sound that emits waves spaced λ apart. The relative speed between the waves and the observer is $v + v_o$, and the frequency with which the waves strike his ear will be

$$f' = \frac{v + v_o}{\lambda}$$

while the actual frequency of emission is

$$f = \frac{v}{\lambda} \quad \text{and} \quad \lambda = \frac{v}{f}.$$

We can again express f' in terms of f for the observer who is either approaching or receding from the source:

$$f' = \frac{v \pm v_o}{v/f} = f \times \frac{v \pm v_o}{v}.$$

The equations above can be combined into the following single expression:

$$f' = f \times \frac{v \pm v_o}{v \pm v_s}.$$

There are rules governing the selection of the + and − signs, but a smattering of common sense will make these rules unnecessary. Approach (of either source or observer) makes the heard frequency higher, and the proper sign can be chosen to make the equation fit the situation. For example, consider car *A* speeding down the highway at 90 ft/s. Ahead of *A* is car *B*, going in the same direction at only 30 ft/s, and *A* sounds his horn before he passes. If the actual frequency of *A*'s horn is 1000 vibr/s, what frequency does B hear? Let us write our equation without regard for sign:

$$f' = 1000 \times \frac{1100 \pm 30}{1100 \pm 90}.$$

The source (car A) is moving toward the observer (car B); this motion, considered by itself, would make the heard frequency greater. Therefore, we must make the denominator of the fraction (which contains v_s) smaller by subtracting the 90. The observer is headed away from the source; this motion, considered by itself, would make the heard frequency less. Therefore, we must make the numerator (which contains v_o) smaller by subtracting the 30. So we have

$$f' = 1000 \times \frac{1100 - 30}{1100 - 90}$$
$$= 1059 \text{ vibr/s}.$$

9-5 Sound Intensity

In perfectly quiet surroundings, the faintest sound you can hear represents a *very* small amount of energy. If you hold up a finger in a room containing barely audible sound, a fingernail (about 1 cm²) will be struck by

FIG. 9-5 Scale of noise, measured in decibels and in watts/cm².

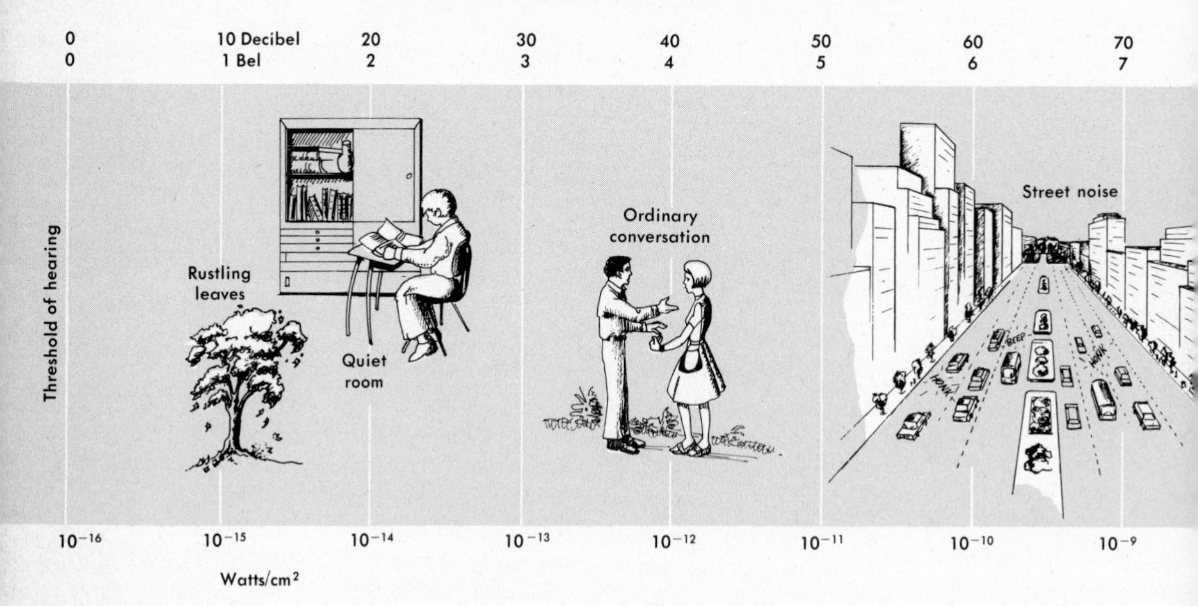

sound waves representing an energy of about 10^{-16} J/s. This can be more briefly described as 10^{-16} W/cm^2. In a fairly (but not exceptionally) noisy factory, the sound might carry energy at the rate of 10^{-7} W/cm^2, a level 10^9 times as great as the barely audible.

Figure 9-5 is a graphic representation of a wide range of sound levels, on a geometric scale, similar to Fig. 1-1, that showed a large range of time intervals and distances. The W/cm^2 scale at the bottom could obviously be extended indefinitely to the left without ever reaching zero. Audiologists have found it convenient to use this powers-of-10 scale, by counting from the threshold of hearing (10^{-16} W/cm^2 or 10^{-12} W/m^2) as zero. From this zero, each added *bel* (named for Alexander Graham Bell) means a multiplication of the sound intensity by 10. In a factory, for example, in which the sound intensity is 10^{-7} W/cm^2, this noise would be designated as 9 bels—that is, 10^9 times the intensity of the hearing threshold of 10^{-16} W/cm^2.

The bel has proved to be a little too large for convenient use, and has been almost completely supplanted by the *decibel* (dB), which, as its name implies, is just one-tenth of a bel. The noise level in the factory described above would be given as 90 dB, as indicated on the scales above Fig. 9-5. When we thus divide each bel into 10 dB on a geometric scale, each decibel must represent multiplication by the tenth root of 10, or 1.26. Doubling the intensity of a sound means an increase of 3 dB—

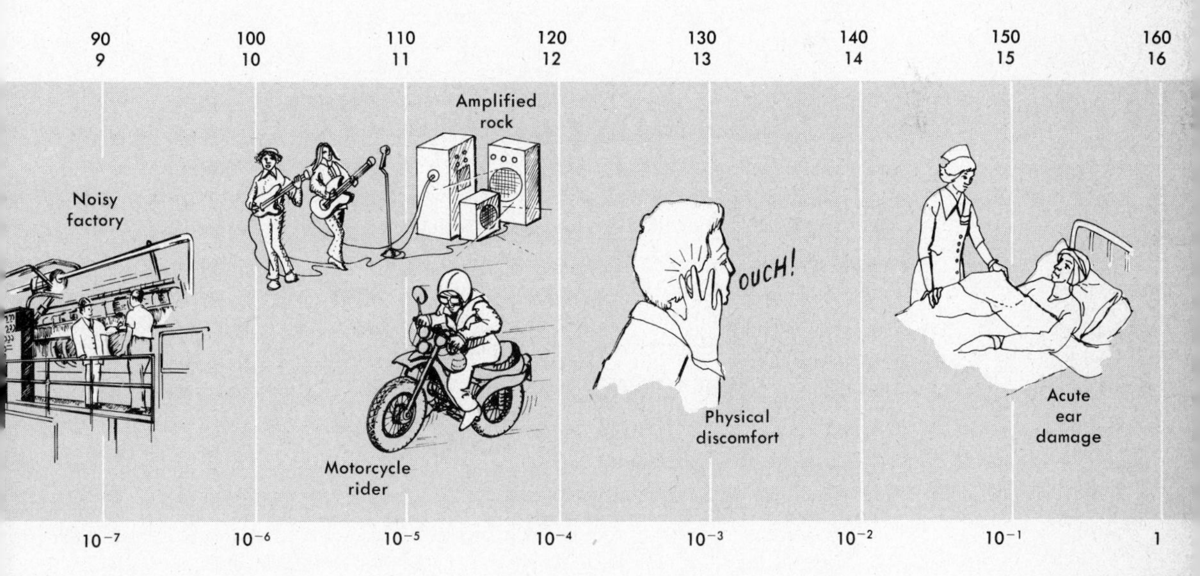

$1.26 \times 1.26 \times 1.26 = 2.00$. A 6-dB increase means a multiplication by 4; 9 dB equals multiplication by 8.

Decibels have provided a convenient scale for measuring the performance of many amplifying devices, not necessarily associated with sound. A 20-dB amplifier, for example, we could at once convert to 2.0 bels, which means amplification by a factor of 10^2, or 100. An amplification of 16 dB we could handle in two parts by breaking the 16 dB into two parts: 10 dB and 6 dB. The 10 dB equals 1 bel, equivalent to a factor of $10^1 = 10$; the remaining 6 dB means multiplication by 4, for a total of $10 \times 4 = 40$.

Questions

(9-1) 1. Lightning is observed to strike a tree on a peak known to be just 1.50 miles away. The sound of thunder is heard 7.1 seconds later. What was the speed of the sound?

2. Derive a simple formula by means of which you can count the time in seconds between a lightning flash and the sound of thunder, and from it determine the distance in miles to the lightning stroke.

3. A man stands some distance from a high cliff and fires a gun. He hears the echo 1.5 s after the gun was fired. How far away is the cliff?

4. A man stands 400 ft in front of a high cliff and fires a gun. How long a time elapses before he hears the echo reflected from the cliff?

5. Through binoculars a man watches a carpenter driving nails at a regular rate of 1 stroke per second. He hears the sound of the blows exactly synchronized with the blows he sees. He hears two more blows after he sees the carpenter stop hammering. How far away is the carpenter?

6. If you stand at one end of a long hallway with a good sound-reflecting surface at its far end, you can clearly hear the echo of a handclap. With a little effort and practice, you can clap your hands rapidly and regularly, so that each clap coincides with the echo of the previous one. Assume you stand 135 ft from the end of the hall, and a colleague counts 120 claps in 30 sec. What is the speed of sound in this hall?

7. A strip of stiff plastic is held against the teeth of a rotating 48-tooth gear, and the sound produced has a frequency of 512 vibr/second. What is the rotational speed of the gear, in rev/min?

8. A siren can be made from a rotating flat disk, which has regularly spaced holes punched through it along a circle concentric with the axis of rotation. An air nozzle is directed against the disk; each time a hole passes the nozzle, a puff of air is released to generate a wave pulse. What frequency sound will be produced by a disk containing 72 holes, and rotating 1800 rev/min?

9. A man in front of a long flight of stairs makes a loud clap by slapping two boards together. The tread of the stairs (the horizontal surfaces one steps on) is 11 in. What frequency is associated with the sound reflected from the stairs?

10. A man stands some distance in front of a long, high flight of stairs, and fires a gun. The stairs have treads (the horizontal surfaces one steps on) that

are uniformly 15 in. deep. Will there be any tone associated with the echo from the steps? If so, what is its frequency?

11. A string on a musical instrument is 22 inches long, and has a fundamental frequency of 272 vibr/s. To what length must the string be shortened (by pressing it against a fret or a fingerboard) to make its frequency 318 vibr/s?

12. A string on a musical instrument is 33 inches long, and has a fundamental frequency of 192 vibr/s. What is the frequency when this string is pressed against a fret to reduce its length to 30.5 inches?

13. (a) What is the fundamental frequency of a tube closed at one end and 8 inches long? (b) What is the frequency of its first overtone? (c) of its second overtone?

14. (a) What is the fundamental frequency of a tube open at both ends and 12 inches long? (b) What is the frequency of its first overtone? (c) of its second overtone?

15. The fundamental frequency of an organ pipe open at both ends is 440 vibr/s. What is the frequency (a) of its first overtone? (b) of its third overtone?

16. The fundamental frequency of a piano string tuned to the note A above middle C is 440 vibr/s. What is the frequency (a) of its first overtone? (b) of its third overtone?

17. What is the length of an organ pipe closed at one end, having a fundamental frequency of 440 vibr/s?

18. What is the length of the organ pipe of Question 15?

(9-2) **19.** Take two "sounds" of the same amplitude, one of 2500 vibr/s and the other an ultrasonic one of 100,000 vibr/s. How many times as much energy does the latter carry than the former?

20. A mechanism for creating 5000-cycle/second sound waves of a certain amplitude in water requires a power of 25 watts. How many watts would be needed to create waves of the same amplitude, but with a frequency of 60,000 cycles/second?

21. The echo of a sonar "beep" is heard 2.50 s later. If the speed of sound through water is 1400 m/s, how far away is the reflecting iceberg?

22. An oceanic depth-sounding vessel surveys the ocean bottom with ultrasonic sonar, of $v = 5020$ ft/s in seawater. What is the time delay in the return of the echo from the bottom, in a trench 6 mi deep?

23. About what is the length of the train of one hundred 10^5-cycle/second sound waves emitted by a bat?

24. Assuming fifty 100-wave bursts per second of sound of $f = 10^5$ cycles/second, about what fraction of the time is the bat (a) emitting sound? (b) listening?

(9-3) **25.** A toy sailboat on a pond moves with a speed of 1 m/s and creates waves whose speed is 30 cm/s. What is the angle included between its bow waves?

26. A ship moving at 30 knots creates bow waves including an angle of 60° between them. What is the speed of the surface waves created by the ship?

27. A jet plane travels at Mach 2 at an altitude of 5000 ft. How far past an observer will the plane be when the shock wave hits him?

28. A bullet travels 510 m/s through the air. What will be the angle between the shock wave and the path of the bullet?

(9-4) 29. At what speed would an observer in a car have to approach a stationary siren in order for its pitch to sound 10 percent higher than it actually is?

30. At what speed would a car have to be approaching an observer for the latter to hear music from the car radio with a pitch 10 percent higher than it actually is?

31. A car, sounding a horn whose frequency is 1200 vibr/s, drives directly toward a vertical cliff at 55 ft/s. What is the frequency of the sound of the horn reflected from the cliff, as heard by the driver of the car? (The car acts as a moving source when the sound is emitted, and as a moving observer for its reflection.)

32. A car has a siren sounding a 2000-vibr/s tone. What frequency will be heard (a) by a stationary observer as the car approaches him at 45 mi/hr? (b) if the car is stationary and the observer approaches it at 45 mi/hr? (c) if the car is going away at 60 mi/hr and the observer chases it at 30 mi/hr?

(9-5) 33. The sound level in a quiet office is 40 dB. (a) What is the sound energy intensity in W/cm^2? (b) How long would it take for 1 erg of energy to be delivered to a microphone that presents an area of 10 cm^2 to the sound?

34. A window 60 cm $\times$ 100 cm is open to a noisy street where the average sound intensity is 80 dB. (a) What is the sound energy intensity in W/m^2? (b) How much sound energy enters the room through this window in an 8-hr day?

35. A noisy office has a sound level of 60 dB, which seriously cuts down efficiency and increases mistakes. What fraction of the present sound energy would have to be kept out and/or absorbed, to bring the level down to 50 dB?

36. The noise level in a hospital room is 40 dB; 20 dB would be much better for its occupants. An architect claims to be able to achieve a 20-dB level quite inexpensively by installing wall covering that will cut the sound energy intensity in half. Is the architect right? Explain.

37. A duet of singers produces 40 dB of sound at one of the seats in an auditorium. It is replaced by an octet. To what decibel rating does this increase the sound at this same seat?

38. At a football game 300 people (visitors) cheer loudly when the home-team quarterback is sacked. This produces 40 dB at the press box. When the home team returns a kickoff 103 yards for a TD, 24,000 people cheer equally enthusiastically. What sound level does this produce in the press box?

10

Temperature and Heat

10-1 Measurement of Temperature

Except for judging by the feel of our skin, which is not a very accurate quantitative guide and is often deceiving, the only way we can measure temperature is to measure the effects temperature changes have on the physical properties of materials. All physical bodies, such as a piece of solid material or a certain amount of liquid, nearly always respond to temperature changes in a simple way: They expand when the temperature rises and contract when it drops. There are many other effects, and most of them are utilized in one way or another to measure temperature: the change in the electrical resistance of wire, the variation with temperature of electric current generated by dissimilar metals joined together, changes in the viscosity or plasticity of materials, changes in the color of the light emitted by very hot bodies, and so on.

The most commonly employed effect, however, is that of expansion. Thermometers used for this purpose are usually constructed in such a way that a small thermal expansion results in a large displacement of an indicator. In thermometers based on the expansion of a liquid, a

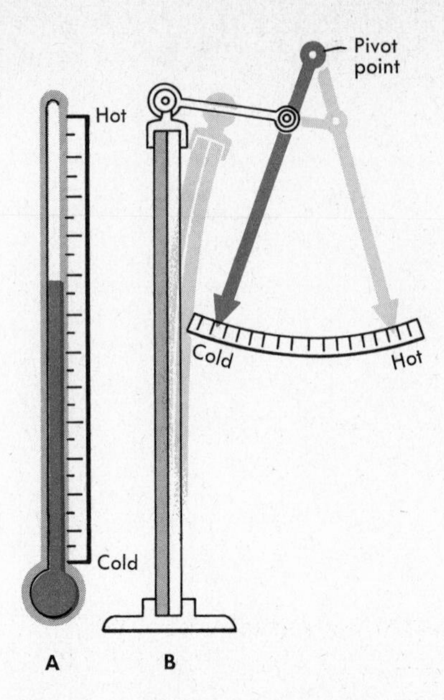

FIG. 10-1 (A) Mercury thermometer; (B) bimetal strip thermometer.

comparatively large amount of the liquid (usually mercury or alcohol) is confined in a bulb, and its expansion causes the excess liquid to rise in a narrow capillary tube (Fig. 10-1A). In thermometers based on the expansion of solids, a double (bimetallic) strip is often used, composed of two metals whose expansions are different when they are heated. For example, in Fig. 10-1B, the darkened strip might be of brass and the unmarked strip (cemented or otherwise securely fastened to the brass strip) could be zinc. Brass expands considerably more than zinc when heated, so as the bimetallic strip is warmed, the greater expansion of the brass forces the strip to curve, as shown, and move an indicator across a scale.

Whatever kind of thermometer we choose, the scale on which the indicator moves must be calibrated in some definite, easily reproducible manner. The two thermometric scales in general use are calibrated at the freezing point and the boiling point of water (at a pressure of 1 standard atmosphere). On the *Centigrade*, or *Celsius*, scale, which is used in scientific work all over the world, the freezing point is 0°C and the boiling point is 100°C. On the *Fahrenheit* scale, used (except for scientific work) in the United States, 32°F is the freezing point and the boiling point is 212°F. Figure 10-2 shows the calibration of a pair of thermometers. With both thermometers immersed in a beaker of wet crushed ice (Fig. 10-2A), we mark the points where the mercury stands as 32 on the Fahrenheit thermometer, and as 0 on the Celsius. Similarly, with the thermometers in steam above boiling water (Fig. 10-2B), we mark

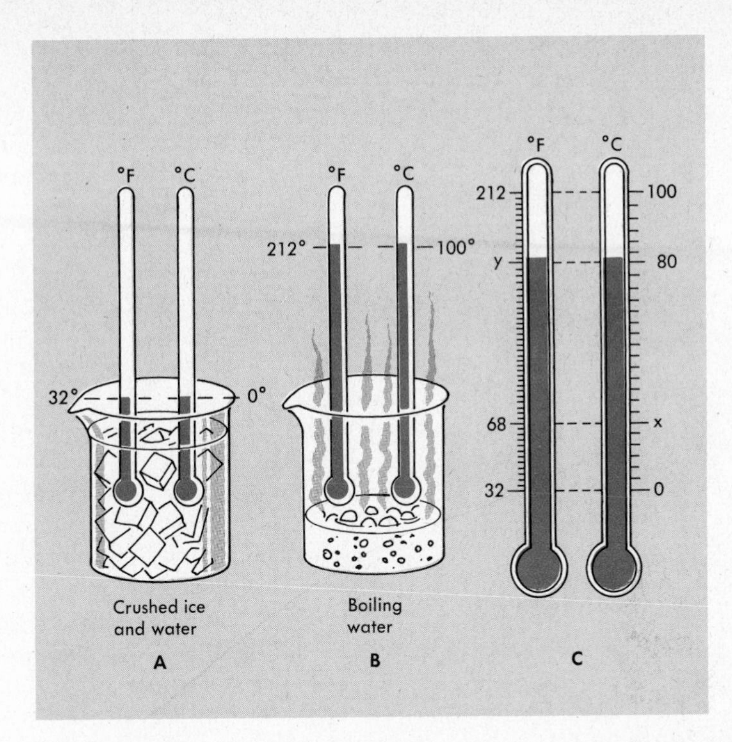

FIG. 10-2 Calibration of Fahrenheit and Centigrade (or Celsius) thermometers.

the points 212 and 100. On the Fahrenheit scale the interval between freezing and boiling is thus divided into 180 degrees (212–32), and on the Celsius scale there are 100 degrees in this same interval. From this it is plain that the Fahrenheit degree is only $\frac{5}{9}$ the size of the Celsius degree.

Any definite Fahrenheit temperature corresponds to some definite Celsius temperature, of course, and vice versa. There are formulas for these conversions, which both authors of this book have gone to great pains *not* to learn. They are too easily forgotten, or remembered wrong; it is considerably more dependable to reason from scratch on each conversion. For example, in Fig. 10-2C, the temperature 68°F is shown. This temperature is $68 - 32 = 36$ Fahrenheit degrees above freezing, which is $36 \times \frac{5}{9} = 20$ Celsius degrees above freezing. Since the freezing point is 0°C, this means that 68°F is the same as 20°C. For a conversion in the other direction, the temperature 80°C is shown. Here we find 80 Celsius degrees above freezing, or $80 \times \frac{9}{5} = 144$ Fahrenheit degrees above freezing. Since freezing is 32°F, 80°C must be $144 + 32 = 176$°F.

10-2 Gas Thermometers

Although all thermometers will behave, generally speaking, in a similar way and give us a clear indication of whether the temperature goes up or down, they will disagree among themselves in smaller details since different materials react somewhat differently to an increase of temperature. Thus, if we mark by 0 and 100 the freezing and boiling points of water on thermometers filled with mercury, alcohol, and water, we will find that, as the temperature increases from zero, the mercury and alcohol columns will rise while the water column will first drop and will begin to rise only after the other two columns have covered about 4 percent of the total distance to the boiling point. Even the mercury and alcohol columns, which are adjusted to show identical values at the two ends of the scale, will not quite agree with each other in between, because they expand at different rates in different temperature intervals. We must therefore look somewhere else for an exact and universal definition of the temperature scale. The solution is provided by gases, since it has been found that ***all gases subjected to heating expand in almost exactly the same way.*** The lower the pressure on the gases, the more nearly alike do all kinds of gases behave, but even at normal atmospheric pressure the differences are very small. This property of gases, which is in contrast to that of solid and liquid materials, is due to the extreme simplicity of the inner structure of gases as compared with the structure of solids and liquids. We can accept as a standard the temperature scale provided by a gas thermometer (Fig. 10-3), regardless of what gas is used to fill it. Having this as a standard scale, we can then properly calibrate any other temperature-measuring instrument.

10-3 Absolute Zero

Let us take a gas thermometer (such as the one shown schematically in Fig. 10-3) and carefully measure the gas volume, first at the boiling point of water and then again when the thermometer is in a slush of ice and water. In defining the Celsius temperature scale, we have agreed to call these temperatures 100°C and 0°C; we can now plot them as points B and F on a graph against our measured volumes (Fig. 10-4). A straight line through these two points can be drawn and extended to left and right to define a scale for measurement over a wide range of temperature.

There is something peculiar that should be noted about the left (low-temperature) side of the graph. We cannot extend the line indefinitely as we can on the other (high-temperature) side because it soon runs into the axis of the graph, indicating a zero volume. It would be ridiculous to extend the line any farther since as far as we know, a gas with a *negative* volume is an idea that has no meaning. So this point, toward which our graph seems to be heading, is called the *absolute zero* of temperature.

The apparent intention of the gas to shrink to zero volume at absolute zero is naturally never fulfilled. All gases liquefy before this point is reached; in fact, even before it begins to liquefy, a gas commences to

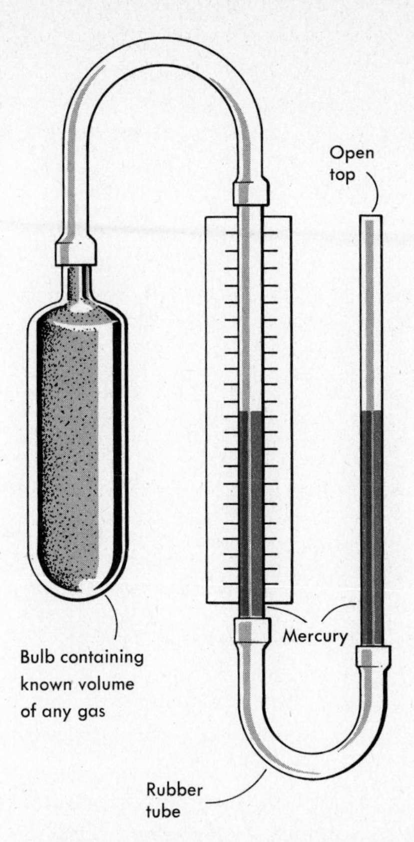

FIG. 10-3 A gas thermometer. The changing volume of the trapped gas is read from the height of the mercury column in front of the scale. In order to keep the mercury level the same on both sides, the right-hand tube is lowered as the gas heats; the gas in the bulb will therefore remain at atmospheric pressure at all temperatures.

deviate considerably from its more regular behavior at higher temperatures. Ingenious experimental procedures and theoretical corrections. however, have revealed that the temperature of absolute zero is −273.15°C. We will not worry ourselves about the odd hundredths of a degree and can be content to refer to this ultimate coldness as −273°C. A temperature scale beginning with 0 at absolute zero is an *absolute temperature scale*. The most common absolute scale uses the same size degree as the Celsius scale and is called the *Kelvin* scale (°K), named for Lord Kelvin, a great nineteenth-century English physicist. Since the Kelvin degree is the same size as the Celsius degree and begins counting 273° lower on the scale, it is apparent that in order to convert a temperature given in °C into °K, we need only add 273.

The concept of an absolute zero is an important one in physics, and we will later interpret it in terms of something more significant than the crossing point of two lines on a graph.

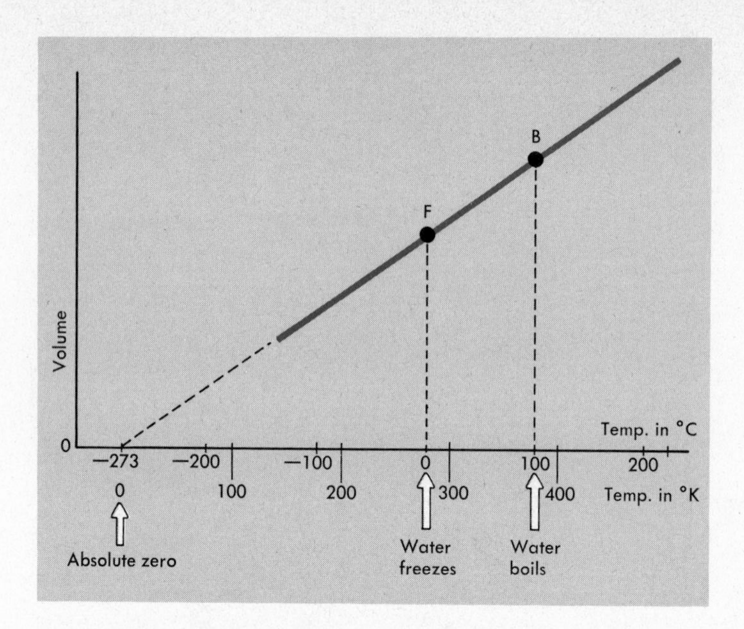

FIG. 10-4 The behavior of a gas at low temperature.

10-4 Pressure in Gases

As was apparent in the discussion of gas thermometers, there are three factors of importance in determining the behavior of gases: temperature, volume, and pressure. We have defined several temperature scales, and the meaning of volume is self-evident. At this point, however, we can well afford to devote a few paragraphs to a closer look at pressure and its effect on gases in general.

Although we seldom pay much attention to it, the air around us and in our lungs and ears is under a very considerable pressure. Earth's atmosphere is a mixture of about 78 percent nitrogen, 21 percent oyxgen, 1 percent argon, 0.03 percent carbon dioxide, plus smaller amounts of many other gases, and a variable amount of water vapor. The total mass of the atmosphere is about 5×10^{15} tons, which amounts to more than one kilogram for every square centimeter of the earth's surface. The weight of the air above our heads was first measured by an Italian physicist, E. Torricelli (1608–1647), using an arrangement shown schematically in Fig. 10-5. He took a long vertical glass tube that was closed on the top, completely filled it with water, and placed it in a water-filled tub. When he opened a faucet at the bottom of the tube, the water level in the tube fell to a height of about 10 m above the water level in the tub, leaving "Torricelli's emptiness"—or *vacuum*, as we call it now—at the upper part of the tube.

Since the volume A in Fig. 10-5 contains nothing but a little water vapor and is completely separated from the atmosphere, it can exert

FIG. 10-5 Torricelli's experiment.

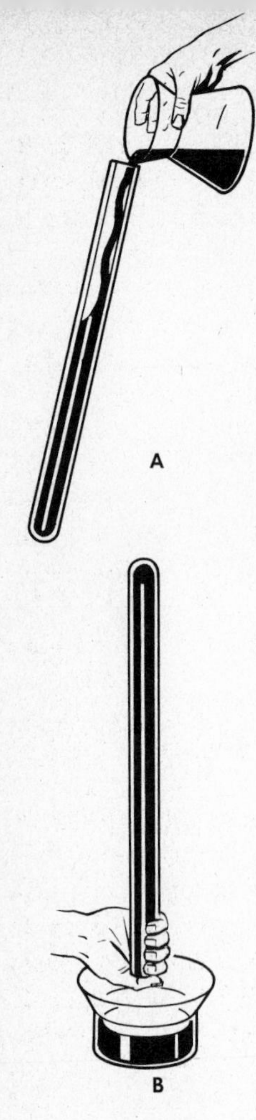

no appreciable force downward against the top of the column of water. So the atmospheric pressure at B, the surface of the water in the tub, is able to force water up the tube without any resistance except that resulting from the height of the water column itself. Hence the column of water (or any other liquid) can be used to measure the pressure of the atmosphere:

$$h \times d = \text{atmospheric pressure}$$

where h is the height of the fluid column and d is its density.

In our version of Torricelli's experiment, with the water standing exactly 10 m above the water surface in the tub, the pressure of the atmosphere in Italy that day must have been

$$1000 \text{ cm} \times 1 \text{ g/cm}^3 = 1000 \text{ g/cm}^2$$
$$= 9.80 \times 10^5 \text{ dynes/cm}^2.$$

This experiment of Torricelli can be repeated in the laboratory much more conveniently by using mercury instead of water.* Figure 10-6 shows how to do this. A glass tube, closed at one end, is filled *completely* with mercury (Fig. 10-6A); put your thumb over the open end, invert the tube, and submerge the open end (thumb and all) in a dish of mercury (Fig. 10-6B). If you are at sea level and the atmospheric pressure is neither abnormally high nor low at the time, when the thumb is removed, the mercury column will fall until its top is about 76 cm above the mercury surface in the dish (Fig. 10-6C). At this height, the pressure at the bottom of the column is exactly balanced by the pressure of the atmosphere against the open surface of the mercury. This simple device is called a *barometer*, and the pressure of the atmosphere is often referred to as *barometric pressure*. Although it is not an actual pressure unit (which must have the dimensions of force/area), the height of a mercury column is such a convenient way of measuring pressure that pressures are often given in terms of "centimeters of mercury" or "millimeters of mercury" (cm Hg or mm Hg, Hg being the chemical symbol for mercury).

Scientists all over the world agree, for example, to use exactly 76 cm Hg as the pressure of a *standard atmosphere*, which is very nearly the average atmospheric pressure at sea level. Using the density of mercury (13.6 g/cm^3) and the average value of 980 cm/s^2 for g, we can readily find that one standard atmosphere is equivalent to a pressure of about 1.013×10^6 dynes/cm^2, or 14.7 lb/in^2. As we rise above the earth's surface (or rather, above sea level) by climbing high mountains or by ascending in a balloon or airplane, less and less air is left above our heads,

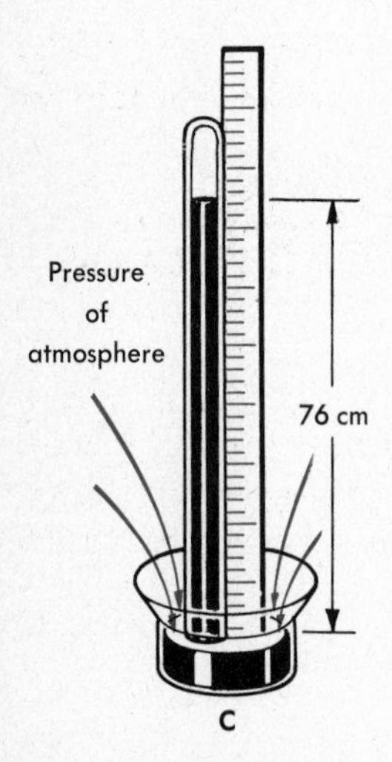

FIG. 10-6 The mercury barometer.

* Since mercury is 13.6 times denser than water, the mercury column will stand only 1/13.6 as high as Torricelli's column of water.

and the atmospheric pressure decreases correspondingly; the amount of air over the top of Mt. Everest is only a third of that above sea level, and climbers are forced to carry oxygen tanks with them in order to breathe.

Among the earliest accurate experiments on gas pressures were those of the Irish physicist Robert Boyle (1627–1691). To repeat one of his experiments, take a closed glass tube with some air trapped in it, and an open tube, and connect them with a mercury-filled rubber tube a couple of meters long, as shown in Fig. 10-7. Start the experiment with the two glass tubes in the relative positions shown in Fig. 10-7A, in which the mercury stands at the same level in both of them. Under these conditions the trapped air is at normal atmospheric pressure. Now move the open tube up until the trapped air is compressed to half of its original volume. You will find that in this case, if you have been careful to keep the temperature constant, the difference in the mercury levels will be about 760 mm (Fig. 10-7B).* Move the open tube higher until the trapped air is squeezed into a third of its original volume and you will find that the difference in the mercury levels is now 1520 mm, or 2 × 760 mm (Fig. 10-7C).

In Fig. 10-7A, the trapped air was subject to atmospheric pressure, i.e., 760 mm of mercury. In Fig. 10-7B, the pressure was increased 760 mm to a total of 2 atms. In Fig. 10-7C, the pressure was that of 3 atms. Since the volumes of trapped air were in the ratios $1:\frac{1}{2}:\frac{1}{3}$, we can conclude that ***the volume of gas at constant temperature is inversely proportional to the pressure to which it is subjected,*** which is the classical *Boyle's law of gases*. Air and other gases ordinarily follow this law quite well, but when they are highly compressed, deviations from it can be observed. This is quite understandable since, in these cases, the density of gas approaches that of liquids, which possess very low compressibility.

The inverse proportionality between volume V and pressure P for a given sample of gas whose temperature does not change can be stated mathematically as

$$\frac{P_1}{P_2} = \frac{V_2}{V_1} \quad \text{or} \quad P_1V_1 = P_2V_2.$$

10-5 The General Gas Law

The relationship between the volume and the temperature of a gas under constant pressure was investigated by Jacques Charles of France in the eighteenth century. Charles's work indicated the proportionality shown by the graph in Fig. 10-4, which we can express mathematically as

$$\frac{V_1}{V_2} = \frac{T_1}{T_2}.$$

---

* It will vary slightly, depending on the atmospheric pressure.

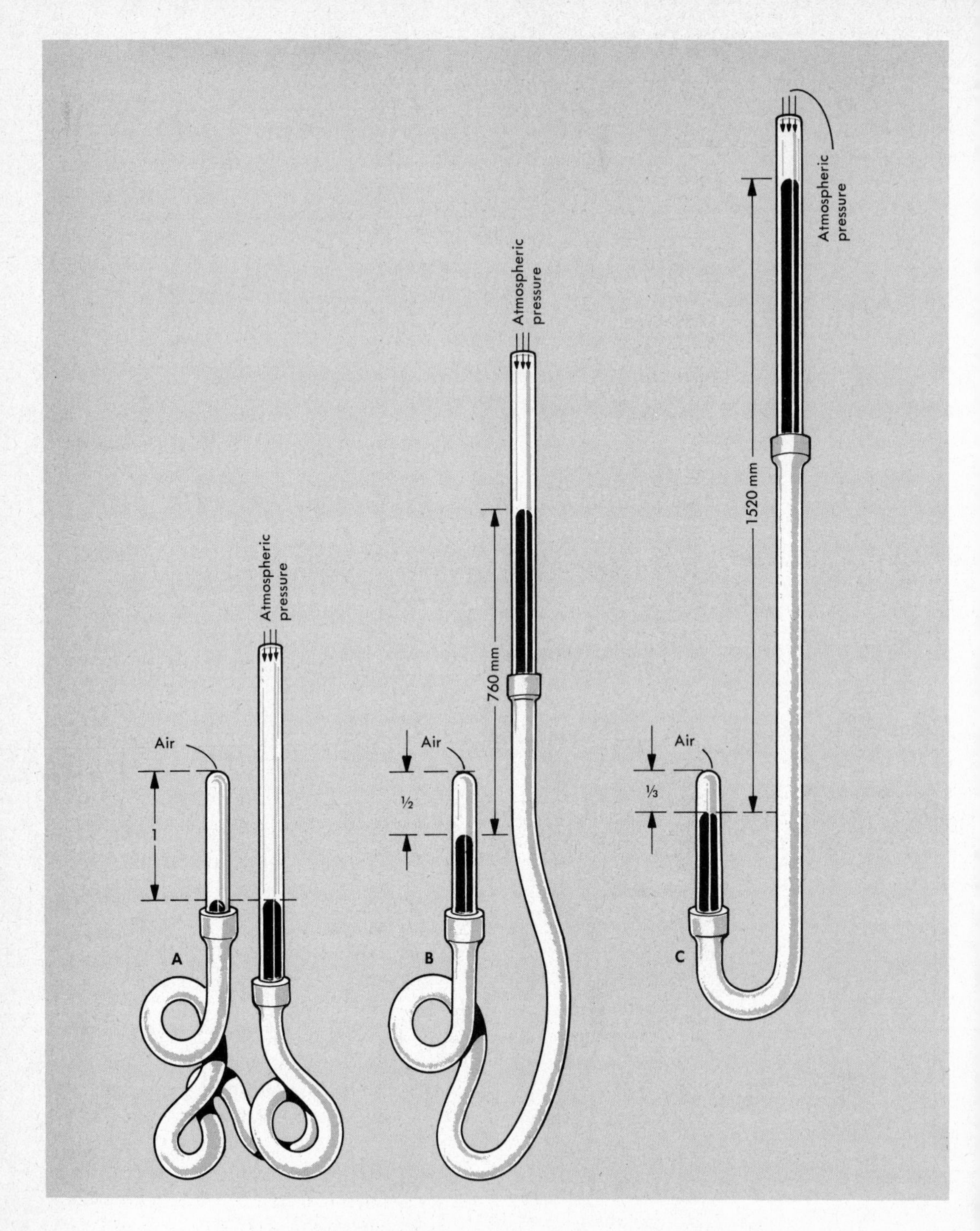

FIG. 10-7 An apparatus for determining the compressibility of gases.

Referring to Fig. 10-4, we can see that this proportion refers to the corresponding sides of the two similar triangles drawn from the points F and B, for example, and that ***the proportion is true only if the temperature used is the absolute temperature.***

Boyle's law and Charles's law can be combined into a single general gas law, which states that for any given and constant mass of gas

$$\frac{P_1V_1}{T_1} = \frac{P_2V_2}{T_2}.$$

We have already mentioned that the *temperatures must be absolute temperatures*; before we try to use this general gas law equation, we must also emphasize that the *pressures must be absolute pressures.*

Everyone is accustomed to seeing and reading pressure gauges; tire gauges and gauges on compressed-air tanks or water tanks are familiar sights. When you have a blowout, the gauge applied to your flat tire reads zero. Does this mean that the inside of your tire is at zero pressure? No, not at all—it only means there is no difference between the pressures inside and outside the tire. In other words, when the gauge reads zero, the air in the tire is at atmospheric pressure, which may be anything from nearly 15 lb/in^2 near the seashore to 10 lb/in^2 in a town in the high mountains.

Let us use the gas law on a problem about tires, which we can assume keep the same volume no matter what the pressure. A man fills his tires to a gauge pressure of 28 lb/in^2 early in the morning (T = 5°C) in a city at sea level (atm pressure = 15 lb/in^2). By afternoon (T = 25°C), he has reached a town high in the mountains (atm pressure = 10 lb/in^2). He checks his tires in this mountain town. What should he expect the gauge to read? Down in the sea-level city, the gauge indicated that the tire pressure was 28 lb/in^2 greater than atmospheric pressure, which was 15 lb/in^2. Thus the *absolute* pressure in the tires was 28 + 15 = 43 lb/in^2; the temperature was 5 + 273 = 278°K. Up in the mountains, the temperature of the air in the tires is 25 + 273 = 298°K. The unknown (but unchanging) volume of the tire is V, and we can write

$$\frac{43V}{278} = \frac{P_2V}{298}$$

or

$$P_2 = 43 \times \frac{298}{278} = 46.1 \text{ lb/in}^2 \quad \text{(absolute)}.$$

What the gauge will indicate is the difference between this 46.1 lb/in^2 inside the tires and the 10 lb/in^2 of the outside atmosphere, or 36.1 lb/in^2.

Actually, no real gas exactly follows the general gas law. Indeed, one way to define the physicists' imaginary *ideal gas* is to say it is one that *does* follow this law exactly. Real gases deviate from it for two reasons. If we look at the law, we see that if the pressure is raised higher and higher (and the temperature is kept constant), the volume must get smaller and approach zero for enormously high pressures. A real gas cannot do this because the gas molecules themselves occupy space. Another factor that enters into the behavior of a real gas is the fact that the molecules actually attract one another slightly, especially when the gas is compressed and the molecules are close together. This attraction helps to squeeze the gas and makes its volume at high pressures a tiny bit smaller than it would be if it were an ideal gas. If one is using a gas thermometer to do very accurate thermometry, these small deviations must be taken into account. However, we do not need to aspire to this hundredth-of-a-degree accuracy in our work here, and so may safely assume that the gases we deal with follow the general gas law.

10-6 Expansion of Solids and Liquids

In contrast to the gases, all of which expand by almost the same amount when heated, solids and liquids are individualists. The amount a solid material will expand when heated is measured by its *coefficient of linear expansion*, generally designated by α. This coefficient gives the fraction by which the length (or width or thickness) of an object will increase for each Celsius-degree rise in temperature. The total increase in length (Δl) that a body experiences will accordingly be the coefficient α multiplied by the length l, times the number of degrees rise in temperature (ΔT). In an equation,

$$\Delta l = \alpha l \, \Delta T.$$

Here are the linear coefficients of expansion for a number of materials:

Material	Coefficient
Aluminum	$\alpha = 25 \times 10^{-6}/°C$
Brass	18×10^{-6}
Ice	50×10^{-6}
Invar*	0.9×10^{-6}
Steel	11×10^{-6}
Platinum	9×10^{-6}
Glass	9×10^{-6}

* Invar is a special nickel–steel alloy designed for use where a low coefficient of expansion is needed.

As an example, consider a steel bridge 200 m long, in a locality where the temperature may vary from $-30°C$ in the winter to a blistering summer heat of $+40°C$. This is a temperature change of 70°C, and from winter to summer the bridge will lengthen by an amount given by

$$\Delta l = 11 \times 10^{-6} \times 2 \times 10^{4} \times 70 = 15.4 \text{ cm}.$$

To allow room for such expansions, many bridges have one end mounted on rollers or some other device that will permit it to expand and contract freely with the changing seasons.

The table above shows that glass and platinum have equal coefficients of expansion. For this reason, platinum wire is often used, in spite of its expense, when the wire must pass through the wall of a piece of glass equipment. Since the two materials will expand and contract by the same amount, there is no danger of the glass cracking, or the wire shrinking from it as the temperature changes.

If the length and width and thickness of a piece of material increase as it is heated, its volume will also increase. The Greek letter *beta* (β) is often used to represent the *volume coefficient of expansion*, which is the fraction of its volume by which a piece of material will expand per Celsius-degree rise in temperature. Consider a rectangular block of material $a \times b \times c$ in size, which has a linear coefficient of expansion α. If we raise the temperature of the block by 1°C, its new dimensions will be $(a + \alpha a) \times (b + \alpha b) \times (c + \alpha c)$, or $a(1 + \alpha) \times b(1 + \alpha) \times c(1 + \alpha)$. The volume becomes $abc(1 + \alpha)^3 = abc(1 + 3\alpha + 3\alpha^2 + \alpha^3)$. The coefficient α is always a small number, and hence α^2 and α^3 are so negligibly small that they can be discarded without producing any measurable error. Our new volume is thus very nearly $abc(1 + 3\alpha)$. The original volume abc has been increased by $3\alpha \times abc$, so the fractional increase is 3α.

The volume coefficient of a solid is, therefore, almost exactly three times its linear coefficient:

$$\beta = 3\alpha.$$

What happens to the hole in a washer when the washer is heated? Figure 10-8A shows an aluminum washer at a temperature of 0°C. When the washer is heated to a temperature of 200°C, we realize that the width (measured as either AB or CD) must increase, and our first inclination is to assume that this increase in width will do two things: It will make the outside diameter (AD) larger; and it will also expand inward to make the diameter of the hole (BC) smaller. But before we come too firmly to this conclusion, we should look at the situation in a little more detail. The width will certainly increase, and at 200°C the increase will be

$$25 \times 10^{-6} \times 1.000 \times 200 = 0.005 \text{ in},$$

thus making the expanded width 1.005 in. (Fig. 10-8B). As far as the outside diameter is concerned, it makes no difference whether the washer has a large hole, a small hole, or no hole at all; its increase will be

$$25 \times 10^{-6} \times 4.000 \times 200 = 0.020 \text{ in},$$

to give an expanded diameter of 4.020 in. To find the new hole diameter, we need only subtract the two widths from the outside diameter, to find

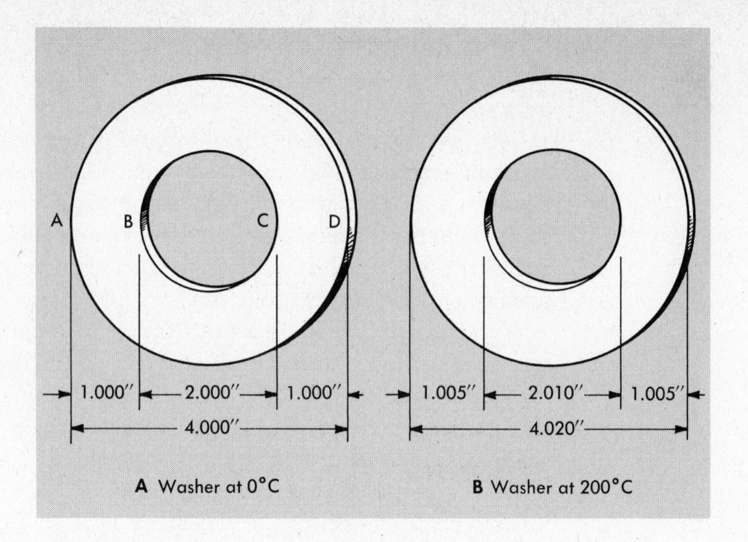

FIG. 10-8 Expansion of a heated aluminum washer.

4.020 − 2(1.005) = 2.010 in. for the expanded diameter of the hole. A little checking will confirm that the hole in the aluminum washer has expanded *exactly as much as though the hole itself were made of aluminum.*

We may argue similarly that a hole inside a block of steel or glass (which is one way of describing a tank or flask) will have the same volume coefficient of expansion as the material surrounding it.

A linear coefficient of expansion for a liquid would be meaningless, but liquids do experience definite changes in volume with changes in temperature. The following list gives the volume coefficients of expansion for some common liquids at ordinary temperatures:

Ethyl alcohol	$\beta = 1.12 \times 10^{-3}/°C$
Benzene	1.24×10^{-3}
Glycerin	0.50×10^{-3}
Mercury	0.18×10^{-3}
Water	0.21×10^{-3}

Suppose a glass flask holds exactly 200 cm³ of mercury, filled to the top, at 20°C. What will happen if we heat the flask full of mercury to 80°C? The volume of the flask will increase by an amount

$$\Delta V_F = \beta \times V \times \Delta t$$
$$= 3 \times 9 \times 10^{-6} \times 200 \times 60 = 0.32 \text{ cm}^3.$$

The mercury itself will also expand:

$$\Delta V_M = 0.18 \times 10^{-3} \times 200 \times 60 = 2.16 \text{ cm}^3.$$

Thus 2.16 − 0.32 = 1.84 cm^3 of mercury will spill over the top of the flask.

10-7 Calorimetry

Even in everyday, nonscientific speech, we make a distinction between "heat" and "temperature," although to many people it is not very clear just what this distinction is. We "add heat" to a body and "raise its temperature," which implies, quite correctly, that heat is a quantity of something, whereas temperature is a property of the body not concerned with how large the body is. We know now that heat is a form of energy—actually, the total kinetic energy of all the molecules of a substance, as we shall see in more detail later. Temperature is a measure of the average kinetic energy of one molecule.

Heat, being energy, can be measured in foot-pounds or joules or ergs. It has been found convenient, however, to define another special unit to measure heat–energy. This special unit is the *calorie* (*cal*), *which is the amount of heat necessary to raise the temperature of one gram of water by one Celsius degree.*

If we take a glass containing, say, 500 g of water at 80°C and mix it with an equal amount of water at 50°C, we will find that the temperature of the mixture will be 65°C, i.e., just halfway between. We have had 500 g of water warming up by 15° from 50° to 65°, which required 500 × 15 = 7500 cal. These calories were furnished by the other 500 g of water, which, in cooling 15° from 80° to 65° ,gave up 7500 cal. The principle of conservation of energy works as well with calories as it does with foot-pounds or ergs.

Different substances require different amounts of heat to raise their temperatures—that is, they have different specific heats. *The specific heat of a substance is equal to the number of calories required to raise the temperature of one gram of the substance by one Celsius degree.* Most substances need considerably less than one calorie to do this. Table 10-1 lists the specific heats of a few common materials.

The specific heat of water is, of course, 1.00, because of the way in which the calorie was defined.

TABLE 10-1 SPECIFIC HEATS OF SEVERAL COMMON SUBSTANCES

Substance	Specific Heat in cal/g-°C
Alcohol	0.58
Aluminum	0.22
Copper	0.093
Ice	0.55
Iron	0.11
Lead	0.030
Mercury	0.033

Some thought will enable us to write an equation that will relate the amount of heat (Q cal), the mass of the substance (m g), the specific heat of the substance (S cal/g-°C), and the temperature change of the substance (ΔT°C). If we take a substance with specific heat S, we know from the definition of specific heat that it will require exactly S cal to raise the temperature of 1 g of the substance by 1°C. So for this special case of 1 g and 1°C, we have $Q = S$. If we have m g, it will take just m times as many calories to change the temperature by 1°C, which we can express by $Q = mS$. If the temperature is to be changed by ΔT°, it will require ΔT times as many calories as it did for 1°C, and we can write as a useful relationship the following:

$$Q = mS\,\Delta T.$$

We should probably mention that for much of the heating and air-conditioning work in Britain and the United States, engineers use a heat unit considerably larger than the calorie. It is a British thermal unit (Btu), which is the amount of heat needed to raise the temperature of 1 lb of water by 1°F. Since this defines the specific heat of water to be 1 in these units, it is apparent that for any substance, S in cal/g-°C = S in Btu/lb-°F.

The unit used by dieticians in figuring the heat or energy content of foods is generally spelled Calorie, with a capital letter, and is equal to a thousand calories. The Calorie is also known as the *kilocalorie* (kcal).

By using the idea that the heat gained by one substance in a mixture must equal the heat lost by another, we can readily solve many problems in *calorimetry*. (Calorimetry means, literally, measuring quantities of heat.) If, for instance, we pour 400 g of mercury at 100°C into 300 g of alcohol at 20°C (in a cup that is insulated so that no heat can enter or leave our mixture from the outside), what will be the final temperature of the mixture? Let us call the final temperature T; T will obviously be somewhere between 100° and 20°C. The mercury will cool from 100° to T°C, a change of $(100 - T)$ degrees. It takes 0.033 cal (the S.H. of mercury) to change the temperature of 1 g of mercury 1 degree; to change 400 g by $(100 - T)$ degrees requires that the mercury lose $0.033 \times 400 \times (100 - T)$ cal. Similarly, to heat 300 g of alcohol from 20°C up to T° requires $0.58 \times 300 \times (T - 20)$ cal. If we set the heat lost to the heat gained, we find

$$0.033 \times 400(100 - T) = 0.58 \times 300(T - 20)$$

$$1320 - 13.20T = 174T - 3480$$

$$187.2T = 4800$$

$$T = 25.6\text{°C}.$$

10-8 Latent Heat

When we place a teakettle on the fire, the temperature of the water gradually rises to 100°C, at which point the water begins to boil. But once the boiling has started, the temperature stays at 100°C until the last drops of water are turned into steam. Although the heat is still flowing into the kettle from the flame, it does not make the water any hotter. What happens to that heat? The answer is, of course, that this heat is used to transform the water into vapor, and measurements show that to do it we must supply 539 calories for each gram of water to be vaporized. This amount of heat is known as the *latent* (*hidden*) *heat of vaporization* and is, of course, different for different substances. Thus to evaporate 1 g of alcohol and 1 g of mercury we need only 204 cal and 72 cal, respectively. The heat absorbed in the evaporation of water plays an important role during hot weather in the cooling of the body through the process of skin perspiration. Indeed, one glass of water evaporated from the surface of the body removes enough heat to cool the entire body by several degrees. Meteorologists use this principle for measuring the relative humidity of air. The apparatus used for this purpose consists of two identical thermometers with the bulb of one of them covered by a wet cloth. This thermometer, because of evaporation, shows a somewhat lower temperature; from the difference between the two readings the weatherman can calculate the rate of evaporation and, consequently, the amount of humidity present in the atmosphere.

A similar phenomenon is encountered when water turns into ice. When the temperature of water comes down to 0°C and the first crystals of ice begin to form, the temperature remains at 0°C until all the water freezes. The *heat of fusion of water* (i.e., the amount of heat that must be taken away from water at 0°C to freeze it, or be given to ice at 0°C to melt it), amounts to 80 cal/g. The heat of fusion of alcohol (which freezes at −114°C) is 30 cal/g, whereas for mercury (freezing at −39°C) it is only 2.8 cal/g. To melt lead (at +327°C), it takes about 6 cal/g, whereas in the case of copper (at +1083°C), the figure is as high as 42 cal/g.

The idea of latent heats can add a certain amount of complication, as well as interest, to calorimetry problems. Suppose we have a calorimeter (an insulated vessel made for doing just such an experiment as follows) containing 500 g of water at 0°C. Into the calorimeter we put 150 g of ice at −20°C and then bubble steam at 100°C into it until the ice is melted and the entire contents are at 60°C. How many grams of steam will it take to do this? The steam passed through at first will, of course, initially condense into water, then cool to 0°C, and finally warm again to 60°C. It will give off just as much heat in cooling from 60° to 0°C however, as it will take up later in going from 0° back to 60°C. Thus we can jump directly from the beginning to the end and say that each gram of steam will give up first 539 cal in condensing into water at 100°C, then 40 cal more to cool to its final 60°C temperature—a total of 579 cal, or $579x$ cal for x g of steam. We can use this released heat energy first to warm the ice to 0°C: $0.55 \times 150 \times 20 = 1650$ cal. The ice must next be melted, which requires $150 \times 80 = 12{,}000$ cal. and gives us $150 + 500 = 650$ g of 0°C water in the calorimeter. To heat this to 60°C will take another $1 \times 650 \times 60 = 39{,}000$ cal. Altogether, then, the x g of steam must contribute $1650 + 12{,}000 + 39{,}000 = 52{,}650$ cal. Thus

$$579x = 52{,}650$$

or

$$x = 90.9 \text{ g of steam.}$$

10-9 Heat Conduction

If we hold one end of an iron rod in our fingers and put the other end in a candle or bunsen burner flame, heat from the flame will flow through the iron into the air around the rod, and also, as we shall notice in a few minutes, into our fingers. Presently, the end we are holding will become so hot that we must drop it. If this experiment is repeated with a copper rod, we must drop it much more quickly than we did the iron. We would still, however, be able to hold a glass rod quite comfortably while the other end is red hot and molten. Apparently, some materials conduct heat much more readily than others; copper is a much better heat conductor than glass.

The rate at which heat energy flows through a rod or slab of material depends on several factors. One obvious factor is the nature of the material, and we can use a standard physicist's scheme to describe the heat conductivity of any material in a quantitative way. Imagine a cube of the material 1 cm on a side (Fig. 10-9A), with one face of the cube kept just 1°C cooler than the opposite face. Heat will flow from the warmer face to the cooler, and the number of calories per second flowing through the cube is the *heat conductivity* or *thermal conductivity* of the material. It is a simple matter to extend this concept to a piece of the material of any size and for any temperature difference. Let us figure a general formula to use on the slab shown in Fig. 10-9B, which is made of a material with a thermal conductivity of k. This means that *if* the slab were 1 cm thick, and *if* it had an area of 1 cm^2, and *if* one side were 1°C warmer than the other, then k cal/sec would flow through it. But it has an area of A cm^2, which considered alone would let A times as much heat flow through. The temperature difference is not 1°C, but ΔT°C, so ΔT times as much heat will flow through because of the greater temperature difference.

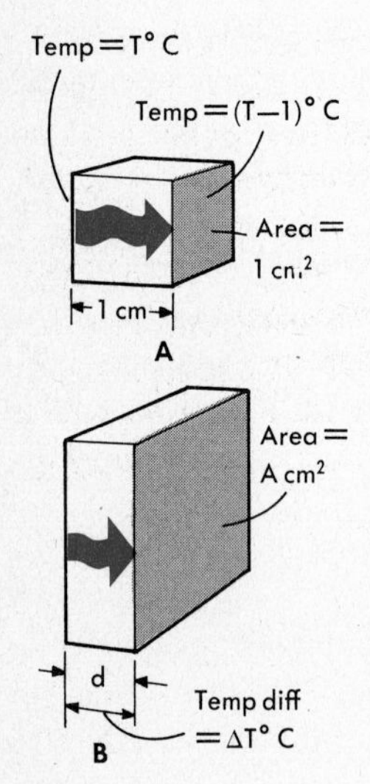

FIG. 10-9 Heat conduction through a slab of material.

The thickness, d cm, will work the other way. Obviously, it is more difficult for heat to flow through a thick slab than a thin one. Experiment, as well as theoretical analysis, shows that the heat flow through a slab of material (other things being equal) varies *inversely* with its thickness. Thus since our slab is d cm thick, only $1/d$ as much heat will flow through it as through the 1-cm thickness.

TABLE 10-2 HEAT CONDUCTIVITY OF DIFFERENT MATERIALS EXPRESSED IN CALORIES PER SECOND FLOWING THROUGH 1 CM², WHEN THE TEMPERATURE GRADIENT IS 1°C/CM

Material	Heat Conductivity (at 18°C)
Silver	0.97
Copper	0.92
Aluminum	0.48
Iron (cast)	0.11
Lead	0.08
Mercury	0.016
Glass	0.0025
Brick	0.0015
Water	0.0013
Wood	0.0003
Asbestos	0.0002
Cotton wool	0.00004
Air	0.00006

The arguments above can all be assembled into a compact equation to give us the rate of heat flow Q/t:

$$Q/t \text{ (cal/sec)} = \frac{kA\,\Delta T}{d}.$$

The heat conductivities of some familiar materials are shown in Table 10-2.

As an example, consider a jar 10 cm in diameter and 20 cm high, made of glass 2 mm thick. The lid of the jar is of copper 1 mm thick, and the water in the jar has been accidentally frozen into solid ice. The jar is put into a tank of 20°C water to thaw out. How long will it take? First, we can find how many calories it will require to melt the ice. The volume of a cylinder is $\pi r^2 h$, or $\pi \times 5^2 \times 20 = 1570\ \text{cm}^3$. Since the density of ice is about 0.92 g/cm^3, this means there are 1450 g of ice to be melted, which will require $1450 \times 80 = 116{,}000$ cal. The area of glass through which the heat will be conducted is $\pi \times 5^2 = 78\ \text{cm}^2$ (bottom) plus $\pi \times 10 \times 20 = 628\ \text{cm}^2$ (sides), for a total of $78 + 628 = 706\ \text{cm}^2$, all 0.2 cm thick. The heat flow through the glass will be

$$\frac{Q}{t} = \frac{0.0025 \times 706 \times 20}{0.2} = 177\ \text{cal/s}.$$

Through the copper lid we have

$$\frac{Q}{t} = \frac{0.92 \times 78 \times 20}{0.1} = 14{,}300\ \text{cal/s}.$$

This gives a total heat flow into the jar of about 14,500 cal/s, and the required time will be 116,000/14,500 = 8.0 s.

Our procedure in working this problem has been perfect; the thermal conductivity values have come from dependable sources that confirm one another, and there are no large arithmetical errors. Yet the answer is ridiculously incorrect; actually, it would probably require an hour or so to melt the ice in the jar. Where have we gone wrong? The 14,500 cal/s rate of heat transfer we computed *is* correct for the instant we put the jar into the bath; the conditions we assumed in working the problem are at this moment correct. In a fraction of a second, however, the outside of the jar becomes covered with a layer of cold water; on the inside the ice is further insulated by the layer of water that has formed from its melting. Our initial conditions no longer hold, and as time goes on and more ice melts, our assumptions become more and more wrong!

A similar condition fortunately holds for windowpanes. If we calculated the loss of heat through glass windows, assuming that the inside of the glass was at a warm 70°F and the outside surface at the 0°F of the winter night, we would be shocked to discover that our entire salary could not pay our heating bill. Actually, there will be less than a 1°F difference between the inside and outside surfaces of the glass; nearly all the insulation comes from a blanket of cool air against the glass on the inside and of warm air on the outside. There are empirical rule-of-thumb multipliers experimentally worked out for use in more complicated procedures that will give approximately correct answers for the conditions common in heating design work. Pure thermal conductivity through walls and partitions is difficult to achieve except by special laboratory procedures.

Nevertheless, although we may not be able to determine the heat transfer exactly, the heat conductivity of various materials plays an important role in all kinds of heat insulation. Since fluffy rock wool and similar materials present 40 times more resistance to the flow of heat than ordinary brick, we can clearly see the advantages of their use for insulating homes. And, since the escape of heat is proportional to the surface of an object, to conserve heat it is advantageous to build houses as compactly as possible—hence the difference in construction styles in southern California and northern Canada. Following the same principle, many animals roll up into a ball when it is cold and stretch out when it is warm.

10-10 Heat Convection

In the case of poor heat conductors, the propagation of heat into the heated body is very slow. For example, it would take hours to heat the water in a teakettle standing on the fire if there were no other heat-carrying processes. In fluids, the propagation of heat is considerably accelerated by the process of convection, which has its basis in the fact

FIG. 10-10 Circulation of water in a teakettle, caused by convection currents.

that heated bodies increase their volume and hence decrease their density. In our teakettle, the water near the bottom is heated by immediate contact with the hot metal, becomes lighter than the rest of the water in the kettle, and floats up, its place being taken by the cooler water from the upper layers (Fig. 10-10). These *convection* currents carry the heat up "bodily," and they mix the water in the kettle so that the tea is ready in almost no time. A similar phenomenon takes place in the atmosphere when, on a hot summer day, air heated by contact with the ground streams up to be replaced by cooler air masses from above. As the air rises to higher and cooler layers of the atmosphere, the water vapor in the air condenses into a multitude of tiny water droplets and forms the cumulus clouds so characteristic of hot summer days. Convection processes are also very important in the life of our sun and stars. In them, the nuclear energy produced in the hot central regions is carried toward the surface by streams of heated stellar gases.

Sometimes the notion of *convective* heat transfer becomes confused with the notion of heat *conduction*. We have seen from Table 10-2, for example, that the heat conductivity of cotton wool is about the same as that of air. Wool, fur, and other materials used to make warm clothes also have about the same degree of conductivity. But if the heat conductivity of air is the same as that of warm clothing materials, why is a naked man less comfortable in cold weather than a man in a fur coat or under a thick woolen blanket? The reason is that the heat is removed from the skin of a naked man not so much by heat conduction into the air as by heat convection: the air warmed by contact with the skin rises and is replaced by more of the cold air. The role of warm clothing materials is to prevent this circulation, to keep the air from moving by trapping it between the numerous interwoven fibers of the materials. If we compress a woolen sweater or a mink coat under a hydraulic press, it will immediately lose much of its ability to keep us warm.

10-11 Heat Radiation

A third enormously important way in which heat energy can be transferred from one body to another is by *radiation*. If you stand outdoors warming yourself at an open fire on a winter day, the heat you receive certainly does not come to you by conduction through the air or the ground, since both of these are cold, and heat flows from you to them rather than the reverse. Neither are you heated by convection, since the hot air over the fire rises into the sky, taking its heat away with it. Just as the bright flames and the glowing coals radiate light, they also send out an even greater amount of radiant heat that travels unimpeded through the air, to be absorbed by your skin and clothing. All the energy we receive from the sun has been radiated in this way across 93 million miles of vacuum. Only a small part of this energy is in the form of light; most of the rest is radiant heat. Light and radiant heat are two forms of electromagnetic radiation, which we shall investigate later in much greater detail.

10-12 Physiological Heat Balance

Homoiothermic (or homeothermic) animals, which maintain a nearly constant internal temperature, must resort to many devices to make themselves independent of their surroundings. This is especially true for man, who lacks a covering of fur to insulate him from the effects of his changing environment.

Consider a 50-kg human whose daily food intake is 2520 Cal or 2520 kcal. The specific heat of the body as a whole can be taken as about 0.8 cal/g-°C = 0.8 kcal/kg-°C. From $Q = mS\,\Delta T$,

$$\Delta T = \frac{2520}{(50 \times 0.8)} = 63^\circ\text{C}.$$

If there were no way to get rid of the heat produced in oxidizing this food, this 63°C rise in temperature would bring her entire body up from its normal 37°C to 37 + 63 = 100°C!

A quantitative discussion of the relative importance of radiation, convection, conduction, and evaporation in maintaining our 37°C is almost impossible, due to the widely varying effects of different types and amounts of clothing, air movement, temperature and humidity, the temperature of walls or other surroundings, etc.

In 22°C surroundings (about 72°F), skin temperature is in the neighborhood of 32°C. Some heat is lost by radiation from the warmer skin to the cooler walls, some by air convection currents over the skin, and some by evaporation. In environments down to 20°C or even lower, there is appreciable perspiration that ordinarily evaporates immediately, so there is no visible moisture; and there is always evaporation of moisture from the surfaces of the lungs.

In the range from roughly 20° to 30°C, body temperature is regulated largely by the fraction of blood that is directed to the surface circulation under the skin. The increase of this surface circulation with rising environmental temperatures raises the skin temperature, which in turn increases the ΔT between skin and surroundings. The larger temperature difference increases all the heat losses mentioned above.

This circulatory adjustment works very well up to about 30°C; above this temperature the body must step up the evaporative losses by increased perspiration, and its resultant evaporation, which carries away 540 cal for each gram evaporated. At still higher temperatures, this mechanism fails—it is useless to perspire more than the surrounding air can evaporate. At this point, humans fall back onto the normal mechanism of the well-insulated, sweatless dog and increase their lung evaporation by panting. This is the end of the line.

At the other extreme of temperature, the body's problem is to conserve heat. Contraction of the surface blood vessels drives the circulation deeper and lowers the skin temperature. This reduces the losses by reducing the temperature difference between skin and surroundings. After this mechanism can no longer keep the rate of heat loss less than its rate of production in the body, we fall back on a useless gesture. "Goose-flesh" or "goose-bumps" results from the contraction of the tiny muscles that make our body hairs stand more nearly erect. This is very useful for furred animals and birds; the insulating layer of hairs or feathers and trapped air becomes thicker, fluffier, and more effective. A million years ago we lost our fur, but its now useless operating machinery still remains. When the heat losses continue in spite of this futile effort, we begin to shiver. The work of muscle contraction converts chemical energy into heat, and the involuntary exercise of shivering increases our internal heat production—the body's last resort, since it cannot further reduce heat losses.

Questions

(10-1) **1.** The melting point of the element sulfur is 113°C. What is this temperature on the Fahrenheit scale?

2. The average normal temperature of the human body is taken to be 98.6°F. What is this temperature on the Celsius scale?

3. The Reaumur temperature scale is almost obsolete, but is still used to some extent in France. On this scale, the freezing point of water is 0°, and the boiling point is 80°. What Celsius and Fahrenheit temperatures correspond to 48°R?

4. Normal room temperature (72°F) corresponds to what temperatures on the Celsius and Reaumur scales? (See Question 3.)

5. A motor is designed to heat up to 50°C above room temperature when running continuously at full load. In how hot a room (in °F) could it be operated if its temperature is not to exceed 200°F?

6. A motor is designed to heat up to 50°C above room temperature when running continuously at full load. What would its temperature be in °F when it is operating in a room at 80°F?

(10-3) **7.** Convert the following temperatures to absolute temperatures (°K): (a) 350°C, (b) 120°F, (c) −250°C, (d) −130°F.

8. Convert the following temperatures to absolute temperatures (°K): (a) 120°C, (b) 1500°F, (c) −30°C, (d) −78°F.

(10-4) **9.** One *bar* is a pressure of 10^6 dynes/cm^2. One standard atmosphere is how many *millibars*?

10. On a mountain, the barometric pressure is 56 cm Hg. Convert this pressure into (a) dynes/cm^2, (b) lb/in^2.

(10-5) **11.** A tire holds 1200 in^3. What volume of air at standard atmospheric pressure would have to be put into the tire to inflate it to a gauge pressure of 40 lb/in^2?

12. A compressor takes 100 ft^3 of air at standard atmospheric pressure and forces it into a tank whose volume is 8 ft^3. What is the gauge pressure of the gas in the tank? (Assume temperature unchanged.)

13. A bottle is tightly sealed at a temperature of 18°C, and is then placed in a 210°C oven. The atmospheric pressure is 13.0 lb/in^2. (a) What is the absolute pressure in the bottle at this high temperature? (b) What is the gauge pressure?

14. A bottle is tightly sealed at a temperature of 70°F, and is then put into a 350°F oven. (a) What is the absolute pressure of the air in the bottle at this higher temperature? (b) What is the gauge pressure? (All this takes place at sea level on an average day.)

15. A compressed-air tank in a filling station has a gauge pressure of 50 lb/in^2 when the temperature is 70°F. A fire in the station raises the temperature of the tank to 250°F. What is the gauge pressure in the tank when hot? (Atm. pressure = 15 lb/in^2.)

16. A man checks his tire gauge pressure as 28 lb/in^2 in the early morning when the temperature is 41°F. In the afternoon, hard driving on a hot pavement has raised the temperature of his tires to 140°F. (a) What is the gauge pressure of his tires, if the atmospheric pressure has not changed? (b) Does it make any difference in your answer whether atmospheric pressure is taken to be 10 lb/in^2 or 15 lb/in^2?

17. A tank is guaranteed safe at 100 lb/in^2 gauge pressure. It is filled with gas to 80 lb/in^2 gauge pressure at sea level and 20°C. If it is taken up in a plane to where atmospheric pressure is 5 lb/in^2, how hot may the tank safely be allowed to get ? (Take sea level atm. pressure = 15 lb/in^2.)

18. A large plastic balloon will hold 10,000 ft^3 when completely filled. It is designed to rise to an altitude of about 10 km, where the pressure will be 22 cm Hg, and the temperature −70°F. How many cubic feet of helium gas at standard atmospheric pressure and 70°F should be put into the balloon if it is to be full when it reaches its designed altitude?

(10-6) **19.** A steel bridge is exactly 250 ft long at 15°C. How long is the bridge when its temperature is 42°C?

20. A steel surveyor's tape is exactly 100 m long under a certain specified tension at 20°C. How long is the tape when the temperature is −10°C?

21. At 15°C the diameter of an aluminum rod is 0.0005 inch too large to fit into a 1-inch diameter hole in a brass plate. (a) Should the rod *and* plate be heated or cooled in order to make a fit possible? (b) At what temperature will the rod fit in the hole?

22. A steel ball 9 cm in diameter has a diameter 0.012 cm too large to permit it to fit into a hole in a brass plate when the temperature is 20°C. (a) Should the ball *and* plate be heated or cooled in order for the ball to fit the hole? (b) At what temperature will the ball fit the hole?

23. A glass flask holds exactly 200 cm^3 at 15°C. What is its capacity when heated to 60°C?

24. A block of aluminum is exactly 10 cm × 10 cm × 10 cm at 0°C. What is the volume of the block when it is heated to 80°C?

25. A 2000-cm^3 aluminum tank is filled with water at 80°C. How much more water can be added when the tank and contents have cooled down to 10°C?

26. A vertical glass tube with the bottom end sealed is filled with mercury to a height of 100 cm, at 15°C. How high will the mercury stand in the tube if the temperature is raised to 35°C?

(10-7) **27.** An insulated tank car holds 15,000 kg of alcohol at 25°C. How much heat must be removed from the alcohol to cool it to 12°C?

28. How much heat is required to raise the temperature of 700 g of mercury from 50° to 122°F?

29. How much copper shot at 95°C must be added to 150 g of water at 25°C in order to heat it up to 39°C?

30. How much heat must be added to 50 kg of ice at −40°C to raise its temperature to −5°C?

31. If 500 g of powdered aluminum at 100°C is added to 200 g of iron filings at 20°C in an insulated cup (calorimeter), what is the final temperature of the mixture?

32. If 300 g of iron shot at 5°C is poured into an insulated cup (called a calorimeter) that contains 500 g of alcohol at 70°C, what is the final temperature of the mixture?

33. In Question 31, we have not taken into account the heat needed to change the temperature of the calorimeter. Suppose it is aluminum, with a mass of 35 g. Refigure Question 31, using this additional information.

34. In Question 32, we have not taken into account the heat needed to change the temperature of the calorimeter itself. Suppose it is made of copper, with a mass of 80 g. Refigure Question 32 using this additional information. (The calorimeter is, of course, always at the same temperature as its contents.)

(10-8) 35. An insulated tank holds 500 kg of water at 20°C. (a) How much water at 100°C must be added to raise the temperature to 45°C? (b) How much steam at 100°C must be added to raise the temperature to 45°C?

36. A cup contains 300 g of very hot coffee at 95°C. (a) How much water at 0°C must be added to reduce the temperature to 75°C? (b) How much ice at 0°C must be added to reduce the temperature to 75°C?

37. In a 75-g copper calorimeter, there are 400 g of crushed ice and 200 g of alcohol at −25°C. If 120 g of steam at 100°C is bubbled through this mixture, what is the final temperature?

38. Suppose 200 g of crushed ice and 600 g of iron shot, all at 0°C, are in a 100-g aluminum calorimeter. If 50 g of steam at 100°C is slowly bubbled through, what is the final temperature?

(10-9) 39. An aluminum rod 40 cm long has a cross-section area of 5 cm^2, and is thermally insulated. One end is kept at −78°C in a bath of dry ice and acetone; the other end is in an insulated container of 100 g of water at 10°C. How long will it take to cool the water down to 0°C?

40. If one end of a thermally insulated copper rod 1 m long and 2 cm in diameter is placed in boiling water and the other end in ice water, how many calories will be transmitted along the rod in 1 min?

41. An insulated vat is divided into two sections by a watertight partition 20 cm × 40 cm made of copper 1 mm thick. On one side is water that is boiling, and on the other side is a mixture of 2 kg of crushed ice kept constantly stirred in water. How long will be required to melt all the ice?

42. A picnic refrigerator is a box 20 cm × 40 cm × 50 cm made of wood 5 cm thick. It contains 5 kg of ice. If we assume the interior temperature stays constant at 41°F while the outside temperature is 95°F, how long will it take for the ice to melt? (There is a drain hole, through which the water leaks out as the ice melts.)

43. Numerous small children have their tongues severely injured by licking a piece of pipe or other metal on a very cold day; licking a piece of wood on the same day would result in no harm. Explain.

44. When we step barefooted out of bed on a winter morning, it is much more comfortable to step on a rug than a wooden floor, although both are at the same temperature. Explain.

11

Heat and Energy

11-1 Mechanical Equivalent of Heat

In an earlier chapter of this book, we saw that friction forces gradually rob mechanical systems of their energy and eventually bring them to a standstill. What is the relation between the mechanical energy lost to friction and the amount of heat produced by it? This question was answered in the middle of the last century by a British physicist, James P. Joule (1818–1889), in his famous experiment on the transformation of mechanical energy into heat. Joule's apparatus, schematically shown in Fig. 11-1, consisted of a water-filled vessel containing a rotating axle with several stirring paddles attached to it. The water in the vessel was prevented from rotating along with the paddles by special vanes attached to the walls of the vessel. The axle with the paddles was driven by a weight hanging from a cord, and thus the work done by the descending weight was transformed by friction into heat produced in the water. Knowing the amount of water in the vessel, Joule could measure the rise of its temperature and calculate the total amount of heat produced; the driving weight and the distance of its descent gave the total

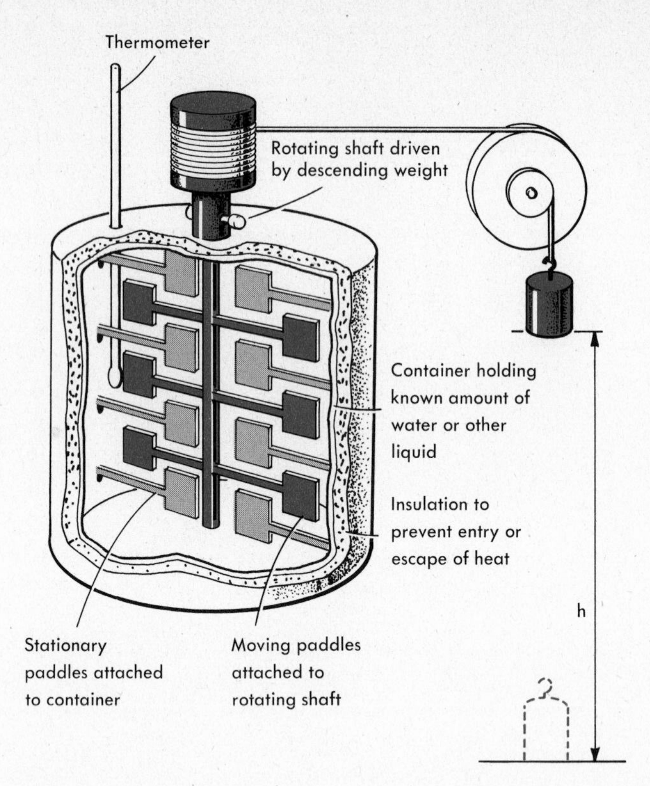

FIG. 11-1 Joule's apparatus for determining the mechanical equivalent of heat.

amount of mechanical work done. Repeating this experiment many times and under different conditions, Joule established that there is a direct proportionality between these two quantities and that "the work done by the weight of 1 pound through 772 feet at Manchester will, if spent in producing heat by friction in water, raise the temperature of 1 pound of water one degree Fahrenheit." In metric units, it means that *one calorie of heat is the equivalent of 4.18 × 10^7 ergs of work, or 4.18 joules.*

Joule's work confirmed the basic idea that was commencing to be seriously considered at the time, namely, that ***heat is energy in the same sense that mechanical energy is, and although one form of energy can be transformed into another (kinetic, potential, heat, electrical, chemical, etc.), the total sum of the energy in any system of bodies or materials remains constant.*** (This assumes, of course, that our "system" does not receive any energy from the outside or lose energy to it.) This law of the *conservation of energy* represents one of the basic pillars of physics and is known as the *first law of thermodynamics.*

The fate of all the mechanical energy on the earth (as well as of electrical and other forms of energy) is to be converted into heat at the exchange rate of 4.18 joules per calorie. The work of winding your watch, which is stored temporarily as the potential energy of the coiled spring, is converted into the kinetic energy of the escapement wheel and gradually converted by friction into heat during the course of the day. The winds are gradually slowed by friction with earth and trees, and their enormous energies also end up as heat.

Suppose 1 joule of work were expended in winding a 30-g watch. If the watch were wrapped in a theoretically perfect heat insulator, how much warmer would the watch be by the time it had run down? The 1 joule of work equals $1/4.18 = 0.24$ cal, which goes into heating the 30 g of steel of which the watch is made. If we take 0.1 cal/g-°C as the specific heat of steel, it will require 0.1 cal to raise the temperature of 1 g by 1°C. The 0.24 cal produced will raise the temperature of 1 g of steel by 2.4°C, and of 30 g, just $\frac{1}{30}$ of 2.4°, or 0.08°C.

This is not very much heat, but we started with very little energy. Braking to a stop, a 4000-lb car traveling 60 mi/hr should give us something more noticeable. The speed of 60/mi hr is 88 ft/s, and the kinetic energy of the car is $\frac{1}{2}mv^2$ or

$$0.5 \times \frac{4000}{32} \times (88)^2 = 484{,}000 \text{ ft-lb.}$$

The mechanical equivalent of heat we know only in terms of joules and calories, so we need to convert our answer from foot–pounds to joules, or newton–meters:

$$484{,}000 \text{ ft-lb} \times \frac{1 \text{ m}}{3.28 \text{ ft}} \times \frac{1 \text{ kg}}{2.2 \text{ lb}} \times \frac{9.8 \text{ N}}{\text{kg}} = 657{,}000 \text{ J}$$

and

$$657{,}000 \text{ J} \times \frac{1 \text{ cal}}{4.18 \text{ J}} = 157{,}000 \text{ cal.}$$

This amount of heat is enough to raise 1570 g of water (about 3 pints) from freezing to boiling, or to raise about 40 lb of steel from room temperature to the boiling point of water.

11-2 Heat and Mechanical Energy

Any amount of mechanical energy can be, and eventually will be, *completely* transformed into heat energy, through the operation of friction, for example. Is this transformation reversible? Can the heat energy contained in a pot of boiling water be *completely* converted into mechanical energy?

A steam turbine, for example, is designed for the sole purpose of converting heat energy into mechanical work. How successful can it be? We can show that even for an idealized perfect engine, this conversion can at best be only partial.

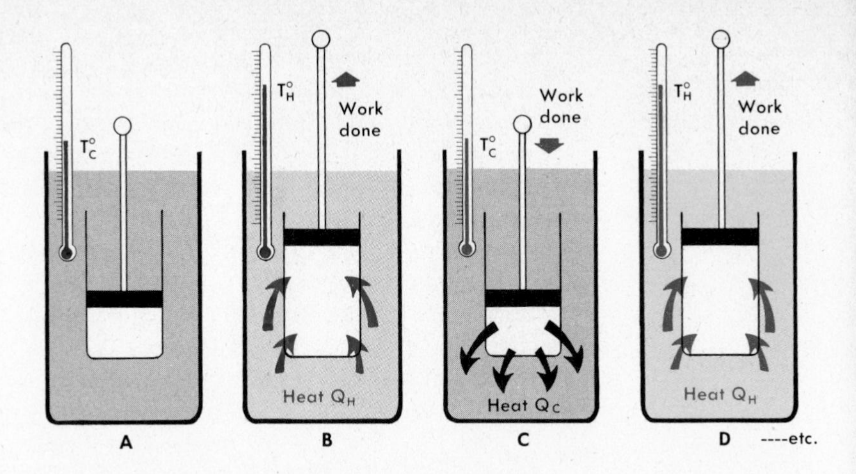

FIG. 11-2 Simple heat engine absorbing heat energy from hot-water tank, and losing heat to cool-water tank.

Figure 11-2A shows a cylinder of gas, with a leakproof and frictionless piston. The cylinder is in a tank of water at the coolest available temperature, T_C. By "coolest available" we mean tap water or water at room temperature, i.e., water that is at the temperature of the environment. A tank of hot water at temperature T_H is also available.

When the cylinder is transferred to this hot tank (Fig. 11-2B), Q_H joules or calories of heat energy flows from the hot water into the cooler gas, thereby raising its temperature and causing it to expand. If we attach the piston rod to some sort of machine, the expanding gas can do mechanical work.

This is half of the working cycle of our heat engine. For the other half, which returns the gas to its original condition, we put the cylinder back into the cool tank (Fig. 11-2C). Here Q_C of heat energy flows from the warm gas into the cool water; the gas temperature drops to T_C, reducing its volume and drawing the piston downward. Again we can have work done by the moving piston rod. The gas is restored to its original temperature, ready to repeat the cycle again and again.

The complete two-part cycle of this primitive engine has done the following:

1. The gas has absorbed Q_H from the hot tank.
2. The gas has rejected Q_C to the cool tank.
3. The gas has done work W.

11-3 Efficiency of a Heat Engine

Figure 11-3 is a schematic diagram of the energies involved in a cycle. From the first law of thermodynamics we know that the energy that goes in must be equal to what goes out:

$$Q_H = Q_C + W.$$

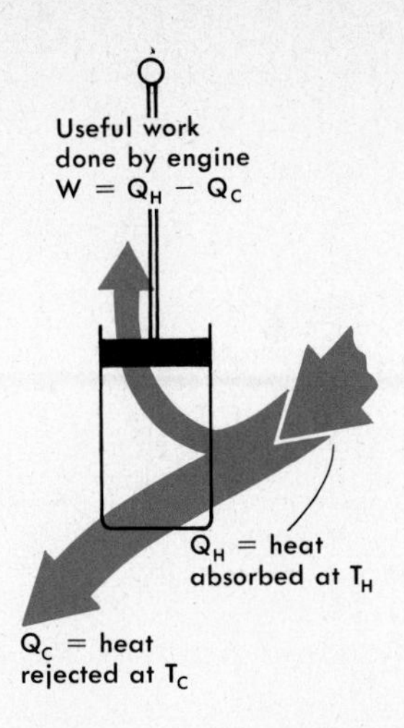

FIG. 11-3 Schematic diagram of the flow of energy in operating a simple heat engine.

The energy input has been Q_H; the useful output has been W. So we have for the efficiency of the engine,

$$\text{eff} = \frac{\text{output}}{\text{input}} = \frac{W}{Q_H} = \frac{Q_H - Q_C}{Q_H} = 1 - \frac{Q_C}{Q_H}.$$

In 1824 a young French military engineer, Nicolas Sadi Carnot, investigated the theory of operation of an idealized heat engine, operating in a somewhat more complex cycle than the one we have considered here. With the use of more mathematics than is appropriate for this course, it can be shown that for such an ideal engine the Q's (heat energies transferred) are exactly proportional to the T's (the *absolute* temperatures at which the energy transfers take place). Thus $Q_C/Q_H = T_C/T_H$, and we can express the efficiency of the ideal engine as

$$\text{eff} = 1 - \frac{T_C}{T_H}.$$

This efficiency of Carnot's engine is the maximum possible. For any other heat engine, real or imaginary—our little cylinder, the steam turbine driving an electric generator, or an internal combustion automobile engine—the actual efficiency can only be less. A large modern electric generating plant sends hot steam into the turbines at about 1000°F (811°K), and the used exhaust steam is in the neighborhood of 100°F (311°K). The maximum possible theoretical efficiency for an ideal Carnot engine would be

$$\text{eff} = 1 - \frac{311}{811} = 1 - 0.38 = 0.62.$$

In a real plant the operation cannot duplicate Carnot's imaginary ideal, and there are other unavoidable heat and friction losses, all of which will bring the efficiency down to perhaps 45 percent. Older, smaller plants would do well to reach 35 percent, and the national overall average efficiency is about 0.33.

Consider a generator plant with an average production of 10 megawatts of electric power at an overall efficiency of 35 percent. This means

$$\text{total heat input} \times 0.35 = 10^7 \text{ W}$$

$$\text{total heat input} = \frac{10^7}{0.35} = 2.86 \times 10^7 \text{ W}.$$

There are two destinies for this total heat input: 1.00×10^7 W of electric power and $2.86 \times 10^7 - 1.00 \times 10^7 = 1.86 \times 10^7$ W of rejected heat, some into the air through the stacks; some into the air from hot pipes, hot machinery, and friction; and most into the cooling water of the lake, sea, river, or cooling tower that serves the same purpose as the "cool tank" of our little cylinder-and-piston heat engine.

11-4 Entropy

If we could transform 100 percent of a given amount of heat into mechanical energy, we would not have to worry about mining coal or drilling for oil and gas. The first law of thermodynamics (which is only the law of conservation of energy) does not say that it cannot be done. An ocean liner could pump in seawater, extract the heat energy from it to drive its propellers, and throw overboard the resulting chunks of ice. An airplane could take in air, turn its heat into kinetic energy, and throw an ice-cold jet out through a nozzle in the rear. But, as Carnot demonstrated, we cannot use the heat content of our surroundings to produce mechanical work any more than we can use the water of the oceans to operate hydroelectric power installations. The potential energy of the water in the oceans is useless because there is no lower water level to which their water could flow; and the heat content of our surroundings is useless because there is no lower temperature region to which this heat can flow.

The big electric generators at Niagara Falls are possible only because water can be drawn from the river above the falls, its potential energy used to drive waterwheels, and then the water discharged at the lower level of the river below the falls. Here the transformations of energy are entirely mechanical but the basic principle applies equally well to heat engines.

Most generating plants (whether the fuel be coal, oil, gas, or fissionable nuclei) are located beside a river or lake in order to have a large cool reservoir into which the necessarily rejected heat can be discharged. A few plants use large cooling towers instead, to discharge the heat into the atmosphere, where the winds distribute it unnoticeably over a wide area.

Obviously, some sort of organization is needed for us to be able to derive useful work from a system. We must have two water reservoirs at different elevations to run a waterwheel; and we must have two heat reservoirs at different temperatures to run a heat engine. If we have two insulated water tanks in a laboratory—one containing 100 gallons of hot water at 95°C, and the other 100 gallons of cold water at 5°C—we can imagine ourselves extracting considerable work from this system by shuttling a properly designed cylinder and piston back and forth between the two tanks until they are finally at the same temperature. If, however, we mix the two tanks together, we shall have merely 200 gallons of 50°C water from which we can extract no useful work (unless, of course, we bring in some more water at a different temperature).

This is the natural way for any transfer of heat energy to take place. If two bodies or materials of different temperature are placed together so that heat can be transferred, the hot body will cool, and the cold body will warm until the final temperature is the same for both. In other words, the natural way of the world is to equalize energy differences—more specifically, differences in temperature.

As far as the first law of thermodynamics is concerned, 200 gallons of 50°C water contains exactly as much heat energy as our two separate tanks at 95°C and 5°C. What is missing is the *organization*, or *orderliness*, of the two separate tanks, carefully segregated with hot water in one tank and cold water in the other.

Scientists have a very useful word—*entropy*—which measures the *disorderliness* or *lack* of organization in a system. If you throw a piece of red-hot metal into a can of water, you have at the beginning a neatly ordered arrangement of the available heat of the metal-plus-water system; the high-temperature material is all in one place, and in the remainder of the can, well separated from the hot metal, is the lower-temperature water. The entropy of the system is relatively low, but in a few minutes this ordered arrangement is lost. The heat energy that was separated into hot material at one place and cool material at another is now randomly and equally distributed, and metal and water are all at the same temperature. The entropy of the system has *increased.*

What of refrigerators, in which, for example, heat is taken from cold water to freeze it into ice, and this heat is discharged into the hot kitchen, to make it still hotter? Figure 11-4 is a diagram of the basic energy transfers in a refrigerator. Compare it to the corresponding diagram for a heat engine shown in Fig. 11-3. The two diagrams are identical, except that the directions on the arrows are all reversed. In the engine, heat flows in its natural direction from hot to cold and a part of it may be converted to useful work. In the refrigerator, work from some outside source must be done on it, to make some heat flow in the abnormal direction from cold to hot.

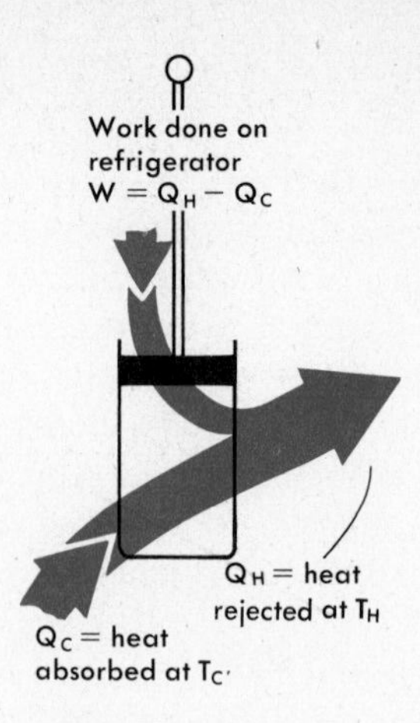

FIG. 11-4 Schematic diagram of the flow of energy in operating a simple refrigerator.

The second law of thermodynamics can be stated mathematically in a number of different ways, all of which have the following consequence, that we could use as another statement of the law itself:

The Second Law of Thermodynamics is that no repeatable process is possible in which entropy does not increase.*

The refrigerator seems to violate this law by transferring energy from cold to hot, but this is only because we have not considered the entire system involved. This includes not only the freezing water and the room, but also the motor and the power plant that generates the electric power to make it run. When we include all this, the negative ΔS of the refrigerator itself is more than offset by a larger positive ΔS in the power plant, so that the total ΔS is positive, as it must always be.

Mathematically, the definition of a change in entropy is very simple:

$$\Delta S = \frac{\Delta Q}{T}$$

where S represents entropy and ΔS is the change in entropy, ΔQ represents the heat added (it is negative if heat energy is lost), and T is of course the absolute temperature at which the ΔQ was either lost or gained. We shall get into trouble if we try to apply this to the hot metal thrown into the can of water; the temperatures of the metal and of the water are constantly changing, which makes calculus necessary for handling the problem. We can, however, think of simple cases in which this difficulty does not arise.

---

*—or at best is unchanged. $\Delta S = 0$ is possible for a cycle of the Carnot engine, or other ideal cycles. No real process, however, can be quite ideal.

Consider, for example, a 100-g ice cube (0°C) thrown into a large container of warm water (40°C). In order to melt, the ice cube must absorb $100 \times 80 = 8000$ cal, and its temperature remains at 273°K until it is all melted. One ice cube will not measurably change the temperature of the water, so we can also consider this to be constant at 313°K. The entropy change of the ice cube is expressed as $\Delta S_i = 8000/273 = 29.3$ cal/°C. For the water (which has lost heat), the change is $\Delta S_w = -8000/313 = -25.6$ cal/°C. Thus the total entropy change in the ice-and-water system is $29.3 - 25.6 = +3.7$ cal/°C. The entropy has increased, as is always true for any spontaneous natural process, if we are careful to include everything that is involved.

Questions

(11-1) **1.** Confirm, by using Joule's quoted figures, that 1 cal $\approx$ 4.18 joules. (This will not come out exactly, as more modern work has shown that Joule's figures, based on relatively crude apparatus, were a little off.)

2. Confirm the statement at the end of Section 1, that 1.57×10^5 cal will heat 40 lb of steel as stated.

3. In a paddle-wheel experiment like Joule's, a 20-kg mass descends through a distance of 1.25 m. After each descent the weight is wound back up and allowed to come down again. This is repeated 24 times. The insulated container holds 2800 g of water. What rise in the temperature of the water could be expected?

4. A 500-g weight is dropped 150 cm to the floor. How many calories of heat are produced?

5. A 1500-kg car traveling 72 km/hr is brought to a stop (whether by a stone wall or the car's brakes makes no difference). How many calories of heat are produced?

6. A 10-g bullet traveling 300 m/s hits a target and is quickly stopped. (a) How much KE is converted into heat? (b) How many calories is this?

7. A 20-kg piece of aluminum is dropped from a building 150 m high. Assume that 0.7 of its energy is converted into heat in the aluminum, and the remaining 0.3 into heat in the ground. (a) How many calories go into heating the aluminum? (b) By how many degrees will the temperature of the aluminum rise?

8. A 10-kg piece of lead is dropped from a building 100 m high. Assume that 0.8 of its energy is converted into heat in the lead, and the remaining 0.2 into heat in the ground. (a) How many calories go into heating the lead? (b) How many degrees will the temperature of the lead rise?

9. An iron bolt is fired at a speed of 300 m/s, by a special gun, into a concrete wall (a common procedure in modern construction practice). Assuming that $\frac{2}{3}$ of the heat produced goes into the iron, find the resulting rise in temperature of the bolt.

10. A lead bullet of unknown mass is fired at a speed of 200 m/s into a tree, in which it stops. Assuming that $\frac{2}{3}$ of the heat produced goes into the bullet and $\frac{1}{3}$ into the wood, find how many degrees the temperature of the bullet is raised.

11. The melting point of lead is 327°C. At what speed would a lead bullet have to be fired into a target in order to completely melt the bullet, if half of the heat produced is absorbed by the bullet? (Bullet is initially at 27°C.) Remember that lead has a heat of fusion.

12. A sack of ice at 0°C is dropped from a plane flying at an altitude of 500 m. What fraction of the ice will be melted by its impact with the ground, if half of the heat produced is absorbed by the ice?

13. A 1500-watt heating coil is immersed in an insulated tank holding 100 kg of ice at −10°C. How long will it take to raise the temperature of the contents to 70°C?

14. A 100-watt lamp is immersed in an insulated container holding 500 g of alcohol at 20°C. How long will it take to warm the alcohol to 50°C?

(11-3) **15.** What is the maximum possible thermal efficiency of an engine whose heat input is at 518°F and which discharges its waste heat at 176°F?

16. What is the maximum possible thermal efficiency (i.e., the fraction of energy supplied that is converted into useful work) that can be obtained from an engine whose heat input is at 400°C and that discharges its waste heat at 60°C?

17. An engine operates between the temperatures 800° and 200°C. By what fraction would its theoretical thermal efficiency be increased if it were found possible to raise its high temperature to 900° C?

18. An engine operates between the temperatures 800° and 200°C. By what fraction would its theoretical thermal efficiency be increased if a redesign lowered its discharge temperature to 100°C?

19. An electric generating plant producing an average output of 50 MW operates at $T_H = 800°C$, $T_C = 50°C$. (a) What is its ideal theoretical efficiency? (b) The plant realizes 70 percent of this theoretical efficiency—what fraction of the heat input is converted to electrical energy? (c) To produce 50 MW, how many joules of heat energy must be provided per second? (d) Cooling is accomplished by a river that carries 10 m^3/s. How much rise in temperature does the heat discharge cause?

20. Consider a nuclear-fueled power plant producing a peak output of 120 MW of electric power. The nuclear reactor must operate at a lower temperature than a fuel-burning boiler can, so T_H is only 400°C, and T_C is 40°C. (a) What is its ideal theoretical efficiency? (b) What fraction of the heat input from the reactor is converted into electric power, if its real efficiency is 78 percent of the theoretical ideal? (c) At its peak output, how many joules per second of heat energy must be supplied by the reactor? (d) Cooling is done in a river that flows 30 m^3/s. What is the temperature rise of the river at peak plant output?

(11-4) **21.** A tank is divided into two compartments by a partition of sheet copper. On one side is water at 80°C; on the other is water at 20°C. In a brief period of time, 1000 cal will flow from the hot side to the cool. What is the total change in entropy during this period? (Assume there is so much water that 1000 cal does not change the temperature appreciably.)

22. A heating coil is immersed in an insulated tank that contains 1000 g of crushed ice mixed with an equal mass of water, all at 0°C. Steam at 100°C is passed through the coil until all the ice is just melted. (a) What has been the entropy change of the steam? (Assume it leaves the far end of the coil as 100°C water.) (b) What has been the entropy change of the ice? (c) What has been the total entropy change?

12

The Molecular Nature of Matter

12-1 The Molecular Hypothesis

Surveying the physical properties of different substances encountered in nature, we find a great deal of variety. Some of the substances are normally solid, melting and turning into gas only at extremely high temperatures. Others are normally gaseous, becoming liquid and freezing only when the temperature drops close to absolute zero. Some liquids are of high fluidity, while others are very viscous. Some substances, generally known as metals, possess a high degree of electric and thermal conductivity, while others, the dielectrics, are very good insulators. Some substances are transparent to visible light while others are completely opaque. . . .

We ascribe all these differences between substances to differences in their internal structure and attempt to explain them quantitatively as well as qualitatively as being due to different properties and interactions of the structural elements of matter. We assume that such seemingly homogeneous substances as air, water, or a piece of metal are actually composed of a multitude of extremely small particles known as *molecules*. All molecules of a given pure substance are identical, and the differences in physical properties between various substances are due to the differences between their molecules. There are as many different kinds of molecules as there are different substances, of which there is indeed an enormous number.

A a solid

The molecules forming any given material body are held together by *intermolecular forces*, which are determined by the nature of the molecule. These forces resist the tendency of internal thermal agitation to break up the molecular aggregates. If intermolecular forces are strong, molecules will be as rigidly cemented together as the bricks in a garden wall (Fig. 12-1A), and the material will remain solid up to very high temperatures. If these forces are comparable to the forces of thermal agitation, they may not be able to hold the molecules rigidly in their places and may permit the molecules to slide more or less freely past each other as if they were grains of fresh Russian caviar (Fig. 12-1B). Thus the substance will keep its volume, but it will take the shape of the container in which it is placed. The viscosity of the liquid will depend on how easily this sliding of molecules can take place, and the substance will become more and more fluid as its temperature and thermal agitation rise. If the intermolecular forces are very weak, the molecules will fly apart in all directions and the material must be kept in a closed container like a bunch of agile flies in a glass jar (Fig. 12-1C). This picture explains the high compressibility of gases, since compression results only in the reduction of the free space in which the molecules are moving. We can get an idea about the amount of free space between the molecules of a gas by comparing the density of the gas with its density in the liquefied state, when the molecules are packed together. For example, the density of atmospheric air, under normal conditions is 0.0012 g/cm^3, and the density of liquid air is 0.92 g/cm^3, or 800 times larger. Since, in the liquid state, the distances between the centers of neighboring molecules are approximately equal to their diameters, the distances in the gaseous state must be $\sqrt[3]{800} = 9.3$ times the diameters. This relation is shown in Fig. 12-2.

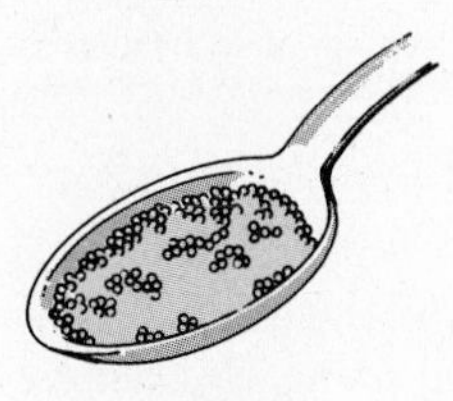

B a liquid

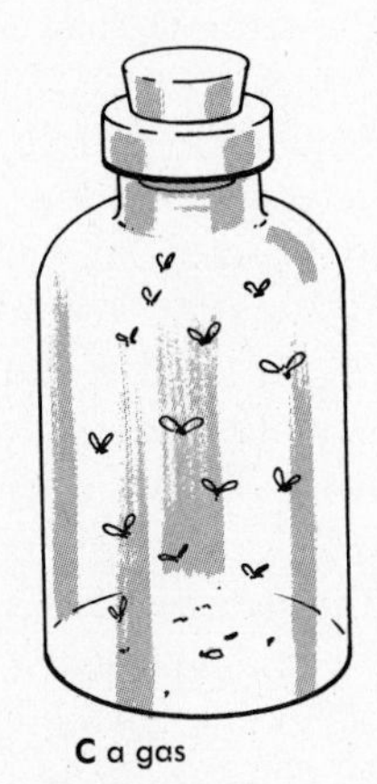

C a gas

FIG. 12-1 Three states of matter.

12-2 Brownian Motion

Although molecules are too small to be seen individually through even the best microscope, their thermal agitation can be noticed by observing the movement of small particles of smoke floating in the air. This phenomenon is called *Brownian motion*, after the British botanist Robert Brown (1773–1858), who in 1827 observed the irregular motions of plant spores floating in water. Such small particles play the role of an intermediary between our familiar surroundings and the world of molecules since they are large enough to be observed microscopically but are

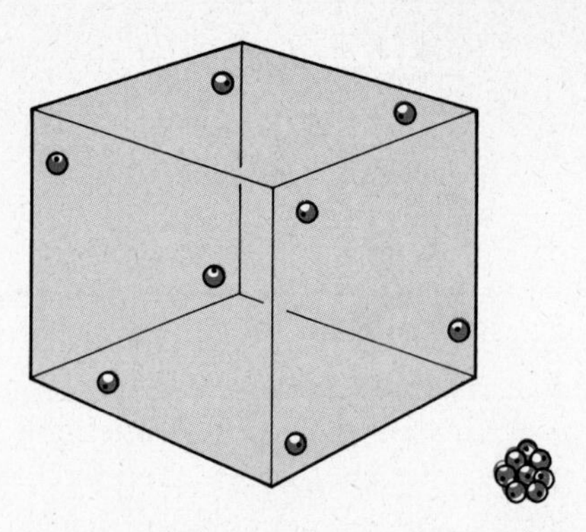

FIG. 12-2 The approximate relative diameters and distances between air molecules in the normal sea-level atmosphere. The blob to the right of the box shows the same molecules in liquid air.

also sufficiently small to be affected by irregular molecular motion. The situation is similar to that of a pilot in a high-flying plane observing a Navy task force in a choppy sea. He cannot, of course, see the waves themselves, and tiny life rafts that follow every movement of the water are invisible to him; at the other extreme, the big aircraft carriers will seem to float without any disturbance at all. But the medium-sized ships, which he can still see, will show a definite roll, and our pilot will know that the sea is rough. In Fig. 12-3, we show the successive positions of a smoke particle, 1 micron ($= 10^{-6}$ m) in diameter, dancing its Brownian dance in atmospheric air. The study of Brownian motion by both Einstein and the French physicist Jean Perrin (1870–1942) led to indisputable proof of the reality of the thermal motion of molecules and gave us valuable information concerning the amount of kinetic energy involved in it.

Figure 12-3 is not an actual detailed chart of the real motion of a smoke particle. As the caption implies, the position of the particle (on a magnified scale) has been recorded at the end of each minute, and then these points have been joined by straight lines. The actual path of the particle between any two points was itself a jagged, irregular one similar to the larger-scale path of Fig. 12-3, but on a scale too small to measure and record. The point of the matter is that if statistical methods are applied to data such as those recorded in the illustration, it is possible to calculate the *average velocity* of the particle during its erratic journey. If the mass of the Brownian motion particle is known, it is then a simple matter to find its *average* kinetic energy, from $\frac{1}{2}mv^2$.

But what is the use of this bit of knowledge, other than a little intellectual exercise? The smoke particle is enormously larger than the molecules of the air in which it is suspended. Its motion, however, is caused by millions of collisions per second with the air molecules. One individual collision will not move the particle appreciably; but

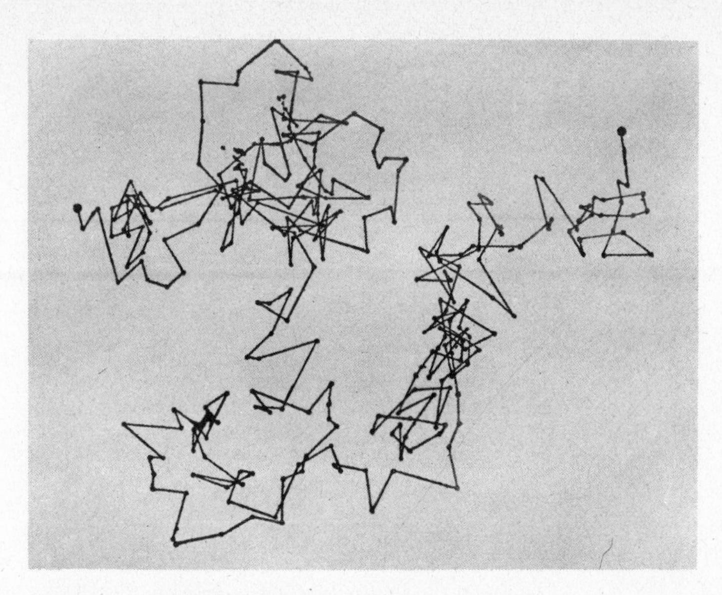

FIG. 12-3 Successive positions of a smoke particle in air, recorded at one-minute intervals (according to J. B. Perrin).

occasionally the number of air molecules striking it on the left side, say, will happen to be greater than the number striking it on the right side, and as a result the particle will move a visible distance to the right. Here again, a complicated statistical analysis of all possible collisions leads to the theorem of the *equipartition of energy*. This theorem states that ***in the random motion of a large number of colliding particles, the average kinetic energies of all particles are the same, regardless of their masses.*** In other words, on the average, $\frac{1}{2}m_{\mathrm{p}}v_{\mathrm{p}}^2$ for the large particle is exactly equal to $\frac{1}{2}m_{\mathrm{M}}v_{\mathrm{M}}^2$ for a molecule of air.

Brownian particles in the neighborhood of 1 micron (μ) in diameter are observed to move at room temperature with an average velocity of 0.65 cm/s, and since their mass is about 5×10^{-13} g, their kinetic energy must be about 10^{-13} erg. Thus, by the equipartition theorem, this value must also represent the kinetic energy of individual air molecules at this temperature.

With an increase of temperature, the intensity of Brownian motion also increases, so that by direct observation of the tiny particles suspended in a gas, or a liquid, we can study how the energy of thermal motion depends on temperature. The results of such experiments performed between 0° and 100°C are shown in Fig. 12-4. The observed points are located on a straight line and, extrapolating it in the direction of lower temperatures, we come to the conclusion that Brownian motion must completely stop at −273°C.

We recall that −273°C is 0°K, the zero of the absolute temperature scale (see Fig. 10-4). From this, and from the straight-line relationship

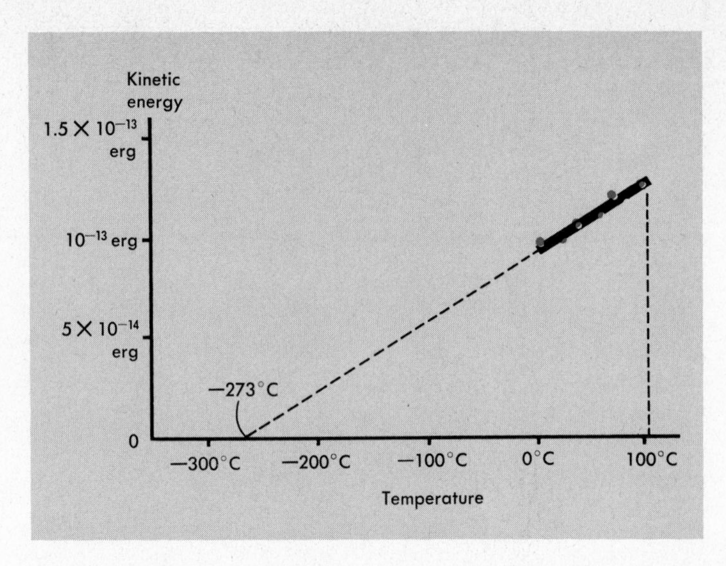

FIG. 12-4 A schematic diagram showing the kinetic energy of Brownian motion measured for temperatures between 0° and 100°C, and extrapolated to lower temperatures.

in Fig. 12-4, we must come to the conclusion that ***the average kinetic energy of the thermal motion of molecules is directly proportional to the absolute temperature.***

As we deduced above from the Brownian motion of smoke particles, the average kinetic energy of an air molecule at room temperature (about 300°K) is about 10^{-13} erg. Now we may imagine taking an air molecule at absolute zero and warming it up to room temperature a degree at a time. Since its 300°K energy is 10^{-13} erg, this means we would have to add $10^{-13}/300 = 3.3 \times 10^{-16}$ erg to raise the temperature of 1 molecule of air by 1°C.

On the other hand, we can deal with air in wholesale quantities, paying no attention to the billions of molecules that it contains. From such bulk experiments, we know that the specific heat of air is 0.16 cal/g/°C. Converting this to ergs, we have

$$0.16 \text{ cal} \times 4.18 \text{ J/cal} \times 10^7 \text{ ergs/J} = 6.7 \times 10^6 \text{ ergs.}$$

It thus takes 6.7×10^6 ergs to raise the temperature of 1 g of air by 1°C. Then, merely by the simple division $6.7 \times 10^6/3.3 \times 10^{-16}$, we learn that 1 g of air must contain 2×10^{22} molecules; or in other words an air molecule has a mass of $1/(2 \times 10^{22})$ = about 5×10^{-23} g. (This, of course, is a rough average for the 80 percent nitrogen and 20 percent oxygen that are the main constituents of air.)

Since 1 g of liquid air occupies a volume of about 1 cm^3 (its density is 0.92, i.e., about 1), the volume of a single air molecule must be about 1 $cm^3/(2 \times 10^{22}) = 5 \times 10^{-23}$ cm^3, and its diameter about $\sqrt[3]{5 \times 10^{-23}} = 4 \times 10^{-8}$ cm.

We can also use the data above to estimate the velocity of molecular motion. At room temperature, we have found the average energy of a molecule to be 10^{-13} erg, and its mass to be not far from 5×10^{-23} g. Then, from KE $= \frac{1}{2}mv^2$,

$$10^{-13} = 0.5 \times 5 \times 10^{-23}\, v^2$$

$$v^2 = \frac{10^{-13}}{2.5 \times 10^{-23}} = 4 \times 10^9$$

$$v = \sqrt{4 \times 10^9} = \sqrt{40 \times 10^8} = \text{about } 6 \times 10^4 \text{ cm/s}$$

$$= \text{about } 0.6 \text{ km/s.}$$

It is interesting to think that all this knowledge of invisible molecules has been derived from measurements made entirely on bulk material such as smoke particles and grams of air. We can summarize the results in Table 12-1:

TABLE 12-1 APPROXIMATE PROPERTIES OF AIR MOLECULES

Mass	5×10^{-23} g
Diameter	4×10^{-8} cm
Velocity (at room temperature)	0.6 km/s

12-3 Molecular Beams

Our faith in the existence of molecules might be a little firmer if it rested on evidence somewhat more direct than that used above. If we make a small hole in the wall of a vessel containing a considerable amount of gas, and the gas escapes out into the surrounding vacuum, the picture will look in general like that of a panicked crowd rushing out of a burning theater. Although the resulting stream of gas will be generally directed away from the opening, individual molecules inside the stream will continue to execute their irregular thermal motion, rushing in all directions and constantly colliding with each other (Fig. 12-5A).

The phenomenon will be different if the density of the gas in the container is so small that the molecules are not likely to collide with one another. In this case, the molecules fly out of the container independently one by one (Fig. 12-5B), and the velocity of the outgoing particles will be determined by the thermal velocity of their molecular motion, corresponding to the temperature of the gas. These *molecular beams*, in which the individual molecules go their own way and do not interact with the others, are very useful for the study of many molecular properties.

The German physicist Otto Stern pioneered in the study of molecular beams and, with his students and followers, developed several ingenious

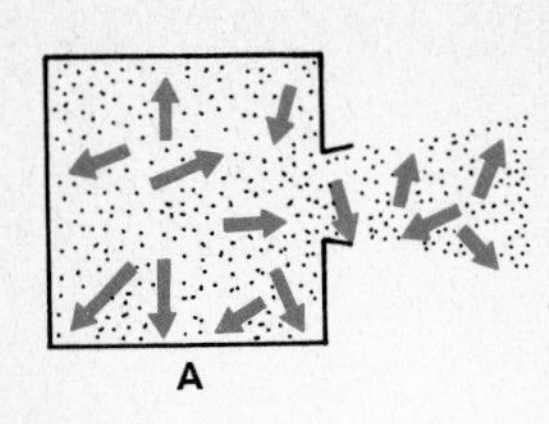

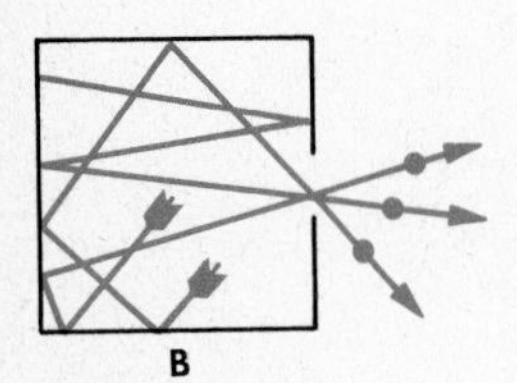

FIG. 12-5 Molecular beams: (A) when the density is high there are many collisions between molecules and the gas flows out in a continuous stream; (B) at very low density there are few collisions and the molecules escape as individuals.

methods for directly measuring molecular velocities. One such piece of equipment is sketched in Fig. 12-6. The entire apparatus is in a high vacuum so that the molecules being studied will not collide with air molecules. The source of the beam is a ceramic cylinder heated to the required high temperature by electric current in a wire wound around its surface. Within the cylinder is placed a small piece of some volatile material (sodium or potassium metals in this particular experiment), which gives rise to a gas of low density and pressure. Individual molecules of the gas fly out through a small opening at the base of the cylinder, and a thin molecular beam is cut out of the divergent stream by a slotted diaphragm.

This thin, parallel-sided beam of molecules falls generally on the outside surface of a rotating metal drum. Once each revolution, however, a thin slit S in the drum passes across the beam, and a small spurt of molecules M passes through to the inside of the drum, as shown in diagram I. Directly opposite the slit in the drum is a glass plate; if the drum were stationary, the molecules would pass across and strike the plate at point A. The drum, however, is rotating rapidly, and during the time the spurt of molecules crosses the diameter of the drum D, the drum has moved so that the flying molecules strike the plate at point B instead of at A (diagram II).

Let us say that the drum is rotating at an angular speed of f rev/s, which is $2\pi f$ rad/s. We recall that the linear speed of any part of a rotating body is $r\omega$, so the linear speed of the glass plate is $(D/2) \times 2\pi f = \pi Df$. In a time interval t, the glass plate will have moved through a distance $d = \pi Dft$, where t will be the time required for the molecules to cross the diameter of the drum. If v is the speed of the molecules, this time will be equal to D/v. Substituting this value for the time into the equation for d, we get

$$d = \pi Df \times \frac{D}{v} = \frac{\pi f D^2}{v}.$$

$$v = \frac{\pi f D^2}{d}.$$

From the above, the experimenter knows that any molecules deposited at a distance d from point A must have traveled with a speed given by the last equation. The values of thermal velocities obtained by this direct method are in perfect agreement with the values obtained by the less direct considerations described in the previous section.

It was found that not all the molecules in the beam struck the plate at the same spot—in other words, that not all the molecules had the same velocity. It is quite easy to measure the relative thickness of the deposit at various distances along the plate and thus to figure the relative number of molecules moving at each different velocity. Although the main bulk of the molecules move at about the velocity that would be

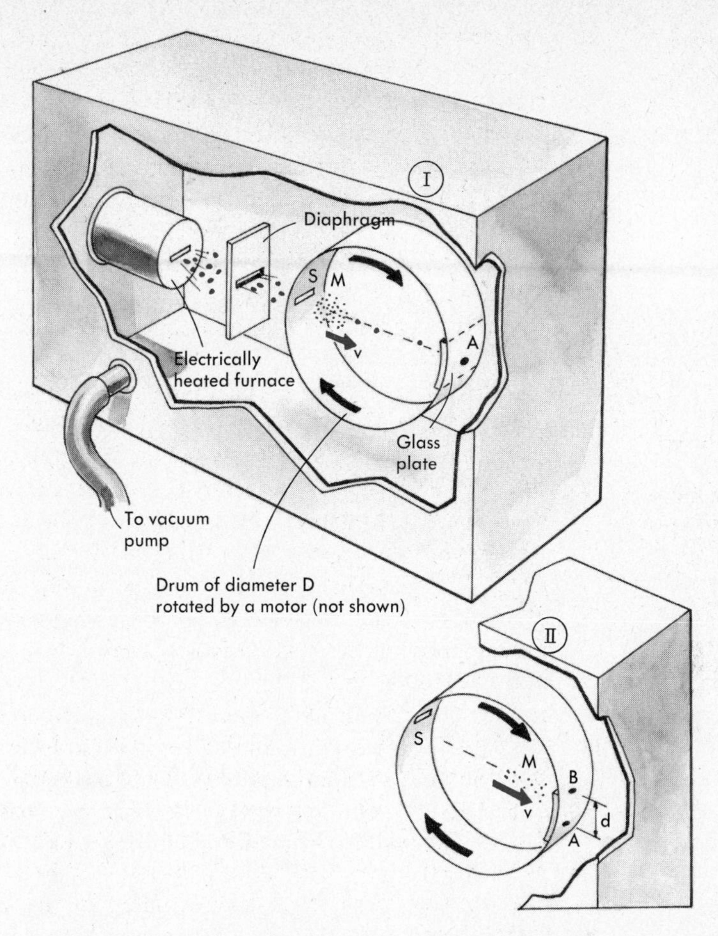

FIG. 12-6 One type of apparatus for measuring the distribution of molecular velocities.

expected from the temperature, there are always some molecules that move considerably slower or considerably faster than the average. Curves showing the distribution of molecular velocities at two different temperatures are given in Fig. 12-7. These deviations from the average result from the statistical nature of molecular motion and the irregularity of molecular collisions that may occasionally almost stop a molecule in its tracks or else send it rushing on with an abnormally high speed. The statistical study of the velocity distribution of the molecules of a gas was carried out in the last century on a purely theoretical basis by the British physicist James Clerk Maxwell, who derived a mathematical expression describing the distribution of molecules of different velocities. The curve obtained by Stern in his experiments stands in excellent agreement with *Maxwell's distribution* of molecular velocities.

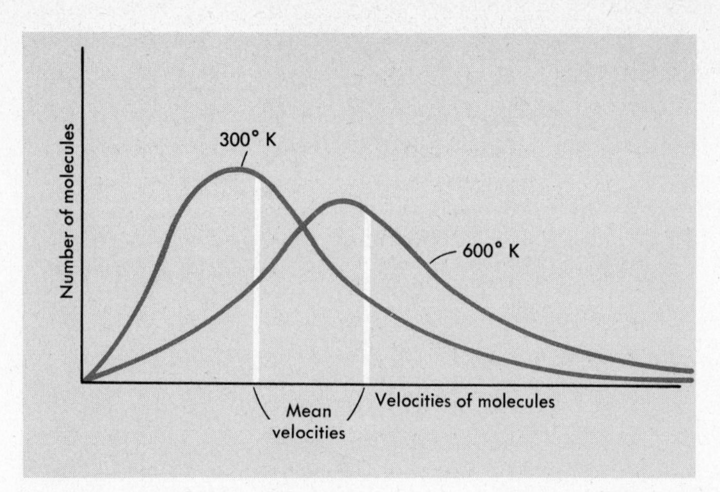

FIG. 12-7 Graph of Maxwell's calculation of the distribution of molecular speeds in a gas at two different temperatures.

12-4 Kinetic Theory of Gases

Consider a closed box (Fig. 12-8) containing a certain amount of gas. The molecules of the gas are rushing ceaselessly in all directions, colliding with each other and bouncing off the walls. The pressure of the gas on the surrounding walls is the result of the continuous bombardment to which the walls are subjected by the onrushing molecules. Suppose first that we keep the temperature of the gas (i.e., the velocity of its molecules) constant but reduce the volume. It is easy to see that in this case the number of molecules hitting a unit area A of the wall will

FIG. 12-8 Molecules of a gas bouncing (reflecting) from an enclosing wall, and thereby exerting a pressure on it. Consider the enclosure to be a cubical box measuring a cm on each side.

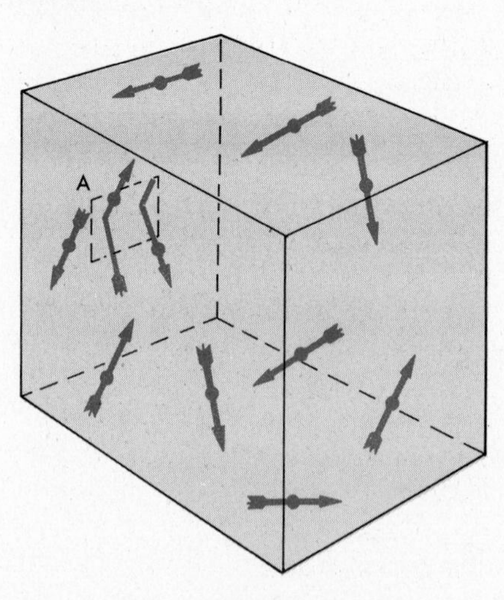

increase in inverse proportion to the decreasing volume. Since the pressure experienced by the walls is proportional to the total number of impacts they receive per unit time, we have here the explanation of the inverse proportionality between the pressure and the volume of gas.

Now keep the volume constant but increase the temperature and thus speed up the molecular motion. This will have two effects:

1. Since the molecules are moving faster, a larger number of them will collide with the wall area per unit time.
2. The impact of each molecule will be more violent because of its greater speed.

Since the pressure depends on both the number of impacts per second and the force of each impact, and since each of these factors is proportional to molecular velocity, the combined result is that the pressure will be proportional to the square of that velocity, i.e., to the kinetic energy of molecular motion, or, what is the same, to the absolute temperature of the gas.

The qualitative arguments given above will be more convincing if our conclusions are backed up with quantitative mathematical arguments as well. Let us put N identical molecules, each of mass m, in the box of Fig. 12-8. They will be moving at many different speeds, and the speed of any molecule will change many times a second as the molecules collide. To make the arithmetic easier, we can imagine replacing these N real molecules with others that are moving with a constant, unchanging speed. What should this speed be? We do not want to change the temperature, so the average KE must remain the same. All the molecules have the same mass, so we need only ensure that the new v^2 of our constant-speed molecules is the same as the average v^2 of all the individual real molecules. So we need only (in imagination, or by the use of complex mathematics) square the speed of each real molecule, average all these squares, and then take the square root of this average (or mean) value. Physicists, chemists, or others interested in molecular behavior, call this the *root-mean-square* velocity, v_{rms}, since it is the square root of the mean square. This is the molecular speed we will be dealing with throughout this chapter, and it will be simpler to use merely the letter v, with the understanding that $v = v_{rms}$.

We can make a further simplification by assuming that at any instant $N/3$ molecules (one-third of the total) are bouncing back and forth vertically between top and bottom, another $N/3$ between the left face and the right face, and $N/3$ between the front and back sides—and all this without colliding with one another. (Actually, since we have assumed the molecules to be perfectly elastic, it makes no difference in the final answer if they *do* collide. If we avoid collisions, however, it helps keep our mental picture a little clearer and simpler.) This seems very unreal and artificial, but a more sophisticated and realistic mathematical analysis brings us to exactly the same final result.

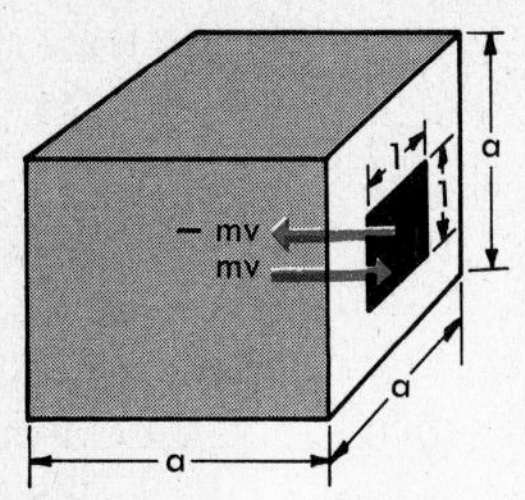

FIG. 12-9 The change in momentum of a gas molecule when reflected from a wall of its container.

Figure 12-9 shows one of the molecules bouncing off the right wall. Its original momentum was mv toward the right; after impact it bounces back with a momentum of mv toward the left, or $-mv$. This gives $2mv$ as the change in momentum for each molecule that strikes the wall. We must now find how many molecules per second strike the wall. To make things as simple as possible, assume (as in Fig. 12-9) the box is a cube a cm $\times$ a cm $\times$ a cm. A molecule going v cm/s will require a/v seconds to travel the a cm from one side of the box to the other and will therefore make v/a collisions per second with the two walls between which it is bouncing. On one wall, the collisions per molecule will thus be $v/2a$ per second. There are $N/3$ molecules bouncing between each pair of walls, so the total number of collisions on one wall is $Nv/6a$ per second.

It has been shown above that each collision produces a momentum change of $2mv$, which means that the total momentum change per second on one wall is $2mv \times Nv/6a = Nmv^2/3a$. The wall has an area of a^2, so the momentum change per unit area per second is

$$\frac{1}{a^2} \times \frac{Nmv^2}{3a} = \frac{Nmv^2}{3a^3}.$$

From our earlier discussions of Newton's second law, we recall that force is equal to the rate of change of momentum, which is just what we have finished figuring. Hence the force on a unit area of wall, which is the pressure, is the last expression we have derived above:

$$P = \frac{Nmv^2}{3a^3}.$$

Let us rearrange this equation to put it in the form

$$\frac{Pa^3}{mv^2} = \frac{N}{3}.$$

The volume of the gas, $V = a^3$. In the denominator, mv^2 is twice the KE of the molecules, which we know to be proportional to the absolute temperature. So mv^2 is also proportional to T, which is a relationship we can write as $mv^2 = kT$, where k is some proportionality constant we will not bother to evaluate. We can substitute these values back in the equation above, to get

$$\frac{PV}{kT} = \frac{N}{3}; \qquad \frac{PV}{T} = \frac{k}{3} \times N = \text{a constant} \times N.$$

Here we have a theoretical justification for the general gas law that we used in Section 5 of Chapter 10. We can also see from this relationship that equal volumes of any gases, if they are at the same temperature and pressure, must contain equal numbers of molecules. A liter of the lightest gas, hydrogen, contains the same number of molecules as a liter of the gas uranium hexafluoride, whose molecules are 176 times more massive.

Another unexpected result can come from our original equation $P = Nmv^2/3a^3$. The total mass of any gas sample is Nm, the number of molecules times the mass of each molecule; and a^3 is the volume. This gives us

$$P = \frac{\text{mass} \times v^2}{3 \times \text{volume}}.$$

Mass/volume is the density of the gas, d, so we have

$$P = \frac{dv^2}{3}; \qquad v = \sqrt{3P/d}.$$

A handbook tells us that the density of air at 1 standard atmosphere (1.013×10^6 dynes/cm^2) and 0°C (273°K) is 1.293×10^{-3} g/cm^3. Substitution of these values gives the rms speed of air molecules under these conditions as

$$v = \sqrt{3 \times 1.013 \times 10^6/1.293 \times 10^{-3}} = 4.85 \times 10^4 \text{ cm/s}.$$

12-5 Surface Tension and Surface Energy

We all are familiar with dewdrops and raindrops, and most of us have seen an escaped bit of mercury gather itself up into elusive droplets. Liquids all show this tendency to assume a characteristic spherical shape in competition with the force of gravity that forces liquids to assume the shape of their containers. A glass can be filled heaping full of water, so that the water level stands slightly above the rim and slopes down toward it at its edge.

These phenomena are examples of *surface tension*, apparent in all liquids. Surface tension can be explained from the molecular point of view by considering that each molecule is subject to attractive forces from its surrounding fellows. Let us consider this from the standpoint of potential energy. As always, we must first agree on what we shall consider to be the level of zero potential energy. As a matter of convenience, let us call the energy zero when a molecule is outside the liquid we shall be considering. Now let us reach inside a drop of liquid with an imaginary pair of impossibly fine tweezers and remove one of the molecules. It is completely surrounded on all sides by other liquid molecules, and to remove it we will have to do work against the attraction of all these neighbors. Since we must do this work to raise its energy to zero, it apparently had a quite large *negative* energy when it was inside the drop.

Having done this, let us repeat the process for a molecule on the outer surface of the drop. This molecule, being on the surface, is *not* surrounded by neighbors on all sides, but only on the bottom. Therefore, less work will have to be done to remove this molecule and bring it up to our agreed-on zero level of energy. And therefore, of course, its energy is *less negative* than the energy of an interior molecule.

So, because it is less negative, we see that a surface molecule has a greater potential energy than an interior molecule. (If you owe $50, you are financially better off than if you owe $100.) We have seen on many other occasions that physical systems come to a state of least energy if they are able to. (This is really a consequence of equipartition of energy and the second law of thermodynamics, although we shall not pursue the reasons any farther.) A truck rolls downhill, a hot body becomes cooler, and so on. And, to reduce the number of its high-energy surface molecules and therefore its total energy, a piece of liquid will do its best (against the pull of gravity) to make its surface as small as possible. The geometrical shape that has the least surface for a given volume is a sphere, so the liquid comes as close as it can to assuming the form of spherical drops.

12-6 Evaporation

The ideas of kinetic theory and of bonds between the molecules of a liquid help to explain the phenomenon of evaporation. We all know that water placed in a pan on a shelf will gradually disappear into the air; if the pan is put on a warm stove, it will evaporate much more rapidly. We must recall that whenever two bodies attract each other, it requires energy to separate them, whether the bodies are a heavy weight and the earth, or a pair of water molecules. If a water molecule happens to be in the top layer of molecules and happens to be moving upward and happens also to have speed enough to break away from the attractive forces of its neighbors, it will then evaporate from the liquid surface and become a water-vapor molecule. Figure 12-10 shows the application to this situation of the Maxwellian distribution of molecular velocities, as illustrated in Fig. 12-7. The velocity v is required for a water molecule to be able to pull free from the attraction of its fellows. At 300°K, only those molecules under the line AC, on the tail of the distribution curve, will have a speed of v or greater. This is a very small fraction of the total area under the 300° curve, and the evaporation rate will be slow. At 350°K, however, the curve is shifted to the right, and at this higher temperature all the molecules under BC will have enough speed to escape. This is a much greater fraction of the total number of molecules, and, as a consequence, the rate of evaporation will be increased a great deal more than would be expected from a mere comparison of temperatures.

We can also see from this molecular explanation of evaporation the reason why evaporation causes a liquid to become cooler. Only the

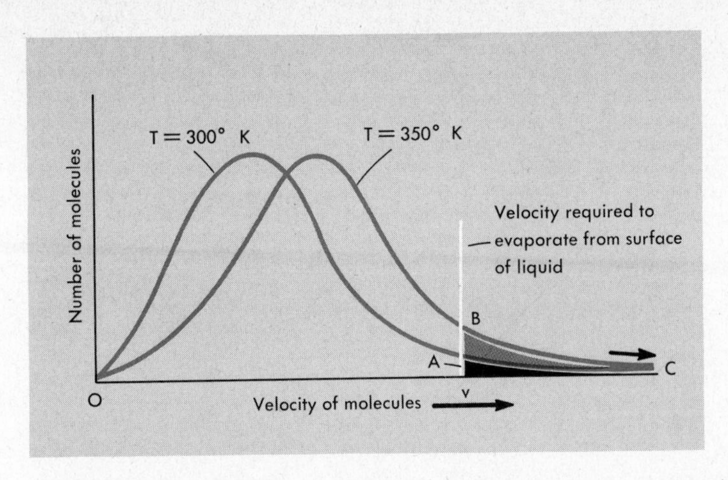

FIG. 12-10 The distribution of velocities in liquids of different temperatures; a slight rise in temperature gives many more molecules the energy needed to evaporate.

fastest-moving and most energetic molecules are able to escape. With these high-energy molecules constantly leaving the liquid, the average energy of those remaining constantly becomes smaller, as long as the evaporation continues. Since temperature is really only a measure of the average kinetic energy of the molecules of a body, the evaporating liquid becomes cooler.

12-7 Diffusion

The phenomenon of diffusion takes place in liquids as well as in gases. If we drop a lump of sugar into a cup of tea and then do not stir it, it will take a very long time, many hours as a matter of fact, until the tea becomes uniformly sweetened. If we *do* wait, we will find that the situation is the same as could have been achieved in a minute by using a spoon. How do the molecules of sugar, without being stirred, travel through the liquid? Well, they just shoulder their way through the crowd of water molecules in the cup as the result of thermal agitation. They do not follow any straight path but execute a "random walk," as illustrated in Fig. 12-11.

It is worth turning back to Fig. 12-3 to note that the motion of a Brownian particle is also a "random walk." The same arguments apply to a smoke particle in collision with air molecules as ot a sugar molecule under bombardment by water molecules. If a person takes N steps all in one direction and the length of each step is L, then obviously he will cover the distance NL away from the point from which he started. If, however, he changes his direction at random with each step, he does not go that far. It can be shown mathematically in this case that the mean distance traveled away from the original point, averaged over many trials, is only $\sqrt{N} \times L$. Thus, since the number of steps taken

FIG. 12-11 A "random walk," in which a person (or other particle) frequently changes his direction by turning at unpredictable angles.

is proportional to time, the distance traveled in a random walk increases only as the square root of time. The situation is illustrated in Fig. 12-12, which shows the diffusion of sugar molecules through the unstirred cup of tea. The phenomenon of diffusion plays an important role in many other processes of physics besides the sweetening of unstirred tea. The principle of the atomic reactor is based on the diffusion of neutrons through the moderator, and the energy quanta produced by thermonuclear reactions in the center of the sun similarly diffuse through the body of the sun until they can fly off freely into space after reaching the surface.

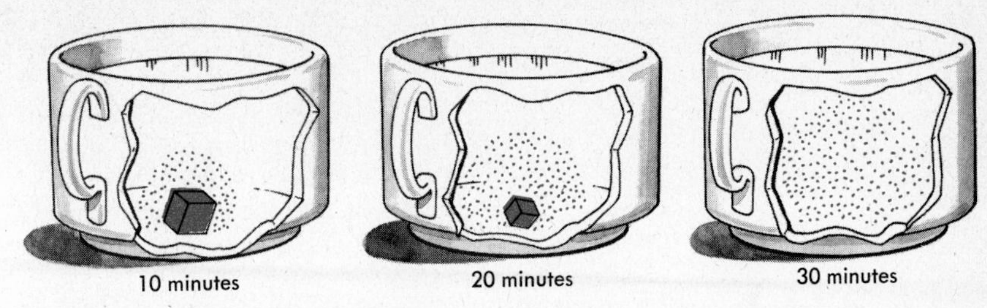

FIG. 12-12 The diffusion of sugar in an unstirred teacup. The distances diffused in 10, 20, and 30 minutes are in the ratio of $\sqrt{1} : \sqrt{2} : \sqrt{3}$, or 1 : 1.41 : 1.73.

Questions

(12-1) 1. The element bromine is a liquid at room temperature with a density of 3.12 g/cm^3. When vaporized (at 60°C) its density is 7.59×10^{-3} g/cm^3. By about how many molecular diameters are its gaseous molecules separated?

2. Liquid sulfur dioxide has a density of 1.4 g/cm^3. In the gaseous state under normal conditions, its density is 2.9×10^{-3} g/cm^3. By about how many molecular diameters are the gaseous molecules separated?

(12-2) 3. The text gives 0.65 cm/s as the average velocity of particles 1 μ in diameter at 27°C. What would be the average speed of particles 8 times this large in diameter, made of the same material, and at at the same temperature?

4. Observation shows that at the same 27°C certain uniform suspended particles have an average speed of 1.95 cm/s, rather than the 0.65 cm/s quoted in the text for particles of 1-μ diameter. What is the diameter of these faster moving particles? (Assume them to have the same density.)

5. By what factor are (a) the kinetic energy and (b) the speed, of the particles in Question 3 multiplied if the temperature is raised to 90°C?

6. Molecules of gas at −73°C have a certain average speed and KE. To what temperature must the gas be heated to (a) double the KE of the molecules? (b) double the average speed of the molecules?

(12-3) 7. A molecular-beam apparatus such as described in the text has a drum 10 cm in diameter that revolves 5400 rev/min. When exposed to a molecular beam of vaporized sodium, the densest deposit on the plate is displaced 0.75 cm from the point opposite the entrance slit. What is the average speed of the vaporized sodium atoms?

8. Consider an oven containing sodium at the same temperature as in Question 7, with a similar rotating drum 14 cm in diameter, rotating 4500 rev/min. How far from point A will the deposit of sodium be densest?

(12-4) 9. A krypton molecule is 2.25 times as massive as a molecule of oxygen. If samples of both gases are at the same temperature, how do their rms speeds compare?

10. A molecule of hydrogen has a mass one-half that of a molecule of helium. How do their rms speeds compare, if both gases are at the same temperature?

11. Box A contains N molecules of a gas at temperature T. Identical box B contains some of the same gas that is in A; its pressure is equal to the pressure in A, but its absolute temperature is $1.5T$. (a) How does the average speed of

the molecules in B compare with those in A? (b) How many gas molecules are in box B?

12. Box A contains N molecules of a gas at temperature T. Box B is identical with A and is at the same temperature; but B contains N molecules of a gas whose molecules are twice as massive as those in A. (a) How does the average energy of the molecules in B compare with those in A? (b) How does their average speed compare? (c) How does the pressure in B compare with the pressure in A?

13. The density of chlorine gas is 3.2×10^{-3} g/cm^3 at 1 atm pressure and 0°C. What is the rms speed of chlorine molecules at this temperature?

14. The density of helium gas at 1 atm pressure and 0°C is 1.78×10^{-4} g/cm^3. What is the rms speed of the helium molecules?

15. Consider a single droplet of 1 g of mercury on a level surface. Now, with a knife or card, divide the droplet into two 0.5-g droplets. (a) Does the mercury now have more or less surface area than it did as a single droplet? (b) Does it now have more or fewer mercury atoms on the surface? (c) Has its surface energy been increased or decreased?

16. Explain why two clean mercury droplets, when pushed into contact, will spontaneously run together to form a single droplet. Will the single droplet formed in this way be a little bit warmer, or a little bit cooler, than the separate pair of droplets?

(12-6) **17.** A pan of water at 27°C is set in the breeze on a warm (27°C) dry day. The water cools to 22°C, after which its temperature does not change. Explain why it does not continue to cool until it is all evaporated away.

18. If a pan of ether (or alcohol, or other relatively volatile liquid) at 27°C is set in the same breeze on the same day as the pan of water in Question 17, will it become cooler than the pan of water? Explain.

13

The Special Theory of Relativity

13-1 Light as Waves

In the nineteenth century, after thousands of years of speculation and groundless argument, experiment finally showed that light was composed of waves. Interference effects left no doubt, and other experimental evidence that we will look at in a later chapter showed that these waves were transverse. It was demonstrated that one color of light differed from another only in its wavelength. Toward the end of the century the mathematical theorist James C. Maxwell (who also derived the distribution of molecular velocities shown in Fig. 12-7) showed that light waves were almost certainly electric and magnetic in nature.

These findings of new knowledge were very satisfactory, but, as is often the case, the answer to one question raised another more difficult question that demanded a further answer.

13-2 The Paradox of Ether

In all our discussions of waves in Chapter 8, the medium through which the waves traveled was of great importance. Whether the medium was a taut wire, the surface of a pond, or the air or other fluid, its ability to transmit the energy of waves depended on its physical characteristics. Something analogous to inertia and elasticity were essential, to give it the ability to temporarily store kinetic and potential energy. Light is transmitted to us from distant stars through what appears to be trillions of miles of an almost perfect vacuum—but the complete nothingness of a vacuum could not have the characteristics that seemed to be needed for the transmission of any sort of waves. So its was concluded that space contained not merely nothing, but was an all-pervading ocean of a hypothetical "ether," through which waves of light could propagate.

Since these waves were without doubt transverse, it was necessary to conclude that the ether was some sort of rigid material, in order to provide the resistance to deformation necessary to transmit transverse waves at the speed of light, which is 3×10^5 km/s. But if the entire space of the universe were filled with this rigid material, how could the stars and sun and planets, including the earth, move through space without any apparent resistance?

Escape from this apparent contradiction was sought in the assumption that ether had properties similar to those of plastic materials, such as sealing wax, which behave as solids under the influence of strong forces acting over a short period of time, but flow as liquids when acted upon by weak but persistent forces such as their own weight. It was argued that in the case of propagating light waves, where the force changes its direction 10^{15} times per second, ether could show the properties of an elastic solid, and that on the other hand, it could flow as a perfect fluid around the bodies of the planets of the solar system since in this case the rotation periods are measured in years. But this "explanation" of the unusual properties of the light-carrying medium remained nothing but words, an no consistent theory of its mechanical behavior was ever worked out. The acid test of the ether hypothesis came late in the nineteenth century and resulted in a complete turnabout of our ideas concerning the nature of light waves and electromagnetic fields.

If it were true that light waves propagate through a jelly-like ether that fills universal space, we should be able to measure our motion through space by observing the effect of that motion on the velocity of light. In fact, since the earth moves in its orbit at a speed of 30 km/s, we would experience an "ether wind" blowing in the direction opposite to our motion in the very same way that a speeding motorcyclist experiences a strong "air wind" blowing into his face. Light waves propagating in the direction of that "ether wind" would move faster, being helped by the motion of the medium, while those propagating in the opposite direction would be slowed down. In theory, one might detect this supposed speed of the earth through the ether by measuring the speed of light coming from stars located in opposite directions in the sky. Such an experiment would not be feasible even now, and near the turn of the century it was unthinkable. So scientists were forced (as they often are) to resort to more indirect methods. In 1887 the American physicist

A. A. Michelson, later aided by E. W. Morley, began a series of experiments designed to measure the earth's velocity through the ether (or, what is equivalent and opposite, the "ether wind" moving past the earth) by comparing light waves traveling in two perpendicular directions.

To better understand the Michelson–Morley scheme, let us first apply the same ideas on more familiar ground. Figure 13-1A shows a ferry connecting the ends of a road, opposite each other across a river, a distance d apart. Let us assume that the ferry moves through the water at a speed c. If the water were quite still, with no current in the river, it would be easy to calculate the time needed to cross the river from A to C and back again: $t = 2d/c$. To go d meters downstream and back (with no river current) would require the same $t = 2d/c$. But if the river is flowing, as indicated, with a current speed v, matters are a little more complicated.

Let us first have the ferry go downstream a distance d, and then return. Downstream, its speed will be $c + v$, and the time needed is $t_D =$

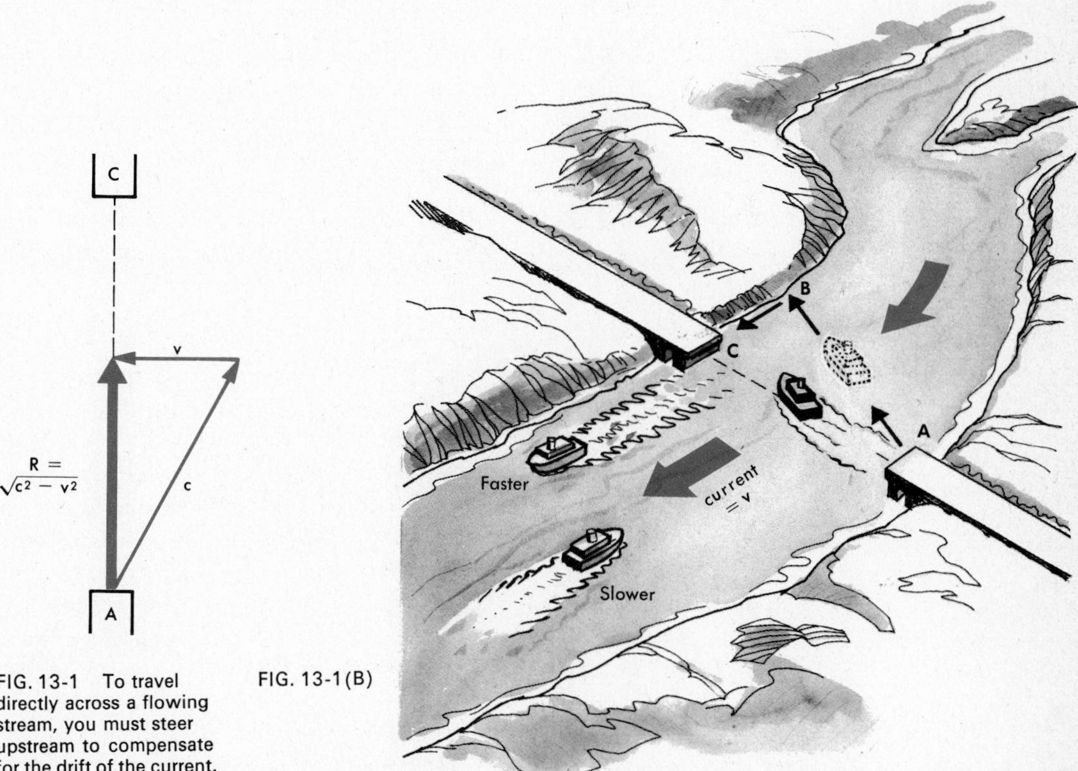

FIG. 13-1 To travel directly across a flowing stream, you must steer upstream to compensate for the drift of the current.

FIG. 13-1(B)

$d/(c + v)$. Returning, the speed of the river current must be subtracted, to give $t_R = d/(c - v)$. The total time is then

$$t_{UD} = t_D + t_R = \frac{d}{c + v} + \frac{d}{c - v} = \frac{d(c - v) + d(c + v)}{(c + v) \times (c - v)}$$

$$= \frac{dc - dv + dc + dv}{(c + v)(c - v)} = \frac{2dc}{c^2 - v^2}$$

$$= \frac{2dc}{c^2(1 - v^2/c^2)} = \frac{2d}{c} \times \frac{1}{1 - v^2/c^2}.$$

Thus, for the trip downstream and back (or upstream and back—the results would be the same) the still-water time $2d/c$ must be multiplied by the fraction shown.

What of the cross-stream trip from A to C and back to A? If the ferry heads from A directly toward C, the current will carry him downstream. To reach C from A he must steer toward a point *upstream* from C; the current will then sweep him downstream enough to put him directly on target. Figure 13-1B shows the velocity vectors involved. To the vector C, his velocity through the water, we must add v, the velocity of the current. Their resultant R will give his actual velocity, $\sqrt{c^2 - v^2}$, directed exactly toward C. We can now find the time needed for this cross-stream round trip:

$$t_{XX} = \frac{2d}{\sqrt{c^2 - v^2}} = \frac{2d}{\sqrt{c^2(1 - v^2/c^2)}}$$

$$= \frac{2d}{c} \times \frac{1}{\sqrt{1 - v^2/c^2}}.$$

This is the same as the expression for the up-and-down-stream time, except that the denominator of the multiplying fraction now has a square root sign on it.

For a numerical example, let us choose a river 150 m wide; the speed of the ferry through the water is 6 m/s, and the river current is 3 m/s.

$$t \text{ (no current)} = \frac{2d}{c} = \frac{2 \times 150}{6} = 50 \text{ s}.$$

$$t_{UD} \text{ (up-and-down)} = 50 \times \frac{1}{1 - 9/36} = \frac{50}{0.75} = 66.7 \text{ s}.$$

$$t_{XX} \text{ (cross-stream)} = \frac{50}{\sqrt{0.75}} = \frac{50}{0.866} = 57.7 \text{ s}.$$

In principle (and with the use of more algebra than it would be convenient to include here), if we could measure the difference between the up-and-down-stream time, and the cross-stream time, we could use this difference to calculate v, the speed of the river current.

This is exactly the scheme that Michelson proposed in order to measure the velocity of the earth through the supposed ether, which was not accessible to more direct methods. We need only to use as c the speed of light through the ether, measured to be very nearly 3×10^8 m/s; v will represent the velocity of the earth through the ether—or, rather, the equal and opposite velocity with which the ether streams past the earth. Figure 13-2 is a schematic sketch of the apparatus used by Michelson and Morley. A light beam from a source S falls on the glass plate P, which is covered with a thin semitransparent layer of silver which reflects half of the beam in the direction of the mirror M_1, and allows half to pass through and continue on to mirror M_2. After being reflected by the mirrors the beams return to plate P. Half of the first beam (from M_1) penetrates the thin silver coating and continues on to the observer's telescope T; half of the second beam (from M_2) is reflected into T by the silver coating. If Fig. 13-2 were correct in representing the "ether wind," the M_1 beam represents the cross-stream trip, and the M_2 beam, the

FIG. 13-2 Michelson's apparatus for trying to measure the speed of the earth's motion through the ether. Instruments were mounted on a heavy stone plate floating on mercury to avoid vibration and warping when the apparatus was rotated.

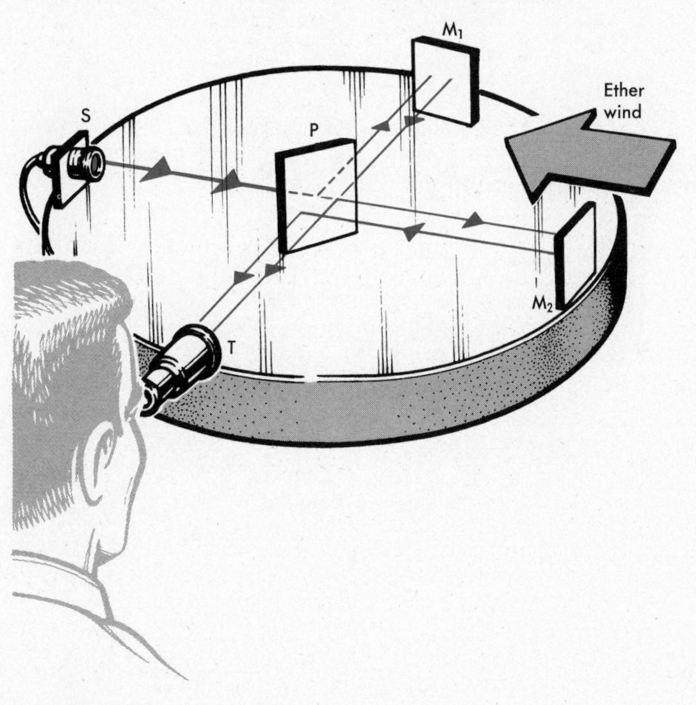

up-and-down-stream trip. Using light instead of a ferryboat, how much time difference could we expect?

Michelson's and Morley's apparatus was actually a little more complicated than Fig. 13-2 indicates. Auxiliary mirrors reflected the beam back and forth several times between P and M_1, and between P and M_2, so that the total travel was equivalent to about 10 m downstream and 10 m back, and an equal 20 m total across the stream and back. For a stationary earth with no "ether wind," the time for either trip would be

$$t = \frac{20 \text{ m}}{3 \times 10^8 \text{ m/s}} = 6.7 \times 10^{-8} \text{ s}.$$

The earth's speed in its orbit is very nearly 30 km/s; the speed of light is 3×10^5 km/s. From these values for v and c, we can figure the value of the fraction $1/(1 - v^2/c^2)$:

$$\frac{1}{1 - 9 \times 10^2/9 \times 10^{10}} = \frac{1}{1 - 10^{-8}}.$$

This fraction is impossible to evaluate on a slide rule, or even on most calculators, and the division would be a long and dreary long-hand job. So let us make use of a very convenient approximation:

If α is much smaller than 1,

$$\frac{1}{1 - \alpha} \approx 1 + \alpha, \quad \text{and} \quad \frac{1}{1 + \alpha} \approx 1 - \alpha.$$

The smaller α is, the more accurate this approximation. In our fraction, $\alpha = 10^{-8}$, and the result will be very accurate indeed.

$$\frac{1}{1 - 10^{-8}} = 1 + 10^{-8}.$$

Thus we have that the *increase* in time required for the beam to go downstream and back in the ether wind is

$$\Delta t_{\text{UD}} = 6.7 \times 10^{-8} \times 10^{-8} = 6.7 \times 10^{-16} \text{ s}.$$

For the cross-stream beam, the multiplying factor is $1/\sqrt{1 - v^2/c^2} = 1/\sqrt{1 - 10^{-8}}$. How do we find the square root of 0.99999999? Here again we can use a very accurate approximation:

If α is much smaller than 1,

$$\sqrt{1 - \alpha} \approx 1 - \tfrac{1}{2}\alpha, \quad \text{and} \quad \sqrt{1 + \alpha} \approx 1 + \tfrac{1}{2}\alpha.$$

Using this approximation, we get

$$\sqrt{1 - 10^{-8}} = 1 - \tfrac{1}{2} \times 10^{-8} = 1 - 5 \times 10^{-9}.$$

And as before,

$$\frac{1}{1 - 5 \times 10^{-9}} = 1 + 5 \times 10^{-9}.$$

So the *increase* in time for the cross-stream beam is

$$\Delta t_{XX} = 6.7 \times 10^{-8} \times 5 \times 10^{-9} = 3.35 \times 10^{-16} \text{ s}.$$

The *time difference* between the up-and-downstream and the cross-stream trips is thus $6.7 \times 10^{-16} - 3.35 \times 10^{-16} = 3.35 \times 10^{-16}$ s. Michelson carefully rotated his apparatus through 90°, so that the paths of the two beams were reversed. This would double the time difference, to give him a difference of 6.7×10^{-16} s between the two positions of his rotated apparatus.

This is a *very* short time difference to try to measure. The best modern oscilloscopes do well to directly measure time intervals a million times larger; and in Michelson's day such instruments had not yet even been imagined. So again Michelson had to make use of an indirect method.

Figure 8-14 on p. 148 showed light and dark bands on a screen, resulting from their differing distances from a pair of slits. In the Michelson apparatus, similar differing distances result from the different cross-stream and up-and-downstream paths. The observer sees an interference pattern that he could expect to shift, or change position, as the stone table rotated.

Our calculated time difference was 6.7×10^{-16} seconds; at the light-speed of $c = 3 \times 10^{10}$ cm/s, this corresponds to a path difference of $6.7 \times 10^{-16} \times 3 \times 10^{10} = 2 \times 10^{-5}$ cm. This is about 0.4 of the wavelength of visible light, and should cause an easily observable and readily measurable shift in the interference pattern.

However, to Michelson's great surprise, no change in the interference pattern could be detected as he rotated his apparatus. This could of course be explained, if by a strange coincidence the earth happened at that time to be stationary with respect to the ether. Figure 13-3A shows how this might occur, if the whole solar system were moving through the ether in just the right direction, at a speed equal to the earth's orbital speed. If this were the case, then six months later (Fig. 13-3B) the earth should be moving through the ether at 60 km/s, to produce a shift twice as large as that calculated above. But six months later, Michelson could still detect no change in the interference pattern as his apparatus was rotated. Over the years, hundreds of other experiments, in dozens of variations, have been performed, with increasingly precise instruments.

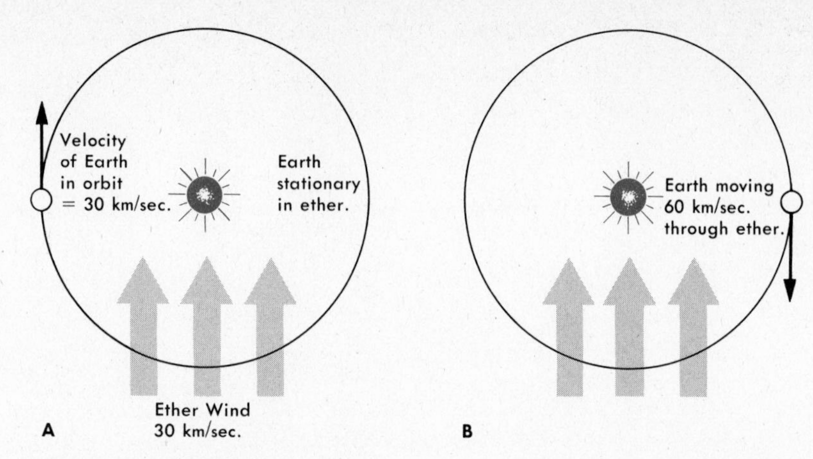

FIG. 13-3 The expected velocity of the earth with reference to the hypothetical ether would change by 60 km/s in six months.

The net result—nothing! It was as though the earth always had zero speed through the ether—or that the speed of light relative to the observer was always the same 3×10^{10} cm/s, no matter what his velocity.

It was suggested that the earth might drag along with it a surrounding volume of ether—but it was shown that this would cause changes in the apparent directions of stars, which were not observed. The Irish physicist G. F. Fitzgerald, and the Dutch physicist H. A. Lorentz both independently suggested that all material bodies (such as metersticks, or Michelson's apparatus) contracted in the direction of their motion through the ether by a factor of $\sqrt{1 - v^2/c^2}$. This would of course always shorten whichever leg of the apparatus was in the up-and-downstream position, just enough to make the time difference zero and explain the absence of shifts in the interference pattern. The circular stone float on which the apparatus was mounted would become elliptical, with the short axis of the ellipse in the direction of the motion through the ether. This distortion would not be measurable, because a meterstick or any other measuring device would be equally shortened.

Lorentz, in 1904, tried to explain this hypothetical contraction in terms of the electric and magnetic forces between atoms moving through the ether. This, and other similar work, however, never led to a satisfactory explanation.

13-3 Einstein's Postulates

In 1905, Albert Einstein, then 26 years old (Fig. 13-4), had been unable to find a job teaching in any of the German schools, and was employed as a patent clerk in Zurich. At that time, Einstein developed a very fundamental theory from the foundations of two postulates:

1. The basic laws of physics are the same for all unaccelerated observers.

FIG. 13-4 Albert Einstein at the age of twenty-six, when his first revolutionary paper on Special Relativity was published.

This includes the laws of electricity, magnetism, optics, etc., as well as those of mechanics. This postulate means that there is no meaning in the idea of an *absolute* velocity, such as that with reference to an imaginary "ether," that Michelson had tried to measure. Only the *relative* motion of one body with respect to another has any significance. As Einstein stated it, the introduction of an ether is superfluous. Since the theory was restricted to unaccelerated observers, it has been called the "restricted" or "special" theory of relativity.

2. The speed of light (in empty space) is the same for all observers, no matter what their velocity or what the velocity of the source of the light.

Thus, instead of trying to explain away the negative results of the Michelson–Morley experiments, this postulate accepts them at their apparent face value.

This second postulate is probably the more difficult one to accept. If our subconscious minds still picture some sort of an ether or other "absolute" space through which light waves are transmitted, and we are moving through it with a velocity v toward a light source, it seems incredible that we should not have $v + c$ for the speed of light. But no—it appears that when we add these velocities, $v + c = c$. On the other hand, if we reject the ether and picture light as some sort of a wave–like projectile emitted from a source approaching us (or we, approaching the source) with a velocity v, it is again incredible that we should not measure the speed of light to be $v + c$. Again, it seems that $v + c = c$.

But this, apparently, is the way the world is, and we must make the best of it. Velocity involves a combination of distance and time; and for any observer, velocity (as *he* measures it) must be equal to distance (as *he* measures it) divided by time (as *he* measures it). If this is true, and if the speed of light is the same c for all observers, then a number of apparently strange results necessarily follow. They can all be demonstrated with only straightforward simple algebra. Because a great deal of algebra is required, however, we will not try to lead the reader through it.

13-4 Relativistic Mechanics

Imagine an observer O who considers himself to be stationary, as he has every right to do. He sees, passing by in the air, a spaceship-laboratory traveling with a velocity v relative to the observer. Through its large windows, O sees an experimenter carrying out some simple laboratory procedures in mechanics. He is equipped with standard masses plainly marked "1 kg," "100 g," etc.; metersticks; and a fine clock on the wall. The name of the manufacturer is visible on all this equipment, and O has no reason to believe they are not proper standards, as marked.

Observer O, however, with his own ingenious equipment (and making proper allowance for the speed of light as it moves to him from the traveler) discovers the following:

O's measurements show that all the traveler's metersticks are too short: instead of 1 meter, they are only $\sqrt{1 - (v^2/c^2)}$ meter long.

O finds the traveler's clock is running slow: a period of 1 minute by the traveler's clock is actually $1/\sqrt{1 - (v^2/c^2)}$ minute by the stationary clock in O's laboratory.

After watching some experiments, O comes to the further conclusion that the traveling kilograms are more massive than they are labeled: their actual masses are $1/\sqrt{1 - (v^2/c^2)}$ kg.

In all fairness, it should be noted that the spaceship experimenter has been closely observing what goes on in O's lab, and comes to exactly the same conclusions: that O's metersticks are too short by the same factor, that O's clock is running slow, and that O's kilograms are really more massive than they are marked to be.

That each of these experimenters came to exactly the same conclusion about the other is just what we should have expected; the relative velocity between the two is what makes the difference, and this is the same for both. (Remember that O, presumably on the earth, who chose to consider himself stationary, was actually spinning several hundred miles per hour with the daily rotation of the earth, about 60,000 mi/hr in his orbiting around the sun, and at about 600,000 mi/hr in the solar system's revolution about the center of our Milky Way Galaxy!)

This little fable is not one that could ever be carried out in actuality. But many pieces of it have been observed to be true, and there is probably nothing in modern science more firmly established than Einstein's Special Theory of Relativity. In later sections on nuclear physics, we shall see much of the evidence.

What is true for metersticks and clocks and mass standards is of course true for all lengths and time intervals and masses aboard any speeding traveler. At ordinary everyday speeds these relativistic changes are too small to be measurable. As an example, let us calculate a few things about an artificial satellite circling the earth at a speed of 11 km/s $= 1.1 \times 10^6$ cm/s. Its rest mass m_0 (i.e., its mass as measured when stationary with respect to the measurer) is, say, exactly 1000 kg, and its rest length l_0 is exactly 5 m. Relativity shows us that a stationary earthly observer would measure its mass and length as

$$m = \frac{m_0}{\sqrt{1 - (v^2/c^2)}}$$

$$l = l_0 \times \sqrt{1 - (v^2/c^2)}.$$

Thus, while the satellite is circling, we would measure its mass (by any means we can imagine) to be

$$m = \frac{10^3}{\sqrt{1 - (1.1 \times 10^6/3 \times 10^{10})^2}} = \frac{10^3}{\sqrt{1 - 1.4 \times 10^{-9}}}$$

$$= \frac{10^3}{1 - 0.7 \times 10^{-9}} = 10^3 \times (1 + 0.7 \times 10^{-9}) = 1000 + 0.7 \times 10^{-6}\,\text{kg}.$$

The measured *increase* in mass is thus only 0.7×10^{-6} kg, or 0.7 mg, an amount certainly not practically measurable for a 1-ton object.

The length of the satellite, as we would measure it, will prove out in the same way

$$l = l_0\sqrt{1 - (v^2/c^2)}$$

$$= 5 \times (1 - 7 \times 10^{-10})$$

$$= 5 - 3.5 \times 10^{-9}\ \text{m}.$$

The shortening would thus turn out to be only 3.5×10^{-9} m $= 3.5 \times 10^{-7}$ cm $= 35$ Å, or about 1 percent of the wavelength of ultraviolet light!

If there were an accurate clock in this satellite, accurately set before it took off, how far off would this clock be when the satellite returned safely to earth circling for, say, 10 days? From our point of view here on earth, the traveling clock will be running slow by a factor of $\sqrt{1 - (v^2/c^2)} = 1 - 7 \times 10^{-10}$. In 10 days there are $10 \times 24 \times 60 \times 60 = 8.64 \times 10^5$ s. The time interval recorded by the satellite clock would then be

$$t = 8.64 \times 10^5(1 - 7 \times 10^{-10})$$
$$= 8.64 \times 10^5 - 6 \times 10^{-4} \text{ s.}$$

That is, it would be slow (according to our stay-at-home standards) by only 6×10^{-4} s.

Figure 13-5 pictures an addition of velocities that our common sense tells us must be right. The Jeep is moving forward at velocity v; the hunter fires forward, the muzzle velocity of the bullet being V. We would have no hesitation in saying the total speed of the bullet (measured with reference to the ground) is $V + v$. When he fires from the rear of the Jeep, the ground velocity of the bullet would similarly be expected to be $V - v$.

But this is not quite the case. If we work from the apparent changes in length and time that we measure in a moving body, we find that the

FIG. 13-5 The velocity of a bullet is affected by the motion of its source *almost* as we would expect to to be.

speed of the forward-fired bullet, measured from the ground, is not $V + v$, but

$$\frac{V + v}{1 + Vv/c^2}$$

where c is as usual the speed of light, measured by *any* observer. Ordinarily Vv/c^2 is so small that the difference would not be measurable. As an example, consider an orbiting laboratory circling the earth at a speed of 7.5×10^5 cm/s (over 16,000 mi/hr). The laboratory fires an experimental projectile forward with a speed (relative to the laboratory) of 7.5×10^5 cm/s. We then have for the speed of the projectile measured from the earth:

$$\frac{7.5 \times 10^5 + 7.5 \times 10^5}{1 + 7.5 \times 10^5 \times 7.5 \times 10^5/(3 \times 10^{10})^2} = \frac{15 \times 10^5}{1 + 6.25 \times 10^{-10}}$$

$$= 15 \times 10^5(1 - 6.25 \times 10^{-10})$$

$$= 15 \times 10^5 - 9.4 \times 10^{-4} \text{ cm/s.}$$

Our intuitive, commonsense answer of 15×10^5 cm/s is too large, but we are off by less than a hundredth of a millimeter per second!

Nuclear and atomic physicists, and cosmologists, however, regularly deal with velocities approaching c, and for them the relativistic outlook is necessary. If, instead of the orbiting laboratory, we take a spaceship traveling at $0.9c$ that fires its projectile forward at $0.8c$, we find for the velocity of the projectile, *not* 1.7 c, but instead:

$$\frac{1.7c}{1 + 0.9c \times 0.8c/c^2} = \frac{1.7c}{1 + 0.72} = 0.988c.$$

No matter how much we add velocities, we cannot exceed c, the speed of light, which thus appears to be an automatically enforced speed limit. In an earlier section we put down $v + c = c$, which at the time seemed a little unbelievable. Let us try it:

$$\frac{c + v}{1 + cv/c^2} = \frac{c + v}{1 + v/c}.$$

If we now multiply numerator and denominator by c, we have

$$\frac{(c + v) \times c}{c + v} = c,$$

again enforcing the speed limit.

Perhaps the most important consequence of the Special Theory of Relativity is one that has not yet been mentioned. Einstein's famous equation

$$E = mc^2$$

has been widely publicized in many ways, and there are few people who cannot recall at least having seen it somewhere. Its application has helped shape the modern world and will no doubt have an equally powerful effect on the future.

Our hypothetical observer O, in calculating from the data observed in the passing spaceship, will be unable to reconcile his observations with the laws of conservation of energy and conservation of momentum unless he adds another item to his list. This item is Einstein's $E = mc^2$, which basically says merely that energy has mass. Although the derivation in Einstein's original work referred only to kinetic energy, Einstein assumed that the relationship must also apply to energy of all kinds. Later experiments have proved him to be right. We now know that not only does energy have mass, but that energy can be converted into mass and mass can be converted into energy.

In the equation, c^2 (the square of the speed of light) is a very large number, which means that a small amount of mass corresponds to a large amount of energy. In the CGS system ($c = 3 \times 10^{10}$ cm/s), one gram of mass is interchangeable with 9×10^{20} ergs of energy. In the MKS system, one kilogram of mass is interconvertible with 9×10^{16} joules of energy.

Because of the large value of the proportionality constant c^2, the mass of ordinary amounts of energy is usually very small. If we burn a kilogram of propane (i.e., combine the propane with oxygen to form water vapor and carbon dioxide gas), the heat produced and conducted or radiated away is about 5×10^7 J. This energy has a mass we can calculate from $E = mc^2$; using $c = 3 \times 10^8$ m/s,

$$5 \times 10^7 = m \times (3 \times 10^8)^2 = 9 \times 10^{16} m$$

$$m = \frac{5 \times 10^7}{9 \times 10^{16}} = 5.6 \times 10^{-10}\ \text{kg}$$

$$= 0.6\ \mu\text{g}.$$

Thus, the water and carbon dioxide formed have a combined mass that would be 0.6 μg less than that of the propane and oxygen that combined to produce them. This is so unmeasurably small a difference that the chemists are quite correct, from a practical standpoint, in emphasizing the conservation of mass in their reactions.

On the other hand, when a certain amount of mass is transformed into energy, the situation is reversed (we now *multiply* mass by c^2), and we obtain a great deal of energy from a very small mass change. Thus, for example, the uranium core of the first atomic bomb lost only about 1 g of its original mass in the course of its transformation into fission products. Being turned into energy, this gram of mass produced an effect equivalent to the explosion of 20,000 tons (20 kilotons) of TNT.

13-5 Space–Time Transformation

Einstein's new laws clearly contradict the classical (commonsense) ideas concerning space and time, so that in accepting these new laws as experimental fact, we are forced to introduce radical changes in our old notions. In his *Principia*, the great Newton wrote:

I. Absolute, true and mathematical time, of itself, and from its own nature, flows equably without relation to anything external.
II. Absolute space, in its own nature, without relation to anything external, remains always similar and immovable.

According to Einstein's views, however, space and time are more intimately connected with one another than it was supposed before, and, within certain limits, the notion of space may be interchanged with the notion of time, and vice versa. To make this statement clearer, let us consider a railroad passenger having his meal in the dining car. The waiter serving him will know that the passenger ate his soup, steak, and dessert in the same place, i.e., at the same table in the car. But, from the point of view of a person on the ground, the same passenger consumed the three courses at points along the track separated by many miles (Fig. 13-6A, B, and C). Thus we can make the following trivial statement: ***Events occurring in the same place but at different times in a moving system will be considered by a ground observer as occurring at different places.***

Now, following Einstein's idea concerning the reciprocity of space and time, let us replace in the above statement the word "place" by the word "time" and vice versa. The statement will now read: ***Events occurring at the same time but in different places in a moving system will be considered by a ground observer as occurring at different times.***

This statement is far from being trivial and means that if, for example, two passengers at the far ends of the diner had their after-dinner cigars lighted simultaneously from the point of view of the dining-car steward, the person standing on the ground will insist that the two cigars were lighted at different times (Fig. 13-6D and E). Since, according to the principle of relativity, neither of the two reference systems should be preferred to the other (the train moves relative to the ground or the ground moves relative to the train), we do not have any reason to take the steward's impression as being true and the ground observer's impression as being wrong, or vice versa.

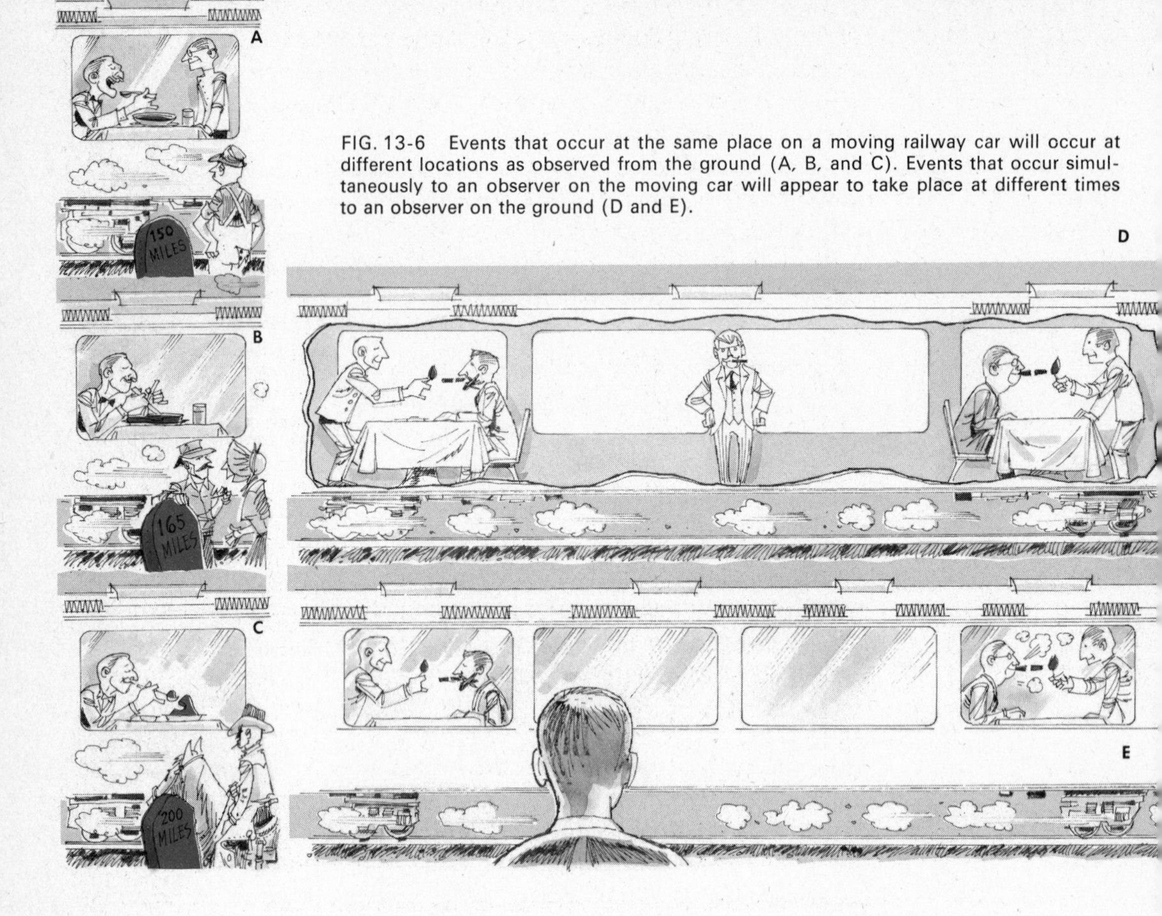

FIG. 13-6 Events that occur at the same place on a moving railway car will occur at different locations as observed from the ground (A, B, and C). Events that occur simultaneously to an observer on the moving car will appear to take place at different times to an observer on the ground (D and E).

Why, then, do we consider the transformation of the time interval (between the soup and the dessert) into the space interval (the distance along the track) as quite natural and the transformation of the space interval (the distance between the two passengers having their cigars lit) into the time interval (between these two events as observed from the track) as paradoxical and very unusual? The reason lies in the fact that in our everyday life we are accustomed to velocities that lie in the lowest brackets of all the physically possible velocities extending from zero to the velocity of light. A race horse can hardly do better than about one millionth of a percent of this upper limit of all possible velocities, and a modern supersonic jet plane makes, at best, 0.0003 percent of it. In comparing space and time intervals, i.e., distances and durations, it is rational to choose the units in which the limiting velocity of light is taken to be 1. Thus, if we choose a "year" as the unit of duration, the corresponding unit of length will be a light-year, or 10,000,000,000,000 km, and if we choose a "kilometer" as the unit of length, the unit of time will be 0.000003 s, which is the time interval necessary for light to cover the distance of 1 km. We notice that whenever we choose one unit in a "reasonable" way (a year or a kilometer), the other unit comes out either too large (a light-year) or too short (3 microseconds) from the point of view of our everyday experience. So, in the case of the passenger eating his dinner on the train, a half-hour interval between the soup and the dessert could result in 200,000,000 mi of distance along the track (time $\times$ c) if the train were moving at a speed close to that of light, and we are not surprised that the actual difference is only 20 or 30 mi. On the other hand, the distance of, let us say, 30 m between two passengers lighting their cigars at opposite ends of the railroad car translates into a time interval of only one hundred-millionth of a second (distance $\div$ c), and there is no wonder that this not apparent to our senses.

The transformation of time intervals into space intervals and vice versa can be given a simple geometrical interpretation, as was first done by the German mathematician H. Minkowski, one of the early followers of Einstein's revolutionary ideas. Minkowski proposed that time or duration be considered as the fourth dimension supplementing the three spatial dimensions and that the transformation from one system of reference to another be considered as a rotation of coordinate systems in this four-dimensional space. Figure 13-7 uses one space axis (the direction of the train's motion) and the time axis. It does not follow Minkowski's more complex scheme, and does not lead to correct answers if

FIG. 13-7 The difference in space and time separations between events, as interpreted by observers moving relative to one another.

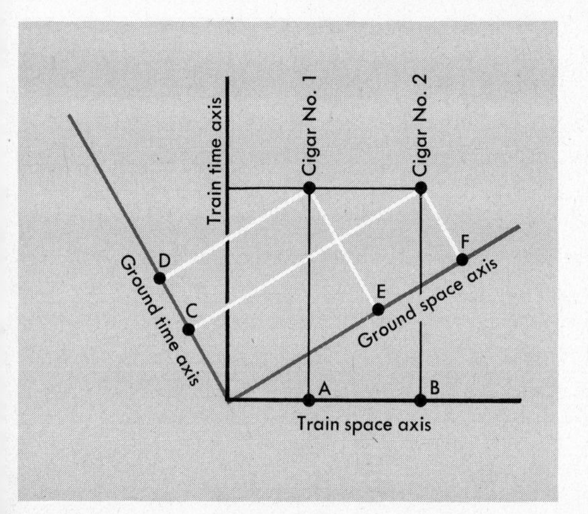

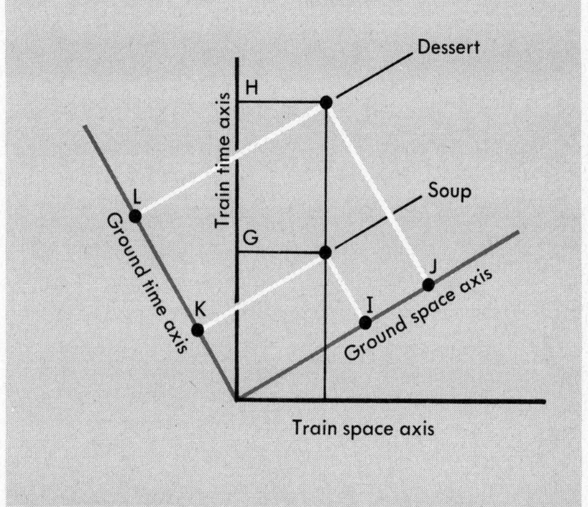

followed in detail, but it does serve to present the basic idea of the interrelationship of space and time. From the point of view of the observer on the train, the time interval between the two cigar lightings is zero, and they are separated in space by the length of the dining car *AB*. To the observer on the ground, there will be a time interval *CD* between the two lightings, and they will also appear to be separated by a shorter distance *EF*. To the train observer, both the soup and the dessert will be eaten in the same place and will be separated by the time interval *GH*. The observer watching from the ground, however, will not agree. He will, of course, say they were eaten in quite different places (*IJ*) and will contend that the time interval between them (*KL*) is shorter than the train observer says.

We see from these diagrams that the appearance of a time interval between two events that were simultaneous in the first system of reference is connected with a shortening of the apparent distance between them as seen from the second system of reference, and vice versa; the appearance of a space interval between two events that were occurring in the same place in the first system shortens the apparent time interval between them as observed from the second. The first fact gives the correct interpretation of the apparent Fitzgerald's contraction of the moving bodies, and the second makes the time in a moving system flow slower from the point of view of the second system. Of course, both effects are relative, and each of the two observers moving with respect to one another will measure the other one to be somewhat flattened in the direction of his motion and will consider the other fellow's watch to be slow.

13-6 Mr. Tompkins

Because these effects become appreciable only when the velocities involved are close to that of light, we do not notice them at all in our everyday snail's-pace life. But we can imagine a fictitious situation that would arise if the velocity of light were much smaller and closer to our everyday experience. This is what happened to Mr. Cyril George Henry Tompkins, who after listening to a popular lecture on the theory of relativity, was transferred, in his dream, to a fantastic city in which the velocity of light was only 20 miles per hour and served as the natural speed limit for its inhabitants.*

> At first, when he found himself on the street of this relativistic city, nothing unusual seemed to be happening around him; even a policeman standing on the opposite corner looked as policemen usually do. The hands of the big clock on the tower down the street were pointing almost

* From G. Gamow, *Mr. Tompkins in Wonderland* (Cambridge: Cambridge University Press, 1939), by permission of the publishers. Certain slight modifications have been made in order to place Mr. Tompkins' dream city in America rather than in England.

FIG. 13-8 The bicyclist seemed to be unbelievably flattened.

to noon and the streets were nearly empty. A single cyclist was coming slowly down the street and, as he approached, Mr. Tompkins' eyes opened wide with astonishment. For the bicycle and the young man on it were unbelievably flattened in the direction of the motion, as if seen through a cylindrical lens (Fig. 13-8). The clock on the tower struck twelve, and the cyclist, evidently in a hurry, stepped harder on the pedals. Mr. Tompkins did not notice that he gained much in speed, but, as the result of his effort, he flattened still more and went down the street looking exactly like a picture cut out of cardboard. Mr. Tompkins felt very proud because he could understand what was happening to the cyclist—it was simply the contraction of moving bodies. "Evidently nature's speed limit is quite low here," thought Mr. Tompkins, "that is why the policeman on the corner looks so lazy; he need not watch for speeders." In fact, a taxi moving along the street at the moment and making all the noise in the world could not do much better than the cyclist, and was just crawling along. Mr. Tompkins decided to overtake the cyclist, who looked a good sort of fellow, and ask him all about it. Making sure that the policeman was

FIG. 13-9 As he speeded up, the city blocks became quite short.

looking the other way, he borrowed somebody's bicycle standing near the curb and sped down the street. He expected that he would be immediately flattened and was very happy about it as his increasing figure had lately caused him some anxiety. To his great surprise, however, nothing happened to him or to his cycle. On the other hand, the picture around him completely changed. The streets grew shorter, the windows of the shops began to look like narrow slits, and the policeman on the corner became the thinnest man he had ever seen (Fig. 13-9).

"By Jove!" exclaimed Mr. Tompkins excitedly, "I see the trick now. This is where the word *relativity* comes in. Everything that moves relative to me gets shorter for me, whoever works the pedals!" He was a good cyclist and was doing his best to overtake the young man. But he found that it was not at all easy to get up speed on this bicycle. Although he was working on the pedals as hard as he possibly could, the increase in speed was almost negligible. His legs already began to ache, but still he could not manage to pass a lamp post on the corner much faster than when he started. It looked as if all his efforts to move faster were leading to no result. He understood now very well why the cyclist and the cab he had just met could not do any better, and he remembered the words in the book on relativity which he had read. It was stated that it is impossible to surpass the limiting velocity of light. He noticed, however, that the city blocks became still shorter and the cyclist riding ahead of him did not now look so far away. He overtook the cyclist at the second turning and, when they had been riding side by side for a moment, was surprised to see that he was quite a normal, sporting-looking young man. "Oh, that must be because we do not move relative to each other," he concluded; and he addressed the young man.

"Excuse me, sir!" he said, "Don't you find it inconvenient to live in a city with such a slow speed limit?"

"Speed limit?" returned the other in surprise, "we don't have any speed limit here. I can get anywhere as fast as I wish, or at least I could if I had a motorcycle instead of this good-for-nothing old bike!"

"But you were moving very slowly when you passed me a moment ago," said Mr. Tompkins. "I noticed you particularly."

"Oh you did, did you?" said the young man, evidently offended. "I suppose you haven't noticed that since you first addressed me we have passed five blocks. Isn't that fast enough for you?"

"But the blocks became so short," argued Mr. Tompkins.

"What difference does it make, anyway, whether we move faster or whether the blocks become shorter? I have to go ten blocks to get to the post office, and if I step harder on the pedals, the blocks become shorter and I get there quicker. In fact, here we are," said the young man, getting off his bike.

Mr. Tompkins looked at the post office clock, which showed half-past twelve. "Well!" he remarked triumphantly, "it took you half an hour to go this ten blocks, anyhow—when I saw you first it was exactly noon!"

"And did you *notice* this half hour?" asked his companion. Mr. Tompkins had to agree that it had really seemed to him only a few minutes. Moreover, looking at his wrist watch he saw that it was showing only five minutes past twelve. "Oh!" he said, "is this post office clock fast?" "Of course it is, or your watch is too slow, just because you have been going too fast. What's the matter with you anyway? Did you fall down from the moon?" and the young man went into the post office.

Continuing his journey down the street he finally saw the railway station. A gentleman obviously in his forties got out of the train and began to move toward the exit. He was met by a very old lady, who, to Mr. Tompkins' great surprise, addressed him as "dear Grandfather." This was too much for Mr. Tompkins. Under the excuse of helping with the luggage, he started a conversation.

"Excuse me, if I am intruding into your family affairs," said he, "but are you really the grandfather of this nice old lady? You see, I am a stranger here, and I never . . ." "Oh, I see," said the gentleman, smiling through his moustache. "I suppose you are taking me for the Wandering Jew or something. But the thing is really quite simple. My business requires me to travel quite a lot, and, as I spend most of my life in the train, I naturally grow old much more slowly than my relatives living in the city. I am so glad that I came back in time to see my dear little granddaughter still alive! But excuse me, please, I have to help her into the taxi," and he hurried away, leaving Mr. Tompkins alone again with his problems.

13-7 Time and the Space Traveler

In the Tompkins story, the relation between the young grandfather and his old granddaughter is, of course, grossly exaggerated, but the fact is that, according to Einstein's theory, such a difference in aging is really expected to occur in the case of relative motion. Thus, if sometime in the future a spaceship were to take off from the surface of the earth to visit other planets of the solar system, or maybe the planetary systems of other stars of the Milky Way, the pilot and the passengers would be relatively younger upon their return than the people of the same original age who had stayed on the earth. This difference might become quite conspicuous if the spaceship were accelerated to velocities close to the velocity of light. It is well-known, however, that human organisms cannot stand strong accelerations and that pilots suffer blackout when their plane makes several g's ($g = 980$ cm/s^2 being the normal acceleration of gravity on the surface of the earth). If we assume that the spaceship is traveling with the comfortable acceleration of 1 g, we find that it takes about a year to approach the velocity of light, when relativistic changes of time rate begin to play any role. (Since a year contains about 3×10^7 s, the acceleration of 980 cm/s^2 will raise the velocity to 98 percent of the speed of light.) For accelerated space trips that last well beyond 1 yr, such as a trip to the nearby star Sirius, which is 8 light-years away, relativistic time changes begin to be quite appreciable. From the point of view of the inhabitants of the earth, the crew of a ship making a round trip to this star comes back just 16 yr after the departure; for the crew itself, these 16 yr will seem only as 9 yr. If, instead of to a nearby star, the spaceship travels with constant acceleration to the center of our own stellar system of the Milky Way and back, it will return 40,000 yr later by the earth's calendar; whereas by its own time reckoning, the trip will take only 30 yr.

Earlier in this chapter, however, it was said that *each* observer would think the clock of the other one was slow. Why, then, could we not equally well argue that when he landed, the traveler would find that the earth clocks were slow and the error would be the other way? To help resolve this argument, we can note that the two are *not* exactly equivalent. Earth has been virtually unaccelerated all the time; the satellite has undergone a takeoff acceleration, a turnaround acceleration, and a landing deceleration. There have been arguments over this question for many years, but now almost all relativists agree that the satellite clock *would* come back reading slow relative to the unaccelerated earth clock. The traveler to Sirius and back *would* return younger than his twin brother who stayed at home.

13-8 The Appearance of Moving Objects

It is not possible for any of us to actually grasp or have a clear mental picture of the speed of light, 3×10^{10} cm/s or 186,000 mi/s. The Tompkins story has the great advantage of reducing the speed of light to 20 mi/hr, which is something our experience qualifies us to deal with.

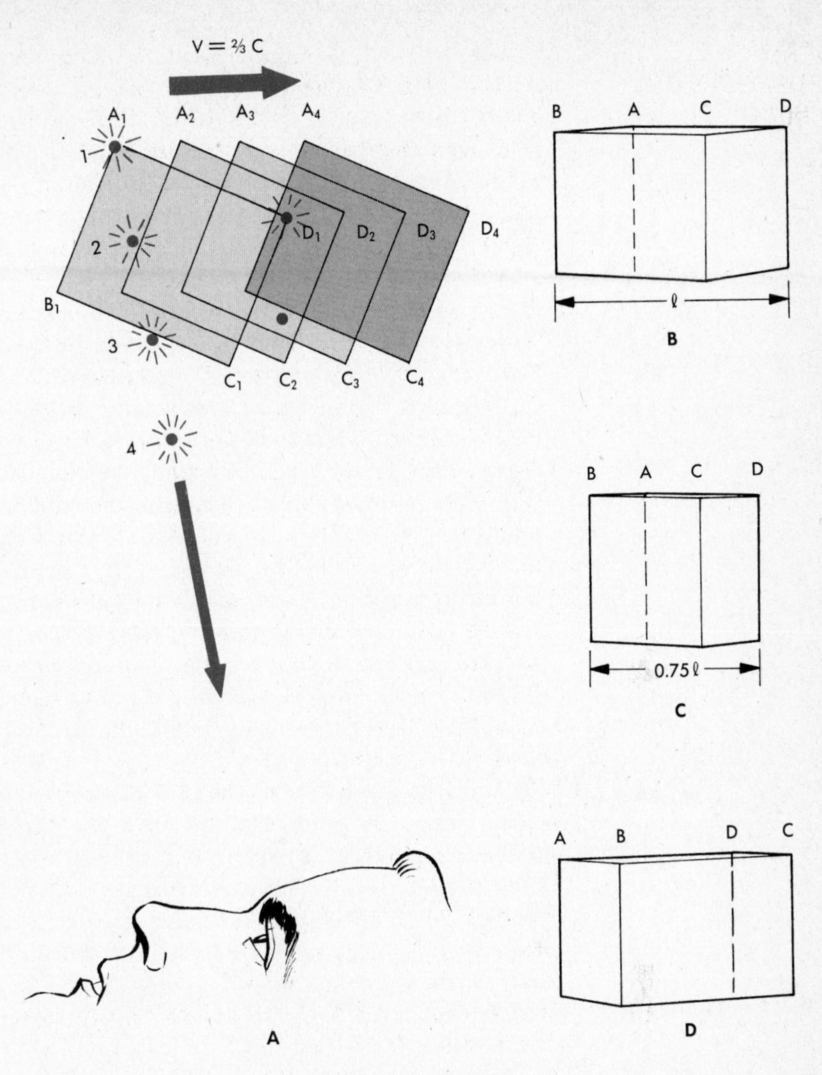

FIG. 13-10 Appearance of a moving block. The observer will not see the actual measured shortening; the block will instead appear to be rotated.

But the story does also have a major defect. It was written in 1939, and it was not until 20 years later than any relativist had considered what we would actually *see* if some familiar object were to whiz by at a very high speed. In 1959, Prof. J. L. Terrell published in the *Physical Review* an analysis of how such fast-moving objects would appear.*

Figure 13-10A shows the top view of a rectangular block *ABCD*

* Also see V. L. Weisskopf, "The Visual Appearance of Rapidly Moving Objects," *Physics Today*, **13** (1960), 24.

moving from left to right at a speed $v = \frac{2}{3}c$, together with an observer. If the block were stationary, the observer would see it as shown in Fig. 13-10B. Now, if the block were moving from left to right at two-thirds the speed of light, we know that the observer would conclude that all dimensions in the direction of motion would be reduced to three-quarters of their stationary values. We might therefore assume (as did all the early relativists) that the moving block would appear as shown in Fig. 13-10C: it has merely been shortened in its direction of motion. This is the assumption on which the artist based Figs. 13-8 and 13-9. No matter what Mr. Tompkins might *figure* about the cyclist and the policeman, however, what he would actually *see* is something quite different.

Imagine a ray of light starting toward the observer from the back edge A when the block is in position 1. This ray could never reach the observer's eye if the block were stationary; but since it is moving at high speed, the results will be different. In some tiny fraction of a second, the light from A_1 heading toward the observer, would have reached point 2 if the block were not in the way; during the same small time interval, the block would have indeed moved out of the way of the light to its second position on the drawing, leaving the ray unobstructed. After another small time interval, the light ray would have reached point 3 and the block (now in its third position) obviously no longer threatens to obscure the light, which is now quite free to proceed on toward the observer's eye.

Similarly, a ray from D_1 has a clear path to the observer from a stationary block. When the block is moving at high speed, however, the ray can never get there. In the same small interval of time, light from D_1 on its way to the observer would have to have reached the point marked by the colored dot; by this time, though, the block would have moved to its second position, where it has clearly intercepted the ray and prevented it from ever reaching the observer. The observer of the moving block would thus not see edge D at all, and edge A would be clearly in view to the left of B.

A mathematical analysis (which we will not inflict on the reader) shows that the moving block would not *appear* to be shortened but would instead look as though it had been rotated, as in Fig. 13-10D. Fundamentally, of course, the cause of this illusion is that light rays that arrive simultaneously at the observer's eye left the moving block at different times. The ray from A, for example, has farther to go than the ray from C and must therefore have started from the block earlier, when it had not yet moved so far to the right.

The calculations can be done in reverse too; if the observer (who sees something like Fig. 13-10D) were to make all the proper allowances for the motion of the block and the differing distances to its various edges, he would finally conclude that the light had emanated from a shortened block such as that in Fig. 13-10C. That is, if he figured back from his observations, he would in this way *measure* the block to be only 0.75 as long as its stationary length.

Thus Mr. Tompkins would actually *see* a cyclist and a policeman of perfectly normal proportions, who would merely appear to be considerably rotated from their actual positions. Figures 13-8 and 13-9 are really all right after all—except that they do not portray what Mr. Tompkins would *see*, but rather what he could *calculate* about what he observed. The astonished expression on his face is still quite justifiable!

Questions

(13-2)

1. A boat that moves at 12 m/sec on still water is on a river whose current flows at 6 m/sec; the river is 3 km wide. (a) How long will it take the boat to cross the river and return? (b) How long will it take it to go 3 km upstream and return?

2. Consider a river 2 mi wide that flows 8 mi/hr, and a motorboat that travels at exactly 24 mi/hr. (a) How long will it take the boat to cross the river and return? (b) How long will it take it to go 2 mi downstream, and then back?

3. Give good approximate evaluations for the following: (a) 1/1.018, (b) 1/0.996, (c) $\sqrt{1.0036}$, (d) $\sqrt{0.988}$, (e) $1/\sqrt{1.00064}$.

4. What are the values of the following, to a good approximation? (a) 1/0.9932, (b) 1/1.0026, (c) $\sqrt{0.9986}$, (d) $\sqrt{1.012}$, (e) $1/\sqrt{0.978}$.

5. Consider a "river" 45 m wide that flows with a current of 60 km/s, and a boat that moves through the water at 3×10^8 m/s. (a) How long will it take the boat to cross the river and return? (b) How long will it take the boat to go 45 m downstream and return? (c) What is the difference in time required for these two trips?

6. Two identical airplanes, each with an airspeed of exactly 900 mi/hr, have a race on a day when the wind is blowing out of the north at 60 mi/hr. Plane *A* flies 300 mi south and returns; plane *B* flies 300 mi east and returns. Which one wins the race, and by how much?

(13-4)

7. A spherical asteroid 10 km in diameter passes the earth with a relative velocity of $0.4c$. What do astronomers measure its diameter to be in the direction of its motion? (This is a purely fictional asteroid—real ones have much smaller speeds.)

8. Imagine a spaceship that has a speed of $0.6c$ relative to the earth. Its rest length is 100 m. What would we measure its length to be as it passed by?

9. Consider the asteroid of Question 7 to be moving at a more realistic relative velocity of 40 km/s. By how much would astronomers observe its diameter to be shortened in its direction of motion?

10. Intercontinental planes will no doubt be traveling at 3600 km/hr in the near future. By how much would such a plane appear to be shortened, as observed from the ground? Take its rest length to be 150 m.

11. A K^+ meson has an average lifetime of about 10^{-8} s. How fast must it be moving to increase this average lifetime by 50 percent?

12. A free neutron (not in an atomic nucleus) has an average lifetime of about 1000 s. How fast must a beam of neutrons be traveling for them to have a lifetime twice this long?

13. By what fraction will the rest mass of a molecule be increased when it is in the 2500-m/s blast of a rocket exhaust?

14. By what fraction is the mass of the earth increased (from the viewpoint of an observer stationary relative to the solar system) due to its 30 km/s orbital speed? What does this amount to, in metric tons? ($m_o = 6 \times 10^{24}$ kg.)

15. The rest mass of a proton is 1.67×10^{-27} kg. What is our measure of the mass of a proton whose velocity is $0.9c$?

16. An electron has a rest mass of 9.11×10^{-31} kg. What is our measure of the mass of an electron traveling at $0.99c$?

17. A beam of ions is shot from the nose of a spaceship at a speed of $0.9c$. The ship is leaving the earth at $0.6c$. What would earthly observers measure the speed of the ions to be?

18. A future spaceship, receding from the earth at half the speed of light, fires from its nose a rocket that travels at half the speed of light with reference to the ship. With what speed does the rocket travel with reference to the earth?

19. In Question 17, the ion beam is directed backward toward the earth. What is the speed of the beam, as observed by scientists on the earth?

20. From the rear of the ship of Question 18, a ray of light is directed backward toward the earth. What will earthly observers find the speed of the light ray to be? (Here, $v = -c$.)

21. If a block of iron (specific heat = 0.1 cal/g/°C) has a mass of exactly 1000 kg at 0°C, by how much will its mass be increased when it is heated up to 100°C?

22. (a) If 50 kg of water at 0°C is frozen into ice at 0°C, is the mass of the ice greater or less than the mass of the water? (b) By how much has the mass changed?

23. The sun loses about 4×10^6 metric tons of mass per second by radiating the equivalent energy into space. What is the wattage of the sun?

24. A nuclear reactor-powered generating plant produces electric power at an average rate of a million watts. The energy is produced by the conversion of the mass of the nuclear fuel. By how much will the mass of the fuel have been reduced at the end of a year?

25. The mass of the solar system (by Earth measurement) is 1.97×10^{30} kg; the entire system is orbiting about the center of our Milky Way galaxy at a speed of nearly 300 km/s. (a) What is the KE of the solar system? (b) What is the mass of this much energy? (c) What is the relativistic increase in the solar system's mass as the result of its speed, judged by an observer stationary in the Galaxy? (d) Does this agree with your answer to part b?

26. The earth's orbital speed is about 30 km/s. (See Question 14.) What is the earth's orbital KE (from KE = $\frac{1}{2}mv^2$)? What mass does this amount of energy correspond to? Does this agree with the answer to Question 14?

(13-6)

27. To the cyclist of Question 28, how many feet long would a city block appear to be?

28. In Mr. Tompkin's city, how long a time (by his own watch) would it take a cyclist to go $\frac{1}{2}$ mi at 15 mi/hr? (Distance measured by counting blocks, which are 16 per mile, measured on the ground.)

(13-7) 29. In the year 2010, a traveler leaves Earth on an exploratory voyage at a speed of $0.99c$. His own accurate clock–calendar tells him that his trip has lasted just 10 yr when he arrives back on Earth. What do the Earth calendars say the year is?

30. A space traveler is 30 yr old when he leaves to explore the galaxy on a ship traveling at 2.4×10^8 m/s. He returns 50 yr later (by Earth calendars). About how old does the traveler appear to be?

14

The General Theory of Relativity

14-1 Acceleration and Gravity

One of the basic postulates of Einstein's Special Theory of Relativity is that it is pointless to speak about absolute motion through space and that only the relative motion of one system with respect to another system can be considered as a physical reality. An observer enclosed inside a windowless vehicle moving with a constant velocity has no way of telling whether he is in a state of motion or at rest, no matter what kind of physical experiments (mechanical, optical, electric, or magnetic) he performs inside his enclosure in order to answer this question.

But what about accelerated motion? When an airplane speeds up along the runway prior to the takeoff, you are pressed to the back of your seat, and it is not necessary to look through the window to know that you are subjected to an acceleration. If the flight is smooth, the conditions inside the airplane are exactly the same as if it were standing at the airport, but any accelerations caused by air currents in bad flying weather are certainly very noticeable to the passengers. Does this mean that, whereas one should not talk about absolute velocities, absolute accelerations have a definite physical meaning?

This problem was attacked by Einstein about 10 years after his original publication of the theory of relativity (now known as the *Special Theory of Relativity*), and its solution (published in 1916) resulted in further theoretical developments now known as the *General Theory of Relativity*. He showed that ***all the physical phenomena that are observed within an accelerated system are identical with those occurring in a resting system placed in a gravitational field.*** To understand this idea, let us consider events taking place in the passenger cabin of a rocket ship traveling through space, far away from any gravitating celestial bodies. (In his original paper, Einstein spoke about a box being accelerated through space by the pull of a rope attached to it.) If the rocket motors are shut off, the ship coasts freely through space with constant velocity, in accordance with Newton's first law. The travelers and all the objects in the ship are "weightless" and float about freely (Fig. 14-1A).

Suppose now that the motors are started so that our rocket ship begins to gain speed. Since the velocity of the ship is now increasing, the things floating in the cabin will continue to move with the same velocity with which the ship was originally coasting through space, and they all will be collected at the rear wall of the cabin and pressed to it by the force of the acceleration (Fig. 14-1B). Realizing the situation, the travelers will rise to their feet and will stand on the rear wall of the cabin as if it were the floor of their laboratory back on the earth (Fig. 14-1C). Being of scientific mind and knowing that the earth is far away, they may try to perform some experiments in order to find out what the difference is between the force acting on them because of the operation of the rocket motors and the force of gravity acting on the earth's surface. Lifting a wooden ball, which for some unknown reason is among the other things in the cabin, one of the travelers (let us call him Dr. A) feels that it presses against his hand as though attracted toward the "floor" of the cabin.

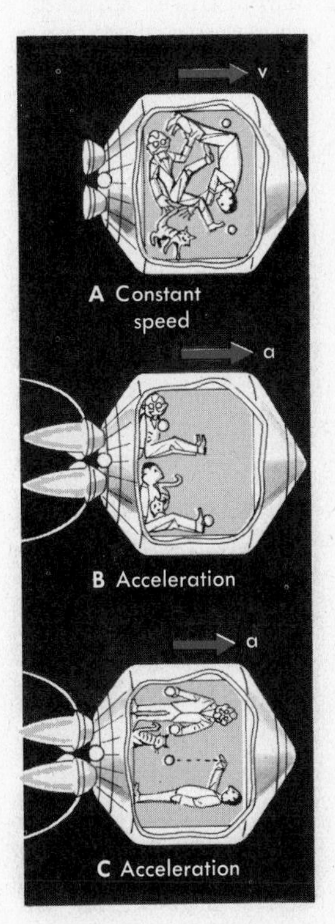

FIG. 14-1 Events taking place in an unaccelerated rocket ship (A); and in an accelerated one (B and C).

"Feels exactly as if I am holding a wooden ball back home," says Dr. A. "If I let it go, it will drop down as it would on the earth."

"Yes, *as if*!" counters Dr. B. "But, you know very well there is no gravity here and that the ball will simply approach the rear wall of our cabin because, once you let it go, it will continue to move with a constant velocity while our rocket is accelerating and gaining speed."

"What's the difference?" objects Dr. A. "It falls just the same. Is there any way to check whether its fall is due to the accleration of our rocket ship or to the gravitational attraction of some very large mass under our feet?"

"Why don't we repeat Galileo's experiment, dropping a light and a heavy ball simultaneously and see if they fall down at the same speed," suggests Dr. B.

"Very well, here is an iron ball; God knows how it got here," says Dr.

A, holding the two balls in his hands. "Now, I let them go!"

You can readily imagine what happens when Dr. A releases the two balls. They will, of course, move *side by side* with the same velocity that the rocket ship had at the moment of their release. The "floor" of the cabin, constantly gaining velocity, would finally overtake them, and hit them both at the same time. From the point of view of the rocket-ship travelers, however, it would seem that both balls accelerate toward the floor and hit it simultaneously.

The principal point of Einstein's paper was that the behavior of material objects in an accelerated system and their behavior in a gravitational field are not only similar to one another but absolutely identical. In other words, ***no matter what kinds of physical experiments we carry out in a closed cabin, we can never find out whether the cabin is resting on the surface of some massive planet or is being accelerated through space far away from any gravitating masses.***

14-2 Gravitational Deflection of Light

One characteristic of a good physical theory is that it not only explains known facts but is also able to make new predictions that can be checked by experiment. Such was the case with Einstein's views concerning the identity of acceleration and gravity. Let us return to our rocket ship and see what would happen if the two scientists tried to perform some optical experiments inside their cabin. One of the basic properties of light is that it propagates along straight lines, so that if we direct a beam of a flashlight at the wall, the illuminated spot will be directly opposite the source. If one repeats this experiment in an accelerated rocket, however, the situation will be rather different, as shown in Fig. 14-2. If the rocket were not accelerated, the beam of light coming from the source on the left would propagate straight across the cabin to produce luminous spots S_1, S_2, and S_3 on the transparent plates of fluorescent glass (No. 1, No. 2, and No. 3) placed across its way. If, however, the rocket is accelerated, the luminous spots will no longer lie on a straight line. The light will take equal intervals of time to cover the distances between source and No. 1, between No. 1 and No. 2, and between No. 2 and No. 3. Because of the accelerated motion of the rocket the displacements of the screens during these time intervals will stand in the ratios $1^2 = 1$, $2^2 = 4$, $3^2 = 9$. Thus the spots S_1', S_2', and S_3', tracing the motion of the light beam will lie on a *parabola* rather than on a straight line. Therefore, the observers in the cabin, considering themselves to be in a gravitational field, will be under the impression that the light beam is deflected by the gravitational field in the same way that a flying bullet is. If the analogy discussed above between the motion of material bodies in an accelerated system and in a gravitational field is more than an analogy and if the two things are really identical, we should expect that ***light rays propagating through any gravitational field should be deflected in the direction of the acting force.***

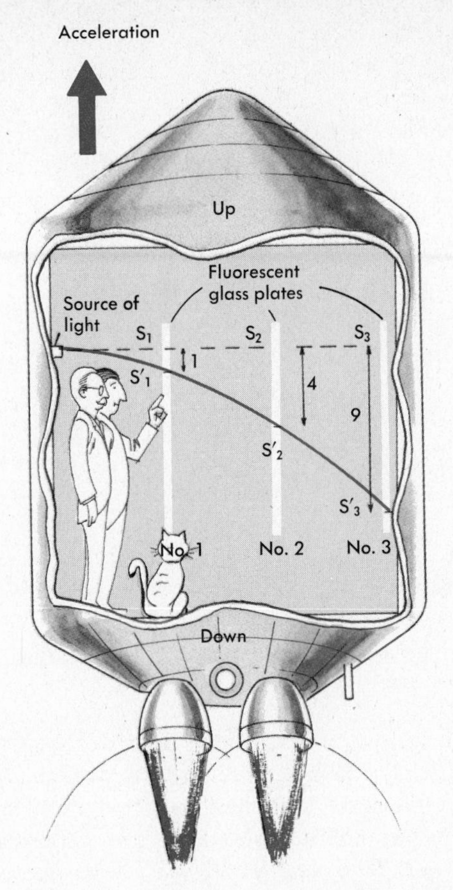

FIG. 14-2 The deflection of a beam of light as observed in an accelerated rocket ship.

This conclusion was first tested in the observation of a total eclipse of the sun in Africa by a British expedition in 1919. These observations proved that light rays from distant stars that pass close to the massive body of the sun are indeed deflected. The situation is shown schematically in Fig. 14-3. If the sun were in some other part of the sky, so that the starlight propagated along straight lines (the dashed lines in Fig. 14-3), an observer O on the earth would find the value θ for the angular distance between the stars. If, however, the rays coming from the two stars were to pass on opposite sides of the sun, their paths would be deflected toward the sun (colored lines), and the observed angle θ' would be larger. Thus the deflection caused by the gravitational field of the sun would make the angle between the stars look larger than it actually is. The test could be made, of course, only during a total solar eclipse, since otherwise the light from the stars would be lost in the brilliant sunshine. The observation of the 1919 eclipse and other later observations confirmed the expected bending of light rays in the gravitational field of the sun.

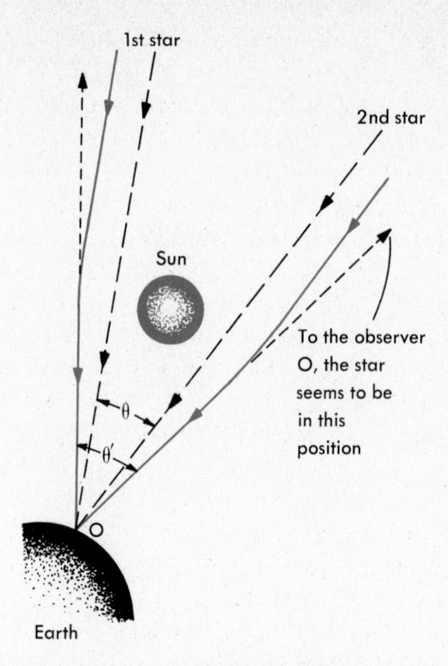

FIG. 14-3 The deviation of light from a star when the light passes close to the sun.

The deflection of starlight as it passes near the sun is very small, however, and is difficult to measure with much precision. In recent years, several other theories (similar to Einstein's General Theory of Relativity, but differing from it in some details) have been proposed. So far, the experimental data from eclipse photographs have not been precise enough to show that any one of the theories is right and the others wrong.

14-3 Other Consequences of General Relativity

Einstein's ideas concerning the nature of the gravitational field also predicted other consequences that could be confirmed by astronomical observations. One of them pertained to the motion of planets around the sun.

The major, i.e., longest, axis of the elliptical orbit of the planet Mercury does not remain always fixed in the same direction, but rather rotates, or *precesses*, about the sun approximately 575 seconds of arc per century (Fig. 14-4). Most of the precession can be explained by purely Newtonian mechanics, applied to the gravitational pulls of the other planets as they, too, revolve around the sun and tug at Mercury as they pass. When all these effects are calculated, 43 seconds/century still remains unaccounted for.

Einstein's General Theory of Relativity predicts results a little different from Newton's, especially for a planet like Mercury, which has a very elliptical orbit and is in a very strong gravitational field close to

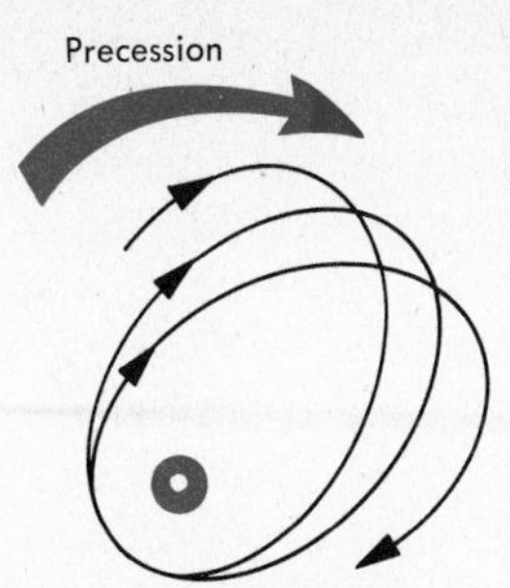

FIG. 14-4 Precession, or rotation of the long axis of an elliptical orbit.

the sun. Einstein's calculations showed that the major axis of Mercury's orbit should be expected to precess at the rate of just 43 seconds/century, even without the pulls of any other planets. The General Theory of Relativity thus gave the answer to a puzzle over which astronomers had been scratching their heads for many decades.

But astrophysicists cannot let well enough alone; it has recently been noted that if the sun were not quite perfectly spherical, it would have an effect on Mercury's precession that had not previously been taken into account. Experiments are now going on to determine whether the rotation of the sun bulges its equator enough to put the predictions of the General Theory of Relativity in question. Eventually, if measurements can be made accurately enough and if we learn more about how the mass of the sun is distributed in its deep, rotating interior, it may be possible to choose between Einstein's General Theory of Relativity and some of the competing not-quite-the-same theories.

Another important consequence of the General Theory of Relativity was that gravity should influence the rate of all physical processes by slowing them down. Thus, on the surface of the moon, where gravity is weaker than on the surface of the earth, a chronometer should gain time with respect to an identical terrestrial chronometer, whereas on the surface of the sun, where gravity is much stronger, it should trail behind. Of course, it is impossible to place a man-made chronometer on the surface of the sun, but fortunately there are natural chronometers already there: the atoms that tick their time by emitting light waves of well-defined frequencies. Thus, in order to see if there is any difference in the clock rate on the surface of the sun and on the surface of the earth, one should compare the frequency of light emitted by identical sources, one on the sun and one on the earth. Light emitted by atoms of different elements provide means for such a study. The vibrations of the atoms in the sun's strong gravitational field should be slower than those of similar atoms on the earth. But this difference is only about two parts in a million and is very difficult to measure accurately. More recent experiments have used the vibrations emitted by the nuclei of atoms (gamma rays), which can be measured with great precision. These, measured at different elevations within the same laboratory, give results that (like the shift in sunlight) agree with Einstein's predictions. Unfortunately, however, all the competing theories—even an extension of Newton's laws—predict the same thing, so that this agreement provides no basis for choosing among them.

Einstein's Special Theory has no competitors, and has been so thoroughly confirmed by experimental evidence that it is now a solid part of the foundations of physics. But the General Theory is still far out along the frontiers of science. Although it is the original theory from which all its present competitors have been derived, the experimental evidence is not yet good enough to say which one of its many modifications, if any, is correct.

14-4 Gravity and Space Curvature

The second great step made by Einstein in his General Theory of Relativity was to associate a gravitational field with a curvature of the four-dimensional space in the neighborhood of gravitating masses. What does it mean that space is curved? The best way to understand the cumbersome notion of curved space is through an analogy with curved surfaces that have only two dimensions and thus can be easily visualized. We know very well the difference between a plane surface, such as the surface of a table, and curved surfaces, such as those of a football or a saddle. Mathematicians distinguish between two kinds of curved surfaces: those with *positive curvature* and those with *negative curvature*. In order to tell which is which, we draw a plane tangent to the curved surface at one point. If the curved surface lies entirely on one side of the plane (Fig. 14-5A), the curvature of the surface is called *positive*. If, on the other hand, the plane and the surface intersect so that one part of the curved surface lies above and another part below the plane (Fig. 14-5B), the curvature is called *negative*. We can easily see that, according to this definition, the surface of a sphere or an ellipsoid has a positive curvature, while the curvature of any saddle-shaped surface is negative.

FIG. 14-5 Surfaces of (A) positive; and of (B) negative curvature. Note that in the case of positive curvature the entire curved surface is on the same side of the tangent plane, while in the case of negative curvature, part of the curved surface is on one side of the tangent plane and part is on the other side.

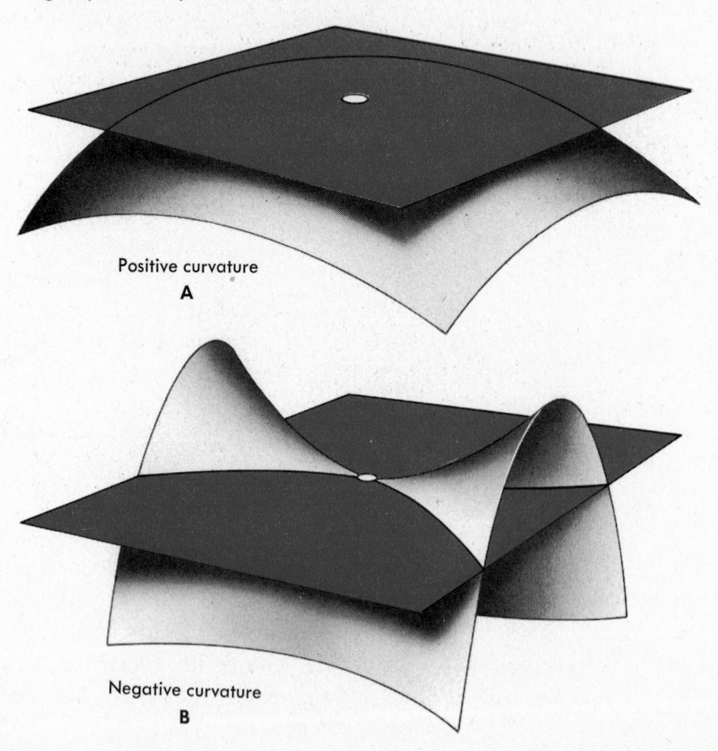

The difference between a surface of positive and one of negative curvature shows in the properties of geometrical figures drawn on them. While figures drawn on a plane surface are subject to the rules of classical Euclidean geometry, it is not so for figures drawn on curved surfaces. Consider, for example, the theorem of Euclidean plane geometry according to which the sum of the angles of a triangle is always equal to two right angles. This theorem does not hold for spherical triangles formed by arcs of great circles connecting three points *A*, *B*, and *C* on the surface of a sphere (Fig. 14-6A) since the surface, so to speak, "bulges up" between the vertices of the triangle. The simplest way to see this is to consider a triangle $A'B'C'$ with one vertex at the pole and two others on the equator of the sphere. Since the meridians forming two sides of that triangle make right angles with the equator, the sum of the angles $A'B'C'$ and $A'C'B'$ is already equal to two right angles and we have in addition the third angle $B'A'C'$ that can be quite large. An opposite result will be obtained in the case of a triangle drawn on a saddlelike surface (Fig. 14-6B). The sum of the angles of a triangle there is smaller than two right angles, since the surface "sinks" between the vertices. It should be noticed that in both the cases above, the sides of the triangle are actually not straight lines in the common sense of the word, but the "straightest" lines that can be drawn on the surface. They actually represent the *shortest distances* between the two given points, and are known in mathematics as *geodesic* lines, or simply *geodesics*. In the geometry of curved surfaces, geodesics play the same role that straight lines do in ordinary plane geometry.

From the simple facts concerning the geometry of surfaces with only two dimensions, we can now generalize for the case of three-dimensional

FIG. 14-6 The sum of the angles of a triangle equals 180° only when the triangle is drawn on a plane surface.

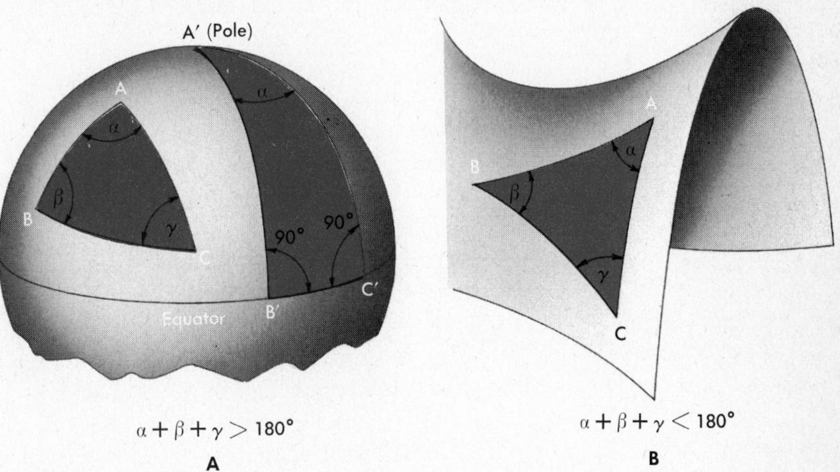

space. Since we are ourselves three-dimensional beings, and hence cannot look at the space we live in *from the outside* as we can look at the surfaces, we are deprived of the ability to visualize curved spaces as easily as we can curved surfaces. What we can do, however, is to say that the three-dimensional space that surrounds us on all sides is curved if its geometrical properties deviate from the laws formulated by Euclid. Thus, if the sum of the angles of a triangle formed by any three points in space is larger than two right angles, we ascribe to the space a positive curvature; if the sum of the angles is less than 180°, the curvature of space is considered to be negative.

In order to give physical meaning to these rather abstract considerations, imagine three astronomers located at three observation points (planets, or even spaceships) around the sun (Fig. 14-7). Each of the three astronomers uses a very accurate instrument for measuring the angles of the triangle *ABC* formed by the three positions in space. If the sun were not inside this triangle, light rays would propagate along "conventional" straight lines (broken lines in Fig. 14-7), and our astronomers would confirm the classical Euclidean statement. The presence of the gravitational field of the sun, however, will cause light rays to be bent (solid lines in Fig. 14-7); adding together their measurements of the three angles, our astronomers will find a value larger than two right angles. Einstein's revolutionary idea was to ascribe the finding of the three astronomers in the hypothetical case above, *not to the deflection of light rays propagating through Euclidean space, but rather to the deviation of the geometry of the space itself from the classical Euclidean rules.* In other words, *light rays always propagate along the "straightest" (geodesic) lines, and the deviation from the rules of Euclidean geometry obtained by measurements carried out in the neighborhood of the sun is due to the curvature of space caused by the presence of the sun's large gravitating mass.* Since the sum of the angles of a triangle is in this case larger than two right angles, the curvature of space in the neighborhood of the sun is to be considered as positive. The idea of replacing the picture of a physical deflection of light rays propagating through a gravitational field by a change in the geometry of space caused by the presence of the gravitating mass turned out to be very helpful to the understanding and mathematical description of the phenomenon of gravity.

14-5 The Curved Space–Time Continuum

In the previous section, we have introduced the idea of a curvature of three-dimensional space caused by the presence of gravitating masses. We have seen, however, that space and time must be unified into a single four-dimensional continuum, each point of which is characterized by four coordinates: x, y, z, and t. Accordingly, the General Theory of Relativity considers the above-described curvature of three-dimensional space caused by the presence of gravitating masses to be the result of a curvature of the four-dimensional space–time continuum

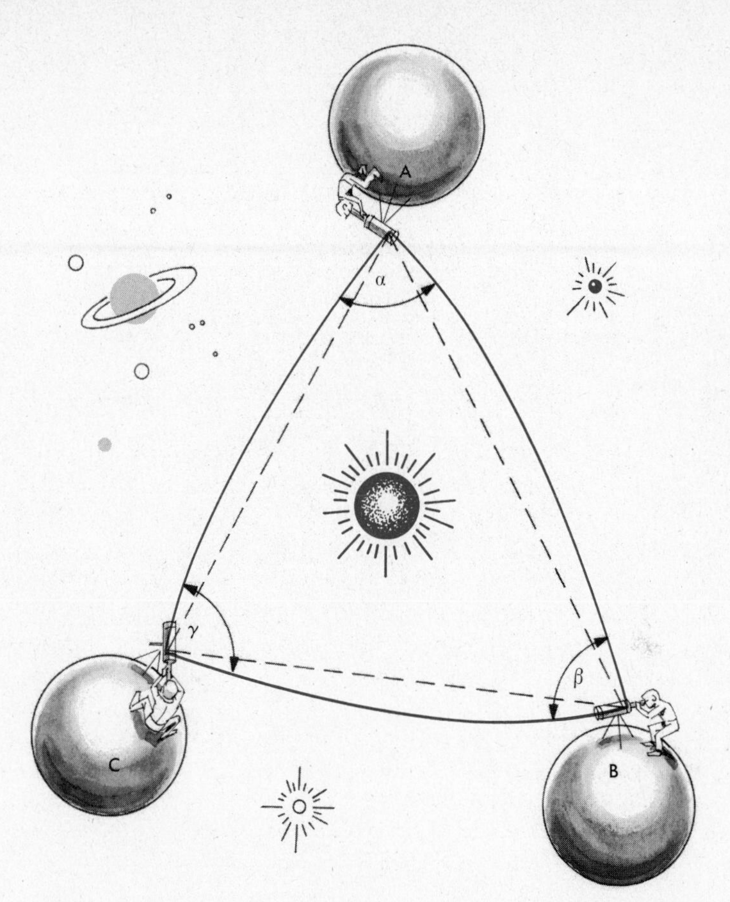

FIG. 14-7 Three observers checking Euclidean geometry for a triangle described around the sun.

itself. This notion can be clarified by considering as an example the rotation of the earth around the sun. Since the earth's orbit lies in a plane, we need retain only the two space coordinates x and y located in this plane and replace the third space coordinate z by t. It is convenient to multiply time by the velocity of light c, so that this coordinate will have the same physical dimensions as the simple space coordinates x and y. As a result, we obtain the picture shown in Fig. 14-8, in which each plane perpendicular to the ct axis gives the position of the earth in its orbit at the time corresponding to the position of the plane. Connecting the consecutive positions of the earth by a continuous line, we obtain a helix that winds around the time axis passing through the sun. Such lines, which if we include the third z coordinate of space would run through the four-dimensional space–time continuum, are known as the *world lines* of the material bodies whose motion they describe.

In our example, the world line of the sun is a straight line perpendicu-

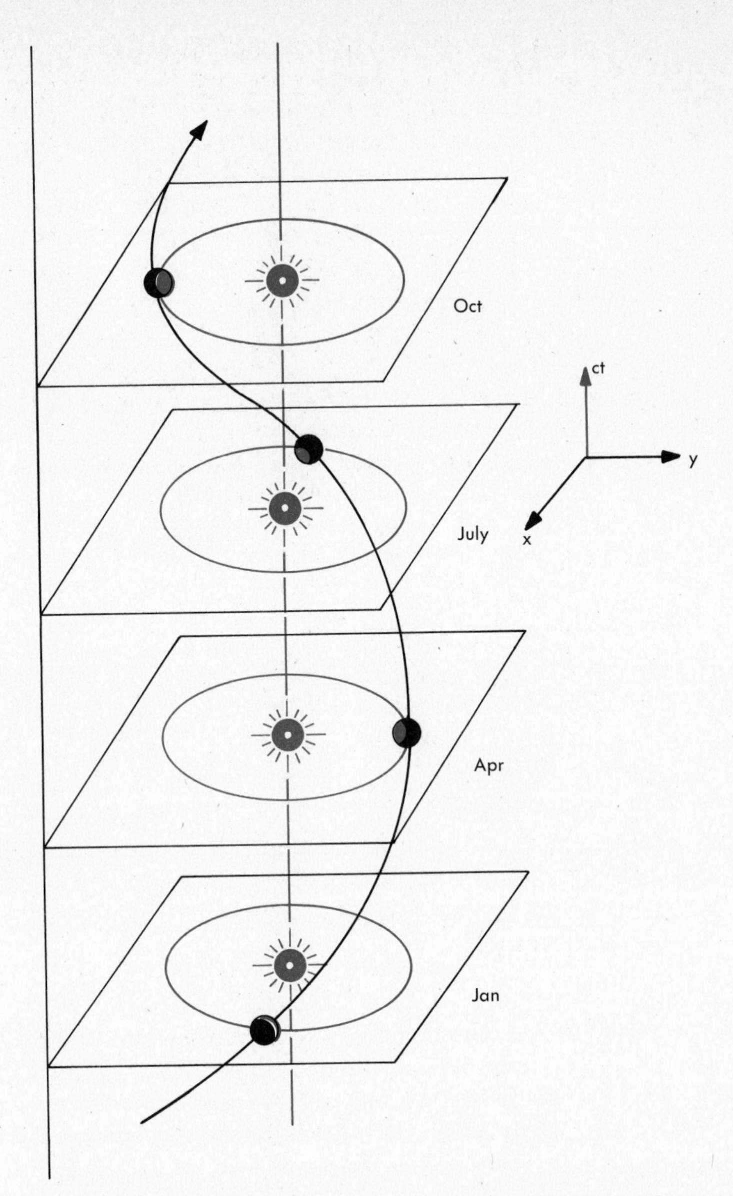

FIG. 14-8 The world line of the earth in its orbital motion around the sun. Only two space dimensions (*x* and *y*) are shown; the third dimension (marked "*ct*," perpendicular to *x* and *y*) represents the solar system's passage through time.

lar to the *xy* plane, whereas the world line of the earth is a helix winding around it. Such would be the situation if we considered the space–time continuum to be subject to the rules of Euclidean geometry. In the General Theory of Relativity, however, the four-dimensional space–time continuum is itself considered to be curved, so that the conventional straight lines of Euclidean geometry must be replaced by geodesic lines in curved four-dimensional space. It was shown by Einstein that the helical world line of the earth shown in Fig. 14-8 is actually such a geodesic, or "straight line" in the curved space surrounding the sun. Thus, just as in the case of the deflection of light rays passing close to the sun, *the orbital motion of the earth can be interpreted, not as being caused by a certain physical force exerted on it by the sun, but as the result of the curvature of space in the sun's neighborhood.*

We can summarize this section by saying that Einstein's General Theory of Relativity gives a geometrical interpretation to the Newtonian theory of universal gravity: *Instead of saying that the propagation of light and the motion of material bodies are compelled to deviate from straight lines by the force of gravity, we say that this motion takes place along the "straightest" (geodesic) lines in a space that is curved by the presence of gravitating masses.*

15

Electrostatics

15-1 Static Electricity

In early experiments with electricity carried out by William Gilbert (1540–1603), personal physician to Queen Elizabeth I, electric charges were produced by such means as "rubbing a galosh against a fur coat" or rubbing a glass rod with a silk handkerchief. We may have noticed that when a passenger in a fur coat slides across a plastic car seat, a spark may jump between his finger and the car door when he touches it. Or we may draw a similar spark from a radiator after walking across a carpet on a dry day. Electricity produced in such ways is *static electricity*, and from early studies of it the first laws of electricity were discovered.

Let us suspend two light metal or metal-covered balls from dry threads a short distance apart, as in Fig. 15-1. Rub a hard-rubber rod with a piece of fur or wool, and touch each ball with the rod; the balls will be seen to repel each other (Fig. 15-1A). They will also repel each other if they are both touched with the fur (Fig. 15-1B). However, if one ball is touched with the rubber rod, and the other with the fur, they will attract each other (Fig. 15-1C). A similar experiment, with exactly

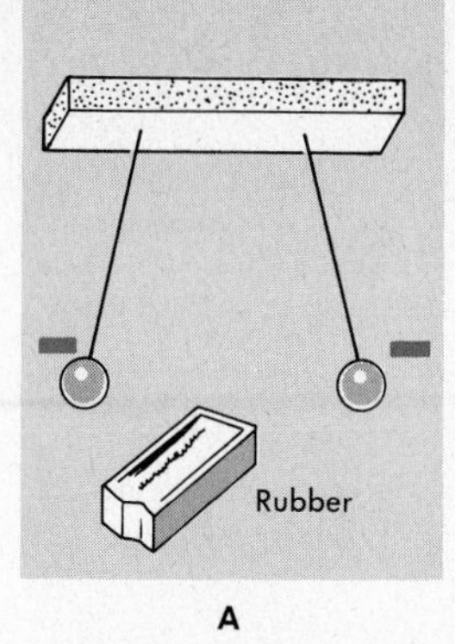

A

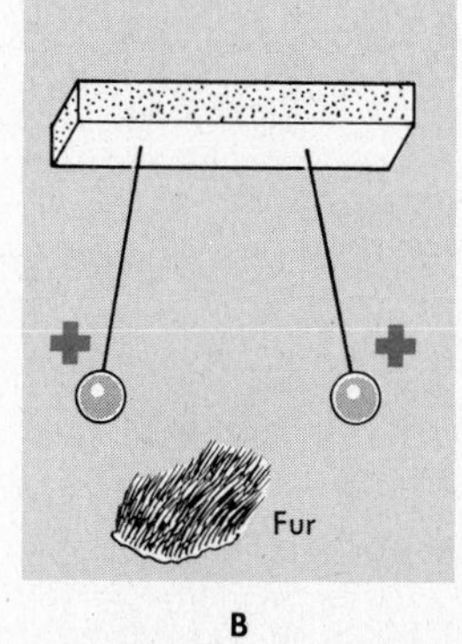

B

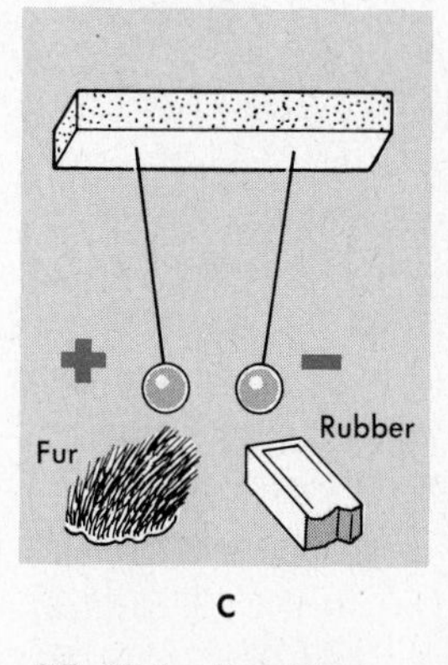

C

FIG. 15-1 Repulsion and attraction between electrically charged bodies.

corresponding results, could be done using a glass rod and a silk cloth. These experiments serve to confirm Gilbert's conclusions that there are two kinds of electric charge, and that ***electric charges of the same kind repel each other, and those of the opposite kind attract each other.***

We need one more experiment to complete this qualitative picture. Touch one of the balls with the hard-rubber rod rubbed with fur; touch the other ball with a glass rod rubbed with silk. The balls attract each other, showing that the charges are opposite. In the late eighteenth century, Benjamin Franklin introduced the terminology of *negative* for the kind of charge on the rubber rod rubbed with fur or wool, and *positive* for that found on a glass rod rubbed with silk.

The French physicist C. A. Coulomb (1736–1806) made quantitative studies of these electric forces, and found that ***the force of attraction or repulsion between two charged bodies is directly proportional to the product of their charges, and inversely proportional to the square of the distance between them.***

A unit for measuring the amount of charge was defined: *the electrostatic unit of charge (esu) is the amount of electric charge that when placed on each of two small bodies 1 cm apart, will cause them to repel (or attract, if they are oppositely charged) each other with a force of 1 dyne.* With this unit of charge defined, we can write Coulomb's law as

$$F = \frac{Q_1 Q_2}{d^2},$$

where the charges Q are measured in esu, the distance in cm, and the force in dynes. This equation has the advantage that the proportionality constant is equal to 1, and we do not even have to write it in. The esu is an inconveniently small unit of charge for most purposes, however, and Coulomb's law, although it was the foundation for all further work in electricity, is not very often of much direct use. In physics and electrical engineering, most work is concerned with other units based on the MKS system, and it is convenient to use a much larger unit of charge known as the *coulomb* (C), which is very nearly 3×10^9 esu. So it will be useful to rewrite Coulomb's law, measuring the charge in coulombs, distance in meters, and force in newtons. To do this we can find the force in newtons between two one-coulomb charges 1 m apart. This will enable us to figure the proportionality constant that will be needed:

$$\begin{aligned} F\text{ (dynes)} &= \frac{Q_1\text{ (esu)} \times Q_2\text{ (esu)}}{(d\text{ (cm)})^2} \\ &= \frac{3 \times 10^9 \times 3 \times 10^9}{10^4} = 9 \times 10^{14}\text{ dynes} \\ &= 9 \times 10^9\text{ N.} \end{aligned}$$

We can now write Coulomb's law:

$$F = \frac{9 \times 10^9 \times Q_1 \times Q_2}{d^2},$$

in which charge is in coulombs, distance in meters, and force in newtons.

As an example, we can calculate the force of repulsion between two electrons 5×10^{-10} m apart. Electrons are tiny particles that are a part of all matter, and which all have an identical negative charge of 1.602×10^{-19} C.

$$F = \frac{9 \times 10^9 \times (1.602 \times 10^{-19})^2}{(5 \times 10^{-10})^2} = 9.24 \times 10^{-10} \text{ N}.$$

15-2 Elementary Atomic Structure

For an understanding of how an electric charge, either + or −, is associated with material bodies, we must take a brief look at the atomic structure of matter.

All matter is made up of tiny particles known as *atoms*. There are only about a hundred different kinds of atoms, and they combine with each other in different ways to form groups called *molecules*. All matter has been found to be composed of atoms or molecules, and some knowledge of how atoms are made will give us valuable information about the behavior of matter.

In 1911, Lord Rutherford in England discovered (by means of ingenious experiments we will discuss farther on) that an atom has a tiny *nucleus* which is positively charged and contains nearly all the mass of the atom. Distributed about the nucleus and revolving about it in orbits* are much less massive negatively charged particles called *electrons*.

In a normal atom, there are exactly as many negatively charged electrons as are needed to neutralize the positive charge of the nucleus, so that the atom as a whole is electrically neutral. This is of course also true of all normal material substances, which are composed of atoms. Figure 15-2 gives a rough idea of how an atom is arranged. The outermost electrons are less strongly bound to the atom than the inner ones, and they are the ones that take part in chemical reactions between atoms and that are responsible for the accumulation of an electric charge on bodies.

Not all atoms or groups of atoms hold their outer electrons equally firmly. The attraction, or affinity, of hard rubber for electrons is greater than that of fur fibers. Accordingly, when a hard-rubber stick is rubbed

---

* The visual picture of electrons revolving in orbits is not quite true, but it will serve as a tentative model.

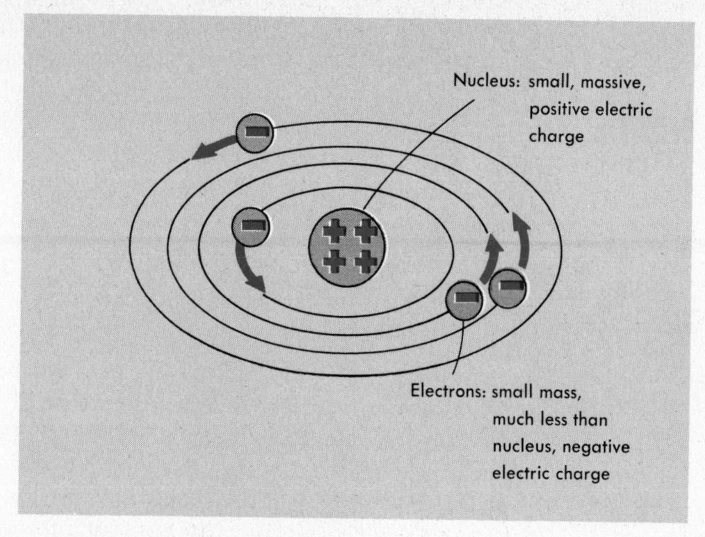

FIG. 15-2 Schematic model of an atom.

with fur, some of the electrons leave the fur and become attached to the rubber. Thus the hard rubber, with an excess of negative electrons, becomes negatively charged; the fur, left without enough electrons to neutralize the positive charges of its atomic nuclei, is left with a net positive charge. If a glass rod is rubbed with a piece of silk, the glass will be found to have a positive charge and the silk a negative charge, indicating that the silk fibers have a greater affinity for electrons than the glass does. We must bear in mind that, in solid materials, only some of the outer electrons may have any freedom to move. The remainder of the atom is firmly fixed in place.

15-3 Conductors and Nonconductors

Let us take a ball made of cork or pith, covered with metal foil, and touch it with a hard-rubber rod previously rubbed with fur or a woolen cloth. Some of the excess electrons on the rod will leave it and transfer to the ball, giving the ball a negative charge. The negative charge on the ball can be confirmed by noting that it is strongly repelled by the rubber rod. (Like charges repel each other.)

Now if you hold in your hand a piece of bare wire, a metal spoon, or a wet stick of wood and touch it to the charged ball, the ball will be found to lose its charge. Apparently, the excess of electrons on the ball is able to flow into your hand and body through the connecting material. If this experiment is repeated, and you hold a piece of dry wood, a glass rod, a tube of paper, or a piece of plastic, the charge will be found to remain on the ball.

This experiment can serve as a basis for separating materials into two general classes: *conductors*, through which a stream of electrons can readily flow, and *insulators*, or *nonconductors*, through which electrons cannot move. In general, metals and water solutions of salts, acids, and alkalis are conductors, whereas nonmetallic solids and oils are non-conductors. Do not take this as a rigorous classification scheme, because all conductors offer some resistance to the passage of electrons and all insulators permit some few electrons to pass through. A number of materials are on middle ground, and are neither good conductors nor good insulators. For many purposes, however, the classification is a useful one.

15-4 Induced Charges

We can be sure a foil-covered pith ball will be electrically neutral if it is touched with a finger or with one end of a wire connected to a water pipe that is in turn buried in the ground. Any excess or deficiency of electrons on the ball will be removed by a flow of electrons to or from your body. Yet such a neutral ball will be attracted by either a positively or a negatively charged rod. We must look for some explanation of this behavior, since we might expect from Coulomb's law, with one of the Q's $= 0$, that a neutral ball would not be affected one way or the other. Figure 15-3 shows what happens in such a situation. The attraction of the positively charged rod draws toward itself some of the mobile electrons in the metal covering of the ball, thus leaving an equal number of the atoms on the distant side of the ball with a net positive charge. The rod consequently attracts the near side of the ball and repels the far side; since the near side is nearer (naturally!), the attraction is stronger than the repulsion, and the net effect is a force of attraction. We need only reverse all the signs in Fig. 15-3 to see that a negatively charged rod would also, and equally, attract a neutral ball.

When the charged rod is removed from the neighborhood of the ball, the excess electrons on the front of the ball return to the rear half and

FIG. 15-3 The attraction of an electrically neutral foil-covered pith ball to a charged rod.

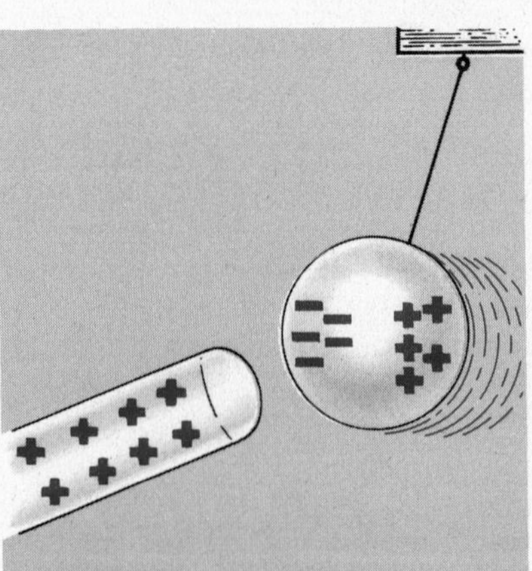

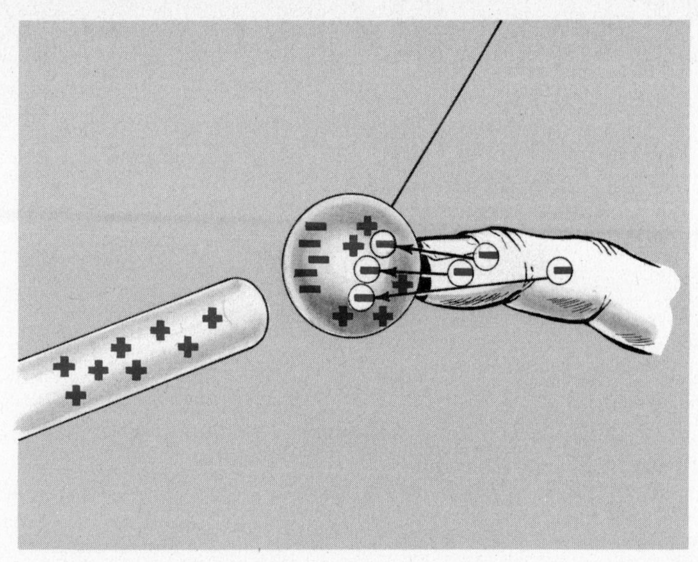

FIG. 15-4 Electrons flow from finger to ball, thus giving the ball a charge by induction.

the ball is again uniformly without charge of either kind. If we touch the ball with a finger *while the charged rod is still near*, however, the situation is changed. Figure 15-4 shows electrons, attracted by both the rod and the electron-deficient half of the ball, flowing into the ball. The finger is then removed, and if the charged rod is taken away, the excess electrons, as a result of their mutual repulsion, distribute themselves evenly over the surface of the ball and give it a uniform negative charge. If the same experiment were repeated with a negative rod, electrons would be repelled from the ball into the finger and the ball would be charged positively.

The attraction of a neutral foil-covered pith ball toward a charged rod was easily explained in terms of the migration of electrons through the conducting foil. However, the charged rod will also attract pieces of dry paper, lint, and hair, which are all good insulators, so that the electron migration explanation will not work. Here we have an example of the *polarization* of atoms in an electric field. We have pictured an atom as consisting of a massive positive nucleus surrounded by a whirling cloud of negative electrons. Normally, the electron cloud is centered on the nucleus, but a nearby charged body causes forces of repulsion and attraction that distort each atom. Figure 15-5A shows a normal atom, with the electron cloud centered on the nucleus. In Fig. 15-5B, the nearby negatively charged rod attracts the nucleus and repels the electrons, with the result that the center of the electron cloud now falls on the far side of the nucleus, so that the attraction for the nucleus is greater than the repulsion of the electrons, and the atom as a whole is slightly attracted toward the rod. This effect is small for any one atom, but there

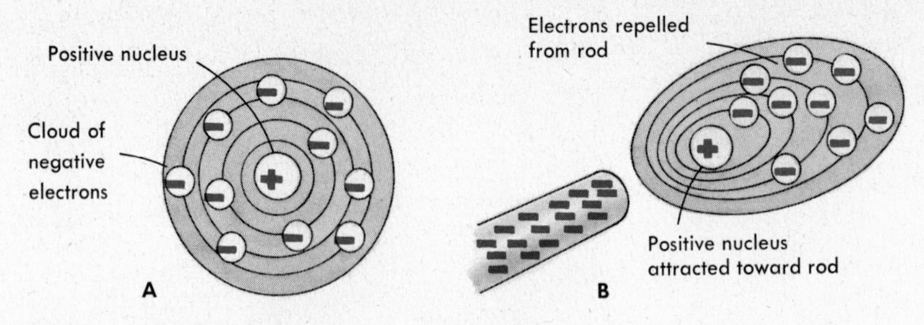

FIG. 15-5 Polarization of an atom in an electric field.

are an enormous number of atoms in a piece of paper or a bit of hair and the total force of attraction may be quite appreciable. You should redraw Fig. 15-5 for yourself, with a positive rod, to demonstrate that this, too, will result in a polarization of the atom in such a way that it will again result in an attraction.

15-5 The Electroscope

For the detection and measurement of electric charge, many different devices have been designed, some of them complicated and expensive electronic instruments. Among the simplest, however, is the *gold-leaf electroscope* (Fig. 15-6). It consists of a metal rod, with a right-angle

FIG. 15-6 The gold-leaf electroscope.

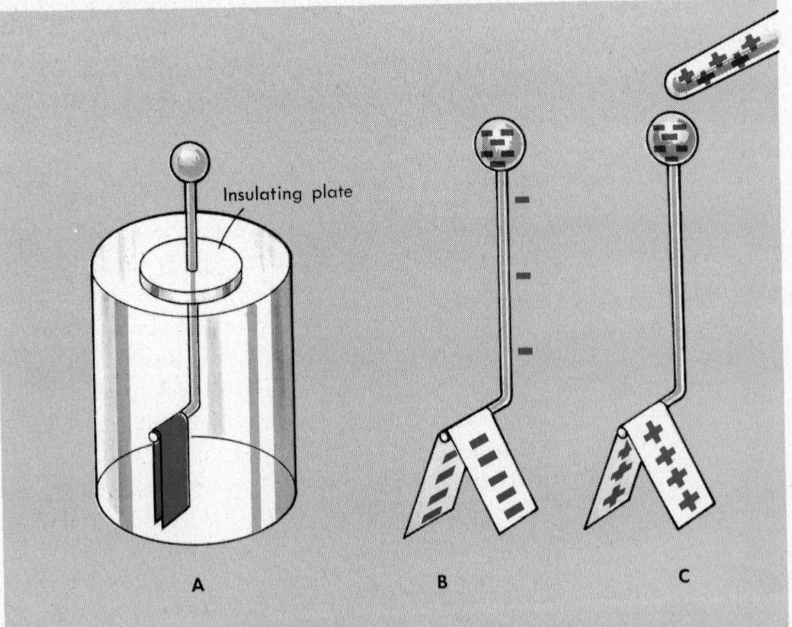

bend at the bottom, and a metal ball, which is generally fastened to the top of the rod. A strip of very thin gold foil is draped over the bottom bend of the rod and cemented to it. The rod is supported by a good insulator in the top of a box that surrounds the fragile gold leaves and protects them from damage. When the electroscope is uncharged, the gold leaves hang straight downward from their own weight (Fig. 15-6A). When the electroscope has been charged, however, by having it touch some charged object, the leaves stand apart (Fig. 15-6B) because of the repulsion of the like charges on each leaf.

Figure 15-6C also shows the leaves standing apart, although the electroscope as a whole has no net charge. The positively charged rod near the knob has attracted electrons from the leaves up into the knob, thus giving the leaves a positive charge so that they repel each other. In this case, when the rod is removed, the electrons will return to the leaves and they will again hang down normally, as in Fig. 15-6A.

But if instead of immediately taking the charged rod away, we first touch the knob, the situation will be the same as that shown in Fig. 15-4: The rod will attract more electrons from the finger into the ball. Now if we first remove the touching finger (so that the electrons cannot flow back) and then take the rod away, the electroscope will be left with a charge *opposite* that of the rod. Charging an electroscope by induction in this way is generally more satisfactory than charging it by direct contact.

15-6 The Electric Field

In considering the mechanical interactions between material bodies, we are accustomed to the fact that such interactions require direct bodily contact. If we want to move an object, we have to touch it with our hand or else have a stick to push it or a rope to pull it. An exception to this need for direct contact is found in the gravitational attraction that all bodies exert on each other. This "action at a distance" bothered Newton when he announced his law of universal gravitational attraction, and he made no attempt to explain *how* gravitational forces were exerted. We say that every mass is surrounded by a *gravitational field* that in some way appears to exert an attractive force on other masses. Einstein has contributed a great deal to our understanding of the gravitational field but there is nevertheless still much mystery connected with forces that act at a distance.

The forces of electrical repulsion and attraction, acting without any apparent connection between charged bodies, are similar examples of action at a distance and can be attributed to the existence of an *electric field* surrounding charged bodies. Without making any attempt to explain its workings, we can nevertheless describe an electric field accurately and understandably in terms of the force it exerts on a *test charge*. The imaginary test charge to be used is a *positive* charge of 1 coulomb. Wherever this test charge is placed in an electric field, it will have exerted on it a force of a certain magnitude and direction.

The strength and direction of the electric field at any point are defined to be the same as the magnitude and direction of the force exerted on our unit test charge when it is placed at that point. When we use the test charge of +1 C, the electric field strength E equals the force F that the field exerts on the test charge, and is in the same direction. This can be made more general by imagining that we use a particle bearing some other charge of Q C. In this case the force will be just Q times as great as for a unit charge, and we can write

$$F = EQ \quad \text{or} \quad E = \frac{F}{Q}.$$

In the MKS system we are using, the field strength will naturally be in newtons per coulomb.

It is easy to derive a general expression for the strength of the electric field caused by a charged sphere, or by a charged particle of any shape that is small enough to be considered a point. Let us take a sphere bearing a charge Q and calculate the field at point A, which is a distance r m from it. We put the imaginary +1 C standard test charge at that point (Fig. 15-7) and figure the force F exerted on it. Coulomb's law gives this as

$$F = \frac{9 \times 10^9 \times Q \times 1}{r^2}.$$

Since we always use +1 C as the test charge, we can omit the 1 from the numerator; and since this force per unit charge is the definition of the field strength E, we have simply

$$E = \frac{9 \times 10^9 Q}{r^2}.$$

FIG. 15-7 Determination of the electric field at point A, at a distance r from a charged particle.

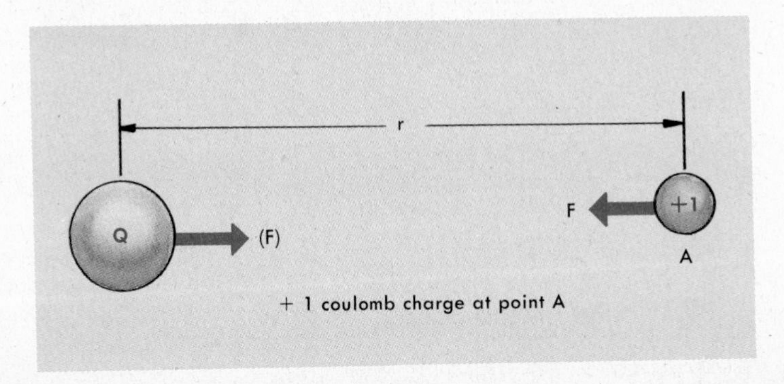

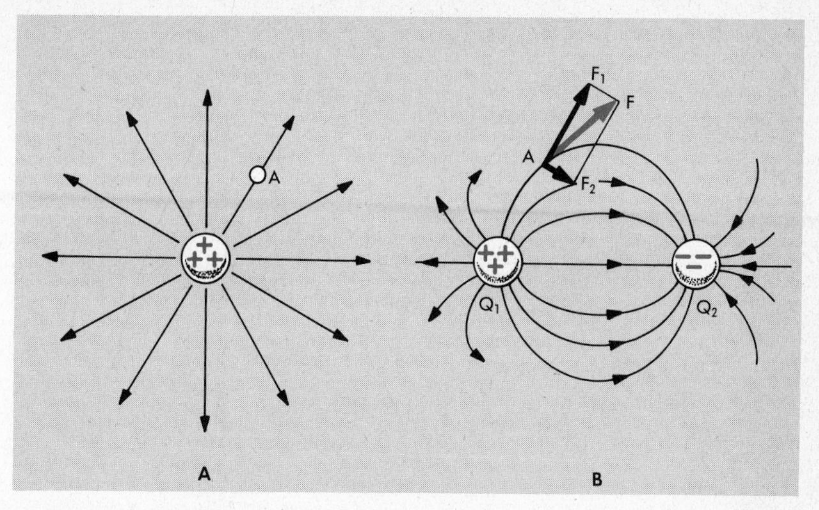

FIG. 15-8 The electric lines of force that map the electric field around charged bodies.

The field can be mapped by drawing lines that at every point run in the direction of the field. Figure 15-8 shows the electric field mapped out with "lines of force" for two simple cases. In Fig. 15-8A, there is a single positively charged sphere, and no matter where the test charge is placed, the force on it will be radially outward away from the central charge in the field we are mapping. Figure 15-8B shows the field resulting from a positive charge Q_1 and a negative charge Q_2 in the same neighborhood. Our +1-C test charge at A will be repelled by Q_1 and attracted to Q_2. The total force on the test charge will be F, the vector sum of F_1 and F_2. The line of force passing through point A, since it is required to give the direction of the field, must be tangent to the vector F.

There is no limit to the number of lines that can be drawn, but it is worth noticing that near the charged bodies, where the field is stronger, the lines are close together; at a distance, where the field is weak, the lines are far apart. Although the drawings show the field only in the two-dimensional plane of the page, the field exists, of course, in three dimensions. It can be shown that in this three-dimensional picture the number of lines of force cutting through a unit area at right angles to the field is exactly proportional to the strength of the field.

Let us take as an example four small charged bodies arranged in a square as shown in Fig. 15-9A. What is the electric field at C, in the center of the square? The field E will be a vector, the resultant, or vector sum, of the four components E_1, E_2, E_3, and E_4 caused by each of the four corner charges. Since C is at the center of the square, its distance from each of the charges is $0.10 \sin 45° = 7.07 \times 10^{-2}$ m (or it could be figured as half the diagonal of the square, $= \frac{1}{2} \times \sqrt{((0.1)^2 + (0.1)^2)} = \frac{1}{2} \times \sqrt{0.02} = 7.07 \times 10^{-2}$ m).

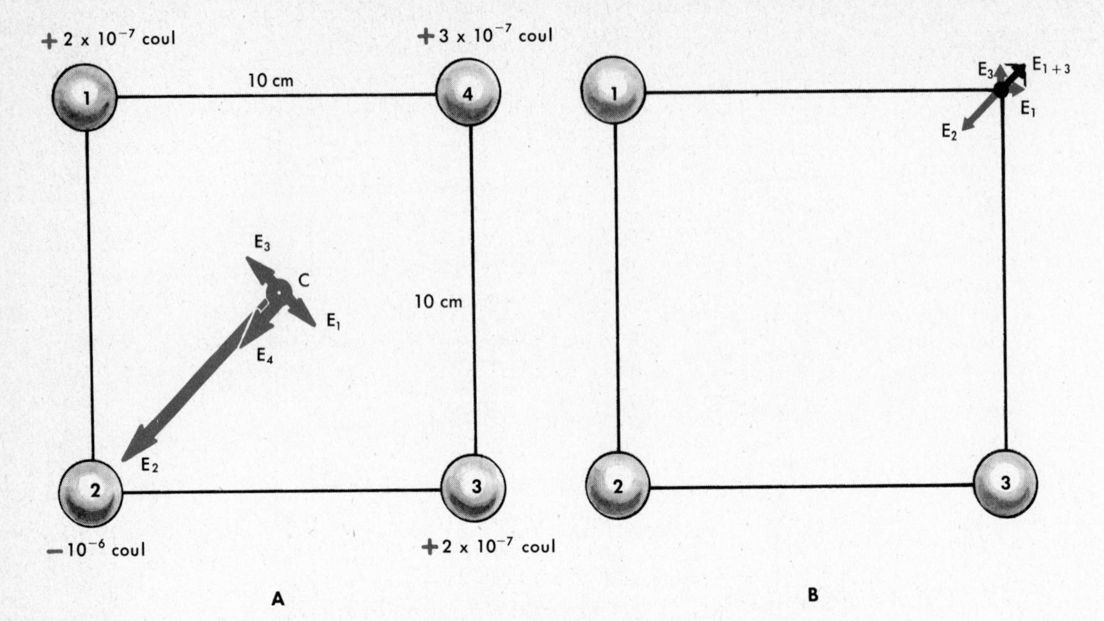

FIG. 15-9 The electric field produced by several charged bodies.

Because of the symmetry of the arrangement, E_1 and E_3 are equal and opposite, and so cancel out. For E_2 we have

$$E_2 = \frac{9 \times 10^9 \times 10^{-6}}{(7.07 \times 10^{-2})^2} = 1.80 \times 10^6 \text{ N/C}.$$

Because Q_2 is negative, the force on the imagined $+1$ charge at C will be an attraction, giving the direction of E_2 as shown in Fig. 15-9A. The field component due to Q_4 is

$$E_4 = \frac{9 \times 10^9 \times 3 \times 10^{-7}}{(7.07 \times 10^{-2})^2} = 0.54 \times 10^6 \text{ N/C}.$$

This E_4 component comes from the repulsion of the positive Q_4, and hence is in the same direction as E_2.

Since E_2 and E_4 are in the same direction, the total field E at point C is simply $1.80 \times 10^6 + 0.54 \times 10^6 = 2.34 \times 10^6$ N/C in a direction pointing toward Q_2. Electric field strength is by definition the *force per +1 coulomb*, so it follows that if we placed a charge of 10^{-7} coulomb at point C, it would experience a force of $2.34 \times 10^6 \times 10^{-7} = 0.234$ N, directed toward Q_2.

We can follow up this idea by finding what force will be exerted on charge 4. Charge 4's own field cannot move charge 4 any more than a man can lift himself from the floor by pulling on his own bootstraps. So let us temporarily remove charge 4 (in imagination, at least), find the field there due to the other three charges, and then put 4 back. Figure 15-9B shows the other three charges and the components of the field that each one produces. Components E_1 and E_3 are perpendicular to each other, and are each $9 \times 10^9 \times 2 \times 10^{-7}/10^{-2} = 1.8 \times 10^5$ N/C. Vectorially, these add to give $E_{1+3} = 1.8 \times 10^5 \times \sqrt{2} = 2.55 \times 10^5$ N/C, at a 45° angle as shown. Directed toward Q_2, $E_2 = 9 \times 10^9 \times 10^{-6}/(0.1414)^2 = 4.50 \times 10^5$ N/C. The sum of all three components of the field (since E_2 and E_{1+3} are along the same line) is $4.50 \times 10^5 - 2.55 \times 10^5 = 1.95 \times 10^5$ N/C toward Q_2. When we replace Q_4 in this field, the force on it will be $1.95 \times 10^5 \times 3 \times 10^{-7} = 5.85 \times 10^{-2}$ N, toward the charge Q_2.

15-7 Electric Potential

Imagine a small body bearing an electric charge of Q C. At any point r meters from this charged body the electric field will have a strength of $E = 9 \times 10^9 \times Q/r^2$. This same point will also have another property called *electric potential. The electric potential at a point is equal to the work needed to bring a +1 coulomb charge to the point from an infinite distance away.*

We do not need to start from the beginning to calculate this amount of work, because we have already done so in Chapter 6 on p. 110, in discussing gravitational potential energy. There we started with a force of gravitational attraction, $F = 6.67 \times 10^{-11} \times m_1 \times m_2/r^2$. The expression for electrical attraction (or repulsion) is a similar inverse-square rerelationship: $F = 9 \times 10^9 \times Q_1 \times Q_2/r^2$.

In the gravitational test case, we might let one of the masses be a 1-kg "test mass," located r cm from another mass M. On p. 110 we found the work necessary to move the 1-kg test mass from its distance of r m to infinity would be simply $6.67 \times 10^{-11} \times M/r$. Conversely, the work needed to bring the test mass from infinity to r would be the negative of this. We could therefore describe the gravitational potential at a point r m from M as $-6.67 \times 10^{-11} \times M/r$. In an analogous treatment of the electrical case, we let Q_2 be a +1-C test charge, and conclude that the electric potential V at a point r m from a charge of Q C is simply $V = 9 \times 10^9 \times Q/r$. The minus sign does not appear here, as it did in the similar expression for gravitational potential. As far as we know now, all mass is alike—there are no $+m$'s and $-m$'s—and these like masses always attract. Like electric charges always repel. If the charge Q were negative, the potential V would be negative.

As an example, take a foil-covered pith ball 3 cm in diameter, with a charge of 2×10^{-9} C. What is the potential 6 cm from its surface? Here, with the charge uniformly distributed over the sphere, its effect will be the same as though all the charge were at the center of the ball,

7.5 cm = 7.5×10^{-2} m from the point where we want to determine the potential. Then

$$V = \frac{9 \times 10^{9} \times 2 \times 10^{-9}}{7.5 \times 10^{-2}} = 240 \text{ J/C}.$$

The potential at the ball's surface, 1.5×10^{-2} m from the center, would be

$$V = \frac{9 \times 10^{9} \times 2 \times 10^{-9}}{1.5 \times 10^{-2}} = 1200 \text{ J/C}.$$

The concept of electric potential is a very important one, and one for which we shall have many uses. Few of these uses will have any apparent connection with the idea of moving a test charge up to a certain distance from a charged body, as we have just done. The procedure of moving up the test charge, however, serves to illustrate and emphasize the important fact that *electric potential is work per unit charge.*

To many of you the unit of potential *joules per coulomb*, may seem a bit strange and unfamiliar. However, nearly everyone is familiar with its other name—the *volt. A volt (V) is a joule per coulomb.*

$$1 \; V = 1 \text{ J/C}.$$

Actually, we are seldom concerned with the absolute potential of a point, that is, with the work needed to bring a unit charge up to the point from infinity. More often, we shall deal with the *potential difference* between two points, which is the work required (either positive or negative) to move a coulomb from one point to the other. When we speak of "a 6-volt battery," we mean that the potential difference between the terminals of the battery is 6 volts, i.e., that a coulomb of charge will do 6 joules of work in flowing from one terminal to the other.

15-8 Capacitance

When we originally defined electric potential, it was associated with the presence of an electrically charged body. The surface of a charged conducting ball must be at some definite potential, because it will take a definite amount of work to bring a unit charge up to the ball. If the charge on the ball were doubled, it would take twice as much work to bring up a unit charge, and the potential of the ball would accordingly be twice as high. We thus see that the charge on a body is proportional to its potential, and the ratio of charge to potential is called the *capacitance* of the body. The basic unit in which capacitance is measured is the *farad* (F), which is a coulomb per volt. That is, if one coulomb of charge added to a body gives it a potential of one volt, it has a capacitance of one farad:

$$C \text{ (farads)} = \frac{Q \text{ (coulombs)}}{V \text{ (volts)}}.$$

The farad is an awkwardly large unit, and capacitance is more often measured in microfarads (μF) or picofarads (pF = 10^{-12} F, often referred to as a "puff." Sometimes the picofarad is written as $\mu\mu$F.)

The capacitance of an isolated body such as an electroscope is very small. If an ordinary dry cell (about 1.5 volts) is connected between the electroscope knob and the ground (Fig. 15-10A), thereby moving electrons from the ground to the knob until the electroscope has a potential of 1.5 volts with respect to the ground, there will be no perceptible movement of the leaves. The Q transferred to the electroscope is so small that it will cause no perceptible repulsion of the leaves.

We can modify the electroscope by replacing the knob with a large flat sheet of metal. On this we put a thin sheet of insulating material and complete the sandwich with another flat metal sheet, as shown in Fig. 15-10B. The battery will pull electrons from the top plate through the battery and into the bottom electroscope plate until the potential difference between the two plates reaches 1.5 volts. The amount of charge transferred will be enormously greater than in Fig. 15-10A, however. The attraction of the positively charged top plate will help bring in and hold the negative charge on the bottom plate. This attraction of the top plate will also prevent much negative charge from accumulating on the gold leaves, and the leaves will hang nearly straight down. The battery can now be disconnected. (The charging of the two plates will take only a

FIG. 15-10 Increasing the charge on an electroscope by using a capacitor.

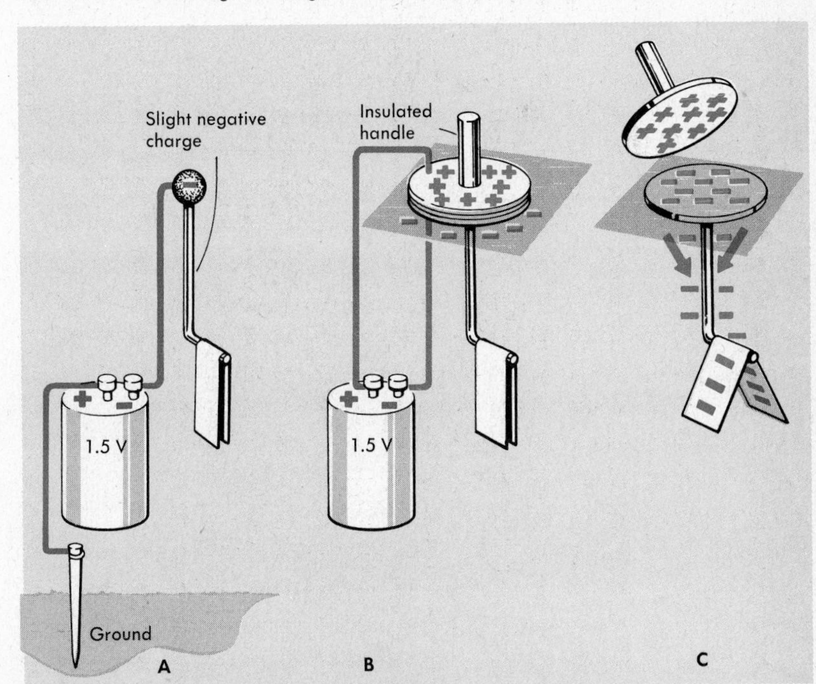

fraction of a second.) When the top plate is removed, the mutual repulsion of the excess electrons on the electroscope will drive some of the charge down into the leaves, which will now stand apart (Fig. 15-10C).

A device such as we have put on the electroscope—two sheets of metal or other conductor, with a sheet of insulating material between them—is called a *condenser* or *capacitor*. The capacitance of the condenser—that is, the amount of electrical charge on its plates divided by the potential difference between its plates—depends on several factors. One such factor is obviously its area; if the plates are made twice as large, the charge held on them will be twice as great. The thickness of the insulating layer between the plates is also important. The closer the plates are to one another, the greater is the amount of charge that is held, because of the increase in the strength of the electric field between the plates as they are brought closer together.

Figure 15-11 shows a pair of parallel charged plates separated by d m and with a potential difference of V volts between them. In the empty space between them will be an electric field with a strength we can call E N/C. This field will be uniform and perpendicular to the plates except at the very edge. In order to find the strength of the field, let us take a small charge of $+q$ C and move it from the negative plate to the positive plate. The work needed to do this is $F \times d$, or Eqd joules. The work can also be found in another way: Since V is by definition the work in joules needed to move 1 coulomb from one plate to the other, then to move q coulombs will require Vq joules. Equating the two expressions for the same work, we find

$$Eqd = Vq$$

or

$$E \text{ N/C} = \frac{V}{d} \text{ V/m}.$$

FIG. 15-11 Charged parallel plates with vacuum (or air) between them.

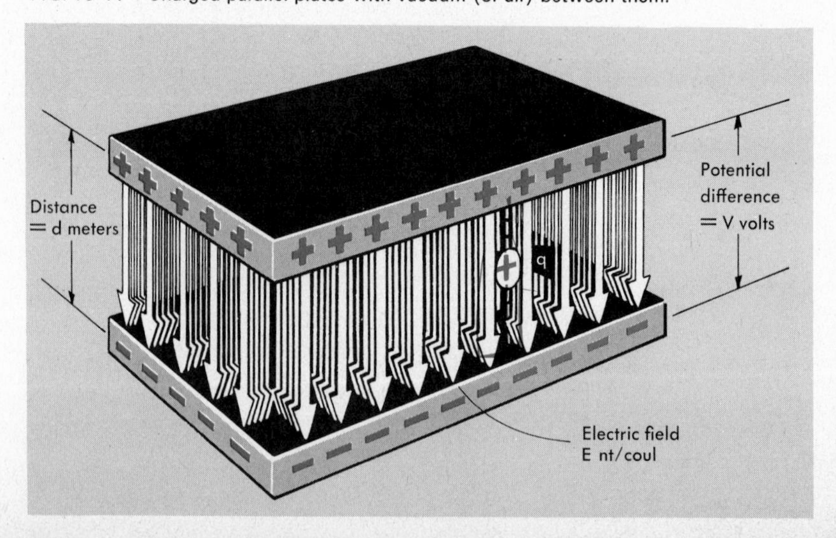

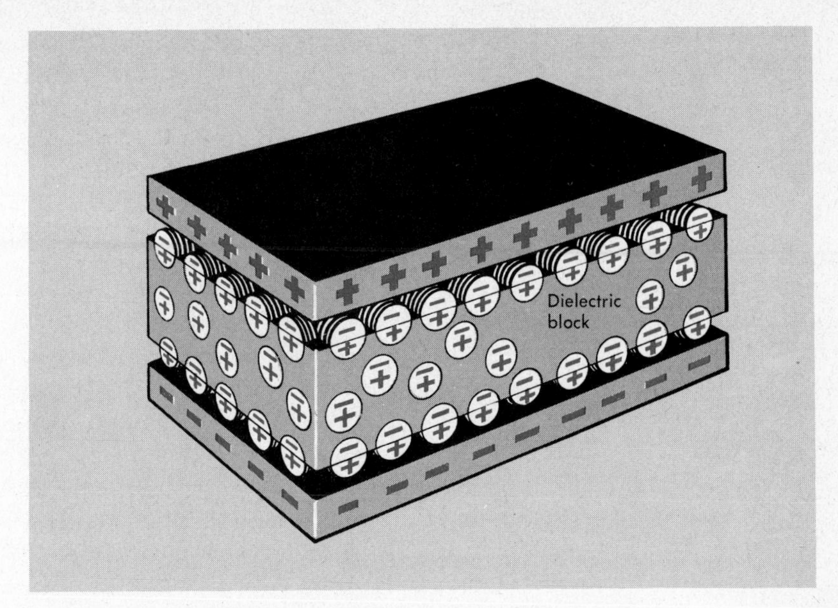

FIG. 15-12 Charged parallel plates separated by a dielectric material.

Thus we see that expressing electric field strength in volts per meter is exactly equivalent to expressing it according to its definition as newtons per coulomb.

If, for example, we have a parallel-plate capacitor made of two metal sheets 10 cm × 10 cm, 2 mm apart, and connect the plates to the terminals of a 45-volt battery, we have for V/d: $45/(2 \times 10^{-3}) = 22{,}500$ V/m. The electric field between the plates is thus 22,500 N/C.

The material between the plates also has a strong influence on the capacitance of a condenser. Figure 15-12 shows a dielectric slab between the plates. (Insulators or nonconductors are known as *dielectrics*.) We have already mentioned the distortion or *polarization* of atoms in an electric field, and in Fig. 15-12 the atoms of the dielectric between the plates are represented schematically in a distorted form. The positive signs represent the nuclei thrust downward by the field between the plates; the negative signs above the nuclei represent the centers of the clouds of electrons that are thrust upward. In the central material of the slab, these atomic distortions average out to no net effect. At the top surface, however, the upward-pulled electrons result in a layer of excess negative charge. Similarly, on the bottom face of the slab, the downward-pulled positive nuclei produce a surface layer of positive charge. These layers of induced charge help maintain the charges on the metal plates and result in a higher capacitance for the condenser. For our purposes, we can define the *dielectric constant* of an insulating material as the factor by which it multiplies the capacitance of a condenser, as compared with the same condenser with a vacuum between the plates. Here are dielectric constants K for a few common dielectrics:

Hard rubber	2.5
Glass	6 to 10
Oil	2 to 2.3
Water (pure)	80
Mica	5.5 to 6.5
Air	1.001

What causes the difference in the dielectric constants for different materials? Why does a sheet of glass between the plates of a capacitor increase its capacitance by a factor of, say, 6, while the same capacitor dipped into oil would have its capacitance only doubled? The answer is, of course, that some atoms or molecules are more readily polarizable than others. In glass, the layers of + and − charge on each side are stronger and better separated than the corresponding layers formed in oil. Water, with its enormous dielectric constant of 80, introduces a new factor, because the water molecule is already polarized, even if it is not in an electric field. One end of a water molecule is positive, and the other negative; in an electric field, the molecule merely rotates (which it can readily do in its liquid state) in response to the electric forces on it, and thereby readily produces strong layers of induced charge on its surfaces.

Putting all this discussion together, we see that the capacitance of a parallel-plate capacitor is proportional to its area A and to the dielectric constant K, and inversely proportional to the distance d separating the plates:

$$C = k \times \frac{KA}{d}.$$

If C is to be in farads, A in square meters, and d in meters, the proportionality constant k can be shown to be 8.85×10^{-12}, which gives us

$$C = \frac{8.85 \times 10^{-12}\, KA}{d}.$$

The capacitor used as an example earlier in this section thus has a capacitance of

$$C = \frac{8.85 \times 10^{-12} \times 1 \times 0.10 \times 0.10^{*}}{2 \times 10^{-3}} = 4.42 \times 10^{-11}\ \text{F}$$
$$= 44.2\ \text{pF}.$$

* If the dimensions were all accurate to one part in a thousand and we wanted an answer to equal accuracy, it would have been necessary to use 1.001 for the dielectric constant of the air between the plates. Since only a very few special capacitors would warrant calculations this precise, the difference between vacuum (1.000) and air (1.001) has been neglected here.

The charge on each of the plates of this capacitor at a potential difference of 45 volts would be

$$Q = CV = 4.42 \times 10^{-11} \times 45 = 1.99 \times 10^{-9}\ \text{C}.$$

If a sheet of hard rubber 2 mm thick were slipped between the capacitor plates, with the battery still connected, V would of course remain the same, but the capacitance and the charge on the plates would each become greater by a factor of 2.5. Suppose, though, that the battery is first disconnected, and then the rubber sheet slipped into position. Since there is no conductor through which electrons can move, the charge on the plates must remain the same. The capacitance is again increased by a factor of 2.5; since $V = Q/C$, the potential difference between the plates drops to 18 volts.

Questions

(15-1) **1.** Two small charged balls, 8 cm apart in air, have charges of +12 and +4 esu. What is the force between them?

2. Two small charged balls are placed 5 cm apart in air. One ball has a charge of +6 esu and the other, −20 esu. What is the force between them?

3. Two small equally-charged balls repel each other with a force of 1.2×10^{-4} N when they are 4 cm apart. How many coulombs of charge is on each ball?

4. Two protons repel each other with a force of 4×10^{-11} N when they are 2.4×10^{-9} m apart. What is the charge on each proton, in coulombs?

5. An electron has a charge of 1.60×10^{-19} C; the charge of an alpha particle is twice as large, and positive. How far apart are an electron and an alpha particle when there is an attraction of 6×10^{-11} N between them?

6. One small ball has a charge of $+2 \times 10^{-9}$ C; another has a charge of -4×10^{-9} C. How far apart must they be to attract each other with a force of 4.5×10^{-5} N?

(15-3) **7.** Two identical metal-covered balls attract each other with a force of 8 dynes when they are 3 cm apart. The balls are now touched together, and their total charge becomes equally divided between them. When replaced at 3 cm apart they are found to repel each other with a force of 1 dyne. (a) What is the explanation for the switch from attraction to repulsion? (b) What is the charge on each ball after touching? (c) What were the charges before touching?

8. Two charged balls are identical, and are covered with a metal coating. They attract each other with a force of 27 dynes when they are 4 cm apart. The balls are touched together, and their total charge divides equally between them. They now repel each other with a force of 9 dynes when placed 4 cm apart. (a) What is the explanation of the change from attraction to repulsion? (b) What is the charge on each ball after touching? (c) What were the charges before touching?

(15-4) **9.** In previous questions, we have tacitly assumed that the conductive balls were small enough in comparison to their distance apart that any migration of charge over their surfaces would not affect our answers. Consider, however, two such balls 0.5 cm in diameter, with a charge of -2 esu on each. Place the balls 2 cm apart, measured center-to-center. Taking the inevitable migration of charge into account, would the actual repulsion between them be greater or less than the 1 dyne we might have expected?

10. Repeat Question 9 for the case in which the charges on the balls are $+2$ and -2 esu. Will their attraction be greater or less than 1 dyne?

(15-5) **11.** Given a neutral electroscope, a piece of silk, and a glass rod, describe two ways to give the electroscope a positive charge, and two ways to give it a negative charge.

12. Given a neutral electroscope, a bit of fur, and a rubber rod, describe two ways to give the electroscope a positive charge, and two ways to give it a negative charge.

13. A hard-rubber rod that has been rubbed with fur is brought near the knob of an already charged electroscope. As it approaches, the leaves come closer together until they hang straight down. When the rod is brought still closer (but without touching the knob), the leaves stand apart again. (a) Explain what has happened during the experiment. (b) Was the charge on the electroscope positive or negative?

14. An electroscope has been charged by touching the knob with a piece of glass that has been rubbed with silk. As a charged object is brought near the electroscope, the leaves come closer together until they hang straight down. When the object is brought still closer (but without ever touching the electroscope), the leaves stand apart again. (a) Explain what has happened during this experiment. (b) Is the charge on the object positive or negative?

(15-6) **15.** Point A is 5 cm west of a particle having a charge of -2.15×10^{-9} C. What is the magnitude and direction of the electric field at A?

16. What is the intensity and direction of the electric field at a point 10 cm north of a small body bearing a charge of $+3.60 \times 10^{-9}$ C?

17. (a) What is the electric field 3×10^{-9} m from an electron? (b) What is the force on an alpha particle at this distance from an electron? (An alpha particle has a positive charge twice as large as the charge on an electron.)

18. (a) What is the electric field 3×10^{-9} m from an alpha particle? (An alpha particle has a positive charge twice as large as the charge of an electron.) (b) What is the force on an electron at this distance from an alpha particle?

19. A particle with a charge of -7×10^{-10} C is 6 cm from another particle with a charge of -63×10^{-10} C. At what point (or points) is the electric field zero?

20. A small body with a charge of $+64 \times 10^{-10}$ C is 12 cm from another small body whose charge is $+256 \times 10^{-10}$ C. At what point (or points) is the electric field zero?

21. Point A is 0.10 m from point B; C is 0.08 m from B and 0.06 m from A. Charges of 9.60×10^{-8} C are at each of the three points. What is the magnitude of the force on the charge at point C?

22. What is the magnitude of the force on an electron that is 3×10^{-9} m from one alpha particle, and 4×10^{-9} m from another, the two alpha particles being 5×10^{-9} m apart? (See Question 18.)

(15-7) **23.** What is the electric potential at point A in Question 15?

24. What is the electric potential at the point described in Question 16?

25. An electron is 5×10^{-9} m from an ion whose charge is $+4.8 \times 10^{-19}$ C. How much work is needed to remove the electron from the vicinity of the ion?

26. An electron and an alpha particle (see Question 18) are 2.7×10^{-11} m apart. How much energy must be expended to separate these two charged particles?

27. (a) In Question 19, is there any point at which the potential is zero? (b) If the 7×10^{-10} C charge is made positive instead of negative, is there now a point (or points) of zero potential on the straight line passing through the charges? If so, where?

28. (a) Is there any point of zero potential in the neighborhood of the charges in Question 20? (b) If the 64×10^{-10} C charge is made negative instead of positive, is there now a zero potential point (or points) on the straight line passing through the charges? If so, where?

29. A charge of -3×10^{-8} C is 20 cm from a $+9 \times 10^{-9}$ C charge. Point A is on the line joining the charges and is 6 cm from the negative charge; point B is on the same line, 3 cm from the positive charge. (a) What is the potential (in volts) at A? (b) What is the potential at B? (c) What is the potential *difference* (in volts) between A and B? (d) How much work would be required to move a charge of -2×10^{-8} C from A to B?

30. A charge of $+10^{-7}$ C is 15 cm from a charge of $-0.2\ \mu$C. (a) What is the potential (in volts) at A, 4 cm from the positive charge, and on the line connecting the two charges? (b) What is the potential B, 6 cm from the negative charge and on the same line? (c) What is the potential *difference* (in volts) between A and B? (d) How much work (in joules) would be required to move a charge of $+1\ \mu$C from B to A?

(15-8) **31.** A capacitor consists of a pair of parallel metal plates with oil (dielec. const. = 2) between them and has a capacitance C. What will its capacitance be if we drain the oil from between them, and make their separation three times what it originally was?

32. A capacitor formed of a pair of parallel metal plates with air between them has a capacitance C. What will be its capacitance if the plates are brought to one-fourth as far apart as they were, and the space between them is filled with mica?

33. A 10^{-11}-F capacitor is made of two flat parallel metal plates 1 cm apart. By connecting them to battery terminals the plates are given a potential difference of 100 volts. (a) What is the charge on each plate? (b) What is the electric field between the plates?

34. A 6-pF capacitor consists of two flat parallel metal plates 8 mm apart. The plates are connected to a 90-volt battery. (a) What is the charge on each plate? (b) What is the electric field between the plates?

35. The battery in Question 33 is disconnected from the plates and then the plates are moved to 5 mm apart. Now, (a) What is the charge on each plate? (b) What is the potential difference between the plates? (c) What is the electric field between them?

36. The plates of the capacitor in Question 34 are moved to 12 mm apart and the 90-volt battery remains connected to the plates. Now, (a) What is the potential difference between the plates? (b) What is the charge on each plate? (c) What is the electric field between them?

37. Battery terminals are briefly connected to the plates of a capacitor, and are then disconnected. Does the electric field between them become stronger or weaker when the air separating the plates is replaced by oil?

38. A parallel-plate capacitor remains connected to a battery. Is the charge on the plates increased or decreased when a sheet of glass is substituted for the air separating them?

39. What is the charge on the plates of the capacitor of Question 33 when the 1-cm space between them has been filled with oil of dielectric constant 2.0? (The battery remains connected.)

40. After the battery of Question 34 is disconnected, the 8-mm space between the plates is filled with oil whose dielectric constant is 2.2. What, then, is the potential difference between the plates?

41. A capacitor consists of two sheets of aluminum foil 3 cm × 10 m, separated by a strip of waxed paper (diel. const. = 2.5) 0.1 mm thick. What is the capacitance, in pF?

42. What is the capacitance (in μF) of a capacitor made of two metal sheets 10 cm × 20 cm, separated by 2 mm of glass whose dielectric constant is 7.5?

16

Electric Currents

16-1 Electric Cells

If we connect a conductor charged with, let us say, positive electricity, to the ground or to a negatively charged conductor by means of a metallic wire (Fig. 16-1A), the conductor will be discharged. For a split second, a current of electrons will flow through the wire. The duration of such an electric discharge is, however, too short for convenient study, and it is desirable to have an arrangement that provides us with a steady current. This became possible after the discovery that steady electric current can be produced by the *electric cell*, invented by the Italian physicist Allessandro Volta (1745–1827). The simplest sort of voltaic cell consists of two plates or rods (called *electrodes*) made of two different materials, such as carbon and zinc, placed in some conducting solution (called an *electrolyte*), such as sulfuric acid (Fig. 16-1B).

The little cells you slip into your flashlight work on the same principle, but the chemical reactions that take place are very complex, and best left to a chemistry course. They all depend, however, on the process of *ionization*. As we have seen in a previous chapter, normal atoms and

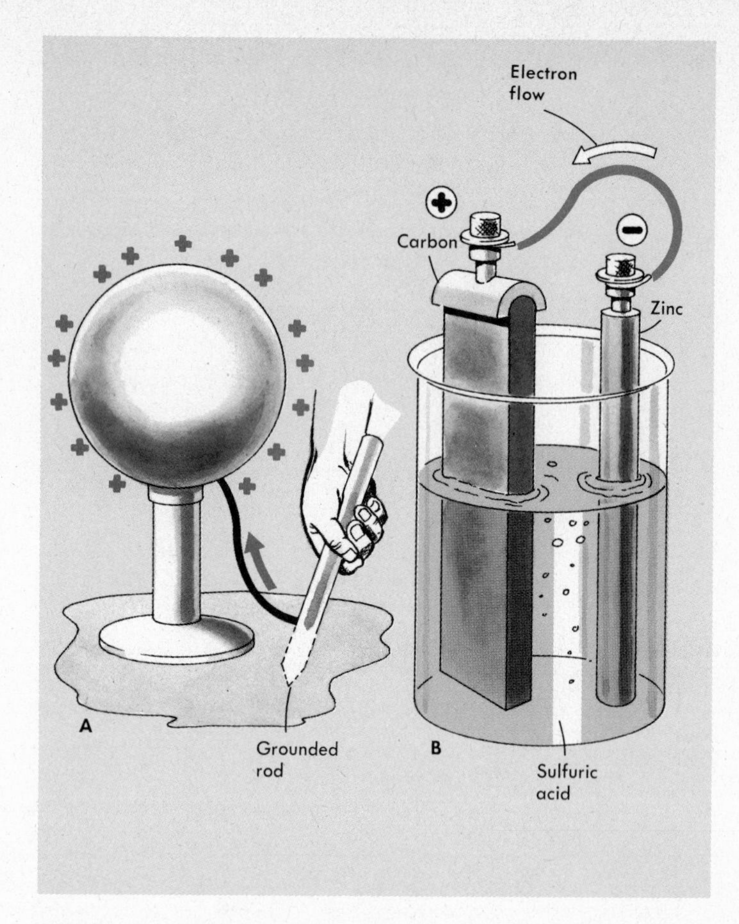

FIG. 16-1 (A) An instantaneous electric current from a charged conductor; (B) a continuous electric current from a chemical cell.

molecules contain equal + and − charges, and are electrically neutral. If we dissolve, say, a spoon of sugar in water, the dissolved sugar molecules remain intact and electrically neutral. Salts behave differently—in solution, the salt molecules split into two parts, and one of these carries with it one or more extra electrons, leaving the other with an equal deficiency of electron charge. We say that salts are *ionized* in solution; that is, they split into electrically charged pieces called *ions*. Many substances—in particular, salts, acids, and bases—ionize when they are dissolved and produce an *electrolyte*, a solution containing ions that are free to migrate through it and thereby conduct a current of electric charge.

In the zinc–carbon-and-acid cell of Fig. 16-1B, zinc metal is not soluble in water. In this acid electrolyte, however, some of the zinc can dissolve, but *not* as zinc atoms. Each zinc atom that goes from the metal rod into the electrolyte leaves two electrons behind and dissolves as a doubly-charged positive zinc ion. This makes the zinc rod more and more negative from the extra electrons that are left behind as the zinc dissolves.

In the meantime there is also activity at the carbon rod. As the electrolyte tends to become positively charged from the presence of the positive zinc ions, it remains neutral by attracting electrons into solution from the carbon, which thereby gets a + charge. This leaves the cell as shown in the drawing: The zinc electrode in negative; the carbon electrode positive.

If we connect the electrodes with a wire, electrons will flow through the wire from zinc to carbon, more of the zinc will go into solution to renew the charge, and so on, until the zinc is all dissolved.

Any pair of dissimilar conducting materials can be put into any electrolyte and will become a cell that will convert chemical energy into electrical energy in a manner similar to the cell described above. These cells will vary widely, however, in the amount of chemical energy released per coulomb of charge. You will recognize this last sentence as another way of saying "these different cells will have different voltages," since "voltage" means the energy in joules per coulomb of charge.

16-2 Electrical Resistance

In dealing with the electric cell, we have spoken of the stream of electrons flowing along the wire connecting the positive electrode of the cell to the negative electrode. This electron flow, or current, could be described as so many electrons per second. This would require the use of enormously large numbers, however, and scientists have found it convenient to invent a special unit to describe the flow of electric charge. This unit, named for a French physicist and early electrical experimenter, André Ampère (1775–1836), is the *ampere* (A). *One ampere is a current of one coulomb per second.*

$$1\ \text{A} = 1\ \text{C/s}.$$

Just as water flows faster through a pipe when the difference in pressure between the ends of the pipe is larger, electric current depends on the difference in electric potential between the ends of the wire. The law discovered by a German physicist, G. S. Ohm (1787–1854), states that for many substances, particularly metals, ***the electric current in a conductor is directly proportional to the potential difference between its ends.*** In other words, every conductor presents some resistance to the passage of an electric current, and to increase a current through it, proportionally more energy per coulomb of charge is required. On the basis of Ohm's law, a unit of electrical resistance has been defined and named the *ohm*. The ohm can be defined as follows: *The resistance of a wire or other conductor is 1 ohm if a potential difference of 1 volt between its ends will cause a current of 1 ampere to flow through it.*

In *metal* conductors, Ohm's law is obeyed, and the electric current I is proportional to the potential difference V and inversely proportional to the resistance of the conductor R. Because of the way the ohm was defined, this relationship can be put into the very brief and convenient form known as Ohm's law:

$$I\,(\text{amperes}) = \frac{V\,(\text{volts})}{R\,(\text{ohms})}$$

or

$$V\,(\text{volts}) = I\,(\text{amperes}) \times R\,(\text{ohms})$$

or

$$R\,(\text{ohms}) = \frac{V\,(\text{volts})}{I\,(\text{amperes})}.$$

Ohm's law does not hold for the current through the electrolytes of cells, for electrical discharge through gases, for vacuum tubes, or for devices that use poorly conducting materials called *semiconductors*. In most electric circuits, however, the current flows through metallic conductors, and we will make extensive use of Ohm's law when we take up electrical circuits.

Electrical conductors come in all shapes and sizes, the most common being a wire, which is merely a long cylinder. A copper wire 0.1 inch in diameter and 1000 ft long will have a resistance of very nearly 1 ohm. (Ω, the Greek capital letter omega, is the recognized abbreviation for the ohm.) Wires of the same size made of other materials will have different resistances. An identical wire made of silver would have a resistance of only 0.9 Ω, and silver thus has a lower resistivity than copper.* Aluminum, because of its lower cost, is being used more and more instead of copper wire, even though its resistivity is 1.6 times that of copper. As a comparison we might consider a wire (or "rod" might be a better term) made of glass, which is an excellent insulator. Such a rod the same size as the 1-Ω copper wire would have resistance of 10^{19} Ω; one made of amber would have 3×10^{22} Ω resistance.

Intuition tells us that a long conductor should have more resistance than a short one, and that a fat, thick conductor should have less resistance than a thin, fine one. Experimental evidence shows this guess to be a good one: Resistance is directly proportional to the length of the conductor and inversely proportional to its cross-sectional area.

16-3 Electrical Circuits

A simple electrical circuit has already been shown in Fig. 16-1, and the flow of charge through the circuit has been discussed qualitatively.

---

* During World War II, when copper was very scarce, many tons of monetary silver were made into heavy electrical conductors for use in industry. After the war the silver was of course replaced with copper and returned.

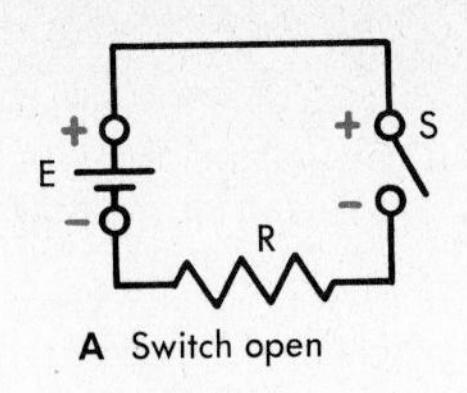

A Switch open

Electron current

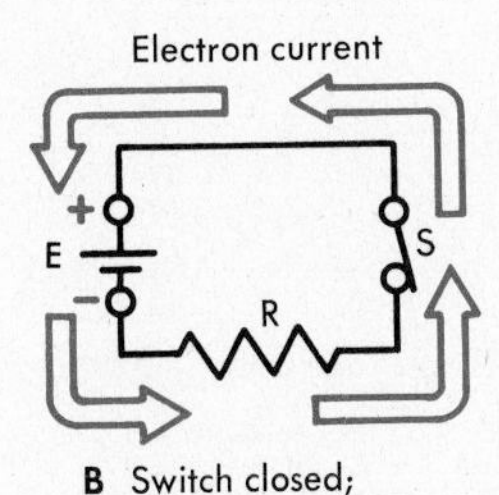

B Switch closed; flow of electrons

"Conventional" current

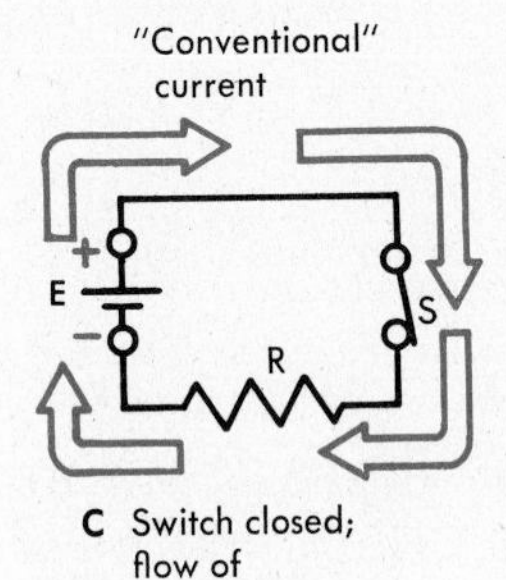

C Switch closed; flow of hypothetical positive charges

A circuit must include some source of potential, or energy (in order to make the charge move), and a path through which the charge can travel. Figure 16-2 shows another electrical circuit. It contains a cell as a source of energy (the cell is symbolized by the pair of parallel lines to the left of the diagram—the accepted convention is that the longer line represents the positive electrode; the shorter line, the negative); a resistance of some sort (symbolized by the zigzag line R—it could be a heater or a light bulb); and a switch.

The letter E by the cell stands for *electromotive force* (emf), which is not a force at all but the electric potential produced by the cell. That is, it is the total energy given each unit of charge as it passes through the cell, as a result of chemical energy being converted into electrical energy. It might well have been labeled V, but since the voltage of the terminals of the cells may be different from its emf, it is well to avoid confusion and reserve E for this use. If R is the total resistance of the circuit and E is the total emf, then the current in the circuit, by Ohm's law, will be $I = E/R$.

In Fig. 16-2A the switch is "open," and there is no complete path, or circuit, around which the electrons can flow. The cell can only pull electrons from the upper part of the circuit and push them into the lower part until the open contacts of the switch have the same potential difference as the terminals of the cell. This will take only a small fraction of a second, and then there will be no further motion of electrons.

The switch has been closed in Fig. 16-2B, and electrons can now flow around a complete circuit. The direction of electron flow will, of course, be from the negative terminal of the cell, where there is an excess of electrons, counterclockwise around the circuit to the positive cell terminal where there is a deficiency of electrons. Unfortunately for the physics student and his instructor, the "conventional" electric current flows in the opposite direction (Fig. 16-2C). This convention of considering the flow to be from the + terminal through the circuit and back to the − terminal was adopted more than a century before electrons were known to exist and is firmly established in the literature of electricity. Actually (on the principle that two negatives are equivalent to a positive) it makes no difference if we consider hypothetical positive charges moving in one direction, or real negative charges moving in the opposite direction. The conventional direction has the advantage, also, that it is more natural to think of a flow from high potential to low than it is to picture the reverse.

16-4 Series Circuits

Figure 16-3 shows a situation a bit more complex. Instead of a single cell, we have a *battery* of five cells connected *in series*. If cells, or resistances, or any other elements in a circuit are in series, it means they are connected together without side branches in such a way that the entire current must flow through each one. The drawing shows not only the

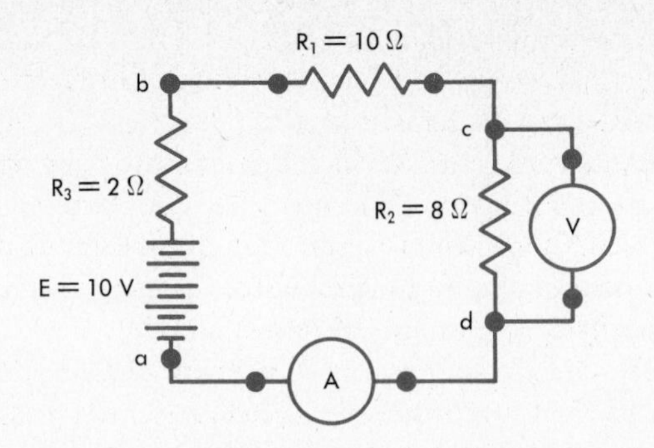

FIG. 16-3 Measurement of potential difference and current in an electric circuit.

cells, but the resistances, connected in series. If each cell has an emf of 2 V, it means that 2 joules of potential energy will be given to each coulomb of charge that flows through the cell. Hence a coulomb of charge, which must flow through each of the five cells in series, will be given 2 joules by each cell, for a total of 10. Thus the total emf of the battery is 10 J/C, or 10 V.

The current passes through resistances of 2 Ω, 10 Ω, and 8 Ω, in series, for a total of 20 Ω. We may summarize the argument above by saying that when emf's or resistances are connected *in series*, we need only add them together:

$$R \text{ total} = \sum R \qquad E \text{ total} = \sum E.$$

The letters *a* and *b* represent the terminals of the battery, and the resistance R_3, lying between the terminals, is the *internal resistance* of the battery. This 2-Ω resistance actually occurs principally within the electrolyte of each of the cells, but for convenience is shown as though it were all gathered together in one place. For the entire circuit, then, we have a total emf of 10 V, a total resistance of 20 Ω, and hence a current of $\frac{10}{20}$, or 0.5 A.

On the drawing are shown an ammeter *A* and a voltmeter *V*. We will not be able to discuss how they operate until later, but even without a knowledge of their operation some things can be said about their use. The voltmeter is connected to the circuit to measure the difference in potential between points *c* and *d* and is not concerned with how much current is flowing. Voltmeters are made with a *very high* resistance, so that only an insignificant trickle of current runs from *c* to *d* through the meter, and the original unmetered circuit is virtually unchanged. Forgetting the rest of the circuit for a moment, we see that the voltmeter is connected to read the potential difference between the ends of a resistance through which a current of 0.5 A is flowing. By Ohm's law we know that this potential difference must be

$$\begin{aligned} V &= IR \\ &= 0.5 \times 8 = 4.0 \text{ volts.} \end{aligned}$$

Because of the way it is figured, we can refer to the potential difference between the ends of a resistance through which a current flows as the "IR drop across the resistance."

Similarly, the IR drop across R_1 is $0.5 \times 10 = 5.0$ V, across R_3 it is $0.5 \times 2 = 1.0$ V. The sum of the three IR drops is $4.0 + 5.0 + 1.0 = 10.0$ V, which is, as it must be, the emf of the battery. The energy given the circulating charge is exactly equal to the energy it expends in heating the resistances through which it flows.

If we were to connect the voltmeter to a and b, the battery terminals, it would not read the emf of the battery, but something less, because some of the energy given to the charge is used up within the battery itself. Work must be done in moving the charge through the internal resistance of the cell. This work, like any work done to overcome resistance, is converted into heat and lost. As we have seen, this internal IR drop inside the battery is $0.5 \times 2 = 1.0$ V, so we would expect the terminal voltage to be $10.0 - 1.0 = 9.0$ V.

In a series circuit, it makes no difference where the ammeter is placed, because the current is the same in every part of the circuit. It is connected in a very different way from the voltmeter, however. It must itself be placed in series in the circuit, so all the current will pass through it. The ammeter must therefore have a *very low* resistance in order for its effect on the circuit to be negligible.

16-5 Parallel Circuits

We have already seen an example of a current that divides into two branches; in Fig. 16-3, the 8-Ω resistance and the voltmeter provide two such branches for the main current to flow through, and are said to be connected *in parallel.* Figure 16-4A shows a circuit in which three resistances are connected in parallel; when the current I reaches point a, it divides into three separate streams, I_1, I_2, I_3, which recombine at b and return to the negative terminal of the battery. We want to compute the value of the one single resistance R (in Fig. 16-4B) that can be substituted for R_1 and R_2, and R_3 and still have the same current I and the same potential difference between a and b.

In the arrangement shown in the drawing, the potential difference between a and b (or, as it is often stated, the potential across a and b) we shall call V. We can now write three simple equations for the separate branch currents:

$$I_1 = \frac{V}{R_1} \qquad I_2 = \frac{V}{R_2} \qquad I_3 = \frac{V}{R_3}.$$

Turning to Fig. 16-4B, we can similarly write for the main current

$$I = \frac{V}{R}.$$

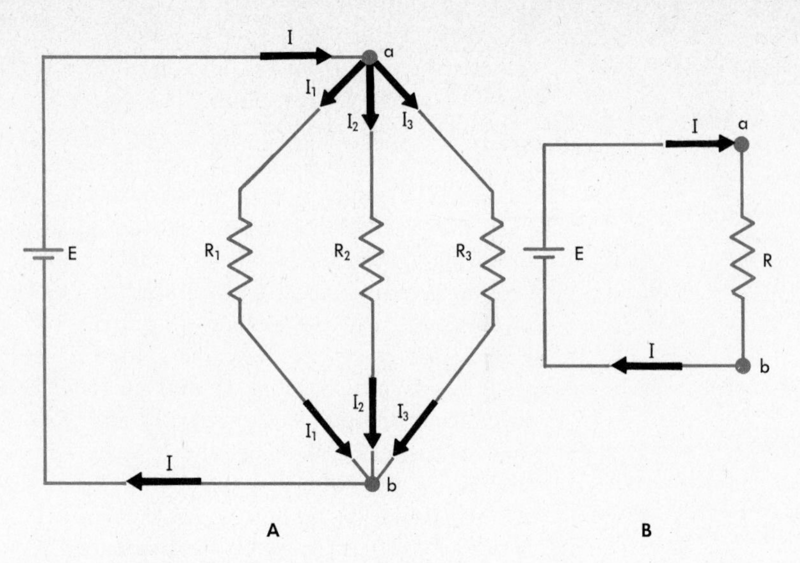

FIG. 16-4 Resistances connected in parallel.

It is apparent that the sum of the three branch currents must equal the main current:

$$I = I_1 + I_2 + I_3$$

or

$$\frac{V}{R} = \frac{V}{R_1} + \frac{V}{R_2} + \frac{V}{R_3}.$$

Dividing this entire equation by V, we get

$$\frac{1}{R} = \frac{1}{R_1} + \frac{1}{R_2} + \frac{1}{R_3},$$

which is the relationship we need to be able to calculate the equivalent resistance R from the values of the resistances connected in parallel. An inspection of the derivation will show that this relationship is applicable not only to three but to any number of resistances in parallel.

Figure 16-5 shows the steps followed in simplifying a complex circuit into an equivalent elementary circuit containing one emf and one resistance. In general, it is a good idea to get rid of any parallel branches by replacing them with the single resistances to which they are equivalent. In Fig. 16-5A, we see there are two such places in the circuit. In the top one, 3 Ω and 6 Ω are in parallel. To find the single resistance with which they can be replaced, we write

$$\frac{1}{R} = \frac{1}{3} + \frac{1}{6} = \frac{1}{2}; \qquad R = 2\,\Omega.$$

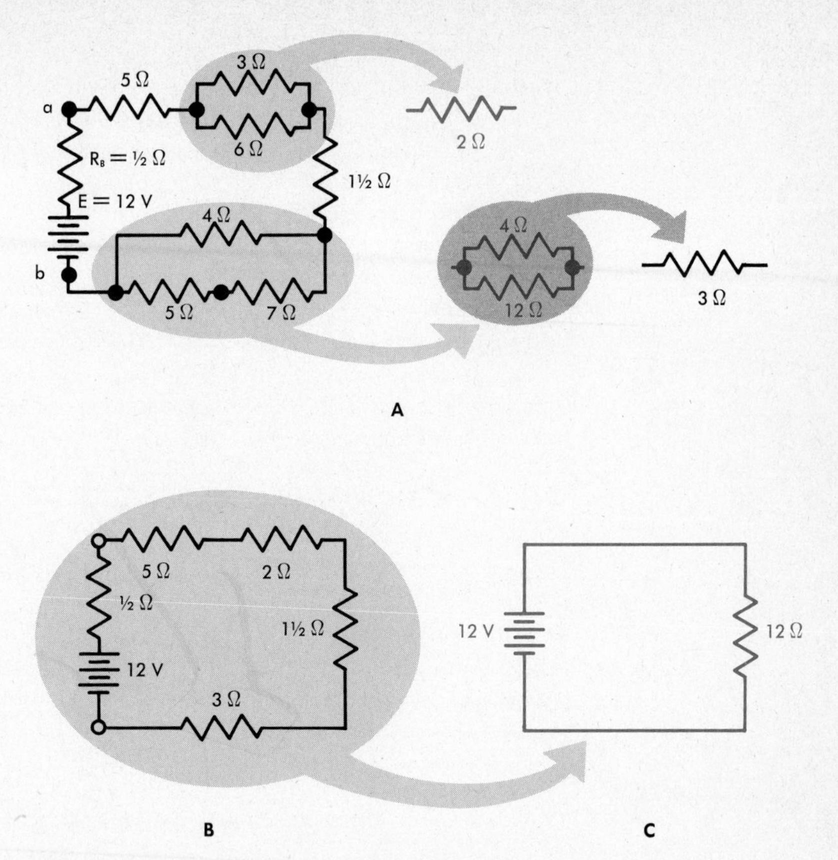

FIG. 16-5 The simplification of a circuit containing parallel branches.

In the bottom circle, the bottom branch consists of 5 Ω and 7 Ω *in series*; these can be immediately combined by simple addition, so that what we have is 4 Ω and 12 Ω in parallel. For this equivalent resistance,

$$\frac{1}{R} = \frac{1}{4} + \frac{1}{12} = \frac{1}{3}; \qquad R = 3\ \Omega.$$

Now, in Fig. 16-5B, each of the two parallel-connected parts of the circuit has been replaced by its single equivalent resistance. From here it is easy merely to add together all these series-connected resistances, and arrive finally at the simplest possible circuit of Fig. 16-5C.

(It should be noted that it is possible to have more complex circuits that may contain elements not unambiguously connected either in series or in parallel with other elements. The straightforward procedure used for Fig. 16-5 may not work in such cases, and a more advanced procedure must be used. We will not deal with such circuits in this book.)

16-6 Electrical Power and Energy

The practical electrical units—volt, ampere, coulomb and ohm—have been defined in such a way as to make the calculation of work and power very simple. In all resistances through which a current flows, electrical energy is converted into heat energy. The filament of an ordinary light-bulb is designed to be heated white-hot by the conversion of electrical energy into heat; the wires leading to the bulb are also slightly warmed by the passage of current. In the 8-Ω resistance of Fig. 16-3, we have calculated a current of 0.5A and a potential difference of 4 V between the ends of the resistance. A current of 0.5 A means that 0.5 C/s travels through the resistance; the 4-V potential difference means that 4 J/C of electrical energy have been lost—in this case, converted into heat. By multiplying together the current in amperes and the *IR* drop in volts, we get the power in watts:

$$0.5 \text{ C/s} \times 4 \text{ J/C} = 2 \text{ J/s} = 2 \text{ W}.$$

In general,

$$P(\text{watts}) = I(\text{amperes}) \times V(\text{volts}).$$

Of course $P = IV$ is not the only expression possible for determining the power in a circuit or in part of a circuit. From Ohm's law, $V = IR$; substituting this in the equation given above, we have

$$P = I \times IR = I^2R,$$

which may be more convenient to use than $P = IV$. Similarly, from $I = V/R$, we can get that

$$P = (V/R) \times V = V^2/R.$$

Since power is always the *rate* at which energy is being used or produced or converted into energy of another kind, the total amount of energy converted in some given time is simply the power multiplied by the time:

$$\text{Work(joules)} = P(\text{watts}) \times t(\text{seconds}).$$

In 5 minutes, a 100-watt heater will produce $100 \times 5 \times 60 = 30{,}000$ joules, or 7180 calories, of heat.

As an example, let us consider the circuit shown in Fig. 16-6A and calculate the power expended in R_1 and in R_2. If we knew either the potential difference across these resistors, or the current through them, the job would be easy. So we should probably begin by simplifying the circuit to find the total current. The first step is to combine R_3 and R_4 for a total $R_{3+4} = 20\ \Omega$. This branch ($R_3 + R_4$) is connected in parallel with R_1. The accidental fact that the drawing shows R_1 on the left side and $R_3 + R_4$ on the right side makes no difference; when the conventional current reaches point *b*, it must divide in two parts that

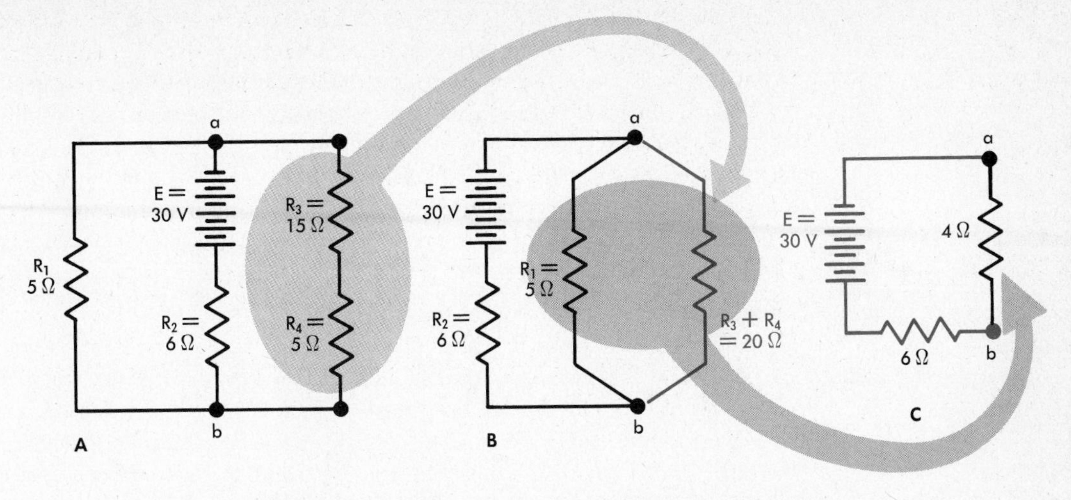

FIG. 16-6 Another example of the simplification of a circuit.

recombine at a to flow back into the battery. This is shown in Fig. 16-6B. It is apparent that the next step should be to replace the parallel resistances of 5 Ω and 20 Ω with a single equivalent resistance:

$$\frac{1}{R} = \frac{1}{5} + \frac{1}{20} = \frac{1}{4},$$

so $R = 4\ \Omega$, as shown in Fig. 16-6C. Now, for the whole circuit, there is a total resistance of $6 + 4 = 10\ \Omega$ and an emf of 30 V, which will result in a current $I = V/R = \frac{30}{10} = 3$ A. The potential difference V_{ab} is just the IR drop through the 4-Ω resistor in Fig. 16-6C, or $V_{ab} = 3 \times 4 = 12$ V. Armed with this knowledge of V_{ab}, we can turn back to Fig. 16-6B and determine the current through R_1: $I_1 = V_{ab}/R_1 = \frac{12}{5} = 2.4$ A. The power expended in R_1 can now be figured: $P = VI = 12 \times 2.4 = 28.8$ W; or, if you prefer, $P = I^2R = (2.4)^2 \times 5 = 28.8$ W. Similarly, the power expended in R_2 can be immediately reckoned as $I^2R = (3)^2 \times 6 = 54$ W.

It is important to remember that the wattage marked on light bulbs and other electrical appliances is not an inherent property of the device but depends on the voltage supply to which it is connected. A 60-watt light bulb is also marked (usually) 120 volts; the manufacturer says, in effect, that *if* the bulb is connected to a 120-volt source of potential, it will convert electrical energy into heat and light at the rate of 60 J/s. Let us figure what its wattage would be if it were connected to a 12-volt battery. From the manufacturer's specifications, we know that the light will draw 60 watts if it is connected to 120 volts. From the relationship $P = IV$, or $I = P/V$, we see that under these circumstances the current will be $\frac{60}{120}$, or 0.5 A. Ohm's law, $R = V/I$, shows that the resistance of the filament must be $120/0.5 = 240\ \Omega$. Assuming that this resistance remains the same, we can see that the current that will flow when the bulb is connected to the 12-volt battery will be $I = V/R = \frac{12}{240} = 0.05$ A. We can now determine the power to be $P = IV = 0.05 \times 12 = 0.6$ W. The value of $I^2R = (0.05)^2 \times 240 = 0.0025 \times 240 = 0.6$ W also, to confirm our previous calculation.

Questions

(16-2)

1. A current of 0.02 A flows through a resistor in a circuit, and the potential difference between the ends of the resistor is 50 V. What is its resistance?

2. A certain resistor is a part of an electric circuit. The potential difference across its terminals is determined by a voltmeter to be 18 V, and a properly connected ammeter shows that a current of 0.6 A flows through it. What is the resistance of the resistor?

3. A uniform wire 2 m long has a total resistance of 10 Ω. A voltmeter indicates a potential difference of 0.40 V between point *A*, 50 cm from one end of the wire, and a point *B*, 25 cm from the other end. What current flows in the wire?

4. A uniform wire 1 m long carries a current of 0.1 A and has a total resistance of 20 Ω. A voltmeter is connected to the wire at two points: 30 cm from one end, and 40 cm from the other. What potential difference does the voltmeter indicate?

(16-3)

5. A current of 2.4 A flows in a circuit containing an emf of 48 volts. What is the total resistance of the circuit?

6. A cell whose emf is 12.0 V is in a circuit whose total resistance is 5.0 Ω. What current flows in the circuit?

(16-4)

7. A battery consists of six cells in series, each cell with an emf of 1.2 V. The battery is in a circuit containing resistances of 1, 6, 3, and 8 Ω connected in series. What current flows through the circuit?

8. Each of the cells in a battery of three cells has an emf of 1.6 volts. This battery is in a circuit that contains resistances of 3, 4, and 5 Ω, all connected in series. What current flows in the circuit?

9. A cell with an emf of 1.5 volts and an internal resistance of 0.1 Ω is connected to four 0.5 Ω resistors in series. (a) What is the total resistance of the entire circuit? (b) What is the current in the circuit? (c) What is the *IR* drop in the internal resistance of the cell? (d) What is the terminal voltage of the cell? (e) What is the potential difference across one of the 0.5 Ω resistors?

10. A battery has an emf of 6 volts and an internal resistance of 1.5 Ω. The external circuit to which its battery is connected contains three 4.5-Ω resistors in series. (a) What is the total resistance of the entire circuit? (b) What is the current in the circuit? (c) What is the *IR* drop through the internal resistance of the battery? (d) What is the terminal voltage of the battery? (e) What is the potential difference across one of the 4.5-Ω resistors?

11. A 6-volt battery has an internal resistance of 0.05 Ω. What resistance connected across the battery will reduce its terminal voltage to 5.8 V?

12. A 12-volt battery has an internal resistance of 0.5 Ω. What resistance connected to the battery will give it a terminal voltage of 11.0 V?

(16-5) 13. Four resistors, two 2 Ω and two 4 Ω, are all connected in parallel. What is the resistance of the single equivalent resistor?

14. Three resistors, 3, 2, and 6 Ω, respectively, are connected in parallel. What is the resistance of the single equivalent resistor?

15. Three resistors of 4, 6, and 12 Ω, respectively, are connected in parallel and to the terminals of a 10-volt battery whose internal resistance is 0.5 Ω. What current flows in each resistor and in the battery?

16. A 4-Ω, a 3-Ω, and a 2-Ω resistor are all connected in parallel, and the combination is connected to the terminals of an 8-V battery of 0.3 Ω internal resistance. What current flows in each resistor and in the battery?

17. Resistances of 3, 7, and 2.5 Ω are connected as shown in Fig. 16-Q17 to a 10-V battery whose internal resistance is 0.4 Ω. (a) What current flows through each resistance, and through the battery? (b) What is the terminal voltage of the battery ($a - b$)? (c) What is the potential difference across $b - c$? (d) Across $a - c$?

18. Resistances of 3, 5, and 6 Ω are connected as shown in Fig. 16-Q18 to a 12-V battery whose internal resistance is 1 Ω (a) What current flows through each resistance, and through the battery? (b) What is the terminal voltage of the battery ($a - b$)? (c) What is the potential difference measured across $b - d$? (d) Across $c - d$?

(16-6) 19. A resistance of 10^4 Ω is connected to a 500-V source of potential. (a) At what rate (in watts) is electrical energy converted into heat in the resistance? (b) How long would it take to produce 1000 J of heat energy?

20. A 12-Ω resistor is connected to a 6-V supply. (a) What is the power dissipated in the resistor? (b) How long will it take to produce 100 J of heat in the resistor?

21. A 10^6-Ω (1 megohm, MΩ) resistor is rated at 0.5 W (i.e., 0.5 watt is the maximum power it can dissipate without overheating). What is the greatest potential difference that should be applied to this resistor?

FIG. 16-Q17

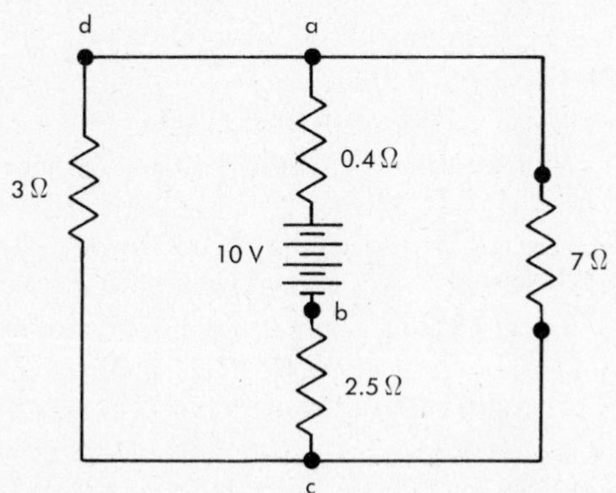

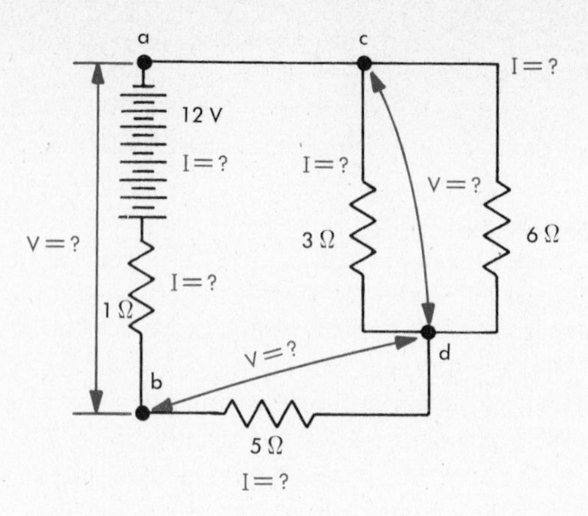

FIG. 16-Q18

22. A 2-Ω resistor can dissipate a maximum of 10 W without overheating. What is the maximum potential difference that can be applied across the resistor?

23. (a) What is the resistance of a 4800-W, 240-V water heater? (b) Assuming the resistance to be unchanged, what would its wattage be if connected to a 120-volt supply?

24. (a) What is the resistance of a 40-watt, 120-volt lamp? (b) Assuming the resistance remains the same, find the wattage of this lamp if it is connected to a 90-volt potential source.

25. An electromagnet is designed to operate at 30 watts from a 12-volt supply but the only available source of potential is 28 V. (a) If it is possible to add an auxiliary resistor to permit the proper operation of the magnet, how should this resistor be connected, and what should its resistance be? (b) What wattage would be dissipated by this auxiliary resistor, and what fraction is this of the power output of the 28-volt supply?

26. A piece of laboratory equipment requires the use of a 6-volt, 30-watt lamp, but the only potential source available is 120 V. (a) Would it be possible to connect a resistor in such a way as to permit the proper use of the lamp? If so, how should the resistor be connected, and what should be its resistance, in ohms? (b) What fraction of the total power consumed would be wasted in heating the resistor?

27. Two small 10-gallon, 120-volt water heaters are rated at 500 W and 1000 W, respectively. These heaters are connected in series across a 120-volt supply. Which heater will be the first to bring its water to proper temperature?

28. A 25-watt, 120-volt lamp and a 100-watt, 120-volt lamp are connected in series across 120 V. Which lamp will burn more brightly?

29. An insulated tank holds 2 gallons of alcohol of density 0.79 g/cm^3. (1 gallon water = 3.79 kg.) The alcohol is heated by submerging a 60-watt lamp in it. How long will it take to raise its temperature from 20° to 50°C?

30. A 4500-watt water heater holds 40 gallons of water at 18°C. (1 gallon = 3.79 kg water.) How long will it take to raise the water temperature to 75°C?

17

Magnetism

17-1 Magnets and Magnetic Fields

The ancient Chinese knew that slender pieces of certain natural iron ores, when suspended by a string, would assume a definite position with one end pointing approximately north and the other approximately south. It is thus apparent from the behavior of the magnetic compass that the earth is surrounded by a *magnetic field* that affects the orientation of lodestones and other magnets, both natural and artificial. The magnetic field that orients the compass needle manifests itself in many other ways; for example, it deflects the beams of electrically charged particles that come to us from the sun, thus producing the magnificent auroras often visible in the polar regions.

We can use the magnetic field of the earth to magnetize steel rods by holding them in the direction of the magnetic field of the earth and hitting them repeatedly with a hammer. The violent impacts shake the tiny particles of the rod and orient them, at least partially, in the direction of the field. As a matter of fact, all steel objects posses a certain small degree of magnetization induced by the terrestrial magnetic field.

If we bring two magnetized steel rods close together, we find that the "homologous" ends, i.e., the ends that pointed the same way during the magnetization process, repel each other and that if one of the rods is turned around, the ends of the rods attract one another.

This behavior shows that a long piece of magnetized material—a splinter of lodestone, a steel bar, or a compass needle—shows its magnetic properties most strongly in regions near its ends, known as the *poles* of the magnet. It also shows that like poles, i.e., poles that point toward the same direction, repel each other and that unlike poles attract. The use of magnets as compasses has given these two sorts of magnetic pole their names: The end that swings toward the geographic north is the *north pole* of the magnet; the other end, pointing in a southerly direction, is called the *south pole* of the magnet. It is interesting to notice (in view of the definitions above and in consideration of the fact that unlike magnetic poles attract) that the magnetic pole of the earth located near its geographical north pole is actually its *magnetic south pole*, and vice versa.

The magnetic field around a magnet can be represented by drawing (or imagining) lines, in very much the same way that we represented the electric field in the neighborhood of charged particles. The direction of the magnetic field at any point is shown by the orientation of a small compass needle placed at the point; the magnetic lines of force drawn in this way will be pointed in the same direction as the north pole of the compass needle, as indicated by the arrows in Fig. 17-1.

Figure 17-2 indicates in a very crude, sketchy manner the idea that magnetic materials are made of atomic particles that are themselves small magnets. In a piece of unmagnetized iron or steel, these particles point equally in all directions (Fig. 17-2A). When the iron or steel is

FIG. 17-1 The magnetic field of a bar magnet.

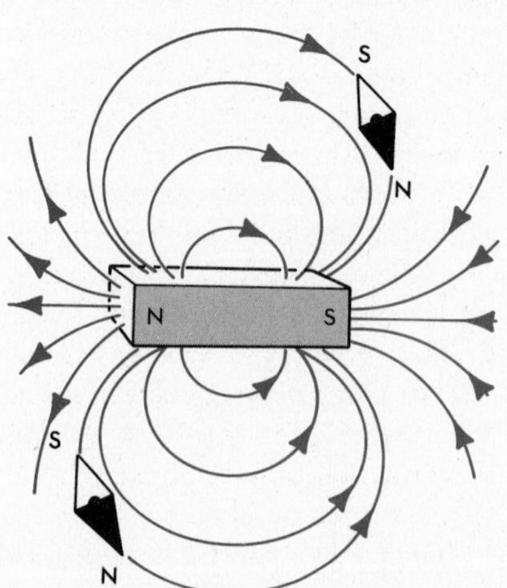

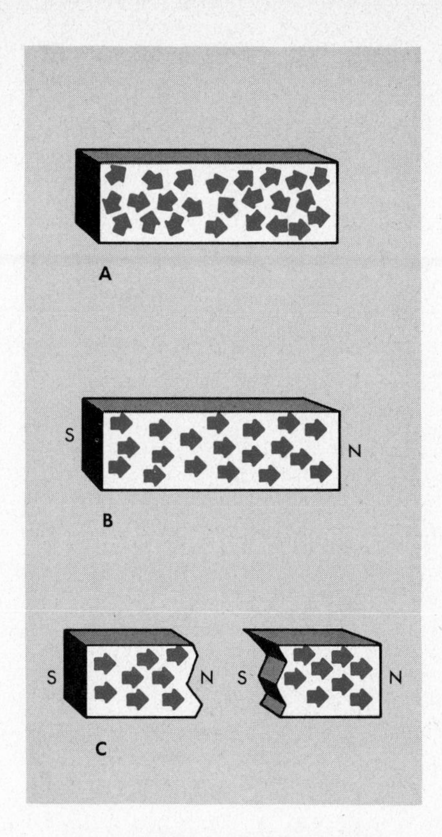

FIG. 17-2 Schematic diagram of alignment of atoms (A) in an unmagnetized bar; (B) in a magnetized bar; (C) in a magnetized bar that has been broken.

strongly magnetized, all the atomic magnets are lined up in substantially the same direction (Fig. 17-2B). In soft iron, the atomic magnets line up very readily when the iron is placed in a magnetic field. If a magnet is covered by a piece of paper or glass, and iron filings are sprinkled on the paper, each filing becomes an *induced magnet* and aligns itself in the direction of the field (Fig. 17-3). Since soft iron magnetizes readily, it also readily loses its magnetism. When it is removed from the magnetic field, the atomic magnets are immediately again thrown into completely random alignment by the jostling of their neighbors.

In hard steel, however, and to an even greater extent in certain special alloys, the atomic magnets, when once aligned, remain aligned until the steel is heated to high temperatures. Permanent magnets are thus made of such materials. Conversely, it requires a strong field to magnetize these materials. In a weak field, as mentioned above, they must be jarred by hammering to permit the atomic magnets to slip into alignment.

It is important to remember that, unlike positive and negative charges, *magnetic poles must always occur in pairs*, and that it is impossible to cut a

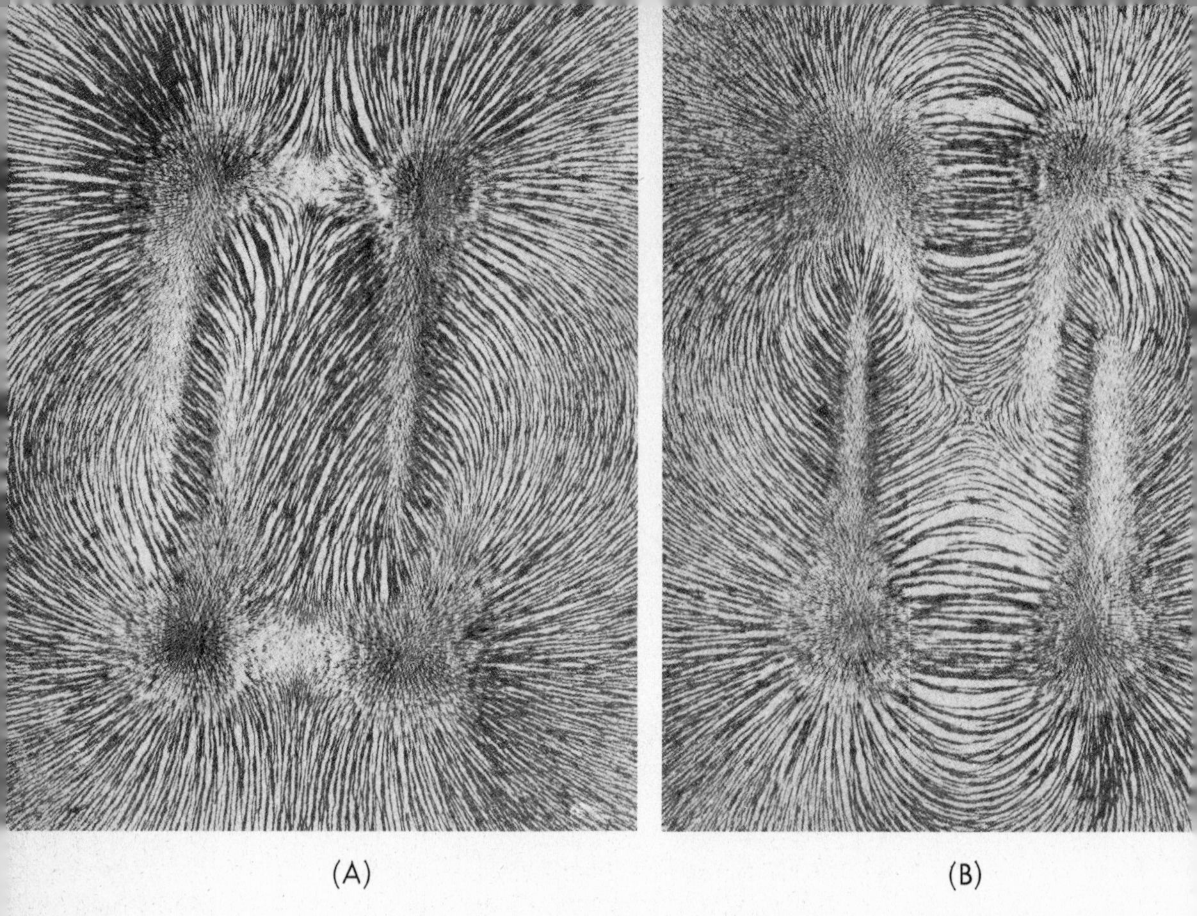

(A) (B)

FIG. 17-3 Iron filings indicating the magnetic field of a pair of bar magnets beneath a sheet of paper. (A) Magnets with like poles opposite each other. (B) Magnets with unlike poles opposite. (*University of Colorado photograph.*)

north or south pole from a magnet and carry it away. If we cut a magnet into two pieces, we will get two smaller magnets, since a new pair of poles will originate at the broken ends. On the basis of the atomic magnetic picture, it is plain to see why north and south poles cannot be separated by breaking a magnet in two. Figure 17-2C indicates the formation of the new pair of poles where the magnet has been broken.

17-2 Currents and Magnetism

The invention of the electric cell and battery, which made it possible to have a continuous flow of electricity, led to the study of various interactions between electric currents and magnets. (*Stationary* electric charges do not interact with magnets at all.) There are several basic laws governing these interactions, all of them discovered early in the nineteenth century. In the year 1820, a Danish physicist, H. C. Oersted (1777–1851), noticed that ***an electric current flowing through a wire deflects a magnetic needle placed in its neighborhood in such a way that the needle assumes a position perpendicular to the plane passing through***

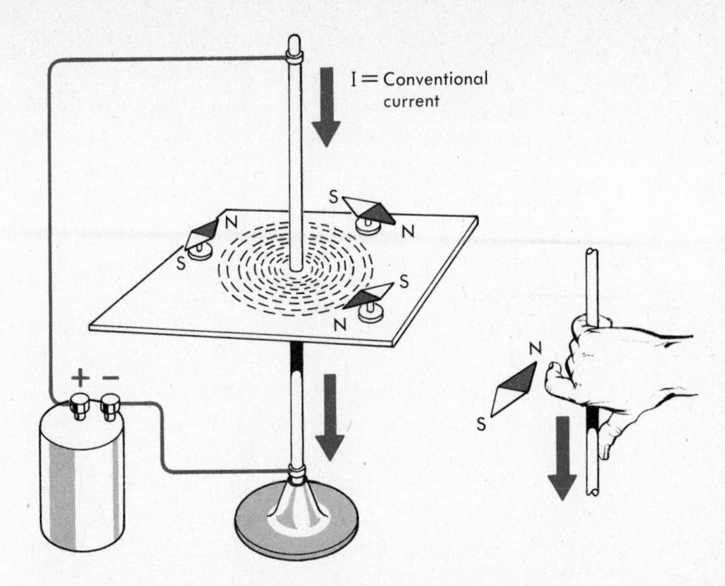

FIG. 17-4 The orientation of magnets (compass needles) in the neighborhood of an electric current. The direction in which the north pole of the needle will point can be found by the following rule: Hold the wire in your right hand, so that the thumb points in the direction of the conventional current; the fingers then point in the direction of the needle's north pole, and hence also of the magnetic field.

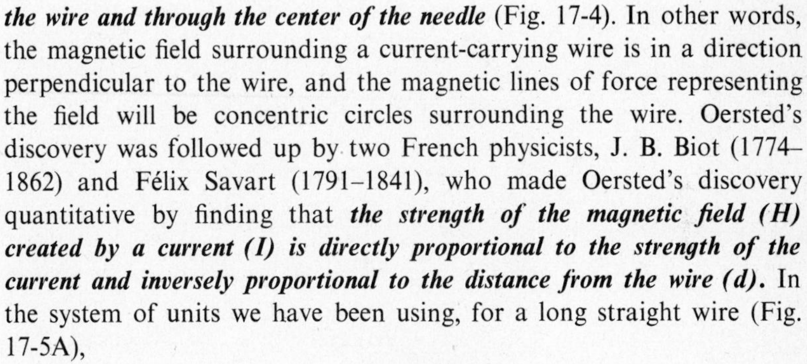

the wire and through the center of the needle (Fig. 17-4). In other words, the magnetic field surrounding a current-carrying wire is in a direction perpendicular to the wire, and the magnetic lines of force representing the field will be concentric circles surrounding the wire. Oersted's discovery was followed up by two French physicists, J. B. Biot (1774–1862) and Félix Savart (1791–1841), who made Oersted's discovery quantitative by finding that ***the strength of the magnetic field (H) created by a current (I) is directly proportional to the strength of the current and inversely proportional to the distance from the wire (d).*** In the system of units we have been using, for a long straight wire (Fig. 17-5A),

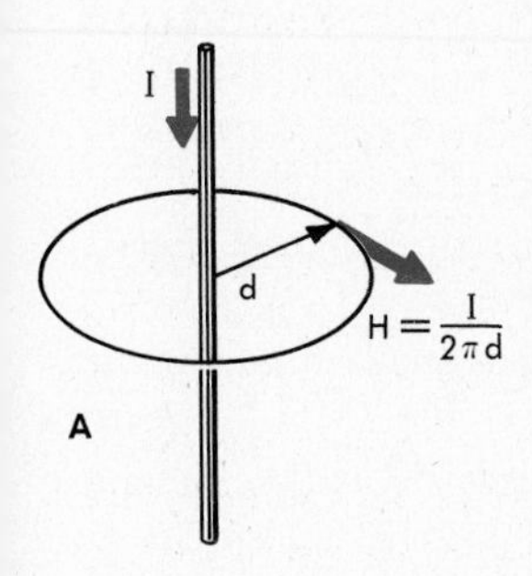

$$H = \frac{I\ (\text{amperes})}{2\pi d\ (\text{meters})}.$$

Thus magnetic field strength is seen to be measured in units of amperes per meter.

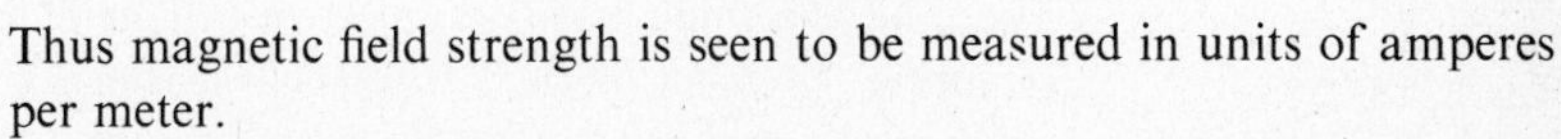

For a single circular loop of wire of radius r m (Fig. 17-5B), the field at the center is

$$H = \frac{I\ (\text{amperes})}{2r\ (\text{meters})}.$$

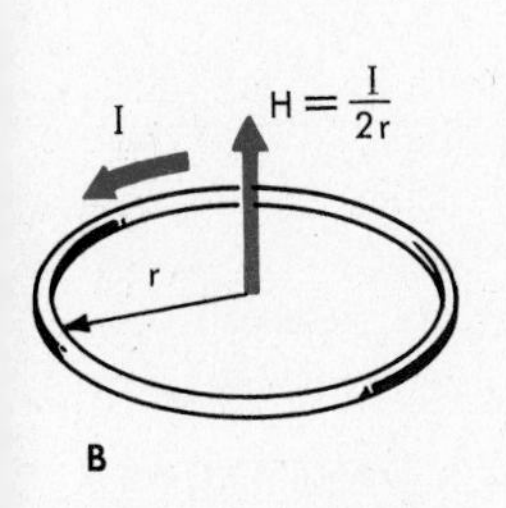

FIG. 17-5 Magnetic fields produced by electric currents.

At first glance, it seems that the causes of the magnetic field observed around a bar magnet and of the identical sort of magnetic field around a current-carrying wire must be entirely different. We can see no current flowing in a bar magnet; it is not connected to a battery or any other source of electric potential, and it can lie on a shelf for decades, quietly maintaining its same field all the while.

Shortly after Oersted's discovery in 1820, André Ampère suggested that in spite of appearances, the causes might actually be the same. He imagined that there might be tiny circulating electric charges within atoms themselves, so that each atom could contain loops of current and therefore itself be a miniature magnet. If this were true, the bar magnet could be explained as the result of the alignment of the atomic electromagnets. This was an inspired guess on Ampère's part, as almost nothing was known about the structure of atoms at that time. We know now that he was fundamentally correct, the only difference being that the field of a bar magnet arises mostly from the spin of its electrons on their axes rather than from their circulation in orbits. But this is a small detail that should not detract from Ampère's credit for the basic idea.

Since the spinning electrons in an atom may be considered to be tiny circulating currents, it is easy to visualize how an atom can actually be a small magnet. This is especially true of some atoms such as those of iron, in which there are more electrons spinning in one direction than in the other.

17-3 Force on a Moving Charge

We mentioned before that a magnet and a *stationary* electric charge do not have any effect on each other. A charged pith ball and a strongly magnetized compass needle will completely ignore one another. An electric current, however, is nothing more than a stream of *moving* charges; and we have seen that these moving charges create a magnetic field whose direction is at right angles to the direction of the current. It should therefore not be too surprising to find that a charge that is *moving* in a magnetic field experiences a force, and that ***the force is at right angles to both the velocity of the charge and the direction of the field.***

Figure 17-6 shows how the direction of the force on the moving charge can be predicted. It shows a *right* hand held flat, the thumb extended at right angles to the other fingers. Orient the hand so the fingers point in the direction of the magnetic field; now turn the hand (keeping the fingers still pointed in the field direction) so the thumb points in the direction the *positive* charge is moving. (If the moving charge is negative, one needs only point the thumb in a direction *opposite* its velocity—a negative charge moving, say, to the left is exactly equivalent to a positive charge moving to the right.) The direction in which you would now push forward with the palm of your hand is the direction in which the magnetic field pushes against the moving charge.

The magnitude of this force is

$$F = qv\mu H$$

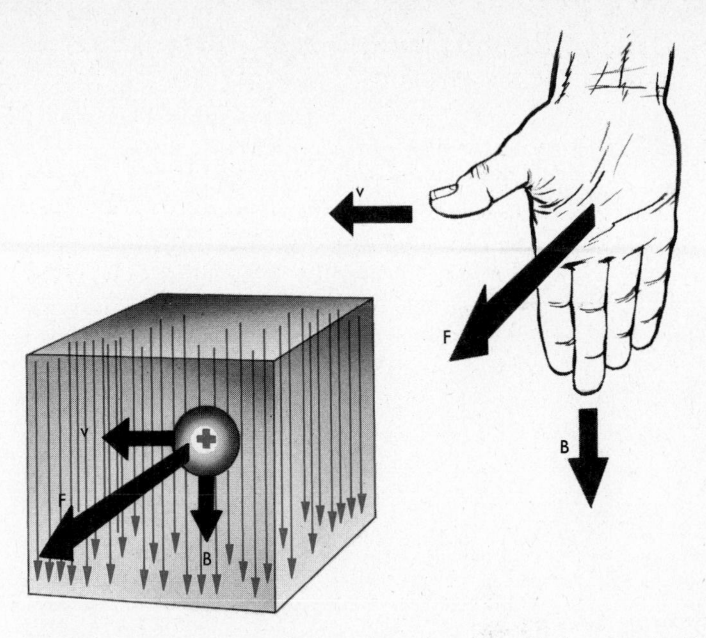

FIG. 17-6 Right-hand rule for determining the relationship between the direction of a magnetic field, the velocity of a charged particle moving through it, and the resulting force on the particle.

with F measured in newtons; q in coulombs; v, the component of the velocity perpendicular to the field, in meters per second; and H in amperes per meter.

The symbol μ (the Greek lowercase letter *mu*) needs some explaining. In the equation above, we have arbitrarily chosen the units of F, q, v, and H to be those with which we were already familiar in many other applications. In order to reconcile all these units, a proper numerical constant is needed. It is as though we had decided to state Newton's second law with force measured in pounds, mass in grams, and acceleration in meters/second^2. If we want to use these units, we cannot write $F = ma$ but must insert the proper conversion factor so that the equation reads $F = 2.24 \times 10^{-4}\, ma$.

The factor μ is called the *permeability* of the material in which the magnetic field is located. In a vacuum (or in air, which for most practical purposes is magnetically the same),

$$\mu = 4\pi \times 10^{-7} = 12.57 \times 10^{-7}.$$

The product μH finds much more use than does H alone; because it appears so frequently, it is given its own private symbol: $\mu H = B$, which has the name *magnetic flux density* or *magnetic induction* and is measured in units of *webers per square meter*, which we will discuss in the next section. Using this symbol B, the expression for the force on a charged particle moving in a magnetic field can be written in the following simpler form:

$$F = Bqv.$$

As an example of the force exerted on a moving charge, imagine an *alpha particle* (charge $+3.20 \times 10^{-19}$ C) 3.0 cm from a long, straight wire, traveling with a velocity of 6×10^5 m/s parallel to the wire. The wire carries a current of 20 A, which flows in the same direction in which the α particle is traveling. What is the force on the particle? (See Fig. 17-7.) We can first find the magnetic field H at a distance of 3 cm, or 0.03 m, from the wire:

$$H = \frac{I}{2\pi d} = \frac{20}{2\pi \times 0.03} = 106 \text{ A/m}.$$

Since the field is in air, the magnetic induction B is $12.57 \times 10^{-7} \times H = 1.33 \times 10^{-4}$ weber/m². It will not be necessary to calculate the component of the velocity perpendicular to the field, because in this problem the velocity itself *is* perpendicular to the field, and we find $F = qvB = 3.20 \times 10^{-19} \times 6 \times 10^5 \times 1.33 \times 10^{-4} = 2.55 \times 10^{-17}$ N. The direction of the force (use Fig. 17-6) is directly in toward the wire. Since this hand rule is designed to use the velocity of a *positive* charge, we

FIG. 17-7 Force on an alpha particle moving in the magnetic field of a current-carrying conductor.

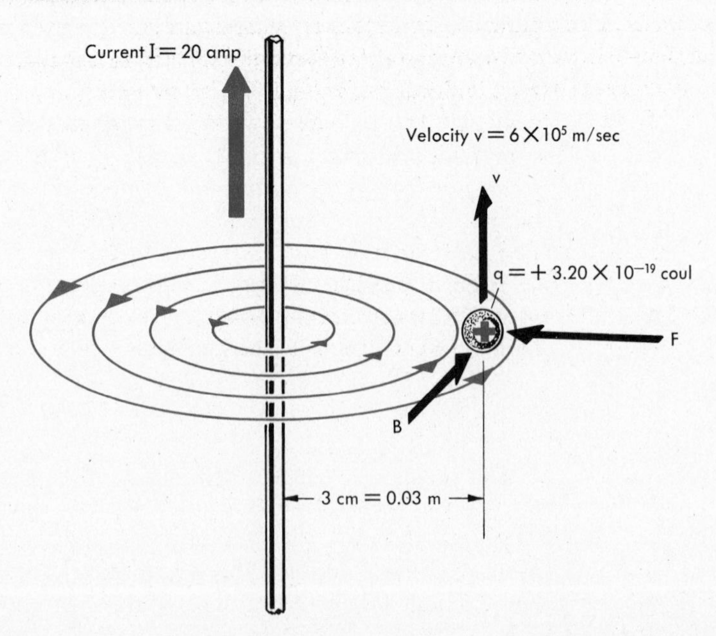

would get an opposite answer for a negatively charged particle moving in the same direction.

17-4 Magnetic Flux

In the preceding section, B has been measured in *webers/square meter*, which serves to introduce the concept of *magnetic flux*. If we want only to show the direction of the field around a magnet or a system of electric currents by sketching the lines of force, there is no law that says how many or how few should be drawn. The concept of lines of force, however, can be made more useful if the lines are made to have a quantitative meaning, in addition to showing merely the direction of the field. To do this, we may put in exactly enough lines so that wherever we want to draw (or, better, imagine) a square meter with its plane perpendicular to the field, the number of lines passing through the square meter is exactly equal to the numerical value of B at that location. Thus B is represented as so many lines per square meter. Where the lines are far apart, the field is weak; where the lines are closer together, the field is stronger (see Fig. 17-1).

Suppose that there is a large barn door facing north and that we have imagined the earth's magnetic field to have been indicated by lines drawn as described in the preceding paragraph. We can then count the number of lines passing through the barn door and say that so many webers of magnetic flux pass through the door. As we have drawn them, each line represents one *weber* (named for a nineteenth-century physicist) of magnetic flux. The magnetic induction B is accordingly measured in units of webers/m^2 and is often quite logically called the *flux density*. The symbol Φ (the Greek capital letter *phi*) is generally used to represent flux. If we have some area A (for simplicity taken perpendicular to the field) through which a total flux of Φ passes, then the average flux density within the area is $B = \Phi/A$ webers/m^2. Conversely, an area A perpendicular to a given magnetic induction B is cut by $\Phi = AB$ webers of flux.

17-5 Solenoids and Electromagnetism

If a current-carrying wire is wound in the form of a helix (Fig. 17-8A), it is called a *solenoid* and can be considered as a large number N of single turns of wire connected in series. It can be shown (with calculus, again!) that the field at the center of a solenoid l meters long is given by

$$H = \frac{NI}{l}$$

and the flux density or magnetic induction is, of course,

$$B = \mu H = \frac{\mu NI}{l}.$$

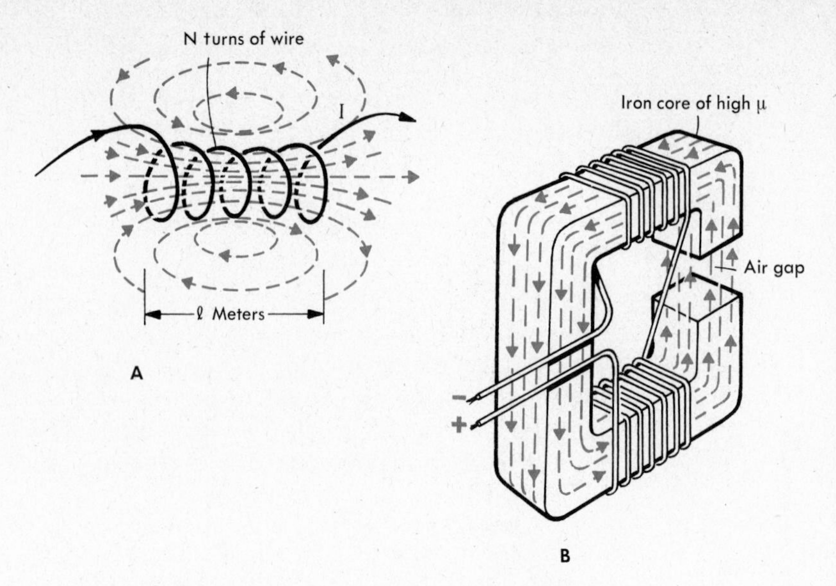

FrG. 17-8 A simple solenoid, and one form of electromagnet.

The flux density can be increased by increasing the current I or by increasing N/l, the number of turns per meter, or by increasing μ.

It is easy to see how one might change the current in a coil or rewind it to change its N/l; but how does one go about changing μ? The permeability factor μ not only takes into account the conversion of units but also the effect of the material in which the field is established. Suppose, for example, a cylinder of iron were slipped inside the coils of the solenoid in Fig. 17-8A. When the current is turned on, the field established by the current will at least partially align the atomic magnets of the iron to point with the field, so that each one contributes a little extra flux in the same direction. The end result of all these atomic contributions is that we may expect to have more flux, and a higher flux density B, with the iron core than without it. The *relative permeability*—that is, $\mu_{\text{material}}/\mu_{\text{vacuum}}$—can be as high as several thousand for some iron alloys and several hundred thousand for other specially manufactured magnetic materials.

Such materials, which markedly increase the flux density, are called *ferromagnetic*; this group, besides iron and many special alloys, also includes the elements cobalt and nickel. For most materials, however, the relative permeability is very nearly 1. Some (like bismuth, relative permeability 0.99983) are *diamagnetic* and make the flux weaker; others (aluminum, relative permeability 1.00002) are *paramagnetic* and contribute very weakly.

Electromagnets are nearly always wound on a ferromagnetic core. With many turns of wire, which are sometimes water-cooled to permit them to carry a high current without overheating, a high flux density

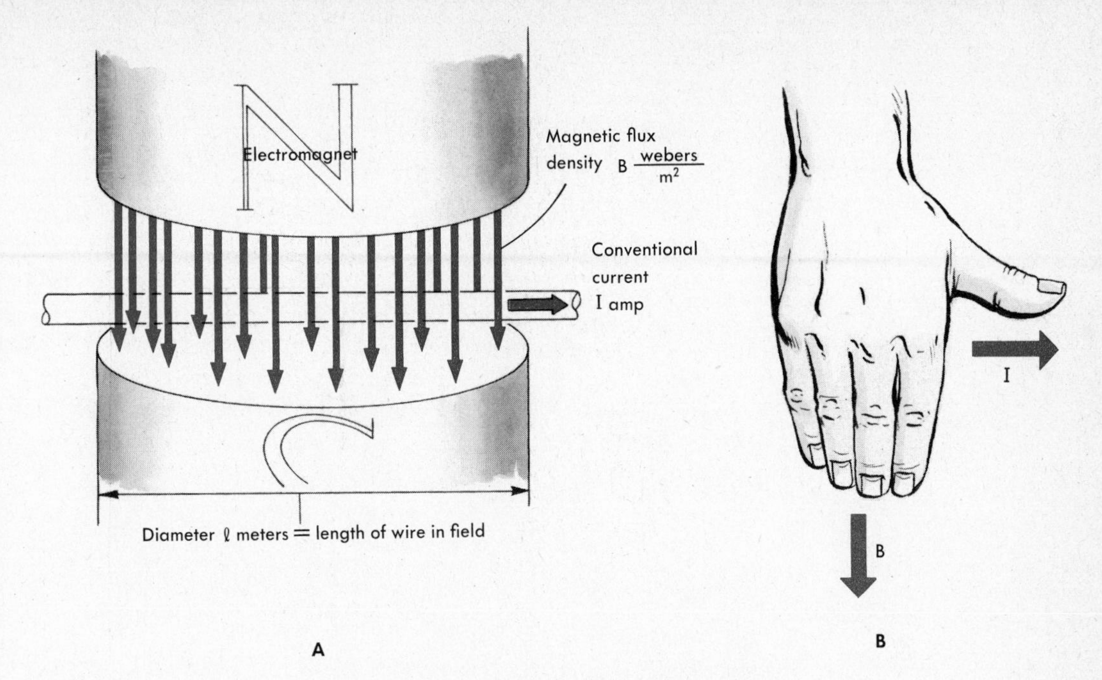

FIG. 17-9 A current-carrying wire in a magnetic field.

can be established. The electromagnet in Fig. 17-8B is merely a solenoid bent into a C, its winding being in two parts.

17-6 Currents in a Magnetic Field

Suppose we place a current-carrying wire in a strong magnetic field, such as the air gap of the electromagnet shown in Fig. 17-8B. Figure 17-9A shows a wire in such a field. It carries a current as shown; since a current consists of moving charges, a force must be exerted on them as they pass through the field. The direction of this force can easily be determined by applying the right-hand rule of Fig. 17-6. The current shown is of course actually a flow of electrons from right to left; this is equivalent to a flow of positive charges from left to right in the direction given for the conventional current. As indicated in Fig. 17-9B, the force is directed away from the observer. The force is actually on the moving charges, but since they are confined within the wire, we can expect that the wire itself will experience a force away from the observer, into the paper.

We know how to calculate the magnitude of the force on a single moving charge, so how can this knowledge be applied to the case of a wire carrying a continuous stream of charges equal to I A? Let us write the expression for the force on a moving charge:

$$F = qvB.$$

Dimensionally, this equation is

$$\text{newtons} = \frac{\text{coulombs} \times \text{meters} \times \text{webers}}{\text{second} \times \text{meter}^2}.$$

Absolutely nothing is changed if we quietly move the "second" in the denominator until it is under the "coulombs":

$$\text{newtons} = \frac{\text{coulombs} \times \text{meters} \times \text{webers}}{\text{second} \qquad \times \qquad \text{meter}^2}.$$

Since a coulomb per second is an ampere, which measures current, the equation becomes

$$F = IlB$$

where l is *the length of wire in the magnetic field.* In the original equation, v was the component of velocity perpendicular to the field; in the equation above, l must therefore be interpreted as the component of the length perpendicular to the field.

Let us make a numerical example of Fig. 17-9: the diameter of the magnet poles is 20 cm and the flux density between them is 1.65 webers/m^2. The wire passes through the center of the field (which means that 20 cm, or 0.20 m, of the wire is in the field) and is perpendicular to it. The current is 30 A. For the force exerted on the wire,

$$F = IlB = 30 \times 0.20 \times 1.65 = 9.90 \text{ N}.$$

(We have already determined that the force will be into the paper.)

17-7 Galvanometer, Voltmeter, Ammeter

One way in which the above force is utilized is in the construction of electric meters of many kinds. The *galvanometer*, besides being an important instrument itself, is the actual working element in most voltmeters and ammeters. There are many kinds of galvanometers, which are designed to measure very small currents and potential differences, but the most common is the d'Arsonval type, in which a very light coil of wire is pivoted in a strong magnetic field (Fig. 17-10).

The direction of the magnetic field, which is defined as the direction in which the north pole of a compass needle would point, always must run from the north pole of a magnet to the south pole, which, in Fig. 17-10, is from left to right. Applying the right-hand rule, we see that the wires of the coil from A to B experience a force toward the reader, out from the plane of the page; the wires from C to D are thrust away from the reader into the page. These forces cause a torque that rotates the coil about its central axis I–I against the restraining torque of a small

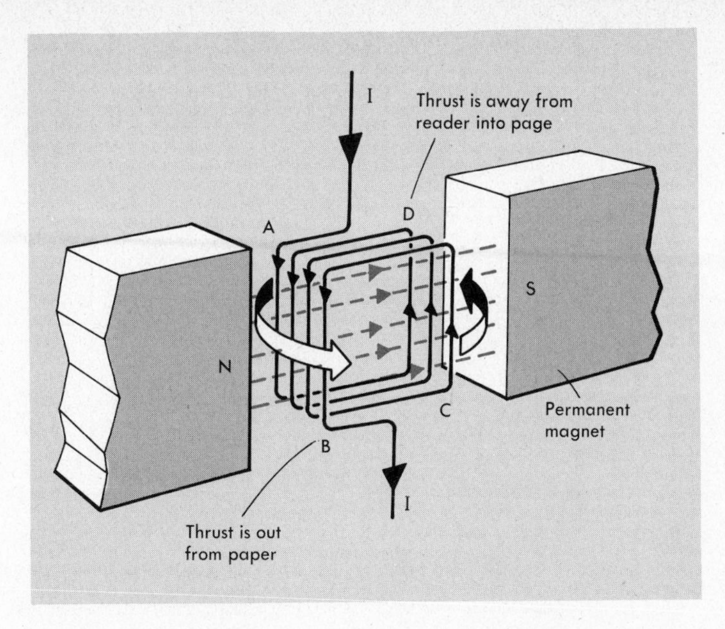

FIG. 17-10 A moving-coil, or D'Arsonval type, galvanometer.

spring not shown in the drawing. The rotating torque is proportional to the current in the coil; hence the greater the current, the greater the angle through which the coil is rotated. In many meters, an indicating needle is attached directly to the coil, but for very sensitive galvanometers a beam of light reflected from a tiny mirror is used instead. Some d'Arsonval galvanometers can indicate currents as small as 10^{-10} A. (There are other more complicated types of current-measuring devices that will give a reading for 10^{-17} A!)

Even the coarsest, most insensitive galvanometer is not adapted to measure large currents or high voltages. If it were so designed, the galvanometer coil itself would add too much resistance when put in series in a circuit as an ammeter and would change the very current it was intended to measure. On the other hand, if it were used directly as a voltmeter, the coil resistance would be so low that considerable current would flow through the galvanometer, thereby changing the *IR* drop it was supposed to measure accurately.

These difficulties can be avoided by using a relatively sensitive galvanometer connected to properly chosen auxiliary resistances. Figure 17-11A shows a high resistance connected in series with a galvanometer. Let us say that the galvanometer coil has a resistance of 50 Ω and that a current of 0.01 A will cause the needle to make a full-scale deflection. What resistance must we put in series with this galvanometer to convert it into a voltmeter that will deflect full scale for 10 volts? This means that when there is a 10-volt potential difference between the

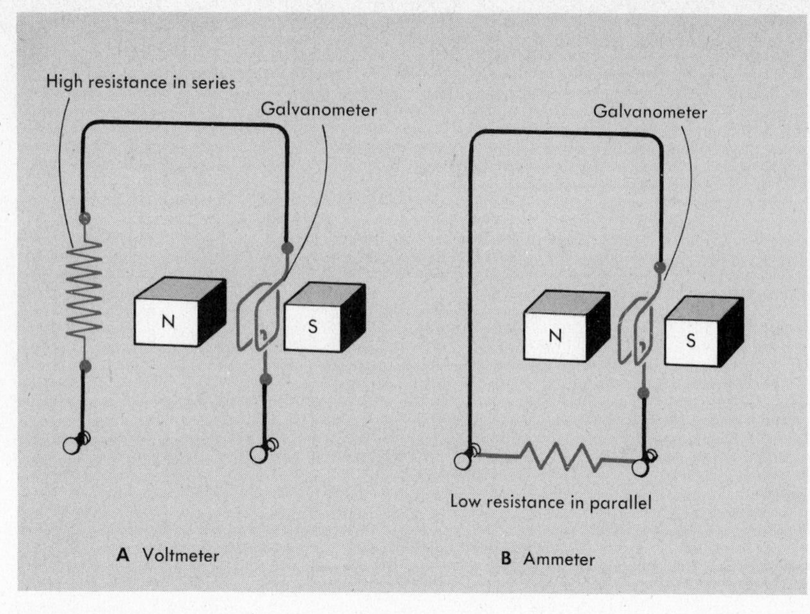

FIG. 17-11 The auxiliary resistances used with a galvanometer to make a voltmeter or an ammeter.

meter terminals, 0.01 A must flow through the coil. Ohm's law gives the required total resistance of the voltmeter:

$$R_T = \frac{V}{I} = \frac{10}{0.01} = 1000\ \Omega.$$

Of this 1000 Ω, 50 Ω are already supplied by the galvanometer coil itself, leaving 950 Ω for the added series resistance.

To make the galvanometer into a 5-A ammeter, i.e., one that deflects full scale for a total current of 5 A, we must provide a *shunt*, or low-resistance path in parallel with the galvanometer (Fig. 17-11B). Of the total 5 A, 0.01 A must flow through the galvanometer and the remaining 4.99 A must go through the shunt. Since the two parallel paths are connected to the same terminals, the potential difference (which is the *IR* drop) across each is the same, and we can say

$$I_S R_S = I_G R_G$$

or

$$4.99 R_S = 0.01 \times 50$$

$$R_S = 0.1002\ \Omega.$$

All the common types of electric motor also operate on this same principle—that a current-carrying conductor in a magnetic field receives a thrust perpendicular to both the field and the current in the direction given by the right-hand rule. In all motors except some special kinds and some toy motors, the magnetic fields are provided by electromagnets.

17-8 Interactions Between Currents

So far we have been talking about the interaction between currents and magnetic fields. What about the interaction between two currents? This was first studied by Ampère, who showed that two wires carrying electric currents flowing in the same direction are attracted to one another, while currents flowing in opposite directions repel each other (Fig. 17-12). This rule for the interaction between currents can be shown to follow from the behavior of current-carrying conductors in magnetic fields. Each wire is in the magnetic field of the other wire; each experiences a force that pulls them apart or pushes them together, depending on the direction of their currents.

As an example, take two long, straight parallel wires M and N, separated by a distance a m (Fig. 17-13). Let us calculate the force experienced by N as a result of its being in the magnetic field of M. At a m from M, the intensity of the magnetic field caused by I_M is

$$H = \frac{I_M}{2\pi a} \quad \text{and} \quad B = \frac{12.57 \times 10^{-7} \times I_M}{2\pi a}.$$

We cannot determine the total resultant force on N, because it is indefinitely long, but for a 1-m length ($l = 1$) of N, the force given by

$$F = BI_N l$$

becomes

$$F = 12.57 \times 10^{-7} \times \frac{I_M I_N}{2\pi a}.$$

FIG. 17-12 The forces acting on two parallel current-carrying wires.

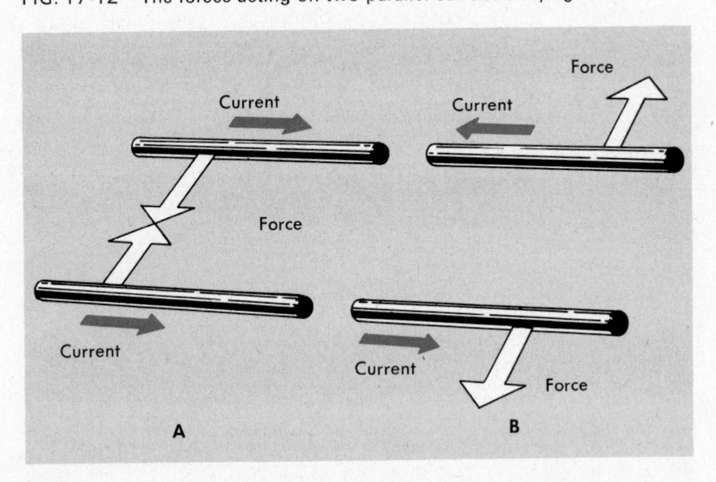

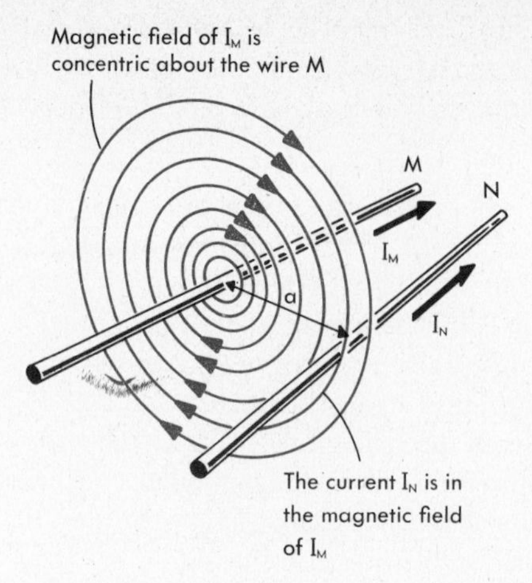

FIG. 17-13 A current-carrying wire in the magnetic field of another parallel current-carrying wire. (*M* is also in the magnetic field of *N*; this has been omitted to avoid complicating the drawing.)

The right-hand rule shows that this force on *N* is in a direction to pull it in toward *M*.

From Newton's third law, we know in advance that if *M* exerts the above force on *N* by means of *M*'s magnetic field, then *N* must exert an exactly equal and opposite force on *M*. A short calculation will convince you that this is indeed true.

If the two currents flow in opposite directions, the magnitude of the force will remain the same, but the right-hand rule tells us that in this case the conductors will repel each other.

As an example of the application of the formula worked out above (known as *Ampère's law*), let us consider two parallel wires 50 cm long in a large electric motor. These wires are separated by a distance of 1 cm, and the insulation between them, being almost perfectly non-magnetic, has the same permeability as air. A short circuit develops in the motor, with the result that a sudden current of 5000 A flows through the two wires (a short circuit is the formation of a very low resistance path for current, generally due to mechanical damage to insulation). What force will tend to pull the wires apart, assuming that the currents run in opposite directions? A direct substitution of numerical values in Ampère's law gives

$$\begin{aligned} F &= 12.57 \times 10^{-7} \times \frac{I_1 I_2}{2\pi a} \\ &= \frac{12.57 \times 10^{-7} \times 5000 \times 5000}{2\pi \times 0.01} \\ &= 500 \text{ N/m}. \end{aligned}$$

Since the length of the wires is 50 cm, or 0.5 m, the force on each wire will be 250 N. A newton is equal to a force of about a quarter of a pound, so the wires will be shoved apart by a force of more than 60 lb.

17-9 Generation of Electric Currents

We have seen that if a wire carrying current is placed in a magnetic field, a force acts on the wire and will move it if it is not fastened in place. In 1831, it occurred to the great British experimental physicist and chemist Michael Faraday, and independently to the American physicist Joseph Henry, that a converse effect should also be observable. That is, if a conductor is moved in a magnetic field, a current should be generated in the conductor. Both men observed that this actually did happen.

Let us arrange an apparatus as shown in Fig. 17-14. It consists of a pair of parallel bare conductors spaced l m apart in a magnetic field of flux density B. Across these is placed another bare conductor making contact with the two parallel wires. A galvanometer is connected to the parallels to complete an electrical circuit and to indicate any current that flows. Now, if the crosswire is moved to the left, the galvanometer needle will deflect in one direction; when the crosswire is moved to the right, the galvanometer deflects in the opposite direction.

Suppose the crosswire is moved to the left a distance of Δs m in

FIG. 17-14 Inducing a current in an electric circuit.

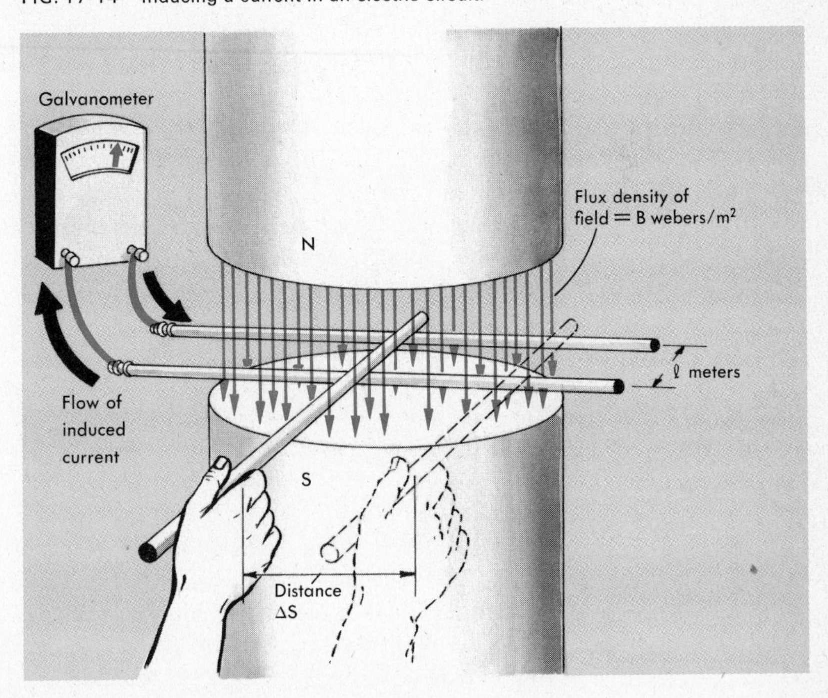

a time Δt; the galvanometer will show us that a current flows in the circuit. A current does not flow unless there is an emf in the circuit; so apparently a potential difference e has been created in the moving wire. Any electric current expends energy while flowing, and the principle of conservation of energy tells us that this electric energy must be supplied by the work we do in moving the crosswire. Assume (ignoring friction and the inertia of the wire) that a force F is required to move the wire; the work done in moving it a distance Δs is equal to $F\,\Delta s$.

The force F is evidently needed because there is a current I flowing through the l m of the crosswire, and the crosswire is in a magnetic field. In the previous section, we found that a force $F = BIl$ would be exerted on the wire; this is the opposing force we must overcome to move the wire, and thus the work done can be expressed as

$$W = F\,\Delta s = BIl\,\Delta s.$$

In Fig. 17-14 there is a complete electrical circuit that encloses a certain amount of magnetic flux. Now as the crosswire is moved to the left, the area of the circuit is reduced and it accordingly encloses less flux. The circuit area is reduced by $l\,\Delta s$ m^2 and as a result it encloses $Bl\,\Delta s$ webers less flux than it did; that is, $\Delta\Phi = Bl\,\Delta s$ webers.

Thus the work done in moving the wire is

$$W = BIl\,\Delta s = I\,\Delta\Phi.$$

All that remains to be done now is to divide both sides of this equation by the time Δt:

$$\frac{W}{\Delta t} = \frac{I\,\Delta\Phi}{\Delta t}.$$

We recognize $W/\Delta t$, work per unit time, as being power, which in an electrical circuit equals the potential times the current. The substitution of eI for $W/\Delta t$ gives

$$eI = \frac{I\,\Delta\Phi}{\Delta t}$$

or

$$e = \frac{\Delta\Phi}{\Delta t}.$$

This equation gives the relationship we have been working toward. It says that whenever the magnetic flux enclosed by a circuit changes, a potential equal to the rate of change of the flux is generated. (Assuming, naturally, that our units are correct, potential will be in volts and rate of change of flux in webers per second.)

A rearrangement gives us another statement that is often useful. Power is also given as force times velocity; that is,

$$P = Fv$$

or

$$Ie = BIlv$$

from which

$$e = Blv.$$

The direction of the emf and the resulting current can be determined by a special application of the rule of Fig. 17-6. In Fig. 17-14, when the crosswire is moved to the left, the hypothetical + charges of the conventional notation are forcibly carried along with the rest of the wire, thus constituting a current to the left. The right-hand rule of Fig. 17-6 shows that these charges will experience a force driving them out of the paper toward the hand holding the crosswire, thus driving the current in a clockwise circuit through the galvanometer.

In science, there is a very general principle known as *the principle of Le Chatelier*. It states that ***whenever we undertake any action to change an existing physical system, the system reacts in such a way as to oppose our action.*** Reflection will show that this principle necessarily follows if we are to have conservation of energy; otherwise perpetual motion machines would be a dime a dozen, and we would be able to create unlimited amounts of energy from any small starting push. Le Chatelier's principle, when applied to the interactions of currents and magnetic fields, is known as *Lenz's law*. Applying Lenz's law to Fig. 17-14, we see that the induced current in the crosswire must flow in such a direction that its reaction with the magnetic field will produce a force acting to the right, opposing the motion of the wire. Our rule shows that the current must flow through the crosswire toward the hand, in the same direction that we had already concluded from a different line of reasoning.

Figure 17-15A shows a coil of 100 turns of wire wound on a circular frame 20 cm in diameter. The plane of the coil is parallel to the earth's magnetic field, which has a flux density of 0.5 gauss. (The gauss is a very commonly used unit for the measurement of flux density. It is derived from the CGS system, and $1\ gauss = 10^{-4}\ weber/m^2$.) The coil is pivoted on an axis OO, perpendicular to the field. If the coil is rotated through 90° about OO in 0.2 s, what average voltage will be generated? As the coil is first oriented, no flux passes through it; after 90° rotation, the flux will pass through the entire area within the coil. This area is $\pi r^2 = 0.0314\ m^2$, and the flux density of the field is 0.5 gauss, or 5×10^{-5} weber/m^2. Therefore, for each turn of the coil, the enclosed flux will change from zero to $0.0314 \times 5 \times 10^{-5} = 1.57 \times 10^{-6}$ weber. This change takes place in 0.2 s, so the induced emf will be $\Delta\Phi/\Delta t = 1.57 \times$

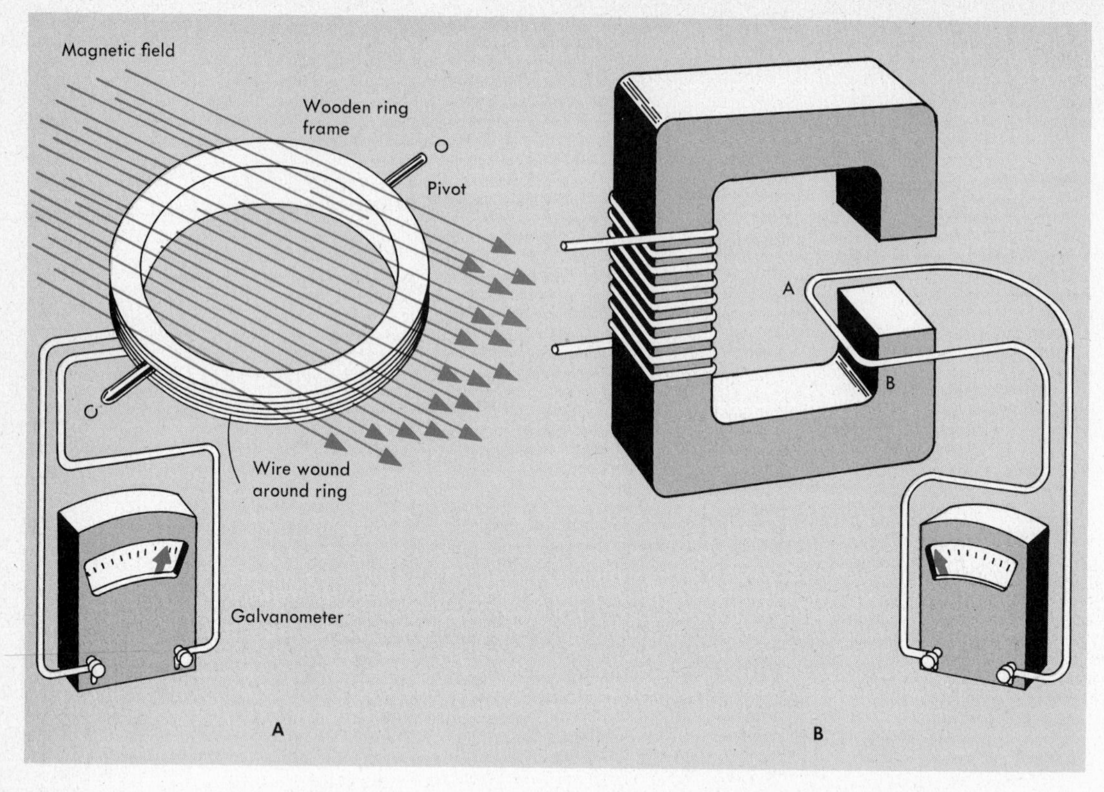

FIG. 17-15 Examples of emf induced in conductors moving in a magnetic field.

$10^{-6}/0.2 = 7.85 \times 10^{-6}$ V. This emf is induced in each one of the 100 turns; hence the total average voltage will be 7.85×10^{-4} V, or 0.785 mV.

Another example is given in Fig. 17-15B. A single wire AB is drawn through the air gap of a magnet at a speed of 1 cm/s. The magnet poles measure 5 cm × 5 cm, and the voltage induced while AB is passing between the poles is 0.05 mV. To find the flux density in the air gap, it will be convenient to use $e = Blv$, which gives us

$$5 \times 10^{-5} = B \times 5 \times 10^{-2} \times 10^{-2}$$

or

$$B = 0.1 \text{ weber/m}^2$$

$$= 1000 \text{ gauss.}$$

(In checking the equation above against the given data, remember that everything had to be converted from centimeters to meters.)

In deriving the relationships

$$e = \frac{\Delta\Phi}{\Delta t} \quad \text{and} \quad e = Blv$$

we used a complete circuit through which a steady current could flow. However, the formulas for induced potential that we derived in this way still hold, even if there is no complete circuit. A single length of conductor moved across a magnetic field will have induced in it the same potential as calculated from the equations above; but since there is no closed path through which it can drive a current, the only effect can be to drive electrons (or hypothetical + charges) to one end or the other of the moving conductor. The result will be that this potential difference will exist between the ends of the conductor.

17-10 Changing Flux

So far all our discussion has been about electric currents that run steadily and smoothly in one direction; however, some very important consequences come from currents that stop and start and change direction. Before we investigate these changing currents, it will be helpful to take another look at the equation $e = \Delta\Phi/\Delta t$.

Although this equation was derived by considering a conductor moving in a magnetic field, the equation holds for *any* situation in which the flux enclosed by a circuit changes. Figure 17-16 shows examples. In Fig. 17-16A, the coil connected to the galvanometer is cut by very little flux; while the magnet is being moved toward it (Fig. 17-16B), the flux passing through the coil is increasing, as is apparent from the fact that the lines of force are closer together closer to the magnet. With increasing flux, the meter needle is deflected to the right. In Fig. 17-16C the magnet is stationary, and although considerable flux passes through the coil, it is no longer changing, and the galvanometer reads zero. If the magnet were withdrawn, the flux would reduce, and the galvanometer would deflect to the left. If this whole experiment were repeated, with the magnet turned so its south pole were nearer the coil, the galvanometer deflections would be reversed in direction.

In Fig. 17-16D, the galvanometer coil is placed near another coil connected to a battery, but the switch is open and there is no current and no flux. In Fig. 17-16E, the switch has been closed; in a tiny fraction of a second, the current and the flux increase to their maximum. During this brief period of growing flux, the needle swings over and immediately returns to zero when the flux has reached its maximum and stops changing. The switch has been opened again in Fig. 17-16F; the current and flux drop to zero; and as the flux rapidly decreases, the needle takes a brief swing, this time to the left. Lenz's law tells us that in the case of the *increasing* flux (Fig. 17-16D and E), the induced current will flow in the direction that will make a flux opposed to that of the battery-connected

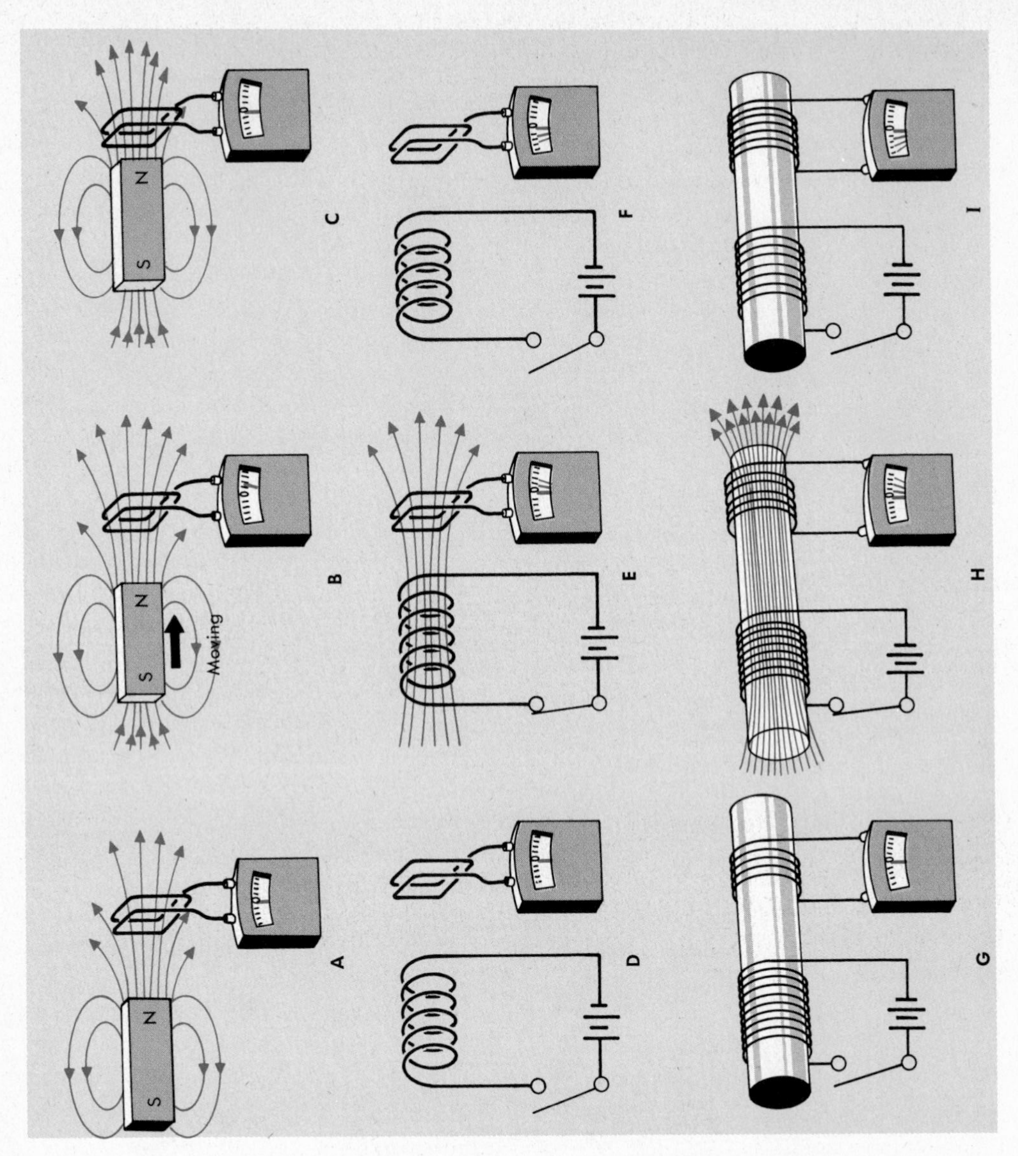

FIG. 17-16 An electric potential is induced when the flux passing through a coil is changing.

coil, in order to oppose its increase. When the switch is opened, the induced current will flow in the direction that will produce flux in the same direction as the flux of the battery-connected coil, in order to keep it from decreasing.

Figure 17-16G, H, and I repeat the procedure of D, E, and F, except that both coils are wound on a piece of soft iron. The permeability μ of the iron is much greater than that of air, so the induced flux is greater, and the galvanometer needle takes a larger swing.

All the discussion above seems a long way from the generation and distribution of the alternating current that is used in nearly all our homes, offices, and factories. It does, however, emphasize the principle that whenever the flux within a coil changes, an emf is induced in the coil, and a current will flow if the coil is a part of a complete circuit. It is this principle that makes the transformer possible, and without the transformer, our present wide distribution of cheap electric power would not be possible.

17-11 Transformers and Alternating Current

A simple *transformer* is shown in Fig. 17-17. It contains a *core*, generally rectangular in shape, made up of sheets of soft iron that magnetizes readily and immediately loses its magnetism when the magnetizing force is removed. On the core are wound two coils, called the *primary* (*P*) and the *secondary* (*S*). There is no fundamental difference between primary and secondary—whichever coil is supplied with power from the outside is called the primary, and the coil that supplies the output or the transformer is the secondary. A transformer can be connected either way, depending on the use to which it is to be put.

If (in Fig. 17-17) the primary is connected to a battery, current will begin to flow in *P*, and it will produce a magnetic flux. The coil is wound on a core of high permeability, which means that the flux is provided with something analogous to a low-resistance path for an electric circuit. Instead of much of the flux going into the surrounding air,

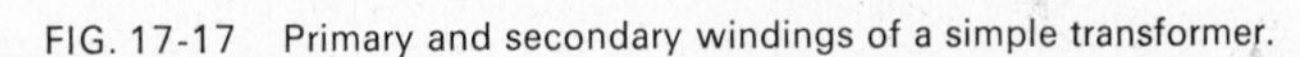

FIG. 17-17 Primary and secondary windings of a simple transformer.

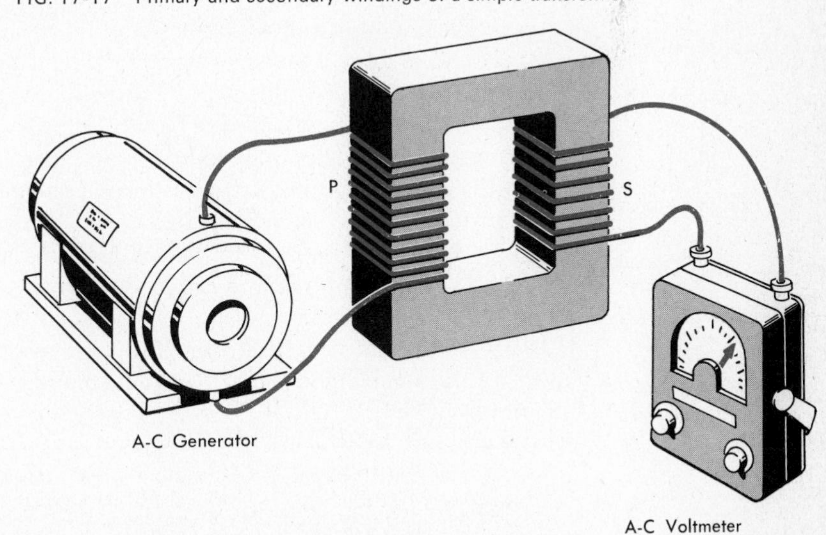

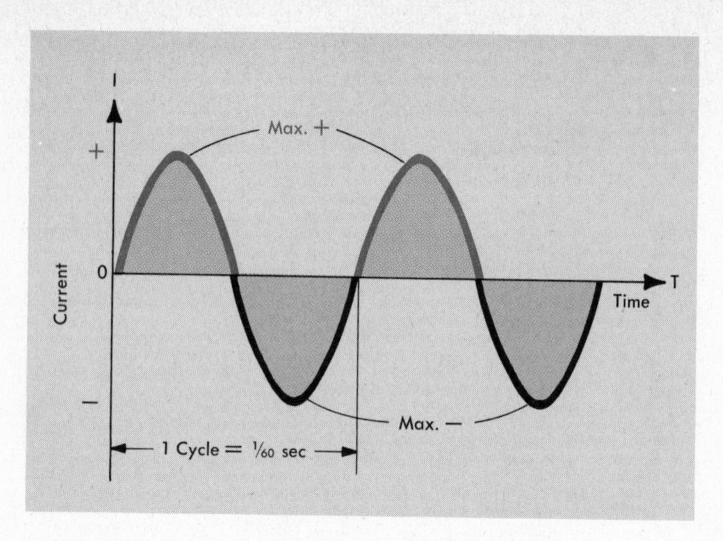

FIG. 17-18 Graph of the changes in an alternating current.

as in Fig. 17-16D, E, and F, almost all the flux will stay within the core and hence also pass through the windings of the secondary. A meter connected to *S* would show a large deflection whenever a battery was connected to, or disconnected from, *P*. It is not necessary, however, that the primary be suddenly connected or disconnected; the emf induced in the secondary depends on the *rate of change* of the flux, and if the primary current undergoes regular periodic changes, so will the flux, and so will the secondary current.

Almost all the residential, commercial, and industrial current in the world today is *alternating current* (ac). This is current that reverses its direction smoothly and regularly many times a second. If you were to analyze* the current flowing through the light bulb in your desk lamp, it would be found to fluctuate in magnitude, and change in direction as shown in the graph of Fig. 17-18. Starting at some instant when the current is zero, it rises to a maximum and then falls to zero and begins flowing in the opposite direction, to reach a maximum again and to fall to zero again—and so on. These described changes constitute one cycle, and in the United States $\frac{1}{60}$ s is required so that the frequency is 60 cycles/second. The smooth rise and fall of the current (and also of the applied potential) follows a *sinusoidal*, or *sine curve*. The graph is the same shape as the sinusoidal transverse wave shown in Fig. 8-5A, generated by a source moving in simple harmonic motion.

* This could not be done with a regular ammeter, which could not follow the rapid fluctuations—nor would you be able to follow its motion if it did. An *oscilloscope*, however, a cousin of the TV tube, has a beam of electrons as its only moving part and is able to draw an accurate graph of rapidly and regularly changing currents or potentials.

Your desk lamp is probably plugged into a standard 120-volt ac outlet. If you were to analyze the changing potential difference of this outlet with an oscilloscope, you would of course find its graph to be of the same sinusoidal shape as the current graph. But you would find the potential difference would rise to a maximum of 170 volts first in one direction and then in the other. This maximum value has been set by the power company because it results (over one or many cycles) in exactly the same heating effect as though a steady direct potential of 120 volts had been applied to a lamp or an iron or a heater.

Meters designed for use on alternating current are always calibrated to read the equivalent steady direct-current values of I or V, and loads and ratings are always given in these same terms.

The sinusoidally varying current I_P in the primary of a transformer causes a sinusoidally varying flux in the core, as shown in Fig. 17-19. At point A the flux is increasing at its maximum rate, and the maximum emf, at a, will be induced in the secondary. At B, the flux is maximum but at this instant is not changing at all (the flux curve is horizontal), so the induced emf at b is zero. As the flux decreases from B to C to D, a secondary emf is induced in the opposite direction. Calculus could show that the graph of the slope, or rate of change, of a sine curve is also a sine curve, so if we supply a sinusoidally varying input to the primary of a transformer, the output of the secondary will also be sinusoidal.

Each turn of the secondary coil will have induced in it the same emf as every other turn. Since all the turns are, in effect, connected in series,

FIG. 17-19 The induction of an alternating secondary emf by an alternating flux in the transformer core.

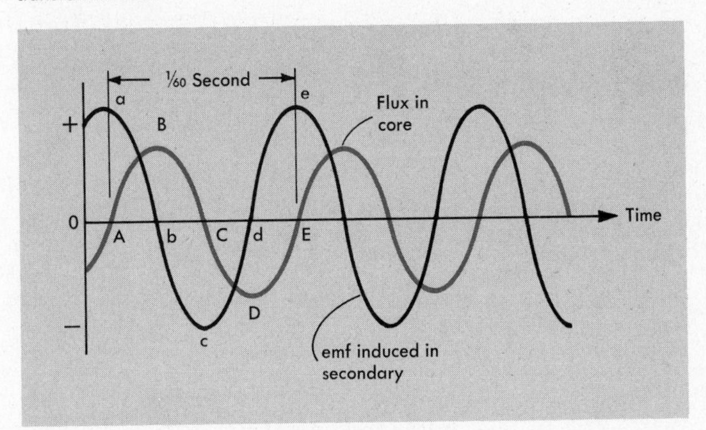

the total emf will be proportional to the number of turns, and the secondary voltage can be made (within certain practical limits) as large or as small as desired by designing the transformer so it has the proper number of turns. The relationship between the primary and secondary voltages and the number of turns on the primary and secondary windings is very simple:

$$\frac{V_S}{V_P} = \frac{N_S}{N_P}.$$

One should not assume that the transformer is a slick way of getting something for nothing! Even though the secondary voltage can be made much higher than the input voltage to the primary, the transformer is a remarkable device that automatically regulates the primary current in such a way that the *power* supplied to the primary is always equal to the power taken from the secondary (plus a small fraction more to make up for the inescapable heating losses). In other words, if the secondary voltage is 10 times the primary voltage, the transformer will see to it that the primary current is 10 times the secondary current, so that the input and output powers remain equal.

The ability of the transformer to change voltage is of enormous practical importance. Let us take as an example a generating plant producing electricity at 2300 volts. The plant must supply 4600 kilowatts or kW, to a nearby town over lines that have a total resistance of 5 Ω. To deliver 4,600,000 watts at 2300 volts would require a current of 2000 amperes, and the power used to heat the transmission wires, I^2R, would be 20,000,000 watts! Obviously, this is a very impractical thing to try to do, but engineers can solve the problem by installing transformers to raise the voltage from 2300 to, say, 69,000 volts. To transmit 4.6×10^6 W at 69,000 V would require a current of only 67 A, and the I^2R losses in the transmission line would be reduced to 22,400 W. This loss would be less than 0.5 percent of the delivered power, and would be quite economical.

In the city, the voltage could be reduced to 2300 for distribution, and every block would have several small transformers to reduce the voltage still further to a less dangerous 120 or 240 volts for use in homes and offices.

Questions

(17-2)

1. What is the value of H at a point 15 cm from a 200-A current flowing in a long straight wire?
2. What is the strength of the magnetic field at a point 20 cm from a long straight wire carrying a current of 60 A?
3. What is the magnetic field strength at the center of a coil of wire consisting of 300 turns, each 12 cm in *diameter*, if the current in the coil is 0.30 A?

4. A flat coil of wire is 8 cm in *diameter*, and consists of 150 turns; a current of 0.5 A flows through the coil. What is H at the coil's center?

5. A flat coil 10 cm in diameter is composed of 50 turns of wire; the magnetic field strength at its center is 120 A/m. What is the current in the coil?

6. At a point 10 cm from a long straight wire, the magnetic field strength is observed to be 20 A/m. What current is flowing in the wire?

(17-3) 7. What is the flux density B at the point described in Question 1?

8. What is the flux density (or magnetic induction) at the point described in Question 2?

9. A horizontal E–W wire carries a westward current. A beam of alpha particles is shot directly downward toward the wire. In what direction will the beam be deflected?

10. A vertical wire carries an upward current. A horizontal beam of electrons is directed toward the wire. In what direction will the beam be deflected?

11. A wire carries a current of 50 A. An alpha particle moves at 3×10^5 m/s parallel to the wire and 20 cm from it, in a direction opposite that of the current. What are the magnitude and direction of the force on the particle?

12. A vertical wire carries a 20-A current flowing upward. An electron (charge -1.60×10^{-19} C) 10 cm from the wire is moving upward with a speed of 5×10^6 m/s. (a) What is the magnitude of the force on the electron? (b) What is the direction of the force?

13. An electron (mass 9.1×10^{-31} kg) travels 10^6 m/s at right angles to a magnetic field whose flux density is 10^{-3} weber/m^2. (a) What is the force on the electron? (b) What is the acceleration of the electron?

14. An alpha particle (mass, 6.65×10^{-27} kg) travels at right angles to a magnetic field, with a speed of 6×10^5 m/s. The flux density of the field is 0.2 weber/m^2. (a) What is the force on the alpha particle? (b) What is its acceleration?

15. Since the acceleration of the electron in Question 13 is always perpendicular to its velocity, the speed of the particle will not be changed by it, although its direction will constantly be changed. (a) Consider the gravitational force on a planet moving in a circular orbit. Do you see a resemblance between the motion of the planet and the motion of the electron? (b) How did we set up the equation from which the radius of the orbit could be calculated? (c) What will be the radius of the path of the electron in the magnetic field of Question 13?

16. An alpha particle (see Question 14) is shot into a magnetic field of $B = 2 \times 10^{-3}$ weber/m^2, with a speed of 8×10^5 m/s at right angles to the field. What is the radius of the particle's circular path?

17. An electron is accelerated from the cathode to the anode of an "electron gun." There is a potential difference of 120 volts between cathode and anode, and the electron flies through a hole in the anode into a region containing a magnetic field that is perpendicular to the velocity of the electron. (a) What is the energy of the electron as it passes through the hole in the anode? (b) With this much kinetic energy, what is the speed of the electron as it enters the magnetic field? (c) What must be the flux density of the field in order to make the electron move in a circle of 6-cm radius?

18. An electron gun (see Question 17) accelerates a stream of electrons through a potential difference of 280 volts; the electrons then enter a uniform

magnetic field perpendicular to their velocity. (a) With what energy do the electrons enter the magnetic field? (b) With what speed do the electrons enter the magnetic field? (c) What is B, if the electron beam is bent into a circle of 8-cm radius?

(17-4) **19.** A circular loop of wire 3 cm in radius is placed with its plane perpendicular to a magnetic field. It is found that there is a flux of 1.5×10^{-5} weber through the loop. What is the flux density of the field?

20. A circular loop of wire 4 cm in diameter is placed with its plane perpendicular to a magnetic field of $B = 3 \times 10^{-2}$ weber/m^2. What is the flux through the loop?

21. In Question 19, the loop is oriented so that its plane makes an angle of 65° with B. What must be the magnetic induction of the field if there is to be the same flux of 1.5×10^{-5} weber through the loop?

22. In Question 20, the loop is oriented so that its plane makes an angle of 70° with B. What, now, is the flux through the loop?

(17-5) **23.** An air-core solenoid 20 cm long and 4 cm in diameter is wound with 5000 turns of wire and carries a current of 0.2 A. What are the magnetic field and magnetic induction within this solenoid?

24. What are the magnetic field and the flux density within an air-core solenoid 10 cm long, consisting of 1200 turns of wire carrying a current of 0.5 A?

25. The current in the solenoid of Question 23 is reduced to 0.05 A and a core (relative permeability = 750 at this value of H) is slipped in the solenoid. What is the resulting flux density?

26. An iron core (relative permeability = 400 at this intensity of magnetization) is slipped in the solenoid of Question 24. What is the flux density, with the current adjusted to be 0.1 A?

(17-6) **27.** A wire in the armature of an electric motor is 20 cm long, and is in (and perpendicular to) a magnetic field whose flux density is 0.1 weber/m^2. What force is exerted on the wire, which carries a current of 30 A?

28. A wire connecting a dry cell to a lamp is 1 m long and carries a current of 0.5 A. It is perpendicular to the earth's magnetic field, which has a strength of 4×10^{-5} weber/m^2. Is there much danger of the wire being ripped loose from its binding posts by the force exerted on the current by the magnetic field?

29. A wire carrying a current of 50 A passes through the center of a cylindrical magnetic field 10 cm in diameter. The wire is perpendicular to the field, and experiences a force of 8 N. What is the magnetic induction of the field?

30. A 10-cm length of wire carrying 20 A is in a magnetic field at right angles to the wire. The wire experiences a force of 0.1 N. What is the flux density of the field?

(17-7) **31.** A certain type of electrometer can measure currents as small as 100 electrons per second. What is this current, in amperes?

32. A current of 10^{-7} A seems quite small. A flow of how many electrons per second constitutes this current?

33. The coil of a galvanometer has a resistance of 300 Ω and a current of 2×10^{-4} A causes the meter to deflect full scale. (a) What auxiliary resistance is needed to convert the galvanometer into a voltmeter that reads 25 volts at full-scale deflection? (b) Show how this resistance must be connected.

34. A galvanometer with a coil resistance of 100 Ω deflects full scale with a current of 10^{-3} A. (a) Sketch how a resistance must be connected to make it into a voltmeter. (b) What must this resistance be if the voltmeter is to read 5 volts at full-scale deflection?

35. The galvanometer of Question 33 is to be converted into an ammeter reading full-scale for a current of 0.5 A. What resistance is needed, and how should it be connected?

36. The galvanometer of Question 34 is to be converted into an ammeter reading full-scale for a current of 2 A. What must be the resistance of the added shunt, and how is it connected?

(17-8) **37.** Two parallel straight wires, M and N, are 5 cm apart, and each is 0.60 m long. The current in M is 30 A; and in N, 5 A in the opposite direction. (a) Calculate the force on M due to the magnetic field of N, and the force on N due to the magnetic field of M. (b) How are the directions of these two forces related?

38. Take two wires, A and B, separated by 0.1 m. A carries a current of 1 A and B, 5 A in the same direction. (a) Calculate the force produced on A by the magnetic field of B, and the force produced on B by the magnetic field of A, per meter of length of wire. (b) How are the directions of these forces related?

39. An aluminium wire 50 cm long has a mass of 60 g and carries a current of 40 A, supplied through flexible leads whose weight is negligible. The wire lies on another wire that is flat on the laboratory bench. What current must flow through the bottom wire in order to repel the top wire to a height of 2 cm above it?

40. A wire weighing 200 g/m carries a current of 50 A, which is fed into the wire by flexible leads whose weight can be neglected. This wire is parallel to, and on top of, a horizontal wire lying on a table. What current must flow through the bottom wire in order to repel the top wire to a height of 3 cm above it?

(17-9) **41.** The flux between the poles of a magnet is 5×10^{-4} weber. A piece of wire is drawn across the flux in 0.2 s. What emf is induced in the wire?

42. A piece of wire is passed through the gap between the poles of a magnet in 0.1 s. An emf of 4×10^{-3} V is induced in the wire. What is the flux between the magnet poles?

43. A small car whose rear hubcaps are 1.60 m apart is driven 120 km/hr. Where the car is, the vertical component of the earth's magnetic field is 0.3 gauss. (a) What is the potential difference between the hubcaps? (b) Could this potential difference be used to light a small lamp by running wires from the lamp to make contact with each hubcap? Explain.

44. A metal airplane with a wingspan of 50 m flies at 360 km/hr. The component of the earth's magnetic field perpendicular to the velocity of the plane is 0.2 gauss. (a) What is the potential difference between the wing tips of the plane? (b) Could you take a lamp, say, and connect it to wires running from each wing tip and have the lamp light up? Explain.

45. A magnet has pole faces 6 cm in diameter, with a flux density of 5000 gauss between them. A piece of fine wire ($R = 10\ \Omega$) is connected to a 30-Ω galvanometer, and then passed through the field in 0.2 s. (a) What average voltage will be induced while the wire passes through the field? (b) What average current will flow through the galvanometer?

46. An electromagnet has square pole faces measuring 10 cm × 10 cm, between which there is a flux density of 3000 gauss. A single conductor of heavy wire is passed across the gap between the poles in 0.5 s. The ends of the wire are connected to a galvanometer whose resistance is 40 Ω. (a) What average voltage will be induced in the wire? (b) What average current will flow through it?

47. A small flat coil 2 cm in diameter that has 120 turns and 40 Ω resistance, is placed in a magnetic field and then withdrawn. The coil is connected to a ballistic galvanometer, a special galvanometer calibrated to indicate the amount of charge passing through it in a single surge. It has a resistance of 50 Ω and indicates that a charge 3×10^{-3} C surged through it when the coil was withdrawn. What is the flux density of the field? (The time taken to withdraw the coil makes no difference. Assume it to be Δt; it should cancel out.)

48. A small flat coil of fine wire is 1 cm in diameter, and has 60 turns and a 12-Ω resistance. The coil is connected to a ballistic galvanometer (see Question 47) whose resistance is 30 Ω, is placed in a magnetic field, and is then withdrawn. As the coil is withdrawn, a charge of 5×10^{-4} C passes through the galvanometer. What is the flux density of the field?

(17-11) **49.** A transformer having 3000 primary turns and 150 secondary turns is supplied from a 240-volt line. The load on the secondary draws a current of 15 A. What are the secondary voltage and the primary current?

50. A transformer has 500 primary turns and 2500 secondary turns. Meters indicate that the load connected to the secondary is operating at 200 V and 8 A. What are the voltage and current supplied to the primary?

51. A pumping station that operates 24 hr/day and 365 days/yr requires 150 kW of power, which is delivered at 4800 volts over lines whose resistance is 5.2 Ω. How much money would be saved per year by transmitting at 12,000 volts instead of 4800? Electrical energy here costs 0.3 cent/kWh. (The kWh, or kilowatthour, is an *energy* unit and represents a power of 1000 watts for an hour. A watt-second is, of course, a joule.)

52. A small factory requires 400 kW and is supplied by lines whose total resistance is 1.2 Ω. How much money would be saved per year by transmitting its power at 11,500 V instead of 2300? Assume the factory operates 8 hr/day for 300 days/yr and that electrical energy costs 0.4 cents/kWh. (The kWh, or kilowatthour, is an *energy* unit and represents a power of 1000 watts for an hour. A watt-second is, of course, a joule.)

18

The Electrical Nature of Matter

18-1 Positive and Negative Ions

In Section 1 of Chapter 16, we considered electric cells designed to convert chemical energy into electrical energy. Similar cells can be operated in reverse, to convert electrical energy into chemical energy.

If we place two electrodes in a container of pure water, and apply a potential difference across the electrodes, practically no current will flow through the cell. Pure water has a high resistance, and so is a fairly good insulator; at room temperature only 1 water molecule in 2×10^8 is ionized into charged particles, which is not enough to conduct any substantial current. When a little of some readily-ionizable material—nitric acid (HNO_3), say, or sulfuric acid (H_2SO_4)—is added, the acid molecules break apart into ions that migrate readily through the water.

Figure 18-1A shows one of such cells in operation. The detailed chemical reactions that take place at each electrode lie outside the scope of a physics text, but we can appreciate the final result, which is to break water molecules (H_2O) into two hydrogen ions, each lacking an electron (H^+), and an oxygen ion holding the two excess electrons (O^{2-}). The

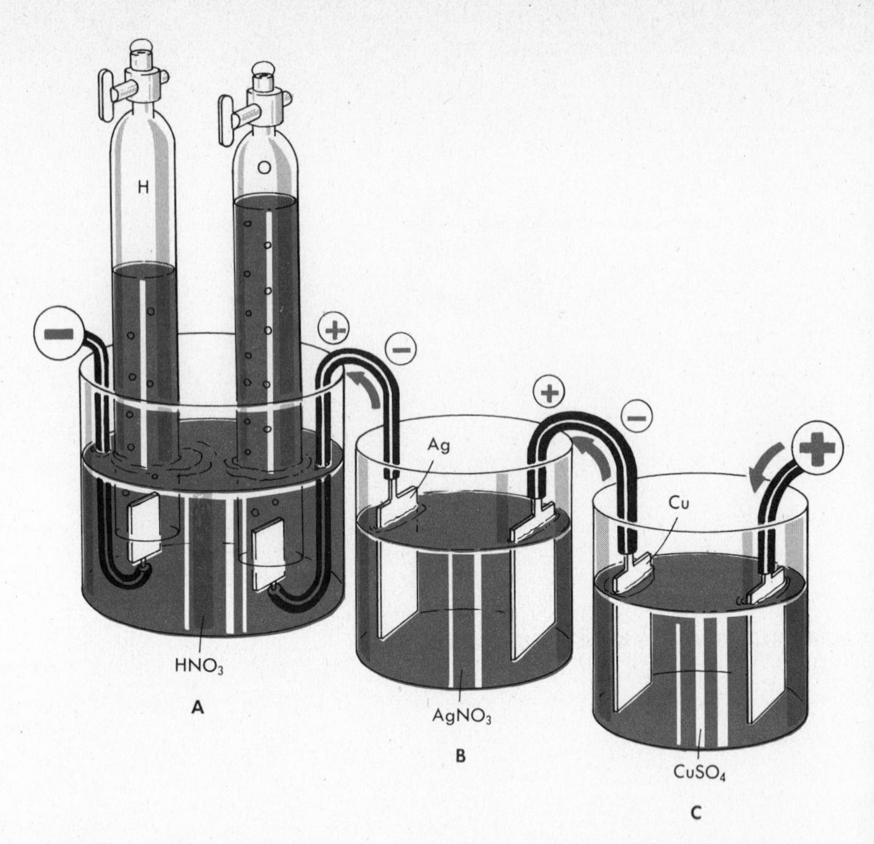

FIG. 18-1 The electrolysis of water solutions of (A) nitric acid, (B) silver nitrate, and (C) copper sulfate by the same current which passes through the three cells in series.

hydrogen ions pick up electrons from the negative electrode to become atoms; the oxygen releases its two excess electrons to the positive electrode. For reasons we can leave to the chemists, the neutral atoms of hydrogen and of oxygen join together in pairs to produce the corresponding molecules H_2 and O_2, which rise in bubbles to displace the liquid in the tops of the collecting tubes. Figure 18-1A shows the volume of the hydrogen to be twice the volume of the oxygen. This is just what we should expect, since the break-down of H_2O produces twice as many hydrogen as oxygen molecules.

In the cell of Fig. 18-1B, instead of the acid of A, a metal salt has been dissolved in the water—in this case, silver nitrate ($AgNO_3$). (It is fairly obvious that the N is the symbol for nitrogen; silver is represented by the symbol Ag.) This readily breaks up in solution into the ions Ag^+ and NO_3^-, which are then driven through the electrolyte in opposite directions by the applied potential. When the Ag^+ ions arrive at the negative electrode, they pick up from it their missing electrons and become neutral silver atoms. This metallic silver deposits as a coating on the electrode, and the process is an example of *electroplating*, which has many useful applications.

Figure 18-1C shows another cell containing copper sulfate ($CuSO_4$) in solution. When this ionizes, the copper leaves two electrons behind to become the Cu^{2+} ion; the SO_4 sulfate group, holding these excess electrons, is the doubly-negative group $SO_4{}^{2-}$. With the passage of an electric current this cell will of course deposit a layer of copper on the negative electrode.

18-2 The Laws of Faraday

Michael Faraday, whose name has already been mentioned in connection with the theory of electric and magnetic fields, was the first to investigate in detail the laws of electrolytic processes. He found first of all that, for each given electrolyte solution, the amount of material deposited at the electrodes is directly proportional to the strength of the electric current and to its duration, or, in other words, that ***the amount of material deposited on the electrodes is directly proportional to the total amount of electric charge which passes through the solution.*** From this *first law of Faraday*, we conclude that each ion of a given chemical substance carries a well-defined electric charge.

Figure 18-1 shows the three electrolytic cells connected in series, so that the current in each must be the same, and the charge transported through each cell must be the same in equal lengths of time. From this reasoning and Faraday's first law, we can conclude that the number of hydrogen atoms in the collector, and of silver atoms on the plated electrode must be equal. However, since the copper lacks *two* electrons, there will be only half as many copper atoms plated out in Fig. 18-1C.

In Faraday's time nobody had any idea of the mass of any single atom, but chemists had determined from many ingenious experiments the *relative* weights of the atoms of many chemical elements. As long as we are dealing with only *relative* weights—the ratio of the mass of one atom to the mass of another—it really makes no difference what we use as a standard. If the ratio of the mass of a hydrogen atom to the mass of an oxygen atom is about 1/16, it is also 2/32, or 1.96/31.36. In the early days of analytical chemistry, hydrogen (the lightest of the elements) was arbitrarily assigned the value of exactly 1. On this scale, the mass of the oxygen atom would be 15.87. Later (primarily because most of the analyses to determine atomic weights were done with oxides of the elements), it was decided to make oxygen exactly 16, which required that hydrogen be raised to 1.008. In recent years, there has been another change. As we will see later, not all oxygen atoms are alike, nor are all carbon atoms. In fact, the atoms of nearly all the elements have several different *isotopes*, which differ from one another only in their masses. The present standard for both physicists and chemists is the C-12 isotope of carbon, which is assigned the value of exactly 12. In other words, the unit of mass in determining the relative masses (or "weights," in common usage) is 1/12 of the mass of one atom of C-12. The periodic table of the elements, Fig. 24-5 on p. 450, gives the masses of the 105 currently known elements relative to this standard.

In the cells in series in Fig. 18-1 we have seen that equal numbers of hydrogen and silver atoms have been collected. If we let the cells run until 1.008 g of hydrogen has been collected, the equal number of silver atoms plated will have a mass of 107.9 g. From the atomic weight of copper, we see that the mass of the same number of copper atoms would be 63.5 g. In the cell of Fig. 18-1C, however, each copper atom carries a double charge, since it has lost two electrons. Accordingly, since the same charge passes through all three cells, there can be only half as many copper atoms plated, or 63.5/2 = 31.8 g.

The number of electrons gained or lost by an ion is one aspect of what the chemist calls *valence*. Hydrogen, silver, and the nitrate group (NO_3) have a valence of 1 (monovalent); copper and the sulfate group (SO_4) have a valence of 2 (*divalent*); while aluminum has a valence of 3 (*trivalent*).

Thus with several electrolytic cells in series, as in Fig. 18-1, for each atom of a monovalent element that is deposited, only $\frac{1}{2}$ of a divalent atom, or $\frac{1}{3}$ of a trivalent atom can be deposited. Chemists call the atomic weight divided by the valence the *equivalent weight*, and *Faraday's second law of electrolysis* states that ***when the same amount of electric charge flows through different electrolytic cells, the amounts of the substances deposited (or liberated) are in direct proportion to their equivalent weights.***

For example, we can place two cells in series (which guarantees that the same amount of charge will flow through each), one cell containing silver nitrate and the other gold chloride (gold is trivalent), and allow current to flow until we have 1.000 g of silver deposited on the cathode of the first cell. At this time, how much gold will have been deposited on the cathode of the other cell? The equivalent weight of silver, since silver is monovalent, is the same as its atomic weight, or 107.9. Gold has an equivalent weight of 197.0/3 = 65.7. Therefore, we can write

$$\frac{\text{weight Ag deposited}}{\text{weight Au deposited}} = \frac{1.00}{x} = \frac{107.9}{65.7}$$

or

$$x = 0.609 \text{ g Au deposited.}$$

It has been found that the passage of 96,500 C of charge (this amount of charge is known as a *faraday*) will deposit a mass, in grams, that is equal to the equivalent weight of any element. (This amount of any element is more formally called a *gram-equivalent weight*; one *gram-atomic weight* is, of course, an amount of substance whose mass in grams equals its atomic weight.) The American physicist Robert A. Millikan (1868–1953) showed with his famous oil-drop experiment (which we will discuss later in this chapter) that the value of a single electronic charge is 1.602×10^{-19} C or 4.80×10^{-10} esu. Thus the passage of 96,500 C means the passage from cathode to anode of $96{,}500/(1.602 \times 10^{-19}) = 6.02 \times 10^{23}$ electrons. These electronic charges are escorted across one at a time by monovalent ions, two at a time by divalent ions, and so on. This leads us to the important conclusion that ***a gram-atomic weight of any element contains 6.02 × 10²³ atoms.*** The number, 6.02×10^{23}, is known as *Avogadro's number*, after the nineteenth-century Italian chemist who first reasoned that gram-molecular weights (the extension of gram-atomic weight to gram-molecular weight should be obvious) of all gases contained the same number of molecules.

18-3 The Passage of Electricity Through Gases

The next step in the study of the electric nature of matter was made by another famous Britisher named J. J. Thomson (1856–1940) (Fig. 18-2). Whereas Faraday studied the passage of electric current through liquids, J. J. (as he was known to his colleagues and his students) later concentrated his attention on the electrical conductivity of gases.

FIG. 18-2 Sir J. J. Thomson (left), discoverer of the electron, and Lord Rutherford, discoverer of the nucleus, discuss problems in the courtyard of the Cavendish Laboratory, Cambridge, England, 1929.

When we walk in the evening along the downtown streets of a modern city, we observe the bright display of neon (bright red) and helium (pale green) advertising signs. Modern offices and homes are illuminated by fluorescent light tubes. In all these cases, we deal with the passage of high-voltage electric current through a rarefied gas—the phenomenon that was the object of the lifelong studies of J. J. Thomson. As in the case of liquids, the current passing through a gas is due to the motion of positive and negative ions driven in opposite directions by an applied electric field. The positive gas ions are similar to those encountered in the electrolysis of liquids (being the positively charged atoms or molecules of the substance in question), and the negative ions in this case are the much less massive singly charged particles that we now know to be electrons.

To study these particles, mysterious at that time, Thomson, in 1897, used an instrument shown schematically in Fig. 18-3. It consisted of a glass tube containing a rarefied gas, with a negative electrode placed at one end of it, and a positive electrode located in an extension on the side. Because of this arrangement the negative ions, which form the *cathode rays* that move from left to right in the drawing, miss the positive electrode and fly on to the right end of the tube. The tube broadens here, and its flat end is covered with a fluorescent material that becomes luminous when bombarded with fast-moving particles. This tube is very similar to a modern TV tube, in which the image is also due to the fluorescence produced by a scanning electron beam. But, in those pioneering days of what we now call "electronics," one was satisfied with much simpler

FIG. 18-3 The passage of electric current, carried by both positive and negative particles through a rarefied gas.

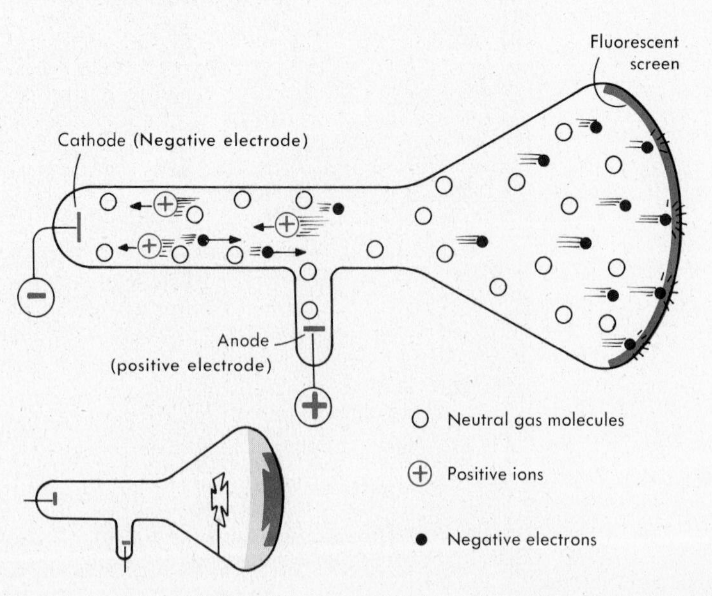

shows; placing a metal cross in the way of the beam, Thomson observed that it cast a shadow on the fluorescent screen, indicating that the particles in question were moving along straight lines, similar to light rays.

18-4 Charge-to-Mass Ratio of the Electron

Thomson's next task was to study the deflection of the beam caused by electric and magnetic fields applied along its path. Indeed, since the beam was formed by a swarm of negatively charged particles, it should be deflected toward the positive one of the parallel plates that produce the electric field shown in Fig. 18-4A. On the other hand, the beam of charged particles should be deflected by a magnetic field directed perpendicularly to its track (Fig. 18-4B) according to the laws of electromagnetic interactions.

The deflection of a particle will depend, of course, on how much force is applied to it. For a charged particle in an electric field, the force

FIG. 18-4 (A) the deflection of negative particles (electrons) by an electric field and (B) by a magnetic field. By combining both deflections in the same tube, the charge-to-mass ratio of the particles can be measured.

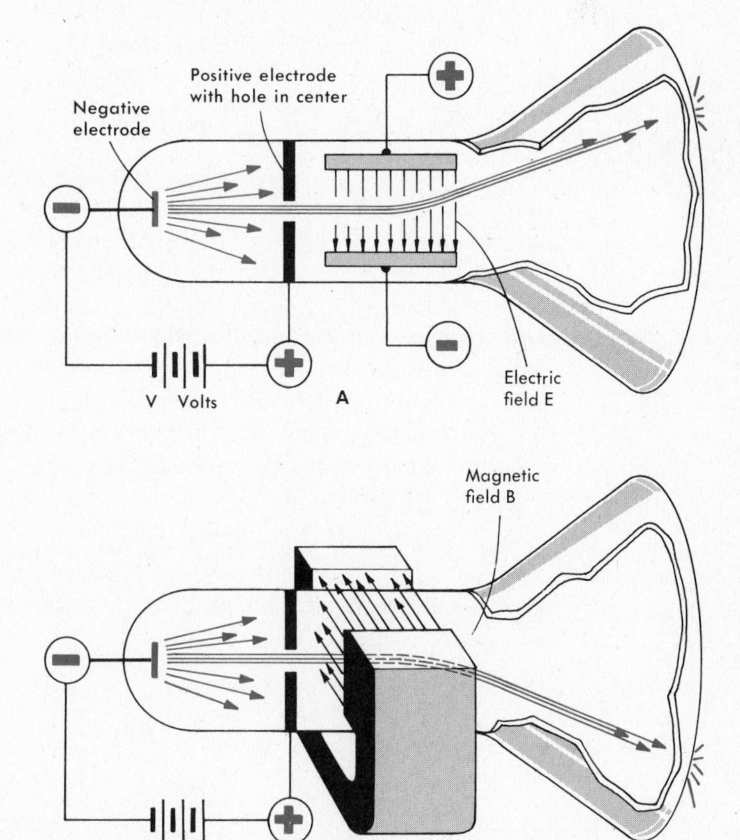

depends only on the particle's charge and on the strength of the field. For a magnetic field, however, the situation is different; a magnetic field has no effect on a stationary charge, but it does exert a force on an electric current, which is nothing more than a stream of *moving* charges. Hence the deflection in this case will depend on the strength of the magnetic field, the charge on the moving particle, and also on its velocity.

By combining the two experiments shown in Fig. 18-4A and B, Thomson was able to obtain valuable information about the little negatively charged particles called "electrons." In a tube equipped with *both* an electric and a magnetic field, at right angles to each other, and each at right angles to the beam of electrons, Thomson adjusted the strengths of the two fields so that the beam of electrons continued straight ahead without any deviation. Under these circumstances, the upward force exerted on the electrons by the electric field must equal the downward force exerted by the magnetic field. An electric field of strength E will exert a force of E N on a charge of 1 C; on an electron whose charge is e, the force will be Ee. In the magnetic field, we have seen that the force on a moving charge is Bvq, which for our electron becomes Bve. If we set these two forces equal, we get

$$Ee = Bve$$

or

$$v = \frac{E}{B}.$$

This equation does not even include the mass or the charge of the electron but does serve to measure the velocity the electrons must have if they proceed undeviated by the combined force of E and B. The experimenter knows the potential difference applied across the electrodes of the tube, which we can call V volts. By the definition of the volt, a charge of 1 C accelerated through this potential difference would have had V joules of work done on it and hence would pass through the center hole with an equal amount of kinetic energy. For an electron with a charge of e C, the energy would be Ve, which we can set down as equal to the $\frac{1}{2}mv^2$ of the electrons:

$$Ve = \tfrac{1}{2}mv^2.$$

When we substitute into this equation the previously determined value of v, we get

$$Ve = \frac{mE^2}{2B^2}$$

from which

$$\frac{e}{m} = \frac{E^2}{2VB^2}.$$

Thus we can measure the charge-to-mass ratio of the electron in terms of the readily measurable quantities E, B, and V that are used in the tube.

Actually, Thomson used a somewhat different method, which was based on measuring the deflection of the electron beam by the electric field alone when the magnetic field was turned off. This leads into complications involving the geometry of the tube that we can avoid by using this simpler method of calculation, which provides equivalent results. Thomson's experiments, and those of later workers, give the value $e/m = 1.759 \times 10^{11}$ C/kg, or 1.759×10^{8} C/g. Unfortunately, he was not then able to solve his equations to determine e, the charge of the electron, because at that time the mass of the electron was not known.

18-5 The Charge and Mass of the Electron

This work of Thomson's paved the way for the work of the celebrated American physicist Robert A. Millikan, who directly measured the charge of the electron by means of a very ingenious experiment illustrated in Fig. 18-5. A cloud of tiny oil droplets was sprayed into the space above the plates, and a small hole in the top plate was uncovered

FIG. 18-5 A schematic diagram of Millikan's experiment for measuring the elementary electron charge.

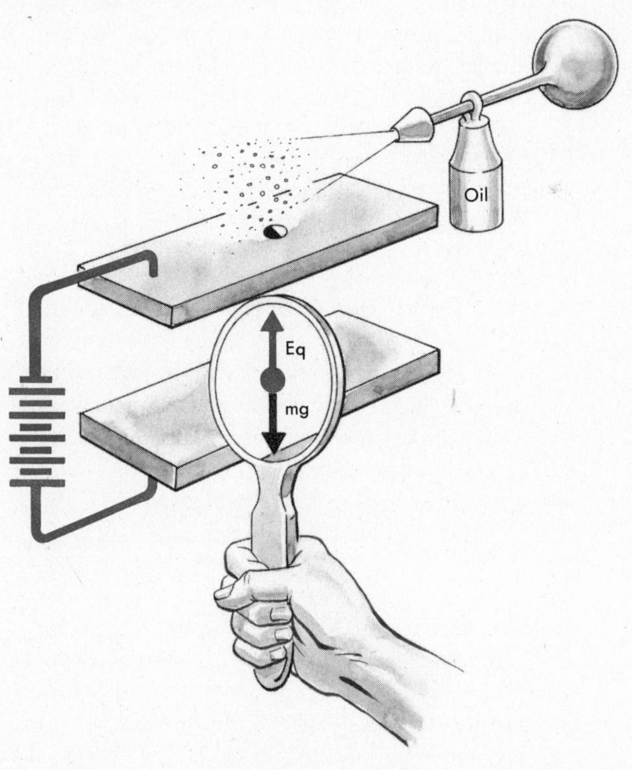

long enough for one of the droplets to drift down through the hole into the space between the plates, where it could be observed through a microscope set into the wall of the vessel. By means of a relationship known as Stokes' law, the weight of a small droplet can be determined from the rate at which it settles downward through the air. Millikan was able to measure the rate of settling with no electric field between the plates and thus compute the weight of the droplet.

Ultraviolet light can drive electrons away from the molecules of objects on which it falls, so by allowing a beam of ultraviolet light to shine between the plates, Millikan was able to cause the droplet to have a slight charge that could change suddenly from time to time as it collided with charged air molecules. By varying the potential applied across his plates, he could adjust the electric field until the droplet would hang motionless, neither rising nor falling. Under these equilibrium conditions, the upward force caused by the electric field was just equal to the weight of the droplet, and thus

$$Eq = mg$$

from which q, the charge on the droplet, could be easily figured.

It turned out that all the charges measured in this way were small *integral* multiples of a certain quantity that was apparently the elementary electric charge, or the charge of an electron. Numerically, Millikan found that the droplets always had a charge of just 1, 2, 3, or some other integer times 1.602×10^{-19} C. His interpretation was that the charge of the droplet must result from an excess (or deficiency) of an integral number of electrons, and that therefore the charge on each electron was 1.602×10^{-19} C.

From Thomson's charge-to-mass ratio and a direct knowledge of the charge on an electron, the mass of an electron can be computed to be

$$\frac{1.602 \times 10^{-19}\ \text{C}}{1.759 \times 10^{11}\ \text{C/kg}} = 9.11 \times 10^{-31}\ \text{kg}.$$

The discovery of the electron as representing a free electric charge and the possibility of its extraction from neutral atoms was the first direct evidence that *atoms are not indivisible particles but complex mechanical systems composed of positively and negatively charged parts.* Positive ions were interpreted as having a *deficiency* of one or more electrons, whereas negative ions were considered as atoms having an *excess* of electrons.

18-6 Thomson's Atomic Model

On the basis of his experiments, J. J. Thomson proposed a model of internal atomic structure (Fig. 18-6) according to which atoms consisted of a positively charged substance (*positive electric fluid*) distributed uniformly over the entire body of the atom, with negative electrons imbedded in this continuous positive charge like seeds in a watermelon.

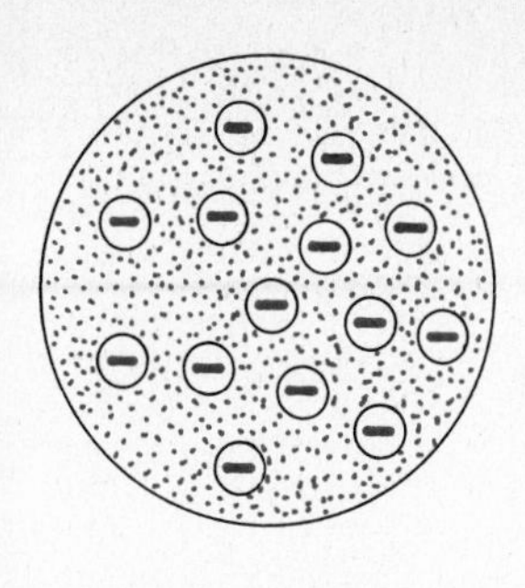

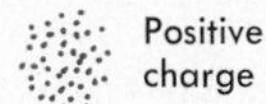
Positive charge

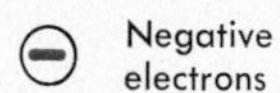
Negative electrons

FIG. 18-6 J. J. Thomson's atomic model, showing negative electrons floating in a ball of positive "electric fluid."

Since electrons repel each other but are, on the other hand, attracted to the center of the positive charge, they were supposed to assume certain stable positions inside the body of the atom. If this distribution were disturbed by some external force, such as, for example, a violent collision between two atoms in a hot gas, the electrons were supposed to start vibrating around their equilibrium positions, emitting light waves of corresponding frequencies.

At this time (1904), it was well-known that each element, when in the form of an incandescent vapor, would emit light of only certain frequencies. Many calculations were made in an attempt to reconcile the theoretical frequencies of vibrations that could be expected from Thomson's electrons with the actually observed frequencies of the light emitted by various elements. All these efforts were failures, and it became clear that there was something radically wrong with Thomson's model. It considered an atom to be a complex structure of positive and negative charge, and thus represented a considerable progress toward the truth; but it was not the true picture of an atom.

18-7 Rutherford's Atomic Model

The correct description of the distribution of positive and negative charges within an atom was made in 1911 by a New Zealander then working at Manchester University in England. This was Ernest Rutherford, who was later made Lord Rutherford for his many scientific achievements. Young Rutherford entered physics during that crucial period of its development when the phenomenon of natural radioactivity had just been discovered, and he was the first to realize that radioactive phenomena represent a spontaneous disintegration of heavy unstable atoms.

Radioactive elements emit three different kinds of rays: high-frequency electromagnetic waves known as γ (gamma) *rays*, beams of fast-moving electrons known as β (beta) *rays*, and α (alpha) *rays*, which were shown by Rutherford to be streams of very-fast-moving helium ions. Rutherford realized that important information about the inner structure of atoms could be obtained by the study of collisions between onrushing α particles and the atoms of various materials forming the target. This started him on a series of epoch-making atomic bombardment experiments that revealed the true nature of the atom. The basic idea of the experimental arrangement used by Rutherford in his studies was exceedingly simple (Fig. 18-7): a speck of α-emitting radioactive material; a lead shield with a hole that allowed a narrow beam of the α particles to pass through; a thin metal foil to deflect or scatter them; and a pivoted fluorescent screen with a magnifier, through which the tiny flashes of light were observed whenever an α particle struck the screen. The screen and magnifier were pivoted, to observe the number of flashes occurring at different angles of deflection. The whole apparatus was evacuated, so that the particles would not collide with air molecules.

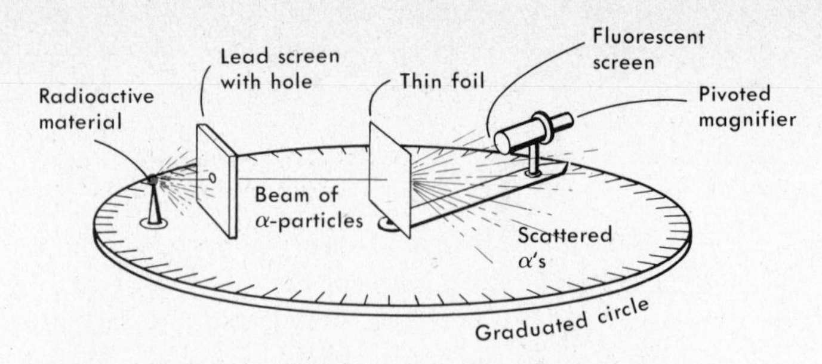

FIG. 18-7 Schematic diagram of the apparatus used by Rutherford in the alpha-scattering experiments that led to his discovery of the atomic nucleus.

How could one expect the α particles to be deflected according to Thomson's model of the atom? The α particles (their structure then unknown) were known to be doubly charged positive ions of helium; they acted as very efficient projectiles in being able to penetrate through thin metal foils at least several hundred atoms in thickness. Mathematical analysis showed that such a positive projectile, after penetrating several hundred of Thomson's spheres of positive charge, would be deflected by electrostatic forces, but the total deflection could not possibly add up to more than a few degrees.

But the results of the experiment, when performed by two of Rutherford's students, Hans Geiger and Ernst Marsden, were very different. Most of the α particles penetrated the foil with very little deflection. An appreciable fraction of them, however, were deflected through large angles—a few were turned back almost as though they had been reflected from the foil. This was a deflection of nearly 180°, and a completely impossible phenomenon according to the Thomson model.

Such large deflections required strong forces to be acting, such as those between very small charged particles very close together. This would be possible, Rutherford reasoned, if all the positive charge, along with most of the atomic mass, were concentrated in a very small central region which Rutherford called the atomic *nucleus*. Now if the α particle were also merely an atomic nucleus, the scattering problem could be treated by an analysis of the repulsion between two mass points that repel each other according to Coulomb's inverse-square law. An α particle penetrating an atom near its edge (Fig. 18-8) would be deflected only a small amount; those passing closer to the nucleus would be repelled with a greater force and deflected through a larger angle. Rutherford, knowing the kinetic energy of his α particles, calculated that they would have to get to within about 10^{-12} cm from the center of the nucleus if they were to be turned back in the direction from which they came. Thus the radius of the nucleus could not be larger than this small distance. A mathematical analysis of the total deflections that an α particle might experience in penetrating a foil was in very close agreement with the experimental data gathered by Geiger and Marsden.

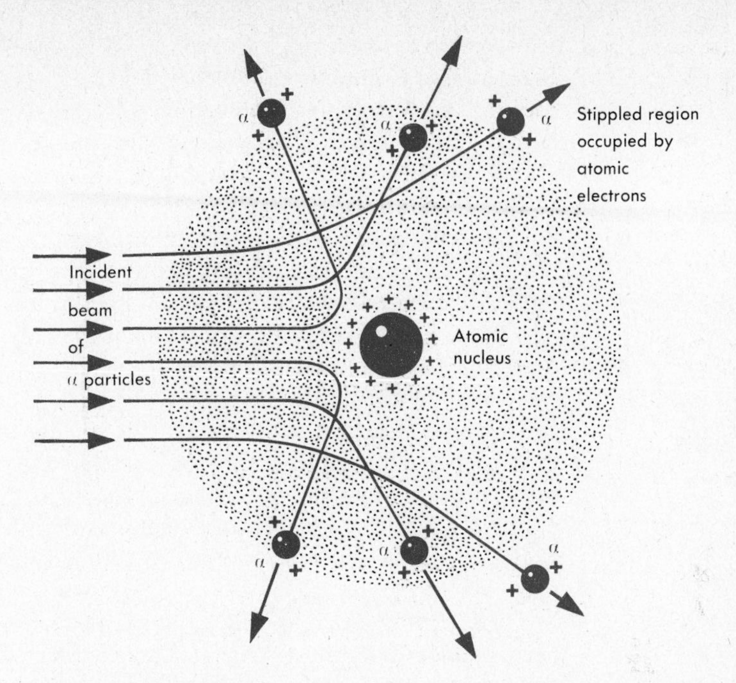

FIG. 18-8 The deflection of alpha particles by the repulsion of the small, massive, positively charged nucleus.

Since the radius of a whole atom is in the neighborhood of 10^{-8} cm, it is apparent that the volume of the atom must be largely made up of electrons arranged around the nucleus in such a way as to occupy this space. Because there would be a Coulomb force of attraction between the positive nucleus and the negative electrons, the two would be drawn together and the atom would vanish unless some provision were made to prevent it. It was suggested that the electrons might be orbiting rapidly around the nucleus, so that the electrostatic attraction would merely provide the necessary centripetal force. Thus in one bold stroke Rutherford transformed the static "watermelon model" of J. J. Thomson into a dynamic "planetary model" in which the nucleus plays the role of the sun and the electrons correspond to the individual planets of the solar system.

Because the mass of an electron is only about 1/7000 of the mass of an α particle, their contributions to the deflection of the α particles in Rutherford's scattering experiments could quite properly be ignored.

18-8 Conduction of Electricity in Solids

We have discussed the passage of an electric current through liquid solutions of acids, bases, and salts and through rarefied gases. In the first case, the current was due to the motion of positively and negatively charged ions, such as Ag^+ and NO_3^-, shouldering their way through the crowd of water molecules. In the second case, we dealt with positively charged ions flying in one direction and free negative charges, or electrons, flying in the opposite direction. But what happens when an electric current passes through solids, and why are some solids (all of them classed as metals) rather good conductors of electricity while the rest of them, known as insulators, pass hardly any electric current at all? Since in solid materials all atoms and molecules are rigidly held in fixed positions and cannot move freely as they do in gaseous or liquid materials, the passage of electricity through solids cannot be due to the motion of charged atoms or atomic groups. Thus the only active electric carrier can be an electron, which, being much smaller than the atoms and molecules forming the crystalline lattice of a solid, should be able to pass between big atoms as easily as a small speedboat can pass through a heavily crowded anchorage of bulky merchantmen. Indeed, this is what takes place in metallic conductors. The high electrical conductivity of these substances is inseparably connected with the presence of free mobile electrons that rush to and fro through the rigid crystalline lattices (Fig. 18-9).

These electrons that belong to no particular atom play an essential part in holding the metallic crystal together. Indeed, without them the positive ions would all repel each other and the crystal would be immediately converted into a puff of vapor. Free as the electrons are to move about, wherever they go, they are surrounded by the positive ions and are attracted by all of them. They thus act as a sort of bonding device that gives the crystals the strength characteristic of most metals.

In nonmetals such as sulfur, each atom holds tightly to all its electrons, and the application of an electric field can cause nothing more than a

FIG. 18-9 The motion of free electrons explains the passage of electric current through metals.

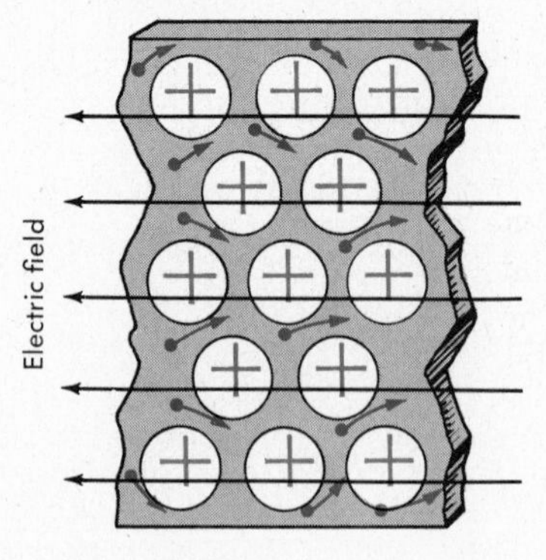

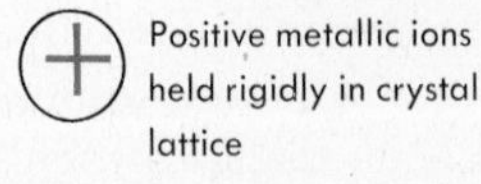

slight deformation (electric polarization) of the atoms forming the crystal lattice. Such nonmetals must therefore in general be classified as non-conductors, or insulators.

18-9 Semiconductors

Some materials, such as the element silicon (Si), cannot be classified as either good conductors or good insulators. The outer electrons of the atoms are more firmly bound than those in true metals; but thermal agitation of the atoms knocks enough of them free to make the material slightly conductive. Such materials are *semiconductors*.

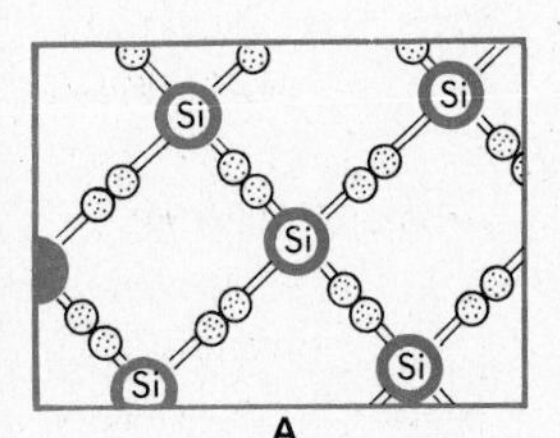

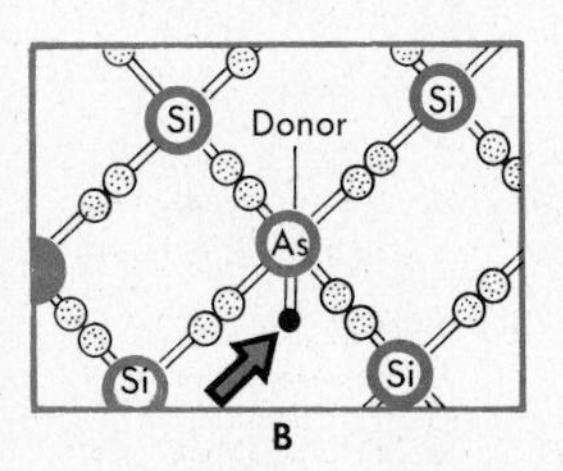

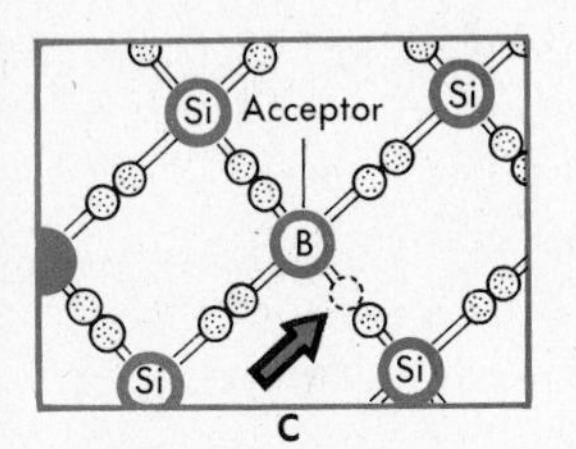

FIG. 18-10 A small proportion of the proper impurities in the crystal structure of a semi-conductor makes it much more conductive.

Figure 18-10A is a very schematic diagram of a regular crystal of pure silicon. Without concerning ourselves with the details, we can see that each atom has four outer electrons (small spotted circles), and each electron joins with an electron from another Si to bond the atoms together in a crystal. All the available electrons are utilized, and only those that are occasionally knocked loose can act to conduct a current.

Now, instead of using pure silicon, let us add a small amount of arsenic (As) to the silicon, and allow this impure silicon to crystallize. Figure 18-10B shows an As atom taking its place in the crystal. Arsenic, however, has five outer electrons; only four can be used in the bonds, so one very loosely bound electron is left free to move in an electric field, and conduct a current, similar to the conducting electrons in metals. This is called an *n-type semiconductor*, since current is carried by the negatively-charged electrons.

In Fig. 18-10C, a small amount of boron (B) has been added as an impurity to the silicon. Trivalent boron has only three outer electrons; it can make bonds with only three of its four Si neighbors, leaving a vacancy, or *hole*, as indicated by the arrow. If we now apply a potential difference across the crystal, to produce an electric field as shown in Fig. 18-11, one of the weakly-bound electrons from a neighboring Si atom to the right or below the boron may move in to fill the vacancy, as in Fig. 18-11A. But in so doing, it of course leaves another vacancy at its original location. This process is repeated, as shown in Fig. 18-11B, C, and D. The net effect is that of a hole moving downward to the right, in the direction of the field, exactly as though it were a positive charge. Such a crystal is a *p-type semiconductor*, because the conduction of current by moving holes is exactly equivalent to conduction by positive charges.

18-10 Semiconductor Rectifiers

One of the many necessary jobs to be done in all sorts of electronic equipment is *rectification*. A *rectifier* converts an alternating potential into direct, which flows always in the same direction in the circuit. Figure 18-12* shows how this can be accomplished with a pair of semi-

* The exaggerated dimensions of Figures 18-12 and 18-14 may give the incorrect impression that these semi-conductor devices consist of separate slabs of the proper type crystal, pressed together. The separate layers are actually formed in place by a complex manufacturing technology, on a thin wafer of silicon less than a hundredth of an inch thick.

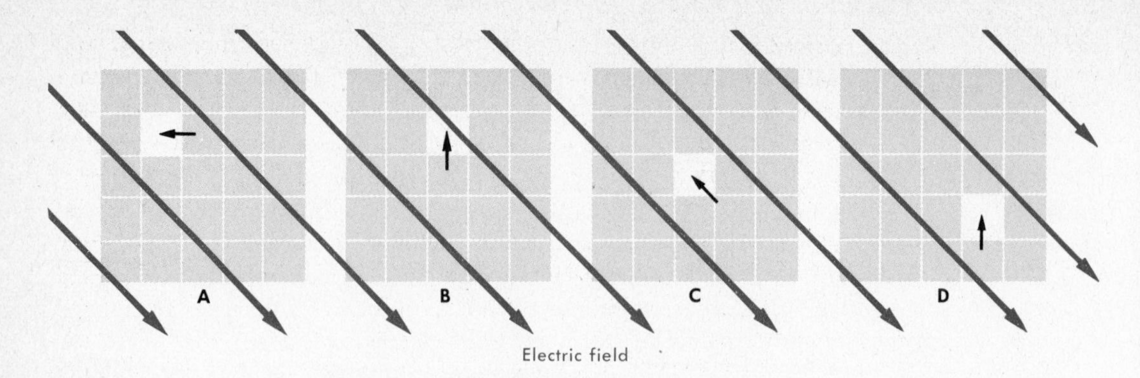

FIG. 18-11 The successive fillings of a "hole" by neighboring electrons (represented by shaded squares) makes the hole move (in this case downward and to the right as the filling electrons move upward and to the left).

FIG. 18-12 The motion of electrons and holes across a *p–n* junction. (A) If the electric field is directed from *p* to *n,* a continuous current flows across the junction. (B) If the electric field is directed from *n* to *p,* no current can flow.

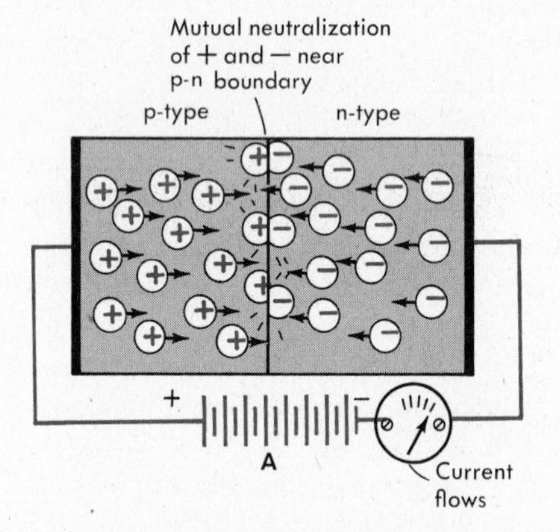

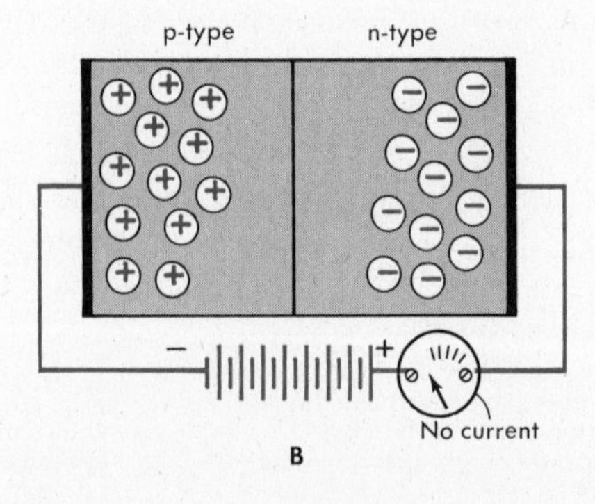

conductors. A *p*-type crystal containing mobile holes is in close contact with an *n*-type, containing mobile electrons. If we now connect a battery as shown in Fig. 18-12A, with the *p*-type side positive and the *n*-type side negative, the holes and electrons are driven toward each other. At the *p-n* boundary there will be mutual annihilation as each hole is filled with an electron. The + and − terminals will continuously provide new holes and new electrons, so that a continuous current flows. In effect, with a small potential causing a relatively large current to flow, we can say that the resistance of this device is small.

With the battery terminals reversed, as in Fig. 18-12B, the result is very different. The holes and electrons are drawn away from the *p-n* boundary, there is no constant mutual annihilation, and practically no current flows in the circuit. Since the potential causes a negligibly small current, we can consider the device to have a very high resistance to a potential in this direction.

Figure 18-13A shows an alternating potential connected across this device, called a *diode* because it has two elements. During one alternation, while the *p*-type is positive, a current flows. During the other alternation, the *p* is negative, the *n* is positive, and little or no current flows. This intermittent current, always in the same direction, is graphed in Fig. 18-13B. The dotted loops, which represent current in the opposite direction that would have flowed through an ordinary wire, have been cut off.

18-11 Transistors

Let us now make a three-element device, as sketched in Fig. 18-14. This is a *transistor*, and its basic function as an amplifier can be easily followed. A *p*-type semiconductor is sandwiched between two *n*-types. (It works equally well to use a *p-n-p* sandwich, and reverse the battery connections.) Consider the left, *n-p*, junction. The battery connections at *e* and *b* are of such polarity that this *n-p* diode element has a *low* resistance, and without the collector element, a small potential would cause a substantial current to flow in the emitter-base portion of the

FIG. 18-13 A *p–n* junction used as a rectifier of an alternating current.

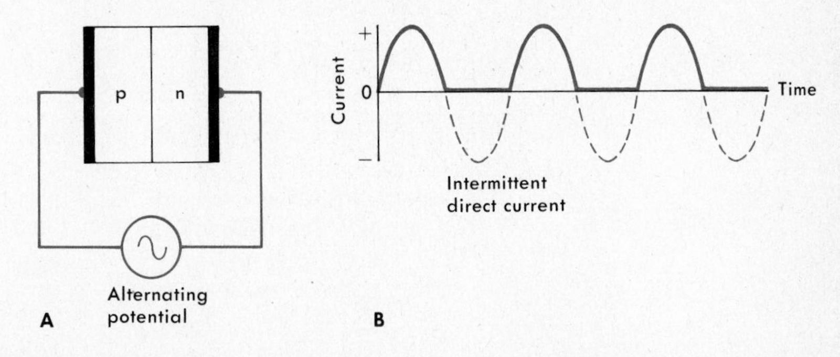

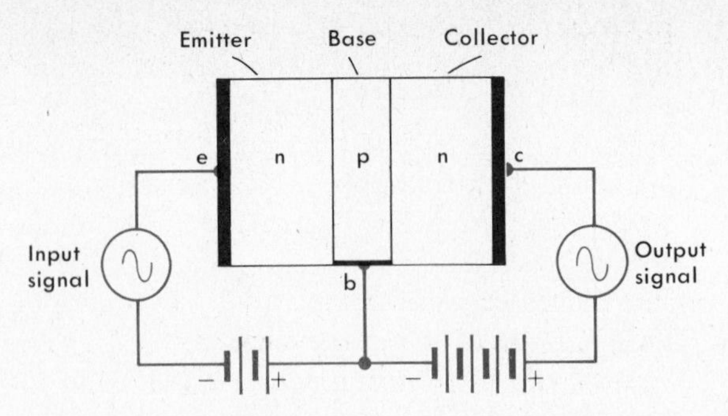

FIG. 18-14 A three-element *n–p–n* transistor used as an amplifier. (A *p–n–p* arrangement will accomplish the same job.)

circuit. Now look at the *p-n* junction between base and collector. The right-hand battery is connected with its + terminal to the *n*-type collector, and our experience with the diode shows that under these circumstances the base-collector *p-n* junction will have a very high resistance. If this were all, we should expect nothing very startling; a large current would flow in the circuit loop through *e* and *b*, and a very small current in the *b-c* loop.

The trick of the transistor lies in the extreme thinness of the base—less than 0.001 inch. Of the electrons entering the base from the emitter, only a few percent ever reach the terminal at *b*; 95 percent or more, in their irregular Brownian-motion-like path toward it, diffuse into the collector through the high-resistance base-collector boundary. Once across this boundary there is no obstacle to their attraction toward the positive terminal at *c*. So we have a nearly uniform current flowing through the whole circuit—from *e*, through the *n-p-n* semiconductors to *c*, through the two batteries, and back to *e*. Only a minor additional trickle is in the left loop—*e*, *n-p*, *b*, and the left-hand battery.

Now let us add an input signal as shown—we might imagine it to be a small alternating potential that will make V_{be} alternately a little more or a little less than its normal direct-current value. This will of course cause a corresponding fluctuation in the electron current flowing through the transistor.

$$\Delta I = \frac{\Delta V_{\mathrm{input}}}{R_{\mathrm{e-b}}}.$$

On the collector, or output, side, the current and its fluctuations are nearly the same, but it flows through the high-resistance *p-n* junction. Hence on this side we have

$$\Delta I = \frac{\Delta V_{\mathrm{output}}}{R_{\mathrm{b-c}}}.$$

From these two expressions for the same ΔI, we find

$$\frac{\Delta V_{\text{input}}}{R_{\text{e-b}}} = \frac{\Delta V_{\text{output}}}{R_{\text{b-c}}}$$

$$\Delta V_{\text{output}} = \Delta V_{\text{input}} \times \frac{R_{\text{b-c}}}{R_{\text{e-b}}}.$$

The input signal is amplified by a factor equal to the ratio of the resistances, which may run as high as 1000 or more.

18-12 From Vacuum Tubes to LSI

Rectifier diodes and amplifying transistors are only two of many types of semiconductor devices. For many decades before their invention and development, their functions were performed by vacuum tubes. Indeed, for a number of specialized operations the vacuum tube is still the only satisfactory answer.

The operation of these tubes, like that of semiconductor devices, depends on the control of a stream of electrons. In the tube, however, the electrons must be emitted from a red-hot electrode (the cathode) and attracted through a vacuum to a positively-charged electrode (the anode). Between cathode and anode are metal grids or coils whose varying potentials serve to control the flow of electrons. There are a number of disadvantages to this system. An overwhelmingly large fraction of the power needed is used merely to heat the cathode to the temperature at which it will emit electrons freely. The disposal of this heat is a major problem that does not exist with transistors. The mechanical structure of the tube—accurately positioned cathode, anode, and grids, all surrounded with an evacuated glass container—makes it necessarily large, fragile, and expensive. The transistor is an inexpensively manufactured small chip of crystal, generally imbedded in a lump of protective plastic. It is remarkably rugged, and its operating lifetime, with no hot filaments to burn out, is indefinitely long.

The miniaturization of semiconductor circuitry has been carried to almost unbelievable lengths, and techniques under development give promise of still further progress. By means of photographic masks to protect or expose the proper areas, *p*- and *n*-type inpurities, conducting metal strips, insulating layers, resistances, etc., can be deposited on or etched away from the surface of a chip of silicon in multiple layers that provide a large number of devices in a very small area, known as *Large-Scale Integration,* or LSI.

Questions

(18-1) 1. What are the ions into which silver nitrate decomposes in solution? What is the charge on each ion?

2. What are the ions into which copper sulfate breaks up when in solution? Give the charge on each ion.

(18-2) 3. Look up lead and zinc in the periodic table, and (a) write their symbols, atomic numbers, and atomic weights. (b) What is the ratio of the mass of a lead atom to the mass of a zinc atom? (c) How does the mass of 3.42×10^{24} atoms of lead compare with the mass of an equal number of atoms of zinc? (d) How does the number of atoms in 20.719 g of lead compare with the number of atoms in 6.537 g of zinc?

4. (a) Write the symbols, atomic numbers, and atomic weights of iron and potassium. (b) How does the mass of an iron atom compare with the mass of an atom of potassium? (c) What is the ratio of the mass of 7×10^{19} iron atoms to the mass of 7×10^{19} potassium atoms? (d) Consider a sample of 55.85 g of iron, and one of 39.10 g of potassium. How do the number of atoms in each of these samples compare?

5. What is the number of atoms in 12 g of lead and the number of atoms in 5 g of zinc?

6. Compare the number of atoms in 5 g of iron to the number of atoms in 20 g of neon.

7. (a) What is an equivalent weight of hydrogen? (b) an atomic weight of silver? (c) an equivalent weight of aluminum?

8. (a) What is an atomic weight of copper? (b) an equivalent weight of silver? (c) an equivalent weight of gold?

9. An electrolytic cell contains a solution of a silver salt and is connected in series with a cell containing a solution of a salt of zinc (divalent). (a) How many coulombs of charge must flow through the cells to deposit 10 g of zinc? (b) In this time, how much silver will have been deposited?

10. An electrolytic cell contains a solution of gold chloride, and is placed in series with another cell containing a solution of a salt of copper (divalent). (a) By the time 0.60 g of copper has been deposited, how much gold will have been deposited on the cathode of the other cell? (b) How many coulombs of charge will have flowed through the two cells?

11. Molten salts (the pure salt, not a solution) are also ionized and can be electrolytically decomposed. Metallic sodium and chlorine gas can be produced by electrolysis of common salt (NaCl, both monovalent). How many pounds of sodium are produced for each pound of chlorine?

12. Most of the copper used to make wire, etc., has been electrolytically refined by depositing it from a copper salt solution (divalent) onto a cathode. What is the cost of electrical energy required per kilogram of copper, if electricity costs the refiner 0.5 cent /kWh? (The cell operates at 0.2 V).

(18-4) 13. What force is exerted on an electron when it is in an electric field of 125 V/cm?

14. An electron is to be given a force of 3.2×10^{-15} N by an electric field. What must be the strength of the field in V/cm?

15. What is the force on an electron whose speed is 10^9 cm/s while moving at right angles to a field whose flux density is 3×10^{-3} weber/m^2?

16. What must be the flux density B, if a magnetic field is to exert a force of 3.2×10^{-10} dyne on an electron traveling 10^9 cm/s at right angles to the field?

17. What potential difference would be needed in the electron gun if the electrons were to be undeviated in passing through an electric field of 270 V/cm, and a perpendicular magnetic field of 8×10^{-4} weber/m^2?

18. In an electron gun, the potential difference between cathode and anode is 300 V. The electron beam passes through a magnetic field of 20 gauss, and a perpendicular electric field, to emerge undeviated. What is the strength of the electric field?

(18-5) **19.** A droplet of oil has four extra electrons. (a) What electric force is exerted on the droplet when it is in an electric field of 220 V/cm? (b) If the drop is motionless in this field, what is its weight?

20. In a repetition of Millikan's oil-drop experiment, the weight of a droplet is found to be 4.60×10^{-15} N. The plates are 1.50 cm apart, and the droplet is motionless when a potential of 72 volts is applied across them. How many electron charges are there on the droplet?

(18-7) **21.** The charge on a nucleus equals the atomic number of the element times the charge of the electron; the charge on a neon nucleus, for example, is $10 \times 4.8 \times 10^{-10} = +4.8 \times 10^{-9}$ esu. (a) What is the potential energy of an alpha particle at the instant it has penetrated to 10^{-12} cm from a gold nucleus and is momentarily stationary? (b) What kinetic energy must the particle have in order to penetrate to this distance from the nucleus?

22. An alpha particle has 1.2×10^{-6} erg of kinetic energy. What is the closest it can come to the nucleus of a silver atom? (See Question 21.)

23. Figure 18-8 is not drawn to scale. In the drawing, the diameter of the nucleus is about 0.3 inch. About what should the diameter of the atom be on this scale, if the drawing were to be in proper proportion?

19

Light Rays

19-1 The Velocity of Light

Light is obviously a form of energy transmitted in some way through space. With your unaided eyes, you can see light that has been traveling from its source for 2 million years to be finally absorbed in the act of stimulating chemical changes in your retinas, which you interpret as "seeing" the Andromeda galaxy.

In order to know that light has taken 2 million years to make this journey, it is apparent that we must have some knowledge of the distance of the Andromeda galaxy (a problem we leave to the astronomers) and also a knowledge of the speed at which light travels.

The first attempt to measure the speed of propagation of light was undertaken by Galileo in a very primitive way. One evening, he and his assistant placed themselves on two distant hills in the neighborhood of Florence, each of them carrying a lantern with a shutter. Galileo's assistant was instructed to open his lantern as soon as he noticed the flash from the one carried by his master. If light were propagating with a finite speed, the flash from the assistant's lantern would have been observed by Galileo with a certain delay. The result of this experiment was, however, completely negative, and we know now very well why. Light propagates so fast that the expected delay in Galileo's experiment must have been about one hundred-thousandth of a second, which is quite unnoticeable to human senses.

The velocity of light was first successfully measured by the Danish astronomer Olaus Roemer, who, instead of observing a helper on a neighboring mountain, studied the moons of the planet Jupiter, thus increasing the distance to be covered by light by a factor of hundreds of millions. Roemer's method is illustrated in Fig. 19-1, which shows the orbits of the earth, Jupiter, and one of Jupiter's moons. Moving around the planet, the moons are periodically eclipsed as they enter the broad cone of shadow cast by Jupiter. Studying these eclipses, Roemer noticed that they occurred ahead of their average schedule when the earth and

FIG. 19-1 Roemer's method for measuring the velocity of light by observing the eclipses of Jupiter's moons.

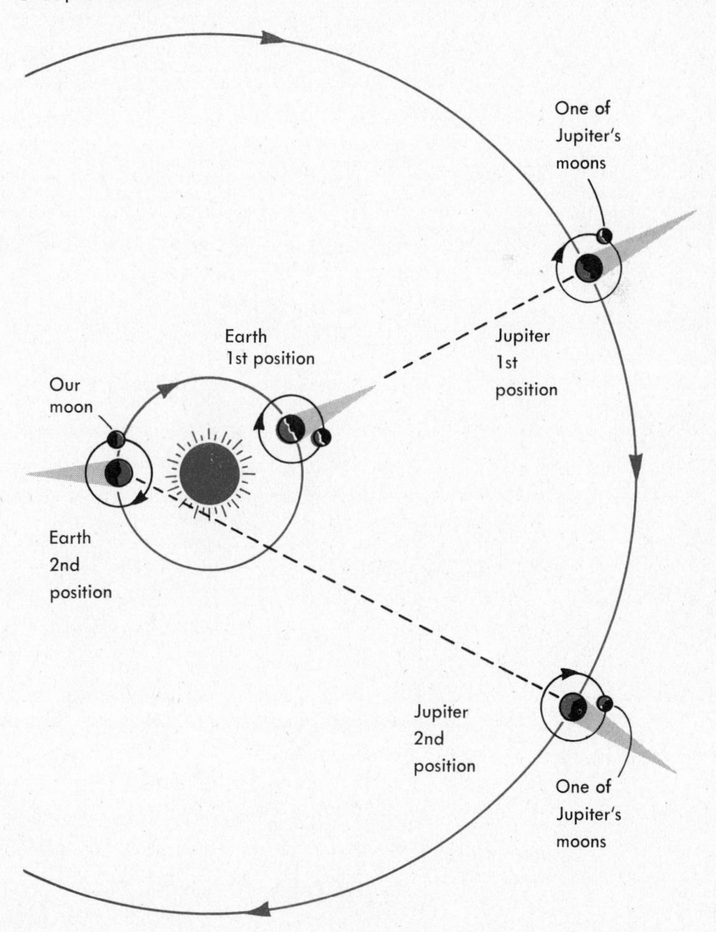

Jupiter were on the same side of the sun (first position) and were delayed in the opposite case (second position). Clocks were not very reliable nor accurate in the seventeenth century, and Roemer's best figures were that the advance and delay in the eclipses were each about 11 minutes. From these observations Roemer correctly concluded that the speed of light was such that it would require 22 minutes to cross the diameter of the earth's orbit, then thought to be about 2.9×10^8 km. This gives a value of 2.2×10^8 m/s for the speed of light. This figure is considerably low, of course, but was nevertheless a remarkable achievement, considering the crude technology of his time. Regardless of its accuracy, it showed that the speed of light was not infinite, as many then believed, but was finite and measurable.

The first laboratory measurement of the speed of light was carried out in 1849 by the French physicist H. L. Fizeau, whose apparatus is shown in basic principle in Fig. 19-2. (Fizeau used lenses to focus the light into a tight beam. These are not shown.) With the 720-tooth wheel stationary, the beam passed through a space between teeth, was reflected from a mirror 8633 m away, and returned between the same teeth to enter Fizeau's eye. When the toothed wheel was set spinning at the proper speed, the outgoing light that had passed between the teeth found its way blocked by a tooth on its return, so that no light reached the observer. This would happen when the time needed for the light to travel $2 \times 8633 = 17{,}266$ m was equal to the time it took the wheel to turn through $1/(720 \times 2) = 1/1440$ revolution. Fizeau found the returning light was completely blocked when his wheel turned at 12.6 rev/s. At this speed, the time for the wheel to turn 1/1440 rev is $1/(12.6 \times 1440) = 5.51 \times 10^{-5}$ s. This gave him for the speed of light, c:

$$c = \frac{1.7266 \times 10^4}{5.51 \times 10^{-5}} = 3.13 \times 10^8 \text{ m/s}.$$

FIG. 19-2 Schematic diagram of Fizeau's method for measuring the velocity of light.

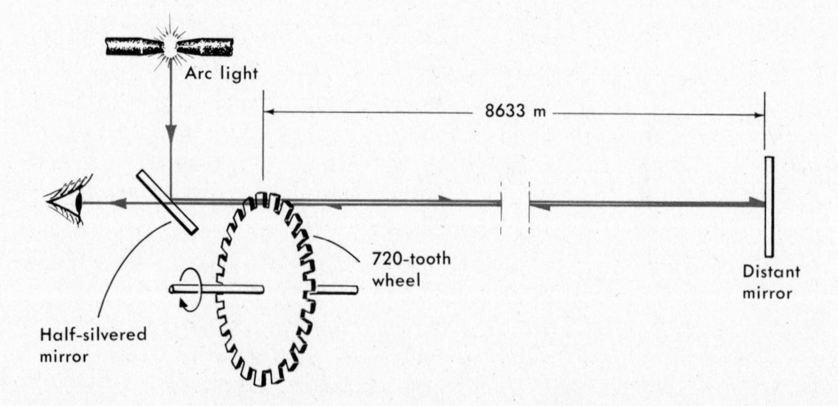

Many experimenters, using many different methods, have continued research on this important constant. The most precise determination to date—from research conducted jointly by the National Bureau of Standards and the University of Colorado—finds the speed of light in vacuum to lie somewhere between 2.99792459 and 2.99792461×10^8 m/s. This was done not by dividing a distance by a time, but by taking advantage of the fact that light and other similar radiation is a wave phenomenon. An accurate measurement of frequency and wavelength then gives the speed of light as $c = f\lambda$. For our work here (in which we generally expect only slide-rule accuracy), we will simplify our arithmetic by using the value $c = 3 \times 10^8$ m/s. (The letter c is universally used to represent the speed of light.)

19-2 Reflection of Light

The study of light is one of the most important parts of physics, since most of our knowledge concerning the world around us is gained through seeing. We learn about the properties of giant stellar systems by means of light that travels for millions of years through empty space to deliver us its message. We learn about the properties of atoms through the light emitted by them, which carries in a hidden form information concerning their internal structure. And, of course, most of the information that we get in our everyday life is also obtained through the medium of light.

Unless we are looking directly at a source of light—the sun, a star, an electric light, a firefly, etc.—what we see is reflected light.

Obviously, reflecting surfaces vary in many respects. The shiny door of a well-polished black car may reflect only 5 percent of the light that falls on it, but this small amount of reflected light presents you with a clear image of your face as you stand before it. A good mirror can do no more, although the image is brighter because it may reflect as much as 85 or 90 percent of the light. A piece of black cast iron and a clean white plastered wall may reflect as much light as the car door and the mirror, but you see no image of your face as you look at these.

The difference, of course, is in the smoothness of the surfaces. The glossy surfaces of glass or enamel or polished metal are smooth and regular, even to a beam of light; paper and plaster are covered with microscopic irregularities from which light is reflected in all directions like Ping Pong balls bouncing from the surface of a rock pile.

Reflection from a glossy surface is *specular* reflection (from the Latin word for "mirror"); the irregular reflection from a comparatively rough surface is *diffuse*. Most of our attention will be to specular reflections, because this reflection behaves in accordance with a simple law.

Figure 19-3 shows a narrow beam of light from a small bright source falling on a mirror. We can imagine such beams from any source, narrow enough to represent by a line—these imagined narrow beams are *rays*, which we will find it useful to draw on many occasions. In Fig. 19-3 we notice a regularity first observed by the ancient Greeks. Draw a

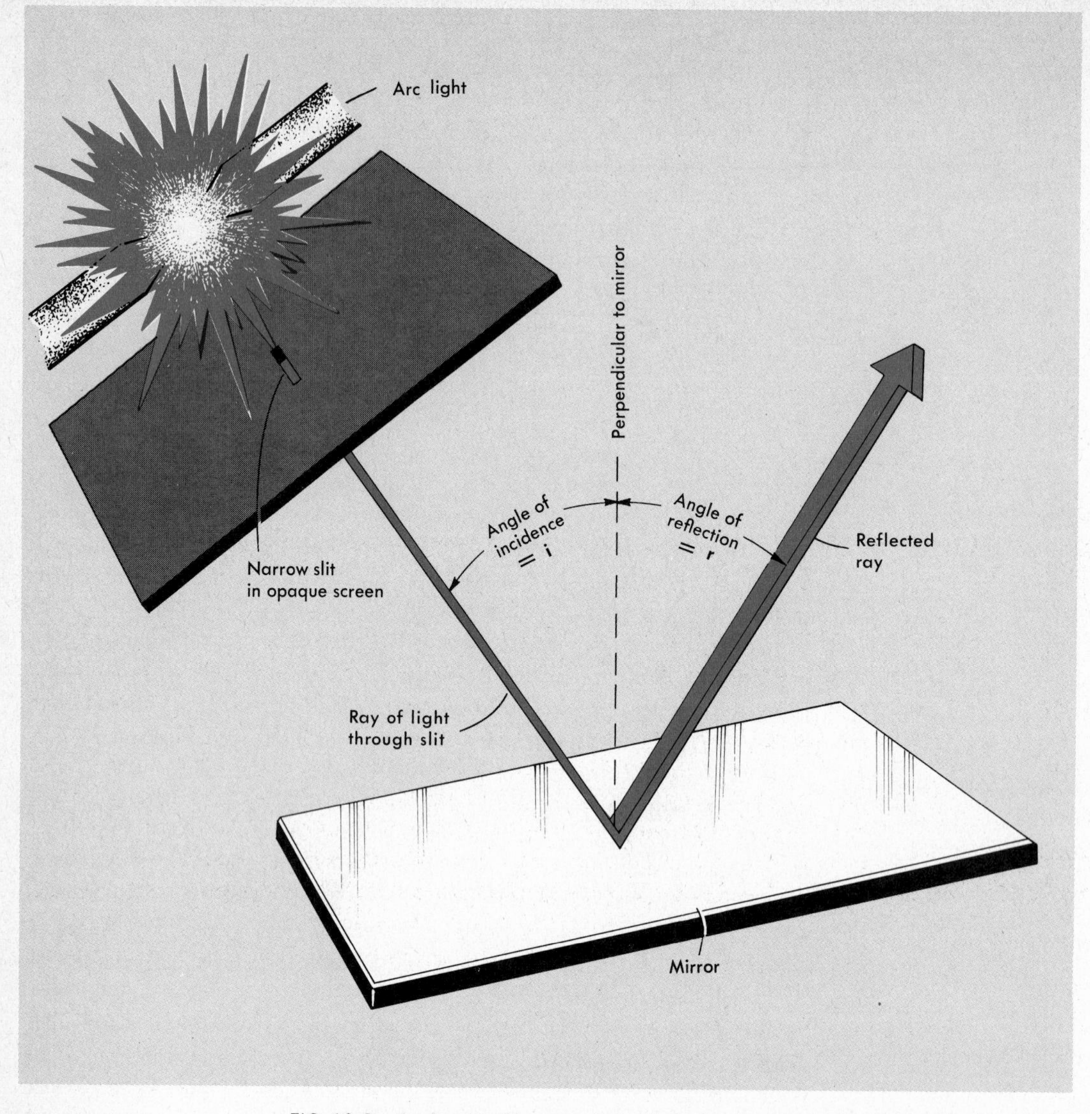

FIG. 19-3 Angle of incidence and angle of reflection for a reflected beam of light.

perpendicular to the mirror where the ray strikes it; the angle between the incoming ray and this perpendicular is the *angle of incidence*. The angle between the reflected beam and the perpendicular is the *angle of reflection*. The law of reflection is that *the angle of reflection equals the angle of incidence*. By applying this single simple law we can analyze how images are formed by various reflective surfaces.

19-3 Plane Mirrors

The mirrors that hang on walls are usually flat, or *plane*, and we all are familiar with the appearance of objects when seen in such mirrors. It

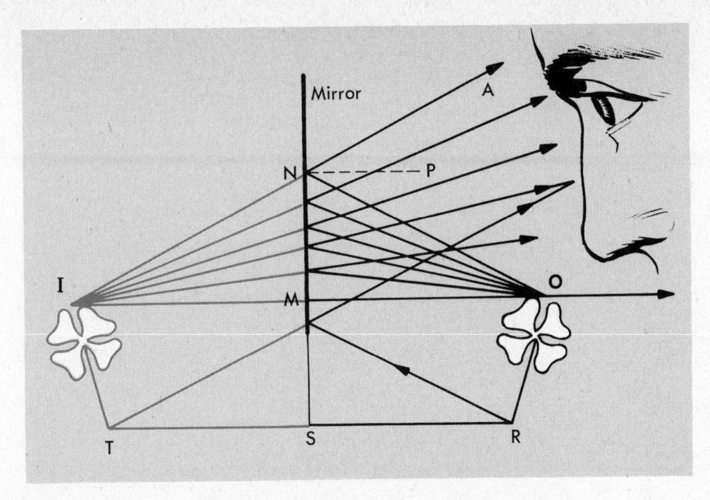

FIG. 19-4 The formation of a virtual image by a plane mirror.

will probably be helpful to our understanding if the preceding statement is rephrased: Instead of "seeing an object in a mirror," it would be clearer to say "seeing the image that the mirror forms of the object." Figure 19-4 shows how the image of an object is formed by a plane mirror. The top of the cloverleaf, *O* (for "object"), is illuminated presumably by the sun; the diffuse reflection of the sunlight from *O* sends out an infinite number of rays in all directions, of which we have shown only a few. *To the eye, it appears as though the reflected rays were all coming from the point I.* Actually, of course, no light ever reaches or passes through *I* (*I* may be in the middle of a thick brick wall), and *I* is in such a case called a *virtual image* of point *O*.

Ray *OM* has been drawn perpendicular to the mirror, and is reflected back in the direction from which it came. At point *N*, *NP* has been drawn perpendicular to the mirror, and for the ray striking *N* (as for all the rays) the angle of incidence (angle *ONP*) equals the angle of reflection (angle *PNA*). The application of a small amount of high school geometry will prove that the triangles *OMN* and *IMN* are equal. From this it is apparent that ***in a plane mirror, the virtual image of an object lies on the line drawn from the object perpendicular to the mirror, and is just as far behind the mirror as the object is in front of the mirror.***

The image of the stem of the clover, and of every point between the top and the stem, could have been constructed in this same way, but we need only draw *RST* perpendicular to the mirror and make *TS* equal to *SR*, in order to locate the image of the stem. It is worth noticing that the eye can see the image of the stem perfectly, even though the mirror does not extend down as far as *S*.

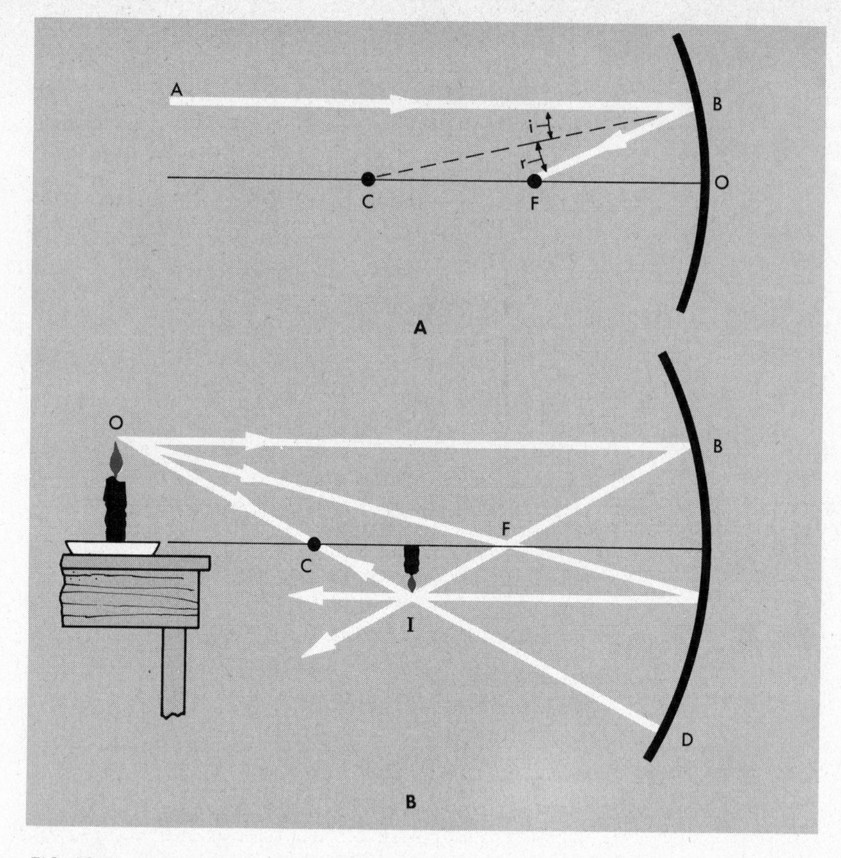

FIG. 19-5 Image formation by a concave mirror.

19-4 Concave and Convex Mirrors

In many optical instruments, from telescopes to microscopes, mirrors that are concave or convex rather than plane are used. The simplest shape for such a mirror is that of a small piece of a polished sphere. Let us see what happens to light that is reflected from a concave spherical mirror, shown in Fig. 19-5A. Point C is the center of the hollow sphere of which the mirror is a part, and is called the *center of curvature* of the mirror. The line CO, from the center of curvature to the center of the mirror, is the *optical axis*. Take a ray of light which comes in parallel to the optical axis and which strikes the mirror at B. A radius of the sphere CB is perpendicular to the mirror, so AB is reflected toward F with its angle of incidence i equal to its angle of reflection r. The reflected ray hits the optical axis at point F, which is called the *principal focus* of the mirror. Since AB is parallel to the optical axis, angle $BCF = i$, and triangle BCF is isosceles, which makes $BF = CF$. If point B is not too far from O, OF and BF are nearly equal, and we can say that OF almost exactly equals half of the radius of curvature CO.

Many other rays from some very distant object could also be drawn—

all parallel to the optical axis and to each other. All these rays would intersect at (or very nearly at) the same point F. The closer the ray AB is to the optical axis CO, the more nearly true is the approximation that $OF = BF$ and that OF equals half the radius of curvature OC. With this restriction in mind, we can define point F as *the principal focus of the mirror; the point at which rays parallel to the optical axis and close to it intersect the optical axis and each other*. The distance OF (generally designated as f) is the *focal length* of the mirror. In all our discussions and examples, we will assume that the mirror itself is small in comparison with its radius of curvature, so that these approximations will be very close to the truth.

Figure 19-5*B* shows the formation of an image by a concave mirror. From point O on the object we can draw an infinite number of light rays that will strike the mirror, and by drawing a radius from C we could construct r equal to i and thus trace the reflection of each. By selecting rays judiciously, however, we can make the job much easier. Select OB, the ray parallel to the optical axis; we know this ray will be reflected through the principal focus F. Another ray whose behavior can be easily predicted is OC; since it travels along a radius of the sphere, it will strike the mirror at D and be reflected back toward C again. Since $i = r$, any reflected ray can also be followed backward; in Fig. 19-5A, a ray from F to B would be reflected parallel to the axis along BA. We can thus draw a third predictable ray from point O: The ray OF will strike the mirror and be reflected back parallel to the optical axis, as shown. These three selected rays intersect at I, and to an observer *it will appear as though the light were emanating from I*. The light rays from O actually do all intersect and cross at I. If we were to put a piece of ground glass at this point, an actual image of O would be focused on it. Such an image is called a *real image*.

In Fig. 19-6, a ray has been drawn from O to M, at the center of the mirror. Angle OMA = angle IMB (why?), so the right triangle OMA is similar to the right triangle IMB. Therefore,

$$\frac{OA}{BI} = \frac{AM}{BM}$$

or

$$\frac{\text{object size}}{\text{image size}} = \frac{\text{object distance}}{\text{image distance}} = \frac{p}{q}.$$

In the drawing are another pair of similar right triangles, OAC and IBC. From these triangles, we find

$$\frac{AC}{BC} = \frac{OA}{BI} = \frac{p}{q}.$$

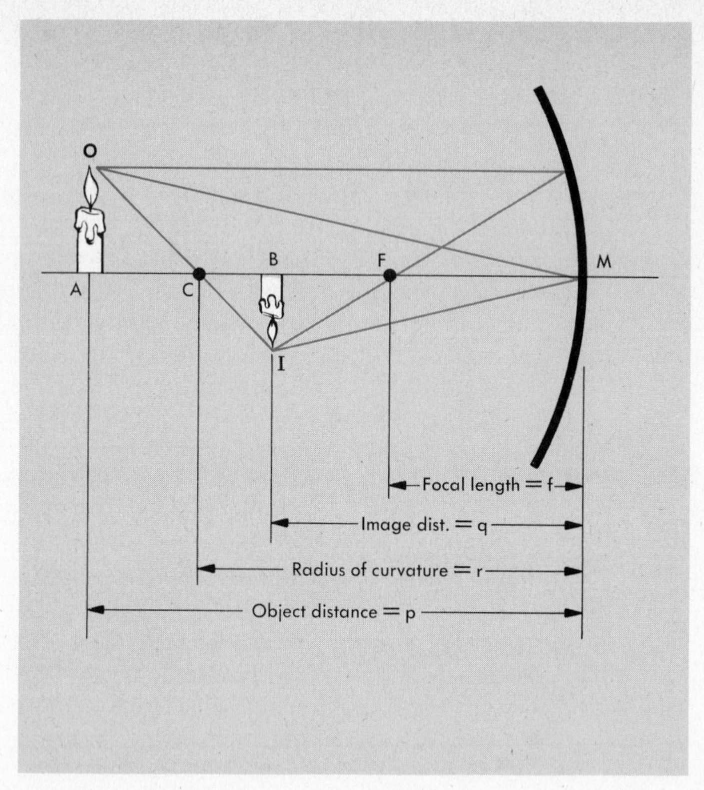

FIG. 19-6 Geometrical factors in the formation of an image by a concave mirror.

We can see from the drawing that AC equals the object distance minus the radius of curvature of the mirror, or $p - r$. Similarly $BC = r - q$. Substituting these values we find

$$\frac{p - r}{r - q} = \frac{p}{q}$$

$$pq - rq = rp - pq$$

$$rq + rp = 2pq.$$

Dividing this last equation by pqr, we find

$$\frac{1}{p} + \frac{1}{q} = \frac{2}{r}$$

and since $r = 2f$, and $2/r = 1/f$, we have

$$\frac{1}{p} + \frac{1}{q} = \frac{1}{f}.$$

As an example, take a concave mirror with a radius of curvature of 24 cm. What can we say about the image of an object 4 cm long that is placed 48 cm in front of the mirror? It is given that $f = 12$ cm and that $p = 48$ cm, so from

$$\frac{1}{p} + \frac{1}{q} = \frac{1}{f}$$

we find

$$\frac{1}{48} + \frac{1}{q} = \frac{1}{12}$$

or

$$\frac{1}{q} = \frac{1}{12} - \frac{1}{48} = \frac{3}{48} = \frac{1}{16}$$

so that

$$q = 16 \text{ cm}.$$

We can now figure the size of the image:

$$\frac{\text{Image size}}{4} = \frac{16}{48}$$

from which the image size equals $4 \times \frac{16}{48} = \frac{4}{3}$ cm long. The image (as you can check with a graphical construction) will be real and inverted.

It is a good idea to adopt a standard way of making sketches of mirror and lens problems. The most common way is to have the light coming in from the left. If you then keep in mind that the example we have just used to derive the general equation relating p, q, and f is the "all +" case, you should have no difficulty with algebraic signs. The center of curvature of the concave mirror lies to the left of the mirror, and we have called its focal length +; hence, for a convex mirror, the center of curvature will be to its right, and f will be negative. In our "all +" case, the object was to the left; so if the object were to the right of the mirror, p would be negative. Similarly, for an image formed to the right of the mirror (opposite to the "all +" case), we would have a negative value for q.

Figure 19-7 shows the graphical construction of the image formed by a convex mirror. The ray OA, parallel to the axis, will be reflected as though it were coming from F. (With a convex mirror, the principal focus F is of course *virtual*, because no ray of light can ever actually reach it.) The ray OB, heading toward the center of curvature along a radius, will be reflected back along the same path. Although the image of the top of the flagpole can be accurately located by the intersection of these two rays alone, we can easily add a third ray: OC is drawn from O toward the principal focus F. It will never reach there, of course, but will

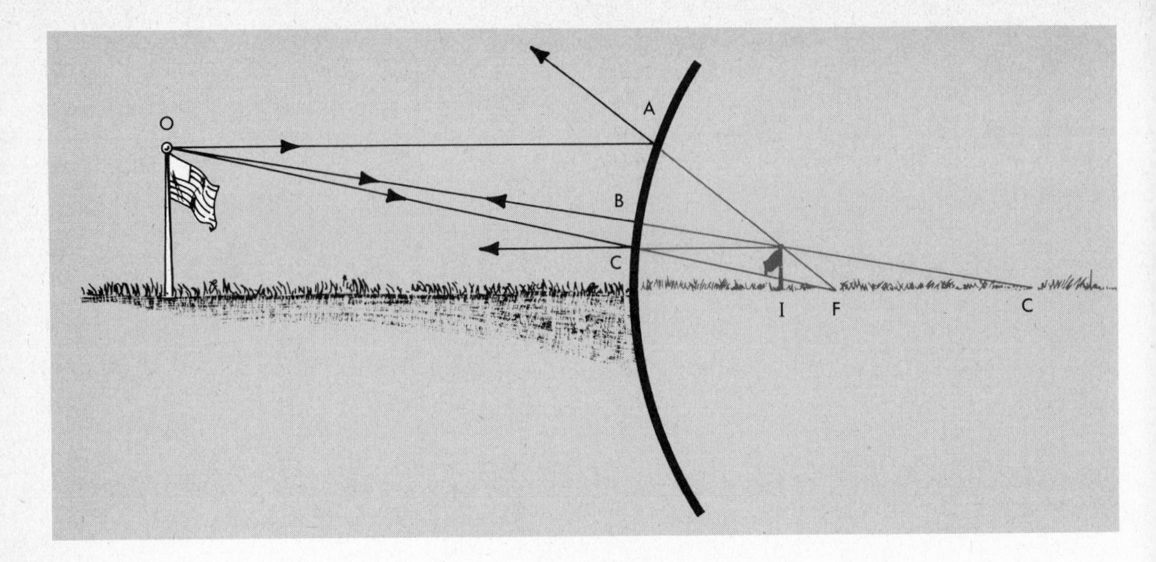

FIG. 19-7 Image formation by a convex mirror.

instead be reflected parallel to the axis and when extended to the right beyond the mirror will also intersect AF and BC at the top of the flagpole image. Thus, to an observer, all three rays (as well as any others we might care to draw) originating at O and reflected by the mirror will appear exactly as though they had come from the virtual image I.

The same equation relating object distance, image distance, and focal length is also applicable to the convex mirror. Suppose an object O were 25 cm in front of a convex mirror with a radius of curvature of -20 cm. The focal length f is then -10 cm, and we have

$$\frac{1}{25} + \frac{1}{q} = \frac{1}{-10} = -\frac{1}{10}$$

$$\frac{1}{q} = -\frac{1}{10} - \frac{1}{25} = -\frac{7}{50}$$

or

$$q = -\frac{50}{7} = -7.14 \text{ cm}.$$

The negative value for q shows that the image is behind the mirror and therefore virtual.

19-5 Refraction

If a beam of light is directed against a pool of water or a piece of glass, its direction will be changed (unless it happens to strike perpendicular to the boundary surface of the water or glass, which is a very special

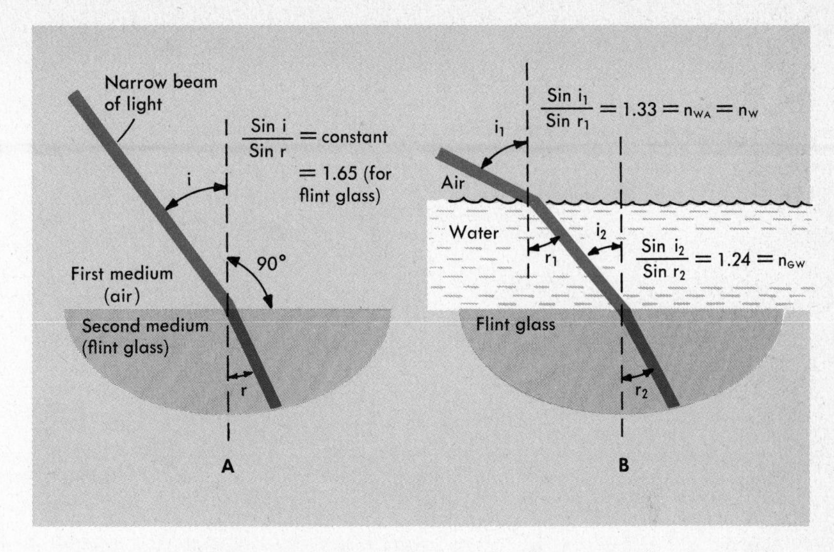

FIG. 19-8 The refraction of a beam of light in passing from one medium into another.

case). Figure 19-8A shows such a beam, which travels first in air and then strikes the surface of a block of flint glass. The angle of incidence i and the angle of refraction r are measured between the ray and the *normal* (i.e., *perpendicular*) to the surface, as shown.

The relationship between i and r is not quite so obvious as the simple $i = r$ law of reflection; it was not until 1621 that it was discovered by the Dutch physicist Willebrord Snell. Snell found that for any ray passing from one medium into another,

$$\frac{\sin i}{\sin r} = n$$

where n is a constant—that is, the ratio $\sin i/\sin r$ is the same for any given pair of media, no matter what the angle of incidence is. This ratio n is called the *index of refraction* of the second medium (into which the ray is refracted) with respect to the first (from which the ray has come). We can use subscripts to indicate the mediums concerned. In the example of Fig. 19-8A we might write $n_{GA} = 1.65$, to indicate that this is the index of refraction of glass relative to air and would apply to the case of a ray in air entering glass.

Ordinarily, if no reference medium is specified, it is assumed to be empty space. Air, however, differs very little from a vacuum in this respect, so that we might say merely that the index of refraction of the flint glass is 1.65. If this is expressed as n_G, without a second subscript, it will always be understood to be with reference to air or vacuum.

TABLE 19-1 REFRACTIVE INDICES OF DIFFERENT SUBSTANCES (for yellow light)

Substance	Refractive Index
Water	1.33
Alcohol	1.36
Glass (flint)	1.65
Diamond	2.42

Table 19-1 gives the index of refraction for a few substances. The index is slightly different for different colors of light; as is customary, these indices are given for yellow light.

If the angle of incidence in Fig. 19-8A were 40°, we would then have

$$\frac{\sin 40^\circ}{\sin r} = 1.65$$

$$\sin r = \frac{\sin 40^\circ}{1.65} = \frac{0.643}{1.65} = 0.390$$

$$r = 23.0^\circ.$$

Figure 19-8**B** shows a situation that is a little more complex; there is a uniform layer of water between the air and the glass. Given the initial angle of incidence i_1, we should have no trouble calculating r_1, the angle of refraction into the water, as the index of water (with respect to vacuum or air) is given in Table 19-1. But the table does not give the index of flint glass with respect to water, which we will need in order to calculate the second refraction. Fortunately (for reasons to be discussed later) there is a simple way out of this difficulty: $n_{GW} = n_G/n_W = 1.65/1.33 = 1.24$.

Now, assuming that $i_1 = 60^\circ$, we can work out the example of Fig. 19-8**B**:

$$\frac{\sin 60^\circ}{\sin r_1} = 1.33$$

$$\sin r_1 = \frac{\sin 60^\circ}{1.33} = \frac{0.866}{1.33} = 0.651$$

and

$$r_1 = 40.62^\circ \quad \text{or} \quad 40^\circ 37' = i_2.$$

For the water-to-glass refraction,

$$\frac{\sin 40.62^\circ}{\sin r_2} = 1.24$$

$$\sin r_2 = \frac{0.651}{1.24} = 0.525$$

$$r_2 = 31.67^\circ \quad \text{or} \quad 31^\circ 40'.$$

It is interesting to consider that if the direction of the beam in Fig. 19-8B is reversed, so that the light passes from glass to water to air, the path of the beam will be identical. The i's will become r's, and the r's will become i's; the indices of refraction will now be of water with respect to glass, and of air with respect to water. These will be the inverse of the indices taken the other way and will both be less than 1.

19-6 Prisms

The prism is a very useful piece of optical equipment, and we will need to speak of it again in later chapters. Figure 19-9 shows a ray passing through a 60° prism, made of glass for which $n = 1.6000$. The angle of incidence of the ray is 45°. Let us trace this ray as it is refracted at both faces of the prism.

$$\frac{\sin i_1}{\sin r_1} = n_{GA} = 1.6000$$

$$\sin r_1 = \frac{\sin i_1}{1.6000} = \frac{0.70711}{1.6000} = 0.44194$$

$$r_1 = 26^\circ\ 14' \quad \text{or} \quad 26.23^\circ.$$

If you need to brush up on your geometry, it might be a good exercise to show that $i_2 = A - r_1 = 60^\circ - 26.23^\circ = 33.77^\circ$.

When the ray emerges,

$$\frac{\sin i_2}{\sin r_2} = n_{AG} = \frac{1}{n_{GA}}$$

and

$$\sin r_2 = \sin i_2 \times n_{GA} = 0.55586 \times 1.60 = 0.88938$$

$$r_2 = 62^\circ\ 48' \quad \text{or} \quad 62.79^\circ.$$

FIG. 19-9 Two refractions of a light ray in passing through a prism.

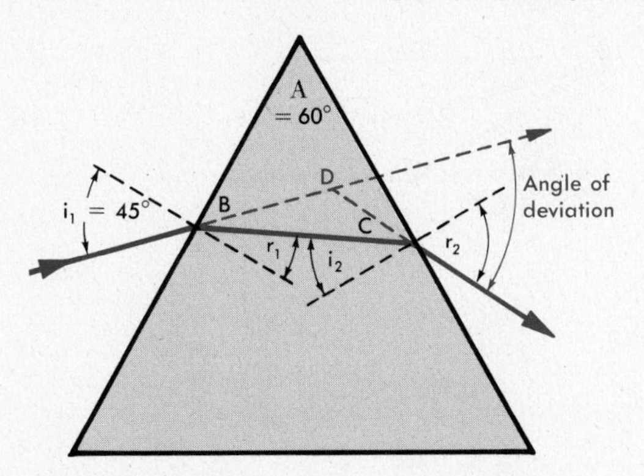

For most applications it is important to know the *deviation* of the refracted ray—i.e., by what angle its direction has been changed. Figure 19-9 shows the incident and emergent rays extended as dotted red lines, to intersect at D. The angle of deviation is shown on the drawing. It is another simple exercise in geometry to show that the angle of deviation $D = i_1 + r_2 - A$. In this case $D = 45.00 + 62.79 - 60.00 = 47.79°$ or 47°47′.

19-7 Total Internal Reflection

For every prism that is used to bend light, there are hundreds, or perhaps thousands, of prisms that are used for an entirely different purpose—to act as mirrors in reflecting light. To see how this is accomplished, look at Fig. 19-10A. There are many kinds of optical glass with indices of refraction (for yellow light) varying from 1.45 to 1.92, but let us assume an index of 1.65 for a ray passing from air to glass. The rays in the drawing pass *from glass into air*, however, so the index is 1/1.65 = 0.606, and we can compute the direction of the refracted rays in air by setting

$$\frac{\sin i}{\sin r} = 0.606.$$

As with all rays when they encounter a change in index of refraction, a part of ray 1 is reflected from the glass–air surface, but most of it is

FIG. 19-10 Total internal reflection.

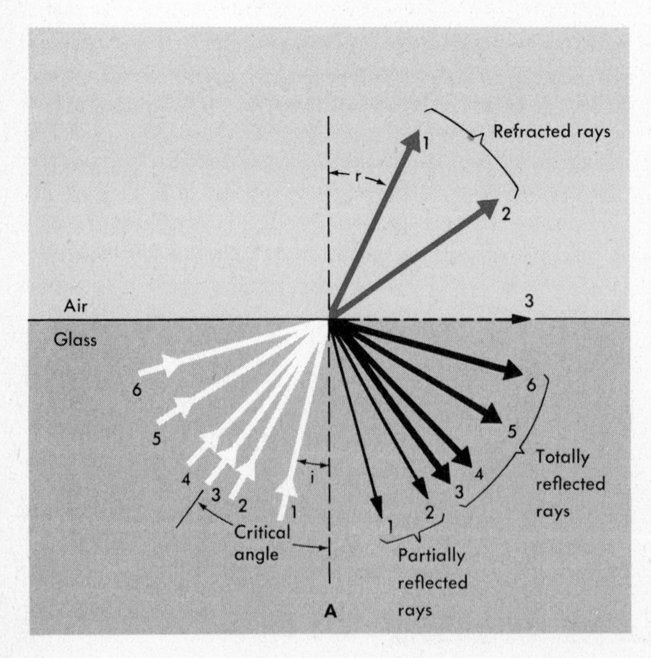

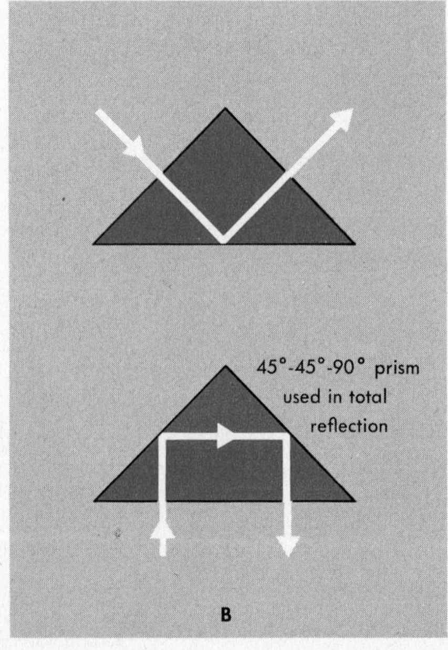

refracted into the air as shown. The behavior of ray 2 is very similar, but ray 3 is quite unusual. If its angle of incidence (for this particular glass) is 37.3°, we can calculate its angle of refraction as

$$\sin r = \frac{\sin 37.3^\circ}{0.606}$$

$$= \frac{0.606}{0.606} = 1$$

and

$$r = 90^\circ.$$

So this particular ray (called the *critical ray*) striking the glass–air surface at this particular angle (called the *critical angle*) cannot really be refracted out into the air at all. If we try to solve the problem for an angle of incidence greater than 37.3° (for this type of glass), we shall find that $\sin r > 1$, which is impossible. Nature also finds it impossible, and the ray will be totally reflected back through the glass, as shown in Fig. 19-10A.

Thus the phenomenon of *total internal reflection* can occur if a ray in a material of higher index of refraction tries to escape into a material whose index of refraction is lower. The ray will be totally reflected if its angle of incidence is greater than the critical angle: $\sin i_{\text{crit}} = 1/n$.

In the 45° prisms shown in Fig. 19-10B, the angle of incidence on the glass–air surface is 45°, which is greater than the critical angle, and the rays will be 100 percent reflected. Besides being a more efficient reflector than any silvered or aluminized surface, there can be no tarnish or corrosion on the glass surface, and it will maintain its 100 percent reflecting ability for an indefinitely long period.

19-8 Lenses

Analogous to concave and convex mirrors are *converging* and *diverging lenses*. These are shown in Fig. 19-11. The converging lens, thicker in the middle than at the edge, bends incoming parallel rays so that they converge to a real focus. The diverging lens, thinner in the middle than at the edge, causes incoming parallel rays to diverge as though they were coming from a virtual focus. As with the mirrors, the distance from a lens to its focal point for parallel rays is called the *focal length* of the lens; for a converging lens the focal length is positive, and for a diverging lens, negative.

With a little juggling about of similar triangles as shown in Fig. 19-11B and following a procedure similar to that which led to a relationship among p, q, and f for spherical mirrors, we can derive what turns out to be the identical relationship applicable to lenses:

$$\frac{1}{p} + \frac{1}{q} = \frac{1}{f}.$$

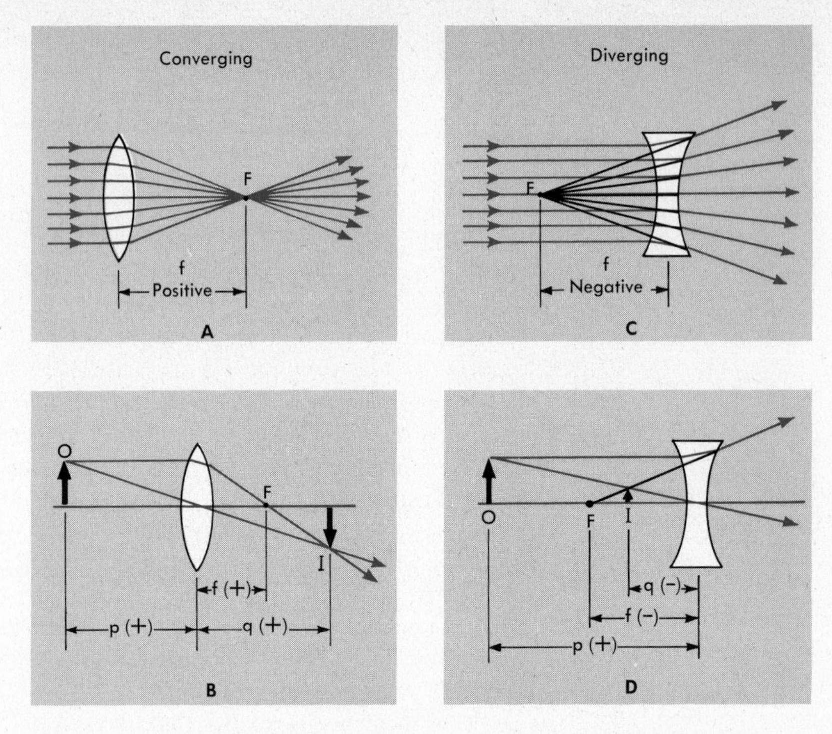

FIG. 19-11 Focal points and image formation by converging and diverging lenses.

In order to establish a convention for + and − signs, look carefully at Fig. 19-11B, which is an "all +" diagram. The light proceeds from left to right; the object distance p, to the left of the lens, is positive, and the image distance q, to the right of the lens, is also positive. Since the lens is converging, f also is positive. Figure 19-11D shows the formation of a virtual image by a diverging lens; f is negative, but since the object is to the left of the lens, p is positive. Since the virtual image is to the left of the lens (on the opposite side from the image in our "all +" Fig. 19-11B), q is negative.

The graphical construction of images is as simple for lenses as for mirrors. In Fig. 19-11B, we can pick two of the infinite number of rays we could draw emerging from O. One ray, parallel to the axis, will be refracted to pass through the principal focus. The other we can draw through the center of the lens. At this point the glass surfaces on both sides of the lens are parallel, and since in this elementary work we must confine ourselves to *thin* lenses, the ray passes through this center point without any deviation. To an observer on the right, the rays look as though they were coming from the real image I, where the rays intersect.

The ratio of image size to object size is, of course, what we call the *magnification*, and a look at the similar triangles in Fig. 19-11B will show that

$$M = \frac{q}{p}.$$

We should note that here, with both p and q positive, M is also positive, and we see that the image is *inverted*. In Fig. 19-11D, since p is positive q is negative, M is negative, and the image is *erect*. This relationship between the sign of M and the orientation of the image is a useful rule.

As an example, let us consider a simple camera with a lens of 50-mm focal length. When a picture is taken of an object at a great distance, the rays from any one point on the object come into the lens almost parallel, and are focused on the film at the principal focus 50 mm behind the lens. For an object 1 m (1000 mm) from the lens, however, the image distance will be somewhat greater. Substituting numbers into

$$\frac{1}{p} + \frac{1}{q} = \frac{1}{f}$$

we get

$$\frac{1}{1000} + \frac{1}{q} = \frac{1}{50}$$

and

$$\frac{1}{q} = \frac{1}{50} - \frac{1}{1000} = 0.020 - 0.001 = 0.019$$

so that

$$q = 52.6 \text{ mm}.$$

Thus, in order that the image of nearby objects may fall exactly on the film, we need to focus the camera by increasing the distance between film and lens.

Again, suppose we want to focus the image of a flag on a wall 10 ft (120 in.) away from it, and suppose we have a converging lens with a focal length of 24 in. Where must the lens be placed? Figure 19-12 illustrates the problem, and to solve it we must note that $q = 120 - p$ (or $p = 120 - q$; either way will lead to the same solutions). Then

$$\frac{1}{p} + \frac{1}{120 - p} = \frac{1}{24}$$

$$\frac{120}{p(120 - p)} = \frac{1}{24}$$

$$p^2 - 120p + 2880 = 0$$

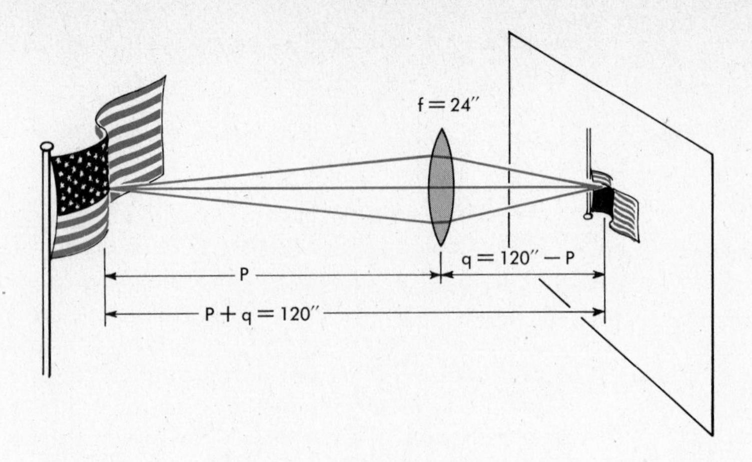

FIG. 19-12 Image formation by a converging lens. The image is reversed left and right and is also upside-down.

and

$$p = \frac{120 \pm \sqrt{14{,}400 - 11{,}520}}{2}$$

$$= \frac{120 \pm \sqrt{2880}}{2}$$

$$= \frac{120 \pm 53.7}{2}$$

$$= 86.8 \text{ in} \quad \text{or} \quad 33.2 \text{ in.}$$

From this, we see that there are two possible locations for the lens. One will give a magnified image, and the other will produce an image smaller than the flag.

19-9 Lens Combinations

Figure 19-13A indicates a problem that at first glance seems quite complicated—where will this system of three lenses form an image of O, and what will be the total magnification? Like many other problems in physics, a seemingly complicated question turns out to be no more than a series of simple questions. Let us take the lenses one at a time. For lens 1, $1/p_1 + 1/q_1 = 1/f_1$ becomes (Fig. 19-13B)

$$\frac{1}{24} + \frac{1}{q_1} = \frac{1}{6}$$

and

$$q_1 = +8.$$

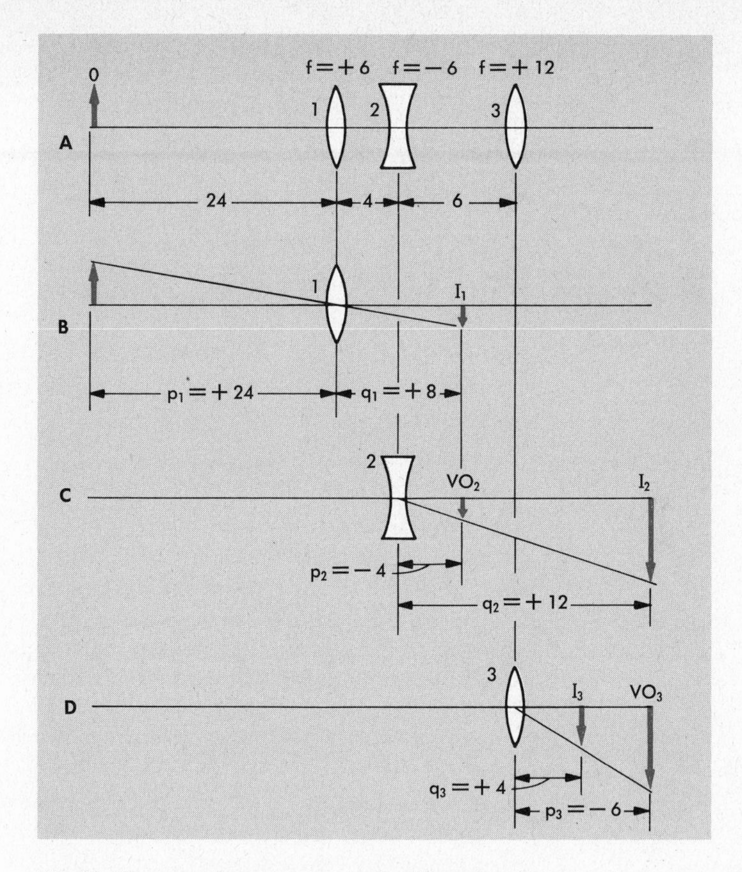

FIG. 19-13 Image formation by a series of lenses.

This image formed by the first lens serves as the object for the second lens, no matter if the image is real or virtual, or whether or not the second lens blocks the rays from ever forming the image at all.

Some adding and subtracting of distances shows that I_1 falls at a distance 4 units to the right of lens 2 (Fig. 19-13C) and becomes a *virtual object* VO_2 for the second lens. Since VO_2 is to the right of lens 2, p is negative, and we have

$$\frac{1}{-4} + \frac{1}{q_2} = \frac{1}{-6}$$

$$q_2 = +12.$$

We then make I_2, the image formed by the second lens, the object for the third lens (Fig. 19-13D) and get

$$\frac{1}{-6} + \frac{1}{q_3} = \frac{1}{+12}$$

$$q_3 = +4.$$

The final image, then, is formed 4 units to the right of the third lens. To find the total magnification, we need only multiply together the magnifications for each of the separate lenses:

$$M = \frac{q_1}{p_1} \times \frac{q_2}{p_2} \times \frac{q_3}{p_3}$$

$$= \frac{8}{24} \times \frac{12}{-4} \times \frac{4}{-6} = \frac{2}{3}.$$

Hence the final image will be $\frac{2}{3}$ the length of the original object, and as is apparent from the drawings and from the sign of the total magnification, will be real and inverted.

Questions

(19-1) 1. How far away is the Andromeda galaxy?

2. In Galileo's experiment on the velocity of light, suppose that (due to the inevitable human reaction time) there was a half-second delay between the assistant's perception of the flash, and his opening of the shutter. If the two experimenters were, say, 2 mi apart, why did Galileo not conclude the speed of light to be 8 mi/sec? What further experiments would have eliminated (and undoubtedly did eliminate) this erroneous conclusion?

3. If one used a distance of 10^4 m and a 300-tooth wheel in an experiment similar to that shown in Fig. 19-2, how fast would the toothed wheel have to rotate, in rev/min?

4. In an experiment like that shown in Fig. 19-2 the distance was 8000 m, the wheels had 360 teeth, and rotated at a speed of 1530 rev/min. What value would this give for the speed of light?

(19-3) 5. A man wants a mirror on a wall, in which he can see his full length. How long must this mirror be? (Consider the man to be a straight line 6 ft tall, with his eye 6 in. down from the top).

6. In Question 5, just how far up from the floor should the bottom of the mirror be placed? Does it make any difference how far the man stands from the wall?

7. Consider two plane mirrors, one on each wall, butted together in a corner of a rectangular room. Place an object near the corner between the mirrors, and sketch the images that would be formed, by both single and double reflections. (*Hint*: an image of an image.)

8. A man has a blue left eye and a brown right eye. When he looks in his shaving mirror, he sees an image that has a brown left eye and a blue right eye. How would his image appear to him if he looked in the mirrors of Question 7, from a distance great enough that he could not see the single reflections?

(19-4) 9. Take a concave mirror of 6-cm focal length. Compute the image distance when an object is placed at the following distances in front of the mirror:

(a) Very large, (b) 48 cm, (c) 16 cm, (d) 12 cm, (e) 6 cm, (f) 4 cm, (g) 2 cm. Which of these images are real, and which are virtual? Which are erect and which are inverted?

10. The same as Question 9, except that the mirror is convex.

11. A concave mirror has a radius of curvature of 24 in. (a) How far from his face should a man hold this mirror if he wishes the image to be formed 12 in. behind the mirror? (b) Will this image be real or virtual? (c) Will it be erect or inverted? (d) If his nose is 3.6 cm long, how long is its image?

12. A concave mirror has a radius of curvature of 36 in., and a man holds it 12 in. away from his face to examine himself. (a) Where is the image of his nose formed? (b) Is this image real or virtual? (c) If his nose is $1\frac{1}{2}$ in. long, how long is its image? (d) Is the image erect or inverted? Construct the image graphically, as well as making analytical solutions.

13. A man holds a *convex* mirror whose radius of curvature is 48 cm. How far from his face should he hold it to make his image appear to be 7.2 cm behind the mirror?

14. A polished reflecting sphere 1 ft in diameter is lying on a lawn. A worm crawls toward it. How far from the surface of the ball is the worm when his image appears to be 2 in. behind the ball's surface?

(19-5) 15. A light ray makes an angle of 60° with the normal to the surface of a cubical flint glass paperweight. What is its angle of refraction on passing into the glass?

16. A ray of light (in air) strikes the surface of a pool of water with an angle of incidence equal to 45°. What is the angle of refraction of the ray into the water?

17. A rectangular tank is half-filled with carbon tetrachloride (n = 1.47), and the top half is filled with water. A ray passing downward through the water strikes the boundary between the liquids at an angle of incidence of 48°. At what angle will it travel in the carbon tetrachloride?

18. In the tank of Question 17, a ray directed upward in the carbon tetrachloride strikes the boundary at an angle of incidence of 28°. What will be its angle of refraction into the water?

19. A ray of light strikes a thick parallel-sided glass plate with an angle of incidence of 50°. (a) What is its angle of refraction in the glass? (b) What angle does it make with the normal to the surface when it emerges into the air again? (The glass has an index of refraction of 1.60).

20. A thick parallel-sided slab of glass (n = 1.60) is submerged in water. A light ray in the water strikes the slab with an incident angle of 40°. (a) What is its angle of refraction in the glass? (b) What is its angle of refraction from the glass out into the water again?

(19-6) 21. The angle between the two refracting faces of a prism is 45°; a ray strikes the first face with an angle of 30° between the ray and the surface, qualitatively similar to Fig. 19-Q23. What is the angle between the ray and the second surface when the ray emerges into the air? (n = 1.600).

22. A ray strikes a 60°-60°-60° prism with an angle of incidence of 45°, qualitatively similar to Fig. 19-Q23. The index of refraction of the prism is 1.500. What is the angle of refraction of the ray into the air after passing through the prism?

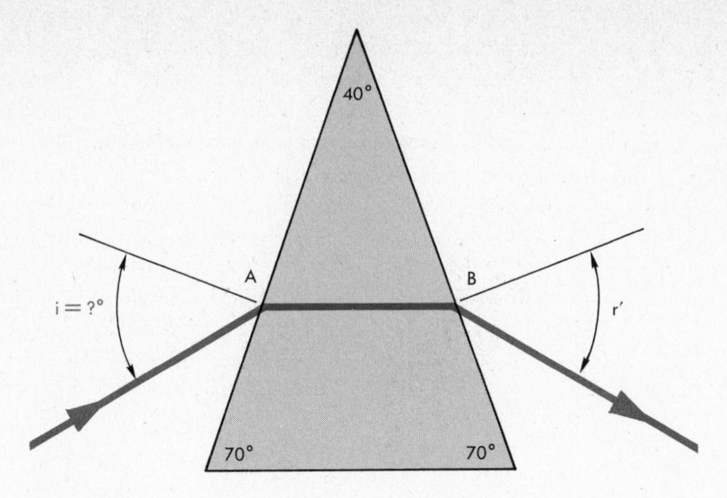

FIG. 19-Q23

23. A ray of yellow light is to pass symmetrically through the prism shown in Fig. 19-Q23; that is, the ray *AB* in the glass is to be parallel to the base of the prism, from which angle r' must be equal angle i. What must be the angle of incidence i if the index of the glass for the yellow light is 1.650?

24. Same as Question 23, with a prism whose angles are 75°-75°-30°.

25. What is the deviation of the ray in Question 23?

26. What is the deviation of the ray in Question 24?

(19-7) **27.** What is the critical angle for total internal reflection for a ray in glass of $n = 1.56$, if (a) air is on the other side of the glass? (b) alcohol is on the other side of the glass?

28. A ray in carbon tetrachloride ($n = 1.47$) is directed upward to the surface of the liquid. What is the critical angle for the ray, if (a) there is air above the carbon tetrachloride? (b) if there is water above it?

29. A 60°-60°-60° prism is made of glass with an index of refraction of 1.414. What must be the angle of incidence of a ray that will be totally reflected at *C*, as sketched in Fig. 19-Q29?

30. A ray is to be totally reflected from the base of a 60°-60°-60° prism, as shown in Fig. 19-Q30. What is the maximum allowable angle of incidence? ($n = 1.500$.)

(19-8) **31.** Same as Question 9, but for a converging lens of 6-cm focal length.

32. Same as Question 9, but for a diverging lens of $f = -6$ cm.

33. A telephoto lens for a small camera has a focal length of 400 mm. How many millimeters must it be moved out from its "infinity-focus" position to focus on an object 5 m distant?

34. The lens of a small camera has a focal length of 35 mm. How many millimeters must it be moved out from its "infinity-focus" position in order to focus on an object 2 m from the lens?

35. In a movie or slide projector, the brightly illuminated film serves as object, and a lens forms its image on the screen. If the image on the screen is

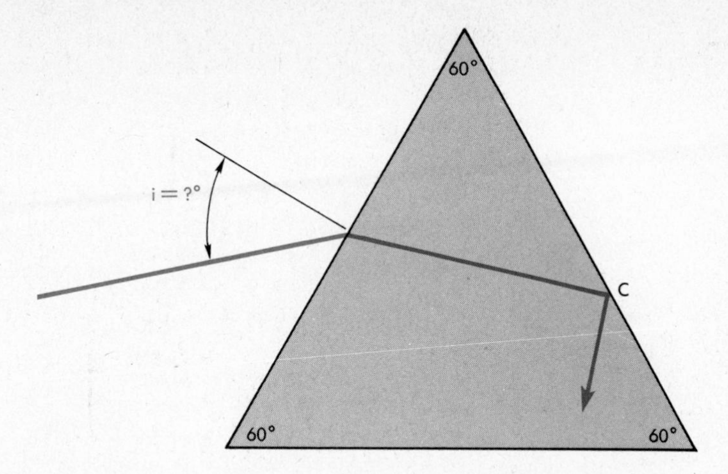

FIG. 19-Q29

to be 16 times as large as the film in linear dimension (it would have an area 256 times as great) when the screen is 12 ft from the projector lens, what must the focal length of the projector lens be?

36. The lens of a slide projector has a focal length of 4 in. If the linear dimensions of the image on the screen are 24 times as large as the film, (a) how far from the film must the lens be and (b) how far from the lens must the screen be placed?

(19-9) **37.** Light from a distant object passes through a converging lens ($f = 8$ in.); 2 in. beyond this lens is a diverging lens ($f = -9$ in.). Where will the image be formed?

38. A diverging lens ($f = -10$ cm) is placed 6 cm beyond a converging lens whose focal length is 12 cm. Where will the image of a distant object be formed?

FIG. 19-Q30

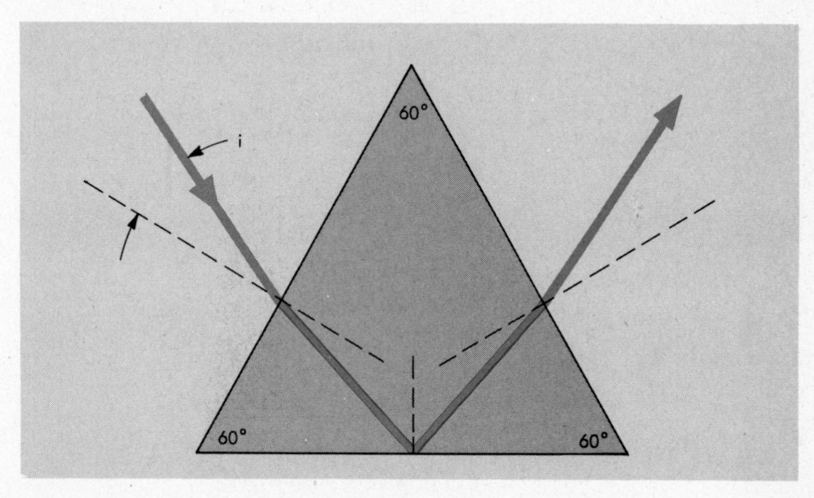

39. Parallel light from a distant object falls on a converging lens ($f = 6$ in.). How far beyond this first lens should another converging lens ($f = 9$ in.) be placed so that the light beyond the second lens is again parallel?

40. Parallel light from a distant object falls on a diverging lens ($f = -10$ cm). How far beyond this lens should a converging lens ($f = 25$ cm) be placed so that the light beyond the converging lens is again parallel?

41. A converging lens, $f = 40$ cm, is immovably fixed 40 cm in front of a screen. It is desired to focus on the screen the image of an object 120 cm in front of this lens. You have another lens of $f = 40$ cm available. Where can you place it in order to achieve the desired results? (There are two answers to this problem.)

42. A converging lens ($f = 8$ in.) is permanently fixed 8 in. from a bright object; 12 in. beyond this lens is a screen upon which a real image of the object is to be formed. You have another converging lens ($f = 6$ in.); where should this second lens be placed in order to form the desired image? (There are two answers to this problem.)

20

Light Waves

20-1 Why is Light Refracted?

Why do light rays change their direction when they travel from air into water or glass? This question has an important bearing on the problem of the nature of light. Sir Isaac Newton believed that a light beam represents a swarm of tiny particles that are emitted from light sources and fly at high speed through space. He visualized the refraction of light rays entering any material medium as being caused by a certain attractive force acting on the particles of light when they cross the surface of a material body (Fig. 20-1A). The vector diagram illustrates Newton's idea of refraction. In it, v_A represents the velocity of the light in air, and Δv is the increase in velocity due to the attractive force at the boundary. The resultant velocity of the particles in glass is given by v_G, which differs from v_A in both magnitude and direction. It is important to realize that, according to Newton's views, the velocity of light in substances with a high refractive index should be larger than the velocity of light in air or in a vacuum, because the force pulling the light particles into the denser materials adds to their velocity.

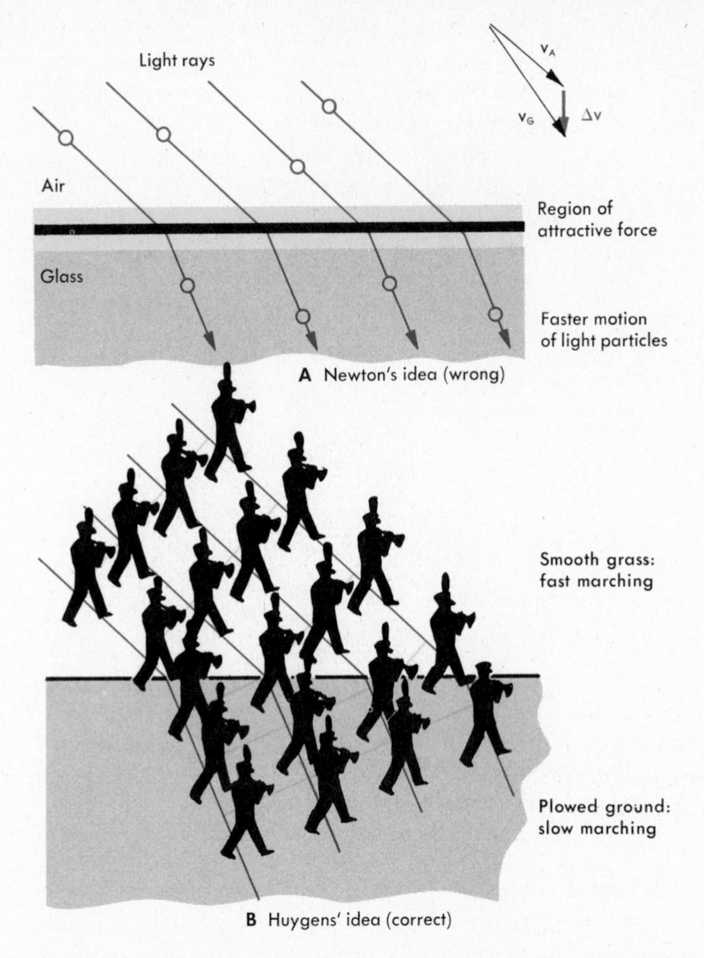

FIG. 20-1 Newton's and Huygens' explanations of the refraction of light.

Newton's views were opposed by the Dutch physicist Christian Huygens (1629–1695), who believed light to be a wave motion in a certain all-penetrating medium ("world ether") and the propagation of light to be similar to the propagation of sound waves through the air. It is interesting to notice that the explanation of the refraction of light based on Huygens' ideas leads to conclusions concerning light velocity in dense media that are directly opposed to those reached by Newton. To understand it in a simple way, let us substitute, for successive waves of light approaching the surface of the glass, successive ranks of a marching band. A part of their practice field, however, has been plowed, and as the ranks of the band cross the boundary from smooth grass to plowed ground, their steps become shorter and their speed is reduced. If they enter the plowed area at an angle, as in Fig. 20-1B, the direction of march will inevitably be changed, and the slower-moving ranks of the band will find themselves closer together.

We can analyze this quantitatively by using Huygens' principle that each point on a wavefront could itself be considered a source of further waves. Figure 20-2 shows a series of wavefronts traveling in the direction of the heavy dashed line. At point A, one of the wave fronts is just entering the glass, and another point B on the same wavefront is only one wavelength away. During the time that B travels to C, the disturbance at A will have traveled in the glass a shorter distance AK, and the wavefront in the glass will be represented by KC. Since BC and AK were covered in the same length of time, the velocities in the two media are in the same ratio as these distances, or

$$\frac{v_A}{v_G} = \frac{BC}{AK}.$$

From the drawing, it is plain that $AK = AC \sin r$ and that $BC = AC \sin i$ so we have

$$\frac{v_A}{v_G} = \frac{AC \sin i}{AC \sin r} = \frac{\sin i}{\sin r} = n_{GA}.$$

We thus see that, according to the wave theory, the index of refraction actually represents the ratio of the velocities of light in the two media concerned. We know, for example, that the speed of light in air or

FIG. 20-2 Wave fronts entering glass from air.

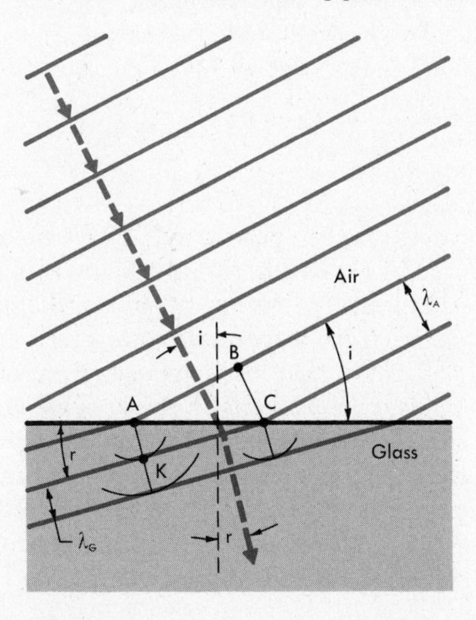

vacuum ($n = 1.00$) is 3×10^{10} cm/s; therefore, in diamond ($n = 2.42$), the speed of light is $3 \times 10^{10}/2.42 = 1.24 \times 10^{10}$ cm/s.

No amount of theoretical reasoning was able to settle the difference between the particle theory and the wave theory. The choice necessarily depended on experimental evidence, and in 1850 the French physicist Jean Foucault used a rotating-mirror modification of Fizeau's experiment to measure the speed of light in water. He found it to be substantially *less* than the speed of light in air. This single crucial experiment settled the century-old uncertainty: Huygens' wave ideas had been right, and Newton's particle theory was wrong.

20-2 Young's Two-Slit Interference

The wave theory for light was satisfactory in explaining both refraction and reflection. (Figure 8-9 shows the formation of an image by the reflection of waves from a plane surface.) The authority of Newton was so powerful, however, that many scientists held to the particle idea until well into the nineteenth century.

In 1802, the English classical scholar, scientist, and engineer, Thomas Young (whose name was given to Young's modulus) performed his two-slit experiment. This experiment, using light rather than surface water waves, was similar to that shown in Fig. 8-13B, in which two openings in a breakwater serve as separate sources to produce an interference pattern.

Light is composed of trains of transverse waves called *photons*, emitted one photon at a time by separate individual atoms in the source. The photons are of course not synchronized with each other, and if we tried to observe interference effects from two different ordinary sources, we could see none at all. The only way to observe light interference in this way is to let the waves of each photon interfere with themselves. In this way, perfect synchronization is guaranteed.

Figure 20-3 is the same as Fig. 8-14, and shows a source of monochromatic light (i.e., light that is all of the same wavelength) illuminating a slit, from which light spreads out to strike two other slits, S_1 and S_2. Each photon that strikes the first slit may be considered to strike S_1 and S_2 simultaneously, and from S_1 and S_2 waves fan out that are accurately in phase with each other. For the sake of clarity, waves have been graphically indicated as traveling along straight rays to points A, B, and C on a screen or a piece of photographic film. Point A is equidistant from S_1 and S_2, so that the waves starting together from the two slits arrive exactly in phase at A, and in this manner reinforce each other to produce a bright illumination. Point B on the drawing was selected so that it is half of a wavelength, $\lambda/2$, farther from S_2 than from S_1. Waves starting out in phase will thus arrive at B exactly out of phase and will annul each other, so that B remains dark. Point C is just λ farther from S_2 than it is from S_1, so when the waves arrive at C, they are shifted one

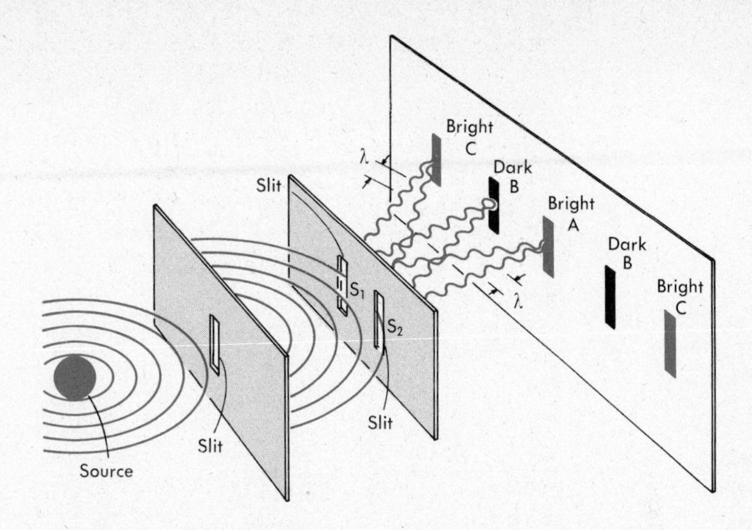

FIG. 20-3 Light and dark bands on a screen, formed by the interference of light coming from two closely spaced slits.

wavelength with respect to each other. This puts them in phase, and *C* is bright.

Such reasoning can be continued and we can make a general statement that any point that is an odd number of half-wavelengths closer to one slit than the other will be dark; any point that is the same distance from both slits, or from which the distances to the slits differ by an integral number of whole wavelengths, will be of maximum brightness. Young's experiments demonstrated these interference patterns, and although there were some holdouts among scientists for several decades, they finally brought the particle theory of light to an end. (This is not quite true; in recent years, light has been found to have some aspects resembling those of particles, but in an entirely different sense than that which Newton had in mind. We will look into this a few chapters further along.)

Interference methods provide a direct way of measuring the wavelengths of light of different colors. If the slits are illuminated with blue light, the pattern of light and dark bands will be much more closely spaced than if they are illuminated with red light. This demonstrates that blue light has a shorter wavelength than red light. The shortest wavelength visible to human eyes is about 4×10^{-5} cm and is beyond the blue color into the violet. The longest wavelength that human beings can see, a deep red, is about 7.5×10^{-5} cm. Because they are so short, wavelengths of visible and near-visible light are often expressed in angstroms (Å) or in nanometers (nm). One angstrom equals 10^{-8} cm, and a nanometer is 10^{-9} m. In these units, we can say that the visible wavelength range is from 4000 to 7500 Å, or from 400 to 750 nm.

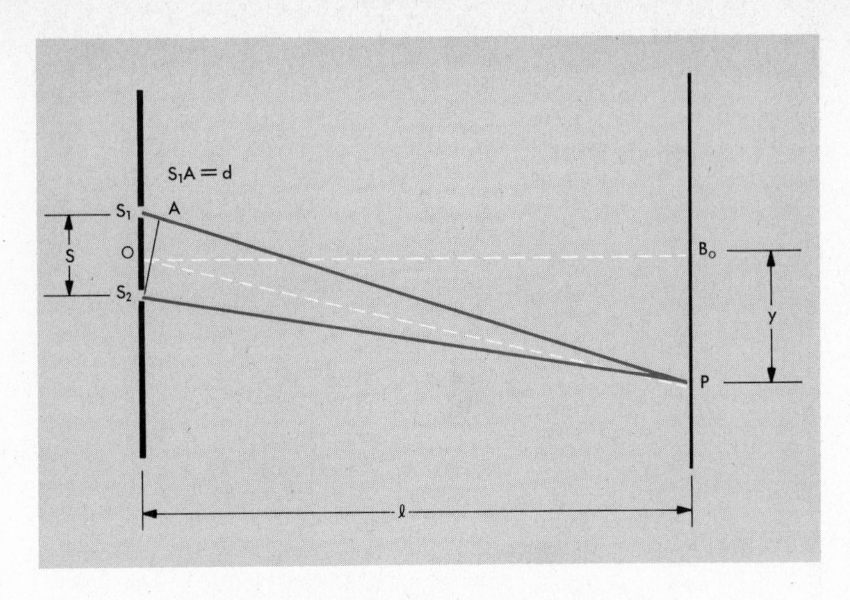

FIG. 20-4 Interference patterns and the wavelengths of light.

As an example of how a pair of slits might be used to determine the wavelength of some particular color of light, let us turn to Fig. 20-4 for the geometry of the arrangement. The two slits S_1 and S_2 are separated by a distance s. They are illuminated from a single slit or other small source of monochromatic light somewhere off to the left of the drawing, which is not shown. If we assume that the incoming waves strike S_1 and S_2 simultaneously, we may mark point B_0 on the screen, directly opposite point O, midway between the slits. Therefore B_0 is equidistant from S_1 and S_2 and no matter what wavelength strikes the slits, it will generate wave trains from S_1 and S_2 that will arrive at B_0 in phase and therefore reinforce each other to produce a bright line there.

Now let us move down (or up) the screen to point P, a distance y from B_0. As we do do, we mark point A on the line S_1P, so that $AP = S_2P$. The path from S_1 to P is longer than the path from S_2 to P by the distance S_1A, which we can refer to by the single letter d. If d is equal to any *integral* number of wavelengths, waves from S_1 and S_2 will arrive *in phase*, and P will mark a location of maximum brightness. If d is equal to any odd number of half-wavelengths ($\frac{1}{2}$, $\frac{3}{2}$. $\frac{5}{2}$, etc.), the waves will arrive exactly *out of phase*, and P will mark a dark line on the screen.

As Fig. 20-4 has been drawn, OB_0 is perpendicular to S_1S_2; OP is perpendicular to AS_2. Elementary geometry (or even elementary common sense reasoning, on which geometry is based) then shows that angles S_1S_2A and POB_0 are therefore equal. The careful geometer will see that triangle S_2S_1A and triangle OB_0P are not quite similar, but he will also agree that the deviation from exact similarity is too small to bother about when s is long in comparison with d, or l is long in comparison with y, as is nearly always the case. Accepting this similarity of triangles, then, we have

$$\frac{d}{s} = \frac{y}{l}.$$

As an example, let us suppose that we have a pair of slits whose separation is 6.1×10^{-3} cm and that they are mounted 25 cm in front of a photographic film. The slits are illuminated by monochromatic light and after a proper exposure, the film shows a series of bright bands spaced 0.21 cm apart. Although the bands are evenly spaced and it makes no difference which pair we choose, let us follow Fig. 20-4, and choose the central bright band B_0 and one of its nearest bright-line neighbors B_1, which will be just $y = 0.21$ cm away. Since B_1 is bright, d is exactly one wavelength, and

$$\lambda = d = \frac{sy}{l} = \frac{6.1 \times 10^{-3} \times 0.21}{25} = 5.1 \times 10^{-5} \text{ cm}$$

or

$$\lambda = 5.1 \times 10^{-5} \times 10^{8} = 5100 \text{ Å}.$$

20-3 Single-Slit Interference

In the preceding section on two-slit interference, the slits were merely assumed to be quite narrow. The necessary width of a slit does, however, produce interference effects of its own, which can be seen readily by allowing light to fall on a single slit.

Figure 20-5 shows a wave train of wavelength λ falling on a single slit whose width is s. Beyond the slit at a distance l is a screen that will be illuminated by the light passing through the slit. In order to find how this illumination will be distributed, let us imagine the slit to be subdivided into a large number of narrower "slitlets"; on the drawing it has not been practical to show more than eight. By Huygens' principle, we can use each slitlet as an independent source of light waves. At the central bright band on the screen, waves from 1 and 8, 2 and 7, 3 and 6, 4 and 5 each have the same distance to travel so they arrive in phase and reinforce. The distance from 1 and 8 is only a fraction of a wave longer than the distance 4 and 5, so all the pairs are also nearly in phase with each other, and the central region on the screen is quite bright.

Now let us consider the point on the screen that is one wavelength farther from 1 than it is from 8. At this point, the waves from 1 will be a half-wavelength behind the waves from 5, so the waves from this pair of slitlets will annul each other. The same is true for the pairs 2–6, 3–7, and 4–8, with the result that this point on the screen will be dark. The same sort of reasoning can be applied to other locations where the path difference is 2λ, 3λ, etc.

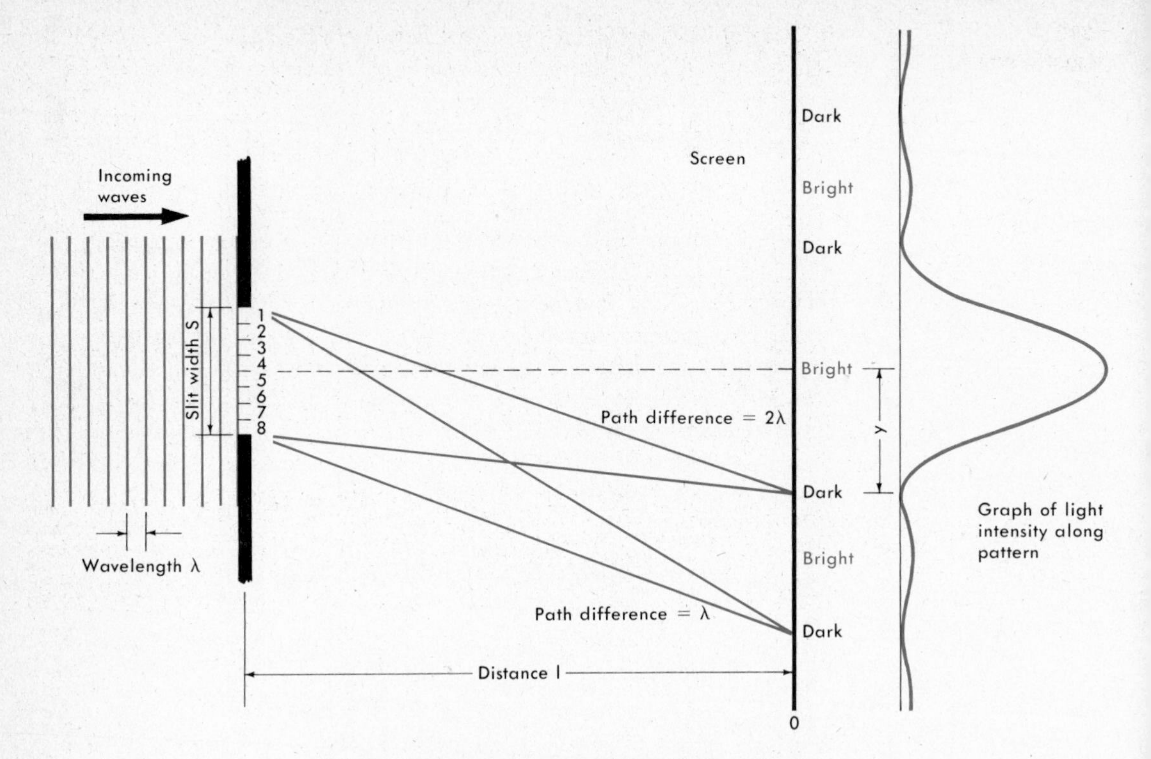

FIG. 20-5 Interference pattern produced by a single slit.

As shown by the graph of light intensity on Fig. 20-5, nearly all the light is confined to the central band. The width of the band is $2y$, and from the same geometry as in Fig. 20-4, we have

$$y = \frac{\lambda l}{s} \qquad 2y = \frac{2\lambda l}{s}.$$

Unless the slits are *very* narrow (in which case little light comes through them), two-slit interference patterns are often a composite of the interference from the two slits, overlaid by the single-slit pattern of the slits themselves. Figure 20-6A shows the interference pattern from a single wide slit; Fig. 20-6B is the wider pattern of a narrower slit. Figures 20-6C and D show the bands of light from a pair of wide and of narrow slits, each pair being spaced equally. The brightness of the bands—or *fringes*—can be seen to vary with the corresponding single-slit effect that overlays them.

20-4 Optical Gratings

The idea of the interference of water waves can be extended to a very long breakwater with many openings in it. This is analogous to the effect of an optical *grating* on light waves. We can make a *transmission grating*

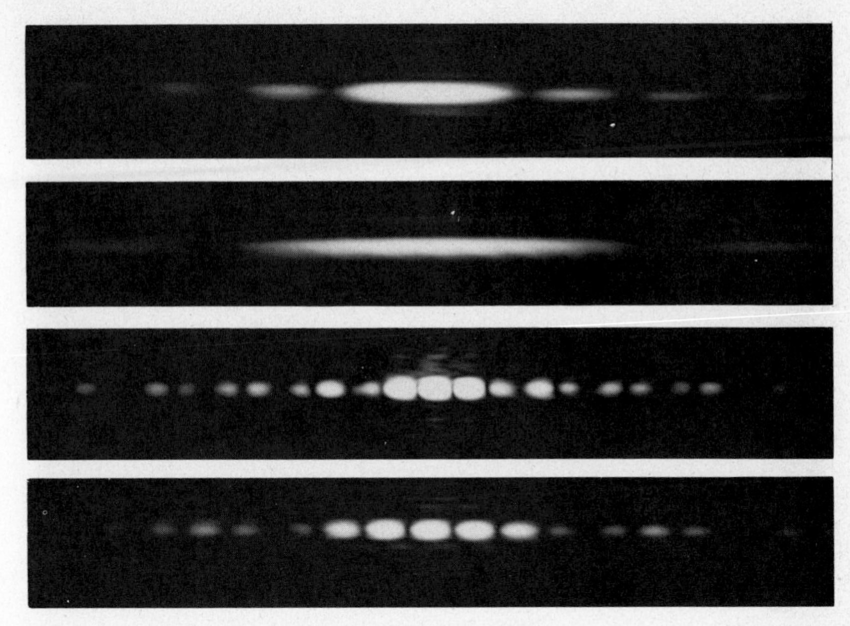

FIG. 20-6 Complex pattern produced by the superposition of single-slit and double-slit interference patterns.

from a piece of glass on which we have scratched a large number of fine parallel grooves. When placed in a beam of light, the glass will transmit, i.e., let pass through, only the light that falls on the smooth unscratched strips between the grooves. A *reflection grating* works in a similar manner by reflecting light from the smooth sides of accurately shaped grooves in the surface of a metal mirror. With modern techniques, we can make gratings of several thousand scratches per millimeter. Such original gratings are very expensive and are ordinarily used for only the most important and delicate research. It is possible to take impressions in plastic from an original grating. These may be nearly as accurate as the original, much cheaper, and satisfactory for many purposes.

Figure 20-7A shows the wavefronts of a beam of light of wavelength λ falling on a transmission grating. The *grating spacing*, which is the distance between lines or scratches, is s. By Huygens' principle, each line of the grating will serve as a source of new waves spreading out to the right of the grating. For the particular wavelength λ, there will be a particular direction, deviated by the angle θ_1, in which the waves from each slit will be exactly one wavelength ahead or behind the waves originating in the adjacent slits. Accordingly, *in this direction*, all the waves from all slits will be in phase with each other.

The direction θ_1, in which the light from each line is just one wavelength λ ahead or behind the waves from adjacent slits, is the direction of the *first-order* spectral line of wavelength λ. In Fig. 20-7A, we can

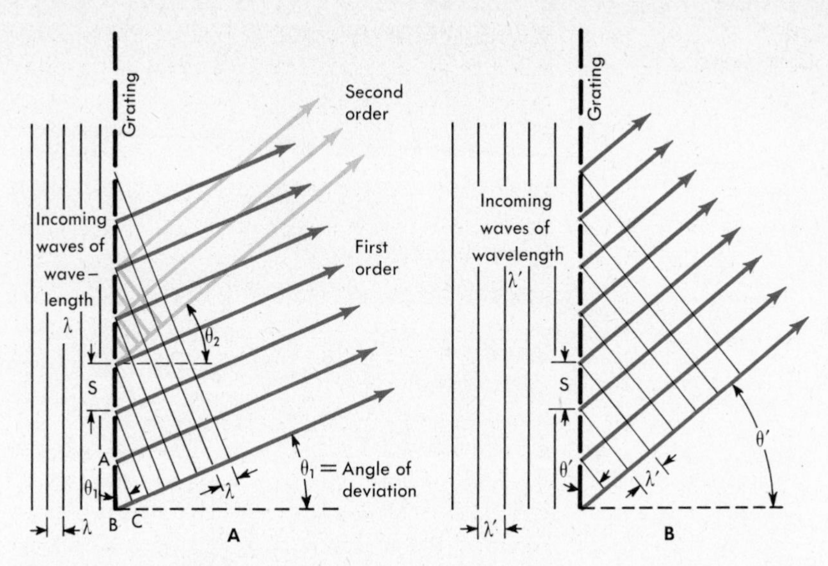

FIG. 20-7 The direction in which a grating reinforces light that falls on it depends on the wavelength of the light. The longer the wavelength, the greater the angle of deviation—the reverse of the behavior of a prism.

see another direction θ_2, in which the waves from each grating line are 2λ ahead or behind the adjacent waves. This is the direction of the *second-order* spectral line of wavelength λ.

The small triangle marked ABC in Fig. 20-7A enables us to derive a simple relationship that determines the angle of deviation θ for any wavelength. Angle BAC equals θ_1, and since the direction of propagation BC is perpendicular to the wavefront AC, angle ACB is a right angle. We can thus write

$$\sin \theta_1 = \frac{BC}{AB}.$$

But $BC = \lambda$, the wavelength of the light, and AB is s, the grating spacing. Accordingly,

$$\sin \theta_1 = \frac{\lambda}{s}.$$

For the second-order direction, it is apparent that

$$\sin \theta_2 = \frac{2\lambda}{s}.$$

Figure 20-7B shows light of longer wavelength striking the same grating. Analysis will show that since λ' is greater than λ, θ' must be greater than θ. In other words, in a grating spectroscope, the longer wavelengths are deviated more than the shorter wavelengths. (The opposite happens when light passes through a prism.)

20-5 Thin Films

If you refer back to Fig. 8-2 and 8-3, you will remind yourself that a transverse wave in a rope is reflected on the *same* side when it meets a junction with a rope of smaller linear mass; at a junction with a rope of larger linear mass (a rigid support is essentially $m_1 = \infty$) the reflection is on the *opposite* side. The same thing is true of light waves when reflected, or partially reflected, from the surface of a medium of lower or higher index of refraction.

Figure 20-8 shows an incoming wave in air striking a thin sheet of some material whose index of refraction is greater than the $n = 1.00$ of air. The wave is partially reflected from the top surface, with the wave reversed. The incoming wave is also partially reflected from the bottom surface; here, because it is about to enter the air, of smaller n, the reflected wave is *not* reversed. As indicated in Fig. 20-8B and C, the reversal of a sine wave is exactly equivalent to a $\lambda/2$ shift, or path difference.

Now let us check the phase relationship of the two reflected waves. The thin sheet or film has a thickness of $\frac{3}{2}$ of the *wavelength in the material*, which is of course $1/n \times \lambda_{air}$. The bottom reflection has had an extra path length of $3\lambda/2 + 3\lambda/2 = 3$ wavelengths, referred to the top reflection. The upper reflection has had a reversal, equivalent to a shift of

FIG. 20-8 Interference between waves partially reflected from top and from bottom of a thin film in air.

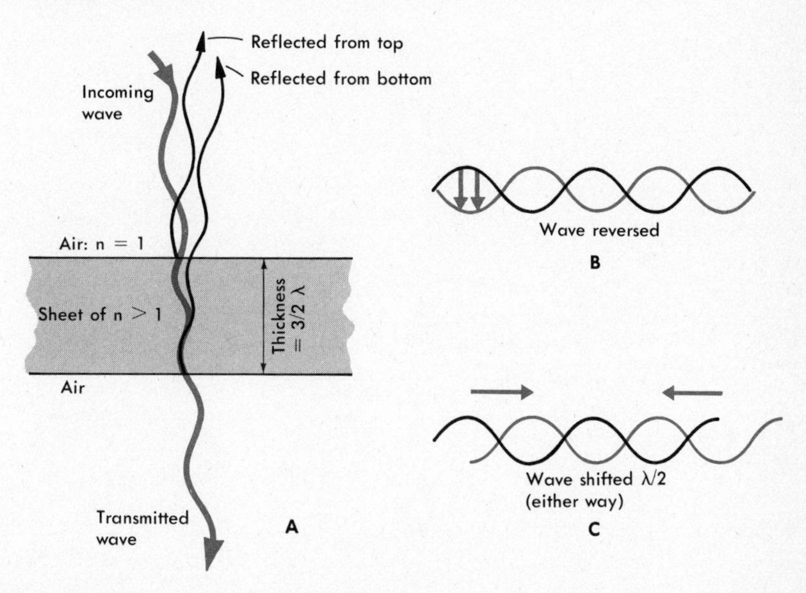

$\lambda/2$, which makes the total path difference $3 \pm \frac{1}{2}$, to put the two reflections exactly out of phase, so that they annul one another. In this case, there will be no perceptible reflection from the thin film. If the film were, say, $5\lambda/4$ thick, the phase difference between the two reflections would be $5\lambda/4 + 5\lambda/4 \pm \lambda/2 = 2\lambda$ or 3λ, which would put them exactly in phase, and we would have a maximum reflection as a result of their reinforcement. We can extend this by saying that if the film (with a material of lower n beyond each surface) has a thickness of any integral number of half-wavelengths, the reflection of that wavelength will be a minimum or zero; if the thickness is any odd number of quarter-wavelengths, the reflection of that wavelength will be a maximum.

Let us see the result of this sort of interference on the light reflected from a similar film of varying thickness—say, a sheet of soapy water across the mouth of a drinking glass lying on its side. The weight of the water will cause the film to drain downward, so that the top becomes thinner and the bottom, thicker. This is sketched in Fig. 20-9. Assume the wedge to be illuminated by white light coming in from right to left. White light is composed of waves of all visible wavelengths; the combined effect of all these colored stimuli on our retinas produces the sensa-

FIG. 20-9 Production of colors from white light reflected from a wedge-shaped film.

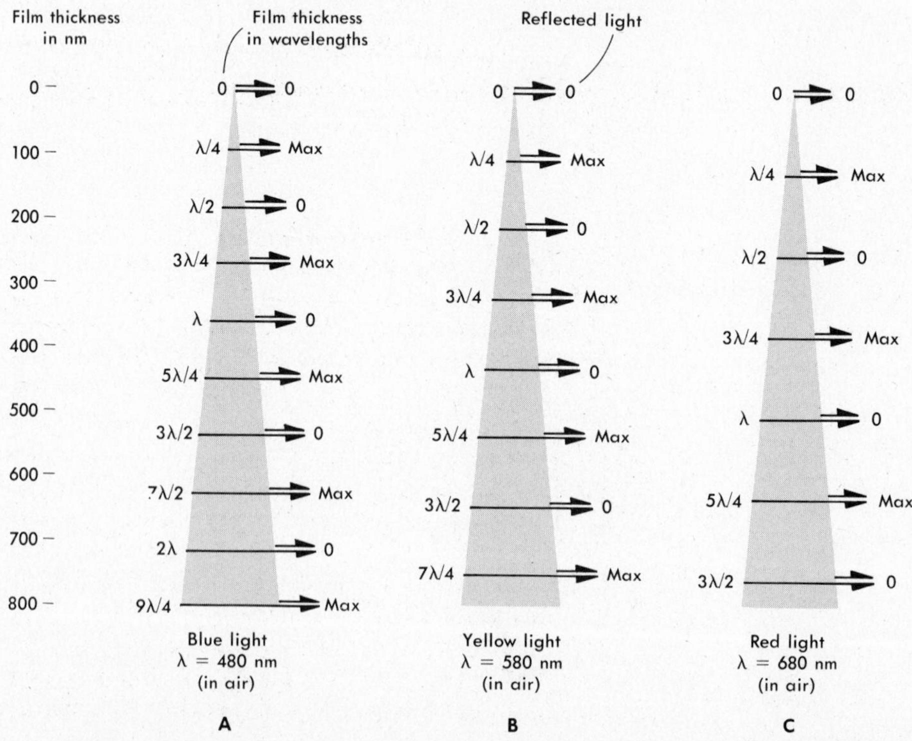

tion of "white", as we will discuss further in a later chapter. From the white light, let us choose the component of $\lambda = 480$ nm, which alone would produce the sensation of "blue." Figure 20-9A gives the thickness of the wedge of film in terms of the corresponding wavelength *in the film*, $480/1.35 = 356$ nm, where 1.35 is the index of refraction of the film. Since air is on both sides, only the first reflection will be reversed. This single reversal ($\pm\lambda/2$) plus twice the film thickness (t) puts the two reflections out of phase at $t = 0, \lambda/2$, etc., where no blue color is reflected. At $t = \lambda/4, 3\lambda/4$, etc., the two reflections are in phase to give a maximum blue reflection.

The drawing also shows the alternating bands of maximum and minimum reflection for $\lambda = 580$ nm (yellow) and $\lambda = 680$ nm (red). Because of their longer wavelength the spacing of these bands is wider. If we look at the reflection of white light from this wedge, we would see a complex mixture not only of the three wavelengths illustrated, but of *all* wavelengths in the visible range (approximately 400 to 750 nm), each with its own different spacing of maxima and minima. Where the tapered film is about 520 to 530 nm thick, we have nearly zero reflection for the blue and red, and nearly maximum for the yellow light, to give the reflection a yellow color at this point.

This effect is often seen when a thin oily film floats on water; the changing colors indicate the changing thickness of the oil film. Soap bubbles also show definite coloration that varies with the thickness of the film. Opals and mother-of-pearl show iridescent colors caused by reflection from the upper and lower surfaces of thin mineral layers. In the animal world, the colors of many feathers (a peacock's tail, for example) and of butterfly wings result from similar reflections from thin plates of nearly transparent material.

20-6 Polarization

All the interference effects we have considered so far could be produced equally well by either transverse or longitudinal waves. Experiments by Thomas Young and others early in the nineteenth century showed that light waves must necessarily be transverse. These experiments dealt with *polarization*, which is not possible for longitudinal waves. Figure 20-10A shows a pair of crystals of the mineral tourmaline (or it could be another of the many minerals that show this same property). About half of the light passes through the first crystal and if the second is properly oriented, if absorbs no further light. This is understandable in terms of the analogy of the slat-back chairs and transverse rope waves shown to the right of the crystals. If either of the tourmaline crystals is rotated through 90°, however, all the light is stopped. It is apparent that the slots in the chair-back allow only vertical components to pass—a wave whose plane was at 45°, for example, would partially pass through the slot as a vertical wave of 0.707 ($\sin 45° = 0.707$) times its original amplitude. Horizontal motion cannot be transmitted. Such crystals as tourmaline have different

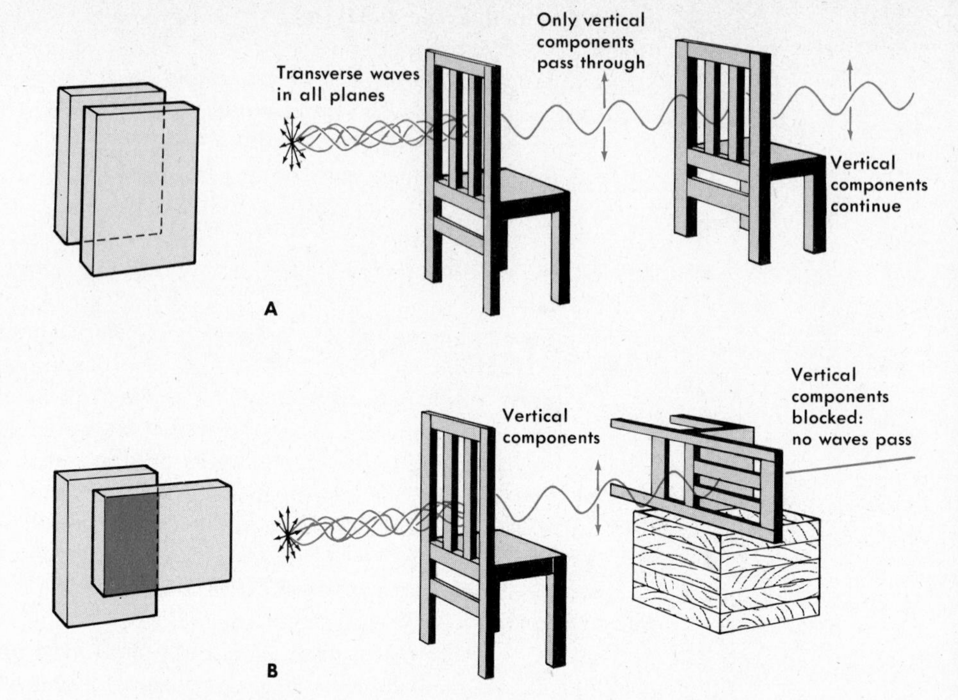

FIG. 20-10 "Polarization" of transverse waves along a rope, as an analog to polarization of light waves by certain crystals.

properties in different directions, so that the components of light waves in one direction are passed; components at right angles to this direction are absorbed. Iodo-quinine sulfate has optical properties similar to tourmaline. The familiar "Polaroid" is composed of billions of tiny needle-shaped crystals of iodo-quinine sulfate imbedded in a sheet of plastic with their long axes all parallel. Light passing through a sheet of Polaroid is almost completely polarized.

Although our eyes cannot detect it, a great deal of light we ordinarily see is also polarized, but by a different mechanism. Figure 20-11 shows a unpolarized ray striking the surface of a nonmetallic, nonconducting material; in this example, glass of $n = 1.60$. A part of the light—generally a few percent—will be reflected at the surface, and the remainder will be refracted into the glass. Unless the ray happens to be exactly perpendicular to the surface, the reflection will be at least partially polarized. When the angle of incidence is *Brewster's angle* (this is $\tan i = n_{GA}$), the reflection is completely polarized, with the waves parallel to the surface, as shown. In the case illustrated, Brewster's angle is the angle whose tangent is $1.60 = 58°$. The refracted ray transmitted into the glass is just partially polarized; it has lost only the reflected vibrations, so it has

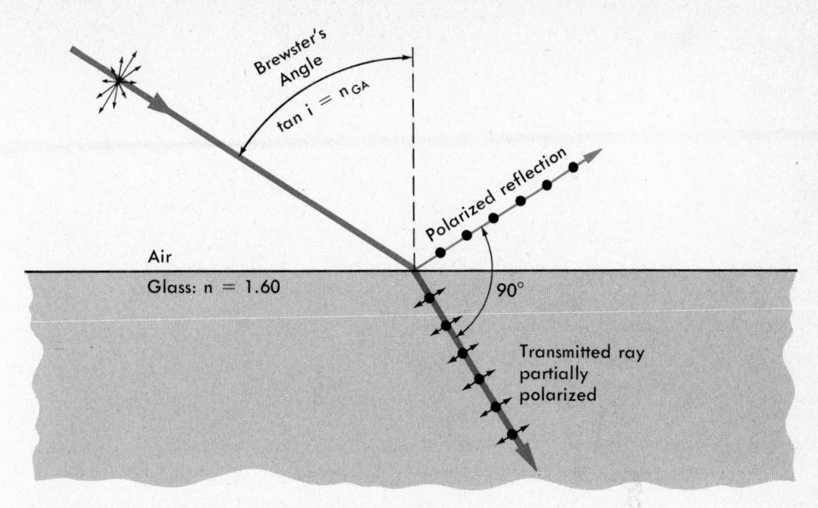

FIG. 20-11 Polarization of light reflected from a non-metallic surface.

only a slight excess of vibrations in the plane perpendicular to the surface.

Sunlight, or artificial light, reflected from water, paper, glossy print, etc., thus contains what we call "glare," which is at least partially polarized. Polaroid sunglasses, with their imbedded crystals properly oriented, will selectively absorb much of this glare.

20-7 The Electromagnetic Spectrum

Michael Faraday, in the course of his far-ranging experimentation, made an odd discovery: Glass, which does not ordinarily polarize rays transmitted through it, does become slightly polarizing when placed in a strong magnetic field. He accordingly speculated that light might be in some way associated with electric and magnetic fields. Faraday was not a mathematician, and was not able to carry his ideas any further.

Twenty years later, however, in 1865, James Clerk Maxwell succeeded in putting Faraday's ideas into mathematical form. His equations, dealing with the relationships between charges and currents and electric and magnetic fields, led to the conclusion that if the proper sort of electromagnetic disturbance were to exist, it would propagate itself through space at a speed of 3×10^8 m/s. This value for the speed was arrived at purely through the relationships between electric and magnetic measurements, and nothing about light had been used in its derivation. Furthermore, Maxwell's theoretical waves were composed of nothing but electric and magnetic fields at right angles to each other and to the direction of their propagation (Fig. 20-12). They were therefore transverse waves, as required to explain polarization.

The agreement between Maxwell's calculations and the properties of

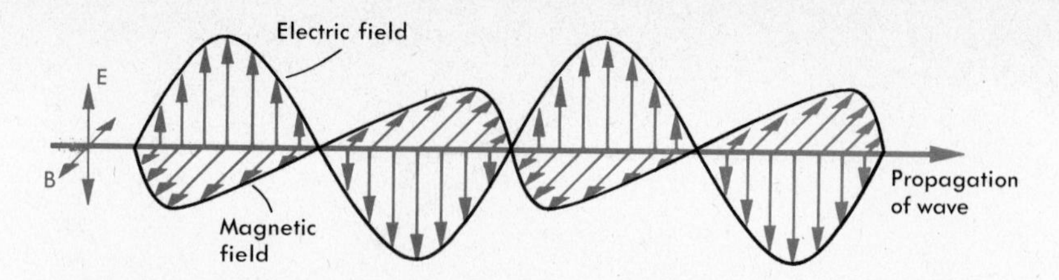

FIG. 20-12 A section of a light wave, or any other electromagnetic radiation, composed of oscillating electric and magnetic fields at right angles to each other, and to the direction of propagation of the wave.

light was too much to pass off as coincidence. Maxwell and his contempories were convinced that light must indeed be electromagnetic waves. But scientists are never satisfied with a theory until it is put in some physical form that can be observed and measured.

Maxwell's equations did more than predict the existence of electromagnetic waves—they also showed that these waves should be generated by accelerated electric charges. In 1887 a German physicist, Heinrich Hertz, built apparatus to actually demonstrate electromagnetic waves. His equipment produced a very-high-frequency alternating current, in which electrons oscillated back and forth millions of times a second, undergoing high accelerations. As predicted by the theory, these accelerating electrons radiated electromagnetic waves that Hertz was able to pick up at considerable distances with a primitive form of radio receiver. Although these radiations were not light, they served to show that Maxwell's electromagnetic waves really existed, and were not just a "paper theory."

Electromagnetic radiation was eagerly studied in the years following Hertz's experiments. As scientists made new discoveries in many fields, new frequency ranges of electromagnetic waves were detected—all traveling in vacuum at the same theoretically predicted speed of 3×10^8 m/s.

Table 20-1 gives some indication of the wide range of these frequencies.

20-8 Colored Light

In Section 5 of this chapter (Thin Films) it was noted that light of $\lambda = 480$ nm gave the visual reaction of *blue*; 580 nm, yellow; and 680 nm, red. It is easy to send a narrow beam of white light against a prism, as in Fig. 20-13. After passing through the prism, the light falls on a screen, on which we see a continuous spectrum of colors, which we can roughly describe as violet, blue, green, yellow, orange, and red. Just where to put the dividing line between blue and green, say, is a matter of subjective

TABLE 20-1 ELECTROMAGNETIC RADIATION

Name	Frequency Range (Hz)*	How Produced
60 cycle	60	The weak radiation from our alternating-current circuits
Radio, radar, and TV	10^4–10^{10}	Oscillating electric circuits
Microwaves	10^9–10^{12}	Oscillating currents in special vacuum tubes
Infrared	10^{11}–4×10^{14}	Outer electrons in atoms and molecules
Visible	4×10^{14}–8×10^{14}	Outer electrons in atoms
Ultraviolet	8×10^{14}–10^{17}	Outer electrons in atoms
X rays	10^{15}–10^{20}	Inner electrons in atoms, and sudden deceleration of high-energy free electrons
Gamma rays	10^{19}–10^{24}	Nuclei of atoms, and sudden deceleration of high-energy particles from accelators

* The *hertz*—abbreviated Hz—means *cycles or vibrations per second.* Its use not only saves time and trouble, but honors a great scientist.

interpretation, and varies with individuals. As a rough guide, the following table serves to give the wavelengths of the colors:

Violet	400–450 nm
Blue	450–500
Green	500–570
Yellow	570–590
Orange	590–620
Red	620–750

20-9 Color in the Sky

From earliest childhood, we all are so accustomed to the blue color of clear sky that it seldom occurs to us to wonder why this should be so. We see the sun, and it is easy to visualize the rays coming from the sun into our eyes. We see the shadows of trees and buildings, and it is also easy to visualize the sun's rays that mark the shadow edges. But in the sky, where there are no such markers to make them visible, it is all too easy to forget that the direct rays of the sun are also passing through every cubic millimeter of the atmosphere, wherever we choose to look.

The sky looks bluest when it is cleaned of dust and smoke, as it often is after a good rainstorm. It is also then very transparent—but *not quite perfectly transparent*! The molecules of the air present tiny obstacles to the free passage of the light. We can picture some of the light bouncing off these molecular obstacles in all directions—in other words, some of

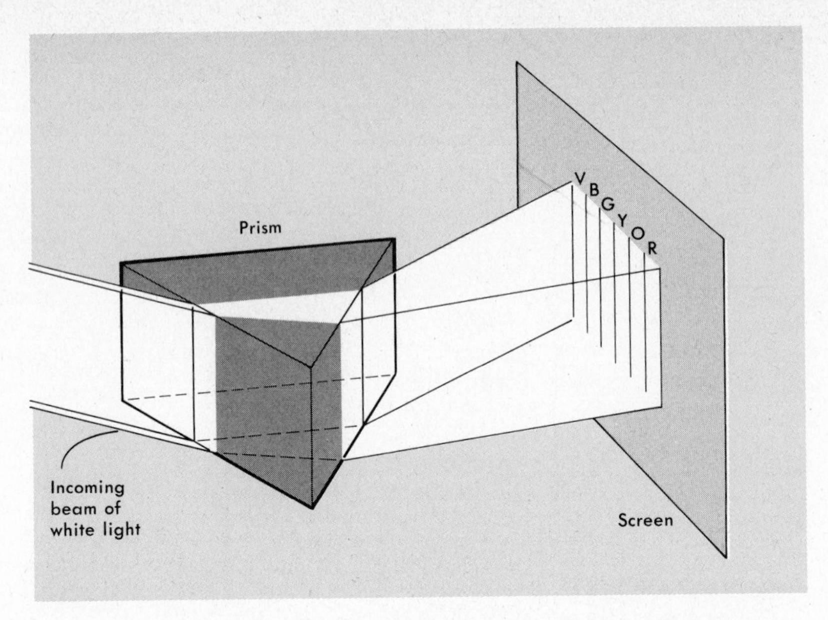

FIG. 20-13 Dispersion of white light into its component visible colors, by a prism.

the incoming light from the sun is *scattered* by the air molecules. Now for reasons that are too mathematical for us to go into here, light of short wavelength is scattered much more than light of long wavelength so that the blue end of the spectrum is more susceptible to this scattering than the red end. Thus, wherever we look in the sky, we see the blue light that has been scattered out of the white sunlight passing through it.

Big particles such as dust and the water droplets that form clouds—enormously larger than the molecules of the air—have very little of this selective effect, and scatter or reflect all colors almost equally. Thus clouds are white, and when the atmosphere is dusty, the blue may be nearly obscured by the overall white skylight.

This selective scattering of the blue end of the spectrum has an effect on the light that comes directly to our eyes. Even at noon, when the sun is most nearly overhead, it appears to be yellowish rather than white. This, of course, is because some of the blue has been scattered out during the passage of the light through the atmosphere over our heads. At sunset or sunrise (or moonset or moonrise) this effect can often be seen in exaggerated form. When the sun is near the horizon, its rays must pass through much more atmosphere to reach us (Fig. 20-14). More of the blue is scattered out of the direct beam; if atmospheric conditions are right, the setting sun may appear to be almost red.

When a rainstorm passes and the sun shines again in the sky, we often see a brightly colored arch against the dark background of the departing

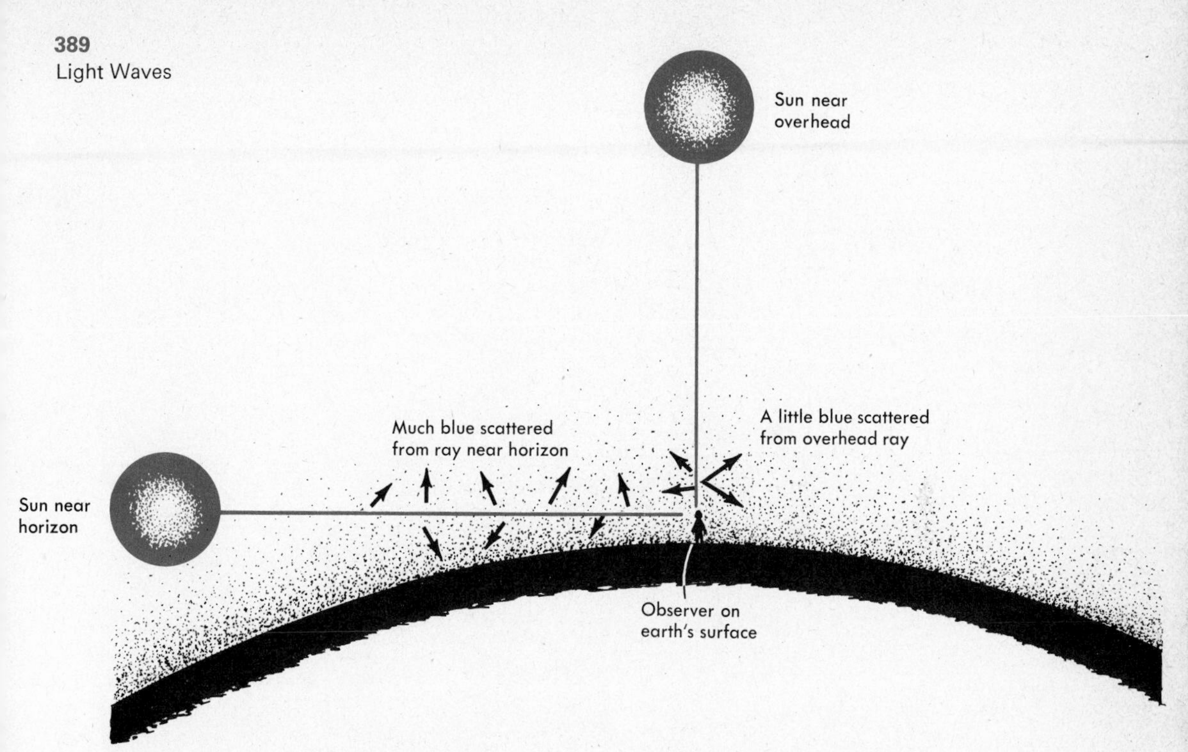

FIG. 20-14 Scattering of blue from the sun's rays by the atmosphere.

clouds opposite to the sun. The rainbow is an optical phenomenon caused by the reflection and refraction of sunlight in the tiny spherical raindrops on which it falls. In order for the direction of the rays of sunlight to be nearly reversed, the rays must enter the raindrops and be reflected from their inner surfaces as indicated on a greatly exaggerated scale in Fig. 20-15. The ray marked W indicates a thin beam of white light coming from the sun. The solid line in Fig. 20-15A shows how the red component is refracted as it enters the drop, is then reflected from the back of the drop, and is refracted again as it leaves the drop to return to the air. This red ray as it leaves makes an angle of very nearly 42° with the direction of the entering white ray.

We have just described the ray SFO, as shown in Newton's own diagram, Fig. 20-15C. The line OP is drawn from the eye directly away from the direction of the sun; if we look 42° away from this line in any direction we shall see the circular arc of red in the sky wherever there is a curtain of water droplets to reflect and refract this color back ot us. Returning to Fig. 20-15A, we see that the violet component (Newton's "most refrangible rays"), emerges from the drop, making an angle of only 40° with the direction of the sunlight. Hence if we describe a

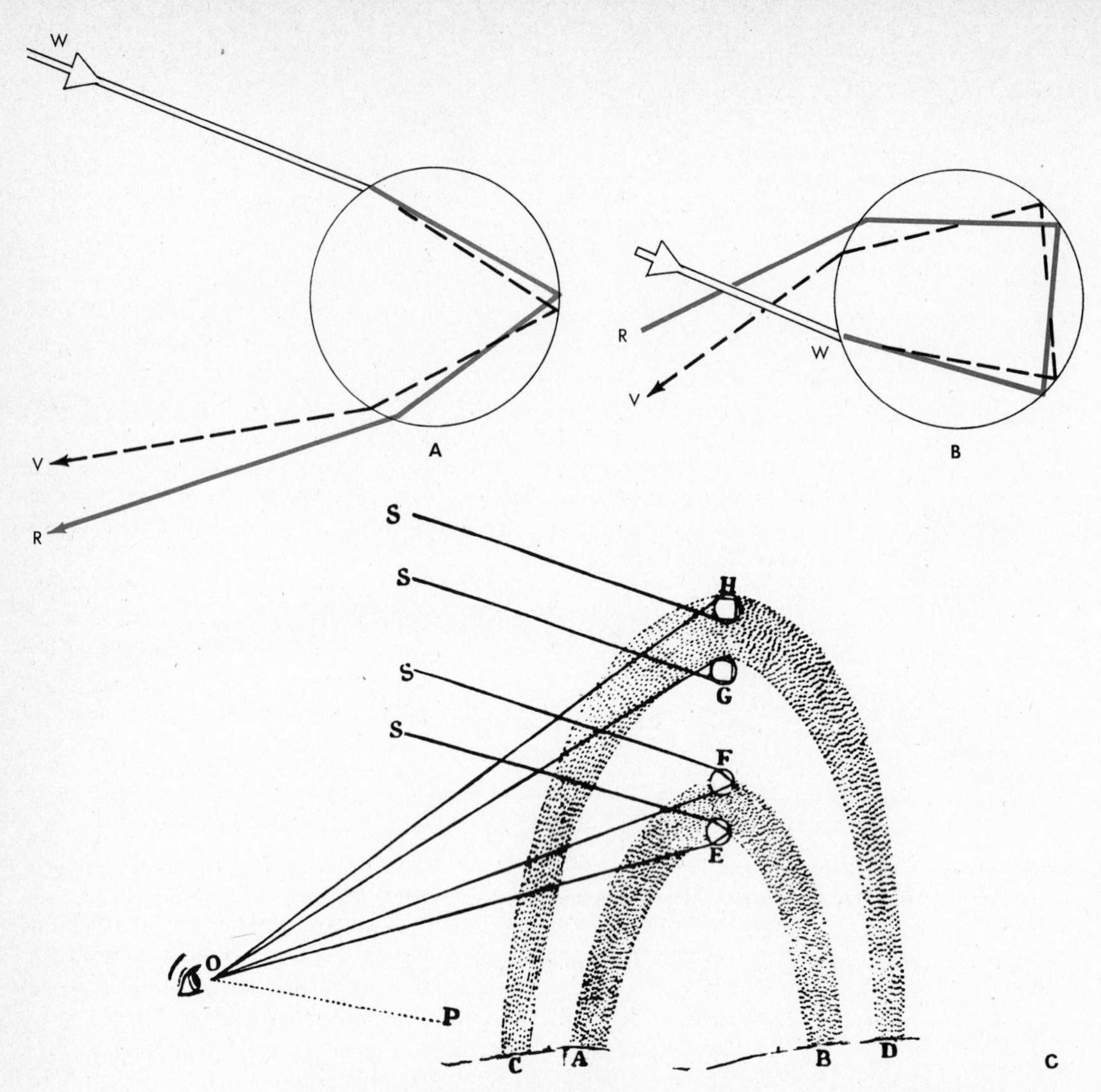

FIG. 20-15 Formation of rainbow colors by refraction and reflection of sunlight within water droplets. Newton's own explanation of a rainbow: "Suppose now that O is the spectator's eye, and OP a line drawn parallel to the sun's rays and let POE, POF, POG, and POH be angles of 40 degr. 17 min., 42 degr. 2 min., 50 degr. 57 min., and 54 degr. 7 min., respectively, and these angles turned about their common side OP, shall with their other sides OE, OF, OG, and OH describe the verges of two rainbows AF, BE, and CHDG. For if E, F, G, and H be drops placed any where in the conical superficies described by OE, OF, OG, and OH, and be illuminated by the sun's rays SE, SF, SG, and SH; the angle SEO being equal to the angle POE, or 40 degr. 17 min., shall be the greatest angle in which the most refrangible rays can after one reflection be refracted to the eye, and therefore all the drops in the line OE shall send the most refrangible rays most copiously to the eye, and thereby strike the senses with the deepest violet color in that region." From Newton's Optics. (Reprinted by G. Bell & Sons Ltd., London, 1931.)

circle in the sky 40° in radius as measured from the point opposite the sun, it will mark the smaller violet arc, lying inside the red one.

Figure 20-15B shows the double reflection within the droplets, which causes the larger secondary bow we can sometimes see.

Questions

(20-1) 1. For the yellow light of a sodium-vapor lamp, a certain glass has $n = 1.75$. (a) What is the speed of this light in the glass? (b) In air, the wavelength of this light is 589 nm. What is its wavelength in the glass?

2. The red light from a hydrogen tube is found to have a wavelength of 405 nm in a piece of special glass. This light has $\lambda = 656$ nm in air. (a) What is the index of refraction of the glass for this light? (b) With what speed does it travel in the glass?

(20-2) 3. A pair of narrow slits, 0.04 cm apart, are 30 cm in front of a photographic film. These slits are illuminated by light of $\lambda = 5893$ Å from another slit. How far apart are the bright bands on the film?

4. A pair of narrow parallel slits are 0.03 cm apart and 20 cm in front of a photographic film. If the slits are illuminated with monochromatic light of $\lambda = 5710$ Å from another narrow slit, how far apart will the resulting bands be on the film?

5. The experimental setup of Question 3 is illuminated by monochromatic radiation of unknown wavelength, and the bands on the film are found to be 0.7 mm apart. What is the wavelength of the radiation? Would it be visible to the eye?

6. The equipment of Question 4 is illuminated by radiation of unknown λ, and the resulting bands (often called *fringes*) on the film are 1.70×10^{-2} cm apart. What is the λ of the radiation? Would it be visible to the eye?

(20-3) 7. Light of $\lambda = 6328$ Å falls on a single slit 0.005 cm wide, as in Fig. 20-5. What is the width of the central bright band, on a screen 2 m behind the slit?

8. As in Fig. 20-5, a beam of light falls on a single slit 10^{-3} in. wide. On a photographic film 90 cm beyond the slit, the central band of the interference pattern is 4.48 cm wide. What is the wavelength of the light?

9. In Question 7, (a) what would the width of the central band be if the slit were twice as wide: 0.010 cm? (b) What would the central band width be if light of two-thirds the wavelength—4219 Å—were used?

10. In Question 8, (a) what slit width would be required to make the central band 1.25 times as wide—5.60 cm? (b) What wavelength of light would produce a central band 5.97 cm wide—i.e., four-thirds its original width?

(20-4) 11. Monochromatic radiation of unknown wavelength is directed against a grating ruled with 13,400 lines/inch. An angle of 20° separates the zero-order reinforcement direction and the first-order direction. What is the wavelength of the radiation?

12. There is an angle of 17° between the reinforcement directions of zero-order and first-order, when radiation is directed against a 12,000 lines/inch grating. What is the wavelength of the radiation?

(20-5) 13. (a) What is the minimum thickness of a soap bubble ($n = 1.35$) at a point where it reflects largely red light ($\lambda_{air} = 6800$ Å)? (b) What other greater thicknesses would also give a maximum reflection of red?

14. (a) What is the minimum thickness of a soap bubble ($n = 1.35$) that will reflect *no* yellow light ($\lambda_{air} = 5800$ Å)? (b) What other greater thicknesses would also reflect no yellow light?

15. Given a spot of *very* thin soap bubble film—i.e., its thickness is a small fraction of the wavelength of light—what would be the appearance of the reflection from this spot, when illuminated with white light?

16. On a plate of dense glass, $n = 1.70$, there is a thin smear of oil, $n = 1.50$. In one spot, the thickness of the smear is a small fraction of the wavelength of light. What would be the appearance of the reflection of white light from this spot?

(20-6) 17. Polaroid sunglasses reduce "reflected glare." Explain how this is done.

18. Given a small piece of Polaroid, how could you use it to determine whether a light beam was polarized or not?

19. Two pieces of Polaroid with their axes at 90° almost completely block the passage of any light through them. What will be the effect of slipping another piece of Polaroid between the first two, with its axis oriented at 45°?

20. It has been suggested that the dangers of night driving could be reduced by reducing the glare of oncoming headlights. Can you propose how this might be done with sheets of Polaroid? (Remember that you must be able to see objects properly illuminated by your own headlights.)

21. If the frequency range shown in Table 20-1 were plotted on a linear (arithmetic) scale from 0 to 10^{24} Hz on a graph 1 km long, how long a piece of this graph would represent visible light?

22. If the frequency range shown in Table 20-1 were plotted on a geometric scale (similar to Fig. 1-1) from 1 to 10^{24} Hz on a graph 1 km long, how long a piece of this graph would represent visible light?

21

Optical Instruments

21-1 Optical Aberrations

In Chapter 19, a lens was considered to have one definite focal length, and we did not concern ourselves with the wavelength of the light used in forming images. At that time it was not necessary to consider the fact that the index of refraction is different for different wavelengths. For all types of glass and plastic, n is larger for the short wavelength blue and violet, and is smaller for the long wavelength red. A single lens, therefore, cannot focus ordinary white light at any point. As shown in Fig. 21-1A, the blue component is refracted more, and so has a shorter focal length than the red component.

This unavoidable defect of a single lens is *chromatic aberration*. Fortunately, it can be almost completely corrected by using a pair of lenses. The *difference* in n for red and blue measures what is called the *dispersion* of the glass. Glasses of different chemical composition have not only different n's, but also have widely varying dispersions. Figure 21-1B shows an *achromatic* doublet. The converging lens alone has strong

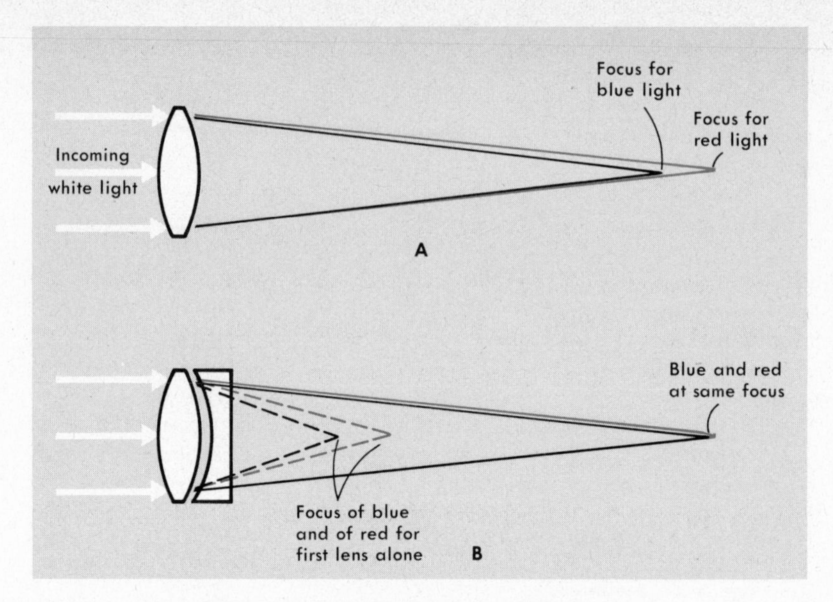

FIG. 21-1 Chromatic aberration and the achromatic doublet lens.

chromatic aberration. The diverging lens is made of a different glass that has a large dispersion. Thus, although it is weaker than the first lens and therefore lets the rays continue to converge, it brings both red and blue to the same focus.

The problem of chromatic aberration does not exist for mirrors, since reflection is not affected by the wavelength. A concave mirror, however, plainly shows another bothersome aberration that is also present in lenses. Figure 21-2A shows a mirror with a spherical concave surface. The rays striking near the edge are reflected to a focus closer to the mirror than the rays in the vicinity of the axis. This imperfection is called *spherical aberration* and is a potential cause of considerable trouble in astronomical reflecting telescopes, whose mirrors must be large in order to collect as much light as possible. This difficulty is solved, however, by making most telescope mirrors parabolic rather than spherical. In such a mirror (which unfortunately is more expensive to make than a spherical mirror and has other aberrations that a spherical mirror does not have), all rays coming in parallel to the optical axis from a very distant object are reflected to intersect the axis at a single focal point (Fig. 21-2B).

This scheme works equally well in reverse; if a small source of light is placed at the focus, a parabolic mirror will reflect it into a cylinder of parallel rays. For this reason, parabolic mirrors are often used on large searchlights.

A parabolic mirror can thus form a perfect image of a distant object that lies on the axis of the parabola; a combination of two or three

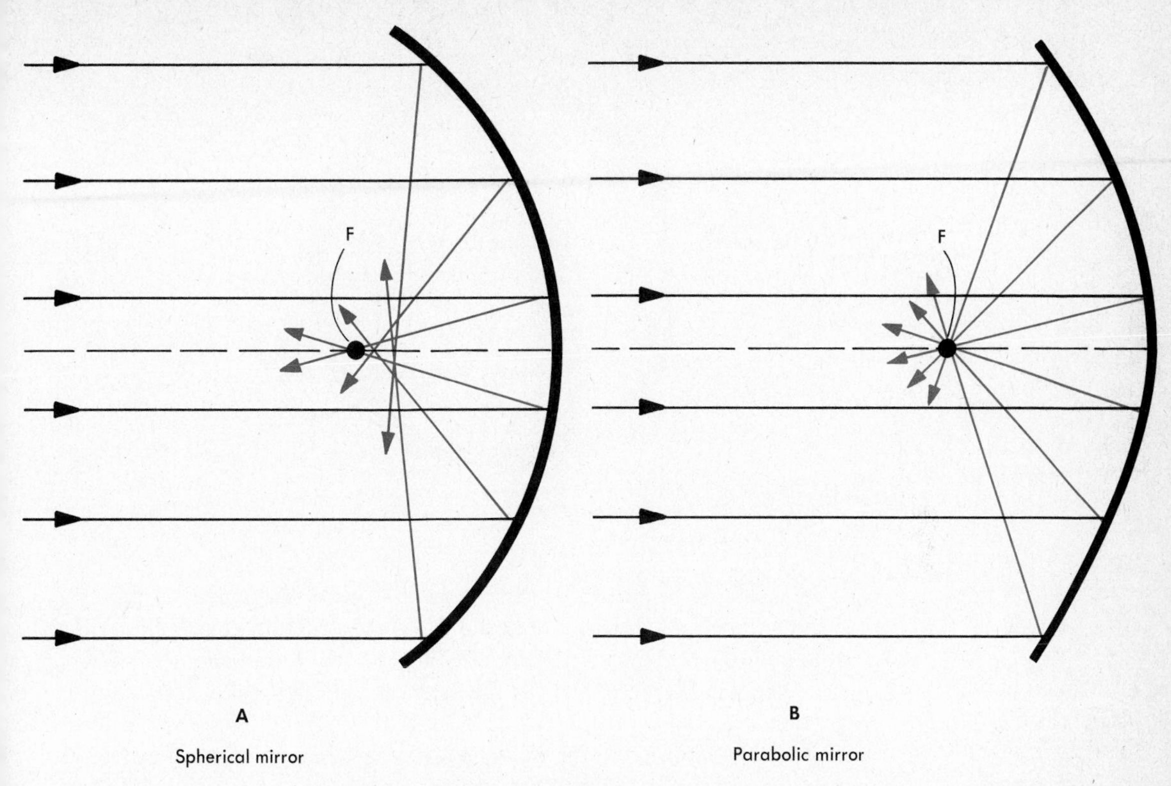

FIG. 21-2 Spherical aberration with a spherical mirror, and its elimination with a parabolic mirror.

lenses can be almost completely corrected for chromatic and spherical aberrations, and so will do almost as well.

This does not mean, however, that a parabolic mirror solves all the optical designer's problems. If the object is not on the optical axis, the image suffers from two *off-axis* aberrations: *coma* and *astigmatism*. In coma, a point on the object produces an asymmetric patch of light for its image, which grows larger and more distorted as the distance from the optical axis increases. For pure astigmatism, an object point is focused as two straight lines perpendicular to each other, and separated by a distance that increases with the distance from the axis. The best image is the circular patch of light midway between the lines.

Lenses also have these same off-axis aberrations, which can be nearly eliminated by a properly designed combination of two, three, or more separate lenses.

The design of any camera, telescope, or microscope is always a compromise in which the designer tries to keep all of these aberrations (plus others that have not been described) so small that they will not cause objectionable defects in the final image.

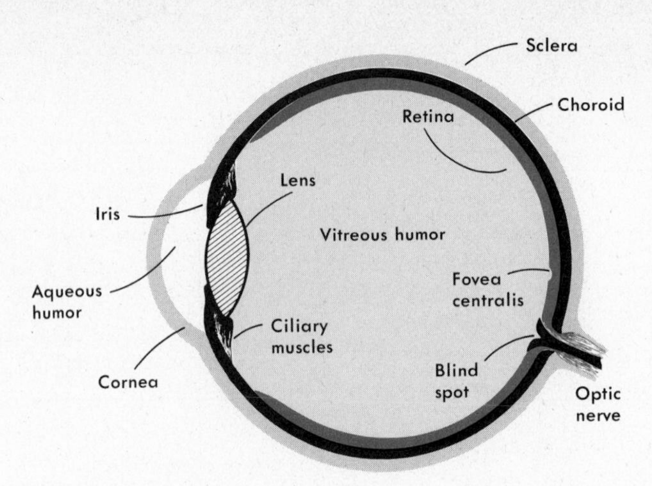

FIG. 21-3 Essential parts of the eye.

21-2 The Eye

The most important and most versatile optical instrument is the eye. Figure 21-3 is a schematic diagram of its construction. The *sclera* is the tough outer membrane to which the muscles that hold and move the eye are fastened. The front part of the sclera is more strongly curved, and is transparent; this transparent section is the *cornea.* The *choroid* layer inside the sclera contains dark pigment so that light penetrating the retina is absorbed, rather than scattered back into the central chamber.

The *retina* is the light-sensitive layer; it is a mosaic of two kinds of cells. The *cones* not only respond to light, but are apparently of three types, each responding most readily to a specific range of wavelengths in the red, the green, and the blue-violet. The *rods* are more sensitive than the cones to dim levels of illumination, but do not distinguish between colors. The *fovea centralis*, near the optical axis, is a small spot a few tenths of a millimeter in diameter, solidly packed with cones. This is the retinal area that provides by far the best discrimination of fine detail and color. When we "look at" something, we swivel our eyes so that its image falls on the fovea. Away from the fovea, the mosaic of light-sensitive cells becomes coarser, and the proportion of rods increases. At the limit of our field of vision, near the edge of the retina, the cells are entirely rods; colors and details cannot be discriminated, although any motion is readily detected.

The *blind spot* is the place where the nerve fibers from the light-sensitive cones and rods are gathered together and leave the eyeball as the *optic nerve*. This small area contains no cones or rods, and is therefore totally blind. The blind spot on every retina is not ordinarily noticeable, as some psychological processing of the incoming data by the brain seems to fill the spot in, to produce a visual effect there similar to its surroundings. Its reality can be easily demonstrated. With both book and head level, hold Fig. 21-4 at arm's length before you and close one eye. If the

R **L**

FIG. 21-4 Chart for demonstrating the "blind spot."

open eye is the right eye, look steadily at the letter **R** and bring the page slowly closer to your face. The **L**, although probably not identifiable, nevertheless remains plainly visible as a dark spot. As you continue to bring the page nearer, however, at a distance of about a foot the **L** disappears, only to reappear again as you bring the page still closer. While it is invisible, its image falls on the blind spot.

The eye is a double-lens system, the first and more important refraction taking place at the surface of the cornea. A second bending of the incoming rays occurs at the transparent capsule marked "lens" in Fig. 21-3. By the combined action of the cornea and the lens, the image of an object at which we look should be formed accurately on the light-sensitive retina. Since we look at objects at varying distances, some adjustment is required to accomplish this.

The image distance q (in $1/p + 1/q = 1/f$) is a fixed quantity in the eye; a normal relaxed eye forms the image of a distant object ($p = \infty$) exactly on the retina. To look at nearby objects with a shorter p, we see that the focal length of the lens, f, must also become shorter. This is accomplished by the ciliary muscles, which act to squeeze the lens and make it more convex. There are no muscles provided to make the lens *less* convex; accommodation appears to depend entirely on making the lens stronger.

The lens of a farsighted eye is not strong enough to focus even distant objects when it is relaxed; thus a farsighted eye must exert a constant squeeze on the lens muscle, and consequently tires easily. The lens of a myopic, or nearsighted, eye is too strong, so that distant objects are focused in front of the retina when the eye is relaxed. It cannot by any muscular effort accommodate for distant objects and hence has a relaxed lens muscle and a blurred image it can do nothing about. In young, normal individuals, the accommodation power of the eye (i.e., its ability to focus at different distances) is very high, and a child can usually accommodate his eyes for clear vision even if the spectacles belonging to a very nearsighted person are put on his nose. With age, the accommodation power of the eye decreases, and its normal focusing abilities very often become defective.

Figure 21-5 shows the focusing of accommodated and relaxed normal eyes, nearsighted eyes, and farsighted eyes. In nearsighted persons, the eye lens is curved too much, so that the image is formed generally in front of the retina. While the eye can still accommodate for good vision of objects located quite close, it cannot possibly bring to the retina the image of distant objects, and they appear blurred. On the other hand, the eye lens of a farsighted person does not have enough curvature, and

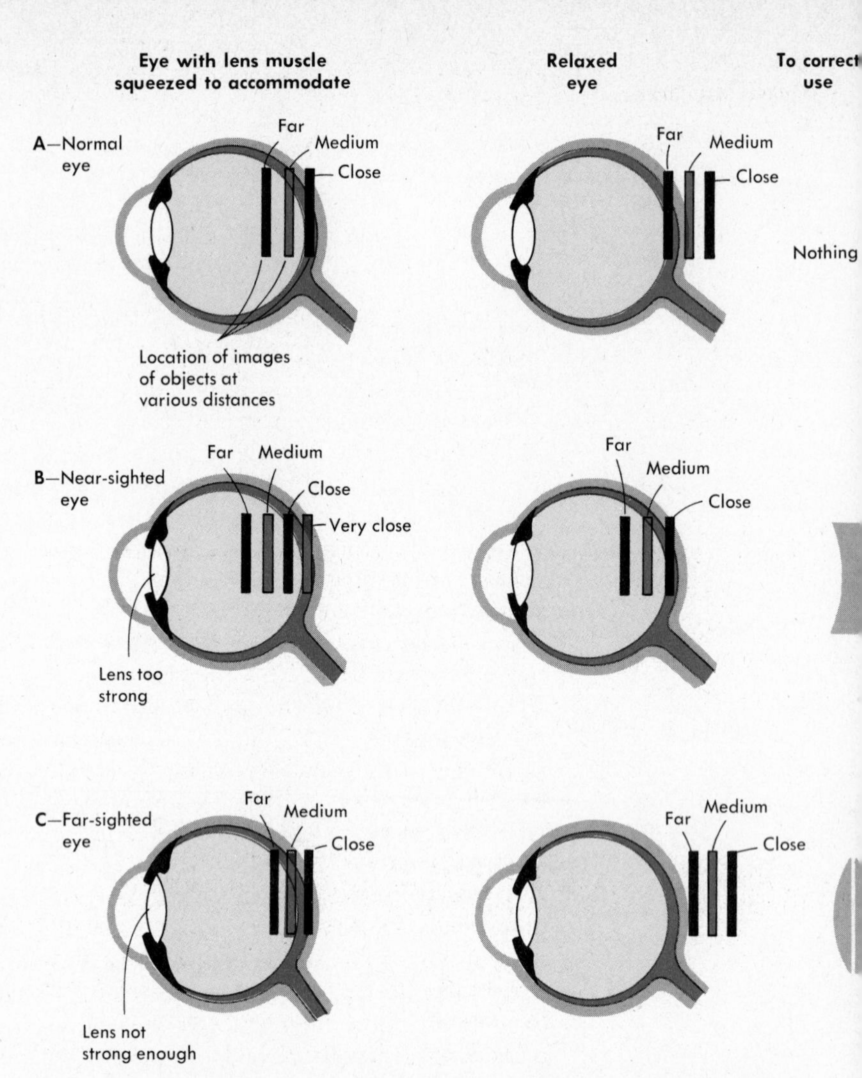

FIG. 21-5 The positions of the image of objects that are at far, medium, or close distances from normal and from abnormal eyes.

the image is formed, on the average, behind the retina. In this case, the eye can still accommodate for the clear vision of distant objects, but fails to do so for objects that are comparatively near. Opticians can help in both cases by prescribing glasses with diverging lenses for nearsighted persons, and converging lenses for farsighted ones.

Take, for example, a nearsighted person who cannot focus clearly on objects more than 36 in. away. What must be the focal length of glasses that will permit him to focus on very distant objects? The problem is a simple one if we rephrase it to ask: What focal-length lens will produce a virtual image 36 in. to the front, where the eye can use it as a virtual object? To answer this we need only write (since $1/p + 1/q = 1/f$)

$$\frac{1}{\infty} + \frac{1}{-36} = \frac{1}{f}$$

$$f = -36\text{-in. focal length.}$$

Consider a farsighted eye that cannot focus on any object closer than 60 cm. What lens will enable it to see clearly a slide rule held at a distance of 20 cm? In this case, the lens must form its virtual image at $q = -60$ cm. The eye can then clearly focus this virtual object:

$$\frac{1}{20} - \frac{1}{60} = \frac{1}{f}$$

$$f = +30 \text{ cm.}$$

Opticians commonly give the *power* of the lens in *diopters.* The diopter power of a lens is merely the reciprocal of its focal length *in meters.* For the lens in the first example above we have $f = -36 \times 2.54 \times 0.01 = -0.91$ m, to give a power of $-1/0.91 = -1.1$ diopters. For the second lens, $1/0.30 = +3.3$ diopters.

The eye–lens system has all the aberrations of artificial lenses, but they are ordinarily not severe enough to cause any noticeable difficulty. Since only the rays very near the axis, which fall on the fovea, are useful in discriminating details, the off-axis aberrations of coma and astigmatism have no effect on vision. The "astigmatism" so commonly referred to as a visual defect is something entirely different. It results from an asymmetry of the surface of the cornea, which is often curved more strongly in one direction than in another. It is corrected by glasses with a similar, but opposite, difference in curvatures.

We see well in sunlight, and reasonably well in moonlight—in these extremes of illumination the light intensity differs by a factor of 10^5 or more. This remarkable range of sensitivity is largely due to the difference between rods and cones—the rods respond to illumination several hundred times smaller than that needed to stimulate the cones. In weak light, this vision by rods must necessarily lack the sensations of color, which are produced only by the less sensitive cones. A flower garden in moonlight is seen only a pattern of grays.

In addition to this fundamental difference between rods and cones, the eye has two other responses to changing light. The first is by means of the *iris* (Fig. 21-3), a colored ring-shaped membrane in front of the lens, visible through the transparent cornea. Automatically functioning muscles change the diameter of the central opening (the *pupil*) as the intensity of illumination changes. In bright sunlight the diameter of the pupil may be as small as 2 mm. When one goes from sunlight into a dimly lighted room, the pupil may expand to an 8-mm diameter. This action provides an immediate adjustment by a factor of $(\frac{8}{2})^2 = 16$.

A slower reaction is that of *dark adaptation*. At low levels of illumination, the supply of light-sensitive chemicals in the rods gradually increases, to reduce the amount of light needed for vision by a factor of a hundred or more. This requires at least a half hour.

21-3 Cameras

In many ways a camera duplicates the parts of the eye and their functions. The camera "lens" may actually contain from three to six or more separate lenses, as needed to correct the various aberrations. Analogous to the eyelid, the camera shutter opens for a predetermined length of time to allow light to enter through the lens and expose the film.

The expansion and contraction of the pupil of the eye to vary the amount of light admitted is a reflex action over which we have no control. The *iris diaphragm* of the camera is a similar mechanism that is under the control of the photographer. Figure 21-6A shows a simplified camera in which this adjustable diaphragm, or *stop*, has been set with its diameter equal to 0.36 times the focal length of the lens. All the light within the red lines is admitted to form the image of a point on the distant object. In Fig. 21-6B the diameter of the stop has been halved to $0.18\,f$.

On almost all cameras the size of the stop opening is indicated on its setting ring or lever by the "f-number:" $f/1.4$, $f/3.5$, etc. *The f-number is the ratio of the focal length to the stop diameter.* In Fig. 21-6A the stop setting, or f-number, is $f/0.36f = 2.8$; in Fig. 21-6B it is 5.6.

Since the $f/5.6$ stop opening has only half the diameter of the $f/2.8$, it admits only a quarter as much light from the object. To give the film the same exposure to light we must open the shutter for four times as long.

FIG. 21-6 Adjustable stop used with a camera lens.

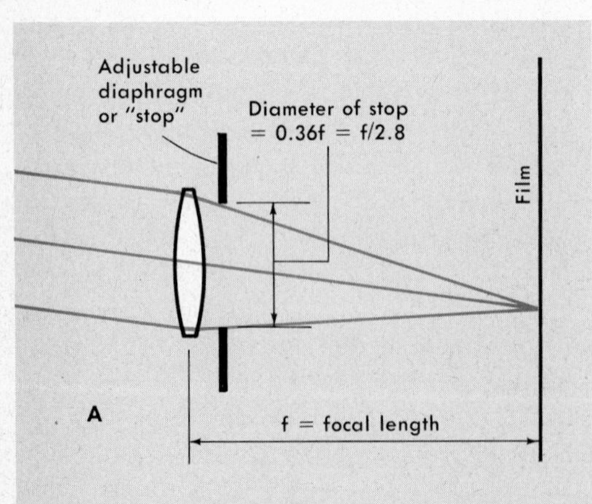

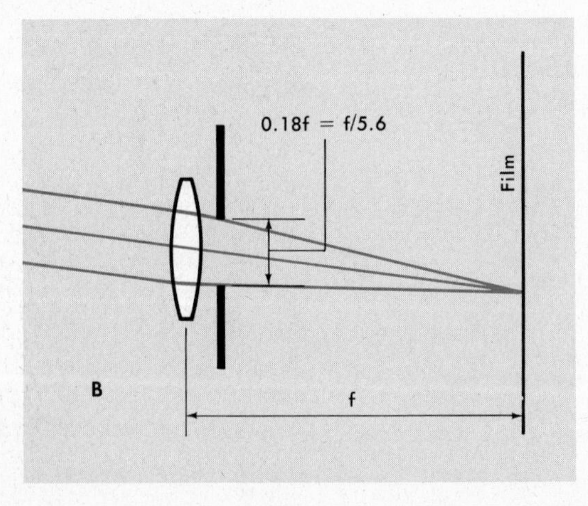

Camera manufacturers ordinarily mark the stop settings so that each differs from the adjacent setting by approximately $\sqrt{2}$; the areas then differ by a factor of 2. A typical camera might have its stop settings marked at 2.8, 4, 5.6, 8, 11.3, 16. As an example, an exposure of $\frac{1}{100}$ s might be the correct exposure time for a certain scene at $f/2.8$. If the photographer increases his f-number by three stops to $f/8$, he would also have to increase his exposure time by $2 \times 2 \times 2$, to $\frac{8}{100} \approx \frac{1}{12}$ s. Alternatively, he could figure his increased time factor as $(8/2.8)^2 \approx 8$.

It should be noted that the photographer does not need to worry about the focal length of his camera. If $\frac{1}{12}$ s at $f/8$ gives a correct exposure for a miniature camera of 50-mm focal length, it will also be correct for a telephoto lens of 200 mm focal length at $f/8$. At $f/8$ the effective lens diameter of the longer lens will be 4 times as great as that of the miniature camera, and will admit 16 times as much light. Its image, however, will also have dimensions 4 times as great, or 16 times greater area. So 16 times as much light is spread over 16 times as much area, giving the same amount of light energy per cm^2, requiring the same exposure.

Larger f-numbers (smaller openings) give a narrower cone of light converging from lens to film, and have several advantages. The effects of any aberrations the lens may have are reduced. Focusing is less critical—if the narrow cone of rays does not converge to a point exactly on the film, each object point will be imaged as a tiny circle of light called the *circle of confusion*. The narrower the cone, the smaller this circle, and the less imperfect the focus.

The photographer's *depth of field* is related to this imperfection of focus. Consider a photographer who sets his stop at $f/2.8$, and focuses* the camera accurately on an object 30 ft distant. As in Fig. 21-6A, $f/2.8$ gives a quite fat cone of light converging on each image point, and a slight error in q will result in a fairly large circle of confusion on the film rather than a point image. Thus only those objects whose distance (p) differs but little from 30 ft will be in acceptably clear focus. However, for a narrow cone of, say, $f/11$, q can be in error by a much greater amount before the circle of confusion becomes too large. Hence objects will still be in acceptable focus at distances considerably greater or less than 30 ft. By increasing the f-number, the photographer has increased his *depth of field*—i.e., the range of distance at which objects will be in satisfactory focus.

To compensate for the advantages of a large f-number, there is a corresponding serious penalty: because less light is admitted, the time of exposure must be longer. This may blur the images of moving objects. And for exposures of $\frac{1}{25}$ s or longer, involuntary muscle tremors make hand-holding unsatisfactory, so that a tripod or other rigid support is necessary.

---

* Focusing is accomplished by moving the lens, to change the lens–film distance (q) so that it is properly related to the object distance (p), in the relationship $1/p + 1/q = 1/f$.

21-4 Resolving Power

In discussing the eye and the camera, we have not considered the wave nature of light. We have drawn rays (in Fig. 21-6, for example) as though they converged to a mathematical point in forming the image of a point on the object. For many purposes, this is a perfectly satisfactory approximation, but when the interference of the light waves is taken into account, we find that this approximation has definite limits.

Figure 20-5 on p. 378 shows the pattern of light on a screen behind a narrow slit illuminated by some distant source. Most of the light is in the central bright band, whose total width is $2y$. We have that

$$\frac{y}{l} = \frac{\lambda}{s}; \qquad y = \frac{\lambda l}{s}.$$

If, instead of a slit of width s, we use a circular opening of diameter $d = s$, we find an analogous circular pattern on the screen, as in Fig. 21-7A. Most of the light falls in the central spot, which has a radius 1.22 times the half-width of the central bright band formed by the slit:

$$r = 1.22y = \frac{1.22\,\lambda l}{d}.$$

This same interference pattern results when a perfect, aberration-free lens of diameter d forms the image of a point object. On a camera

FIG. 21-7 (A) Diffraction pattern images of two separated point sources. (B) Diffraction pattern images separated by the Rayleigh criterion. (*University of Colorado photograph.*)

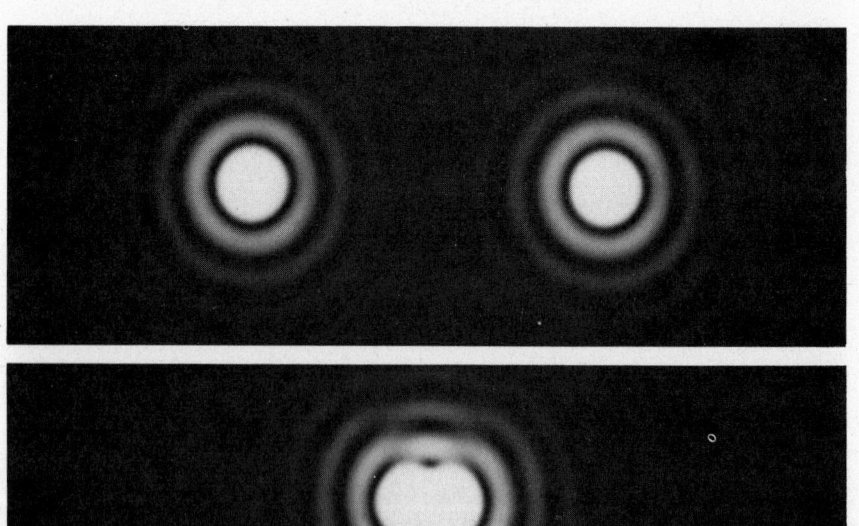

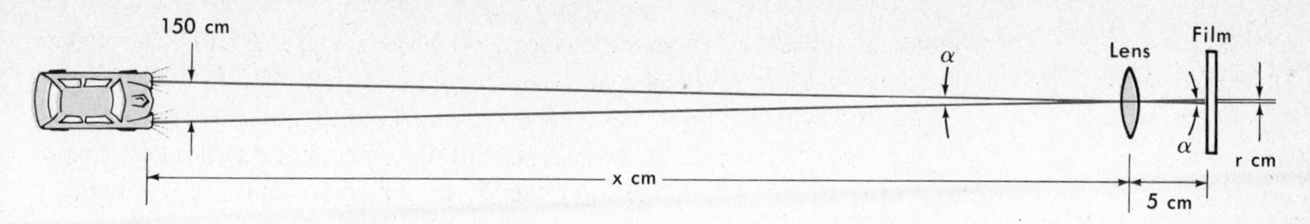

FIG. 21-8 Resolution of two point objects by a camera.

film, say, how close together can two such patterns be, if we are to be sure that they represent two object points rather than one? There is no way to derive a precise answer to this subjective question. In 1874, however, the English physicist Lord Rayleigh suggested a quite simple criterion: Two patterns are barely distinguishable as separate if the central maximum of one falls on the first dark ring of the other. This is shown in Fig. 21-7B; the two patterns are separated by the radius of the central spot $1.22\lambda l/d$.

Consider a photographer whose camera lens has a focal length of 50 mm, set at $f/3.5$. It is night, and in the distance he sees an approaching car whose headlights are 150 cm apart. How close must the car be before his camera can *resolve* the two headlights—i.e., show them to be two sources, rather than one?* Figure 21-8 sketches the situation. As indicated by the undeviated rays passing through the center of the lens, the angle α separating the object points is the same as the angular separation of their images. For resolution of the image points, we have that their separation must be $r = 1.22\lambda l/d$. It is given that $l = 5.0$ cm; at $f/3.5$, the effective diameter of the lens $d = 5.0/3.5 = 1.4$ cm; for the yellowish light of headlights we can take $\lambda = 6.5 \times 10^{-5}$ cm. So

$$r = \frac{1.22 \times 6.5 \times 10^{-5} \times 5.0}{1.4} = 2.8 \times 10^{-4} \text{ cm}.$$

For a small angle, this gives

$$\alpha = \frac{2.8 \times 10^{-4}}{5.0} = 5.6 \times 10^{-5} \text{ radian}$$

$$= \frac{5.6 \times 10^{-5} \times 180}{\pi} = 3.2 \times 10^{-3} \text{ degree}$$

$$= 3.2 \times 10^{-3} \times 3600 = 12 \text{ seconds.}$$

* We will assume that the graininess of the film is fine enough that the resolving power is determined entirely by the lens.

This—12 seconds—is the *resolving power* of this lens at $f/3.5$. It can show as separate, two object points 12 seconds apart. To find the separation x in the problem, we have that $150/x = 5.6 \times 10^{-5}$ rad, and

$$x = \frac{150}{5.6 \times 10^{-5}} = 2.7 \times 10^6 \text{ cm}$$

$$= 27 \text{ km, or about 17 mi.}$$

21-5 Resolving Power of the Eye

Like any other optical instrument, the eye is limited in its ability to resolve, or see as separate, points that are close together. The diffraction pattern of the eye lens is certainly one limiting factor. Under bright sunlight, the pupil may close down to 2.0 mm, or 0.20 cm; the effective focal length of the eye is about 1.7 cm; the wavelength to which the eye is most sensitive is 5.5×10^{-5} cm. From this, Rayleigh's criterion gives us

$$r = \frac{1.22 \times 5.5 \times 10^{-5} \times 1.7}{0.20} = 5.7 \times 10^{-4} \text{ cm.}$$

This represents an angular separation of

$$\frac{5.7 \times 10^{-4}}{1.7} = 3.4 \times 10^{-4} \text{ rad}$$

$$= 3.4 \times 10^{-4} \times \frac{180}{\pi} \times 60 = 1.2 \text{ minutes.}$$

If, then, the only factor to be considered were the diffraction pattern formed by the eye-lens on the retina, we should be able to distinguish two points separated by an angle of 1.2 minutes.

The structure of the retina, however, gives us something else to consider. On the fovea, the cones are tightly packed together in a mosaic of light-sensitive cells. In order to register two separate light spots, a minimum of three cones are needed: two outer ones stimulated by the light, and a third between them in relative darkness. Thus, the mosaic can distinguish light-spots separated by two cell diameters, measured center-to-center of the outer cones. The diameter of a foveal cone is about 2.9×10^{-4} cm, which gives a resolution of 5.8×10^{-4} cm. Expressed in angular terms, this is

$$\frac{5.8 \times 10^{-4}}{1.7} \times \frac{180}{\pi} \times 60 = 1.2 \text{ minutes.}$$

giving the same resolving power as that determined by the interference pattern formed by the lens. Nature is indeed economical—a finer mosaic of cells in the human eye would serve no useful purpose.

21-6 The Simple Magnifier

The most commonly used optical instrument is the simple magnifying glass. As we bring a small object up closer and closer to one eye, it appears larger and larger; if there were no limit to this process, magnifying glasses would be unnecessary. There is a limit to the focusing ability of the eye, however; for a normal adult, 10 in., or 25 cm, is about as close as possible for an object that is to be examined comfortably for any extended period of time. One way of considering the simple magnifier is to think of it as making the eye lens stronger, so that the object may be held closer to the eye and still be focused as a clear image on the light-sensitive retina at the back of the eyeball.

Figure 21-9 shows the use of the simple magnifier, or magnifying glass. If we place the object slightly closer to the magnifier than its focal length, a virtual image is formed at the closest distance for distinct vision, which is assumed to be 10 in. or 25 cm. The magnification is the ratio of the size of this virtual image to the size of the object at the same distance. To determine the magnification, we have the known f of the magnifier, and we know that $q = -25$ cm. From this,

$$\frac{1}{p} - \frac{1}{25} = \frac{1}{f}$$

$$\frac{1}{p} = \frac{1}{f} + \frac{1}{25} = \frac{25 + f}{25f}$$

$$M = \frac{q}{p} = q \times \frac{1}{p} = \frac{25(25 + f)}{25f} = \frac{25 \text{ cm}}{f \text{ cm}} + 1.$$

FIG. 21-9 The simple magnifier.

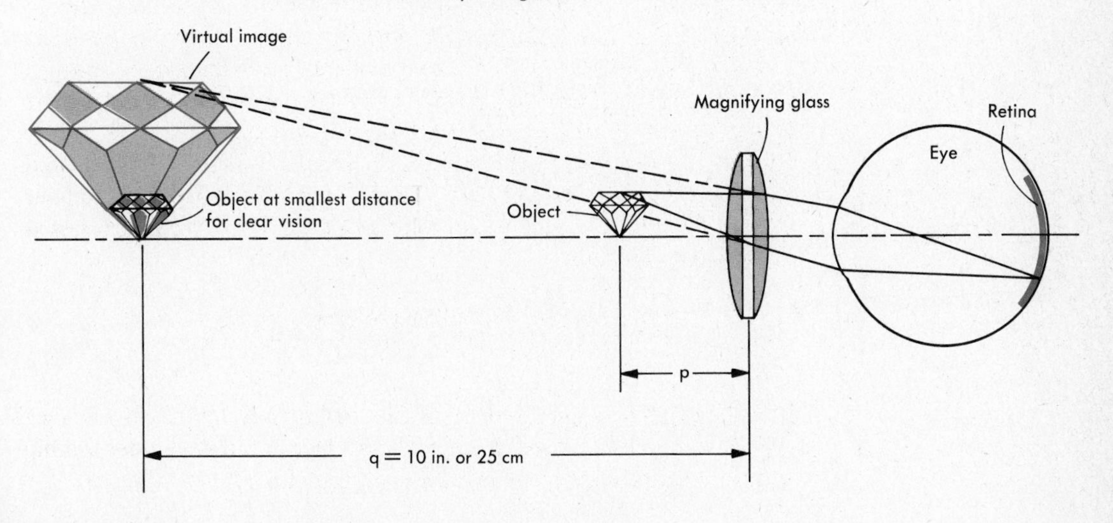

If we had taken all our dimensions in inches, it would have come out

$$M = \frac{10 \text{ in.}}{f \text{ in.}} + 1.$$

21-7 Telescopes

The ordinary refracting telescope has two elements: the *objective*, which receives the light from the object; and the *eyepiece*, which serves as a magnifier to examine the real image formed by the objective. Figure 21-10 shows a very simplified version. Objectives are almost always two lenses, in order to reduce chromatic and other aberrations. Eyepieces contain from two to six separate lenses.

For the magnification of telescopes, comparing the actual size of the image and the object would be meaningless, and the idea of *angular magnification* is used instead. In Fig. 21-10 rays are shown entering the objective from two distant objects, which we can call the "red star" and the "black star," to identify each with the color of the rays on the drawing. In the sky, the two stars are separated by the angle α, and, beyond the objective, the angle between the undeviated rays through the center of the lens is also α. At the focus of the objective (where we could photograph them on a piece of film if we cared to) the two star images are separated by a distance d, so that $\alpha = d/f_o$ rad.

Through the eyepiece, the images are seen to be separated by the angle β. It is not possible to make a direct geometrical construction of these rays after they are refracted by the eyepiece—*but* if the objective had been a little larger, we would have had rays passing undeviated through the center of the eyepiece lens. Figure 21-10 shows these hypothetical rays as dashed lines from the images. The eyepiece is adjusted so the star images are at its focus; thus all the rays from either image point emerge from the eyepiece parallel to each other—in Fig. 21-10

FIG. 21-10 Angular magnification of a simple telescope.

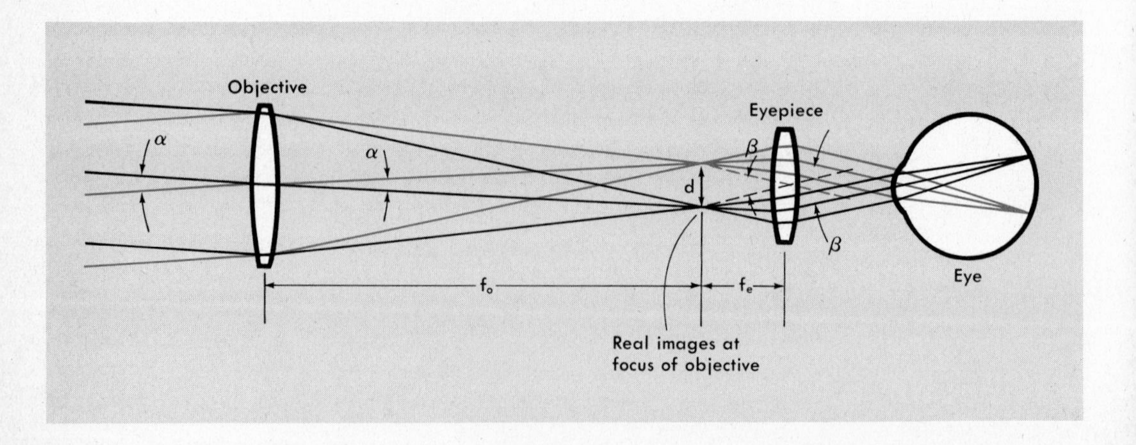

they have been drawn parallel to the hypothetical dashed rays. To the eye, the black star and the red star are separated by the angle β, which we can evaluate approximately as $\beta = d/f_e$ rad. For the angular magnification this gives

$$M = \frac{\beta}{\alpha} = \frac{d/f_e}{d/f_o} = \frac{f_o}{f_e}.$$

For example, consider a telescope whose objective has a focal length of 35 in., used with an eyepiece of $f_e = 0.5$ in. Look through this telescope at a pair of stars whose actual separation is 19 minutes. $M = 35/0.5 = 70$, and to the observer the stars would look as though they were $19 \times 70 = 1330' = 22°10'$ apart.

The simple telescope of Fig. 21-10 has one drawback that does not bother astronomers. Seen through the telescope, the black star appears *below* the red star, rather than above it, as it actually is; that is, the image is inverted.

Figure 21-11 shows three refracting telescopes that give erect images and are thus suitable for ordinary nonastronomical use. The Galilean type is so named because Galileo used a diverging eyepiece on his first astronomical telescope. Since objective and eyepiece are only $f_o - f_e$ apart, it has the advantage of being short. It has a small angular field of view, however, and now finds use only in low-magnification, inexpensive telescopes, often mounted in pairs to make binoculars, ordinarily referred to as "field glasses." Figure 21-11B shows an added lens that inverts the image formed by the objective, so that the world is oriented properly when seen through the eyepiece. The penalty for this advantage is in the added length and weight of the instrument. Figure 21-11C shows one of the side-by-side telescopes of "prism binoculars." These turn the image right-side-up and put the left and right sides back where they belong, by reflecting the rays four times from the surfaces of 45° prisms. This "folding" of the rays has the added advantage of making the telescopes shorter and less cumbersome.

A real image can be formed by a concave mirror, as well as by a converging lens, and the majority of astronomical telescopes use such mirrors as their objectives. Figure 21-12 shows three of the many types of reflecting telescope. The most common is the Newtonian (Fig. 21-12A). A small flat mirror at 45° to the axis reflects the converging cone of rays to the side, where a photographic plate may be placed at the focus, or the image examined through an eyepiece. A reflecting telescope has no chromatic aberration, but the parabolic mirror of the Newtonian is subject to off-axis aberrations; the diagonal secondary mirror blocks some light, and, with its necessary supports, also produces additional diffraction effects in the image.

The Cassegrain reflector (Fig. 21-12B) requires a hole in the center of the parabolic primary mirror, and a small convex secondary mirror. It has all the minor defects of the Newtonian, mentioned above, but

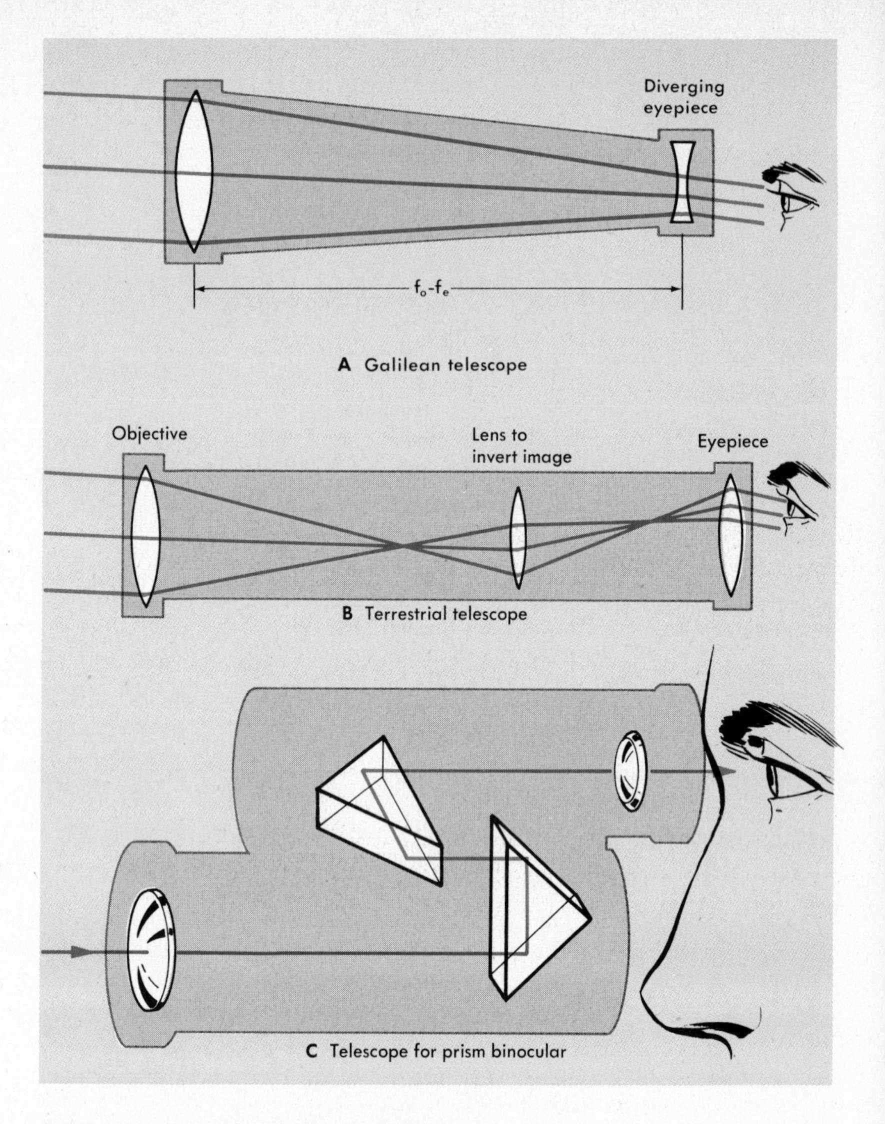

FIG. 21-11 Three types of refracting telescope.

these are balanced by advantages. It is convenient to have the eyepiece (or a photographic plate) on the axis of the telescope just below the mirror. The Cassegrain also gives a long focal length in a short telescope. The convex secondary mirror reflects the converging rays from the primary, so that they converge much less rapidly, in a narrower cone. On the schematic diagram of Fig. 21-12B, the effective focal length is about three times the length of the telescope.

A spherical primary mirror has none of the off-axis coma and astigmatism that limit the performance of the parabolic mirror—but spherical

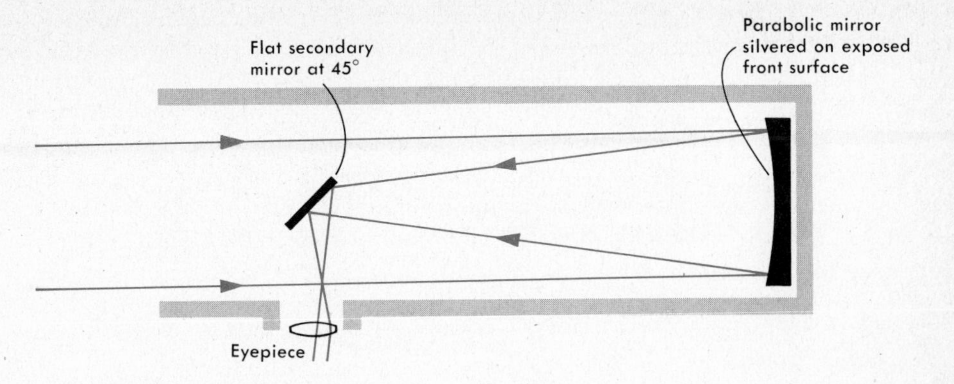

A Newtonian Reflector

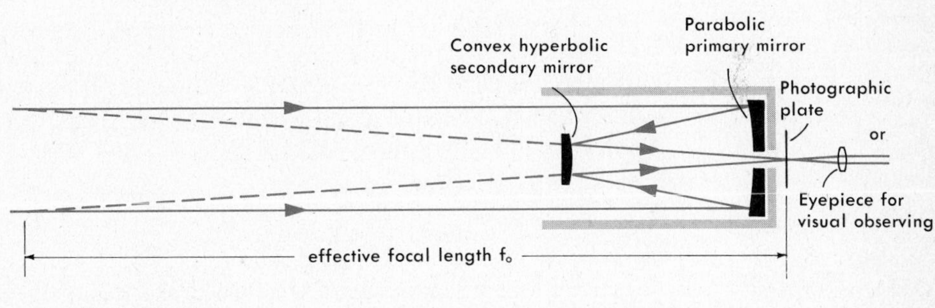

B Cassegrain

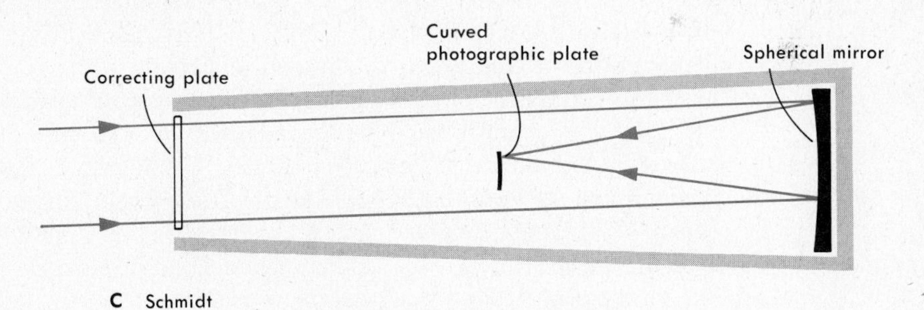

C Schmidt

FIG. 21-12 Three types of reflecting telescope.

aberration prevents its direct use in a telescope. The Schmidt telescope uses a correcting plate at the center of curvature of the mirror to eliminate the spherical aberration. This is a *very* weak, thin lens with a nonspherical curvature, that corrects the spherical aberration of the mirror. With no off-axis aberration, the Schmidt can photograph a wide angle of the sky; but it is a long instrument (twice the focal length of the mirror) and the photographic plate must be curved to a spherical surface.

The interference of light waves puts a limit on the ability of any telescope to resolve the details of an object, as with an eye or a camera.

Let us consider a telescope of $f_o = 35$ in. The star images formed by this objective are really diffraction patterns given by the same $r = 1.22\lambda l/d$ we used before. Assume the diameter of the objective (d) is 2.5 in., and we already have that $f_o = 35$ in., to give us l. For visual use, $\lambda = 5.5 \times 10^{-5}$ cm, which must be converted to inches to match the other dimensions: $\lambda = 5.5 \times 10^{-5}/2.54 = 2.2 \times 10^{-5}$ in. We are probably most interested in the minimum angular separation α that this telescope can resolve. This is $\alpha = r/l$ rad, or $\alpha = 1.22\lambda/d$. Then

$$\alpha = \frac{1.22 \times 2.2 \times 10^{-5}}{2.5} = 1.07 \times 10^{-5} \text{ rad}$$

$$= 1.07 \times 10^{-5} \times \frac{360}{2\pi} \times 60 \times 60 = 2.2 \text{ seconds.}$$

Using this objective in a telescope of any design, one could not be sure if a pair of stars closer than 2.2 seconds were really a pair, or only one. These same limitations apply to both refractors and reflectors.

21-8 Microscopes

In its basic principle, the microscope is similar to the telescope. The objective forms a real image of the object, and this image is examined with the eyepiece. Figure 21-13 sketches the optics of a microscope; the basic difference from a telescope is that the small object is very close to the objective, so that $p \approx f_o$. The total magnification is the product of the magnification of the objective and the magnification of the eyepiece. The distance (q) from the objective to its real image is fixed by the design of the microscope, and is fairly well standardized at 180 mm. On such a microscope, if we use an objective of $f_o = 4$ mm and an eyepiece of $f_e = 25$ mm, we have that q/p for the objective is approximately $M_o = \frac{180}{4} = 45$. (p is not exactly f_o, but the difference is very small.) We have used $M = 25/f_e$ (in cm) $+ 1$ as the magnification of a simple magnifier like an eyepiece, when the virtual image is 25 cm from the lens or eye. Most microscopists prefer to focus so that the virtual image is at infinity—that is, the rays from any image point enter the eye all parallel, so that a *relaxed* eye will bring them to a focus on the retina. For this, he must sacrifice a little magnification: The $+1$ is lost, and $M_e = 25/f_e$ (in cm). Thus the total magnification $M = M_o \times M_e = (\frac{180}{4}) \times (25/2.5) = 450$.

Microscopes, like other optical instruments, are limited in the detail they are able to resolve. In telescopes, the rays from the distant object point are parallel as they enter the objective; in a microscope the rays from an object point (say, the head of the worm in Fig. 12-13) diverge strongly. This has an effect on the interference of the light waves; and for other reasons too mathematical to consider here, it is more realistic to omit the factor 1.22 we have used before. So, for a microscope we must

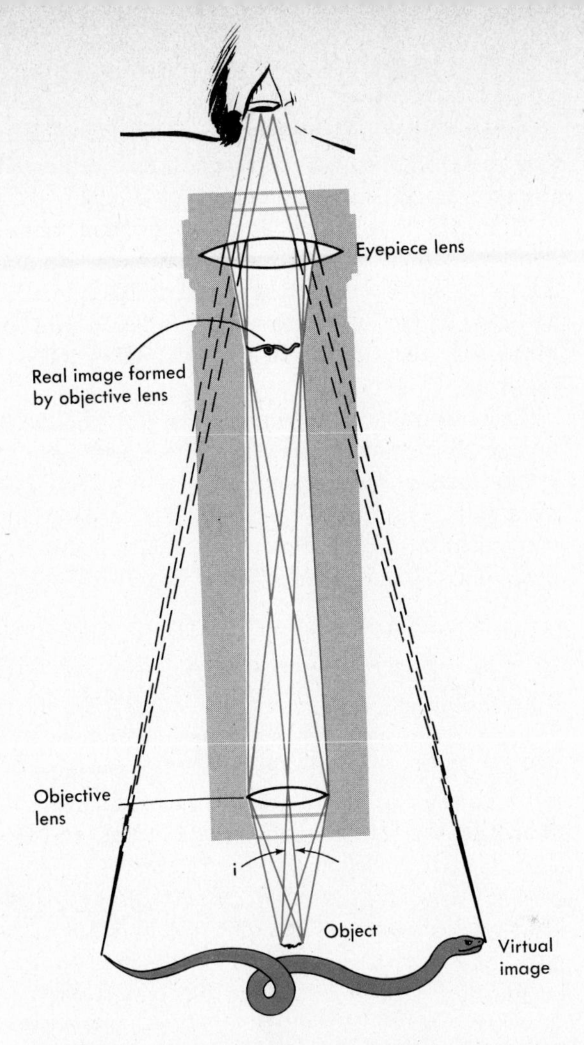

FIG. 21-13 The paths of light rays and the formation of images in a microscope.

use a different expression for the minimum separation between distinguishable details in the object:

$$r = \frac{\lambda}{2n \sin i},$$

where n is the index of refraction of the medium between object and objective, and i is half the angular size of the objective as seen from the object (see Fig. 21-13).

The expression $n \sin i$ is called the *numerical aperture* of the objective. Its numerical aperture and its focal length or magnification (q/p) are generally marked on each objective. For low- and medium-power objectives, air lies between object and objective, so $n = 1$. Many high-magnification, short f_o objectives, however, are designed to be used with a drop of oil filling the space between the slide and the objective lens. For these *oil-immersion* objectives, r is decreased, and the resolution thereby improved, by the larger n of the oil.

Consider a microscope whose oil-immersion objective is marked "90X-NA 1.30," used with a 10X eyepiece. The total magnification is of course $90 \times 10 = 900$. The smallest detail it can resolve is $\lambda/2n \sin i = \lambda/2\text{NA} = 5.5 \times 10^{-5}/(2 \times 1.30) = 2.1 \times 10^{-5}$ cm. This assumes visual use, with the wavelength λ approximately that to which the eye is most sensitive.

One obvious way to increase the resolving power is by using illumination of shorter wavelength. For this reason, photomicrographs (photographs taken through a microscope) are often taken by short-wavelength ultraviolet. Later, we will look at the electron microscope, whose high magnification and resolution are made possible by using wavelengths thousands of times shorter than visible wavelengths.

21-9 Spectroscopes

We have already noted (Fig. 19-9) that the index of refraction of glass is different for light of different wavelengths. This fact has enabled scientists to analyze the wavelengths present in light from different sources, and by this means to probe into the structure of atoms and the nature and composition of stars. One device for such analysis is the *prism spectroscope*, shown in Fig. 21-14. Figure 21-14A shows a spectroscope without its

FIG. 21-14 The principle of the prism spectroscope.

A

B Monochromatic light

C Light composed of two colors

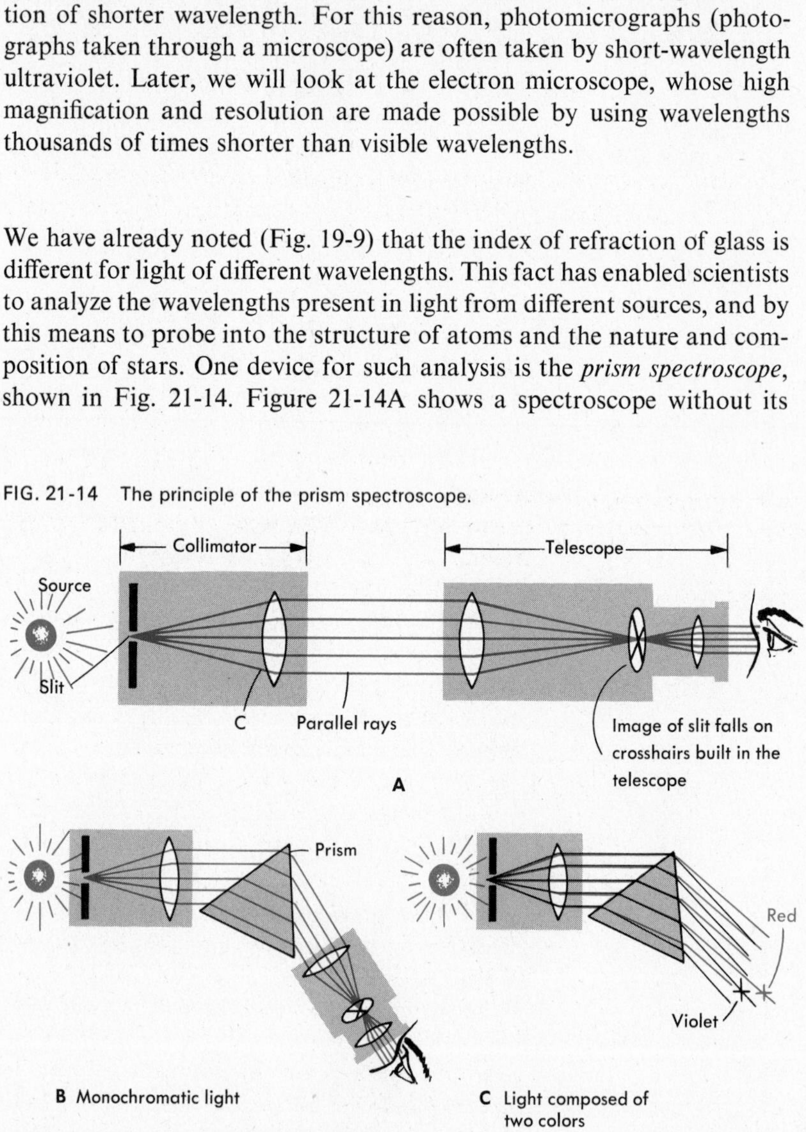

prism. Light from a source that is to be investigated falls on a narrow slit; the slit is placed at the principal focus of a lens C ("C" for *collimator*, a device to make light rays or any other things parallel), so that after being refracted by C the rays from each point on the slit emerge parallel. These parallel rays fall on a telescope objective that forms an image of the slit at its principal focus. The image of the slit is examined through an eyepiece, as with any telescope. The telescope pivots on a graduated circle, and in order to align it properly it has cross hairs built in it at the focus of the objective. The telescope can then be moved until the slit image falls exactly on the cross hairs when observed through the eyepiece.

Now, if we place a prism (usually 60°-60°-60°) between the collimator and the telescope (Fig. 21-14B), the light will be deviated, and we will have to swing the telescope around to make the slit image fall on the cross hairs again. If the light is all one single wavelength (*monochromatic*), there will be only one slit image. However, if (as in Fig. 21-14C) the light is a mixture of two definite wavelengths that we can call red and violet, the violet will be deviated more by the refraction of the prism, and we will see two separate slit images. To avoid confusion in the drawing, only the cross hairs of the telescope have been shown; in the position marked V they are on the violet image, and in position R, on the red image.

Light such as that from an ordinary incandescent lamp contains *all* wavelengths, and viewing this light through a spectroscope we see an infinite number of slit images side by side. These overlapping images result in a *continuous spectrum* ranging from the deepest red our eyes can see, to the extreme violet. In sunlight certain wavelengths are missing, so if sunlight illuminates the slit we see a continuous spectrum crossed by many dark lines, each dark line representing the slit image that is missing because that particular wavelength has been absorbed in the sun's own atmosphere. Such spectra are *dark-line absorption spectra*. A light such as a neon sign will be seen as a series of bright lines of many colors, most of them red and orange. This indicates that the neon gas emits light composed of only certain specific wavelengths. For each wavelength the slit image will be seen as a bright line of the corresponding color. This type is a *bright-line spectrum*. In a later chapter we will see the reasons why different sources emit different types of spectrum.

Most spectroscopes now use gratings rather than prisms for separating the various wavelengths present in a light source. They serve the same purpose as prisms in deviating light of different wavelengths through different angles that can then be accurately measured. The simple transmission grating spectroscope is sketched in Fig. 21-15A. The principle of its operation has already been discussed in Chapter 20, Section 4, and the directions in which the reinforcement occurs are illustrated in Fig. 20-7.

The opaque strips separating the slits of a transmission grating necessarily waste a great deal of the incoming light by reflecting or scattering it back. The *reflection grating* (Fig. 21-15B) is much more

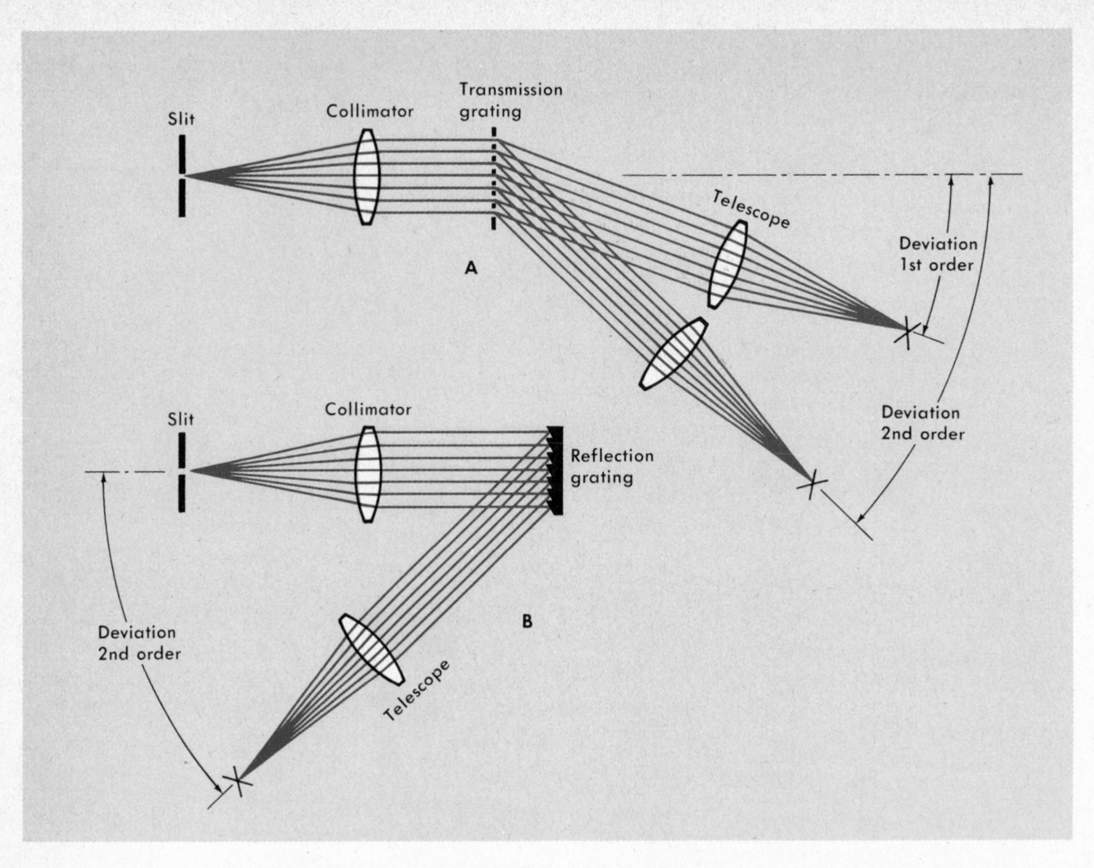

FIG. 21-15 Transmission and reflection gratings as used in a spectroscope.

efficient. It is generally ruled with an accurately shaped diamond point, in the soft aluminum of a mirror surface. Each reflecting strip serves the same purpose as one of the slits in a transmission grating, with little light wasted.

Most spectroscopy is done photographically, as we might with the instruments of Fig. 21-15 by placing a strip of film where the cross-hairs are indicated, at the focus of the telescope objective. This makes it a spectro*graph* rather than a spectro*scope*. Most research spectrographs carry complications a step further by ruling the reflection grating on the surface of a concave mirror. The collimator and telescope are no longer necessary; the grating itself serves a double purpose in also focusing the slit images directly on a curved strip of photographic film.

Questions

(21-2) **1.** From your experimentation with Fig. 21-4, does the blind spot lie noseward from the fovea, or to the outside of it?

2. Measure the distance from eye to page (Fig. 21-4) when the image of the vanishing letter is about centered on the blind spot. From this and whatever other data you need, find the approximate distance between fovea and blind spot. (The focal length of the eye is about 1.7 cm.)

3. A nearsighted man cannot focus clearly on objects farther away than 60 cm. What focal length lens will enable him to see clearly objects at a great distance?

4. A farsighted man cannot focus clearly on objects closer than 48 in. away What focal length lens will permit him to read a paper at a distance of 18 in?

5. A nearsighted person wears glasses of $f = -50$ cm. Through these glasses he has perfect distant vision but cannot focus clearly on an object closer to his eyes than 25 cm. How close can he bring an object and still see it clearly if he takes his glasses off?

6. An elderly man who has lost nearly all ability to accommodate wears bifocal glasses. The upper part, through which he looks at distant objects, has a focal length of $+50$ cm. What is the focal length of the lower part of his glasses, through which he can read a book held 30 cm from his eyes?

7. Following are the focal lengths of some lenses. Give their strength or power, in diopters: (a) 20 cm, (b) -200 cm, (c) 5 in., (d) -25 mm.

8. Lenses have these focal lengths: (a) -50 mm, (b) 20 in., (c) -40 cm, (d) 300 cm. Give their powers in diopters.

(21-3) 9. A lens has a focal length of 85 mm. What is the effective diameter of the used portion of the lens with a stop set at (a) $f/4.5$? (b) $f/12.7$?

10. A lens of $f = 135$ mm is set at (a) $f/5.6$; (b) $f/16$. What is the effective diameter of the lens in each of these?

11. How does the amount of light admitted at $f/4.5$ compare with that admitted at $f/12.7$?

12. How does the amount of light admitted at $f/5.6$ compare with that admitted at $f/16$?

13. A camera with a 50-mm focal length has a built-in exposure meter that calls for $\frac{1}{250}$ s at $f/2.8$. The photographer wants to photograph the same scene with another camera of $f = 105$ mm at $f/4.5$. What should the length of the exposure be?

14. A built-in exposure meter on a camera of $f = 85$ mm calls for an exposure of $\frac{1}{100}$ s at $f/3.5$. To photograph the same scene with another camera with an $f = 200$ mm lens at $f/8$, what exposure would be needed?

15. In photographing a football line from the sidelines, a photographer accurately focuses on the center. In order to picture both ends as clearly as possible, which would be better, $\frac{1}{400}$ s at $f/2.8$, or $\frac{1}{100}$ s at $f/5.6$? Why?

16. A tripod-mounted camera is used to photograph a garden on a still day. Assuming the camera to be focused on plants in the middle distance, would it be better to use $\frac{1}{12}$ s at $f/12.7$ or $\frac{1}{100}$ s at $f/4.5$? Explain.

(21-4) 17. A perfect lens 10 cm in diameter, $f = 60$ cm, is used to photograph a group of stars, which can be considered to be point objects with $\lambda = 6 \times 10^{-5}$ cm. (a) What is the *diameter* of the central spot on these star images? (b) What is the diameter of the central spot of the star images with another lens, of 2-inch diameter and $f = 12$ in? (c) Rewrite the formula for r given in the text, using the f-number of the lens.

18. A bright blue-white star emitting light of $\lambda = 5 \times 10^{-5}$ cm is photographed with an optically perfect lens 10 cm in diameter, and 150 cm focal length. (a) What is the *diameter* of the central spot of the star image? (b) Restate the expression for r given in the text, using the f-number of the lens. (c) The Yerkes Observatory telescope lens is 40 inches in diameter, and has a focal length of 63.5 ft. It is used for star photography like any camera lens. What is the diameter of the central spot on a star image formed by this lens (at $\lambda = 5 \times 10^{-5}$ cm)?

19. What is the minimum separation (in seconds) of a double-star pair that can be resolved by the telescopic camera of Question 17?

20. What is the minimum separation in seconds of a pair of stars that can be resolved by the Yerkes refractor? (See Question 18(c).) Take $\lambda = 6 \times 10^{-5}$ cm.

(21-5) **21.** A referee in the end-zone holds up two fingers. How far apart must the fingers be if they are to be distinguished as separate by an observer at the other end of the field (100 yards distant)?

22. In order to be distinguished, how far apart must two specks be, on a photograph held at a distance of 25 cm from the eyes?

(21-6) **23.** What is the focal length (a) in centimeters and (b) in inches, of a simple magnifier that has a magnification of 6 (often expressed as "6-power" or "6X")?

24. What is the focal length (a) in centimeters and (b) in inches, of a 10X simple magnifier? (See Question 23.)

(21-7) **25.** An astronomer uses an eyepiece whose focal length is 32 mm on a telescope with an objective focal length of 2.0 m. What is the angular magnification?

26. What is the angular magnification of a telescope whose objective has a focal length of 48 in., with an eyepiece whose focal length is 0.5 in?

27. An $f/8$ telescope objective has a diameter of 15 cm. What is the focal length (in mm) of an eyepiece to give a magnification of 60?

28. An eyepiece is to give a magnification of 40 when used with an $f/15$ objective 4 inches in diameter. What is its focal length (in inches)?

29. (a) What is the angular resolution of the telescope objective in Question 27? (b) Using the eyepiece specified in Question 27, how far apart would a pair of barely resolvable stars appear to be? (c) Would the observer's eye be able to resolve them? (d) What focal-length eyepiece would enable an observer to take full advantage of the resolving power of the objective? What would the resulting magnification be?

30. (a) What is the angular resolution of the objective of Question 28? (b) Using the eyepiece specified in Question 28, how far apart would a pair of stars appear to be, if they were barely resolvable by the objective? (c) Would the observer's eye be able to distinguish these as separate stars? (d) If the observer wants to be able to see all the detail possible with this objective, what is the maximum length of the eyepiece that he should use?

(21-8) **31.** What is the magnification of a microscope with an objective of $f_o = 2.0$ cm, using an eyepiece of $f_e = 5.0$ cm?

32. A microscope has an objective whose focal length is 0.75 cm; the focal length of the eyepiece is 2.5 cm. What is its total magnification?

33. Details of about 1 μ ($= 10^{-6}$ m) are to be examined with a microscope. An $f_o = 4$ mm, NA $= 0.85$ objective is used, with a 5X eyepiece. (a) Can the objective resolve detail this fine? (b) What is the magnification of this microscope? (c) At the nominal least distance of distinct vision (25 cm), what is the angle between two points 1 μ apart? (d) Under the magnification of part b, would the microscopist's eye be able to distinguish these details?

34. A high-power oil-immersion objective is marked 90X, NA 1.30. (a) What is the separation of the finest detail resolvable by this objective? (b) What is the angle between such barely resolvable points at the 25-cm nominal least distance of distinct vision? (c) What total magnification would be required, to make the apparent angle the 1.2 minutes resolvable by the eye? (d) Would the greater magnification of a shorter-focus eyepiece make smaller detail observable?

(21-9) **35.** A grating with 4000 lines/cm is used in a spectrograph to examine light of $\lambda = 4960$ Å. (a) Through what angle from its zero-order position must the spectroscope be rotated to see the first-order line? (b) the second-order line? (c) the third-order line?

36. A grating spectrograph is used to examine light of $\lambda = 6200$ Å. The grating has 5000 lines/cm. The zero-order image will of course be seen when the collimator and telescope are in line. From this position, (a) through what angle must the telescope be turned to see the first-order line? (b) Is it possible to see the second-order line? If so, through what angle must the telescope be turned? (c) Same as part b, for third-order.

37. One component of the light emitted by incandescent hydrogen is the F line, $\lambda = 4861$ Å. (a) What is the highest order F line that can be seen through a spectroscope using a grating of 5600 lines/cm? (b) Will this line appear to be of the same color in each order?

38. The "C" line is a monochromatic component of the light emitted by incandescent hydrogen gas and has a wavelength of 6563 Å. (a) What is the highest order C line that can be observed with a spectroscope using a grating with 6000 lines/cm? (b) Will this particular line appear to be of the same color in each order?

22

The Energy Quantum

22-1 Light Emission by Hot Bodies

We know that in order for them to emit enough light to be seen, material bodies must be heated above a certain temperature. Hot radiators (100°C) emit radiation that we can feel with our cold hands in the winter but none that our eyes can see. The heating elements of an electric range (at a low heat of about 750°C) glow with a faint reddish light that can be seen only if the kitchen is not too brightly illuminated. The filament of an electric light bulb (about 2300°C) emits an intense yellow light. We can thus see qualitatively that the intensity of the radiation emitted by a heated body increases as its temperature increases and that the wavelength of the most intense radiation shifts with increasing temperature from the red toward the blue end of the spectrum. Figure 22-1 shows how the intensity of radiation varies with wavelength at different temperatures.

Studies of experimentally determined curves such as these led to two fundamental laws governing radiation from hot solids and liquids.

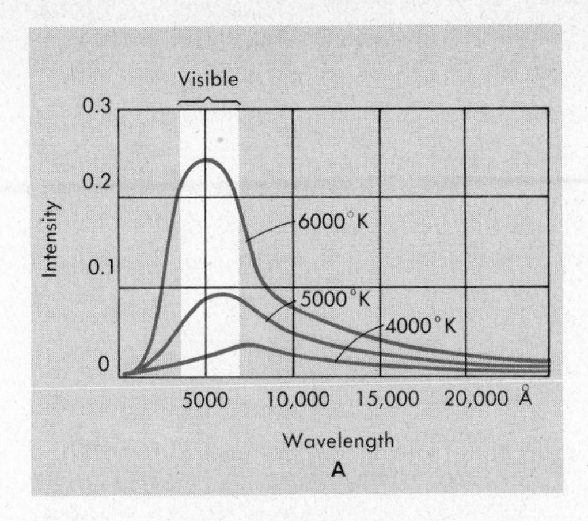

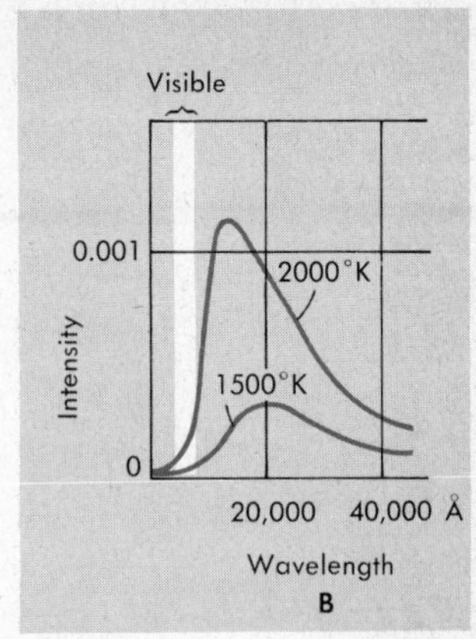

FIG. 22-1 Intensity distribution in the emission of radiation by bodies at different temperatures. The upper curve in (A) corresponds to the surface temperature of the sun—a large fraction of the total energy is in the visible part of the spectrum. The upper curve in (B)—notice that the scale is different from that of (A)—shows only a small fraction of visible radiation, most of which is red. In the 1500°K curve in (B) practically all the radiation is in the invisible infrared.

Wien's law: The wavelength of maximum intensity is inversely proportional to the absolute temperature of the emitting body.

In other words, the hotter the body, the shorter the wavelength. Wien's law can be put in quantitative numerical form by using the relationship $T\lambda_{\max} = 0.29$ cm-°K. For the 100°C radiator (373°K) we find that $\lambda_{\max} = 0.29/373 = 78 \times 10^{-5}$ cm or 78,000 Å. This wavelength at which the radiator radiates most intensely is 10 times as long as the longest deep-red wavelength of visible light; it should not be surprising that it does not glow in the dark!

The Stefan-Boltzmann law: The total rate (per unit area) of emission of energy of all wavelengths is directly proportional to the fourth power of the absolute temperature.

Quantitatively, this is $E = 5.67 \times 10^{-8}\ T^4$ W/m². Our 100°C radiator can again serve as an example. For this relatively cool body, $5.67 \times 10^{-8} \times (3.73 \times 10^2)^4 = 1100$ W/m² or 0.11 W/cm². Note that this total rate of emission, calculated for different temperatures, is proportional to the areas under the different temperature curves in Fig. 22-1.

22-2 Infrared and Ultraviolet Radiation

Just as in acoustical phenomena a human ear can hear only the sounds within a certain frequency (or wavelength) interval, so the human eye can see only the light within narrow limits of frequencies (or wavelengths). Radiation with wavelengths *longer* than that of red light (about 7.5×10^{-5} cm) is known as *infrared radiation*. It is also often called "heat radiation," since it is emitted from hot bodies (such as a room radiator) that are not yet hot enough to be luminous. In fact, heat rays are emitted by all material bodies no matter how low or high their temperatures are, but, according to the Stefan–Boltzmann law, their intensity falls very rapidly with the temperature.

Ultraviolet radiation has wavelengths *shorter* than the blue-violet end of the spectrum (about 4×10^{-5} cm), and this radiation becomes more and more important with the increasing temperature of the emitting body. While the ordinary electric light bulb (at 2300°C) does not emit any ultraviolet radiation to speak of, the sun, which has a surface temperature of about 6000°K, emits an appreciable amount of ultraviolet. The atmosphere is nearly opaque to wavelengths shorter than about 3000 Å (3×10^{-5} cm), but enough ultraviolet radiation of longer wavelength gets through to sunburn or tan the exposed parts of the human skin. Some stars have a surface as hot as 30,000°K. At such high temperatures, their most intense radiation is at a wavelength less than 1000 Å, far into the invisible ultraviolet.

22-3 The Ultraviolet Catastrophe

During the last decade of the nineteenth century the British physicist and astronomer Sir James Jeans tried to find a theoretical explanation for the observed distribution of energy among the different wavelengths radiated by hot bodies (Fig. 22-1). Thirty years earlier Maxwell had based his gas-law calculations on the assumption that the kinetic energy of gas molecules was (on the average) distributed equally among them, which led to his successful explanation of the distribution of molecular velocities (Fig. 12-7). Jeans worked from a similar assumption that the energy of radiation would be equally distributed among all the possible wavelengths.

To approach this idea in a simple way, he imagined what might happen to radiation enclosed in a cubical box whose walls were perfect mirrors. Figure 22-2A shows two opposite walls with standing waves reflecting between them. As with a vibrating string or the vibrating air columns in an organ pipe, the only waves that can persist are those with nodes at the walls, so the separation l must contain an integral number of half wavelengths. Figure 22-2A sketches some of the possible standing waves—there is no limit to the integral number of half-waves possible, and the drawing could be continued on to the right toward an infinite number.

The black dots represent particles of coal dust or some other hypothetical absorbers that can absorb some of the energy at one wavelength and reradiate it at another. This presumably goes on until all the radiation energy is divided equally among all of the infinite number of possible standing waves. This would naturally give each mode of vibration an infinitely small share of the total. To make the point, however, the artist has taken the liberty of giving each mode a small but finite energy in

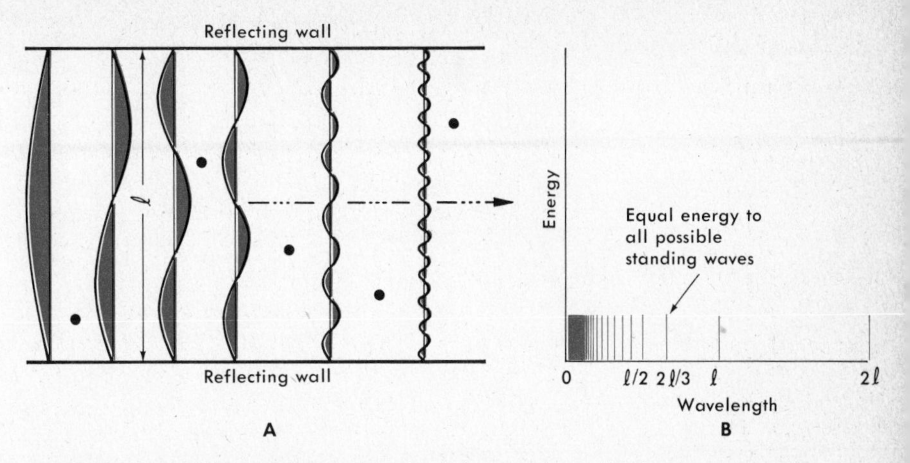

FIG. 22-2 Some of the infinite number of standing waves possible between two walls of "Jeans's cube." Part B shows the distribution of energy with wavelength, if each standing wave received an equal share of the total energy.

Fig. 22-2B. We can see that practically all the energy will be concentrated in the shortest imaginable wavelengths, where the different λ's are bunched infinitely close together.

This result, predicted by the classical laws of physics, is *not* what happens. If it were, the radiation emitted by glowing coals in the fireplace would all be converted to invisible but deadly γ rays before it could leave the grate. Other physicists worked on the problem, but with equally unsatisfactory results. The situation was known as "Jeans's paradox" or the "ultraviolet catastrophe." The problem was not to be solved until a radically new and different concept was introduced.

22-4 The Birth of the Energy Quantum

Just before the end of the last century, in Christmas week, 1899, at a meeting of the German Physical Society in Berlin, the German physicist Max Planck (1858–1947) presented his views on how to save the world from the perils of Jeans's ultraviolet catastrophe. His proposal was as paradoxical as Jeans's paradox itself, but it was much more helpful. According to Planck's view, the rays of the sun that pour into a room through the windows or the light of a table lamp do not represent a continuous flow of light waves, but rather a stream of individual *photons* (Fig. 22-3). A photon is a unit of electromagnetic radiation (you might imagine it as being a finite train of waves) having a certain wavelength and frequency, and having also a certain definite amount, or *quantum*, of energy.

To each photon frequency there corresponds a definite quantum of energy, and it is just as nonsensical to talk about three-quarters of a quantum of green light as it would be to talk about three-quarters of an atom of copper. Planck proposed that photons of different frequencies

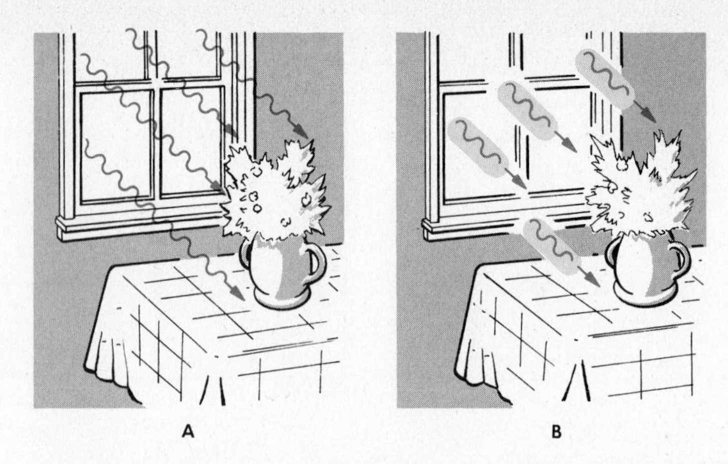

FIG. 22-3 (A) The old view of light as continuous wave trains whose amplitude increases with increasing intensity of the light; (B) the new view which shows light as a stream of "photons," each having a certain "quantum" of energy. The number of photons per second determines the intensity of the light.

carry different quanta of energy, and that ***the energy of a photon is directly proportional to its frequency.*** Writing f for the frequency of the photon and E for the quantum of energy it carries, we can express Planck's assumption as

$$E = hf$$

where h is a universal proportionality constant known as *Planck's constant*, which has the value of 6.63×10^{-27} erg-s or 6.63×10^{-34} J-s.

How does Planck's assumption of light quanta help to remove the troubles of the ultraviolet catastrophe? To understand this, let us look further into the consequences of the basic assumption that $E = hf$, that is, that radiant energy such as light flies about in packets of energy, the sizes of which are proportional to the frequency of the radiation. The long wavelength waves of radio have low frequencies and hence their quanta of energy are small. Visible light, with frequencies a billion times greater, comes in quanta whose energy is also a billion times greater. Energy must be absorbed and emitted in whole quanta, exactly —no fractional parts of quanta are allowed.

The difference between the energy demands of longwave (low-frequency) and of shortwave (high-frequency) radiation has a very important effect on the application of the equipartition principle. Planck's solution of the problem can be considered to take the form of a probability distribution: The vibrations with high demands have a very small chance of having their demands satisfied; low-frequency radiation, which asks but little, has a very good chance of getting it. In other words, both ends of Planck's energy distribution curve (Fig. 22-4) approach zero; at the short-wavelength, high-frequency end, the radiation has prac-

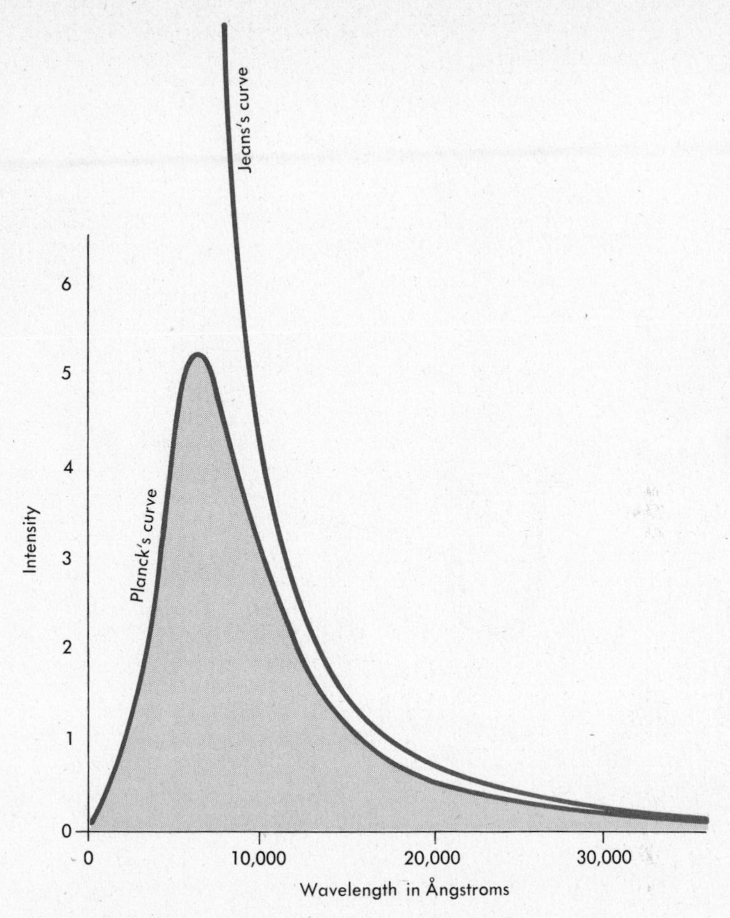

FIG. 22-4 Jeans's curve (wrong) and Planck's curve (right). These curves were theoretical attempts to derive the experimental curves shown in Fig. 22-1.

tically no chance of receiving a great deal; and at the long-wavelength end, the radiation stands a very good chance of receiving practically nothing. So instead of looking like the original Jeans's curve, the distribution curve obtained by Planck took a much more reasonable shape, in perfect agreement with the experimentally determined curves of Fig. 22-1. Further studies of the problem enabled Planck to derive from his theory the thermal radiation laws of Wien and Stefan–Boltzmann.

22-5 The Puzzle of the Photoelectric Effect

A few years after Max Planck introduced the notion of quanta in order to circumvent the difficulties of Jeans's ultraviolet catastrophe, a new and, in a way, much more persuasive argument for the existence of these packets of radiant energy was put forward by Albert Einstein. Einstein based his argument on the laws of the *photoelectric effect*, i.e., the ability

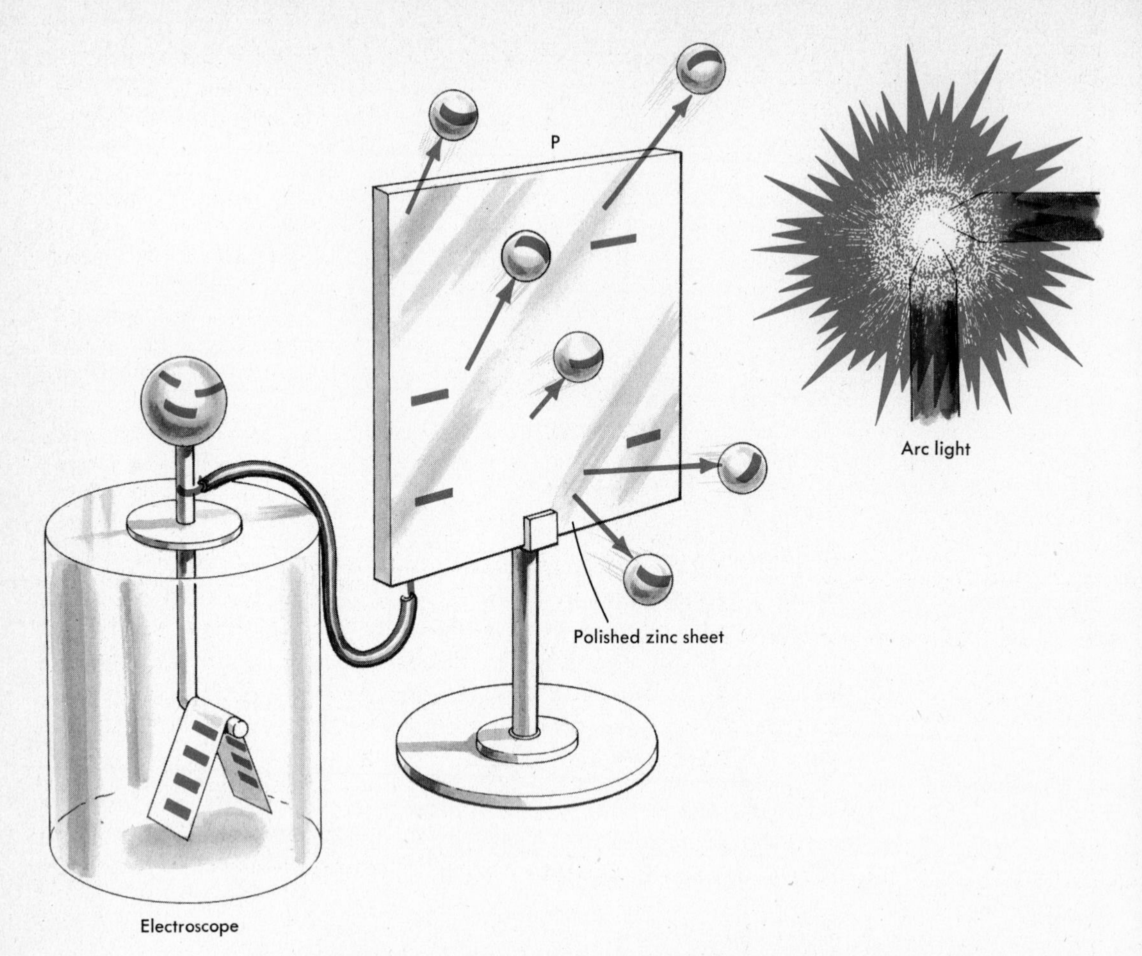

FIG. 22-5 A zinc plate losing electrons when exposed to ultraviolet radiation from an arc light.

of various materials to emit free electrons when irradiated by visible or ultraviolet light. An elementary arrangement for the demonstration of the photoelectric effect is shown in Fig. 22-5. A freshly sandpapered piece of zinc P is attached to an electroscope and given a *negative* charge. If light from an electric arc is allowed to fall directly on the zinc plate, the electroscope leaves will come together, showing that the plate has lost its charge. The closer the arc light is to the plate, the more rapidly will the charge be lost; conversely, as the experiment is repeated with the arc removed to greater and greater distances, the charge will be lost more slowly. However, we find that if a sheet of ordinary glass is put between the arc light and the zinc, the zinc will retain its negative charge, even if the arc is brought very close. Also, we find that if the zinc is originally given a *positive* charge, the arc light will have little apparent effect on the rate at which the charge is lost.

More careful experiments were performed, in which monochromatic radiation was used and in which the numbers and kinetic energies of the ejected electrons were carefully measured. These experiments gave results according to the following rules:

1. For a given frequency of incident radiation, the number of electrons produced by the photoelectric effect is directly proportional to the *intensity* of the radiation, while the velocity (or energy) of the emitted electrons remains the same, no matter what the intensity.
2. When radiation of different frequencies is used, the energy of the emitted electrons increases with increasing frequency. The graph of the kinetic energy of the emitted electrons against the radiation frequency is a straight line (Fig. 22-6).
3. When different metals are used, the graphs of Fig. 22-6 are different lines, but they all have the same slope. With radiation of a frequency less than a certain *threshold frequency* (different for different metals), *no* electrons will be emitted, no matter how great the intensity of the radiation.

These straightforward observations of the photoelectric effect presented great difficulties for the understanding of electromagnetic radiation from the classical point of view. According to classical theory, a propagating wave of light or any other electromagnetic radiation is essentially a varying electromagnetic field, and the strength of the electric and magnetic forces in this field increases with increasing intensity. If the ejection of electrons from metal surfaces is due to the electric forces of

FIG. 22-6 The kinetic energy of emitted photoelectrons plotted against the frequency of the illuminating radiation, for three different metals.

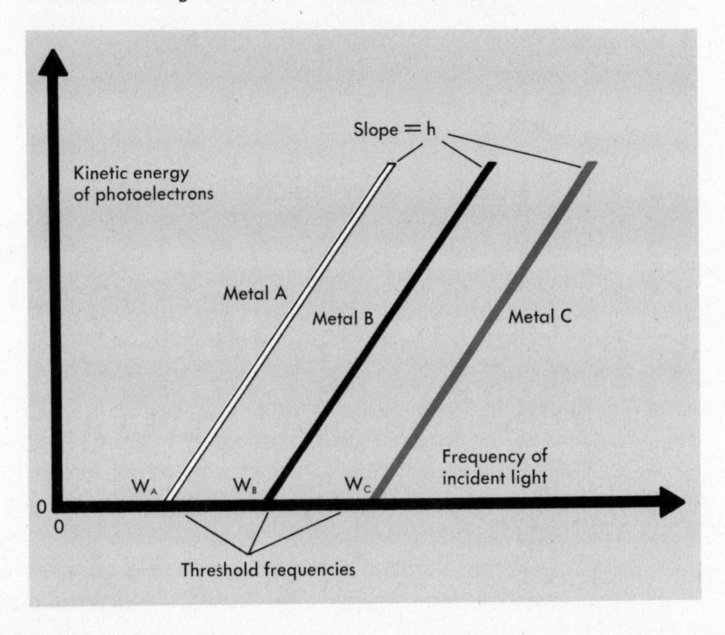

the incident waves, then the energy with which the electrons are ejected should increase with increasing radiation intensity, just as seashore bathers are thrown off their feet more violently by big waves than by small ones. This quite natural conclusion does not fit the experimental facts, however, since an increase of the intensity of the radiation increases only the number of ejected electrons and does not affect their velocity or energy at all.

All this experimental evidence, said Einstein, could be quite satisfactorily explained by Planck's new idea of energy quanta. A certain amount of energy W_{Zn} is required to pull an electron loose from the attraction of the atoms in a zinc plate. According to the old classical theories, the energy of light or ultraviolet radiation spread out in spherical waves, so the amount of energy an electron could absorb from one tiny spot on such a spreading wavefront would be negligible. Einstein argued, however, that the old classical picture does not represent what actually happens. The entire energy of the photon is absorbed in one bite by a single electron. Planck's relationship $E = hf$ tells how much energy will be in a quantum of any given frequency f.

For the zinc plate, W_{Zn} is greater than the energy associated with a photon of visible light, so that no matter how much visible light shines on the plate, no electron will receive enough energy to break loose. This is the situation with a sheet of glass screening the arc light—ordinary window glass shuts out the invisible but highly energetic ultraviolet radiation. With the glass removed, the ultraviolet radiation from the arc, being of higher frequency than visible light and hence of proportionally higher energy, is absorbed by the electrons in the plate. The energy of a photon of ultraviolet is greater than W_{Zn}, so the electrons can escape, carrying any leftover excess energy with them in the form of kinetic energy. With the plate negatively charged, the departing electrons are repelled, and the plate gradually loses its charge. A positively charged plate, however, will attract the electrons back as quickly as they escape, so there is in this case no loss of charge.

For a general energy relationship, we need to consider three terms: hf, which is the entire energy of the quantum absorbed by the electron; W, the energy required to pull the electron free from the surface; and $\frac{1}{2}mv^2$, the kinetic energy of the electron as it leaves. Simple consideration of the conservation of energy gives us

$$hf - W = \tfrac{1}{2}mv^2.$$

Now, if we graph f against $\frac{1}{2}mv^2$, as in Fig. 22-6, we should expect a straight line. (Remember that h and W are both constants.) Planck's constant h determines the slope of the line, which will therefore naturally be the same for all metals. The bottom ends of the lines on the drawing represent electrons with zero kinetic energy—i.e., electrons just barely pulled away from the attractions of their neighbors.

For an electron to be pulled off without anything left over for kinetic energy, the hf of the absorbed quantum must equal W, and the frequency at which this occurs is the *threshold frequency*. The threshold frequencies and threshold wavelengths for some representative metals are as follows:

Platinum	$\lambda = 1980$ Å	or	$f = 1.51 \times 10^{15}$ Hz
Silver	$\lambda = 2640$ Å	or	$f = 1.13 \times 10^{15}$ Hz
Potassium	$\lambda = 7100$ Å	or	$f = 4.22 \times 10^{14}$ Hz

In his classical paper on this subject, published in 1905, Einstein indicated that the observed laws of the photoelectric effect can be understood if, following the original proposal of Max Planck, one assumes that ***electromagnetic radiation propagates through space in the form of individual energy packages called photons, and that on encountering an electron, such a photon communicates to the electron its entire energy.***

This revolutionary assumption explains quite naturally the observed fact that an increase of the intensity of light leads to an increase of the number of photoelectrons, but not of their energy. More intense light means that more light quanta of the same kind will fall on the surface per second, and since a single photon can eject one and only one electron, the number of electrons must increase correspondingly. On the other hand, by decreasing the wavelength of incident light we increase the frequency and, consequently, the amount of energy carried by each individual photon, so that in each collision with a free electron in the metal these photons will communicate to it a correspondingly larger amount of kinetic energy.

As an example, let us investigate a little further the behavior of the zinc plate in Fig. 22-5. From a handbook, we can find that for zinc, W (called the photoelectric *work function*) is about 6.4×10^{-12} erg. This is the energy it must absorb if an electron is to be just barely pulled away, with nothing left over for kinetic energy; and this energy must be absorbed from a single photon. From $E = hf$, it can be determined that for a photon to have this much energy it must have the frequency

$$f = \frac{E}{h} = \frac{6.4 \times 10^{-12}}{6.63 \times 10^{-27}} = 0.96 \times 10^{15} \text{ Hz.}$$

For the corresponding wavelength, we go to $v = f\lambda$ and find its wavelength to be $\lambda = 3 \times 10^{10}/0.96 \times 10^{15} = 3.1 \times 10^{-5}$ cm, or 3100 Å. This is well beyond the visible into the ultraviolet. Radiation of wavelengths longer than this will have energy quanta too small to pull electrons free, and none will be emitted, no matter how high the intensity; shorter wavelengths will have more than enough energy, and the excess will result in kinetic energy of the electrons emitted.

FIG. 22-7 Dr. Arthur H. Compton, who visualized a collision between a photon and an electron as analogous to a collision between elastic balls. The photon shown has less energy after its collision, and a correspondingly longer wavelength.

22-6 The Compton Effect

The Planck–Einstein picture of individual energy packages, or light quanta, forming a beam of light and colliding with the electrons within matter, intrigued the mind of an American physicist, Arthur Compton (Fig. 22-7) who, being of a very realistic disposition, liked to visualize collisions between photons and electrons as similar to those between the ivory balls on a billiard table. He argued that, in spite of the fact that the electrons forming the planetary system of an atom are bound to the central nucleus by attractive electric forces, these electrons would behave almost as if they were completely free if the photons that hit them carried sufficiently large amounts of energy. Suppose that a black ball (electron) is resting on a billiard table and is bound by a thread to a nail driven into the table's surface and that a player, who does not see the thread, is trying to put it into the corner pocket by hitting it with a white ball (photon) (Fig. 22-8). If the player sends his ball with a comparatively small velocity, the thread will hold during the impact and nothing will come from this attempt. If the white ball moves somewhat faster, the thread may break, but in doing so it will cause enough disturbance to send the black ball in a completely wrong direction. If, however, the kinetic energy of the white ball exceeds, by a large factor, the work necessary to break the thread that holds the black ball, the presence of the thread will make almost no difference, and the result

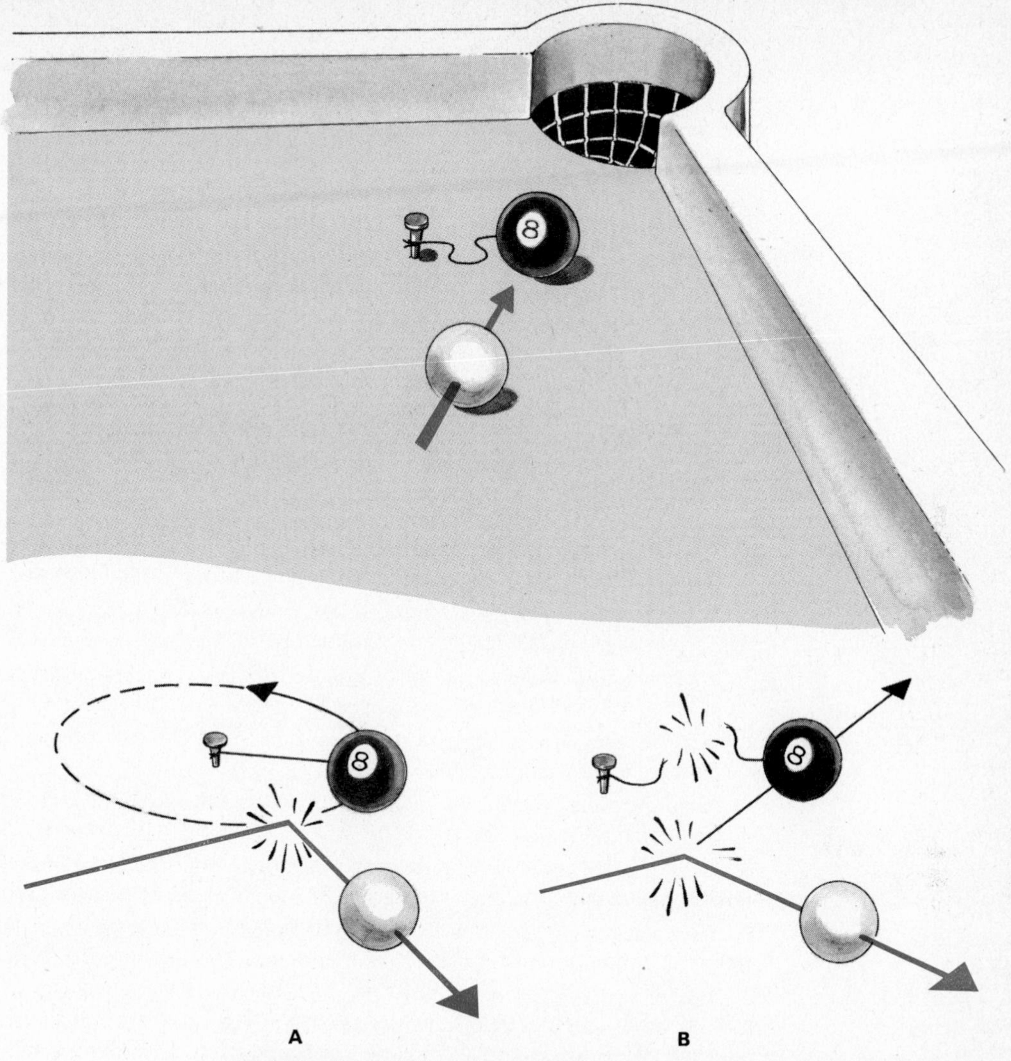

FIG. 22-8 A ball tied by a thread (black) is struck by a moving ball (white). If the white ball has little energy, the effect of the thread is important (A); if the white ball has high energy, the result will be nearly the same as though the black ball had not been tied (B).

of the collision between the two balls will be about the same as if the black ball were completely unbound.

Compton knew that the binding energy of the outer electrons in an atom is comparable to the energy of the photons of visible light. Thus, in order to make the impact overpoweringly strong, he selected for his experiments the energy-rich photons of high-frequency X rays. The result of a collision between X-ray photons and (almost) free electrons can be indeed treated in very much the same way as a collision between two billiard balls. In the case of an almost head-on collision, the black ball (electron) will be thrown at high speed in the direction of the impact, while the white ball (X-ray photon) will lose a large fraction of its energy. In the case of a side hit, the white ball will lose less energy and will suffer a smaller deflection from its original trajectory. In the case of a mere touch, the white ball will proceed almost without deflection and will lose only a small fraction of its original energy. In the language of light quanta, this behavior means that in the process of scattering, ***the photons of X rays deflected through large angles will have a smaller amount of energy and, consequently, a longer wavelength.*** The experiments carried out by Compton confirmed the theoretical expectations in every detail, and thus gave additional support to the hypothesis of the quantum nature of radiant energy.

Questions

(22-1) **1.** What is the wavelength of the most intense radiation from the surface of (a) the star Antares, at about 2600°C? (b) the star Sirius, at about 9700°C?

2. What wavelength is most intense in the radiation into space from (a) the surface of the earth at about 27°C? (b) the surface of Mercury, at about 327°C?

3. What must be the temperature of a plate that is to emit its most intense radiation at 1.2×10^4 Å?

4. What is the surface temperature of a star that emits its most intense radiation at a wavelength of 2000 Å?

5. About how many W/m^2 are radiated from (a) the surface of Sirius? (b) the surface of Antares? (See Question 1.)

6. How does the rate of energy emission per square mile from the surface of earth compare with rate from the surface of Mercury? (See Question 2.)

7. The temperature of an electric heater element is 1200°K when it draws 200 watts. What must its wattage be in order to raise its temperature to 1800°K?

8. The voltage applied to a lamp filament is adjusted until its temperature is 1500°K, at which point the filament draws 25 W. The voltage is now increased until the filament draws 50 W. What is the filament temperature now?

(22-2) **9.** Would radiation of the following wavelengths be called ultraviolet, visible, or infrared? (a) 5.1×10^{-7} m, (b) 312 mμ, or 312 nm; (c) 3×10^{-6} cm.

10. Given radiation of these wavelengths: (a) 8400 Å; (b) 630 nm or 630 mμ; (c) 6.3×10^{-8} m. Classify each as ultraviolet, visible, or infrared.

(22-4) **11.** What is the energy of a quantum of (a) a radio wave whose frequency is 1490 kHz; (a) an infrared photon of $\lambda = 2 \times 10^4$ Å; (c) a gamma-ray photon, $\lambda = 6 \times 10^{-3}$ Å?

12. What is the energy of a quantum of (a) a radio wave, frequency 870 kHz? (b) green light, $\lambda = 5500$ Å? (c) an X ray, $\lambda = 0.6$ Å?

13. What is the wavelength of a photon whose quantum of energy is 3×10^{-15} erg?

14. If a photon is to have an energy of 10^{-12} erg, what must be its wavelength?

(22-5) **15.** How much energy is needed to pull an electron away from a platinum plate?

16. How much energy is needed to pull an electron away from a sheet of silver?

17. A sheet of silver is illuminated by monochromatic ultraviolet radiation of $\lambda = 1810$ Å. (a) How much work is needed to free an electron? (b) How much energy is carried by each incident photon? (c) What is the maximum energy a photoelectrically-emitted electron can have in this experiment?

18. Yellow light ($\lambda = 5890$ Å) falls on a potassium surface. (This must be done in a vacuum to avoid oxidation of the potassium by the air.) (a) How much work is needed to pull an outer electron free from the potassium? (b) How much energy is carried by a photon of this yellow light? (c) With what kinetic energy are electrons ejected from the potassium surface?

19. If photoelectrons are to be emitted from a potassium surface with a speed 6×10^7 cm/s, what frequency of radiation must be used?

20. What frequency of radiation must fall on a silver surface in order that the photoelectrons may be emitted with a speed of 10^8 cm/s?

21. An X-ray photon of $\lambda = 2.00$ Å is scattered by an electron in a Compton interaction. The electron flies away with a kinetic energy of 3×10^{-10} erg. What is the wavelength of the scattered photon?

22. The wavelength of a photon is 0.080 nm before it is scattered by an electron in a Compton interaction. The wavelength of the scattered photon is 0.087 nm. What is the energy of the scattered electron?

23

The Bohr Atom

23-1 Bohr's Postulates

When Rutherford (at that time plain Ernest Rutherford, not yet Sir Ernest or Lord Rutherford) was at the University of Manchester performing his epoch-making experiments that demonstrated the existence of the atomic nucleus, a young Danish physicist named Niels Bohr (1885–1962) came to work with him. Bohr was impressed by Rutherford's new atomic model, in which electrons revolved around the nucleus in much the same way that planets revolve around the sun.

Rutherford's original model, however, had troubles that prevented its general acceptance. It was unable to make any predictions about the light that an atom would emit; and more serious than this was its conflict with the accepted laws of electromagnetic theory. Maxwell, in his work on electricity, had showed that an accelerated electric charge would emit energy in the form of electromagnetic radiation—and this theoretical prediction had been proved true by the experiments of Hertz. An electron revolving rapidly around a nucleus must have a continual centripetal acceleration, and this acceleration would cause a continuous loss of energy by radiation. Bohr calculated that this emission of radiation would cause the electrons in an atom to lose all their energy and fall into the nucleus within a hundred-millionth of a second! Since matter composed of atoms exists permanently, as far as we know, there was obviously something wrong here. Bohr's conclusion was that the conventional classical laws of physics must be wrong, *at least when applied to the motion of electrons within an atom.*

It is always easier to say that something is wrong than to find a way to make it right that will be in agreement with experimental evidence. Bohr's proposed solution was so unconventional that he kept the manuscript locked in his desk for almost two years before deciding to send it in for publication. When the paper appeared in 1913, it amazed and shocked the world of contemporary physics.

Bohr, in defiance of the well-established laws of classical mechanics and electrodynamics, proposed that the following rules must hold:

I. ***Of all the infinite number of mechanically possible orbits for an electron revolving about a nucleus, only a few are permitted. These are the orbits in which the angular momentum of the electron is an integral multiple of $h/2\pi$.***

II. ***While circling around these permitted orbits, the electrons do not emit any electromagnetic radiation, even though conventional electrodynamics holds that they should.***

III. ***Electrons may jump from one orbit to another, in which case the difference in energy between the two states of motion is radiated as a photon whose frequency is determined by the quantum rule $\Delta E = hf$.***

These rules allowed Bohr to interpret the heretofore unexplained regularities observed in the spectra of hydrogen and other hydrogen-like atoms, and to construct a consistent theory of atomic structure.

It had been known for many years that chemical elements, when in the form of an incandescent vapor, emitted radiation of only certain specific frequencies. Each element has its own pattern of spectroscopic lines, some of them very complex. The spectrum of hydrogen is especially simple—it has only four lines in the visible range, as shown in Fig. 23-1.

FIG. 23-1 Part of the spectrum of hydrogen atoms, in or near the visible range, indicating the color of the emitted frequencies.

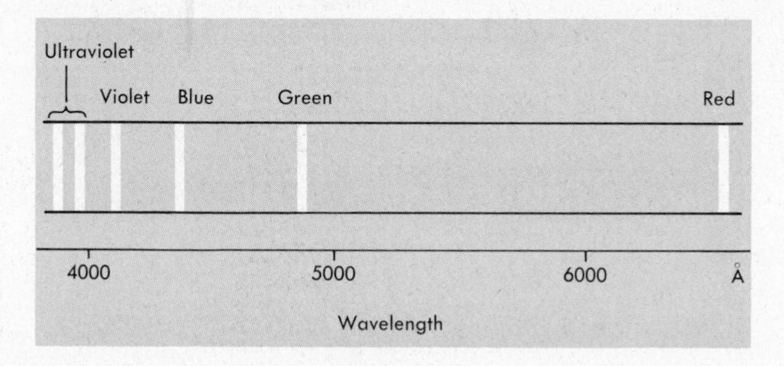

In 1885 a German schoolteacher, J. J. Balmer, discovered that the frequencies of these visible lines could be very accurately given by a simple formula:

$$3.29 \times 10^{15}\left(\frac{1}{2^2} - \frac{1}{3^2}\right) = 3.29 \times 10^{15}(\tfrac{1}{4} - \tfrac{1}{9}) = 4.57 \times 10^{14} \text{ Hz}$$

$$3.29 \times 10^{15}\left(\frac{1}{2^2} - \frac{1}{4^2}\right) = 3.29 \times 10^{15}(\tfrac{1}{4} - \tfrac{1}{16}) = 6.17 \times 10^{14} \text{ Hz}$$

$$3.29 \times 10^{15}\left(\frac{1}{2^2} - \frac{1}{5^2}\right) = 3.29 \times 10^{15}(\tfrac{1}{4} - \tfrac{1}{25}) = 6.91 \times 10^{14} \text{ Hz}$$

$$3.29 \times 10^{15}\left(\frac{1}{2^2} - \frac{1}{6^2}\right) = 3.29 \times 10^{15}(\tfrac{1}{4} - \tfrac{1}{36}) = 7.31 \times 10^{14} \text{ Hz.}$$

The actual frequencies of these lines, determined by measuring the wavelength in a grating spectrometer, and then using $f = c/\lambda$, are 4.569×10^{14}, 6.168×10^{14}, 6.908×10^{14}, and 7.310×10^{14} Hz.

In 1906, a Harvard spectroscopist, Lyman, discovered that the spectrum of hydrogen also included a series of lines in the ultraviolet, and that their frequencies could be given by a similar formula:

$$f\text{ (Lyman)} = 3.29 \times 10^{15} \times \left(\frac{1}{1^2} - \frac{1}{n^2}\right)$$

in which n took on successive integral values of 2, 3, 4, ...

Two years later, a German spectroscopist, Paschen, found another hydrogen spectrum series in the infrared, with frequencies

$$f\text{ (Paschen)} = 3.29 \times 10^{15} \times \left(\frac{1}{3^2} - \frac{1}{n^2}\right)$$

in which n took integral values of 4, 5, 6, ...

Before Bohr's work, the multiplier 3.29×10^{15} was completely empirical—it was merely the number needed to make the answers come out right, and had no theoretical justification. It was one of the jobs of Bohr's theory to show *why* the differences between the squares of these peculiarly simple fractions entered into the calculations of the frequencies of the radiations, and why the multiplier had the specific value 3.29×10^{15}.

23-2 Bohr's Orbits

The centripetal force that keeps the moon curving in its nearly circular orbit around the earth is of course the gravitational attraction between the two bodies. For an electron orbiting in a hydrogen atom, the necessary centripetal force is the electrical attraction between the negative electron and the massive, positively-charged proton that is the atom's nucleus. The amount of charge is the same for both particles; we can refer to it as e C. They are separated by the radius of the electron's orbit, r m. Coulomb's law gives us that their electrical attraction in newtons is $F = kQ_1Q_2/d^2 = ke^2/r^2$, where k has the numerical value 9×10^9. This attraction *is* the centripetal force mv^2/r, so we can write

$$\frac{ke^2}{r^2} = \frac{mv^2}{r}; \qquad r = \frac{ke^2}{mv^2}.$$

This is a perfectly good equation, but it contains *two* unknown quantities: r and v. We need another equation containing r and v, which we get from Bohr's first postulate, or rule, about the angular momentum of the electron in its orbit. Angular momentum is $I\omega$, and for a small particle like an electron the moment of inertia I is by definition mr^2. The angular velocity ω we can express as v/r. This gives for the electron's angular momentum $I\omega = mr^2 \times (v/r) = mvr$. According to Bohr, this was permitted to be *only* some integral multiple of $h/2\pi$, where h is Planck's constant. Thus

$$mvr = \frac{nh}{2\pi}; \qquad r = \frac{nh}{2\pi mv},$$

in which Bohr's *quantum number n* can be any integer. That is, $n = 1, 2, 3$, or any other whole number, each number determining one of the permitted orbits.

We now have two independent expressions for r; since they both equal r, they must be equal to each other:

$$\frac{ke^2}{mv^2} = \frac{nh}{2\pi mv}$$

$$v = \frac{2\pi ke^2}{nh} \quad \text{or} \quad \frac{1}{n} \times \frac{2\pi ke^2}{h}.$$

This gives the speed of the electron in terms of only known quantities. We can substitute this for v in either of our expressions for r, to find r in terms of known quantities. Let us take the second one:

$$r = \frac{nh \times nh}{2\pi m \times 2\pi ke^2} = n^2 \times \frac{h^2}{4\pi^2 kme^2}.$$

Does this give a reasonable value for the size of a hydrogen atom? Let us try it, with $n = 1$. Planck's constant $h = 6.63 \times 10^{-34}$ J-s;

$k = 9 \times 10^9$; m, the mass of an electron, is 9.11×10^{-31} kg; and the electron charge e is 1.602×10^{-19} C. These values give

$$r_1 = \frac{(1)^2 \times (6.63 \times 10^{-34})^2}{4\pi^2 \times 9 \times 10^9 \times 9.11 \times 10^{-31} \times (1.602 \times 10^{-19})^2}$$

$$= 5.3 \times 10^{-11} \text{ m} = 0.53 \text{ Å}.$$

This gives an atomic diameter of about 1 Å, which seems reasonable in comparison with the 4-Å diameter of the larger air molecule, estimated in Chapter 12 by an entirely different approach.

Larger orbits are also possible when n is given values of 2, 3, 4, etc. It is not necessary to refigure these orbits using all the constants included above. Everything is the same except the n's, and r is proportional to n^2:

$$r_2 = 4r_1, \qquad r_3 = 9r_1, \qquad r_4 = 16r_1, \qquad \text{etc.}$$

23-3 Electron Energies

Since the frequency (and wavelength) of an atom's radiation is related to energy through $E = hf$, it will probably be useful to calculate the energy of the electron in its permitted orbits. This must of course include both kinetic and potential energies. Kinetic energy can be readily figured, using the value of v determined in the last section:

$$\text{KE} = \tfrac{1}{2}mv^2 = \frac{m}{2} \times \left(\frac{2\pi ke^2}{nh}\right)^2$$

$$= \frac{2\pi^2k^2me^4}{n^2h^2}, \quad \text{or} \quad \frac{\pi^2k^2me^4}{n^2h^2} \times 2.$$

It was shown in an earlier chapter that the electric potential at a point r m from a charge of Q_1 C is $V = kQ_1/r$, and that the potential energy of another charge Q_2 placed at that point is $\text{PE} = kQ_1Q_2/r$. These expressions were based on the arbitrary assumption of PE = 0 at an infinite separation; for our charges of $+e$ and $-e$ the PE will be negative. But we are concerned only with *energy changes*, and these will be the same no matter what zero reference we select. So, on this basis, we have

$$\text{PE} = \frac{-ke^2}{r} = -ke^2 \times \frac{4\pi^2kme^2}{n^2h^2}$$

$$= \frac{-4\pi^2k^2me^4}{n^2h^2}, \quad \text{or} \quad \frac{\pi^2k^2me^4}{n^2h^2} \times (-4).$$

The expressions for KE and PE are identical except for the multipliers written at the end of each. The total energy, KE + PE, is thus

$$\frac{\pi^2 k^2 m e^4}{n^2 h^2} \times (-2)$$

$$= -\frac{1}{n^2} \times \frac{2\pi^2 k^2 m e^4}{h^2}$$

$$= -\frac{1}{n^2} \frac{\times 2\pi^2 \times (9 \times 10^9)^2 \times 9.11 \times 10^{-31} \times (1.602 \times 10^{-19})^4}{(6.63 \times 10^{-34})^2}$$

$$= -\frac{1}{n^2} \times 2.18 \times 10^{-18} \text{ J.}$$

Note that this energy is a *negative* number because of the way PE = 0 was chosen. For $n = 1$ it is -2.18×10^{-18}; for $n = 2$ it is only a fourth as large, or -0.545×10^{-18}; and so on. Since E_2 is a *smaller negative number* than E_1, it follows that the electron in its $n = 2$ orbit has *more* energy than in $n = 1$, and so on. As the quantum number n is increased, the electron energy becomes greater. At $n = \infty$, the energy of the electron has increased to zero.

23-4 Radiation and Energy Levels

Bohr's third postulate stated that when an electron jumps from one level to another level of lower energy, the energy difference ΔE is radiated as a photon of energy $hf = \Delta E$. Let us consider as an example an electron in a hydrogen atom, that jumps from the $n = 4$ level to $n = 2$.

$$\Delta E = E_4 - E_2$$

$$= -\frac{1}{4^2} \times 2.18 \times 10^{-18} - \left(-\frac{1}{2^2} \times 2.18 \times 10^{-18}\right)$$

$$= -\tfrac{1}{16} \times 2.18 \times 10^{-18} + \tfrac{1}{4} \times 2.18 \times 10^{-18}$$

$$= 2.18 \times 10^{-18} \times (\tfrac{1}{4} - \tfrac{1}{16})$$

$$f = \frac{\Delta E}{h} = \frac{2.18 \times 10^{-18}}{6.63 \times 10^{-34}} \times \left(\frac{1}{4} - \frac{1}{16}\right)$$

$$= (3.29 \times 10^{15}) \times 0.1875 = 6.17 \times 10^{14} \text{ Hz.}$$

Thus we see above that Bohr's work gave a convincing theoretical explanation for the puzzling multiplier 3.29×10^{15} that had to be put in as a purely empirical factor by Balmer, Lyman, and Paschen.

It is easy to go from this specific example to a general formula covering all the possible lines in the hydrogen spectrum:

$$f = 3.29 \times 10^{15} \times \left(\frac{1}{n_1^2} - \frac{1}{n_2^2}\right)$$

in which n_1 can be any integer, and n_2 can be any integer greater than n_1. For $n_1 = 1$ we obtain the Lyman series; for $n_1 = 2$, the Balmer series; for $n_1 = 3$, the Paschen series; for $n_1 = 4, 5, 6, \ldots$, we get other series of lines of generally lower frequency and longer wavelength in the infrared. (The worked-out example above for the jump from $n = 4$ to $n = 2$ is the second line in the Balmer series.) Figure 23-2 is a diagram of the hydrogen energy levels for different n's, and of the energy differences for the various series.

A question naturally arises about how the electron in the hydrogen atom could get up to a higher energy level in order to jump back and emit energy. Obviously, the electron can get in a higher energy orbit only by absorbing energy. This absorbed energy may come from collisions if the gas is heated to a high temperature. It may come from the energy of an electric spark or cathode-ray tube discharge, or *it may arise from the gas absorbing, from radiation falling on it, those same frequencies that it is able to emit.*

Since an atomic electron can occupy only those particular energy levels (or orbits) that we have just calculated, the energy differences between levels are the same whether the electron emits radiation or

FIG. 23-2 A schematic diagram of Bohr's hydrogen atom model, showing the origin of different series of lines in the hydrogen spectrum.

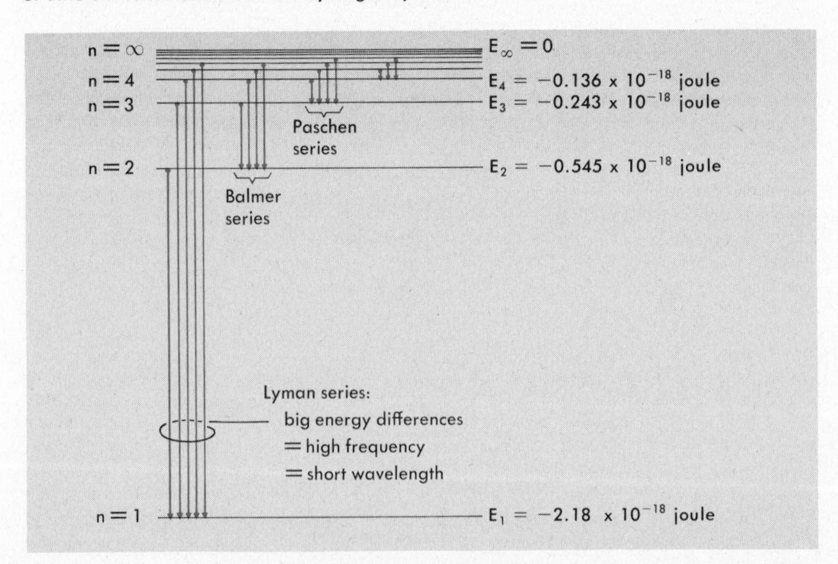

absorbs it. It cannot absorb a quantum that would bring it to an energy between permitted levels, and it cannot absorb a part of a quantum from a passing photon; it must be the whole quantum or nothing. Thus the frequencies of the absorbed photons must be exactly the same as the frequencies of the photons that the atom can radiate.

23-5 Successes and Limitations of the Bohr Theory

As far as we have seen, the Bohr theory of the hydrogen atom seems to be a very successful one. But does this highly artificial picture of light emission by a hydrogen atom really make any sense? Haven't Bohr's postulates been specially adjusted so as to lead in the end to the empirically established Balmer's formula? Certainly they were! But this is exactly how a new theory is usually introduced in physics. Newton introduced the notion of universal gravity in order to interpret the observed motion of the moon around the earth and of the planets around the sun, and in the very same way Bohr introduced his three postulates pertaining to electron motion in an atom, and light emission by "jump" processes in order to interpret the observed laws of hydrogen line spectra. However, the criterion for the validity of any new theory in physics is not only that this theory should give a correct interpretation of previous observations but that it also *predict* things that can be later confirmed by direct experiment.

To a limited extent, Bohr's original theory was able to do this. Other spectral series were discovered in the infrared, just as predicted. The theory was also successfully applied to other one-electron systems with great accuracy. (Ionized helium is such a system. Helium has two electrons, but if one of these is ionized off, the resulting He^+ is very like H, except that its nucleus has a charge of $2e$. Doubly ionized lithium, Li^{2+}, is another.)

With the development of highly precise spectroscopy, however, complications began to present themselves. A Balmer series line, for example, is not a single line, but *six* lines, separate and distinct, but all of very nearly the Bohr frequency. A part of these complications was resolved by the German physicist A. Sommerfeld and is illustrated in Fig. 23-3. While retaining, unchanged, the first of Bohr's orbits, Sommerfeld added one elliptical orbit to Bohr's second orbit, two elliptical orbits to Bohr's third orbit, etc. Although the elliptical orbits added by Sommerfeld had different geometrical shapes, they nevertheless corresponded to *almost* the same energies as Bohr's circular orbits. A long, narrow ellipse, however, brings the electron close to the nucleus, where (by Kepler's second law) its speed must be much higher than its speed in a circular orbit of about the same energy. When relativistic corrections were applied, Sommerfeld found that a part of the splitting of the Bohr lines was successfully explained.

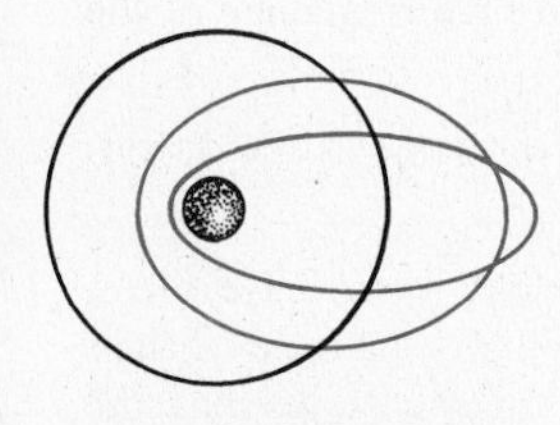

FIG. 23-3 Circular and elliptical atomic orbits of nearly identical energy, according to Sommerfeld.

For further explanation, it was necessary to consider each electron as a little ball of charge, spinning rapidly on its own axis (like the daily rotation of the earth) and contributing an additional effect not originally included by Bohr.

All these elaborations of the original theory were well and good, and would still permit the atom to be pictured as a system of definite ball-like charged particles revolving in orbits about the massive central particle of the nucleus. This became a difficult picture to imagine, however, when it was found that some "orbits" must have *zero* angular momentum. The only orbit that could have zero angular momentum would have been an oscillation back and forth along a straight line penetrating to the center of the nucleus. This seemed a most unlikely situation.

And although the Bohr model worked well for the main features of the hydrogen atom, it would not give correct answers for atoms with more than one electron. The theory could not be adjusted or patched up to make it yield the frequencies actually observed in even the two-electron atom of helium.

As we shall see, this picture had to be abandoned. It is still, however, a great milestone in scientific achievement, and one that is still very useful. It showed that Planck's quantum ideas were a necessary part of the atom and its inner mechanism; it introduced the idea of quantized energy levels and explained the emission or absorption of radiation as being due to the transition of an electron from one level to another. As a model for even multielectron atoms, the Bohr picture is still useful. It leads to a good, simple, rational ordering of the electrons in larger atoms and qualitatively helps to predict a great deal about chemical behavior and spectral details. Even the most abstract theorist, although he now ignores the Bohr model in his calculations, still *imagines* the atom as composed of electrons in various orbits and finds this picture to be a very useful thing.

Questions

(23-1) 1. Figure 23-1 shows the first two of the theoretically infinite number of lines in the ultraviolet that Balmer was not able to observe. Calculate the frequency of the one of longer wavelength.

2. Of the pair of UV lines shown in Fig. 23-1, calculate the frequency of the one of shorter wavelength. (See Question 1.)

3. Show that the dimensions of Planck's constant are the same as the dimensions of angular momentum.

4. Since both the electron and the proton in a hydrogen atom have mass, there is also a gravitational attraction between them. What is the ratio of their gravitational to their electrical attraction? Do you think this justifies the gravitational force having been neglected in computing the Bohr orbits? (The mass of a proton is about 1840 times the electron mass.)

(23-2) 5. What is the radius of the fourth permissible orbit in a hydrogen atom?

6. What is the radius of the third quantum orbit of the electron in a hydrogen atom?

7. The nucleus of the lithium atom has a charge of $+3e$, and the atom has three orbital electrons. Consider a doubly-ionized lithium atom that has only one electron left, making it similar to the hydrogen atom. What is the radius of its smallest ($n = 1$) orbit? The derivation would be the same as that for hydrogen, except that at the beginning, instead of $e \times e = e^2$, you have $e \times 3e$. With this knowledge, you can skip at once to the end and make the necessary change in the expression for hydrogen.)

8. The atomic number of helium is 2; i.e., its nucleus has a charge of $+2e$, and it has two electrons. Consider a singly ionized helium atom, with only one electron left. What is the radius of its smallest ($n = 1$) orbit? (See Question 7.)

(23-3) 9. What is the total energy of a hydrogen electron in its fifth ($n = 5$) orbit?

10. An electron is in its $n = 6$ orbit in a hydrogen atom. What is its total energy?

11. (a) What is the kinetic energy of a hydrogen electron in its $n = 5$ orbit? (b) What is its potential energy? (It is not necessary to work this out laboriously. The solution is easily found from the answer to Question 9.)

12. What is (a) the KE, and (b) the PE, of a hydrogen electron in its $n = 6$ orbit? (See Questions 11 and 10.)

13. Give the PE, KE, and total energy of an electron in the $n = 3$ orbit of ionized helium. (Do this the easy way, from the given value of E_1 for hydrogen. Questions 7 and 11 may contain helpful remarks, if needed.)

14. A doubly ionized lithium atom (atomic number 3) has an electron in its $n = 4$ orbit. What are its PE, KE, and total energy? (Questions 7 and 11 may help in doing this the easy way, starting from the known value of E_1 for hydrogen.)

(23-4) 15. A hydrogen atom makes a transition from $n = 4$ to $n = 3$. (a) What is the wavelength of the emitted photon? (b) Is this photon visible? (c) Of which spectral series is this photon a member?

16. (a) What is the wavelength of the photon emitted when a hydrogen electron jumps from $n = 2$ to $n = 1$? (b) Is this a photon of visible light? (c) What is the name of the series of which this spectral line is a member?

17. Doubly ionized lithium ions (see Question 7) produce a spectrum similar that of hydrogen atoms. Transitions from higher n to $n = 2$ will give a spectral series analogous to the Balmer series of hydrogen. How will the frequencies of these lines compare to the Balmer frequencies?

18. A singly ionized helium ion (see Question 8) has a spectrum qualitatively similar to the spectrum of atomic hydrogen. Transition from higher n to $n = 1$ will produce a series analogous to the Lyman series of hydrogen. How will their frequencies compare with those of the hydrogen Lyman series?

19. Consider a sample of hydrogen in which most of the atomic electrons are in their $n = 1$ state. White light passed through this gas will show absorption lines where some frequencies have been absorbed. Which series will show the stronger absorption, the Balmer or the Lyman?

20. The lower dense layers of a star's atmosphere emit a continuous spectrum of all frequencies; in passing through the rare upper layers of the star's atmosphere, frequencies are absorbed characteristic of the elements in the upper atmosphere, which always contains much hydrogen. In the spectrum of cool stars (up to 5000°K) the Balmer absorption lines are very weak. They become darker in stars of higher temperature up to about 10,000°K. In hotter stars, the Balmer lines become weaker again. Explain.

24

The Structure of Atoms

24-1 Quantum Numbers

Having become acquainted with the possible orbital motions of the one hydrogen electron, as permitted by the Bohr–Sommerfeld rules, we can now tackle the question of the structure of the multielectron atoms.

Bohr introduced the *principal quantum number n*. In his model, this was done by quantizing, or assigning only certain specific values to, the angular momentum of the electron; however, this did not last long. We still keep n as the principal quantum number, but it now refers to the general energy level of the electron, which is also quantized, or allowed only certain values, as we have previously calculated.

The second quantum number l, the *orbital quantum number*, came into the picture as a result of Sommerfeld's introduction of elliptical orbits. This quantum number, rather than n, is the measure of the angular momentum of the electron, which depends on the ellipticity of the orbit. A circular orbit has the maximum angular momentum; an oscillation along a straight line penetrating the nucleus has a zero angular momentum. It was found (through the analysis of spectra of various elements) that l could have only integral values ranging from 0 to $n - 1$. For $n = 1$, for example, l can have only the value 0; for $n = 3$, l can be 0, 1, or 2. For the hydrogen atom, the shape of the orbit has little effect; except for the small relativistic correction, the energies are all the same for a given value of n. In larger atoms, though, the effect may be considerable. Figure 24-1 is a schematic diagram of a sodium atom, which has 11 electrons and whose nucleus therefore has a charge of $+11e$. The eleventh electron, when in a circular orbit, is constantly shielded from the nucleus by the negative charges of the 10 inner electrons. In an elliptical orbit, it penetrates this shield and spends part of each orbital revolution close to the strong attraction of the less well-shielded nucleus. Here, the electron has a *larger negative* PE, and therefore a lower total energy than it has with a more nearly circular orbit of larger l. Hence electrons of different l's can have quite different energies, even though n may be the same.

Spectroscopists had observed that many normally single spectral lines split up into several lines when the atoms of the gas emitting them were in a strong magnetic field. It was obvious from this that something else was quantized; this turned out to be the direction, or orientation, of the angular momentum.

Involved quantum–mechanical arguments, which must be left to a more advanced course, show that the original Bohr–Sommerfeld concepts were oversimplified. The angular momentum of the electron in its

FIG. 24-1 Schematic diagram of a sodium atom: the energy of the outermost eleventh electron depends on whether its orbit is circular or elliptical.

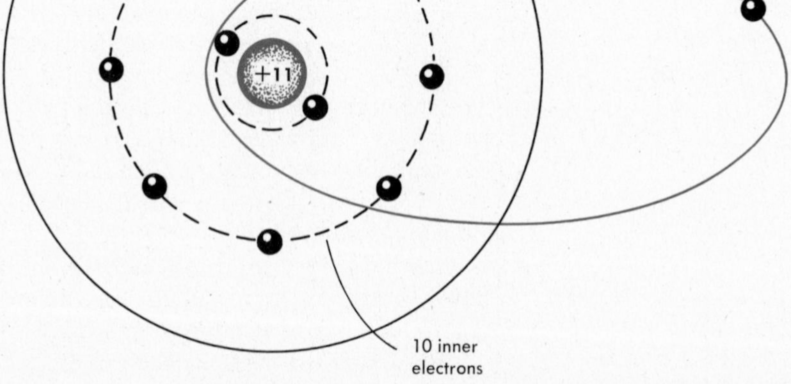

orbit turns out to be $\sqrt{l(l+1)} \times h/2\pi$, rather than the integral multiple of $h/2\pi$ first assumed. The angular momentum vector (perpendicular to the plane of the orbit) cannot point in just any direction—it can take only those orientations in which its *component* in the direction of the magnetic field is an integral multiple of $h/2\pi$.

We may look on the orbiting, negatively charged electron as being a little electric current that produces a tiny magnetic field perpendicular to the plane of its orbit. This magnetic field (technically, its *magnetic moment*) is directly proportional to the angular momentum; so it must therefore also be quantized in the direction of the magnetic field. The integral multiplier m (often written as m_l) is the *magnetic quantum number*.

It takes work to turn a bar magnet into various orientations with a magnetic field, and therefore each different orientation will have a different potential energy. Similarly, the orbiting electron will have slightly different energies in each of its permitted orientations, from a minimum at $m = -l$ to a maximum at $m = +l$. These added energy levels account for a part of the splitting of spectral lines in a magnetic field.

As Fig. 24-2A shows, when $l = 1$, m can be -1, 0, or $+1$, each one with a slightly different energy, the energy differences being proportional to the strength of the field. Figure 24-2B illustrates the seven different m's possible when $l = 3$. In general, m can take on $2l + 1$

FIG. 24-2 Vectors representing the orientation of an electron's magnetic moment in a magnetic field. Only those directions are allowed that give an integral component (+, −, or 0) in the direction of the field.

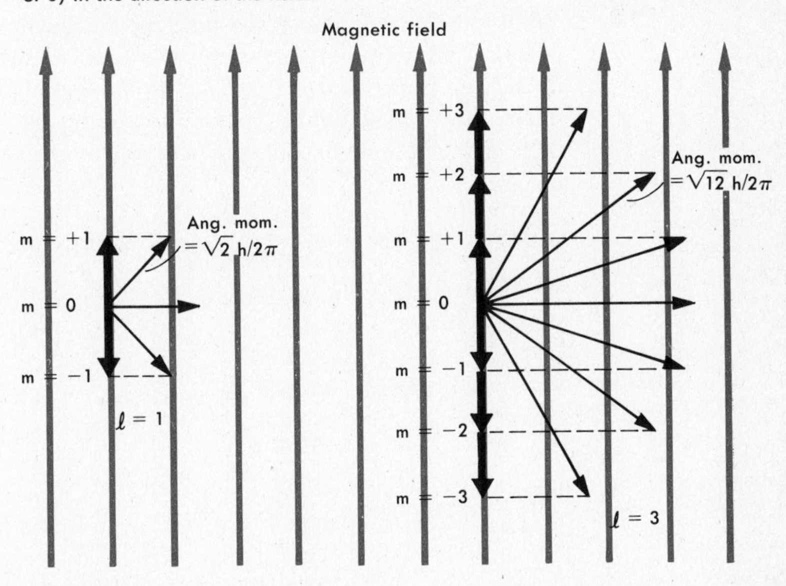

different values. (When $l = 0$, m can of course have only the single value $m = 0$.)

There was also a further line splitting not accounted for that was explained by the concept of *electron spin*. Still holding to the picture given by the Bohr model, we can imagine the electron as a spinning ball of mass and charge, which would itself have a quantized angular momentum and magnetic moment, regardless of its orbital motion. Following quantum-mechanical arguments analogous to those for l, the spin angular momentum and the quantized magnetic moments come out as in Fig. 24-3. The *spin quantum number*, s, (often called m_s), can have but two values: $-\frac{1}{2}$ and $+\frac{1}{2}$; $s = 0$ is not allowed.

This gives a total of four quantum numbers: n, l, m, and s—a far cry from the simple n first introduced by Bohr. It should be noted that the quantization of m and s exists whether or not the atom is in an external magnetic field—the field serves only to separate the energy levels and make them apparent in the line splitting of the spectrum.

24-2 The Exclusion Principle

We have seen that in the case of the hydrogen atom, the electron puts itself in the lowest energy level available; if it is excited into a higher energy, it at once falls back to $n = 1$, either all in one jump or in several steps, emitting its excess energy as radiation as it does so. Is this also the case for the multielectron atoms?

If it were, the normal, unexcited state for any atom would find all its electrons—105 of them in the artificially produced element hahnium! —packed together in an overcrowded ring in the first of the Bohr orbits. With atoms of increasing atomic number, this ring would become more and more crowded because it would have to accommodate more electrons, and also because the ring would become smaller due to the stronger attraction of the increasingly positive nucleus.

If this were true, the size of atoms would decrease rapidly with increasing atomic number, and a lead atom would be much smaller than one of aluminum. Experiment tells us this is not so; although atomic

FIG. 24-3 The only two possible orientations of the spin vector.

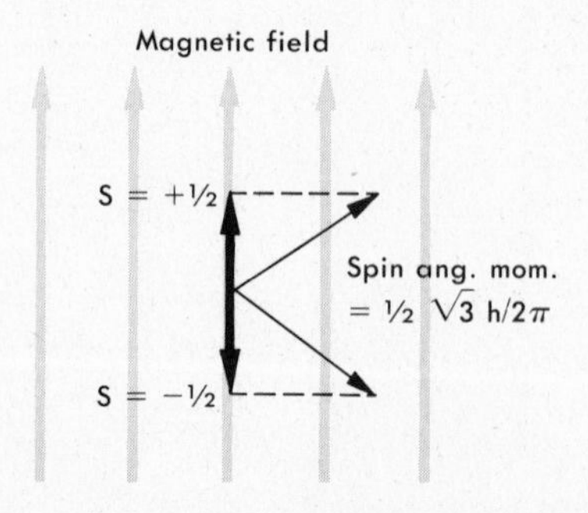

sizes show periodic variations, they remain roughly the same for the entire list of elements. To avoid this contradiction, and to allow for an interpretation that would explain the observed frequencies in each element's spectrum, some new restriction was necessary. This was proposed by the Austrian physicist Wolfgang Pauli in 1924. It is known as the *exclusion principle*, and says that ***in any atom, no two electrons can have the same identical set of the four quantum numbers.***

24-3 Electron Shells and the Periodic Table

With the help of the quantum numbers and the exclusion principle, we are in a position to find the pattern of the electrons in the elements. The element immediately following hydrogen is helium, the atom of which contains two electrons. If these two electrons spin in opposite directions, both can be accommodated on the first (circular) Bohr's orbit, as shown in Fig. 24-4. The quantum numbers for helium's two atoms are $n = 1$, $l = 0$, $m = 0$, $s = -\frac{1}{2}$; and $n = 1$, $l = 0$, $m = 0$, $s = +\frac{1}{2}$. These two electrons are all that can be accommodated in $n = 1$, because l and m can have only the value 0, and we have used up both s's.

All the electrons with the same n are what is called a *shell*; there can be a maximum of $2n^2$ electrons in a full shell. For historical reasons dating back to the early days of spectroscopy, the $n = 1$ shell is known as the K shell, the $n = 2$ shell is the L shell, and so on.

The next element is lithium, with three electrons. (It is the number of electrons that gives each element its *atomic number*. Since atoms are electrically neutral, the atomic number also gives the positive charge of the element's nucleus.) Because there is no more room in the K shell ($n = 1$), the third electron must be content with a place in the L shell ($n = 2$). There it will fit the energy level (or Bohr orbit, if we care to hold to this picture) $n = 2$, $l = 0$, $m = 0$, $s = -\frac{1}{2}$. The fourth electron of beryllium finds the lowest vacant level to have the same quantum numbers except for the spin, which must be reversed to $s = +\frac{1}{2}$ if it is to fit in the same subshell with the third electron. (A subshell is a group of atoms having the same l.) Since there are $(2l + 1)$ possible values for m, and each m may have two different spins, a subshell can hold $2(2l + 1)$ electrons.

The added electrons for the next six elements (boron, carbon, nitrogen, oxygen, fluorine, and neon) fill the $l = 1$ subshell, which also completes the $n = 2$ shell. So, considering the next eight elements in the atomic number sequence (sodium through argon), these go into the $n = 3$ shell, filling the $l = 0$ and $l = 1$ subshells.

In this $n = 3$ shell, $l = 2$ is permitted, and after argon (Z = atomic number = 18), we might expect that the last electron of the next element, potassium, Z = 19, would go into the $n = 3$, $l = 2$ subshell. But this is *not* what happens. Because of the lower energy of the more elliptical orbits of smaller l, $n = 4$, $l = 0$ represents a *lower energy* than $n = 3$, $l = 2$; so the outermost electrons of potassium and calcium are placed

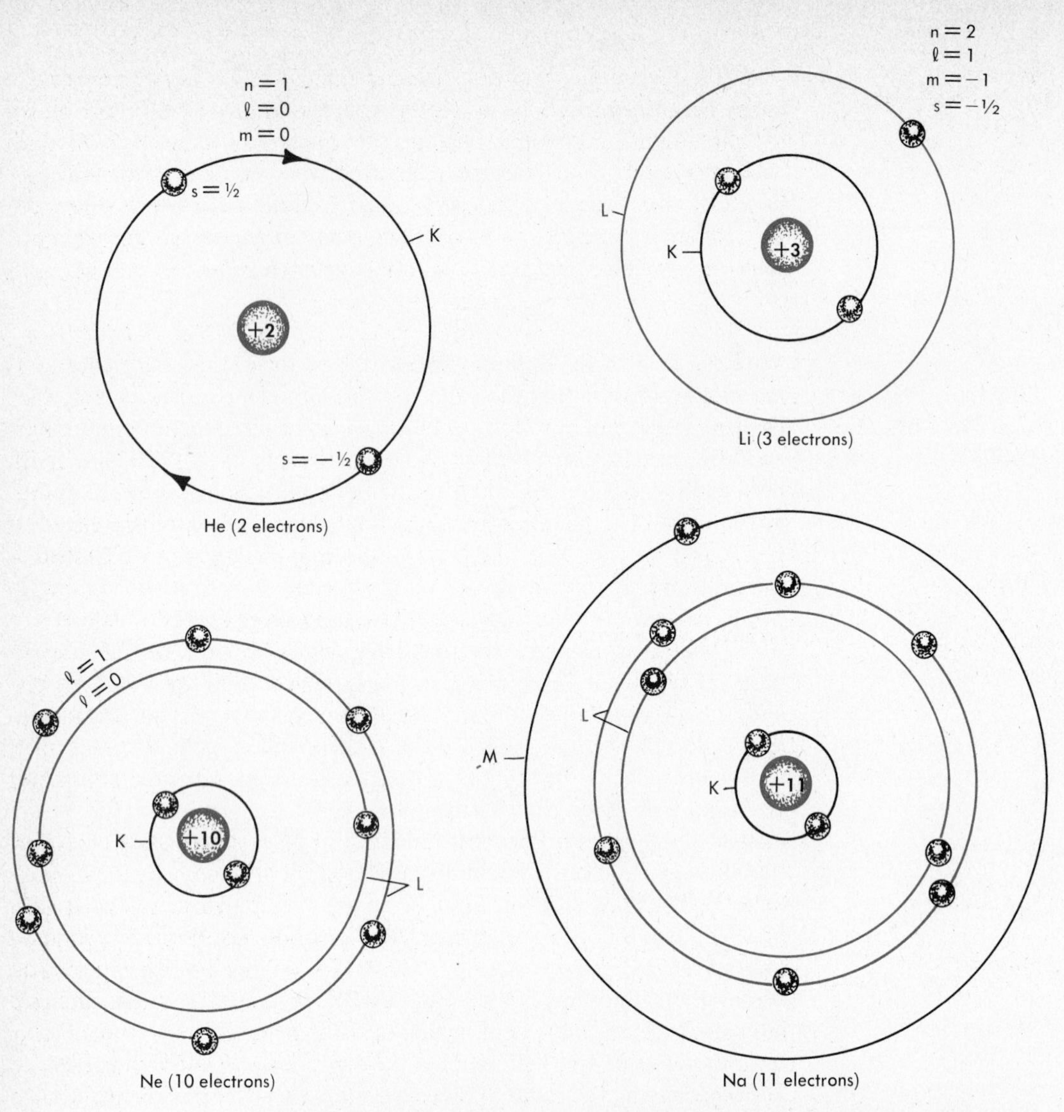

FIG. 24-4 How the electron shells and subshells are filled in atoms of increasing atomic number. As the atomic number increases, the size of the shells becomes smaller because of the increased attraction of the greater nuclear charge. The result is that all atoms have roughly the same size, no matter how many electrons they contain.

in this subshell of lowest available energy. The next 10 elements—scandium (Z = 21) through zinc (Z = 30)—go back to fill the vacant $n = 3$, $l = 2$ subshell. The interrupted filling of the $n = 4$ shell is then continued, and its $l = 1$ subshell is completed by the last electron of krypton (Z = 36); then back to start the $n = 5$ shell, and so on.

In this sequence, we can see that there is a sort of periodic regularity as we check through the list of elements in atomic-number order. Similar periodic regularities in chemical characteristics were noticed by nineteenth-century chemists when the elements were arranged in the order of their atomic weights. But the important step of actually arranging the elements into a periodic table, or chart, was not made until 1869, by the Russian chemist Dmitri Mendeleev (1834–1907). At that time, the systematic electron structure of atoms was not even suspected, and Mendeleev was further handicapped by the fact that the list of chemical elements then known was rather incomplete.

Driven by a strong belief that there *must be* a regular periodicity in the natural sequence of elements, Mendeleev made the bold hypothesis that the deviations from the expected periodicity in his list were due to the failure of contemporary chemistry to have discovered some of the elements existing in nature. Thus, in constructing his table, he left a number of empty spaces to be filled in later by future discoveries, and in certain instances he reversed the atomic-weight order of the elements in order to comply with the demands of the regular periodicity of their chemical properties.

Mendeleev thought his needed reversals of atomic-weight order were evidence that some of the atomic weights had been erroneously determined. This was not the case, however, and we can see now that his difficulty was caused by the fact that the fundamental property of atoms that dictates their chemical behavior and the order in which they should be arranged is *not* their atomic weights (which are only incidental), but their atomic numbers.

Such a table in a more modern and complete form is shown in Fig. 24-5. The columns of elements running up the page are *periods*, and include elements whose outer electrons are in the same shell—or, what is the same thing, have the same principal quantum number n. The rows of elements running across the page are more important to the chemist. Up from the bottom of the page, the first two rows, and the last six rows at the top, include the eight major chemical *groups*, whose members have similar chemical behavior. In Group I, at the bottom, for example, we have hydrogen* and the alkali metals, all of low density and chemically very active, and each with one outer $l = 0$ electron. Their chemical behaviors are very similar, as we might expect, because chemical behavior is controlled largely by the outer electrons. The members of Group II are also all chemically similar, as each has two outer electrons. The other groups likewise have quite similar behavior, on to Group VIII, the so-called "noble gases" that resemble each other in forming no stable chemical compounds with other elements, or even with themselves.

---

* Hydrogen seems out of place here, because this gas does not appear to be at all metallic. But there is good theoretical reason to believe that at very high pressures this ordinary gas crystallizes to form a solid with the characteristics of metals. Astronomers, in fact, believe the cores of the giant planets Jupiter and Saturn are composed of this so-called "metallic hydrogen."

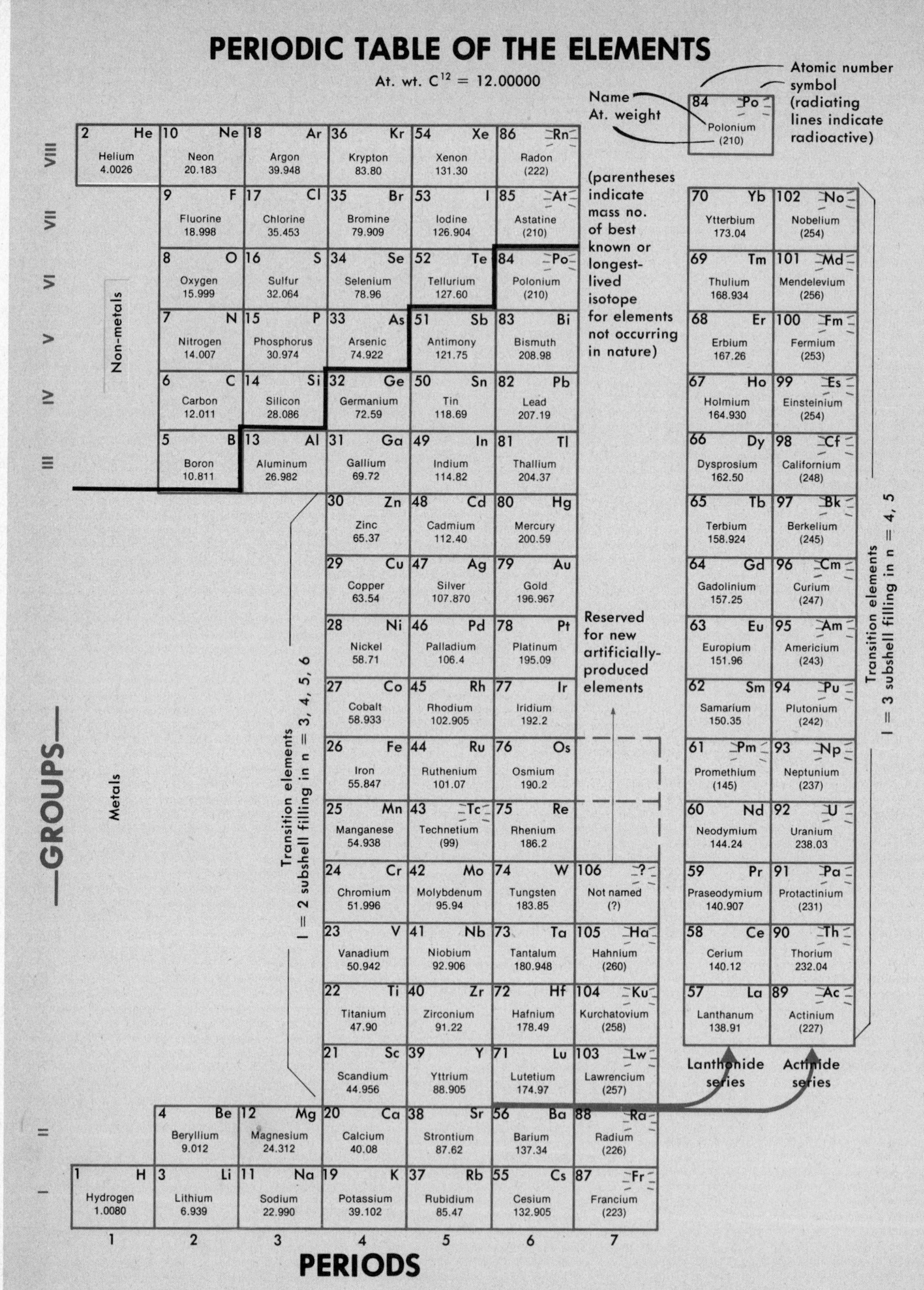
PERIODIC TABLE OF THE ELEMENTS
At. wt. C12 = 12.00000
Name
At. weight
84 Po Polonium (210)
Atomic number
symbol
(radiating lines indicate radioactive)
(parentheses indicate mass no. of best known or longest-lived isotope for elements not occurring in nature)
VIII
VII
VI
V
IV
III
II
I
Non-metals
Metals
—GROUPS—
2 He Helium 4.0026
10 Ne Neon 20.183
18 Ar Argon 39.948
36 Kr Krypton 83.80
54 Xe Xenon 131.30
86 Rn Radon (222)
9 F Fluorine 18.998
17 Cl Chlorine 35.453
35 Br Bromine 79.909
53 I Iodine 126.904
85 At Astatine (210)
8 O Oxygen 15.999
16 S Sulfur 32.064
34 Se Selenium 78.96
52 Te Tellurium 127.60
84 Po Polonium (210)
7 N Nitrogen 14.007
15 P Phosphorus 30.974
33 As Arsenic 74.922
51 Sb Antimony 121.75
83 Bi Bismuth 208.98
6 C Carbon 12.011
14 Si Silicon 28.086
32 Ge Germanium 72.59
50 Sn Tin 118.69
82 Pb Lead 207.19
5 B Boron 10.811
13 Al Aluminum 26.982
31 Ga Gallium 69.72
49 In Indium 114.82
81 Tl Thallium 204.37
30 Zn Zinc 65.37
48 Cd Cadmium 112.40
80 Hg Mercury 200.59
29 Cu Copper 63.54
47 Ag Silver 107.870
79 Au Gold 196.967
28 Ni Nickel 58.71
46 Pd Palladium 106.4
78 Pt Platinum 195.09
27 Co Cobalt 58.933
45 Rh Rhodium 102.905
77 Ir Iridium 192.2
26 Fe Iron 55.847
44 Ru Ruthenium 101.07
76 Os Osmium 190.2
25 Mn Manganese 54.938
43 Tc Technetium (99)
75 Re Rhenium 186.2
24 Cr Chromium 51.996
42 Mo Molybdenum 95.94
74 W Tungsten 183.85
106 ? Not named (?)
23 V Vanadium 50.942
41 Nb Niobium 92.906
73 Ta Tantalum 180.948
105 Ha Hahnium (260)
22 Ti Titanium 47.90
40 Zr Zirconium 91.22
72 Hf Hafnium 178.49
104 Ku Kurchatovium (258)
21 Sc Scandium 44.956
39 Y Yttrium 88.905
71 Lu Lutetium 174.97
103 Lw Lawrencium (257)
4 Be Beryllium 9.012
12 Mg Magnesium 24.312
20 Ca Calcium 40.08
38 Sr Strontium 87.62
56 Ba Barium 137.34
88 Ra Radium (226)
1 H Hydrogen 1.0080
3 Li Lithium 6.939
11 Na Sodium 22.990
19 K Potassium 39.102
37 Rb Rubidium 85.47
55 Cs Cesium 132.905
87 Fr Francium (223)
Reserved for new artificially-produced elements
Transition elements
l = 2 subshell filling in n = 3, 4, 5, 6
70 Yb Ytterbium 173.04
102 No Nobelium (254)
69 Tm Thulium 168.934
101 Md Mendelevium (256)
68 Er Erbium 167.26
100 Fm Fermium (253)
67 Ho Holmium 164.930
99 Es Einsteinium (254)
66 Dy Dysprosium 162.50
98 Cf Californium (248)
65 Tb Terbium 158.924
97 Bk Berkelium (245)
64 Gd Gadolinium 157.25
96 Cm Curium (247)
63 Eu Europium 151.96
95 Am Americium (243)
62 Sm Samarium 150.35
94 Pu Plutonium (242)
61 Pm Promethium (145)
93 Np Neptunium (237)
60 Nd Neodymium 144.24
92 U Uranium 238.03
59 Pr Praseodymium 140.907
91 Pa Protactinium (231)
58 Ce Cerium 140.12
90 Th Thorium 232.04
57 La Lanthanum 138.91
89 Ac Actinium (227)
Lanthonide series
Actinide series
Transition elements
l = 3 subshell filling in n = 4, 5
1
2
3
4
5
6
7
PERIODS

INDEX TO PERIODIC TABLE

Alphabetical by Name					
Actinium	Ac	89	Mendelevium	Md	101
Aluminum	Al	13	Mercury	Hg	80
Americium	Am	95	Molybdenum	Mo	42
Antimony	Sb	51	Neodymium	Nd	60
Argon	Ar	18	Neon	Ne	10
Arsenic	As	33	Neptunium	Np	93
Astatine	At	85	Nickel	Ni	28
Barium	Ba	56	Niobium	Nb	41
Berkelium	Bk	97	Nitrogen	N	7
Beryllium	Be	4	Nobelium	No	102
Bismuth	Bi	83	Osmium	Os	76
Boron	B	5	Oxygen	O	8
Bromine	Br	35	Palladium	Pd	46
Cadmium	Cd	48	Phosphorus	P	15
Calcium	Ca	20	Platinum	Pt	78
Californium	Cf	98	Plutonium	Pu	94
Carbon	C	6	Polonium	Po	84
Cerium	Ce	58	Potassium	K	19
Cesium	Cs	55	Praseodymium	Pr	59
Chlorine	Cl	17	Promethium	Pm	61
Chromium	Cr	24	Protactinium	Pa	91
Cobalt	Co	27	Radium	Ra	88
Copper	Cu	29	Radon	Rn	86
Curium	Cm	96	Rhenium	Re	75
Dysprosium	Dy	66	Rhodium	Rh	45
Einsteinium	Es	99	Rubidium	Rb	37
Erbium	Er	68	Ruthenium	Ru	44
Europium	Eu	63	Samarium	Sm	62
Fermium	Fm	100	Scandium	Sc	21
Fluorine	F	9	Selenium	Se	34
Francium	Fr	87	Silicon	Si	14
Gadolinium	Gd	64	Silver	Ag	47
Gallium	Ga	31	Sodium	Na	11
Germanium	Ge	32	Strontium	Sr	38
Gold	Au	79	Sulfur	S	16
Hafnium	Hf	72	Tantalum	Ta	73
Hahnium	Ha	105	Technetium	Tc	43
Helium	He	2	Tellurium	Te	52
Holmium	Ho	67	Terbium	Tb	65
Hydrogen	H	1	Thallium	Tl	81
Indium	In	49	Thorium	Th	90
Iodine	I	53	Thulium	Tm	69
Iridium	Ir	77	Tin	Sn	50
Iron	Fe	26	Titanium	Ti	22
Kurchatovium	Ku	104	Tungsten	W	74
Krypton	Kr	36	Uranium	U	92
Lanthanum	La	57	Vanadium	V	23
Lawrencium	Lw	103	Xenon	Xe	54
Lead	Pb	82	Ytterbium	Yb	70
Lithium	Li	3	Yttrium	Y	39
Lutetium	Lu	71	Zinc	Zn	30
Magnesium	Mg	12	Zirconium	Zr	40
Manganese	Mn	25			

Alphabetical by Symbol					
Ac	Actinium	89	Mg	Magnesium	12
Ag	Silver	47	Mn	Manganese	25
Al	Aluminum	13	Mo	Molybdenum	42
Am	Americium	95	N	Nitrogen	7
Ar	Argon	18	Na	Sodium	11
As	Arsenic	33	Nb	Niobium	41
At	Astatine	85	Nd	Neodymium	60
Au	Gold	79	Ne	Neon	10
B	Boron	5	Ni	Nickel	28
Ba	Barium	56	No	Nobelium	102
Be	Beryllium	4	Np	Neptunium	93
Bi	Bismuth	83	O	Oxygen	8
Bk	Berkelium	97	Os	Osmium	76
Br	Bromine	35	P	Phosphorus	15
C	Carbon	6	Pa	Protactinium	91
Ca	Calcium	20	Pb	Lead	82
Cd	Cadmium	48	Pd	Palladium	46
Ce	Cerium	58	Pm	Promethium	61
Cf	Californium	98	Po	Polonium	84
Cl	Chlorine	17	Pr	Praseodymium	59
Cm	Curium	96	Pt	Platinum	78
Co	Cobalt	27	Pu	Plutonium	94
Cr	Chromium	24	Ra	Radium	88
Cs	Cesium	55	Rb	Rubidium	37
Cu	Copper	29	Re	Rhenium	75
Dy	Dysprosium	66	Rh	Rhodium	45
Er	Erbium	68	Rn	Radon	86
Es	Einsteinium	99	Ru	Ruthenium	44
Eu	Europium	63	S	Sulfur	16
F	Fluorine	9	Sb	Antimony	51
Fe	Iron	26	Sc	Scandium	21
Fm	Fermium	100	Se	Selenium	34
Fr	Francium	87	Si	Silicon	14
Ga	Gallium	31	Sm	Samarium	62
Gd	Gadolinium	64	Sn	Tin	50
Ge	Germanium	32	Sr	Strontium	38
H	Hydrogen	1	Ta	Tantalum	73
Ha	Hahnium	105	Tb	Terbium	65
He	Helium	2	Tc	Technetium	43
Hf	Hafnium	72	Te	Tellurium	52
Hg	Mercury	80	Th	Thorium	90
Ho	Holmium	67	Ti	Titanium	22
I	Iodine	53	Tl	Thallium	81
In	Indium	49	Tm	Thulium	69
Ir	Iridium	77	U	Uranium	92
K	Potassium	19	V	Vanadium	23
Kr	Krypton	36	W	Tungsten	74
Ku	Kurchatovium	104	Xe	Xenon	54
La	Lanthanum	57	Y	Yttrium	39
Li	Lithium	3	Yb	Ytterbium	70
Lu	Lutetium	71	Zn	Zinc	30
Lw	Lawrencium	103	Zr	Zirconium	40
Md	Mendelevium	101			

In the center of the table are the *transition elements*. Here, as we progress up the page in order of atomic number, electrons are being added in buried shells beneath the surface of the atom. In Period 6, for example, as we go from bottom to top in the transition elements, electrons are being added to the $n = 5, l = 2$ subshell. In this same Period 6, elements $Z = 57$ to 70 (the lanthanides) are added in the $n = 4$, $l = 3$ subshell.

Figure 24-6 sketches the energy levels in the subshells of each shell in a schematic, qualitative way. In this, the standard chemists' notation for the subshells has been used:

s-electrons: $l = 0$; subshell holds 2 electrons
p-electrons: $l = 1$; subshell holds 6 electrons
d-electrons: $l = 2$; subshell holds 10 electrons
f-electrons: $l = 3$; subshell holds 14 electrons.

The notation "4*d*," for example, represents the subshell of $n = 4$, $l = 2$. The number of electrons in that subshell is often written as a superscript, resembling an exponent. We might write the entire electronic structure for the element zirconium (Zr; Z = 40) in this way: $1s^2$, $2s^2$, $2p^6$, $3s^2$, $3p^6$, $4s^2$, $3d^{10}$, $4p^6$, $5s^2$, $4d^2$. (Follow this on Fig. 24-6!)

Some of the subshell levels overlap somewhat, as the energies are affected by added electrons. For example, niobium (Nb; Z = 41) we should expect to be identical to Zr, except for the final notation, which would put a third electron in the 4*d* shell to give $4d^3$. This is *not* quite the case, however. With the added forty-first electron, the 5*s* subshell *loses* one, and the 4*d* subshell *gains* two, giving $5s^1$, $4d^4$. There are several other such irregularities deviating from the scheme of Fig. 24-6 —it is enough for us to recognize that they are there, and leave the details to the analytical chemists.

It may be worth noting that each period begins with the two $l = 0$ *s*-electrons of the period number, and ends with the filling of the $l = 1$ *p*-subshells of the same number.

24-4 Spectra of Multielectron Atoms

The single electron of hydrogen has only a relatively small number of energy levels, which are easily calculated by the Bohr theory. For this reason, it has a simple spectrum whose regularities were apparent. The 104 other and larger atoms have spectra that are much more complex and whose fundamental regularities are not at all apparent. Nevertheless, each element when vaporized has its own set of frequencies that it is able to emit or absorb, in a pattern as characteristic of that element as a set of fingerprints is of an individual human.

These characteristic spectra are produced by the outer electrons of the elements. But since each outer electron is influenced not only by the nucleus but by other electrons as well, the arrangement of possible energy levels can be very complicated, and the differences between

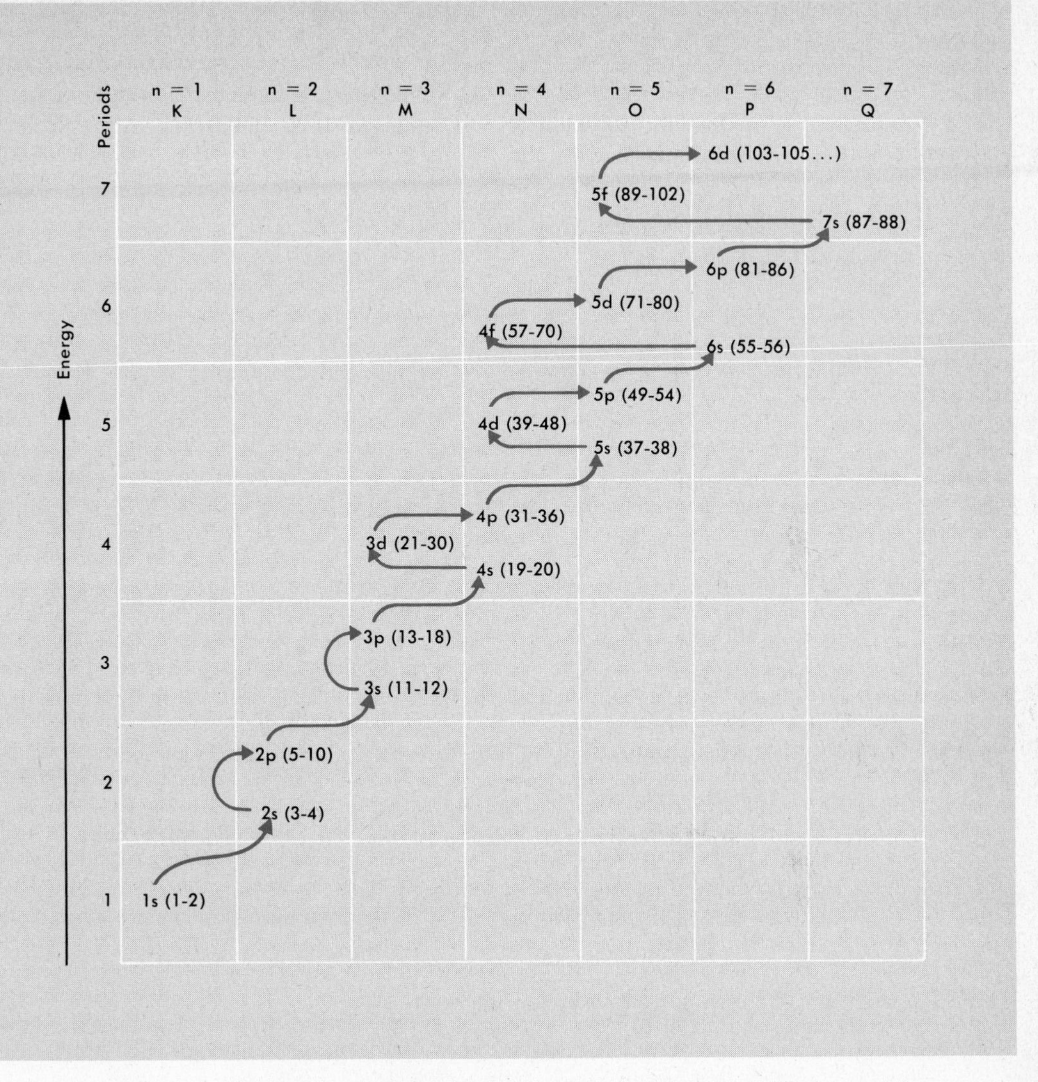

FIG. 24-6 Qualitative diagram of atomic energy levels, as affected by n and l.

energies (which control the frequencies emitted or absorbed) more complicated still.

The spectra of some elements (iron, for example) may have thousands of lines in the visible and near-visible range.

The fact that each chemical element emits or absorbs a set of spectral lines characteristic of that particular element is one of the bases of spectral analysis. In addition, the spectroscopist also deals with the spectra of molecules, relatively stable groups of atoms, which contribute their own energy levels and spectral lines. Analysis of its spectrum (emission or absorption) can tell not only of what atoms or molecules a substance is composed, but in what relative amounts they are present. This is enormously useful in many industrial fields.

Without the spectroscopic analysis of the light from distant stars and gas clouds, astronomy would still be back in the nineteenth century. In a later chapter we will look in more detail at the contributions of spectroscopy to our knowledge of the universe.

24-5 Lasers

So far, we have pictured only two steps in the production of the spectrum of an element. An outer electron (the *only* electron in the case of hydrogen) absorbs exactly enough energy from a passing photon, from a collision, or from an electric impulse, and is thereby raised to some permitted higher energy level. Then, after a brief waiting period, generally about 10^{-8} s, the electron *spontaneously* drops back to its lowest energy level, either in one jump or perhaps in several if there happen to be intermediate levels along the way, emitting one or more photons as it does so.

The only role we have assigned to the passing photon is to give up its entire energy in exciting the electron to a higher energy level. However, photons can and do play another part in this scheme. During the time the excited electron is pausing to wait for its spontaneous fallback down to a lower level, it is very susceptible to having this fallback *stimulated*, or induced to take place sooner than otherwise, by interaction with a photon of the proper energy.

Imagine a population of excited atoms: In some the outer electron is in its lowest energy level E_1; some others are excited to the second level E_2; and some are in the E_3 state. Now consider a passing photon whose energy hf is equal to $E_3 - E_2$. We have already noted one effect it may have—it may interact with an electron in the E_2 state and give up its energy to raise the electron to E_3. But it may also interact with an electron in the E_3 state, to stimulate the electron to drop back at once and emit its $hf = E_3 - E_2$ photon. What makes the laser possible is that ***this emitted photon travels in exactly the same direction as the stimulating photon, and is exactly in phase with it.*** This is very different from a spontaneously emitted photon, which may be in any direction, and whose phase is not necessarily in step with that of any other photon.

Ordinarily this effect is not noticeable, because there are generally more atoms in the lower energy states, so the fate of most photons of the proper energy is to be absorbed in exciting the electrons. One of the tricks that makes a laser work is to provide more electrons in a state of higher energy. One type of laser uses a rod of synthetic ruby crystal—made of colorless aluminum oxide plus a fraction of a percent of chromium atoms. An intense flash of white light excites most of the chromium atoms to a higher-energy state that is *metastable*—that is, the electrons remain at this level for the enormously long average time (atomically

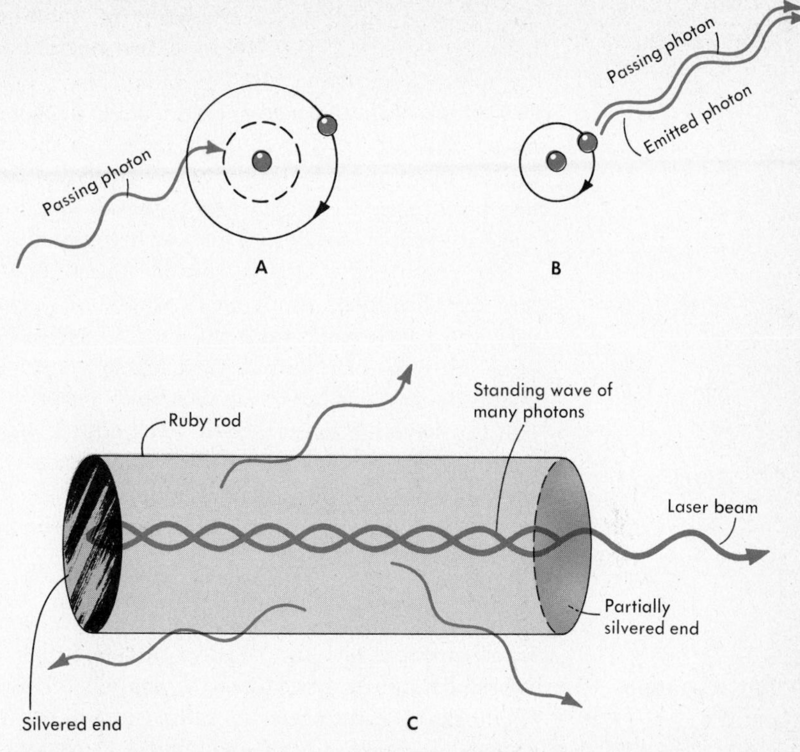

FIG. 24-7 (A) The atomic electron, in a high energy level, is ready to be stimulated into emission by a passing photon of the proper energy. (B) The electron has dropped back to a level of lower energy, emitting a photon exactly duplicating the stimulating photon. (C) Photons not parallel to the axis of the rod leak out of the cylindrical sides. The intense standing wave, composed of many coherent photons, can be started by one photon exactly in the axial direction.

speaking) of about 10^{-3} s before spontaneously dropping back to a level of lower energy.

The earliest of these spontaneously produced photons stimulate the production of others; each one is exactly in phase with the stimulating photon, and traveling in exactly the same direction (Fig. 24-7A and B). If this were all, though, it would be only an interesting experiment, rather than an increasingly useful tool. To produce a laser of this type, the synthetic ruby is made in the shape of a cylinder, with its ends exactly flat and exactly parallel. These ends are now silvered—one completely, to reflect as perfectly as possible, and the other only partially, so that a fraction of the incident photons may escape.

An early spontaneous photon that is of just the right wavelength and is exactly perpendicular to the silvered ends, will be reflected back and forth between them to set up a standing wave (Fig. 24-7C). This stimulates the emission of other photons, which follow it exactly in both phase and direction, and these stimulate others, and so on until the standing wave grows to enormous intensity. Photons from this standing wave that leak through the partially silvered end constitute the actual laser beam.

This is a beam of *coherent* light. All the photons are in phase with each other, and all are traveling in the same direction. Such a coherent beam will spread very little. It can be focused by lenses or mirrors to a tiny point of concentrated energy that is able to burn through steel, to make microscopic welds, or to sear living tissues in bloodless surgery.

The ruby laser ("laser," from the initial letters of "***L***ight ***A***mplification by ***S***timulated ***E***mission of ***R***adiation") emits its great energy in very short pulses. Between pulses, another flash of light is needed to again raise the electrons to their higher energy levels. Most gas lasers, operating on the same basic principle, have a much less powerful output, but have the advantage of being able to emit their coherent beams continuously, rather than in intermittent pulses. Recent research, however, has made possible the development of gas lasers whose continuous beams rival, or even exceed, the intensity of the intermittent flashes from crystal lasers.

24-6 Continuous Spectra

Jeans's trouble with the "ultraviolet catastrophe" and Planck's solution for the difficulties arose from the analysis of radiation from hot bodies. If you examine the spectrum radiated from the hot tungsten filament of an incandescent lamp, you will find it to be continuous—*all* frequencies are present, and there are no lines to be seen. Iron vaporized in an electric arc emits a very definite pattern of many lines, by which the element iron can be identified. But a crucible of incandescent molten iron or a red-hot piece of solid iron will emit a continuous spectrum that cannot be distinguished from the emission from gold or tungsten or any other solid or liquid substance at the same temperature.

In a gas at relatively low pressure, the atoms are separated far apart, and the energy levels available to the outer electrons come from the atom itself, without disturbance from the outside. These energies have definite values, and the spectrum of billions of similar undisturbed atoms will include only certain definite frequencies. In solids or liquids, however, or even in compressed gases, the atoms are crowded close together. The energy levels are constantly and unpredictably being distorted by the electric fields of neighbors, who come and go and interfere continuously and randomly. Under such circumstances, all frequencies are possible and the spectrum is continuous.

The spectrum of the sun is an interesting combination. First accurately observed and measured by the German physicist Joseph Fraunhofer (1787–1826), the solar spectrum consists of a continuous spectrum crossed by many thousands of dark lines, each one indicating a frequency that is missing (or at least reduced in intensity) from the sun's radiation. The sun is composed entirely of hot gas, and its density rapidly increases

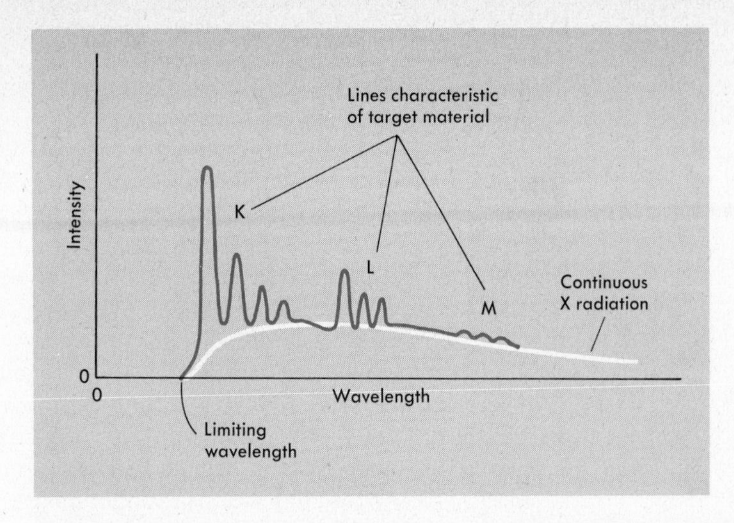

FIG. 24-8 The sun—a sphere of gas without a definite boundary. The photosphere is the layer of gas that emits the sunlight we see; some frequencies of this emission are absorbed in the chromosphere.

with depth. In the deep and denser layers of the solar atmosphere, gas atoms are crowded so close together that they emit a continuous radiation of all frequencies. The depth at which this continuous emission occurs is the *photosphere* (Fig. 24-8), which is what appears to us to be the "surface" of the sun as we observe it. In coming from the photosphere to our instruments, this radiation passes through thousands of miles of solar atmosphere of lower density, in which the atoms can act as individuals. In this less-dense region (the *chromosphere*), the atoms of each element present absorb their own characteristic frequencies. These absorbed frequencies we see as the dark *Fraunhofer lines*.

Analysis of these lines shows that, apart from the large predominance of hydrogen and helium gases, the chemical composition of the sun is identical with that of our earth. By using the methods of spectral analysis, astronomers have been able to learn a great deal not only about the sun, but also about the chemical composition of the planetary atmospheres and the outer envelopes of distant stars.

24-7 X-Ray Spectra

Corresponding to the single electron of the hydrogen atom, the outermost electrons of other atoms also have orbits of higher energy to which they can jump when excited by heat or strong electrical fields or by the absorption of radiation of the proper frequency and energy. In dropping back to their normal levels, these electrons radiate frequencies characteristic of the atoms to which they belong. As in the hydrogen atom, these outer-electron frequencies are in the range of visible light, infrared, or ultraviolet. X rays are likewise produced by electron jumps from one energy level to another, but the energy differences radiated away are enormously greater than those associated with outer electrons, and the resulting X rays are of very high frequency.

An X-ray tube is similar in principle to a cathode-ray tube, and electrons emitted from the cathode are accelerated through a vacuum by a potential of many thousands of volts, to strike against a metal anode. There are two ways in which this electron bombardment may cause high-frequency radiation to be emitted. First, the electron may be rapidly decelerated when it strikes the target. We know from Maxwell's work that any accelerated charge will emit electromagnetic radiation, and these electrons (since they are not in atomic orbits) are no exception. The deceleration of the free bombarding electrons is not quantized, so the resulting photons they emit can have any and all frequencies less than a certain limit. This limit comes from the principle of conservation of energy; the decelerating electron cannot emit a photon whose quantum of energy is greater than the kinetic energy of the electron. If, for example, the electrons are accelerated in the X-ray tube through a potential difference of 20,000 V, each electron will have an energy of $E = Vq = 2 \times 10^4 \times 1.6 \times 10^{-19} = 3.2 \times 10^{-15}$ J, or 3.2×10^{-8} erg. From Planck's $E = hf$, we see that the highest possible frequency that can be emitted is $f = 3.2 \times 10^{-15}/6.63 \times 10^{-34} = 4.8 \times 10^{18}$ Hz. This corresponds to a wavelength of $\lambda = 3 \times 10^8/4.8 \times 10^{18} = 6.2 \times 10^{-11}$ m, or 0.62 Å. This is shown as the "limiting wavelength" in Fig. 24-9.

Sometimes, however, a bombarding electron may actually knock an electron out of one of the inner shells of an atom of the target, or anode. When a heavy metal atom thus loses an inner electron, the vacancy is filled by one of the outer electrons falling down to take its place, and the energy differences between inner and outer electron shells are enormous,

FIG. 24-9 Schematic diagram of an X-ray spectrum.

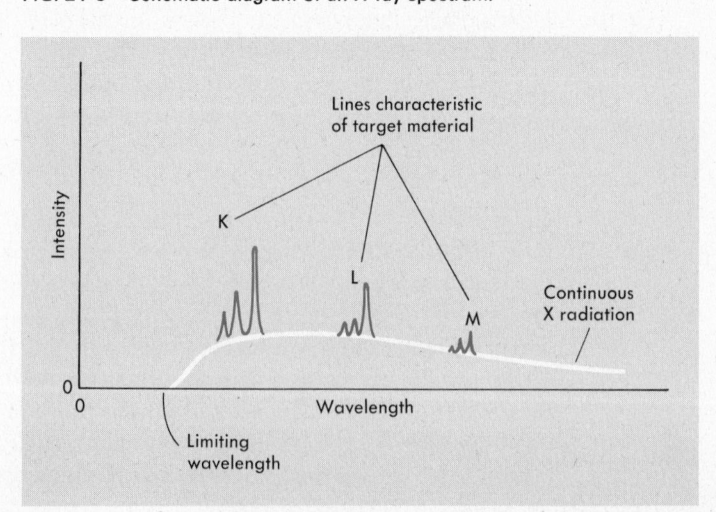

particularly for the atoms of heavy metals. These large energy differences, by the $E = hf$ relationship, are radiated away as energetic photons of very high frequency.

These photons of definite energy are shown as the spikes on the graph in Fig. 24-9. The series of lines marked K are caused by vacancies in the innermost K shell being filled by electrons from the L, M, and other outer shells. The lines marked L are emitted by electrons from outer shells falling into L-shell vacancies, etc. Each element has its own particular schedule of energy differences, and thus each element has a characteristic X-ray spectrum by which it can be identified.

Questions

(24-1)

1. (a) If $n = 4$, what values can l have? (b) What is l if there are shown to be five different values for m?

2. (a) For $n = 3$, what are the possible values for l? (b) For $l = 3$, what are the possible values for m?

3. In some systematic tabular form, list all the possible different sets of quantum numbers for $n = 3$.

4. List in some tabular form each individual different set of four quantum numbers possible for $n = 2$.

(24-2)

5. The volume of a gram-atomic weight of iron (at. wt, 55.8; density, 7.9 g/cm^3) is $55.8/7.9 = 7.1$ cm^3. Since an atomic weight contains 6×10^{23} atoms, the volume of an iron atom is $7.1/6 \times 10^{23} = 1.2 \times 10^{-23}$ cm^3. What is the approximate volume of an atom of (a) aluminum (at. wt, 27; density, 2.7 g/cm^3)? (b) lead (at. wt, 207; density, 11.4 g/cm^3)?

6. Compare the volumes of an atom of lithium, the lightest metal (at. wt, 6.94; density, 0.53 g/cm^3) and an atom of osmium, the densest metal (at. wt, 190; density, 22.5 g/cm^3).

(24-3)

7. List the four quantum numbers for each of the electrons in an atom of lithium ($Z = 3$).

8. List the four quantum numbers for each of the electrons in an atom of beryllium ($Z = 4$).

9. What are all the possible values of l for electrons in the M shell?

10. What are all the possible values of l for electrons in the N shell?

11. Tabulate in *spdf* notation all the electrons in an atom of phosphorus.

12. In *spdf* notation, list all the electrons in an atom of potassium.

13. Refer to the periodic table: What elements have (a) $n = 2$ and $l = 1$? (b) $n = 3$ and $l = 0$?

14. In the periodic table, Fig. 24-5, find the elements that have (a) $n = 2$ and $l = 0$; (b) $n = 3$ and $l = 1$.

15. Give n and l for (a) elements 71–80, (b) elements 31–36, (c) elements 57–70.

16. Give n and l for (a) elements 39–48, (b) elements 89–102, (c) elements 103–105.

(24-7) **17.** High-energy X rays of $\lambda = 0.08$ Å are to be produced by a special tube. What is the minimum potential difference that must be applied between its cathode and anode?

18. In an X-ray tube, there is a potential difference of 12,500 V between the cathode (which emits the electrons) and the anode (target). What is the wavelength of the highest-frequency X rays that this tube can emit?

25

The Wave Nature of Particles

25-1 De Broglie Waves

Although Bohr's theory of atomic structure was successful in explaining a large number of known facts concerning atoms and their properties, the three fundamental postulates underlying his theory remained quite inexplicable for a long time. The first step in the understanding of the hidden meaning of Bohr's quantum orbits was made by a Frenchman, Louis de Broglie, who tried to draw an analogy between the sets of discrete energy levels that characterize the inner state of atoms and the discrete sets of mechanical vibrations that are observed in the case of violin strings, organ pipes, etc. "Could it not be," de Broglie asked himself, "that the optical properties of atoms are due to some kind of standing waves enclosed within themselves?" As a result of these considerations, de Broglie came out with his hypothesis that *the motion of electrons within the atom is associated with a peculiar kind of waves which he called "pilot waves."* According to this unconventional view, each electron circling around an atomic nucleus must be considered as being accompanied by a standing wave that runs around and around the electronic orbit. If this were true, the only possible orbits would be those whose lengths are an integral multiple of the wavelength of the corresponding *de Broglie wave.*

Figure 25-1 shows the application of de Broglie's idea to the first three orbits of the Bohr hydrogen atom. In order to have n complete wavelengths (λ_n) fit into the circumference of the nth orbit, the following relationship must be true:

$$n\lambda_n = 2\pi r_n.$$

In Bohr's theory of the hydrogen atom, we had

$$r_n = \frac{n^2h^2}{4\pi^2kme^2}.$$

Substituting this into the preceding equation gives

$$n\lambda_n = \frac{n^2h^2}{2\pi kme^2}$$

$$\lambda_n = \frac{nh^2}{2\pi kme^2} = \frac{h}{m} \times \frac{nh}{2\pi ke^2}.$$

The last multiplier in the equation above may have a vaguely familiar look. If we turn back to Bohr again, we find it to be the inverse of the electron velocity:

$$v_n = \frac{2\pi ke^2}{nh}.$$

So, in the expression for λ_n, instead of multiplying by $nh/2\pi ke^2$, we can divide by v_n to find the very simple equation

$$\lambda_n = \frac{h}{mv_n}.$$

The relationship above states de Broglie's fundamental hypothesis, that ***the wavelength of the wave associated with a moving particle is equal to Planck's quantum constant divided by the momentum of the particle.***

Although we have justified this by considering an electron in a Bohr atom, if electrons are really accompanied by these mysterious de Broglie waves while moving along the circular orbits within an atom, the same might be true for the free flight of electrons as observed in free electron beams. And, if the motion of electrons in the beams is associated with some kind of waves, we should be able to observe the phenomena of *interference* and *diffraction* of electron beams in the same way that we

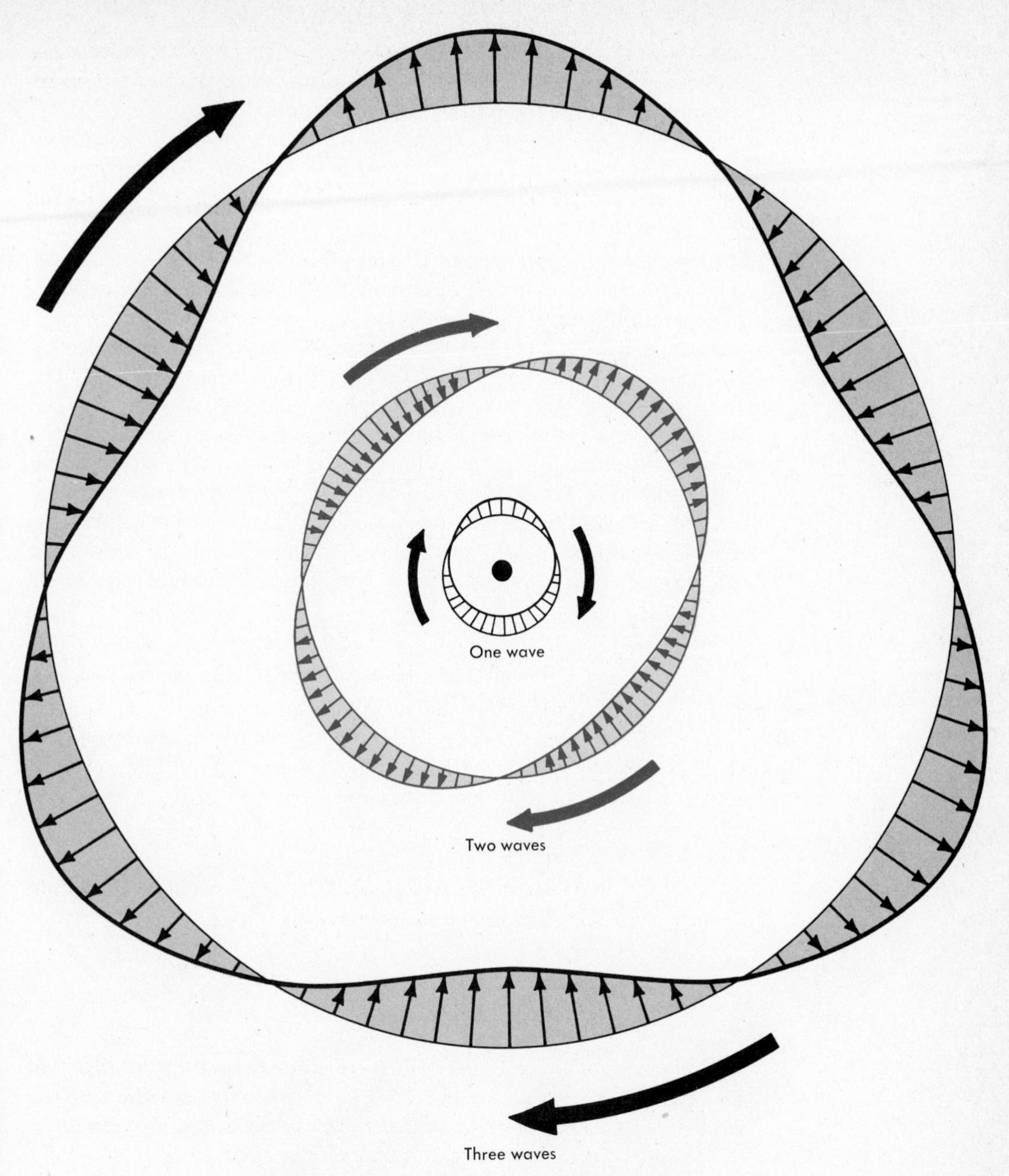

FIG. 25-1 De Broglie's waves as applied to the first three orbits in Bohr's hydrogen atom model.

observe these phenomena in beams of light. For a stream of moving electrons, we assume for the wavelength of the waves the same formula that applied to orbital electrons within the atom:

$$\lambda = \frac{h}{mv}.$$

For the electron beams used in laboratories, this wavelength comes out to be much shorter than that of ordinary visible light and is comparable, in fact, with the wavelengths of X rays, i.e., about 10^{-8} cm.

Fortunately, the crystal spectrograph had already been developed by the British physicists, father and son W. H. and W. L. Bragg, for the study of X rays. The finest optical gratings that it is possible to make have a line spacing of several thousand angstroms, and are useless for studying the diffraction patterns of radiation whose wavelength is in the neighborhood of 1 Å. Nature, however, has provided suitable gratings. We have seen that the diameter of atoms is a few angstroms, and in the regular crystals of metals, salts, and other substances, the atoms are arranged in perfect rows and layers separated by this distance (Fig. 25-2).

FIG. 25-2 A beam of X rays being reflected by the gratinglike molecular layers in a crystal lattice. (*Courtesy Scientific American.*)

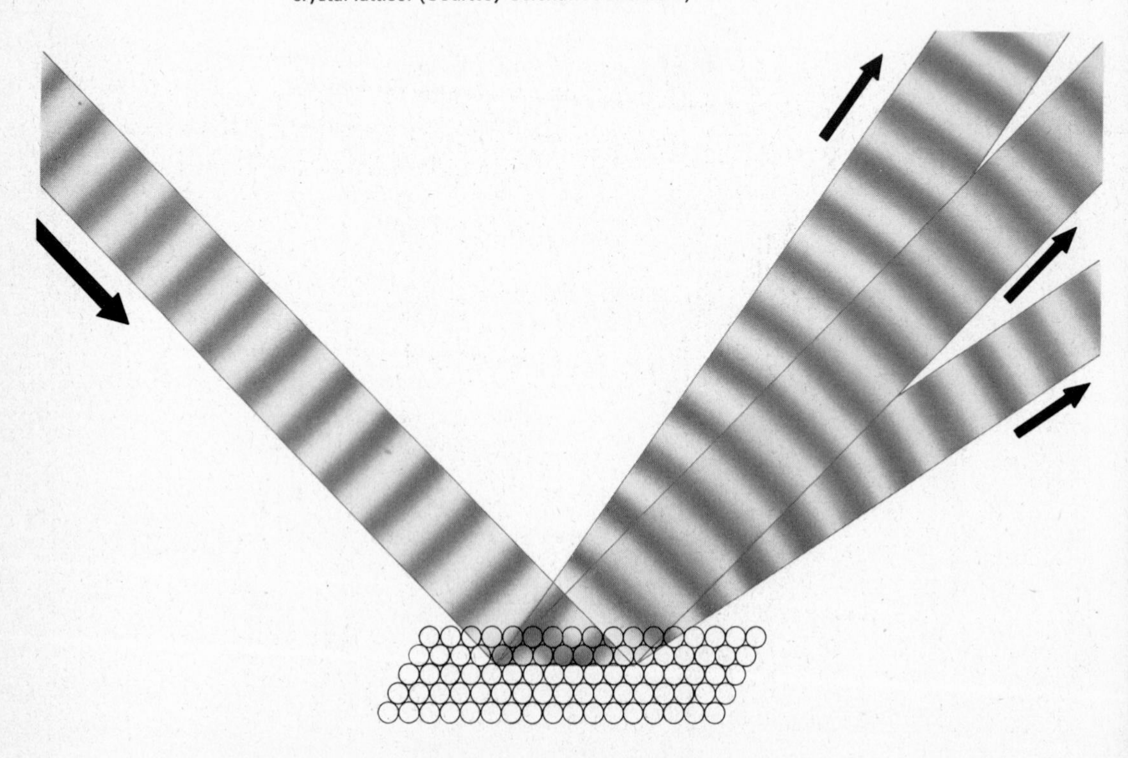

Incoming radiation of short wavelength is reflected, or scattered, by these regularly spaced crystal atoms, to be in phase and reinforced only in certain definite directions. Their behavior is similar to that of the longer waves of light when falling on a ruled reflection grating, as indicated in Fig. 21-15B. The resulting interference patterns are of course more complex than this, because the crystal is equivalent to three gratings acting simultaneously in three dimensions.

Figure 25-3A shows the pattern created by an X-ray beam passing through a layer of aluminum filings. The X rays, if they were diffracted by the regular array of atoms in a *single crystal*, would have produced a pattern of dots on the photographic film. The filings, being composed of a great number of crystals oriented in every possible direction, give the same pattern we would have if the single-crystal dots were rotated about the center line of the beam; all the dots the same distance from the center would combine to result in a circle.

Figure 25-3B is the pattern produced by a beam of electrons passing through a piece of thin aluminum foil. Electrons are not penetrating enough to pass through a layer of filings, but aluminum foil accomplishes

FIG. 25-3 (A) Diffraction of a beam of X rays by a layer of aluminium filings. Wavelength = 0.71A. (B) Diffraction of a beam of electrons by thin aluminium foil. Wavelength = 0.50A. (The actual photographs have been enlarged by different amounts to make the resulting patterns the same size in this reproduction.) (*Photos from the Film Studio, Education Development Center.*)

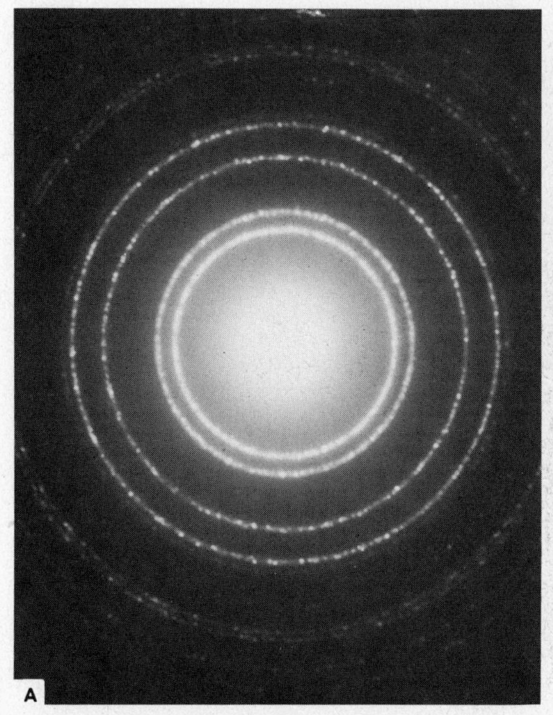

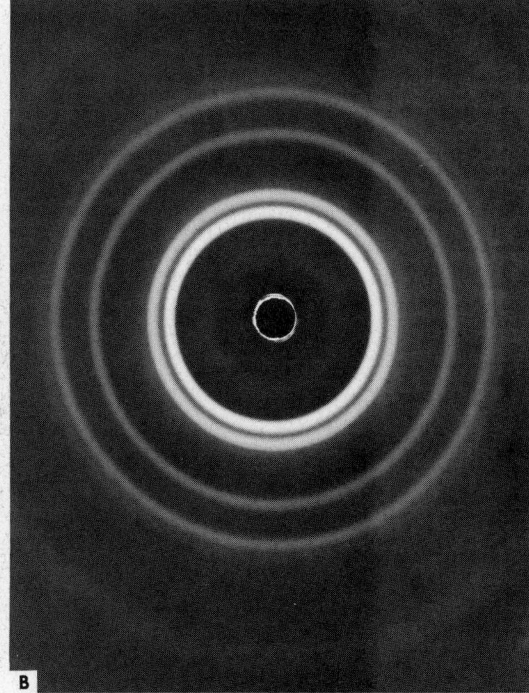

the same thing. Although it does not look to be, the foil (like the filings) is made up of a great number of individual crystals oriented in every direction. The resulting electron-beam diffraction pattern is a duplicate of the one produced by the X rays.

As an example, let us consider a beam of electrons accelerated in an "electron gun" through a potential difference of 200 V. Since 200 V is 200 J/C and since the electron has a charge of 1.6×10^{-19} C, each electron gains an energy $E = Vq = 200 \times 1.6 \times 10^{-19} = 3.2 \times 10^{-17}$ J. This is in the form of the electron's kinetic energy, so we have $\frac{1}{2}mv^2 = 3.2 \times 10^{-17}$ J, and $v^2 = 2 \times 3.2 \times 10^{-17}/9.11 \times 10^{-31} = 7.0 \times 10^{13}$. The electron speed is then $v = \sqrt{70 \times 10^{12}} = 8.4 \times 10^6$ m/s. Now we can use de Broglie's relationship to find the wavelength associated with these moving electrons:

$$\lambda = \frac{h}{mv} = \frac{6.63 \times 10^{-34}}{9.11 \times 10^{-31} \times 8.4 \times 10^6}$$

$$= 8.7 \times 10^{-11} \text{ m}, \quad \text{or} \quad 0.87 \text{ Å}.$$

This is a wavelength very suitable for showing diffraction by most crystals.

A few years later, similar experiments were repeated, using a molecular beam of sodium atoms (Fig. 25-4) instead of an electron beam, and it was found that diffraction phenomena exist in that case, too. Thus it became quite evident that, in material particles as small as atoms and electrons, the basic ideas of classical Newtonian mechanics should be radically changed by introducing the notion of waves associated with material particles in their motion.

The once definite distinction between waves and particles seems to have broken down. There are many sorts of interference experiments in which light shows itself to be unquestionably a wave phenomenon; yet in the photoelectric effect it concentrates all its energy on a single electron, as though it were a bulletlike particle. And now electrons and atoms, so surely particles, behave in some experiments as though they were waves.

The wave aspect and the particle aspect seem to be so mutually contradictory that it is quite natural to ask which one is "really" correct for a beam of light or an electron. The modern physicist will say that neither one is "really" correct. We are trying to make the submicroscopic world of the photon and the atom fit models we imagine as being analogous to tiny bullets and tiny ripples on water tanks. The world of the atom and the photon cannot be described in the same terms we use to describe the behavior of the macroscopic world of matter in bulk. That we cannot is attested to by the wave–particle dilemma and the contradictions of a similar nature we run into when we try. The physicist has mathematical equations whose solutions will give the correct answers,

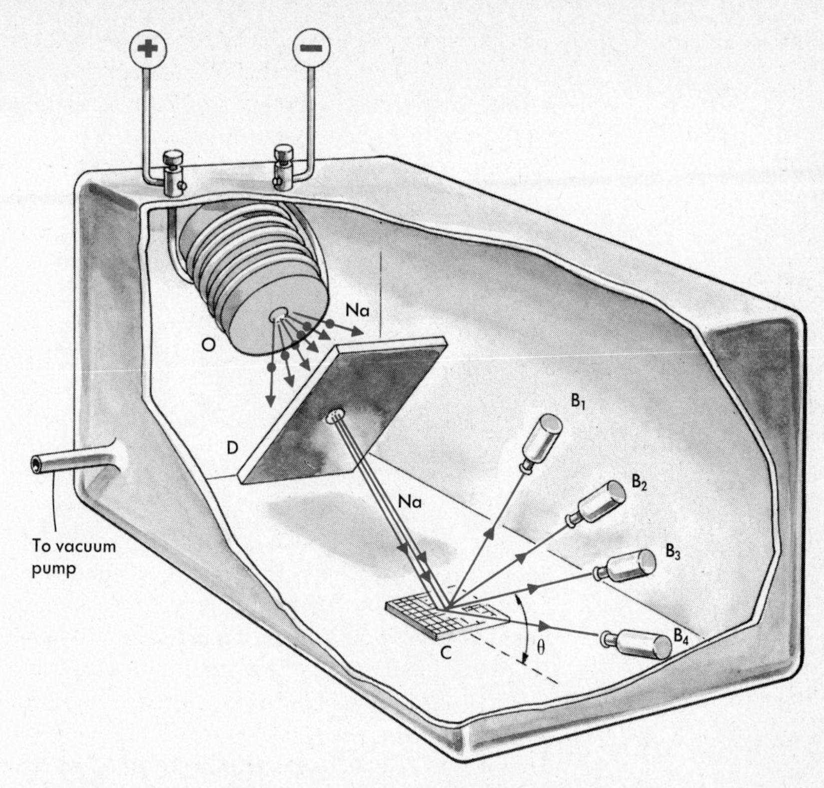

FIG. 25-4 O. Stern's demonstration of the diffraction of atoms. Some of the atoms from the oven O pass through the small hole in the barrier *D*, from which they emerge as a narrow beam. The atoms reflected from the crystal are trapped and measured in the collectors *B*. The results show a strong maximum in the direction required by the ordinary law of reflection, plus a number of other secondary maxima corresponding to a diffraction pattern.

whether it be wave or particle that is involved; but to these equations no model or picture is connected. We must learn either not to ask questions about which model is "really" correct or not to think of any model at all. We quite plainly cannot have it both ways. An intricate mathematical method for handling this kind of problem was worked out by an Austrian physicist, E. Schrödinger, and it represents the subject matter of an important but rather difficult branch of modern theoretical physics known as *wave mechanics*, or *quantum mechanics*.

25-2 The Electron Microscope

In Section 8 of Chapter 21, it was noted that the ability of a microscope to resolve fine detail is limited by the wavelength of the light used to illuminate the object and form its image. This limited resolution makes magnification of more than 1000X, or at most, 2000X, useless in an optical instrument using visible light or ultraviolet.

The reality of "electron waves" has made the *electron microscope* a routine tool. Let us figure the wavelength of electrons that have been accelerated through a potential difference of 15,000 volts:

$$E = Vq = 1.5 \times 10^{4} \times 1.6 \times 10^{-19} = 2.4 \times 10^{-15} \text{ joule.}$$

$$\text{KE} = \tfrac{1}{2}mv^{2} = 2.4 \times 10^{-15}$$

$$v = \sqrt{2 \times 2.4 \times 10^{-15}/9.11 \times 10^{-31}}$$

$$= 7.26 \times 10^{7} \text{ m/s.}$$

$$\lambda = h/mv = 6.63 \times 10^{-34}/(9.11 \times 10^{-31} \times 7.26 \times 10^{7}) = 10^{-11} \text{ m}$$

$$= 0.1 \text{ Å.}$$

This is tens of thousands times shorter than the wavelength of visible light, but so far the electron microscope is not able to distinguish details this fine. The numerical aperture is necessarily much smaller than that of an optical microscope, and there are technological difficulties that prevent focussing electron beams as precisely as a lens system can focus light. In spite of these difficulties, magnifications of 100,000X can be reached—100 times the magnification of an optical microscope.

Of course, speeding electrons cannot be refracted by glass or plastic lenses; instead, their paths are bent by electric or magnetic fields. Figure 25-5 is a schematic diagram of an electron microscope. Its "magnetic lenses" are ring-shaped coils in iron cases, carefully designed to provide a magnetic field of accurately varying intensity and direction, which will refract streams of electrons in the same way that a converging lens refracts light rays.

25-3 The Uncertainty Principle

Now things seem to be going from bad to worse. First, we had Bohr's "quantized orbits" that looked like railroad tracks along which the electrons were running around the atomic nucleus. Then these tracks were replaced by mysterious "pilot waves" that were supposed to provide "guidance" for the electrons in their orbital motion. It all seemed to be against common sense, but, on the other hand, these developments of the quantum theory provided us with the most exact and most detailed explanation (or description) of the properties of atoms—their spectra, their magnetic fields, their chemical affinities, etc. How could it be? How could such a picture, nonsensical at first sight, lead to so many positive results? Here we have to repeat what was already said in connection with Einstein's theory of relativity. Modern physics extends its horizons far beyond the everyday experience upon which all the "commonsense" ideas of classical physics were based, and we are thus bound to find striking deviations from our conventional way of thinking and must be prepared to encounter facts that sound quite paradoxical to our

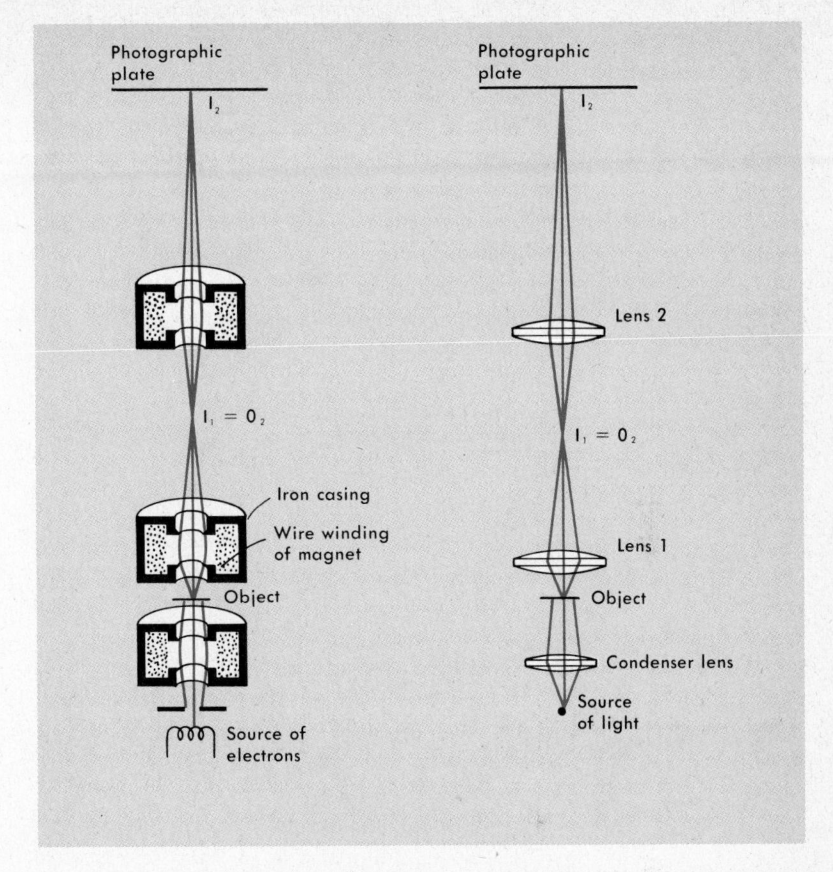

FIG. 25-5 In an electron microscope, beams of short wavelength electrons are "refracted" by magnetic fields. The principle of magnification is the same as that of an optical microscope with glass lenses, shown at the right.

ordinary common sense. In the case of the theory of relativity, the revolution of thought was brought about by the realization that space and time are not the independent entities they were always believed to be, but are parts of a unified space–time continuum. In the quantum theory, we encounter the nonconventional concept of *a minimum amount of energy*, which, although of no importance in the large-scale phenomena of everyday life, leads to revolutionary changes in our basic ideas concerning motion in tiny atomic mechanisms.

Let us start with a very simple example. Suppose we want to measure the temperature of a cup of coffee but all we have is a large thermometer hanging on the wall. Clearly, the thermometer will be inadequate for our purpose because when we put it into the cup it will take so much heat from the coffee that the temperature shown will be considerably less than that which we want to measure. We can obtain a much better result if we use a small thermometer that will show the temperature of the coffee and take only a very small fraction of its heat content. The smaller the thermometer we use for this measurement, the smaller is the disturbance caused by the measurement. In the limiting case, when the thermometer is "infinitely small," the temperature of the coffee in the cup will not be affected at all by the fact that the measurement was carried out. The commonsense concept of classical physics was that this is always the case in whatever physical measurements we are carrying out, so that we can always compute the disturbing effect of whatever gadget is used for the measurement of some physical quantity and find the exact value we want. This statement certainly applies to all large-scale measurements carried out in any scientific or engineering laboratory, but it fails when we try to stretch it to such tiny mechanical systems as the electrons revolving around the nucleus of the atom. Since, according to Max Planck and his followers, energy has "atomic structure," *we cannot reduce the amount of energy involved in the measurement below one quantum*, and making exact measurements of the motion of electrons within an atom is just as impossible as measuring the temperature of a demitasse of coffee by using a bulky bathtub thermometer! But, whereas we can always obtain a smaller thermometer, it is absolutely impossible to find less than one quantum of energy.

A detailed analysis of the situation indicates that the existence of minimum portions (quanta) of energy prevents our being able to describe the motion of atomic particles in the conventional way, by calculating their successive positions and velocities. Quantum mechanics shows us that the position of a particle and its momentum can be known only within certain limits. These limits are negligibly small for large-scale objects, but become of great importance in the submicroscopic world of atoms and atomic particles.

The uncertainty in our knowledge of the coordinate x and the momentum p of any particle can be expressed by writing $x \pm \Delta x$ and $p \pm \Delta p$, which means that all we can say is that the particle is located somewhere between $x - \Delta x$ and $x + \Delta x$, and the momentum of the particle lies somewhere between $p - \Delta p$ and $p + \Delta p$. The German theoretical physicist Werner Heisenberg showed in 1927 that these uncertainties are related by the expression

$$\Delta x \times \Delta p = \frac{h}{2\pi}.$$

If there is no appreciable relativistic change in mass, $p = mv$, and we can rewrite the uncertainty principle as

$$\Delta x \times \Delta v = \frac{h}{2\pi m}.$$

Let us apply this to a particle with a mass of 1 mg. Then, in CGS units,

$$\Delta x \times \Delta v = \frac{6.63 \times 10^{-27}}{2\pi \times 10^{-3}} \approx 10^{-24},$$

which means that if we know the position of the particle to within $\pm 10^{-12}$ cm, we are permitted by nature to know the particle's velocity only to within 10^{-12} cm/s. Obviously, such small uncertainties are of absolutely no significance when we deal with milligrams—or even micrograms—or larger particles.

When we deal with atomic particles, however, the case is different. Let us consider an $n = 1$ electron in a hydrogen atom (Fig. 25-6). We cannot at any instant know just where in its orbit the electron is—it might be on the left side or the right side, as shown. Our uncertainty in its position is $\pm r$, which equals $\pm h^2/4\pi^2 mke^2$. Similarly, we cannot know if it is moving upward or downward with its velocity $\pm v$. We thus have an uncertainty of its momentum equal to $\pm m\,\Delta v = \pm 2\pi mke^2/h$. Putting these together, we find

$$\Delta x \times \Delta p = \frac{h^2}{4\pi^2 mke^2} \times \frac{2\pi mke^2}{h} = \frac{h}{2\pi}.$$

These uncertainties are so large as to make the classical Bohr picture of definite particles in definite orbital motion completely invalid. De Broglie's waves and Heisenberg's uncertainty principle led physicists to a new way of describing the behavior of atomic "particles."

25-4 Waves of Probability

When light was shown to have wave properties, the nineteenth-century physicists at once asked, "waves in what?" In order to answer this question, they invented the "ether," with all of its contradictory properties, only to have it taken away from them by Einstein.

FIG. 25-6 Uncertainty of position, and uncertainty of velocity (and hence of momentum) of an electron in a hydrogen atom.

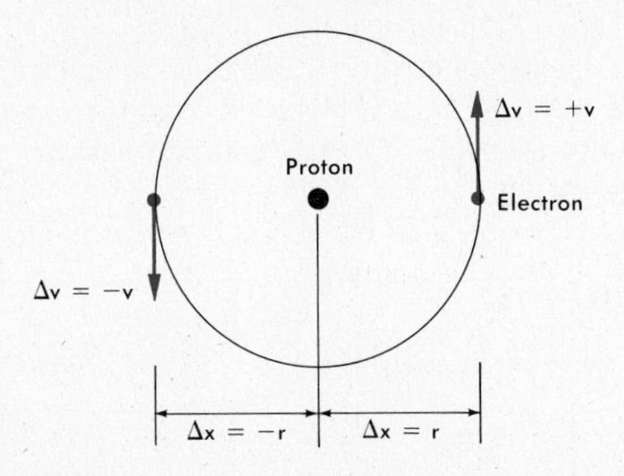

It was not necessary to invent a medium for the de Broglie waves but it was necessary to invent an *interpretation* of them. It turns out that the waves have no real material characteristics but are purely a measure of *probability*. And since the waves are an inescapable attribute of every material particle, we must regretfully conclude that we can never fully know the momentum of a particle, or its location, but can at best only say that some certain momentum is more probable than another or that the particle is more likely to be here than there.

Fundamentally, this was the basic difficulty with Bohr's atomic model. It pictured an atomic electron as a particle with a definite location, and a definite velocity and momentum at all times. If a modern physicist were forced to make some sort of a model, he might picture the electron of a hydrogen atom, say, as a pattern of standing waves. These standing waves are not confined by definite boundaries, as are those of a vibrating violin string or the electromagnetic waves in Jeans's imaginary reflecting cube. They are instead confined by the inverse-square Coulomb force of attraction between the nucleus and the electron's negative charge. Since this force extends out more and more weakly without limit, there is a *very* small chance that any given electron may be anywhere in the universe. Calculations show, however, that in the hydrogen atom in its lowest-energy $n = 1$ state, the probability waves, which can be pictured as spreading out in a sort of cloud of varying density, are most dense at a distance of 0.53 Å from the nucleus. This is where the wave mechanics calculations say the electron is most likely to be, and this is exactly the radius of Bohr's first hydrogen orbit!

In spite of this agreement, there is a great conceptual difference between the Bohr model and the quantum mechanics interpretation.

In the Bohr model, there was no real reason for saying that angular momentum must come in discrete units of $h/2\pi$, except that this is what *must* be assumed if the answers are to come out right and in agreement with the experimental spectral data. In the solutions of Schrödinger's probability-wave equation, no assumptions of this sort are necessary, and the quantum numbers n, l, and m come out with their allowed integral values as a part of the answer. The probability cloud takes on different shapes and dimensions to represent the higher-energy excited states, but still remains a diffuse cloud.

If the shape and density of this cloud are all we can ever know about an electron (or, if indeed, this is all there *is* to know about it), then the uncertainty principle follows naturally as an inescapable, built-in characteristic of how the world is made.

Questions

(25-1) 1. The speed of an electron is 1.2×10^6 m/s. What is its de Broglie wavelength?

2. What is the wavelength of an electron moving with a velocity of 2×10^7 cm/s?

3. With what speed must an electron travel in order to have a de Broglie wavelength of 10^{-8} cm?

4. Interference effects show that the wavelength associated with a beam of electrons is 3.2×10^{-8} cm. What is the velocity of the electrons in the beam?

5. What wavelength is associated with a hydrogen electron in its $n = 3$ orbit?

6. What is the wavelength associated with the hydrogen electron in its $n = 2$ orbit?

7. (a) What is the kinetic energy of an electron that has a de Broglie wavelength of 10^{-8} cm? (b) Through what potential difference must it have been accelerated?

8. (a) What is the kinetic energy of the electrons in Question 4? (b) Through what potential difference must the electrons have been accelerated in order to have this much energy?

9. A proton and an alpha particle have the same kinetic energy. (a) How do their speeds compare? (b) How do their momenta compare? (c) How do their de Broglie wavelengths compare? ($m_\alpha = 4m_p$.)

10. A proton and an electron are given the same kinetic energy ($m_p = 1836m_e$). (a) How do their speeds compare? (b) How do their momenta compare? (c) How do their de Broglie wavelengths compare?

11. What wavelength is associated with a beam of protons (see Question 10) accelerated through a potential difference of 75 volts?

12. What wavelength is associated with a beam of electrons accelerated through a potential difference of 75 volts?

13. Potassium (at. wt, 39.1) vaporizes at a temperature of 1047°K, and the atoms of potassium vapor at this temperature have an rms speed of about 510 m/s. What is the wavelength of a potassium atom at this speed?

14. Vaporized sodium (at. wt, 23) has a temperature of about 1165°K, which gives the atoms of vapor an rms speed of about 700 m/s. What is the de Broglie wavelength of a sodium atom at this speed?

(25-3) 15. A 1-g bullet is fired with a speed of 300 m/s, which the experimenter knows is accurately determined to within 0.01 percent. What restriction does the uncertainty principle put on the determination of the bullet's position at any time?

16. Take one of the droplets in Millikan's oil-drop experiment and assume it to have a mass of 10^{-12} g. If it were possible to determine its position at any moment with an accuracy of $\pm 10^{-3}$ cm, what must be the irreducible minimum uncertainty as to its speed at that moment?

17. A beam of protons (see Question 10) has a speed of 5×10^7 cm/s, with an uncertainty of 0.1 percent. What is the least possible uncertainty in the location of a proton?

18. Electrons in a beam have a speed of 5×10^7 cm/s, with an uncertainty of 1 part in 10^3. What is the least possible uncertainty the experimenter may have in the position of any electron?

19. A proton (see Question 10) is in imagination confined in small box 1 μ wall-to-wall. (a) What is our uncertainty as to its location? (b) What is our minimum uncertainty about its speed? (c) Can the proton ever be definitely known to be at rest within the box? Explain.

20. An electron is placed in the box of Question 19, to replace the proton. How does our uncertainty about its speed compare with the uncertainty of our knowledge about the proton's speed?

26

Radioactivity and the Nucleus

26-1 Discovery of Radioactivity

The discovery of radioactivity, like that of many other unsuspected aspects of physics, was purely accidental. It was discovered in 1896 by the French physicist A. H. Becquerel (1852–1908), who was interested at that time in the phenomenon of fluorescence, which is the ability of certain substances to transform ultraviolet radiation that falls on them into visible light. In one of the drawers of his desk, Becquerel kept a collection of various minerals that he was going to use for his studies, but, because of other pressing matters, the collection remained untouched for a considerable period of time. It happened that in the drawer there were also several unopened boxes of photographic plates, and one day Becquerel used one of the boxes in order to photograph something or other. When he developed the plates, he was disappointed to find that they were badly fogged as if previously exposed to light. A check on the other boxes in the drawer showed that they were in the same poor condition, which was difficult to understand, since all the boxes were sealed and the plates inside were wrapped in thick black paper. What could have been the cause of this mishap? Could it have something to do with one of the minerals in the drawer? Being of inquisitive mind, Becquerel investigated the situation and was able to trace the guilt to a piece of uranium ore labeled "Pitchblende from Bohemia." One must take into account, of course, that at that time uranium was not in vogue as it is today. In fact, only a very few people had ever heard about this comparatively rare and not very useful chemical element.

But the ability of uranium ore to fog photographic plates through a thick cardboard box and a layer of black paper rapidly brought this obscure element to a prominent position in physics.

The existence of penetrating radiation that can pass through layers of ordinarily opaque materials as if they were made of clear glass was a recognized fact at the time of Becquerel's discovery. In fact, only a year earlier, in 1895, a German physicist, Wilhelm Roentgen (1845–1923), discovered, also by sheer accident, what are now known as X rays, which could penetrate equally well through cardboard, black paper, or the human body. But, whereas X rays could be produced only by means of special high-voltage equipment shooting high-speed electrons at metallic targets, the radiation discovered by Becquerel was flowing steadily, without any external energy supply, from a piece of uranium ore resting in his desk. What could be the origin of this unusual radiation? Why was it specifically associated with the element uranium and, as was found by further studies, with two other heavy elements, thorium and actinium?

The early studies of the newly discovered phenomenon, which was called *radioactivity*, showed that the emission of this mysterious radiation was completely unaffected by physical and chemical conditions. One can put a radioactive element into a hot flame or dip it into liquid air without the slightest effect on the intensity of the radiation it emits. No matter whether we have pure metallic uranium or its chemical compounds, the radiation flows out at a rate proportional to the amount of uranium in the sample. These facts led the early investigators to the conclusion that the phenomenon of radioactivity is so deeply rooted in the interior of the atom that it is completely insensitive to any physical or chemical conditions to which the atom is subjected.

Becquerel's discovery attracted the attention of the Polish-born Marie Sklowdowska. She was at that time a graduate student in chemistry at the Sorbonne in Paris, and soon married the brilliant young physicist Pierre Curie. The Curies suspected that the strong radioactivity of certain uranium ores was due to other highly radioactive elements present in the ore in small quantities. After a long and tedious series of chemical separations performed on large quantities of uranium ore, in 1898 they succeeded in isolating small samples of two new highly radioactive elements. These were named *polonium* (in honor of Marie Curie's native country) and *radium*.

26-2 The Nature of the Nucleus

It is obvious to us now that the mysterious radiation emitted by certain elements of high atomic number is an activity of the atomic nucleus. Madame Curie and her contemporaries, however, did not have the advantage of our more modern knowledge. (Remember that her separation of polonium and radium took place 13 years before Rutherford's scattering experiments first showed that atoms even have a nucleus!)

Rather than grope with our ancestors through decades of speculation until the neutron was discovered in 1932, we can help our understanding more by taking advantage of our superior historical position. In a later chapter, we will look at the nucleus and its behavior in more detail; for our present purpose, it will be enough to merely sketch some of its main features.

In the Bohr atom, we identified the hydrogen nucleus to be a single *proton*—a particle having a mass of 1.673×10^{-27} kg, about 1836 times the mass of an electron. Its charge is the same as the charge on an electron (1.602×10^{-19} C), but is positive.

But what of the nuclei of the heavier elements? It was suspected in the early days that they were made up primarily of aggregations of protons. Oxygen, for example, with a mass about 16 times the mass of a hydrogen atom, was thought to have a nucleus of 16 protons. Since its positive charge is only $8e$, it was assumed that the oxygen nucleus also contained 8 electrons to neutralize the proper part of the proton charge. Quantum mechanics, however, showed that the de Broglie waves of these supposed nuclear electrons could not possibly fit in the small volume of a nucleus. As we shall see, the discovery of the *neutron* in 1932 solved this uncomfortable problem. The neutron has a mass slightly greater than the proton (1.675×10^{-27} kg) and has no electrical charge.

Although there is still a great deal that we do not understand about nuclei, we do know that for most practical purposes we can consider them to be aggregations of protons and neutrons. The oxygen nucleus, with a charge of $+8e$, must contain 8 protons; to make up the rest of the required mass, it also contains 8 neutrons. Armed with this basic knowledge, we are in a better position to understand much of the early experimental work.

26-3 Canal Rays and Isotopes

While the study of cathode rays in Thomson's tube led to the discovery of electrons, the study of *canal rays*, which are streams of positively charged gas ions, was also very helpful to the understanding of the inner nature of the atom. The apparatus Thomson used for the study of canal rays was a modification of the tube used for determining the e/m ratio and is shown in Fig. 26-1. A small amount of gas is left within the tube, and when a swiftly moving electron collides with a gas molecule, an electron is likely to be knocked off, making the molecule into an ion with a positive charge. The mass and electric charge of these positively

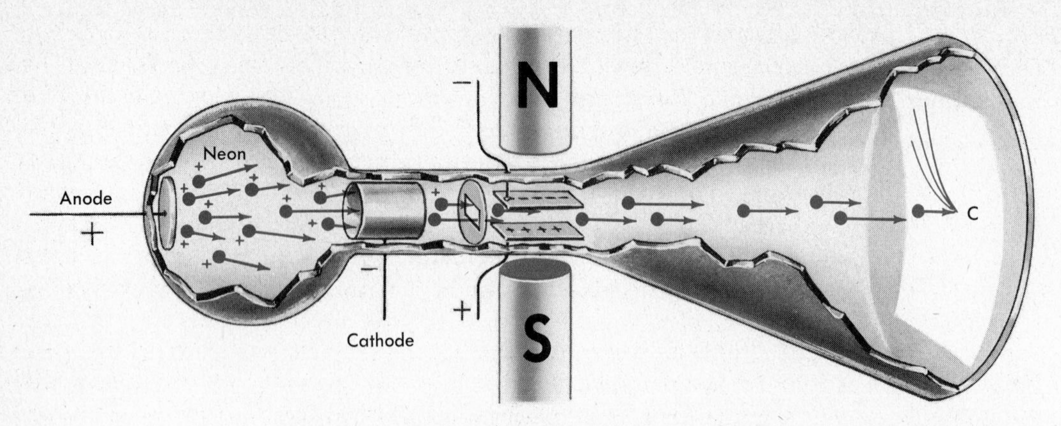

FIG. 26-1 The apparatus that led to the discovery of isotopes. Positive neon ions were accelerated by an electric field and formed a thin beam after passing through a slit. The beam was deflected by an electric field and a magnetic field, and fell on a fluorescent screen at the end of the tube. If all of the neon ions had the same mass, the trace on the screen would have been a single parabola (different points on the parabola corresponding to different velocities). Actually, there were three different parabolas, corresponding to masses of 20, 21, and 22.

charged canal rays can be analyzed by deflecting them in parallel electric and magnetic fields.

In his e/m work, Thomson used electric and magnetic fields perpendicular to each other, so that the two deflections could be made to add up to zero. If you go back and check over that section, you will see that his approach was possible only because all the electrons in his beam had the same velocity. This was because all of them originated on the cathode, and all were accelerated through the same total potential difference between cathode and anode.

With the canal rays, composed of positive ions, the situation was different. The neutral neon atoms he used were unaffected by the electric field between cathode and anode. Their ionization by the energetic electrons streaming out from the cathode was caused by collisions that could take place almost anywhere in the field. The result was that some positive ions would be accelerated through only a small part of the field, some through nearly all of it, and others in between. The beam of canal rays therefore consisted of a stream of positive neon ions with a wide range of speeds.

Although we will not stop to do it, it is not difficult to show that such a stream of ions, all of the same charge and mass but with different velocities, will leave a trace in the shape of a parabola. The mass of the ions can be computed quite accurately from the geometry of the parabola.

In measuring the mass of the particles forming canal rays in a tube filled with neon gas, Thomson expected to confirm the chemical value of the atomic weight of neon, which was known to be 20.18. Instead of this value, however, he found only 20.0, which was considerably lower, and well beyond the limits of possible experimental error. The discrepancy was explained when Thomson found later that the beam of ions passing through the magnetic and electric fields was not deflected as a single beam, but was split into three branches, as shown in Fig. 26-1. (Actually, the third branch was discovered later by Thomson's co-worker, F. W. Aston, using improved apparatus.) The particles in the main branch, containing over 90.5 percent of all the neon ions, had a mass value of 20.0; the other fainter branch contained about 9.2 percent, and had a mass of 22.0. And a still fainter branch contained 0.3 percent of mass 21.0.

This was very remarkable! Here Thomson had found three kinds of neon atoms, *identical in chemical nature and having the same optical spectra, but different in mass.* On top of this, the mass values were almost exactly integral numbers. Ordinary neon, then, was actually a mixture of three different neons, and the chemical weight was just the average weight of this mixture.

The different types of neon were called *isotopes* of this element, which means in Greek "same place" and refers to the fact that all the neons of different weight occupy the same place in the table of elements. We can confirm this average weight by taking the weighted average from Thomson's data:

$$\begin{aligned} 0.905 \times 20 &= 18.10 \\ 0.092 \times 22 &= 2.02 \\ 0.003 \times 21 &= \underline{0.06} \\ \text{Average mass} &= 20.18. \end{aligned}$$

We can understand now just what the difference is between the three isotopes of neon. Since the atomic number of neon is 10 ($Z = 10$), we know that each atom must have 10 protons in its nucleus. If this number were *not* 10, it would have a different number of atomic electrons and would be a different element, rather than neon. To make up the observed masses, the isotope of mass 20 must have $20 - 10 = 10$ neutrons in its nucleus; the neon-21 must have $21 - 10 = 11$ neutrons; and the neon-22, 12 neutrons.

There is a fairly standardized system of notation for indicating different isotopes. In front of the letters of the chemical symbol of the element, a subscript gives the *atomic number* (Z), which is the number of protons in the nucleus. After the symbol, the *mass number* is placed as a superscript. The mass number is merely the total number of protons plus neutrons in the nucleus—or, more simply, the total number of *nucleons*, a term used to include both protons and neutrons. Thus, for our three isotopes of neon, we could write ${}_{10}Ne^{20}$, ${}_{10}Ne^{21}$, and ${}_{10}Ne^{22}$. (The prefix subscript 10 is not necessary; if the element is neon, this number *must* be 10 and it is therefore redundant. Nevertheless, for convenience, it is often used.)

Thomson's original crude apparatus has been improved by Aston,

A. J. Dempster, and K. T. Bainbridge; and the modern *mass spectrograph* can determine the relative masses of isotopes with great accuracy.

Further studies have shown that almost every element represents a mixture of several isotopes. While in some cases (as in gold and iodine), one isotope accounts for 100 percent of the material, in many other cases (as in chlorine and zinc), different isotopes have comparable abundances. The isotopic composition of some of the chemical elements is shown in Table 26-1.

TABLE 26-1 ISOTOPIC COMPOSITION OF SOME ELEMENTS

Atomic Number	Name	Isotopic Composition Percentage Shown in Parentheses
1	Hydrogen	1 (99.985); 2 (0.015)
6	Carbon	12 (98.9); 13 (1.1)
7	Nitrogen	14 (99.64); 15 (0.36)
8	Oxygen	16 (99.76); 17 (0.04); 18 (0.20)
17	Chlorine	35 (75.4); 37 (24.6)
30	Zinc	64 (48.89); 66 (27.81); 67 (4.07); 68 (18.61); 70 (0.62)
48	Cadmium	106 (1.215); 108 (0.875); 110 (12.39); 111 (12.75); 112 (24.07); 113 (12.26) 114 (28.86); 116 (7.58)
80	Mercury	196 (0.15); 198 (10.02); 199 (16.84); 200 (23.13); 201 (13.21); 202 (28.80); 204 (6.85).

26-4 Alpha, Beta, and Gamma Rays

Armed with a little twentieth-century knowledge, we can better understand some of the problems uncovered by the research of earlier years. Becquerel and his followers were puzzled to find that the radiation from their impure mixtures of radioactive elements consisted of three different components.

It is quite easy (in principle, at least) to separate these three types of radiation when they are emitted from a small piece of material containing a mixture of radioactive elements. If we drill a small hole in a block of lead, which is a good absorber of radiations of all kinds, and place a speck of radioactive material at the bottom of the hole, a narrow, well-defined beam of radiation will be emitted from the top of the hole (Fig. 26-2). If this beam is passed through a strong electric field between a pair of parallel plates, the single beam will be separated into its three components, as shown.* The same separation will follow if, instead of

* Figure 26-2 is somewhat deceptive. The deflection of α's and β's cannot be measured by application of the same field, due to the very much greater mass of the alpha particle. A field (either electric or magnetic) that deflects α's satisfactorily winds the electron paths into tight curls; a field satisfactory for the β's leaves the α's practically undeviated.

an electric field, the beam is passed through a strong magnetic field perpendicular to the plane of the drawing. Lacking knowledge of just what these radiations were, the experimenters named them simply alpha (α), beta (β), and gamma (γ) radiation, from the first three letters of the Greek alphabet.

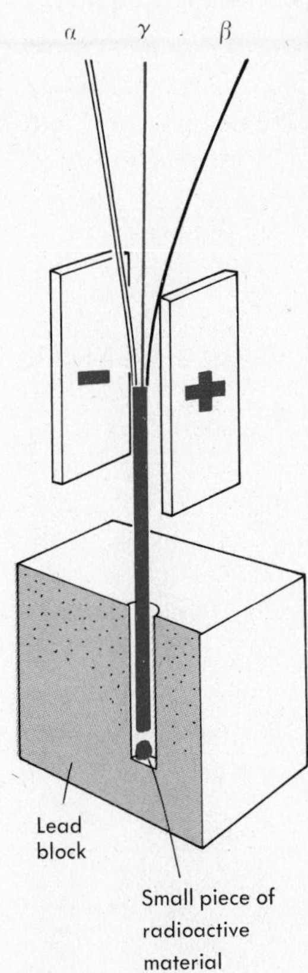

FIG. 26-2 The separation of alpha, beta, and gamma radiation by passage through an electric field. (The same sort of separation can also be accomplished by a magnetic field.)

From their behavior in electric and magnetic fields, it was apparent that the α radiation consisted of positively charged particles; the β radiation, of negatively charged particles; and the γ radiation was either neutral particles or electromagnetic radiation. They were all soon identified.

Alpha rays were found (by Rutherford) to consist of the fast-moving nuclei of helium atoms, now called α particles. From the atomic weight and atomic number of helium, we see that the helium nucleus must contain 2 protons and $4 - 2 = 2$ neutrons, making 4 nucleons altogether. Thus, when an unstable nucleus ejects an α particle, it loses 4 nucleons, and its *mass number is reduced by 4.* And, since 2 of these nucleons are protons, its *atomic number is reduced by 2,* and it of course then becomes another element!

Beta rays were found to be nothing more than energetic, fast-moving electrons. No one has ever been able to detect any difference between these electrons emitted from nuclei and the ordinary atomic electrons with which we are already familiar. But we have said that an electron cannot exist inside a nucleus; how, then, can a nucleus emit something it did not contain in the first place? The answer to this is that the electron is manufactured as it is emitted. The mass of an electron, 9.11×10^{-31} kg, can be produced (via Einstein's $E = mc^2$) from $9.11 \times 10^{-31} \times (3 \times 10^8)^2 = 8.20 \times 10^{-14}$ J of energy. To account for the electron's negative charge, we may think of the neutral neutron as containing both $+e$ and $-e$ of charge. If the $-e$ is carried away by the electron, then the nuclear particle that was a neutron becomes a nucleon with a charge of $+e$—in other words, a neutron has been changed to a proton. Thus, when a β particle is emitted, the *mass number is unchanged* and the *atomic number is increased by 1.*

Gamma rays are the pure energy of electromagnetic radiation and are emitted together with some α particles and some β particles. They do not change the number of nucleons, and they have no charge. Therefore, *the mass number and the atomic number are both unchanged.*

26-5 Families of Radioactive Elements

Radioactivity, as observed by Becquerel, turned out to be a composite effect that was due to the presence of a large number of radioactive elements. In fact, studies by the British physicist Soddy and his famous collaborator Rutherford showed that this mixture contained over a dozen individual elements.

In the principal uranium family, which also includes radium, uranium plays the role of the head of the family, and being very long-lived, produces numerous children, grandchildren, great-grandchildren, etc. The genealogy of this family is shown in Fig. 26-3. The nucleus of an atom of $_{92}U^{238}$ emits an α particle and is transformed into another element that we can temporarily call X. Since the α particle carried away 2 protons and 2 neutrons for a total of 4 nucleons, we see that X must have an atomic number of $92 - 2 = 90$ and a mass number of $238 - 4 = 234$. The periodic table shows that the element of atomic number 90 is called thorium (Th). This transformation of an isotope of uranium into an isotope of thorium can be simply recorded as a nuclear equation:

$$_{92}U^{238} \rightarrow {_2He^4} + {_{90}Th^{234}}.$$

Since, in all nuclear reactions, charge (i.e., atomic number) is conserved, and since the number of nucleons (mass number) is conserved, it is apparent that the superscripts and subscripts must separately add up to the same values on both sides of the arrow.

Now the $_{90}Th^{234}$ is also unstable and radioactive, and it emits a β particle to become something else we may call Y—not to be confused here with the symbol for the element yttrium! The emission of this electron does not change the number of nucleons, so Y must have a mass number of 234. But the emission of the negative electron is the same thing as adding a positive charge, so that the atomic number of Y must be 91. This atomic number identifies Y as an isotope of protactinium (Pa), and we have

$$_{90}Th^{234} \rightarrow {_{-1}\beta^0} + {_{91}Pa^{234}}.$$

After a total of 7 α emissions and 6 β emissions, we arrive at a polonium atom, which emits an eighth α particle and turns into an atom of lead (Pb), with atomic number $92 - (8 \times 2) + (6 \times 1) = 82$, and mass number $238 - (8 \times 4) = 206$. The nuclei of Pb^{206} are stable and no further radioactive transformations take place.

Genealogically speaking, the thorium and actinium families are very similar to that of uranium and terminate with stable lead isotopes Pb^{208} and Pb^{207}, respectively. It may also be mentioned that apart from these radioactive families, which include the heaviest elements of the periodic system and are transformed by a series of intermittent α and β decays into isotopes of lead, there are also a few isotopes of lighter elements that go through a one-step transformation. These include samarium (Sm^{148}), which emits α rays and turns into stable Nd^{144}, and two β emitters, potassium (K^{40}) and rubidium (Rb^{87}), which turn into stable isotopes of calcium (Ca^{40}) and strontium (Sr^{87}).

26-6 Decay Energies

The velocities of α particles emitted by various radioactive elements range from 0.98×10^9 cm/s for Sm^{148} up to 2.06×10^9 cm/s for Th^{223}, which correspond to kinetic energies of from 3.2 to 14.2×10^{-6} erg.

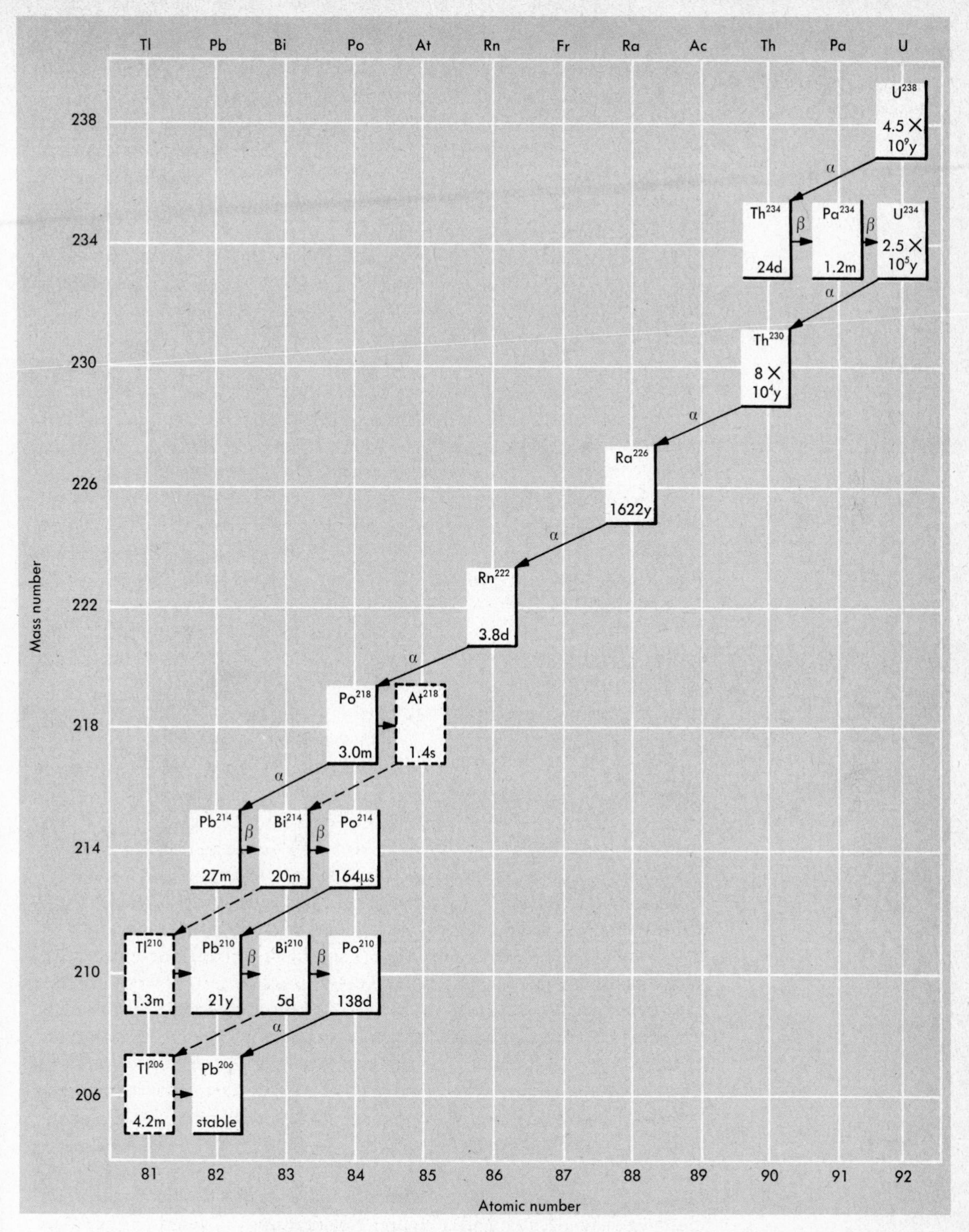

FIG. 26-3 The uranium-238 family. The numbers in the squares give the half-lives in years (y), days (d), hours (h), or seconds (s). Notice that near the end of the decay chain there are several alternative paths shown by dashed lines, which are taken by some of the decaying atoms.

The energies of β particles and γ photons are somewhat smaller but of the same general order of magnitude. These energies are enormously higher than the energies encountered in ordinary physical phenomena. For example, the kinetic energy of atoms in thermal motion at such a high temperature as 6000°K (surface temperature of the sun) is only 1.25×10^{-12} erg, i.e., several million times smaller than the energies involved in radioactive decay.

In speaking about the energies liberated in radioactive transformations, nuclear physicists customarily use a special unit known as the *electron volt* (eV). This unit is defined as *the energy gained by a particle carrying one elementary electric charge* (no matter whether it is an electron or any singly charged positive or negative ion) *when it is accelerated through an electric field with a potential difference of one volt.* Thus the electrons accelerated in J. J. Thomson's tube, with 5000 volts applied between the anode and cathode, acquire by this definition an energy of 5000 eV. On the other hand, the energy of a doubly charged oxygen ion, O^{2+}, accelerated through the same potential difference will be 10^4 eV, since the electric force acting on the ions, and consequently the work done by it, is twice as large. Remembering that the value of an elementary charge on an electron, proton, or any singly charged ion is 1.602×10^{-19} C, and that a volt is 1 J/C, we have that

$$1 \text{ eV} = 1.602 \times 10^{-19} \times 1 = 1.602 \times 10^{-19} \text{ J}, \quad \text{or} \quad 1.602 \times 10^{-12} \text{ erg}.$$

Also in common use is the MeV, which stands for "million electron-volts" or "mega-eV."

$$1 \text{ MeV} = 10^6 \text{ eV} = 1.602 \times 10^{-13} \text{ J} = 1.602 \times 10^{-6} \text{ erg}.$$

26-7 Half-Lifetimes

The process of natural radioactive decay is ascribed to some kind of intrinsic instability of the atomic nuclei of certain chemical elements (especially those near the end of the periodic table), which results from time to time in a violent breakup and the ejection from the nucleus of either an α particle or an electron. The nuclei of different radioactive elements possess widely varying degrees of internal instability.

Whereas in some cases (such as U^{238}) radioactive atoms may remain perfectly stable for billions of years before they are likely to break up, in other cases (such as Po^{214}) they can hardly exist longer than a small fraction of a second. The breakup process of unstable nuclei is a purely statistical process, and we can speak of the "mean lifetime" of any given elements in just about the same sense as insurance companies speak of the mean life expectancy of the human population. The difference is, however, that, whereas in the case of human beings and other animals the chance of decaying (i.e., dying) remains fairly low up to a certain age and becomes high only when the person grows old, radioactive atoms have the same chance of breaking up no matter how long it has been since they were formed. Automobiles and bodies and light bulbs wear out, and with the passage of time become more and more nearly ready for the junk-pile. Atoms and their nuclei, however, are not affected by any wear and tear associated with the aging process. A Ra^{226} formed only this morning by the α-decay of Th^{230} (see Fig. 26-3) has exactly the same chance of decaying into Rn^{222} as one formed a million years ago.

The number of radioactive atoms that decay per unit of time is proportional to the number of atoms available, but is quite independent of the age of these atoms.

The time period during which the initial number is reduced to one-half is known as the *half-life* of the element. At the end of two half-lives, only a quarter of the original amount will be left; at the end of three half-life periods, only one-eighth will be left, etc. This decay is shown graphically in Fig. 26-4. Since the amount present is halved for each successive half-life, there is a simple way to express the amount of a decaying element that is left after any number of half-lives.

If we start out with an amount N_0 of some radioactive material, after n half-lives have passed, there will be left an amount N:

$$N = N_0 \times (\tfrac{1}{2})^n.$$

The gas radon, for example, has a half-life of 3.8 days. If we start out with 5 mg of radon, how much will be left after a month? A month is $30/3.8 = 8$ half-lives, and

$$N = 5 \times (\tfrac{1}{2})^8$$
$$= 5 \times \tfrac{1}{256} = 0.02 \text{ mg left.}$$

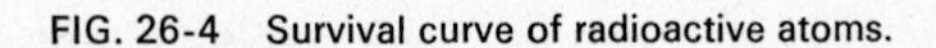

FIG. 26-4 Survival curve of radioactive atoms.

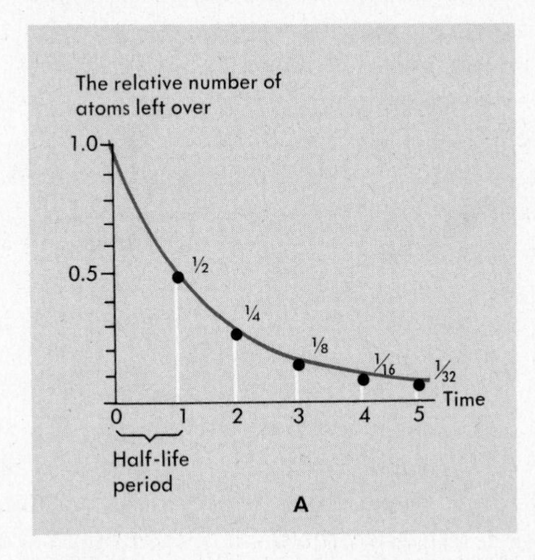

As we have seen above, various elements possess widely different lifetimes (Fig. 26-3). The half-life of U^{238} is 4.5 billion years, which accounts for its presence in nature in spite of the fact that all atoms of both stable and unstable elements may have been formed about 5 billion years ago. The half-life of radium is only 1622 yr, and the 200 mg of radium separated in 1898 by Marie and Pierre Curie now contains only 193 mg. The short-lived atoms of Po^{214} exist, on the average, for only 0.0001 s between the moment they are formed by β decay of Bi^{214} and their subsequent transformation into Pb^{210} atoms.

26-8 Uranium–Lead Dating

The decay of radioactive elements and its complete independence of physical and chemical conditions gives us an extremely valuable method for estimating the ages of old geological formations. Suppose we pick up from a shelf in a geological museum a rock that is marked as belonging to the late Jurassic era, that is, to the period of the earth's history when gigantic lizards were the kings of the animal world. Geologists can tell approximately how long ago this era was by studying the thicknesses of various prehistoric deposits and by comparing them with the estimated rates of the formation of sedimentary layers, but the data obtained by this method are rather inexact. A much more exact and reliable method, based on the study of the radioactive properties of igneous rocks, was proposed by Joly and Rutherford in 1913 and soon became universally accepted in historical geology. We have seen above that U^{238} is the father of all other radioactive elements belonging to its family and that the final product of all these disintegrations is a stable isotope of lead (Pb^{206}).

The igneous rock of the Jurassic era that now rests quietly on a museum shelf must have been formed as a result of some violent volcanic eruption of the past when molten material from the earth's interior was forced up through a crack in the solid crust and flowed down the volcanic slopes. The erupted molten material soon solidified into rock that did not change essentially for millions of years. But if that piece of rock had a small amount of uranium imbedded in it, as rocks often do, the uranium would decay steadily, and the lead resulting from that decay would be deposited at the same spot. The longer the time since the solidification of the rock, the larger would be the relative amount of the deposited lead with respect to the leftover uranium. Thus, by measuring the ratio of U^{238} to Pb^{206} in various igneous rocks, we can obtain very exact information concerning the time of their origin and the age of the geological deposits in which they were found.

Similar studies can be carried out by using the rubidium inclusions in old rocks and measuring the ratio of leftover rubidium to the deposited strontium. This method has an advantage over the uranium–lead method, because we deal here with a single transformation instead of the long sequence of transformations in the uranium family. In fact, one of the members of the uranium family is a gas (radon) and might partially diffuse away from its place of formation, thus leading to an underestimation of the age of the rocks.

26-9 Carbon-14 Dating

Apart from the natural radioactive elements mentioned above, which are presumably as old as the universe itself, we find on the earth a number of radioactive elements that are being continuously produced in the terrestrial atmosphere by cosmic-ray bombardment. Cosmic rays consist mostly of protons plus α particles, and in smaller quantities the nuclei of more massive atoms. These high-energy particles, traveling at very nearly the speed of light, shower down constantly from all directions onto the earth's atmosphere. Their origin is not yet clear, but they may start from violent supernova explosions of distant unstable stars, to be further accelerated by passage through the weak but vast magnetic fields of space. Wherever they come from, their collisions with the atoms of our upper atmosphere give rise to the formation of showers of new particles, some from the shattered debris of the collisions and some created new from pure energy. Among these particles are many neutrons, which may themselves cause further changes. One of these changes is the reaction between a high-energy neutron and the nucleus of a common atom of atmospheric nitrogen (N^{14}). This produces an unstable isotope of carbon (C^{14}) and a proton:

$$ {}_0n^1 + {}_7N^{14} \rightarrow {}_1H^1 + {}_6C^{14}. $$

These constantly replenished C^{14} atoms are soon oxidized by the atmospheric oxygen to become incorporated into the molecules of atmospheric carbon dioxide. Since plants use atmospheric carbon dioxide for their growth, this radioactive carbon is incorporated into each one, making all plants—and the animals that eat the plants—slightly radioactive throughout their lives.

As soon as a tree is cut or falls down and all its metabolic processes stop, no new supply of C^{14} is taken in, and the amount of radioactive carbon in the wood gradually decreases as time goes on. Carbon-14 is a β emitter, and, as we can see, reverts back to ordinary nitrogen in the process:

$$ {}_6C^{14} \rightarrow {}_{-1}\beta^0 + {}_7N^{14}. $$

Since the half-life of C^{14} is 5700 yr, the decay will last for many millennia; and by measuring the ratio of C^{14} to the ordinary C^{12} in old samples of wood, we are able to estimate dates of origin. The studies in this direction were originated by an American physicist, Willard Libby (1908–   ) and are playing the same role in the exact dating of ancient human history as the measurement of the uranium–lead ratio in the dating of the history of our globe. The measurement of C^{14} radioactivity in old samples of wood is a very delicate matter. In the atmosphere, and in the growing plant, the ratio C^{14}/C^{12} is about 10^{-12}; after each half-life, it becomes only half this much. So, to find the relative amount of C^{14} in an old sample by measuring its emission of β-electrons requires not only a sensitive counter, but heavy shielding as well. The ever-present cosmic-ray electrons must be kept from the counter—and the experimenter himself has a much higher C^{14} concentration than the sample he is studying. Figure 26-5 gives a few examples of the measured and

FIG. 26-5 The carbon-14 radioactivity of various old objects as a function of their age, as measured by Dr. Willard Libby.

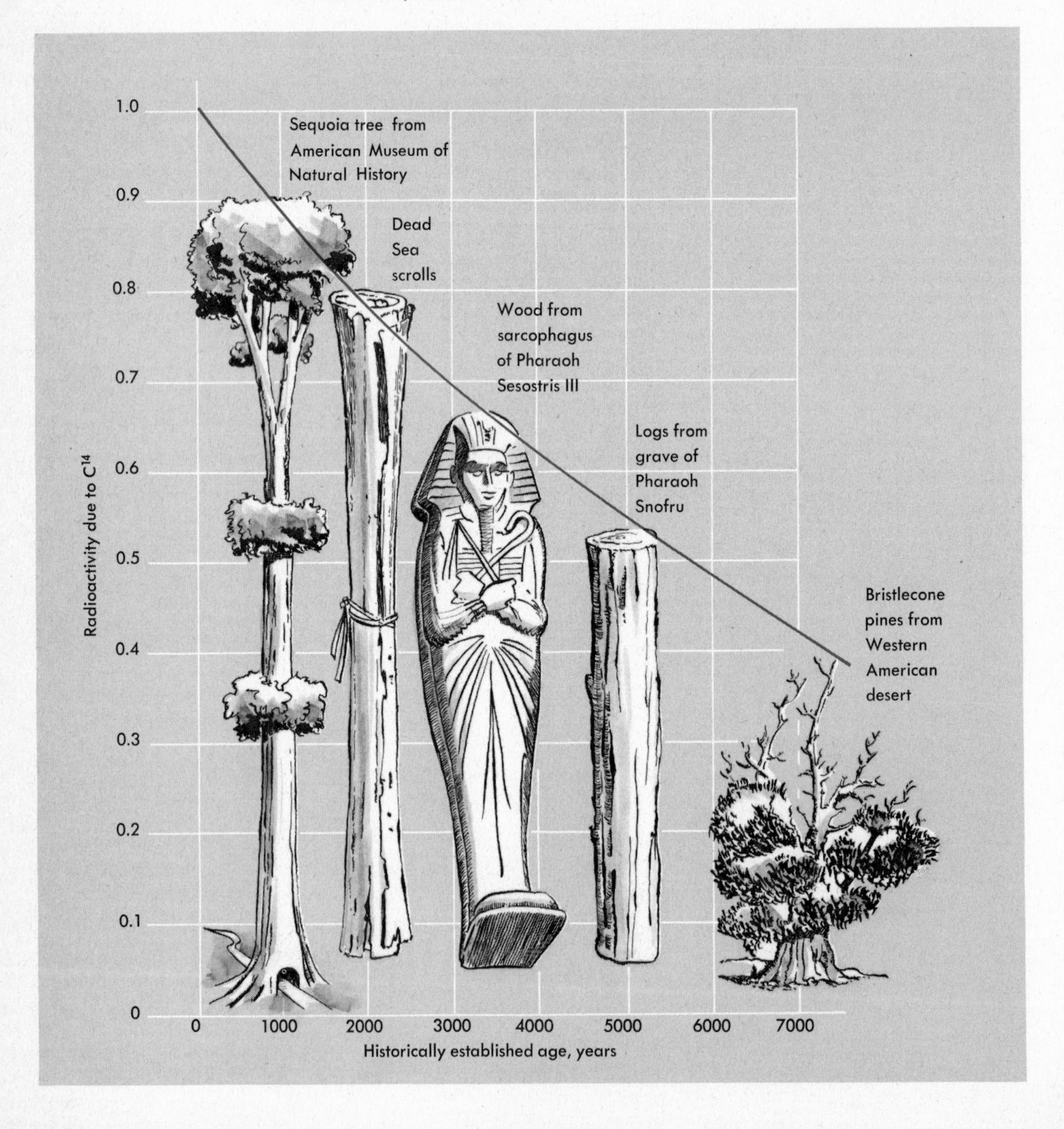

expected concentrations of C^{14} in various wood samples of known age.

Measurements of the C^{14} content of trees felled by ice-age glaciers have established that the last glaciation of the northern United States was much more recent (about 11,000 yr ago) than had been previously supposed.

26-10 Tritium Dating

Another method of dating by the use of radioactive materials, which was also worked out by Libby, utilizes the radioactivity of tritium, the heavy unstable isotope of hydrogen with atomic weight 3. Tritium is produced in the terrestrial atmosphere by the action of cosmic radiation and is carried to the surface by rains. Tritium also decays by β emission, which converts it into ${}_2He^3$. Tritium's half-life is only 12.5 yr, however, so that all age measurements involving this isotope can be carried out only for comparatively recent dates. It seems that the most interesting application of the tritium dating method is in the study of the movements of water masses, both in ocean currents and in underground waters, since by taking samples of water from different locations and from different depths, we can tell by their tritium content how long ago this water came down in the form of rain.

Samples of old water are more difficult to collect than samples of old wood, and Libby resolved this problem by analyzing the tritium content in wine of different vintages, originating in different countries. The unpleasant part of this task is that an entire case of fine wine has to be used for each measurement and is rendered undrinkable in the process. But the agreement with the expected tritium content was in all cases excellent, as demonstrated in Fig. 26-6.

Questions

(26-2) 1. Give the number of neutrons and protons in the nuclei of calcium, aluminum, radium, and selenium.

2. How many protons and how many neutrons are there in the nuclei of the following elements: carbon, bismuth, scandium, and gold?

(26-3) 3. Singly ionized ions of sodium form a canal ray originating in a 300-volt tube. What is their maximum energy, in eV and J?

4. What is the maximum possible energy for a doubly ionized neon ion that is part of a canal ray from a tube operating at 500 volts? Give the answer in eV and J.

5. From the data of Table 26-1, calculate the atomic weight of zinc.

6. From data in Table 26-1, calculate the atomic weight of chlorine.

7. List, using standard notation, the isotopes of nitrogen and chlorine.

8. Write, in standard notation, the isotopes of oxygen and zinc.

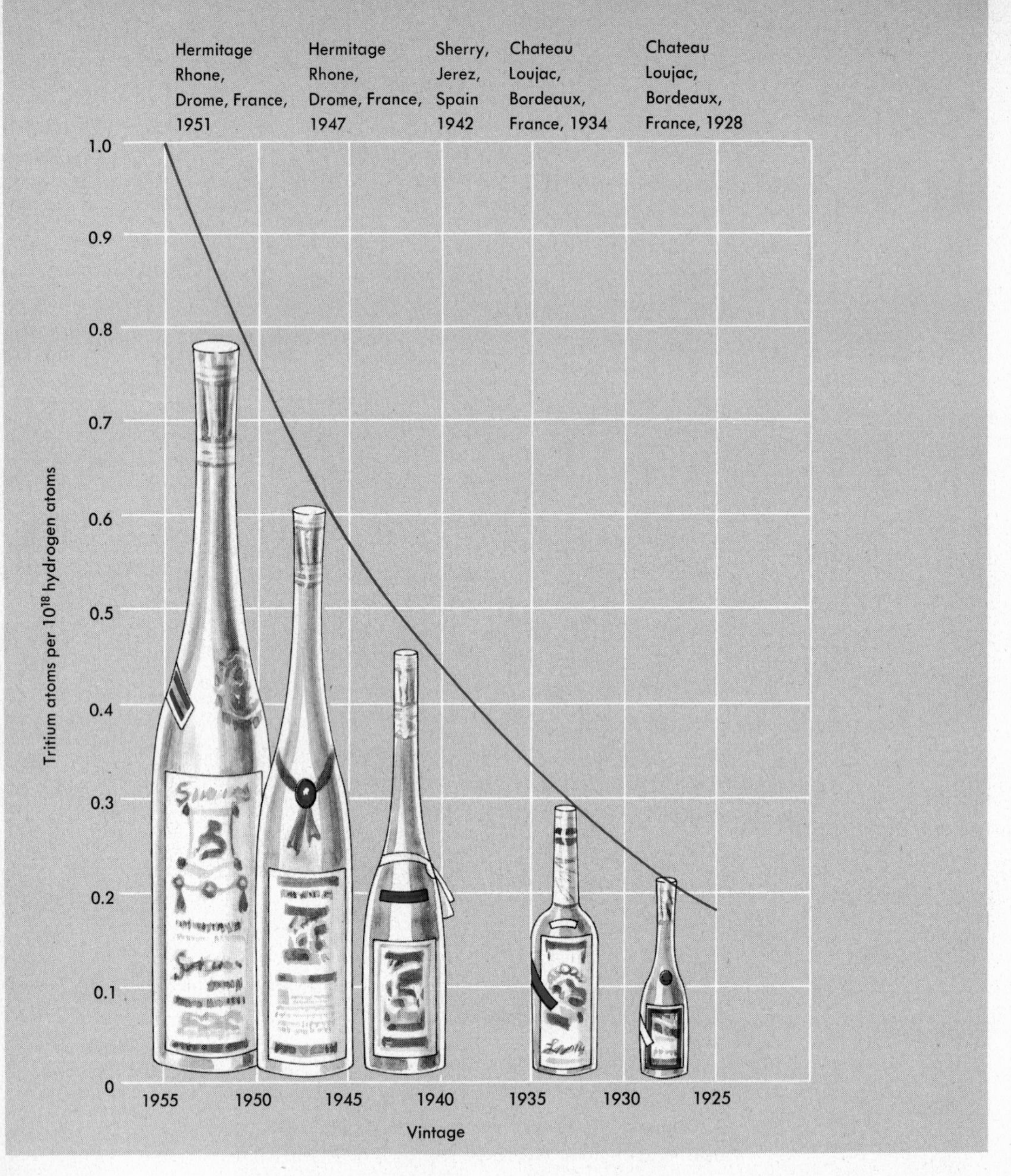

FIG. 26-6 The agreement between the calculated and the measured relationship between age and tritium activity in various wines.

(26-4) **9.** The following unstable isotopes decay by emitting an alpha particle. Write the proper designation of the isotopes into which this decay transforms them: (a) ${}_{94}Pu^{236}$, (b) ${}_{91}Pa^{226}$, (c) ${}_{88}Ra^{221}$.

10. The following isotopes are known to be unstable, and emit alpha particles. Give the designation (atomic number, chemical symbol, and mass number) of the isotopes into which they are transformed: (a) $_{89}Ac^{225}$, (b) $_{90}Th^{223}$, (c) $_{92}U^{238}$.

11. The isotopes listed below decay by emitting beta particles. Into what isotopes does this decay transform them? (a) $_{82}Pb^{211}$, (b) $_{83}Bi^{210}$, (c) $_{94}Pu^{243}$.

12. The following unstable isotopes decay by beta emission. Designate the isotopes into which they are transformed: (a) $_{88}Ra^{225}$, (b) $_{86}Rn^{223}$, (c) $_{83}Bi^{215}$.

(26-5) **13.** Along the chain of decays starting with $_{90}Th^{232}$, $_{84}Po^{216}$ is formed, which emits an alpha particle to become lead. Write a nuclear equation for this decay.

14. In the uranium decay series, radium ($_{88}Ra^{226}$) is formed, which emits an alpha particle to become radon (Rn). Write a nuclear equation for the transformation, showing the atomic number and mass number for this isotope of radon.

15. An unstable isotope of chlorine (Cl^{36}) often decays by β emission. Write a nuclear equation describing this transformation.

16. Potassium-40 often decays by β emission. Write the nuclear equation for this transformation.

17. Chlorine-36 sometimes decays by K-capture, which means that instead of emitting a particle, the nucleus absorbs, or captures, one of the electrons in the atom's K-shell. Write a nuclear equation for this particular Cl^{36} decay.

18. Potassium-40 sometimes decays by K-capture (see Question 17). Write the nuclear equation for this transformation.

19. Another decay series does not exist in nature, but begins with the artificially produced $_{93}Np^{237}$. The series contains 8 α decays and 4 β decays. What is the designation of the stable isotope at the end of this series?

20. Similar to the uranium family is that of thorium, which starts with $_{90}Th^{232}$ and ends with stable $_{82}Pb^{208}$. (a) How many alpha emissions are in this decay series? (b) How many beta emissions?

(26-6) **21.** In the β decay of Pa^{234}, gamma rays of 0.043-MeV energy are also emitted. (a) What is the energy of these photons in ergs? (b) What is their wavelength, in cm and in Å?

22. In the decay of radium into radon by alpha emission, gamma rays with wavelengths of 6.52×10^{-10} cm are also emitted. (a) What is this wavelength in Ångstrom units? (b) What is the energy of these gamma-ray photons, in MeV?

23. What is the energy in eV, ergs, and joules, of an alpha particle that has been accelerated through a potential difference of 75,000 volts?

24. What is the energy (in eV, ergs, and joules) of an electron that has been accelerated through a potential difference of 5×10^5 volts?

25. A beam of alpha particles and a beam of ions of singly ionized neon are each accelerated through a potential difference of 3.1×10^4 volts. How do their kinetic energies compare?

26. Take an electron and a doubly ionized nitrogen atom (N^{2+}), and accelerate each from rest through a potential difference of 5000 volts. How will their kinetic energies compare?

(26-7) **27.** Assuming that it were to remain undisturbed, calculate how much of Madame Curie's 200 mg of radium would be left in the year 8378 A.D.

28. Radon-222 has a half-life of 3.8 days. If we begin with 10.24 μg of this radon isotope, how much is left after 38 days?

29. Strontium-90 (produced in appreciable quantities in atmospheric atom bomb tests) has a half-life of 28 years. How long a time is needed for the initial Sr^{90} radioactivity to fall to 3 percent of its original intensity?

30. An atomic bomb test produces a small amount of radioactive fission product that has a half-life of 4 months. How much time must elapse before the radioactivity of this particular product is reduced to less than 1 percent of its original intensity?

31. In the U^{238} decay series, Pa^{234} is produced by the β decay of Th^{234}, which has a half-life of 24 days. If 16 μg of pure Th^{234} were put in a sealed container for 72 days, (a) how much Th^{234} would remain? (b) would you expect to find 14 μg of Pa^{234} in the container? (c) would you expect there to be any Pb^{206} present?

32. The parent of radium in the uranium series is ${}_{90}Th^{230}$, which has a half-life of about 8×10^4 yr. Suppose we were to set a gram of this pure thorium isotope aside for 160,000 yr: (a) How much Th^{230} would be left? (b) Would we now have 0.75 g of radium? (c) Would analysis show that some Pb^{206} was also present?

(26-9) **33.** Charcoal from a long-buried ancient campfire has a C^{14}/C^{12} ratio about 12.5 percent as great as that found in modern wood. What is the approximate date of the ancient campfire?

34. A piece of wood taken from an archaeological excavation has a ratio of C^{14} to C^{12}, which is about one-fourth as large as the ratio in recently cut trees. How old is the piece of wood?

(26-10) **35.** In the year 2234 A.D. the interstellar exploration ship Enterprise III finds an abandoned and previously unknown colony on the planet Glogu. The planet's orbit had been drastically shifted, and there was no record remaining of the length of the Gloguan "year." A sample from the lake providing the colony's water supply showed a count of 900/s of the 0.0181-MeV electrons characteristic of tritium decay. From old stores, water bottles were found dated in the "years" 52 and 38. Samples prepared in the same way from these bottles give counts of 113/s and 28/s, respectively. (a) What was the length of the Gloguan "year?" (b) On what earth date was the planet colonized?

36. A diver inspecting a sunken ship finds in one of the cabins an unopened bottle of whiskey. The radioactivity of the whiskey due to tritium was found to be only about 3 percent of that measured in a recently purchased bottle marked "7 years old." How long ago did the ship sink? (The whiskey on the ship was moonshine, presumably made, bottled, and sunk in the same year.)

27

Artificial Nuclear Transformations

27-1 Bombarding Atomic Nuclei

After Rutherford became completely persuaded that the radioactive decay of heavy elements was due to the intrinsic instability of their atomic nuclei, his thoughts turned to the possibility of producing the artificial decay of lighter and normally stable nuclei by subjecting them to strong external forces. True enough, it was well-known at that time that the rates of radioactive decay are not influenced at all by high temperatures or by chemical interactions, but this could be simply because the energies involved in thermal and chemical phenomena are much too small as compared with the energies involved in nuclear disintegration phenomena. Whereas the kinetic energy of thermal motion (at a few thousand degrees) as well as the chemical energy of molecular binding are of the order of magnitude of only 10^{-12} erg, the energies involved in radioactive decay are of the order of 10^{-6} erg, i.e., a million times higher. Thus, in order to have any hope of a positive outcome, the light stable nuclei must be subjected to a much stronger external agent than just a high temperature or a chemical reaction, and the bombardment of light nuclei by high-energy particles ejected from unstable heavy nuclei was the natural solution of the problem.

Following this line of reasoning, Rutherford directed a beam of α particles emanating from a small piece of radium against nitrogen gas and observed, to his complete satisfaction, that besides the α particles that passed through the nitrogen and were partially scattered in all directions, there were also a few high-energy protons that were presumably produced in collisions between the onrushing α projectiles and the nuclei of nitrogen atoms. This conclusion was later supported by cloud chamber photographs, as we shall discuss in the next section. We can express this reaction by the following nuclear equation:

$$_7N^{14} + {}_2He^4 \rightarrow {}_8O^{17} + {}_1H^1.$$

Following this original success, Rutherford was able to produce the artificial transformation of other light elements such as aluminum, but the yield of protons produced by α bombardment rapidly decreased with increasing atomic number of the target material, owing to the increase in electrostatic repulsion of the α particle by the greater + charge of the larger nuclei, and he was not able to observe any ejected protons for elements heavier than argon (atomic number, 18).

27-2 Photographing Nuclear Transformations

The study of nuclear transformations was facilitated by the ingenious invention of still another Cavendish physicist, C. T. R. Wilson. This device, known as the "Wilson chamber" or *cloud chamber*, permits us to obtain a photograph showing the tracks of individual nuclear projectiles heading for their targets and also the tracks of various fragments formed in the collision. It is based on the fact that whenever an electrically charged fast-moving particle passes through the air (or any other gas), it produces ionization along its track. If the air through which these particles pass is saturated with water vapor, the ions serve as the centers of condensation for tiny water droplets, and we see long, thin tracks of fog stretching along the particles' trajectories. The scheme of a cloud chamber is shown in Fig. 27-1. It consists of a metal cylinder C with a transparent glass top G and a piston P, the upper surface of which is painted black. The air between the piston and the glass top is initially saturated with water (or alcohol) vapor, generally by a coating of moisture on top of the piston. The chamber is brightly illuminated by a light source S through a side window W. Suppose now that we have a small amount of radioactive material on the end of a needle N, which is placed near the thin window O.

The particles ejected by the radioactive atoms will fly through the chamber, ionizing the air along their paths; however, the positive and negative ions produced by the passing particles recombine rapidly into neutral molecules. Suppose, however, that the piston is pulled rapidly

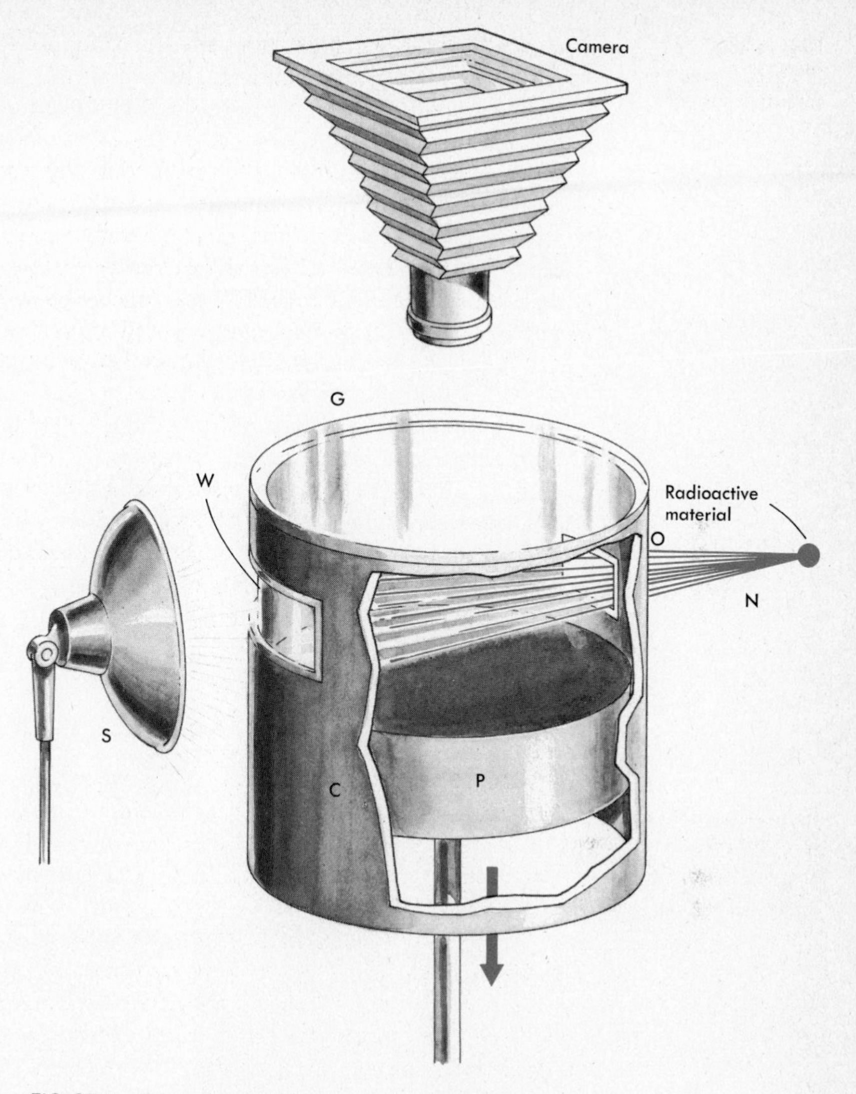

FIG. 27-1 Schematic diagram of a simple cloud chamber.

down for a short distance. The rapid expansion of the air will cause it to cool, and the already saturated air now becomes supersaturated with moisture that will condense into water droplets. In order to condense, however, the droplets need centers of some sort around which to form. The natural condensation of raindrops takes place on dust particles, tiny salt crystals, or ice crystals. In *cloud seeding*, airplanes scatter minute crystals of silver iodide to encourage the condensation of rain droplets.

In the cloud chamber, however, there is no dust, and the droplets condense on the ions (as the next best thing) that have been formed along the path of the speeding particles. Thus the tracks of the particles that passed by just before, or just as, the piston was pulled down will show up as trails of microscopic water droplets. The tracks of α particles are quite heavy, since the massive, doubly charged α particle ionizes the air strongly. The track of a proton is less strongly marked, and along the path of an electron, the ions and hence the droplets are much more sparse. In much of the present cloud chamber work, an intense magnetic field is created within the chamber, so the charged particles are deflected into curved paths. By measuring the curvature shown on the photographs, we can compute the speed and energy of the particles.

Figure 27-2 is a classical cloud chamber photograph taken in 1925 by P. M. S. Blackett that shows the collision of an incident α particle with the nucleus of a nitrogen atom in the air that fills the chamber. The long, thin track going almost backward is that of a proton ejected in that collision. It can be easily recognized as a proton track because protons are four times lighter than α particles and carry only half as much electric charge; therefore they produce fewer ions per unit length of their path than α particles. The short, heavy track belongs to the nucleus $_8O^{17}$ formed in the process of collision.

FIG. 27-2 The first cloud chamber photograph of a nuclear disintegration, taken by P. M. S. Blackett. The diagram to the right of the photo explains the tracks involved in this nuclear reaction.

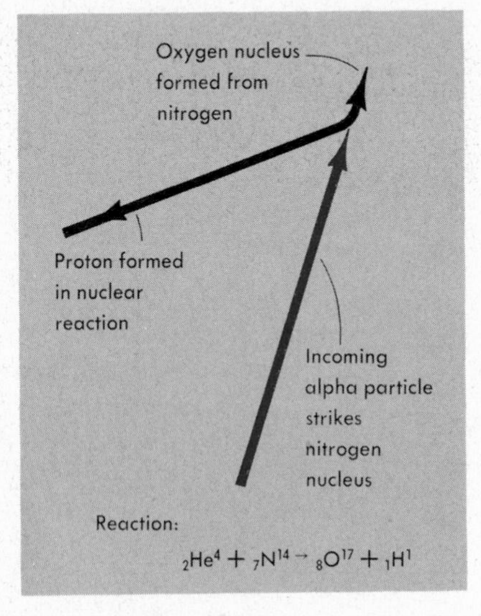

27-3 Other Trackers and Counters

In recent years, the *bubble chamber* has been developed to supplement the work of the cloud chamber. Although the general principle of its operation is the same as that of the cloud chamber, the bubble chamber is filled with a liquid (often propane or, more recently, liquid hydrogen) that is kept exactly at its boiling-point temperature. A slight expansion will reduce the pressure on the liquid, and bubbles of vapor will form on the ions that have been produced in the liquid by passing particles. The bubble chamber has a great advantage when the collision events to be observed are relatively rare. In the closely packed atoms of a liquid, many more nuclear collisions will occur than in a gas, and the observer will stand a much better chance of photographing what he is looking for than he would with a cloud chamber. Figure 30-5 shows a photograph of tracks of particles in a bubble chamber.

The *spark chamber* consists of a stack of flat metal plates spaced a few centimeters apart and separated by insulators. Alternate plates are connected together and a high potential difference is maintained between each plate and its adjacent neighbors (Fig. 27-3). The voltage between plates is not quite enough to cause a spark to jump between them. But an incoming high-energy nucleus or other particle ionizes the air or other gas in its passage, making a low-resistance path along which sparks immediately jump and are photographed.

Besides their obvious application in the conventional photography of cloud-, bubble- and spark-chamber tracks, photographic emulsions are used directly in recording the paths of charged particles. Special thick emulsions are used, often in stacks; the photosensitive chemicals in them mark the trail of ions left by the particle in its passage. Figure 30-4 shows such a record.

Counting particles is often as important as observing their tracks. Probably the best-known device for this purpose is the *Geiger* (or

FIG. 27-3 Schematic diagram of spark chamber.

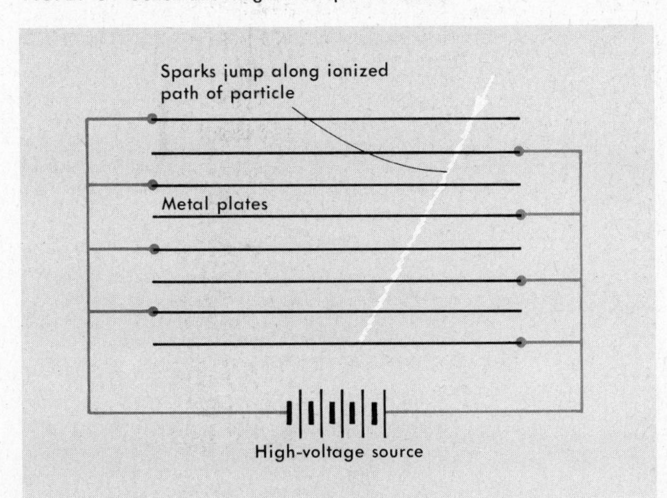

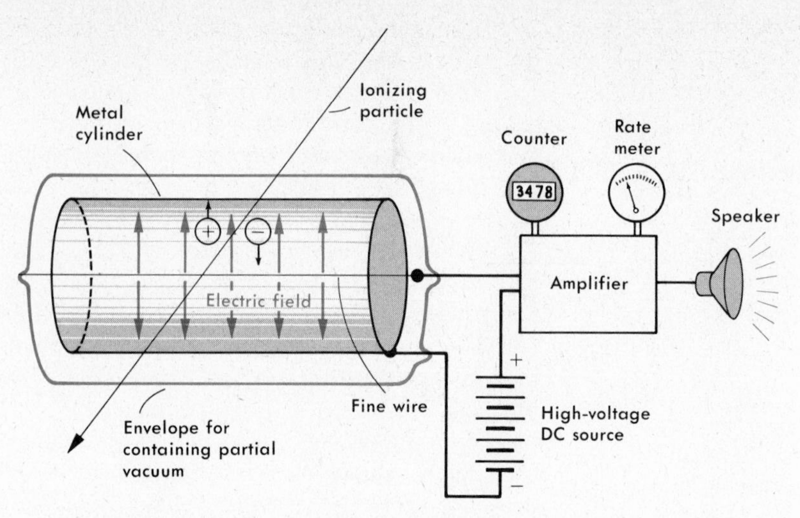

FIG. 27-4 Basic operation of a Geiger (or Geiger-Muller) counter.

Geiger–Muller) *counter* (Fig. 27-4). Basically, it consists of a thin metal cylinder with a fine wire along its axis, in a partial vacuum. An energetic particle entering the cylinder ionizes one or more of the gas molecules into an electron and a massive positive ion. The high potential difference between the cylinder and the central wire creates a strong electric field that immediately accelerates the ions in opposite directions. These accelerated ions create more ions by colliding with gas molecules—these create still more, and so on, so that the electrodes each receive a shower of electric charges. The resulting small surge of current through the external circuit is amplified to operate a counter, a rate meter (which shows the number of surges occurring per second), or a loudspeaker that produces a click for each particle entering the tube.

We have already encountered a *scintillation counter* in its simplest form: In Rutherford's experiments that led to his discovery of the atomic nucleus, the scattered α particles struck a transparent screen coated with zinc sulfide. The resulting flashes of light were painstakingly counted by an observer watching through a magnifier for endless hours. More modern variations use fluorescent crystals or special plastic blocks as detectors. The light flashes in these detectors fall on sensitive surfaces which emit photoelectrons which are then amplified and automatically counted.

27-4 Early Particle Accelerators

In the early days of nuclear physics, before the development of the counters and trackers now available, the only massive projectiles available were the α particles of naturally occurring radioactivity. It was apparent that protons would be even more effective as projectiles. Since

FIG. 27-5 Sir John Cockcroft (left) and author Gamow discuss the possibility of breaking up the nucleus by using artificially accelerated protons. (Cambridge, England, 1929.)

they have only half the charge of α particles, the repulsion of the target nucleus would be only half as great and they would have a correspondingly better chance of penetrating to their targets. At the urging of Lord Rutherford, John Cockcroft (see Fig. 27-5) and E. T. S. Walton designed something similar to a high-voltage electron gun. Using this new device to accelerate protons to an energy of several hundred thousand eV, Cockcroft and Walton in 1932 directed their projectiles against a lithium target and observed the first nuclear transformation caused by artificially accelerated particles:

$$_3Li^7 + {}_1H^1 \rightarrow 2\,{}_2He^4.$$

A boron target struck by the proton beam gives the following reaction:

$$_5B^{11} + {}_1H^1 \rightarrow 3\,{}_2He^4.$$

Figure 27-6 shows the tracks of the three newly formed α particles as they fly apart.

At about the same time, R. Van de Graaff at Princeton University in the United States designed an electrostatic accelerator generally referred to by his name as a *Van de Graaff*. It is based on the classical principle that because of the mutual repulsion of like charges any charge given to a conductor will be distributed entirely on its outer surface. Thus, if we take a hollow spherical conductor with a small hole in its surface, insert through this hole a small charged conductor attached to a glass stick,

FIG. 27-6 A cloud chamber photograph (by P. Dee and C. Gilbert), showing the three alpha particles (arrows) resulting from a boron nucleus struck by an artificially accelerated proton.

and touch the inside surface of the sphere (Fig. 27-7A), the charge will spread out to the outer surface of the big sphere.

Repeating the operation many times, we would be able to transfer to the large conductor any amount of electric charge, and raise its potential as high as desired (or, at least, until the sparks start jumping between the conductor and the surrounding walls).

In the Van de Graaff machine (Fig. 27-7B), the small charged ball is replaced by a continuously running belt that collects electric charges from a source at the base and deposits them on the interior surface of the large metallic sphere. The high electric potential developed in this process is applied to one end of an accelerating tube in which the ions of different elements are speeded up to energies of many millions of electron volts.

27-5 The Cyclotron

In the meanwhile, E. O. Lawrence at the University of California worked on a different idea: Instead of accelerating particles by means of one

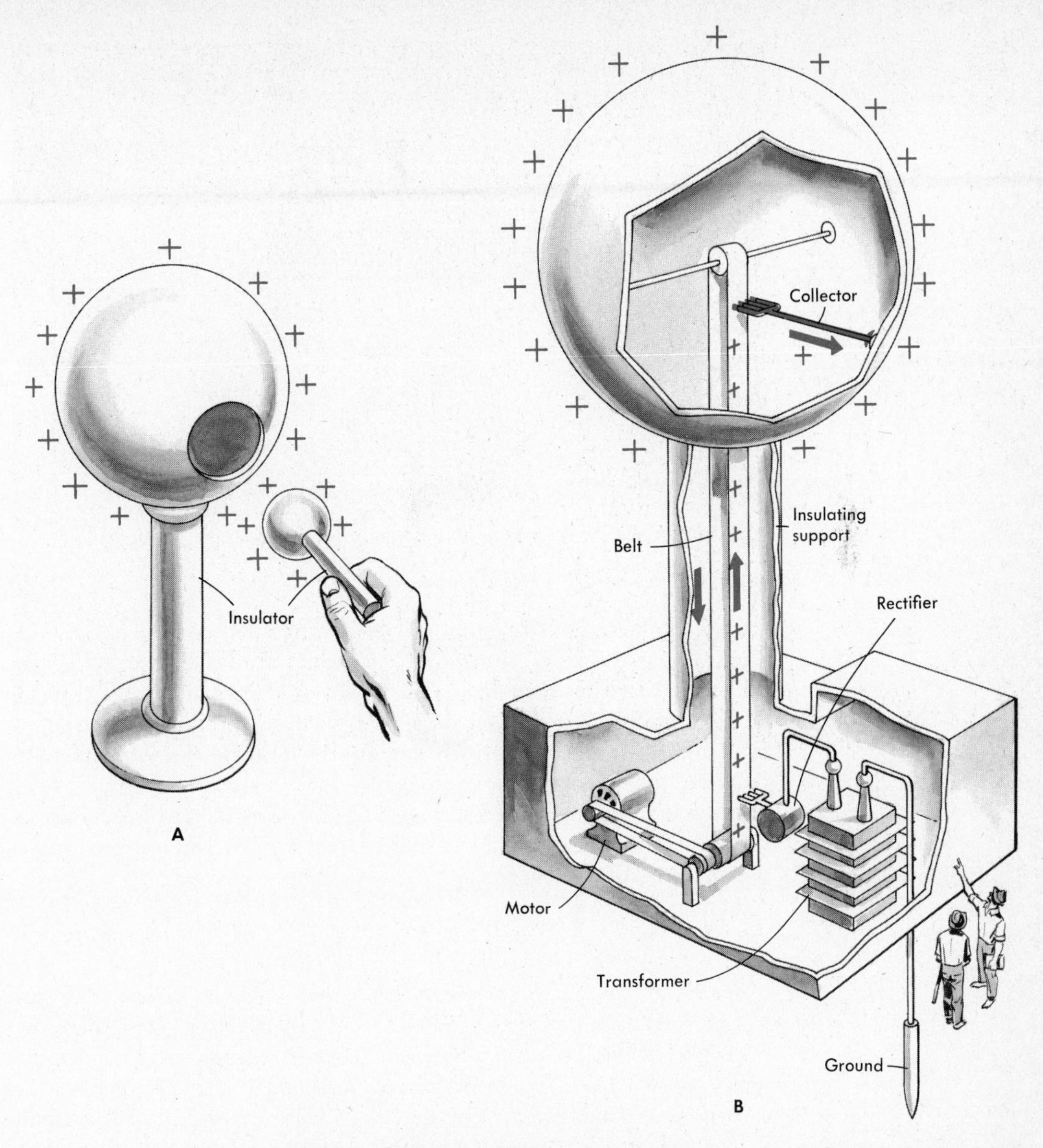

FIG. 27-7 The basic principle (A), and the application of this principle in Van de Graaff's high-voltage electrostatic generator (B).

large potential difference, let them travel in circular paths and apply a smaller, more manageable potential difference many times. This is the principle of the *cyclotron*, illustrated in Fig. 27-8. It consists essentially of a circular metal chamber cut into halves, C_1 and C_2, and placed between the poles of a very strong electromagnet. The half-chambers C_1 and C_2 (called "dees" because of their shape) are connected to a

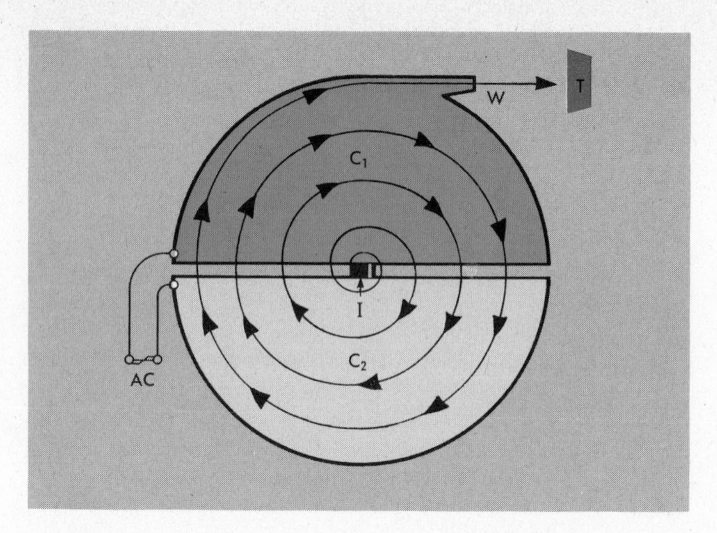

FIG. 27-8 Schematic diagram of the principle of the cyclotron.

source of alternating potential of comparatively high voltage, so that when C_1 is positive, C_2 is negative, and vice versa. The entire device is, of course, in a high vacuum so that air molecules will not interfere with the acceleration of the particles. These particles, protons for example, are projected into C_1 by an ion gun I that works on the principle of a canal-ray tube. Because the protons (or other ions bearing charge q) are moving at right angles to a strong magnetic field B, they experience a sideways force that curves them into a circular path. The magnetic force Bqv provides the centripetal force

$$Bqv = \frac{mv^2}{r}.$$

The time required for the proton to complete a semicircle within C_1 is the length of the path πr divided by the velocity v:

$$t = \frac{\pi r}{v}.$$

From the centripetal force equation above, we can determine that

$$v = \frac{Bqr}{m}.$$

This value for v, substituted into the expression for t, gives us

$$t = \frac{\pi rm}{Bqr} = \frac{\pi m}{Bq}.$$

It is perhaps surprising to find that the time required for the ion to traverse a semicircle in one of the dees is not influenced by the speed of the ion nor by the radius of the path. This gives the secret of the cyclotron's success.

When the ion leaves C_1 and enters the space between the dees, C_1 has been made positive and C_2 negative, so that the space contains an electric field accelerating the ion toward C_2. (We must bear in mind that on the inside of either of the dees, there can be no electric field, so the only force here is that produced by the magnetic field.) Within C_2, then, the ion moves faster but describes a larger semicircle, so the time needed remains the same. On leaving C_2, the ion finds that the polarity of the dees has automatically been reversed, and so on. Every time it crosses the gap, a properly timed electronic oscillator has reversed the electric field, so the ion eventually receives a high energy from many small pushes rather than from one large one. Gathering speed, the ions move along an unwinding spiral trajectory and will finally be ejected through the window W in the direction of the target T. The field may oscillate millions of times per second, and at every oscillation a fresh squirt of ions is injected at I. In this way, the ions strike the target in little bunches millions of times a second, which amounts to almost a steady stream.

The largest existing cyclotron is in the radiation laboratory of the University of California. It has an accelerating circular box 60 inches in diameter and produces artificial α beams with an energy of 40 MeV.

27-6 Other Particle Accelerators

Another look at the expression for the time a proton or other ion spends traversing a semicircle will reveal one drawback of the simple cyclotron:

$$t = \frac{\pi m}{Bq}.$$

The time (which is connected with oscillator frequency by the relation $f = 1/2t$) is seen to depend on m, the mass of the accelerated particle. For bullets or other particles traveling at ordinary speeds, this would not concern us, since m is practically constant. We know, however, from Einstein's Special Theory of Relativity that mass increases as particle speeds begin to approach the speed of light. So, if we try to speed up particles too much in a cyclotron, they will, in the outer spirals, be going so fast that their mass increase will throw them out of step with the oscillating field.

To eliminate this problem and permit particles to be accelerated to higher energies, many ingenious variations have been proposed and built: the synchrotron, the synchrocyclotron, the betatron, the proton synchrotron, and others. In general, these machines use changing oscillation frequencies or changing magnetic field strengths, or both. We will not try to look into the details of their operation. Particles may be accelerated to nearly the speed of light and to energies measured in billions of electron volts.

In these larger accelerators the particles travel in a path of constant radius in an evacuated tube curved in a circle. A series of electromagnets spaced along this path gives the necessary centripetal acceleration.

When nuclear scientists try for higher and higher energies with these circular-path accelerators, they encounter another source of trouble. The necessary centripetal acceleration of the circling ions causes them (in accordance with Maxwell's equations) to radiate some of their energy as electromagnetic radiation. As added energy is given the particles, their increased centripetal acceleration causes most of it to be radiated away, thus defeating the purpose of the machine. There is one way to avoid this: Centripetal acceleration $a_c = v^2/r$. Thus, for some given energy, the acceleration may be made less by making the radius of the path larger. This has led to the construction of some enormous accelerators, which have the handicap that cost increases rapidly with increasing size.

These difficulties have given a renewed interest to the linear, or straight-line, accelerator. No huge magnets are needed when the particle path is straight. Stanford University now has a linear accelerator 2 miles long, in which electrons are accelerated to energies of 20 GeV (giga electron volt $= 10^9$ eV). An electromagnetic wave is sent down a huge conducting tube with a speed only a trifle less than the speed of light. The electrons are pushed along by the traveling field of this wave like surfers riding the moving water wall of an ocean wave.

At present the world's largest accelerator is that of the National Accelerator Laboratory at Batavia, Illinois. It is actually four accelerators in series (Fig. 27-9), each one sending the particles into the next with a good running start. Designed primarily for protons, it has achieved energies of 300 GeV.

In Geneva, Switzerland, the European Council for Nuclear Research (CERN) is constructing an even more advanced and more powerful synchrotron that is expected to produce 400-GeV protons by 1978.

Questions

(27-1) 1. Write the following in the form of a nuclear equation: When an alpha particle strikes a boron-10 nucleus, a nitrogen isotope is produced, with the emission of a neutron.

2. Write a nuclear equation for the following reaction: Beryllium-9 is bombarded with alpha particles, and carbon-12 and neutrons are produced.

3. Write the following nuclear reactions in the form of equations: (a) A certain isotope, under proton bombardment, emits an α particle and becomes ordinary sulfur. (b) Lithium-6, struck by a neutron, becomes an α particle and something else.

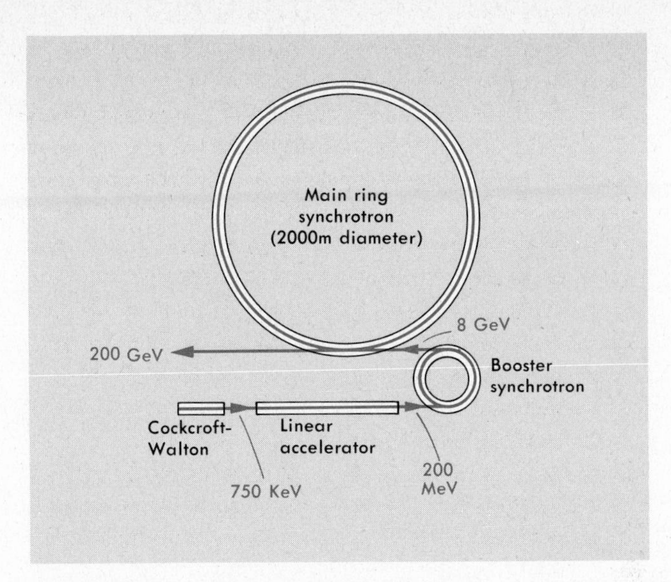

FIG. 27-9 The four-stage accelerator at the U.S. National Accelerator Laboratory in Batavia, Ill.

4. Write nuclear equations for the following reactions: (a) Ordinary sodium, when bombarded with a proton, emits an α particle and becomes something else. (b) Tritium decays into something else by emitting a β particle.

5. Complete the following reactions by giving the complete designation of the missing particle: (a) $Al^{28} \rightarrow e^- + ?$ (b) $Al^{27} + He^4 \rightarrow P^{30} + ?$ (c) $Al^{27} + H^2 \rightarrow Si^{28} + ?$ (d) $P^{30} \rightarrow e^+ + ?$ (e) $Al^{27} + n \rightarrow Na^{24} + ?$

6. Complete the following reactions by giving the complete designation of the missing particle: (a) $B^{11} + H^1 \rightarrow 3(He^4) + ?$ (b) $Al^{27} + H^2 \rightarrow Mg^{25} + ?$ (c) $N^{14} + H^2 \rightarrow 4(?)$ (d) $Al^{27} + H^2 \rightarrow Al^{28} + ?$ (e) $Al^{27} + H^1 \rightarrow \gamma + ?$

(27-2) **7.** An α particle passes into a cloud chamber with a speed of 5×10^6 m/s. The chamber is in a magnetic field $B = 0.35$ weber/m^2, perpendicular to the path of the α. What is the radius of curvature of the track?

8. A magnetic field $B = 0.20$ weber/m^2 is maintained in a cloud chamber. A proton traveling 3.2×10^6 m/s enters perpendicular to the field. What is the radius of curvature of its path?

9. In a cloud chamber, a proton leaves a curved track of $r = 6.1$ cm. It is traveling at right angles to a magnetic field of 0.15 weber/m^2. (a) What is the speed of the proton? (b) What is its energy, in joules or ergs? (c) in eV?

10. A cloud chamber is in a magnetic field of 0.30 weber/m^2. An α particle entering the chamber perpendicular to the field leaves a track whose radius of curvature is 8.9 cm. What is the energy of this α particle, in eV?

(27-4) **11.** What energy, in eV, MeV, ergs, and joules, is acquired by a proton accelerated through a potential difference of 1.50×10^5 V?

12. An α particle is accelerated in a 350,000-V Cockcroft–Walton accelerator. What is its energy in eV, MeV, ergs, and joules?

(27-5) **13.** A cyclotron has a magnetic field intensity of 6000 gauss (0.6 weber/m^2), and has an alternating potential difference of 5×10^4 V between its dees. Protons are injected. How many *revolutions* must the protons make before obtaining a final energy of 4 MeV? ($m_P = 1.67 \times 10^{-27}$ kg.)

14. Deuterons are injected into a cyclotron whose magnetic field intensity is 8000 gauss (0.8 weber/m^2). The alternating potential difference between the dees is 4×10^4 V. How many *revolutions* do the deuterons make before achieving an energy of 3 MeV? ($m_D = 3.34 \times 10^{-27}$ kg.)

15. How long does it take one of the protons of Question 13 to complete a semicircular trip through one of the dees?

16. How long does it take the deuteron of Question 14 to traverse a semicircle within one of the dees?

17. What is the frequency of the alternating potential applied to the dees of the cyclotron in Question 13?

18. What is the frequency of the alternating potential applied to the cyclotron dees of Question 14?

19. What is the speed of a 4-MeV proton?

20. What is the speed of a 3-MeV deuteron?

21. What is the required diameter of the dees and magnet poles of the cyclotron of Question 13?

22. What is the required diameter of the dees and magnet poles in the cyclotron of Question 14?

(27-6) **23.** A particle has a rest mass of m_0. By what fraction is its mass increased if accelerated to a speed of $0.75c$? By what fraction would the period of its revolution in a cyclotron be increased?

24. If a cyclotron had been mistakenly designed to accelerate particles to half the speed of light, by what fraction would the time required to pass through a dee be increased, as compared with the time required at nonrelativistic speed?

25. In the Stanford linear accelerator, electrons have been given an energy of 24 GeV. (a) What is the rest mass of an electron, in MeV? (a) At 24 GeV, by what factor does the relativistic mass-energy exceed the rest mass of the electron?

26. (a) What is the rest mass of a proton, in MeV? (b) If a proton is given an energy of 200 GeV, how many times greater than its rest mass is its relativistic mass-energy?

28

The Structure of the Nucleus

28-1 Nuclear Particles

We have mentioned the early ideas holding that the nucleus was made up of protons and electrons—ideas that were rejected in the 1920s when quantum mechanics began its development. Rutherford, in 1920, without any definite experimental evidence to support him, had suggested a massive, electrically neutral nuclear particle, and had even given it a name, the *neutron*.

Actual evidence was not forthcoming, however, until 1930, when a German physicist, W. Bothe, noticed that the bombardment of beryllium by α particles from polonium gave rise to a very peculiar radiation of high penetrating power. Bothe believed that this radiation, unaffected by magnetic fields, was composed of high-energy photons. But in 1932, James Chadwick, a colleague of Rutherford, proved that the radiation must be composed of neutral particles of about the same mass as the proton. Rutherford had already provided the name for them, and it could now be seen that Bothe's original experiments could be written as

$${}_4Be^9 + {}_2He^4 \rightarrow {}_6C^{12} + {}_0n^1.$$

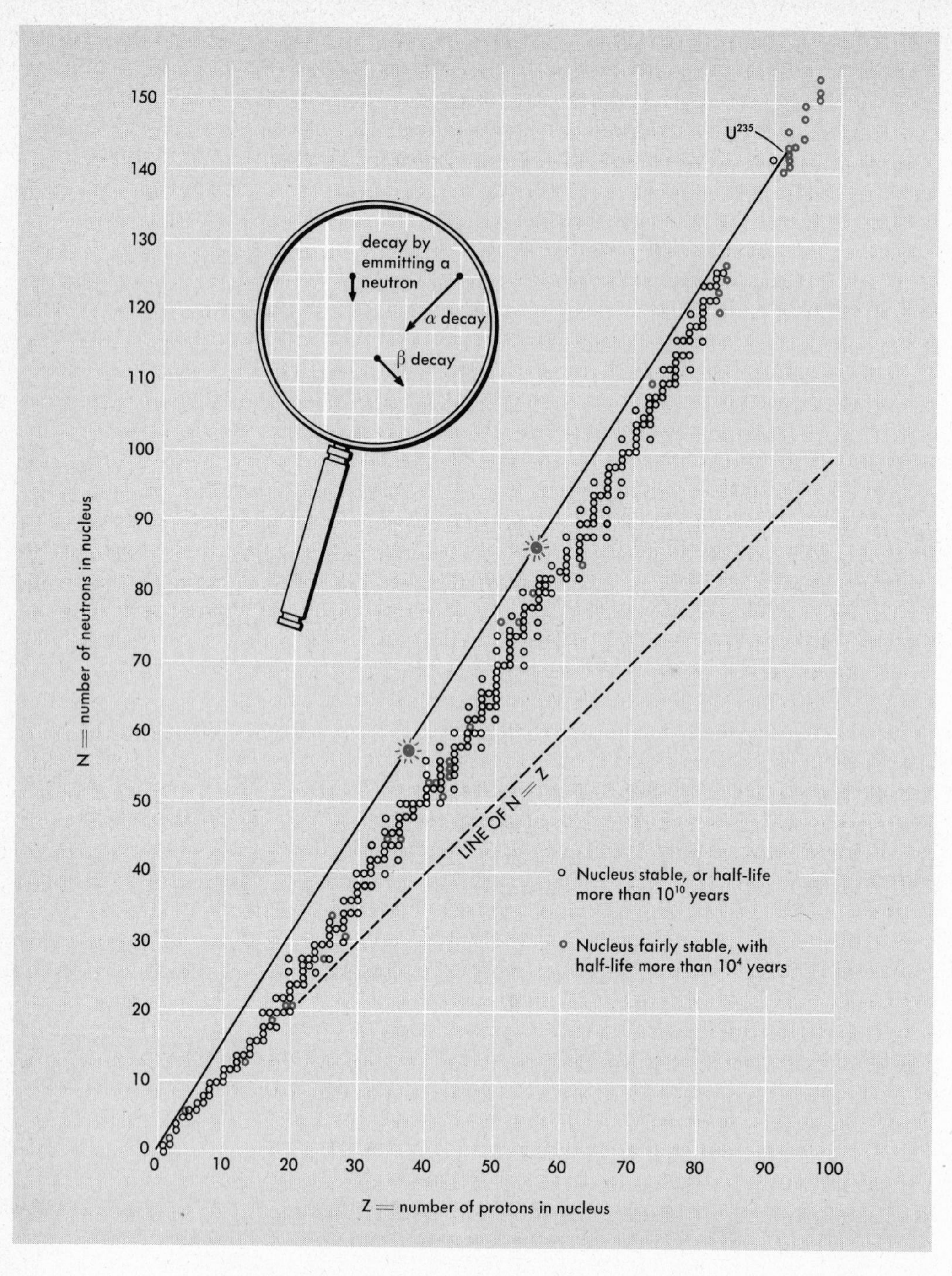

FIG. 28-1 Chart of number of neutrons (N) and number of protons (Z) in nuclei of stable or long-lived nuclei.

With this information experimentally confirmed, it was easy to assign to each isotope of each element the number of protons and neutrons contained in its nucleus. We have already used Z to indicate atomic number, which equals the number of protons in the nucleus. The mass number is customarily called A, which equals the number of protons plus the number of neutrons. For N, which we can use for the neutron number, we then have $N = A - Z$.

Thus, in 1932, physicists engaged in studying the atom and its nucleus had only three particles to consider: the proton, the neutron, and the electron. Their masses are now well-known and are, in amu (atomic mass unit, one-twelfth of the mass of the C^{12} isotope of carbon):

$$\begin{aligned} \text{mass of neutron} &= 1.00867 \text{ amu} \\ \text{proton} &= 1.00728 \\ \text{electron} &= 0.00055. \end{aligned}$$

It is an easy job to go through a list of stable or long-lived isotopes of all the elements, and from their mass numbers (A) and atomic numbers (Z) figure the protons and neutrons in each nucleus. Figure 28-1 shows N and Z for this list. For the lighter elements, the numbers of neutrons and protons are about equal, but with increasing mass the proportion of neutrons becomes greater and greater. In ${}_{92}U^{238}$, for example, $N/Z = 1.59$; for ${}_{8}O^{16}$, $N/Z = 1.00$.

28-2 Mass Defect and Nuclear Binding Energy

It might be expected that the mass of an atom or nucleus would be equal to the sum of the masses of its constituent particles. Let us try this for the O^{16} atom, containing 8n, 8p, and 8e:

$$\begin{aligned} 8 \text{ neutrons} &= 8 \times 1.00867 = 8.06936 \\ 8 \text{ protons} &= 8 \times 1.00728 = 8.05824 \\ 8 \text{ electrons} &= 8 \times 0.00055 = \underline{0.00440} \\ & \qquad\qquad\qquad\quad 16.13200 \end{aligned}$$

It was necessary to add in the 8 electrons because atomic weights always include the weight of the atomic electrons in their values. Since the number of atomic electrons is always equal to the number of protons in the nucleus, it may be more convenient, in figuring nuclear discrepancies, to use the hydrogen atom instead of the proton, as this automatically includes the proper number of electron masses. For oxygen, we could have written

$$\begin{aligned} 8 \text{ neutrons} &= 8 \times 1.00867 = 8.06936 \\ 8 \text{ hydrogens} &= 8 \times 1.00783 = \underline{8.06264} \\ & \qquad\qquad\qquad\quad 16.13200 \end{aligned}$$

The mass of an O^{16} atom, however, is 15.99491 amu, which leaves a discrepancy of $16.13200 - 15.99491 = 0.13709$ amu. For the principal isotope of iron, ${}_{26}Fe^{56}$, the results are similar:

$$\begin{aligned}
30 \text{ neutrons} &= 30 \times 1.00867 = 30.26010 \\
26 \text{ hydrogens} &= 26 \times 1.00783 = \underline{26.20358} \\
& \qquad\qquad\qquad\quad 56.46368
\end{aligned}$$

which is to be compared with the value of 55.9349 for the atomic weight of Fe^{56}. Here we find that the atomic weight of Fe^{56} is 0.5288 amu smaller than the combined masses of its components.

The explanation of this *mass defect* lies in Einstein's $E = mc^2$, and might perhaps have better been called an "energy defect." Since the mass of, say, the O^{16} nucleus is 0.13709 amu less than the combined masses of the 8 neutrons and 8 protons that make it up, it must therefore have the equivalent of 0.13709 amu less energy than the 16 nucleons considered separately. It is easy to see (at least qualitatively) that this must be so. The separated nucleons have great potential energy because of the strong attractive nuclear forces between them. Or we may look at the process in reverse, and imagine ourselves picking the nucleus apart. Every nucleon that is removed from it will have to be dragged away against the strong attraction of its neighbors. When the 16 nucleons (for O^{16}) have been separated in this way, the work we had to do will be just equal to the added potential energy of the separated nucleons. This work, or energy, is the *binding energy* of the nucleus. For the oxygen nucleus, we see that this total binding energy is equivalent to a mass of 0.13709 amu.

It is convenient to be able to convert this mass into other units. The mass of Avogadro's number of C^{12} atoms is taken as exactly 12 grams. Then

$$\begin{aligned}
1 \text{ amu} &= \frac{1}{12} \times \frac{12}{\text{Av. No.}} \\
&= \frac{1}{6.022 \times 10^{23}} \\
&= 1.660 \times 10^{-24} \text{ g} \quad \text{or} \quad 1.660 \times 10^{-27} \text{ kg.}
\end{aligned}$$

This is readily converted to joules by $E = mc^2$, and since 1 eV = 1.602×10^{-19} J, we have the following tabulation:

TABLE 30-1

1 amu	= 931 MeV	= 1.492×10^{-10} joule	= 1.660×10^{-27} kg
1.074×10^{-3} amu	= **1 MeV**	= 1.602×10^{-13} joule	= 1.783×10^{-30} kg
6.70×10^{9} amu	= 6.24×10^{12} MeV	= **1 joule**	= 1.113×10^{-17} kg
6.02×10^{26} amu	= 5.61×10^{29} MeV	= 8.99×10^{16} joule	= **1 kg.**

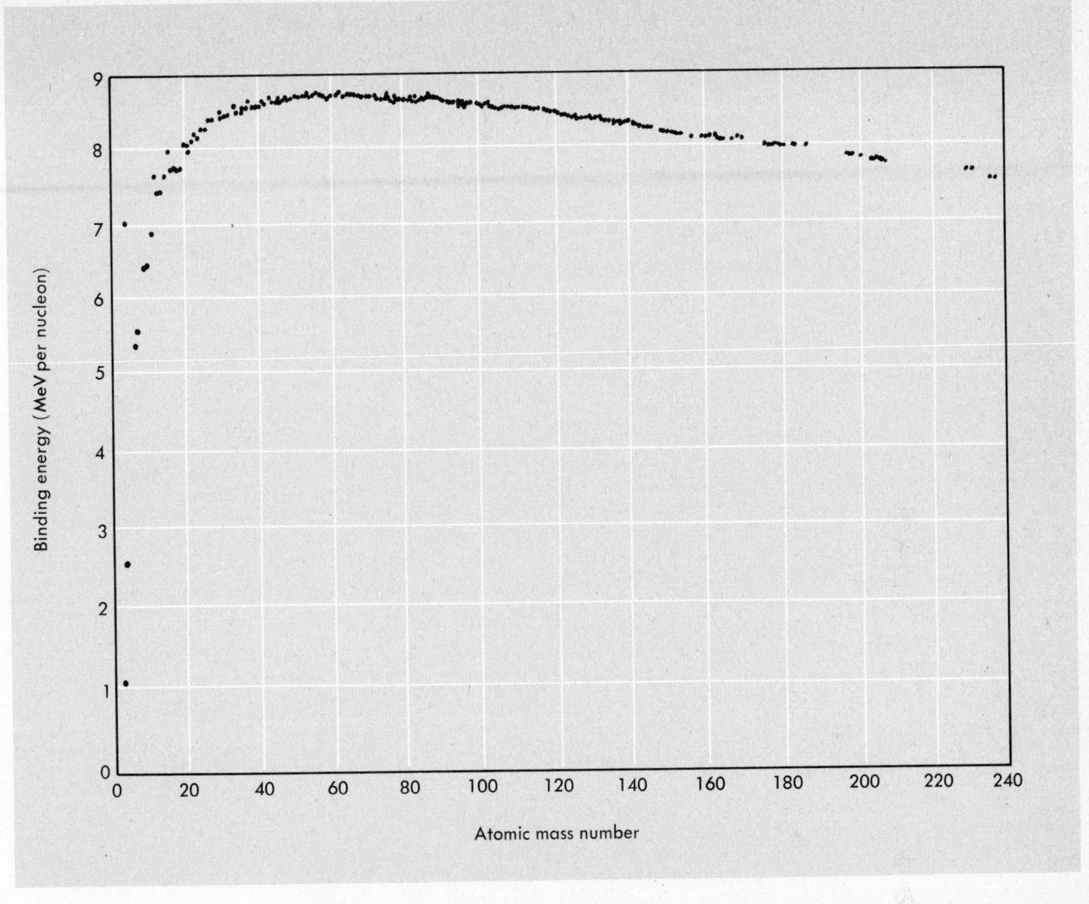

FIG. 28-2 The nuclear binding energies, per nucleon, of stable and long-lived nuclei.

Figure 28-2 shows the binding energy *per nucleon*, plotted against mass number A for a representative sample of elements. This binding energy per nucleon, relatively small for the lighter elements, increases with mass number to a maximum at about A = 60 and then decreases steadily for still more massive nuclei. As we shall see, the relationship between binding energy per nucleon and the mass number, as illustrated in this graph, is what makes it possible to release energy and generate power by nuclear fission and fusion.

28-3 Mass Defect and Nuclear Reactions

The exact knowledge of the atomic weights of the isotopes permits us to evaluate the energy balance of various nuclear reactions, since the mass equivalent of the liberated or absorbed nuclear energy must enter into the equation of the conservation of mass during the transformation.

Thus, in the case of Rutherford's original reaction,

$$_7N^{14} + {_2He^4} \rightarrow {_8O^{17}} + {_1H^1}.$$

The sums of the masses of the atoms entering the reaction and those resulting from it are, respectively,

$$\begin{array}{lr} _7N^{14} & 14.00307 \\ & + \\ _2He^4 & 4.00260 \\ \hline & 18.00567 \end{array} \qquad \begin{array}{lr} _8O^{17} & 16.99914 \\ & + \\ _1H^1 & 1.00783 \\ \hline & 18.00697 \end{array}$$

The combined mass of the reaction products is larger than the combined mass of the atoms entering into the reaction, by 0.00130 amu. Thus, to make the reaction an *equation*, we must add 0.00130 amu = 1.21 MeV of energy on the left side, meaning that the bombarding α particle must have 1.21 MeV more KE than the O^{17} nucleus and the proton on the right side.

The reaction $_3Li^7 + {_1H^1} \rightarrow 2{_2He^4}$, observed by Cockcroft and Walton, gives a different result:

$$\begin{array}{lr} _3Li^7 & 7.01601 \\ & + \\ _1H^1 & 1.00783 \\ \hline & 8.02384 \end{array} \qquad \begin{array}{lr} _2He^4 & 4.00260 \\ & + \\ _2He^4 & 4.00260 \\ \hline & 8.00520 \end{array}$$

In this case, 0.01864 amu = 17.4 MeV must be added to the right side, indicating that energy is released, and that the two α particles produced have 17.4 MeV more KE than did the bombarding proton.

28-4 The Liquid-Drop Model of the Nucleus

Figures 28-1 and 28-2 present nuclear information calculated directly from simple experimental facts. For this it was not necessary to have any knowledge of how the nucleus was put together. Of course it is not possible to describe the small world of particles in terms of the large-scale phenomena we can see and touch. But tentative models—the wave picture of light, the Bohr atom, etc.—can often be helpful if we recognize they are only crude analogies and do not represent any "real" truth.

In 1936 Bohr suggested that the nucleus might be like a droplet of dense liquid, composed of the subdroplets of the neutrons and protons it contained. This is analogous to our picture of a small water droplet composed of molecules.

Clearly, the forces holding the neutrons and protons together cannot be electrical—the neutrons have no charge, and the mutual repulsion of the positively charged protons would pull a nucleus apart. We must conclude that there is another force acting between all nucleons, holding them together by the attraction of each nucleon for its immediate neighbors, as the intermolecular forces hold a real liquid droplet together. As a first approximation, consider that each nucleon in a nucleus is surrounded by others on all sides. If we now reach in and remove one nucleon, its binding energy is the work we must do to pull it outside, away from the nucleus, against the attraction of its neighbors. This will be the same for each nucleon, no matter what the size of the nucleus. Figure 28-3 shows this as *U*, "uncorrected binding energy" per nucleon, running in a straight horizontal line across the graph.

There are at least two corrections to make. Some of the nucleons are on the surface, and are therefore *not* surrounded by attracting neighbors on all sides. It will take *less* work to pluck these nucleons away, as discussed in Chapter 12, Section 5, in connection with the surface energy and surface tension of a real liquid droplet. So the "uncorrected" line *U* will have to be reduced to take this into account. The surface-to-volume ratio of the droplets becomes smaller as the drop—or nucleus—becomes larger; a large nucleus has a small fraction of its nucleons on the surface; for a small nucleus the fraction is larger. In Fig. 28-3 the line *S*, "surface correction," shows this reduction, the correction large for the small nuclei, and getting smaller as A increases.

The other obvious correction is due to the mutual electrostatic repulsion of the protons. It will need less work in our imaginary experiment to withdraw a proton than a neutron, because the repulsion of the other protons will be helping us. If the number of protons is, say, doubled,

FIG. 28-3 Qualitative graphical analysis of the curve of binding energy per nucleon.

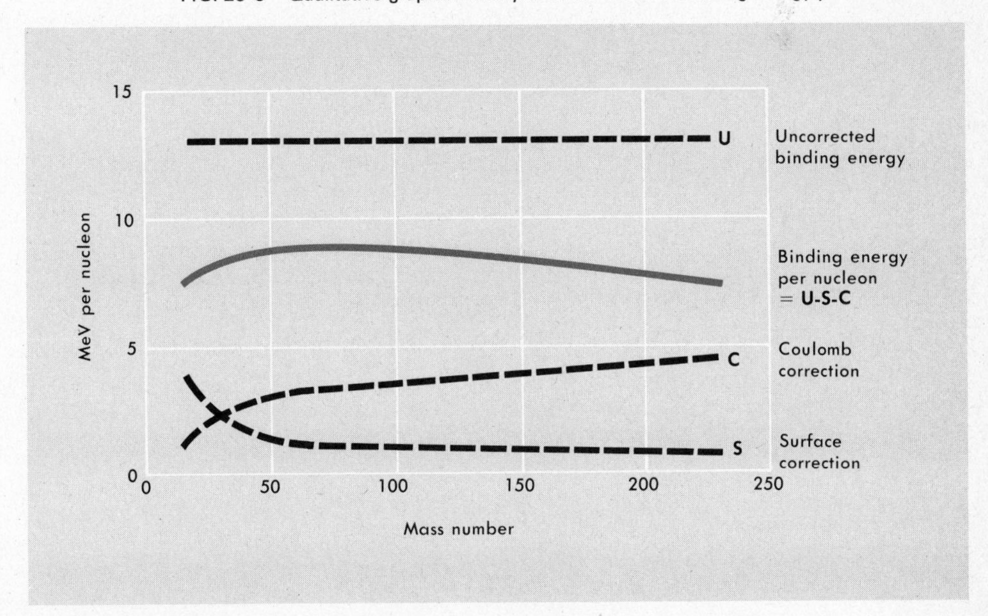

there will be twice as many repulsions contributing to the reduction so this average correction per nucleon should become larger with larger A. The fact that the fraction of protons becomes less as A increases (Fig. 28-1), and the fact that in a larger nucleus the distance to the outer protons becomes greater, combine to make a quantitative calculation difficult; but the general trend of the curve is as shown marked C, "Coulomb correction," in Fig. 28-3. When the total binding energy, $+U - S - C$, is plotted, its shape is similar to that of Fig. 28-2.

28-5 The Nuclear "Shell" Model

The liquid-drop model leaves some unanswered questions. Why are Z and N roughly equal for light nuclei, with N somewhat larger in heavy nuclei? It would seem that the more neutrons and the fewer protons, the more tightly bound and stable a nucleus would be, because of the reduction of the mutual repulsion of the protons. Yet there is no helium isotope $_2He^{35}$, or an oxygen $_8O^{113}$. Why not?

In 1949 two papers (Maria Goeppert–Mayer in the United States and J. H. Jensen in Germany) independently proposed a "shell model" for the nucleus. We have seen the Bohr atomic model, with its various shells and subshells of electrons. Nuclear particles also fit into a similar but more complicated scheme, with quantum numbers analogous to those of the electrons in the outer atom; and the Pauli exclusion principle works as effectively for nuclei as for atoms. All electrons are alike, however, and there is only one sequence of energy levels to accommodate them all. But for protons and neutrons, although the nuclear binding force makes no distinction between them, each has its own set of energy levels.

Figure 28-4 sketches a crude representation of the independent

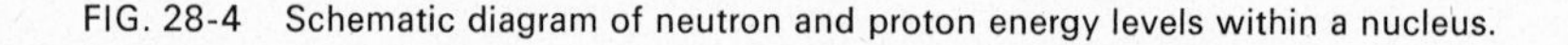

FIG. 28-4 Schematic diagram of neutron and proton energy levels within a nucleus.

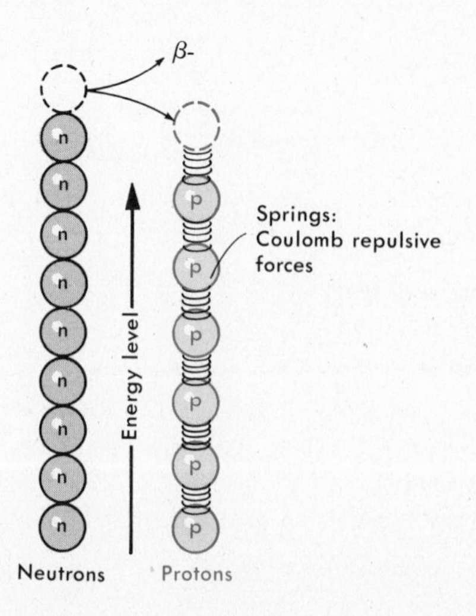

neutron and proton energy levels in the nucleus. The neutrons we might imagine to be lead balls piled one atop another—the proton pile is similar, but between each pair of protons is a spring, representing the electrostatic repulsion between them. This makes their spacing a little larger than that of the neutrons. The imaginary nucleus represented in the drawing is not stable—i.e., there is a way it can achieve a lower total energy. If one of the neutrons were to become a proton, and occupy the top level of the proton pile, the total energy of the nucleus would be reduced—and this it does by emitting an electron in β decay, as shown. For an excess of protons (whose energy levels are more widely spaced than those of the neutrons) equalization can be approached by removing two neutrons and two protons in an α decay. Thus in a radioactive decay chain (as in Fig. 26-3) we see an irregular alternation of α and β decays, each overshooting the line of stable nuclei until a stable configuration is reached.

Nuclear energy levels are of course much more complex than those in the simple stacks of lead balls in Fig. 28-4; both protons and neutrons have shells and subshells similar to those of atomic electrons. In the study of nuclear behavior it was noticed that nuclei with 2, 8, 20, 28, 50, or 82 protons, or 2, 8, 20, 28, 50, 82, or 126 neutrons, were in general quite stable and inert. In exposing these nuclei to bombardment by neutrons, the chances are extraordinarily small that they will change their configuration by absorbing one of the neutrons. As an example, the element tin ($_{50}Sn$) has 10 stable isotopes ranging from Sn^{112} to Sn^{124}, each with 50 protons. There are 6 stable nuclei, each containing 50 neutrons: $_{36}Kr^{86}$, $_{37}Rb^{87}$, $_{38}Sr^{88}$, $_{39}Y^{89}$, $_{40}Zr^{90}$, and $_{42}Mo^{92}$. We might look on this series of nuclear "magic numbers," as they are called, as being analogous to the series $Z = 2, 10, 18, 36, 54$, and 86 in the shell structure of atoms. These mark the filling of the p subshells, and identify the stable, inert noble gases. In a similar way the nuclear shell model has been able to predict stability and other characteristics of nuclear behavior.

Questions

(28-1) **1.** Calculate the neutron/proton ratio in the nuclei of (a) K^{39}, (b) Pd^{106}, (c) Xe^{131}.

2. What is the ratio of the number of neutrons to the number of protons in the nucleus of (a) Sc^{45}, (b) Cd^{112}, (c) Bi^{209}?

(28-2) **3.** The mass (atomic weight) of Li^7 is 7.01601 amu. (a) What is the binding energy of Li^7? (b) How much energy in MeV would be liberated if one could make a Li^7 nucleus out of neutrons and protons?

4. (a) What is the binding energy of the alpha particle? (b) How many MeV of energy would be liberated if one were to make an alpha particle out of neutrons and protons? (Mass of $_2He^4$ atom $= 4.00260$ amu.)

5. What is the binding energy per nucleon of Li^7? (See Question 3.)

6. What is the binding energy per nucleon of Be^9 (atomic weight, 9.01219)?

7. What is the binding energy per nucleon of Zn^{64}? (Mass = 63.9291.) Give answer in amu and MeV.

8. What is the binding energy per nucleon of Fe^{56}? Give answer in amu and MeV.

9. What is the binding energy per nucleon of Pb^{208}? (Mass = 207.9766.)

10. What is the binding energy per nucleon of U^{238}? (Mass = 238.0508.)

(28-3) 11. $N^{14} + He^4 \rightarrow H^1 + O^{17}$. (For Questions 11 through 16): (a) What is the mass (or energy) difference between the two sides of the reaction equation, in amu and in MeV? (b) Does this difference represent an excess of energy that the bombarding particle must have in order to conserve energy, or an excess of energy acquired by the product particles? Following are some atomic masses not given in the text: Be^9 = 9.01219, B^{10} = 10.01294, B^{12} = 12.01437, Ne^{20} = 19.99244, Na^{23} = 22.98977, S^{32} = 31. 97207, Cl^{35} = 34.96885.)

12. $Be^9 + He^4 \rightarrow C^{12} + n$ (See Question 11.)

13. $B^{10} + n \rightarrow Li^7 + He^4$ (See Question 11.)

14. $S^{32} + He^4 \rightarrow H^1 + Cl^{35}$ (See Question 11.)

15. $Be^9 + He^4 \rightarrow H^1 + B^{12}$ (See Question 11.)

16. $H^1 + Na^{23} \rightarrow Ne^{20} + He^4$ (See Question 11.)

29

Large-Scale Nuclear Reactions

29-1 The Discovery of Fission

Early in January, 1939, the German radiochemists Otto Hahn and his associate Fritz Strassmann published the results of bombarding uranium with neutrons. Their results were difficult to interpret, as they had confidently expected to produce "transuranic" elements—i.e., elements beyond uranium in the periodic table. There was evidence that elements in the middle range of atomic weights had been produced by the bombardment, but Hahn, primarily a chemist, was reluctant to make a definite claim to this effect because such a conclusion would be "at variance with all previous experiences in nuclear physics."

The previous year, Hahn's long-time co-worker, the Jewish Lise Meitner, had left Germany and the growing power of Hitler to join Bohr in Copenhagen. Here she and her nephew Otto Frisch found definite confirmation in chemical analyses and cloud-chamber photographs that uranium, when bombarded by neutrons, could split into two medium-sized fragments, with the release of enormous amounts of energy. Bohr immediately reported this to meetings of physicists in the United States. In the months that followed, workers in laboratories all over the world rushed to discover more about this new phenomenon of nuclear fission.

29-2 Neutrons, Uranium, Energy

Naturally-occurring uranium consists almost entirely of two isotopes: about 0.7 percent is U^{235} and 99.3 percent is U^{238}. Experiments soon indicated that nearly all the fissions were of the rarer U^{235}, and confirmed that an enormous amount of energy—about 200 MeV—was released in each fission. This is millions of times more energy than that given off by any chemical reaction.

For the source of this energy, let us turn back to the binding energy curve of Fig. 28-2 on p. 511. The dot representing U^{235} indicates a binding energy of about 7.55 MeV per nucleon; since it contains 235 nucleons, this gives a total binding energy of $7.55 \times 235 = 1770$ MeV. There is no way to predict exactly how the nucleus will split when it fissions, but the most probable division is into pieces containing about 40 percent and 60 percent of the original nucleus. Assuming that the U^{235} nucleus fissions into fragments of this size, we have parts whose mass numbers are $235 \times 0.4 = 95$, and $235 \times 0.6 = 140$. Figure 28-2 gives values of about 8.6 and 8.3 MeV per nucleon for these, for a total of $8.6 \times 95 + 8.3 \times 140 = 1980$ MeV. This indicates a release of $1980 - 1770 = 210$ MeV of energy for each fissioning nucleus, in agreement with more direct experimental evidence.

What sort of fragments are the $A = 90$ and $A = 140$ parts into which the uranium nucleus fissioned? How might we expect them to behave? In Fig. 28-1 on p. 508, one of the circles represents U^{235}, with an n/p ratio of $143/92 = 1.55$. Anywhere along the straight line drawn from U^{235} to the origin of the graph, the n/p ratio will be the same. Presumably the two fragments will contain the same relative amounts of neutrons and protons, and will lie on this line, where they have been indicated by "radioactive" circles. Both lie well above and to the left of the band of stable nuclei, and are accordingly very unstable. In the "magnifying glass" are arrows showing the changes resulting from different modes of decay. An α decay should carry the nucleus still farther from the stable region, and therefore does not occur. The emission of an electron by β decay *does* occur, as it brings the fragment closer to stability. More importantly, neutrons are also emitted as another way to reduce the too-high n/p ratio. So, after emitting a series of electrons and neutrons to form intermediate nuclei of various half-lives—some long and some short—the fragment eventually becomes a stable isotope of one of the middle-weight atoms.

29-3 Uranium and Neutron Bombardment

The nuclei of uranium and of transuranic elements (those of higher Z than uranium) contain so many protons that it requires little disturbance to allow the repulsive Coulomb forces to break them in two. However, alpha particles, protons, and any other positive particles are so strongly repelled that they cannot penetrate the nucleus to damage it. The relatively tiny mass of the electron makes it equally ineffective. Neutrons—relatively massive and without electric charge—are the ideal projectiles to cause these large nuclei to fission.

One might expect to find that the higher the neutron's speed and energy, the more effective it would be at causing fission. But this is not necessarily the case—the neutron requires time to react with the nucleus. In some circumstances a too-fast neutron may merely pass through the nucleus and come out the other side without having any effect. Table 29-1 tabulates in a very rough way the effects of neutrons of different energies on U^{235} and U^{238} nuclei.

The differences between the reactions of U^{235} and U^{238} form the basis of the technology needed to release the energy of fission, whether suddenly for an explosive effect, or gradually and under control for the operation of power-producing nuclear reactors.

29-4 Chain Reactions

There is nothing mysterious or even unusual about "chain reactions." A sheet of paper is quite stable in the presence of the oxygen in the atmosphere. A match held under a corner of the sheet, however, will add enough thermal energy for the oxygen to be able to combine with the material of the paper in a chemical reaction that releases a large amount of heat. This released heat in turn enables the neighboring paper to combine with oxygen, giving off more heat, which . . . , and so on until the entire paper is consumed.

If we substitute "neutrons" for "heat energy," we have a similar situation with the fission of uranium nuclei. Consider a piece of U^{235}. There are always a few neutrons in the air, with the same average kinetic energy as the molecules with which they have been colliding. One of

TABLE 29-1

Neutron energy	U^{235}	U^{238}
High energy	Little effect	Causes fission
Medium energy (a few MeV)	Little effect	Neutron absorbed
Low energy (thermal neutrons, less than 1 eV)	Causes fission	Little effect

these low-energy (about 0.025 eV at room temperature) "thermal" neutrons will soon enter the uranium and cause the fission of a U^{235} nucleus, a job at which thermal neutrons are particularly effective, as indicated in Table 29-1. Each fission will produce two or three neutrons on the average, each having a few MeV of kinetic energy. This is too much energy for them to be very effective in causing U^{235} fission, but after a number of collisions this energy is reduced to the point where they are again effective fissioners, so that each one is capable of causing another fission, each of which produces two of three more neutrons to produce two or three more fissions . . . , and so on, as indicated in Fig. 29-1. (All this is happening, remember, on a time scale measured in microseconds or picoseconds.)

29-5 Critical Size

What has been described above might be a nuclear fission bomb, commonly called an *atom bomb*. As described, however, it might not work—the size of the piece of U^{235} has an important influence. In Fig. 29-1, of the 34 fission neutrons shown produced, 11 of them, or 32 percent, have escaped through the surface and are lost as far as causing further

FIG. 29-1 A nuclear chain reaction developing in a piece of fissionable material with a "branching ratio" of 2—i.e., each neutron-induced fission produces two more neutrons. The reaction starts at 1 by a single neutron from the outside. After seven successive generations in this particular piece, seven neutrons remain inside the piece, and eleven have been lost through the surface.

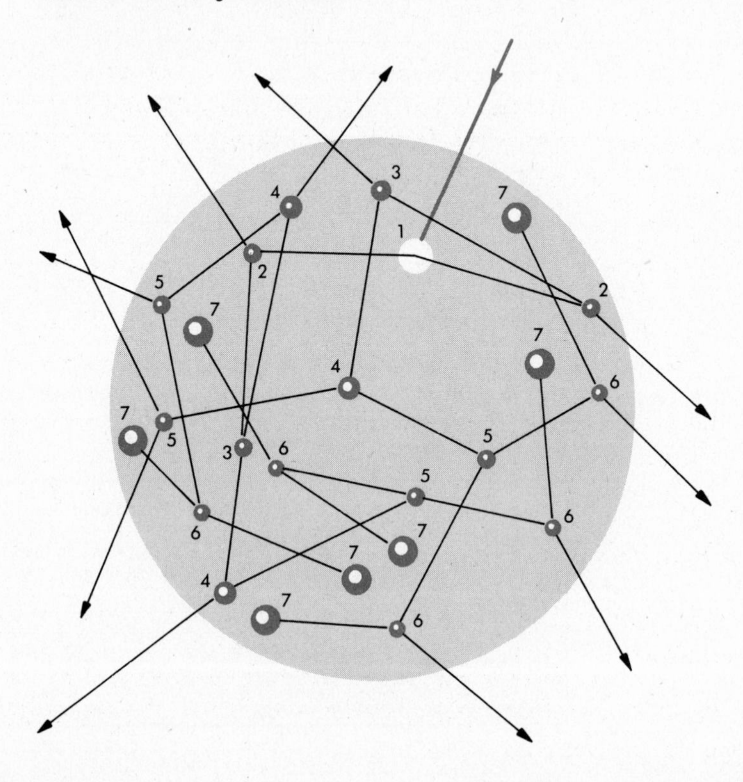

fissions is concerned. To exactly sustain a nuclear chain reaction such as this, without the rate of producing fissions either growing larger or fading away, *exactly one* neutron from each fission must cause another fission. In the hypothetical case shown, *more* than one of them does produce further fissions, so the rate of fissioning would rapidly increase.

The neutrons emitted by fission may have four different fates awaiting them:

1. They may lose energy by collisions until they are able to cause another fission of U^{235}.
2. In a uranium mixture containing U^{238}, they may be absorbed by a U^{238} nucleus. This is a *very* important event that we will discuss later; but it does eliminate the neutron as a fission producer.
3. Absorption by other materials present, including previously formed fission fragments.
4. They may be lost by escaping through the surface.

In the U^{235} bomb we are imagining, we assumed (and this is *much* more easily assumed than accomplished) that U^{238} is not present, so we can neglect item 2, and assume that item 3 is also negligibly small. This leaves only two alternatives: A fission neutron will either cause another fission or it will escape through the surface and be lost.

In order to make the losses through the surface as small as possible, we must make the surface (through which the escapes take place) as small as possible in comparison to the volume (in which the fissions occur). The larger the piece of uranium, the smaller the surface-to-volume (S/V) ratio becomes. Consider a cube measuring d cm on each side. Its volume is d^3; its surface area is $6d^2$, giving $S/V = 6/d$; the S/V ratio varies *inversely* with its size. A cube 0.1 inch on a side has $S/V = 6/0.1 = 60$; for a 10-inch cube, 100 times larger, $S/V = 6/10 = 0.6$—100 times smaller. Although we have figured this for cubes, the same inverse relationship between size and S/V holds for any shape.

Thus, if our imaginary U^{235} bomb is too small, neutron losses through its surface will prevent a chain reaction. As we make it larger and larger, S/V becomes smaller, and eventually we reach the *critical size*, at which exactly one of the two or more neutrons will be retained to carry on the chain reaction. For a piece larger than critical size, losses will be less, so that *more* than one of the neutrons produced in each fission will be able to cause another fission, and the chain reaction will grow rapidly, to produce an explosion. Figure 29-2 shows a primitive bomb in which the two separated pieces are each slightly less than the critical size. When the conventional explosive brings them together, the resulting piece is much larger than critical and the fission chain reaction rapidly builds up, to release explosive energy until the entire device is vaporized and dissipated into the atmosphere.

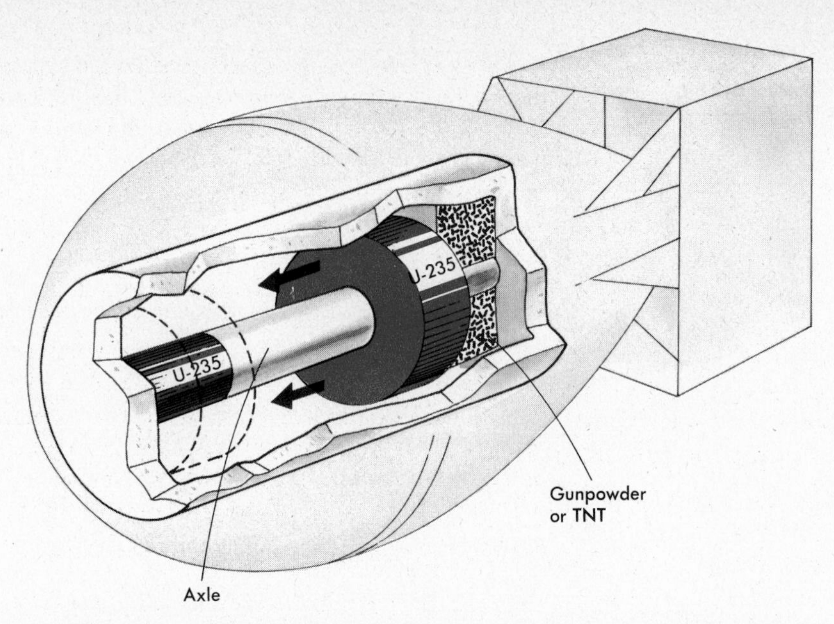

FIG. 29-2 The principle of the "gun-type" atomic bomb. A conventional explosive drives a U-235 or plutonium cylinder along the axle until it surrounds another piece of fissionable material forming the far end of the axle, thereby producing a single fissionable mass of more than critical size.

29-6 The First Nuclear Reactor

In the early prebomb days of fission research, it was not even known if it would be possible to sustain a chain reaction to produce either an explosion or a steady flow of controlled heat energy. It was decided to build a research reactor, to learn as much as possible about the many unknown aspects of the fission process.* This was done in great secrecy in a squash court under the University of Chicago's Stagg Field Stadium, directed by the Italian–American physicist Enrico Fermi.

The two principal isotopes of uranium are chemically identical and differ in their masses by only a little more than 1 percent. It was obvious that the problem of separating them in large quantities would be a lengthy and difficult one. It was therefore necessary to design a reactor that could operate with the naturally-occurring mixture of isotopes. No self-sustaining chain reaction is possible in a single piece of such uranium, no matter how large its size. Most of the neutrons produced by fission would at once be absorbed by the U^{238} nuclei—there are nearly 140 of them for each U^{235} in the natural mixture. But—if they could be slowed down to the thermal energy range, they would merely bounce off the U^{238}'s until they struck a U^{235} nucleus to cause fission. The best way to slow them down is by collisions with light nuclei in a material referred to as a *moderator*.

---

* Germany was also occupied with similar projects—a substantial portion of the Kaiser Wilhelm Institute in Berlin, as well as the uranium mines of German Bohemia, were sealed off under the tightest possible security.

From considerations of conservation of momentum and energy, it can be shown that when a particle collides with another particle much less massive than itself, it is slowed down very little and loses scarcely any energy (recall the lack of effect that electrons have in deflecting α particles). At the other extreme, if a particle collides with another particle much more massive than itself, it bounces back with a speed and energy that are little changed. To be most effective, then, in helping the neutron lose energy by collision, the moderator atoms should be light atoms, comparable in size to the neutron, and should not absorb neutrons.

Of course, from the purely mechanical point of view, ordinary hydrogen (or some hydrogen-rich material, such as plain water) would be the best moderator. Its nuclear protons, having almost the same mass as the neutrons, would absorb the maximum possible energy in each collision and would be most effective in slowing the neutrons. Unfortunately, however, some of the collisions would result in the absorption of the neutron, and its loss from the fission chain:

$$ {}_0n^1 + {}_1H^1 \rightarrow {}_1H^2. $$

This heavy isotope of hydrogen, ${}_1H^2$, is such a common material in nuclear research that it has even been given its own name—*deuterium*—and its own chemical symbol D. Its nucleus, a proton and a neutron, is called a *deuteron*. About 1 H atom in 7000 is actually a D atom, so deuterium is relatively plentiful. "Heavy water," D_2O, is a fine neutron moderator, and can be separated from H_2O by a long and tedious series of repeated electrolysis or distillation; because of the smaller mass and greater mobility of its molecules, H_2O distills and electrolyzes a little more readily than D_2O. Fermi's group decided to use the readily available element carbon, rather than heavy water, whose production would present another long technological problem to be solved.

A neutron can travel several centimeters through uranium before it strikes a nucleus (recall the small size of the nucleus compared to the size of the atom). The experimental reactor used balls of natural uranium and uranium oxide surrounded by bricks of highly purified carbon in the form of graphite.

The balls fit into cavities machined in the bricks, and on the floor of the squash court layers of graphite and uranium were put down to form a pile of ever-increasing height. (This and other early nuclear reactors were called "piles" because of this manner of construction). Figure 29-3 shows the pioneer reactor under construction.

Although they are not clearly shown in the photograph or in the later drawing (Fig. 29-4), the structure contained cavities and channels for instruments of many kinds, and slots into which cadmium rods could be pushed to varying depths. Cadmium (as well as boron and a number of other materials) is a strong absorber of neutrons, and such *control rods* serve to regulate the net rate of neutron production. By pulling the rods farther out, or by pushing them in, the chain reaction could be prevented from dying out, or from steadily increasing until the resulting increase in temperature might damage the reactor.

The uranium balls were made small enough so that nearly all of any fission-produced neutrons would pass through them into the surrounding graphite. Here they could collide to their hearts' content with carbon nuclei and thereby reduce their energies to the point where they could no longer be absorbed appreciably by the U^{238}. Eventually, these low-energy neutrons, following a Brownian-motion path, would diffuse back into one of the uranium balls, where they would be almost certain to cause the fission of the first U^{235} nucleus they encountered.

On December 2, 1942, after the fifty-seventh layer had been placed, critical size was reached, and the first self-sustaining nuclear chain reaction was accomplished. Ten days later the pile was operating smoothly at its maximum allowed power production of 200 watts and was furnishing much of the data needed for further work on both bombs and power production.

29-7 Fission Fuels

So far, our discussion has assigned U^{238} the role of a useless, unwanted neutron absorber that provided no benefits, and many difficulties. By making a huge pile with an enormous volume of moderator, Fermi's group was able to overcome these difficulties, and sustain a modest U^{235} chain reaction. The information gained by this research reactor was of great importance and essential to any further progress. But a nuclear bomb was the first-priority project at that time—it was a life-and-death race to beat Hitler's Germany to this goal.

FIG. 29-3 The only photograph (taken in November, 1942) made during the construction of the first nuclear reactor. Here the nineteenth layer (almost covered up) is formed by graphite blocks studded with spheres of uranium metal and uranium oxide. The even-numbered layers consisted only of graphite blocks without uranium. The pile became critical on December 2, 1942, after 57 layers had been laid down. (*Courtesy Argonne National Laboratory.*)

FIG. 29-4 A drawing of the first nuclear reactor, constructed in a squash court under the west stands of Chicago University's Stagg Field in 1942. (*Courtesy U.S. Army Engineers.*)

As we have seen, a large chunk of nearly pure U^{235} was essential to make a bomb, and this required the separation of U^{235} from the more than 99 percent of U^{238} present in the natural mixture. Since the two isotopes are identical chemically, it was necessary to use a physical separation method that would be able to take advantage of the slight difference in their masses.

One obvious method was to use a *mass spectrograph*, referred to in Chapter 26, Section 3. Figure 29-5 shows the scheme of operation of a primitive model of such an instrument. Uranium, in the form of the gaseous compound uranium hexafluoride (UF_6) is ionized by an electric discharge. As some of the ions pass through a slit in a partition, an electric field accelerates them toward a second slit, from which they emerge, all with the same energy KE $= Ve$, to enter a magnetic field. Since both the $U^{235}F_6$ and the $U^{238}F_6$ have the same KE, the less massive $U^{235}F_6$ must have a slightly higher velocity as well as a smaller mass, and is therefore deflected by the magnetic field into an arc of smaller radius. The two isotopes are thus separated, and can be collected separately. This method was used, with a battery of thousands of spectrographs running 24 hr a day. The disadvantages were many: Each

FIG. 29-5 Schematic diagram of separation of uranium isotopes in a mass spectrograph type separator.

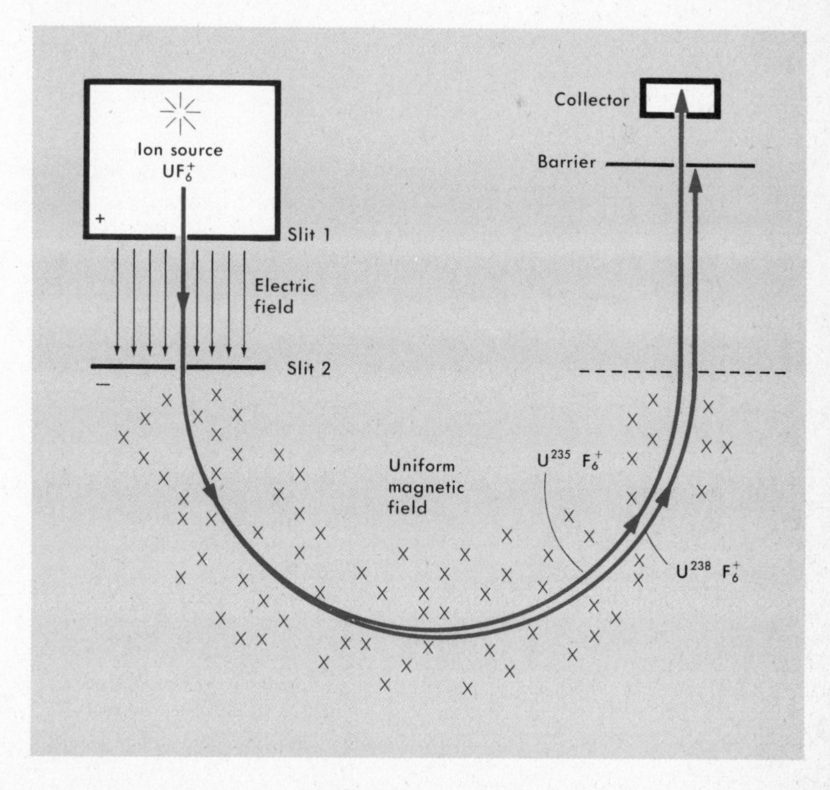

spectrograph could produce no more than a fraction of a milligram a day; enormous amounts of electrical energy were needed to operate the magnets, with a huge cooling system required to take away the I^2R heat from them.

So most of the separation was done with a diffusion system, in which UF_6 gas passes through a porous membrane separating regions of high and low pressure. Since the rate at which molecules diffuse through the pores of the membrane depends on their speed, slightly more $U^{235}F_6$ molecules will pass through to the low-pressure side. Thus at each passage through the porous membrane, the relative concentration of U^{235} will be increased; after thousands of such passages, U^{235} can be nearly completely separated from the U^{238}. By these laborious methods, enough U^{235} was accumulated to make a bomb.

Meanwhile, another important development was taking place. In Table 29-1, there is a brief notation "neutron absorbed" for the reaction between a medium-energy neutron and a U^{238} nucleus. This absorption, which we have so far treated as a disadvantage to be avoided at all costs, has a very remarkable consequence.

$$ {}_{92}U^{238} + n \rightarrow {}_{92}U^{239} \qquad \text{(half-life, 24 min).} $$

The unstable uranium-239 quickly undergoes two β decays:

$$ {}_{92}U^{239} \rightarrow {}_{-1}\beta^0 + {}_{93}Np^{239} \qquad \text{(half-life, 2.4 days)} $$

$$ {}_{93}Np^{239} \rightarrow {}_{-1}\beta^0 + {}_{94}Pu^{239} \qquad \text{(half-life, } 2.4 \times 10^4 \text{ yr).} $$

The plutonium-239 produced in this way is very similar to uranium-235 in its behavior under neutron bombardment. In Table 29-1, its listing would be the same as that in the U^{235} column. As a nuclear fuel, it is the equal of U^{235}.

In this way, a neutron absorbed by U^{238}, although it loses its place in maintaining the chain reaction, does actually produce another fissionable atom of nuclear fuel. And plutonium, being chemically different from uranium, can later be separated out by chemical means, rather than the lengthy physical processes needed to separate isotopes of the same element.

29-8 Nuclear Fission Power Reactors

Fission reactors can be built for several purposes. Many, like the first 1942 pile, are purely for research. Their designs are many and varied, depending on the purpose of the research, and operating efficiency may be a very secondary consideration.

Public interest, however, is directed toward large reactors designed to produce heat, to generate steam, to operate turbines, to drive electric generators. Figure 29-6 is a *very* simplified sketch of a typical nuclear

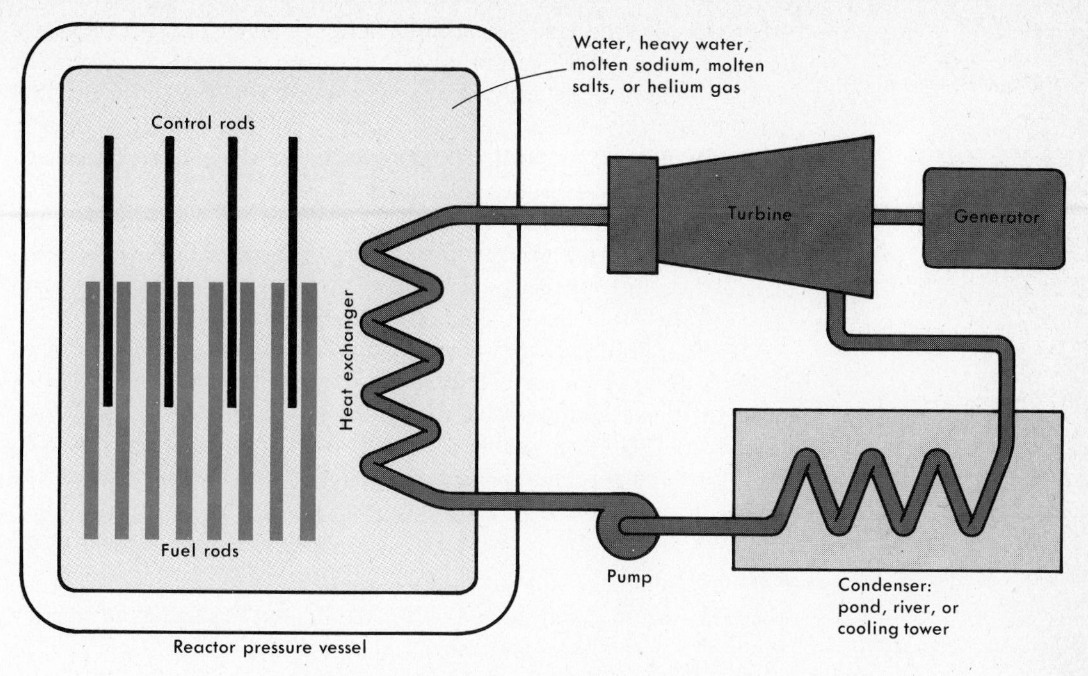

FIG. 29-6 Basic elements included in a typical nuclear generator plant.

generating plant. The fuel is normal natural uranium enriched by the addition of a few percent of U^{235}, enclosed in stainless steel or other special metal tubing.

Most of the power reactors now in operation use ordinary water as the circulating medium to receive the heat generated by the reactor. In these, the water serves a double purpose—it is also the moderator. By using enriched fuel, such a system can afford the small loss of neutrons that occurs in the $H + n \rightarrow D$ reaction. Some reactors omit the secondary heat exchanger sketched in Fig. 29-6. Steam is then fed directly into the turbine from the pressure vessel that contains the reactor. The steam and water, however, are somewhat radioactive from fission fragments that penetrate the metal casing of the fuel rods. Thus, although the overall system is simpler, extraordinary precautions must be taken with the turbine to guard against the possibility of steam leakage.

The fuel rods must be taken out and replaced by new ones every year or so. By that time, the U^{235} content has been appreciably lowered and the fuel has become contaminated with fission fragments, some of which absorb neutrons. By remote-control handling, the highly radioactive old rods are chemically processed. The recovered uranium (with its now reduced U^{235} content) is ready to be made into new rods after more U^{235} is added; some of the U^{238} will have been converted to Pu^{239}, which can be used for fuel enrichment or for bombs; the very radioactive fission fragments must be disposed of in some way.

The disposal of these fission fragments presents a problem that has not yet been solved in a really satisfactory way. The used fuel rods are stored for months or years before reprocessing, to allow the shortest-lived fission products to decay to lower radiation levels. Chemical treatment then separates out the reusable uranium and plutonium. The residue, "hot" in both the radioactive and thermal sense, is at present dissolved in water and stored for several years in large underground tanks while its radioactivity and temperature gradually decrease.

In time, all this radioactive liquid must be converted to solid form, as insoluble compounds, or incorporated into insoluble ceramic pellets or concrete. It must then be stored for centuries as its radioactivity gradually dies down. This will probably eventually all be done in caverns in thick beds of salt or gypsum, or in other geological formations in which the chances for pollution of groundwater are as nearly zero as possible.

29-9 Breeder Reactors

It is very easy to describe the fuel for the present generation of nuclear reactors as natural uranium "enriched by the addition of a few percent of U^{235}." But early difficulties in separating U^{235} from U^{238} are still with us, and the separation is still accomplished by diffusion of UF_6 through porous membranes in great plants that consume enormous amounts of power to operate their many pumps and refrigerators. New centrifuge designs will no doubt allow these to be replaced by much more efficient plants, in which UF_6 is passed through a series of high-speed centrifuges in which the gas at the circumference contains a higher proportion of the heavier $U^{238}F_6$, while the lighter $U^{235}F_6$ is somewhat more concentrated near the axis. It is still a costly process—in terms of both money and energy—and a ton of uranium metal must still be mined and refined to obtain 14 pounds of the fissionable U^{235}.

The *breeder reactor* is designed not only to produce heat for power generation, but to create more fuel than it consumes. This is accomplished by the reactions given in Section 7, in which fissionable Pu^{239} is produced from U^{238}. The most common isotope of thorium (Th^{232}) reacts similarly to neutrons, and produces the fissionable U^{233}:

$$_{90}Th^{232} + n \rightarrow {}_{90}Th^{233} \qquad \text{(half-life, 22 min)}$$

$$_{90}Th^{233} \rightarrow \beta^- + {}_{91}Pa^{233} \qquad \text{(half-life, 27 days)}$$

$$_{91}Pa^{233} \rightarrow \beta^- + {}_{92}U^{233} \qquad \text{(half-life, } 1.6 \times 10^5 \text{ yr).}$$

Thorium is about three times more plentiful than uranium, and is considerably cheaper. The U^{233} it produces fissions in the same way as U^{235} and Pu^{239}. The operation of breeder reactors can thus actually manufacture new nuclear fuel from the otherwise useless and relatively plentiful supply of both U^{238} and Th^{232}.

In order for a reactor to function as a breeder, the neutrons produced in fission must operate on a very tight budget. (1) From each fission, on the average, exactly one neutron must of course cause another fission to maintain the reaction. (2) Some neutrons will be absorbed by the control rods, or escape from the reactor, or will otherwise be lost as far as productive reactions are concerned; it will be difficult to make this below 0.2 neutron/fission. (3) At least one neutron must be absorbed by U^{238} or Th^{232} to produce another atom of fissionable fuel, to replace the one used up in (1). Anything more than this will represent a net gain —more fissionable fuel will be reclaimed from the spent fuel rods than was put in them when they were made.

Since items (1) and (2) are nearly constant, it is apparent that any increase in the average number of neutrons per fission will go to increase (3) and hence the amount of new fuel produced. Table 29-1 must not lead to the conclusion that there is a *sudden* change in the reaction to be expected as the neutron energy increases from the thermal energy range. Increased neutron energy makes fission less probable, but increases the number of neutrons produced per fission; it also increases the probability of producing a new fissionable fuel atom through absorption by U^{238} or Th^{232}. Breeder reactor design calls for a very careful balance among all of these factors.

The few breeder reactors now in use are relatively small, and still experimental in nature. Most of them are "fast" breeders, a term that refers *not* to the breeding rate but to their use of fast, more energetic neutrons to cause fission, with more neutrons per fission as a result. Some are "thermal" breeders, referring to the lower-energy neutrons, more effective for Th^{232}–U^{233} fuel. These experimental breeder reactors will provide much of the design data needed for construction of the larger reactors to follow. These will not only produce more fissionable fuel than they consume but will operate at higher temperatures to give efficiencies of 40 percent or more, the equal of the best coal-, oil-, or gas-burning plants now operating.

29-10 Fusion Reactors

Let us refer again to the nuclear binding energy curve of Fig. 28-2. In fission the high-A parent nucleus has a relatively low binding energy per nucleon; the lower-A fission fragments have a considerably higher binding energy. The difference is the energy released by the fission.

There is another way, evident on the graph, to go from low binding energy per nucleon, to high. If it were possible to take two light nuclei of, say, A = 20, and cause them to fuse together to make a single nucleus of A = 40, we would also be going in the proper direction to release energy. The A = 20 nuclei have a binding energy of about 8.0 MeV/nucleon. In taking them apart into their constituent neutrons and protons we would have to expend $2 \times 20 \times 8.0 = 320$ MeV. When the separate nucleons were reassembled into a single nucleus of A = 40, the energy released would be about $40 \times 8.6 = 344$ MeV, to give a net gain of $344 - 320 = 24$ MeV. This is the process called *fusion*, in which light nuclei combine to make a heavier one.

In the sun and in all the other stars, nature produces enormous amounts of energy from nuclear reactions in which fusion, rather than fission, is the energy source. As we will discuss in more detail in Chapter 33, nearly all this energy comes from the simplest possible fusion reaction, in which four hydrogens combine to form a helium.

A closer look at Fig. 28-2 will give an idea of the energetic possibliities of such fusion reactions. The dot at A = 1 and zero binding energy represents the single proton of H, which is bound to no other nucleon. Above it at 1.1 MeV/nucleon is deuterium, H^2 or D; tritium (H^3 or T) is at 2.6 MeV/nucleon. With much more binding energy than any of these, at 7.1 MeV/nucleon, to the left of the main curve, is helium, He^4.

In combining 4 hydrogens to make a helium, we thus have a release of about 7 MeV/nucleon. In uranium fission there is about 200 MeV/235 nucleons = less than 1 MeV/nucleon. $4H \rightarrow He$ is indeed an energetic reaction.

This is accomplished in a number of steps in the center of the sun where the particles are held together at enormous pressure and a temperature of about 20,000,000°K. One place where high pressure and multimillion-degree temperatures are available is in the immediate vicinity of an exploding nuclear fission bomb. For this reason the so-called "hydrogen bomb" (actually, it is rumored, now made largely of lithium hydride, LiH) is extremely simple in principle. Essentially, all that needs to be done is to surround a fission bomb with a layer of material of low atomic number. When the fission bomb explodes, it provides an environment in which the light nuclei are forced to come together and release still more energy of nuclear fusion.

Such a procedure, however, is far from satisfactory for operating a power plant. Intensive work is being carried out by all countries that have the necessary financial and scientific resources, in an effort to find some practical way of maintaining a steady, controllable flow of energy from nuclear fusion reactions.

In fusion experiments in which beams of particles from accelerators are directed against light-element targets, the yield of the desired reactions is very low. The charged particles, such as protons or α particles, rapidly lose their energy in the ionization of the target material, and only a very few of them (about 0.01 percent) have a chance to collide with another nucleus before having spent their energy in merely tearing off atomic electrons. Such experiments are extremely valuable for the study of nuclear properties, but are worthless for the purpose of large-scale nuclear energy liberation.

The simple $4H \rightarrow He$ reaction used by stars is not at present seriously considered as a source of fusion energy. For a given particle energy, probability of a fusion reaction in a H–H collision is very small; it is much greater for a deuterium–deuterium (D–D) collision and greater still for deuterium–tritium (D–T). These last schemes are therefore receiving all the current experimental attention.

Deuterium is plentiful, cheap, and easy to separate from the common H^1 hydrogen. In all the waters of the oceans, about 1 hydrogen in 6500 is this H^2 isotope. If the deuterium in a single cubic meter of water were to be put through the D–D fusion process, it would release energy equal to the burning of 2000 barrels of oil. One cubic kilometer of water would yield about as much energy as all of the worlds' known oil reserves. With 1.4×10^9 km^3 of water available to us, this energy source will last longer than the world will be habitable.

Tritium, however, is naturally present in such small quantities that it must be artificially produced. Exposing lithium to neutron bombardment in a reactor does this job:

$$n + Li^6 \rightarrow He^4 + T.$$

Since lithium is not plentiful, and only 7 lithium atoms in 100 are the Li^6 isotope, the T–D reaction can only be expected to buy a century or so of time, if it is needed, while the means to use the basic D–D reaction is being perfected.

It is easy to give a compact *beam* of deuterons the kinetic energy corresponding to millions or billions of degrees; it is a much more difficult problem to give them the *random* heat motions that lead to collisions between them. A gas composed of charged particles—in this case D^+ and e^-—is called *plasma*; the problem is to confine the hot plasma until enough collisions take place to produce a profitable amount of nuclear fusion energy. (By "profitable" is meant that more energy must be produced than is expended in operating the reactor.) A simple material container is out of the question, because the plasma would at once be cooled to the temperature of the confining walls. Since charged particles are subjected to strong forces when they move rapidly through a magnetic field, the most common experimental approach has been to confine the plasma by ingeniously designed magnetic fields. The problem is a difficult one, though, and has not yet been satisfactorily solved. But steady progress is nevertheless being made, and even though the goal is still distant, many researchers believe that magnetic field confinement is the method that will lead to final success.

But another very different scheme has also been proposed and worked on intensively during the past few years. Specially designed lasers are already capable of emitting enormous pulses of energy in brief bursts, and research is steadily improving their performance. The idea, as schematically sketched in Fig. 29-7, is to use fuel in the form of millimeter-sized frozen pellets of deuterium (or deuterium–tritium). As each

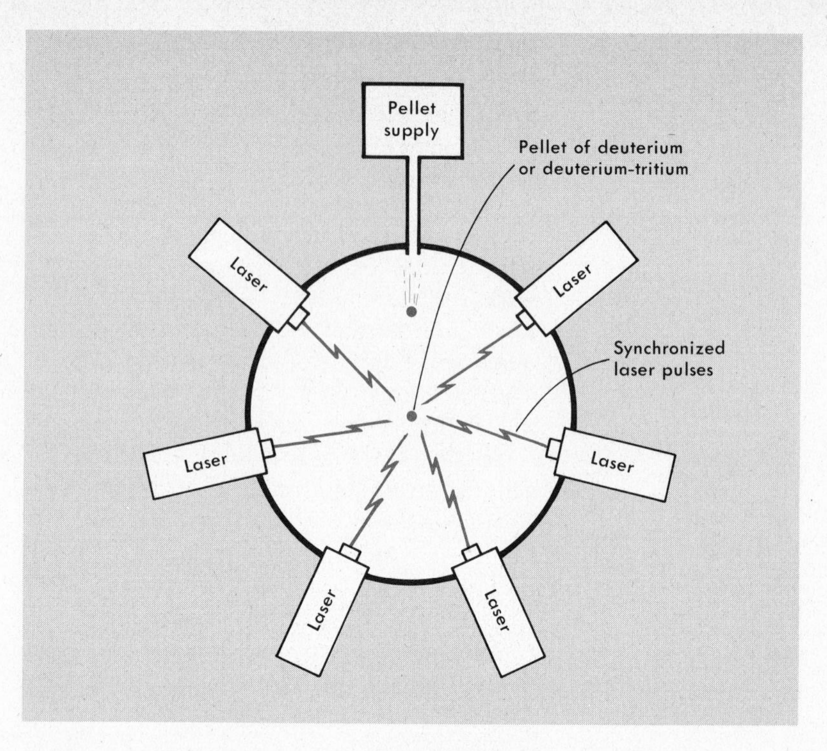

FIG. 29-7 One experimental method for inducing D-T fusion by laser beams.

falling pellet reaches the center of a spherical enclosure, a number of accurately synchronized laser pulses strike it simultaneously. This causes practically instantaneous vaporization; the inertia of the gas keeps it from expanding outward as fast as it is formed, with the result that high pressures, densities, and temperatures are achieved at the center of the pellet. Research in this direction is becoming rapidly more encouraging. Regardless of its success, this method will probably always have the disadvantage of being a series of explosions whose energy must be absorbed as heat in a surrounding layer, probably of molten metal.

There are very few who doubt that the scientific and technological problems of fusion energy will eventually be solved. But a number of major breakthroughs are still needed, and it cannot be predicted when they will be achieved. The time needed may be measured in years, in decades, in centuries, or millenia. In the meanwhile, the end of our petroleum resources lies just over the horizon; U^{235} as used in our present nuclear reactors might the stretch available time to a century or two; the coming generation of breeder reactors will extend our fuel supplies of U^{238} and Th^{232} for thousands of years. By that time, all but the most dedicated pessimists will argue that we will *surely* be able to tap the inexhaustible supply of deuterium, and our energy supply will be assured forever. We will also be largely relieved of the problem of disposing of the radioactive waste products of fission—fusion produces only a little tritium, which will be largely reused, and a minimal amount of radioactive material caused by the neutron bombardment of the reactor structure itself.

Questions

(29-2) 1. Another nuclear fuel, U^{233}, is produced in breeder reactors. From Fig. 28-2, make an estimate of the energy released if a U^{233} fissioned into two parts in the mass ratio of 0.45:0.55.

2. Breeder reactors produce the fissionable Pu^{239}; suppose a Pu^{239} fissions into two fragments of 32 and 68 percent of its mass. From Fig. 28-2, estimate the energy released in this fission.

3. Assume that (to the nearest integer) the protons and neutrons in the U^{233} divide in the proportion given in Question 1. (a) What are the isotopes initially formed by this fission? (b) How many electrons and neutrons would be emitted in converting the smaller fragment to the stable ${}_{44}Ru^{102}$? (c) The larger fragment into the stable ${}_{52}Te^{74}$?

4. In the Pu fission of Question 2, assume that both protons and neutrons divide in the same given ratio (to the closest integer, of course). (a) What are the isotopes initially produced by this fission? (b) Into what stable isotope does the smaller fragment decay after emitting 3 neutrons and 2 electrons? (c) Into what stable isotope does the larger fragment decay after emitting 2 neutrons and 2 electrons?

(29-3) 5. Compare the effects of medium-energy neutrons, such as the majority of those emitted in fission, on U^{235} and U^{238}.

6. Compare the effects of low-energy thermal neutrons on U^{235} and U^{238}.

(29-5) 7. Consider a mouse, and a much larger rat of the same shape. If they are equally well insulated by fur, and have the same body temperature, which will have to eat more per gram of body weight? Explain why.

8. A cubical metal box, each side measuring 4 ft, contains 64 40-watt lamps, evenly spaced 1 ft apart. Another similar box, containing 216 equally spaced 40-watt lamps, is a 6-ft cube made of the same material. (a) Which box will have the warmer surface? (b) In the 6-ft cube, with what wattage should we replace all the 40-watt lamps, in order to make both boxes have the same surface temperature?

(29-6) 9. Consider a direct, head-on, perfectly elastic collision between a moving particle of mass m and a stationary particle of mass $2m$. (a) What fraction of its kinetic energy would m lose in this collision? (b) How many such collisions would be needed to reduce the KE of m to less than 5 percent of its initial value?

10. Same as Question 9, except that the target particle has a mass of $10m$.

(29-7) 11. The operation of specialized mass spectrographs for separating the principal isotopes of uranium depends on the difference in the e/m ratios, and use the ions of UF_6^+. (a) Which has the larger e/m, $U^{235}F_6^+$, or $U^{238}F_6^+$? (b) What is the ratio of their e/m's?

12. In the diffusion separation of U^{235} and U^{238}, which depends on the relative speeds of the molecules involved, the gaseous compound UF_6 (uranium hexafluoride) was used. At any given temperature, (a) which molecules have the higher speed, $U^{238}F_6$ or $U^{235}F_6$? (b) What is the ratio of their speeds?

13. Nuclear bombs are customarily rated in terms of their explosive effect, as compared with the chemical explosive TNT. A 1-kiloton bomb liberates the same energy as 1000 tons of TNT ($= 5 \times 10^{12}$ J). The earliest nuclear fission bombs were rated at 20 kilotons. (a) How many ergs of nuclear energy were released? (b) How many MeV? (c) How many U^{235} or Pu^{239} atoms must have fissioned? (d) About how many grams of uranium were actually fissioned in a 20-kiloton bomb? (e) About how much mass was actually converted into energy?

14. How long would it take a 1-GW generating plant to produce as much energy as a 50-kiloton nuclear bomb?

(29-9) **15.** A generating plant produces an average of 8 GW-hr of electrial energy per day, at an efficiency of 32 percent. (a) How much thermal energy must be supplied per day? (b) How much heat is dissipated in the cooling water or into the atmosphere via cooling towers per day?

16. Refigure Question 15 for an efficiency of 40 percent. By what fraction does this change parts a and b?

30

Particles, Particles!

30-1 The Positron: Antielectron

In the early days of nuclear physics, the materials considered necessary to make the universe comprised a very short list: electrons, protons, and photons. It was still a relatively simple world when the neutron was added to this list in 1932. But three years earlier, in 1929, the purely theoretical picture had been complicated by a British physicist, P. A. M. Dirac (Fig. 30-1), who was busy trying to reconcile the basic principles of the quantum theory with those of Einstein's theory of relativity. On the basis of very abstract theoretical considerations, Dirac came to the conclusion that in addition to the "ordinary" electrons that revolve around atomic nuclei, or fly in beams through TV picture tubes, there must also exist an incalculable multitude of "extraordinary" electrons distributed uniformly throughout what one usually calls empty space. Although, according to Dirac's views, each unit volume of vacuum is packed to capacity with these extraordinary electrons, their presence escapes any possible experimental detection. The "ordinary" electrons studied by physicists and utilized by radio engineers are those few excess

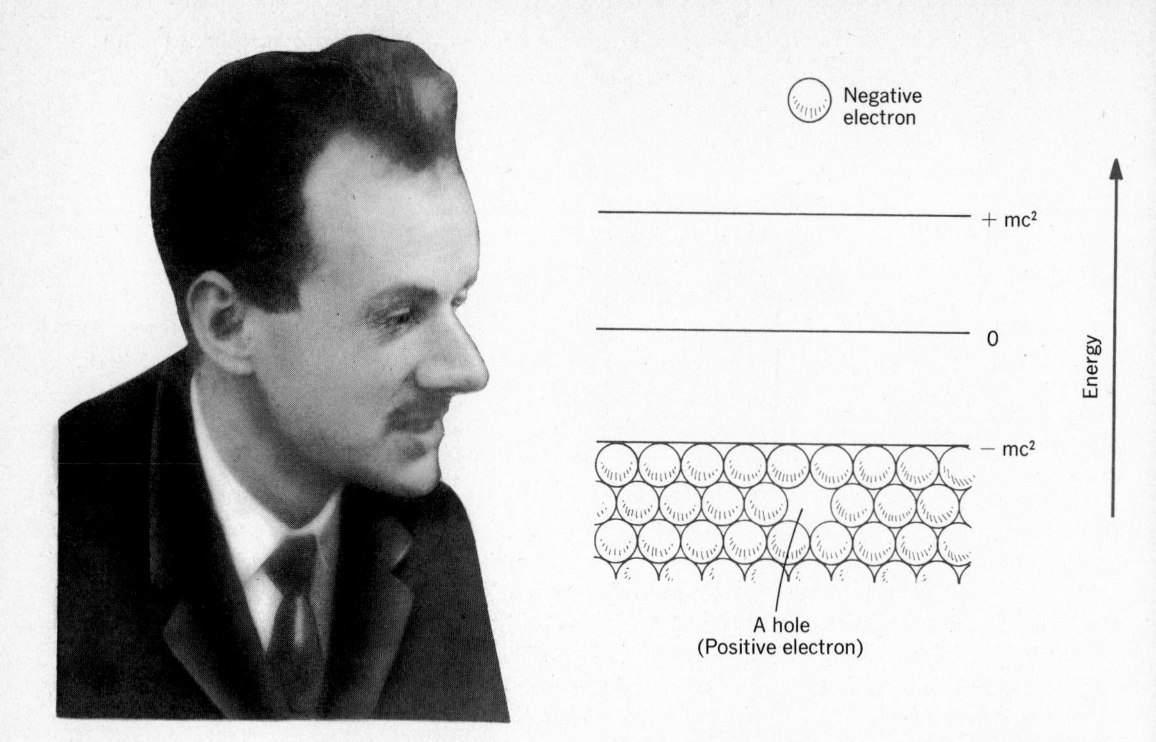

FIG. 30-1 Dr. P. A. M. Dirac, who conceived the idea that "empty" space is actually tightly packed with electrons of negative mass that are inaccessible to any physical observation. We can observe an electron only when it is raised into the region of positive energy (above, on the right). The removal of an electron from the continuous distribution leaves a "hole" (below, on the right), which represents a positive electron.

particles that cause an overflow of Dirac's ocean of "extraordinary" particles (Fig. 30-1, right), and they thus can be observed individually. If there is no such overflow, nothing can be observed, and we call the space empty.

In addition to not being individually observable by any physical means, these extraordinary electrons possess, according to Dirac, a *negative mass* and accordingly a negative $E = m_0c^2$ corresponding to their rest mass, and (if they are moving) a negative kinetic energy. The minimum energy (either positive or negative) that the electron can have is the energy of its rest mass. It is thus impossible for an electron to exist between these two levels, as shown in Fig. 30-1.

Because of their uniform distribution, the extraordinary electrons forming Dirac's ocean are invisible to observation, but what happens if one of these particles is absent, leaving in its place an empty hole? (See Fig. 30-1). This hole in the uniform distribution of particles of negative charge and mass represents the *lack of a negative charge*, which is equivalent to the *presence of a positive charge*, and the *lack of a negative mass*, which is the same as having our ordinary *positive mass*. Thus the electrical instruments used in our physical laboratories would register this hole as a positively charged particle with the same numerical value of charge as an ordinary electron, but with the opposite sign, and the same ordinary mass as an electron. This is similar to the notion of holes in the uniform distribution of electrons in semiconductors, which led to a successful explanation of their properties. But, whereas in that case the notion of a hole can be readily visualized on the basis of an ordinary picture of the electric nature of matter, Dirac's holes belong to a much more abstract concept. His infinite ocean of negative-mass, negative-energy electrons brings back in a sense something like the discredited "ether" but in an entirely new and different way.

Dirac's paper was not taken too seriously until three years later. Then in 1932, an American physicist, Carl Anderson, confirmed by direct observation the existence of positively charged electrons corresponding to the "holes" predicted by the Dirac theory. These particles, carrying a positive charge and a positive mass equal to those of the electron, are now well-established, and are called *antielectrons* or *positrons*.

In an attempt to visualize the meaning of Dirac's abstract mathematics, our picture of an infinite sea of negative-mass, negative-energy electrons seems ridiculously unreal. But there are measurable effects not easily brushed aside. Consider a "real" electron imbedded (as it must be) in Dirac's ocean. The electron would repel the negative-mass (but still negatively charged) electrons in its vicinity, with the same effect as though it were surrounded by a small region of positive charge, and the positive region in its turn surrounded by a more negative layer where the negative-mass electrons must be pushed somewhat closer together. This phenomenon is called "vacuum polarization" and would have certain measurable effects.

One of these effects is a very slight distortion of the electron energy levels in the hydrogen atom, measured in millionths of an eV. This predicted energy difference was accurately measured and confirmed in 1947 by the English physicists Lamb and Retherford. The magnetic moment of an electron would also differ slightly from what it would be in a perfect vacuum, and this difference, too, has been accurately measured, in confirmation of Dirac's theory. As an additional mark in its favor, the theory brings in the concept of electron spin as a necessary outcome. Unlike the quantum numbers n, l and m, which appear automatically in the solution of Schrödinger's equation, the idea of the spin quantum number s had previously been an empirical factor added without any theoretical justification, in order to explain certain details of spectral lines.

So, in spite of its weird unrealness, we should not be too hasty in raising our eyebrows at the idea of Dirac's infinite "ocean." Perhaps it appears ridiculous only because we are trying to picture a model for something that is not picturable in everyday terms.

30-2 Pair Production and Annihilation

From what has been said, we can conclude that in order to form a positron, we have to remove a negative electron from its place in Dirac's ocean. When this electron is removed from the uniform distribution of negative electric charge, it becomes observable as an ordinary negatively charged particle. Thus *the positive and negative electrons always must be formed in pairs*. We often call this process *the "creation" of an electron pair*, which is not quite correct because the pairs of electrons are not created from nothing but are formed at the expense of the energy spent in carrying out the process of their formation. According to Einstein's $E = mc^2$, the energy necessary to produce two electron masses is $E = 2 \times 9.11 \times 10^{-28} \times (3 \times 10^{10})^2 = 1.64 \times 10^{-6}$ erg, or 1.02 MeV. Thus, if we irradiate matter with electromagnetic radiation whose photons have this much energy or more, we should be able to induce the formation of pairs of positive and negative electrons. The electron pairs discovered by Anderson were produced in atmospheric air, and also in metal plates placed in a detecting cloud chamber, by the high-energy γ radiation associated with the cosmic rays that fall on the earth from interstellar space. Figure 30-2 pictures such a pair production. The cloud chamber is in a magnetic field that causes the newly born electron and positron to curve in opposite directions.

The opposite of the "creation" of an electron–positron pair is the "annihilation" of both the electron and the positron when the pair collide. According to Dirac's picture, the annihilation process occurs when an ordinary negative electron, which moves "above the rim" of the com-

FIG. 30-2 A cloud-chamber photograph of two electron pairs produced in a metal plate by a high-energy cosmic ray photon. (*Photo by Dr. Carl Anderson, California Institute of Technology.*)

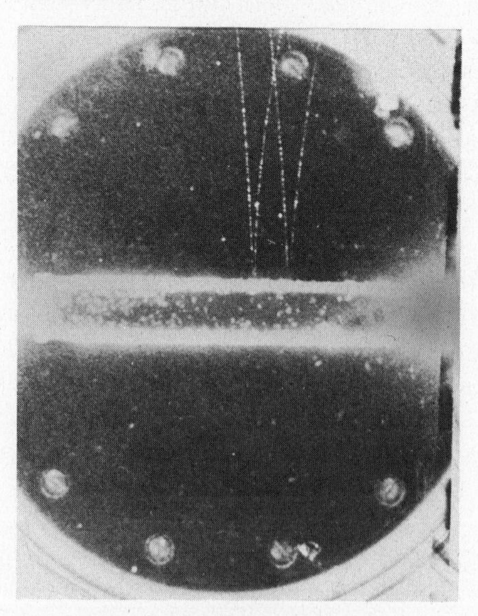

pletely filled Dirac's ocean, finds a "hole" in the distribution and falls into it. In this process the two individual particles disappear, giving rise to two photons of γ radiation with a total energy equivalent to the vanished mass, radiating from the place of encounter. We can readily see that *two* photons are necessary by considering conservation of momentum. Imagine the mutual annihilation of a slow-moving electron and a slow-moving positron, whose total momentum is almost zero. A single photon would have a large momentum in one direction; for the photon momentum to be zero, there must be two photons in opposite directions

30-3 Antiprotons and Antineutrons

After the experimental confirmation of Dirac's theory of antielectrons, physicists were interested in finding the *antiprotons* that should be the particles of proton mass carrying a negative electric charge, i.e., *negative protons*. Since a proton is 1840 times heavier than an electron, its formation would require a correspondingly higher input of energy. It was expected that a pair of negative and positive protons should be formed when matter is bombarded by atomic projectiles carrying not less than 4.4 GeV of energy. With this task in mind, the Radiation Laboratory of the University of California at Berkeley and the Brookhaven National Laboratory on Long Island, New York, started construction of the gigantic particle accelerators—*Bevatron* on the West Coast and *Cosmatron* on the East Coast—that were supposed to speed up atomic projectiles to the energies necessary for the proton-pair production. The race was won by the West Coast physicists, who announced in October, 1955, that they had observed negative protons being ejected from targets bombarded by 6.2-GeV atomic projectiles.

The main difficulty in observing the negative protons formed in the bombarded target was that these protons were expected to be accompanied by tens of thousands of other particles also formed during the impact. Thus the negative protons had to be filtered out and separated from all the other accompanying particles. This was achieved by means of a complicated labyrinth formed by magnetic fields, narrow slits, etc., through which only the particles possessing the expected properties of antiprotons could pass. When the swarm of particles coming from the target (located in the bombarding beam of the Bevatron) was passed through this labyrinth, the experimentalists were gratified to observe the expected particles coming out at a rate of about one every 6 minutes. As further tests showed, the particles were genuine negative protons formed in the bombarded target by the high-energy Bevatron beam. Their mass was found to have a value of 1840 electron masses, which is the mass of ordinary positive protons.

Just as the artificially produced positive electrons are annihilated in passing through ordinary matter containing a multitude of ordinary negative electrons, negative protons are expected to be annihilated by encountering positive protons in the atomic nuclei with which they collide. Since the energy involved in the process of proton–antiproton annihilation exceeds, by a factor of almost 2000, the energy involved in an electron–antielectron collision, the annihilation process proceeds much more violently, resulting in a "star" formed by many ejected particles.

The proof of the existence of negative protons represents an excellent example of an experimental verification of a theoretical prediction concerning properties of matter, even though at the time of its proposal the theory may have seemed quite unbelievable. It was followed in the fall of 1956 by the discovery of *antineutrons*, i.e., the particles that stand in the same relation to ordinary neutrons as negative protons do to positive ones. Since in this case the electric charge is absent, the difference between neutrons and antineutrons can be determined only on the basis of their mutual annihilation ability.

30-4 The Elusive Neutrino

Early studies of β decay indicated that there was something wrong with the energy balance involved. Alpha particles emitted by a given radioactive element carry a definite amount of energy characteristic of the particular isotope that emitted them. The cloud-chamber photo of Fig. 27-2 on p. 496 shows α particles emitted from a speck of material containing two different radioactive elements. Two different ranges for the emitted α particles show quite clearly; some travel to the top of the photograph before they come to rest, while the other group, with less energy, manages to go less than half this distance. Beta particles (electrons), however, emitted from a single radioactive isotope, show a wide energy spread, from almost zero up to a definite high-energy limit. Since (as confirmed by α decay) the energy liberation in a particular nuclear decay should be some particular definite amount, it seemed that β decay was violating the principle of conservation of energy. And if this were not enough, let us look at the decay of tritium into He^3:

$$_1H^3 \rightarrow {}_2He^3 + {}_{-1}e^0.$$

Not only do the emitted electrons have a wide range of energies but the spins of the particles do not add up right. For reasons we will not go into, the spins of the H^3 and the He^3 are both known to be $\frac{1}{2}$, as is that of the electron. If we arrange the spin vectors on the right side of the reaction either parallel or antiparallel, they will add up to 1 or to 0, neither of which is equal to the $\frac{1}{2}$ on the left side. Since spin is in effect a measure of angular momentum, we see a violation of not only conservation of energy, but of conservation of momentum as well!

Pauli, in 1932, suggested that another undetected particle was emitted with the electron, and carried with it the missing energy and spin. The basic theory of this undetected particle was worked out in 1936 by Fermi, who also gave it the name *neutrino*, meaning "little neutral" in Italian.

Like the photon, the neutrino has zero charge and a zero rest mass, so that it can exist only when traveling at the speed of light. The photon, however, reacts readily with matter, while the neutrino is very reluctant to react at all. The thickest shield is no more effective in stopping neutrinos than a barbed-wire fence is in stopping a swarm of mosquitoes. It would take a shield of solid lead many light-years thick to reduce a neutrino beam to half of its original intensity.

Although its actual detection was seen to be nearly impossible, the neutrino was firmly believed in by almost all physicists. It was much more satisfactory to believe in such an invisible ghost than to accept the alternative, which was to concede the violation of the principles of conservation of energy and momentum.

It was very reassuring when direct experimental evidence of the neutrino was achieved in 1956. Frederick Reines and Clyde Cowan of the Los Alamos Scientific Laboratory designed apparatus as shown in Fig. 30-3, to trap these elusive particles. The expected reaction was

$$\text{p} + \text{neutrino} \rightarrow \text{n} + \text{e}^+.$$

The "target" tanks were filled with a solution of cadmium chloride in water; in the scintillator tanks was an organic liquid that emits a flash of light when it absorbs a γ photon. Photoelectric cells surrounding the tanks detected these flashes. The positron emitted in the reaction is at once annihilated with an electron, with the production of two γ photons of definite energy, and in opposite directions. The neutron is absorbed by a cadmium nucleus (recall the cadmium control rods in reactors), which then emits γ radiation of characteristic energy. Thus, the absorption of a neutrino by a proton is marked by a nearly simultaneous emission of this group of gamma rays of definite energies.

The apparatus was placed near a nuclear reactor at the AEC Savannah

FIG. 30-3 Double-sandwich arrangement of target and scintillator tanks used in detecting neutrino reactions.

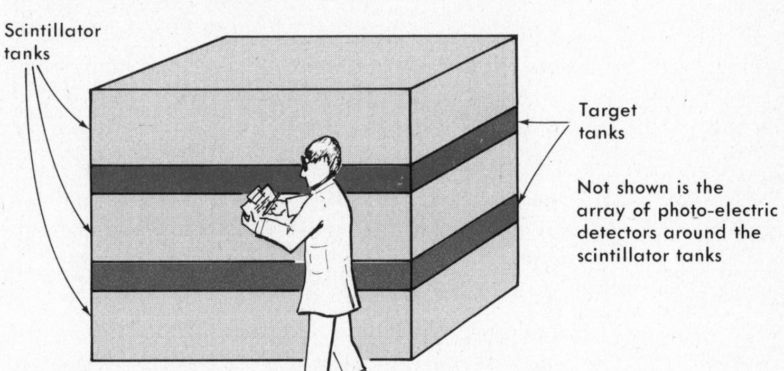

River installation in Georgia. Since the operation of a large reactor and the subsequent decay of its fission products includes an enormous number of β rays, it must also emit an equal number of neutrinos. Other radiation from the reactor was absorbed by heavy shielding that had little effect on the neutrinos. The experimenters were well repaid for their efforts when their counters indicated the expected pattern of γ rays was being emitted at the rate of several per hour. Confirming experiments have since been performed with bigger and more elaborate apparatus, and the neutrino has now taken its place with the other well-established particles.

30-5 Exchange Forces and Mesons

The next member to enter the growing family of auxiliary nuclear particles was also born as the result of purely theoretical considerations. In 1935, a Japanese theoretical physicist, Hidekei Yukawa (known as "Headache" Yukawa to students who struggle with his mathematics), proposed a new particle that would account for the strong forces binding neutrons and protons together in the nucleus. The idea that a force between two particles could be explained by the introduction of a third particle does not seem to make much sense at first glance. Although the mathematical ideas underlying this idea are quite formidable, a crude analogy will help emphasize the basic simplicity of the concept. If you and a friend stand a few feet apart and throw a heavy medicine ball rapidly back and forth, the total effect will be the same as though there were a force of repulsion between you. If each of you stood on a wheeled platform, you would be quickly pushed apart; as the distance between you increased, the time intervals between throwing and catching the ball would likewise increase. This would mean that the rate of change of momentum, and thus the force between you, would become smaller as the distance grew larger.

A force of attraction is not quite so clearly pictured. However, if you imagine yourself grabbing the medicine ball out of your friend's hands, and him then grabbing it from yours, and so on, the result will be a force of attraction between you. The repellent and attractive forces described above could be termed *exchange forces*, since they arise from the exchange of the medicine ball.

Yukawa's work gave an answer to the question: How big a medicine ball do we need to explain the short-range forces of attraction between the nucleons in a nucleus? Scattering experiments (similar to those of Rutherford in his discovery of the nucleus, but with high-energy particles of corresponding short de Broglie wavelength) show that the strong attractive nuclear forces do not appear unless the bombarding particle is within about 1.4×10^{-13} cm from a target nucleus. The hypothetical exchange particles would no doubt travel at high speed within the nucleus, but still substantially less than the speed of light. Let us suppose their speed is something like $2c/3 = 2 \times 10^{10}$ cm/s. This would require a time of $t = d/v = 1.4 \times 10^{-13}/2 \times 10^{10} = 7 \times 10^{-24}$ s for the particle to travel a distance equal to the range of the force.

We must now return to the uncertainty principle, which we have used as $\Delta p \times \Delta x = h/2\pi$. The principle can take another form, relating uncertainty in energy (or its equivalent mass), and uncertainty in time:

$$\Delta E \times \Delta t = h/2\pi.$$

This can be interpreted as meaning that nature has no objection to a violation of mass (or energy) conservation, provided that the violation does not last too long. Suppose our exchange particle is suddenly created—it must last 7×10^{-24} s if it is to travel a distance equal to the range of the nuclear force. Its mass can then be

$$\Delta E = \frac{h/2\pi}{\Delta t} = \frac{6.6 \times 10^{-27}}{2\pi \times 7 \times 10^{-24}} = 1.5 \times 10^{-4} \text{ erg.}$$

Converting this to grams via $E = mc^2$ gives the mass of the exchange particle as

$$m = \frac{E}{c^2} = \frac{1.5 \times 10^{-4}}{9 \times 10^{20}} = 1.7 \times 10^{-25} \text{ g.}$$

This is roughly about 200 times the mass of an electron. It was Yukawa's suggestion that such particles might continually appear and disappear within a nucleus to account for the strong binding forces.

No particle of this approximate mass had ever been observed. But a few years later, C. D. Anderson in the United States discovered just such particles in cloud-chamber photographs of the results of bombardment by the high-energy particles of "cosmic radiation", constantly falling on the earth from space. These particles, now called *muons*, carry one electron charge, either + or − (μ^+ and μ^-) and have a mass approximately equal to $207m_e$. It should be noted that the particles observed by Anderson were "real," being made from some of the energy of the cosmic-radiation protons; Yukawa's particles, made from energy on a *very* short-term loan, cannot be observed outside their nucleus, and are called "virtual."

It was at once assumed that muons were Yukawa's predicted particles, but further investigation showed this could not be true. The exchange particles, in order to exert their forces, must be readily absorbed by nucleons. Muons are very reluctantly absorbed—those produced in the atmosphere by cosmic radiation penetrate far into the earth, and can still be measured in deep mines and at the bottom of deep lakes.

The actual particle was discovered in 1947. It is called the *pion*, and has either +1 or −1 electron charges or may be uncharged (π^+, π^0, π^-). Although discovered in the study of cosmic radiation collisions, they are now produced in large numbers in the targets of the beams of high-energy accelerators, and have been thoroughly studied. No one doubts that pions are Yukawa's predicted particles, which are primarily responsible for the nuclear binding forces. They are unstable, and a charged pion decays into a muon and a neutrino:

$$\pi^{\pm} \rightarrow \mu^{\pm} + \nu$$

The uncharged π^0 becomes a pair of γ photons. The pion is thus the parent of the muon, which is itself unstable and decays into an electron and a pair of neutrinos. Figure 30-4 shows this decay scheme of $\pi \rightarrow \mu \rightarrow$ e.

FIG. 30-4 A series of nuclear events in a photographic emulsion. A cosmic-ray proton strikes an atomic nucleus (upper left) and produces a burst of many different fragments. One of the fragments, a pion, travels to the right edge, where it decays into a muon and a neutrino. (Neutrinos leave no tracks in cloud chambers, emulsions, or any other similar devices.) In the lower left corner the muon decays into two neutrinos and an electron which then travels to the right. (*Photograph by Dr. E. Pickup, National Research Council, Ottawa, Canada.*)

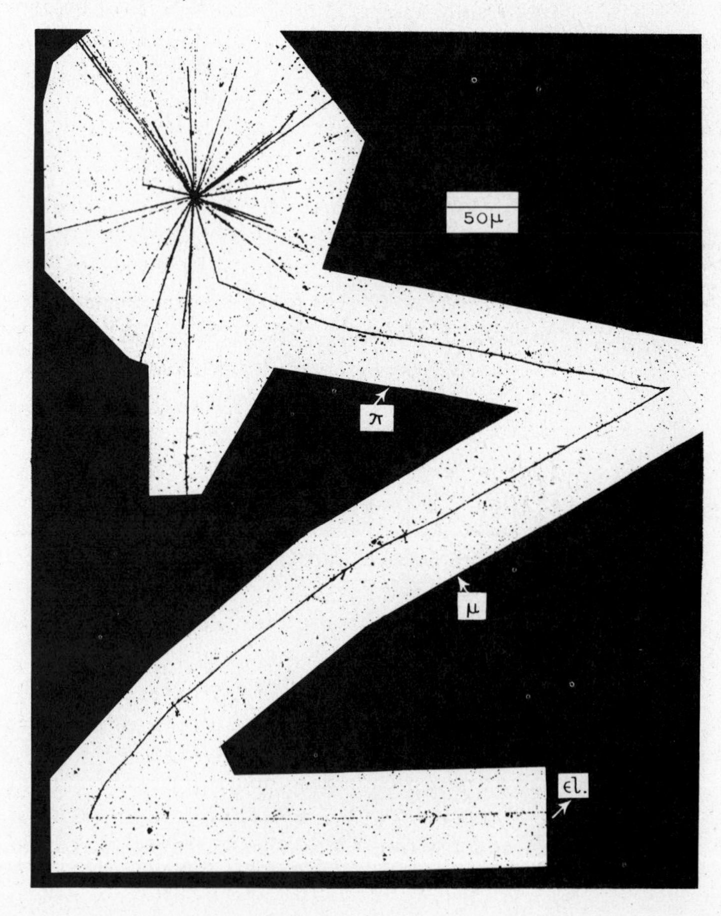

Table 30-1

	Particle		Anti-particle	Rest Mass in Electron Masses	Half-Life of Free Particle
photon		γ		0	stable
e-neutrino	ν_e		$\bar{\nu}_e$	0	stable
μ-neutrino	ν_μ		$\bar{\nu}_\mu$	0	stable
electron	e^-		e^+	1	stable
muon	μ^-		μ^+	207	2.2×10^{-6} s
pion	π^-		π^+	273	2.6×10^{-8} s
neutral pion		π^0		264	1.8×10^{-16} s
proton	p^+		p^-	1836	stable
neutron	n		$\bar{n}$	1839	~15 min

Table 30-1 tabulates the particles we have encountered so far.

In the table, only two particles, the photon and the uncharged pion, have no antiparticles—alternatively, each of these can be considered to be its own antiparticle. The neutrino has expanded into four. Experimental evidence has shown neutrinos to be of two types—one involved in any decay or reaction that involves an electron; the μ-type appears whenever a muon is involved. The decay of a muon into an electron therefore requires two neutrinos, one of each type. (There are conservation laws that we will not go into, determining whether it is the neutrino or the antineutrino that is included in any specific decay or other reaction.) We are accustomed to thinking of the neutron as a stable particle that lasts forever—and this is true when the neutron is tightly bound within a nucleus. A free neutron, however, outside a nucleus, decays with a half-life of 12 to 17 min. (The determination of this half-life is a very difficult experimental job, and it is not yet accurately known.)

$$n \rightarrow p^+ + e^- + \bar{\nu}_e.$$

30-6 Elementary (?) Particles

The 16 particles listed in Table 30-1 include the most common and the most important, but it is only the beginning. A host of other particles have been discovered, measured, and classified: the kaons, or K mesons; lambda (Λ), sigma (Σ), xi (Ξ), omega (Ω) particles, with half-lives near 10^{-10} s; and a growing myriad of very short-lived (about 10^{-23} s) particles called *resonances*. Figure 30-5 shows bubble-chamber tracks of some of these.

It is becoming increasingly difficult to think of all these as being "elementary" particles. It is (on a smaller scale) as though we knew nothing of atoms, and therefore considered each of the millions of different kinds of molecule as being something fundamental or elementary in itself. The way the known particles can be classified into groups in terms of various of their properties suggests that these patterns could be constructed by only three different and more fundamental particles with

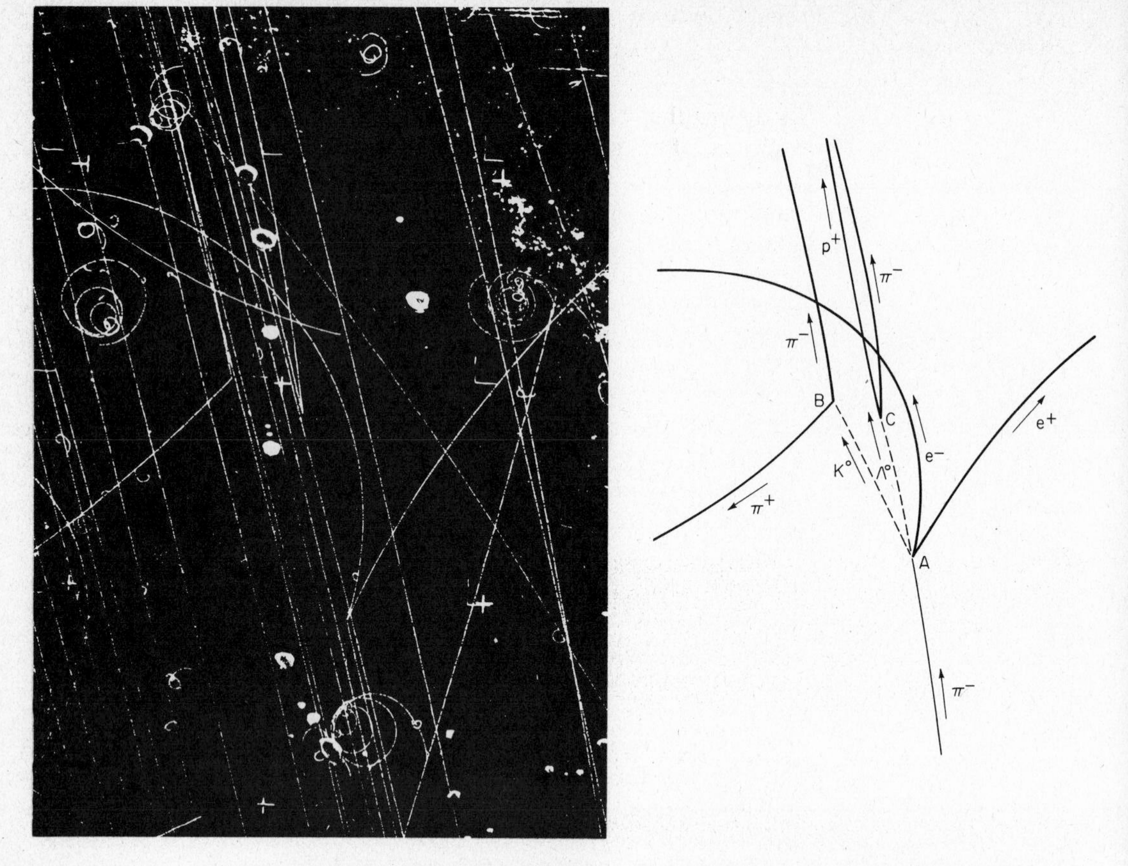

FIG. 30-5 Most of the bubble trails in this bubble chamber photograph (left) are caused by particles merely passing through and by torn-off electrons. The photograph, however, also shows one unusual sequence of events. As indicated by the diagram on the right, which shows the relevant tracks in the photograph, a π^- meson collides with a proton in the chamber liquid at point A. These two particles, plus the energy of the meson, produce a Σ^0 and a K^0. Almost immediately the Σ^0 decays into a Λ and an electron pair. These two decays have taken place so rapidly that the electron pair shown appears to be coming from point A. The K^0, which produces no ionization, decays at point B into a pair of π mesons. The Λ^0, likewise leaving no trail, decays at point C into a proton and a π^-. The diagram below shows the decay scheme.

$$\pi^- + p^+ \xrightarrow{(A)} \Sigma^0 + K^0$$

$$\Sigma^0 \xrightarrow{(A)} \Lambda + e^+ + e^-$$

$$\Lambda \xrightarrow{(C)} \pi^- + p^+$$

$$K^0 \xrightarrow{(B)} \pi^- + \pi^+$$

The entire chamber is in a strong magnetic field perpendicular to the plane of the picture. Notice that this causes positively charged particles to curve to the right, and negatively charged particles to curve to the left. The amount of curvature depends on the speed of the particle, as well as its mass. The photograph was taken by the Lawrence Radiation Laboratory of the University of California, and is reproduced with permission.

fractional electric charges, together with their antiparticles. These hypothetical particles, called "quarks" by Nobel laureate Murray Gell–Mann, so far exist only as speculations. They are being searched for among the enormous number and variety of particles created by the collisions of beams from the high-energy accelerators. And while the search goes on, theoretical considerations have suggested that there should be *four* different sorts of quark rather than just three. At this writing, none has been found.

At the beginning of this chapter we could picture the nucleus as an aggregation of balls—red balls, say, for the protons and black for the neutrons—rather like a small bunch of two-color grapes. We must now change the picture, so that each nucleon is represented by a small core, surrounded by a cloud of virtual pions eternally coming and going as they are created from nothing and quickly vanish into nothing again. Recent scattering experiments with very high-energy and thus very short-wavelength particles, suggest that the results can be better interpreted if each nucleon has several tiny cores or scattering centers rather than one. Is this evidence of quarks, or perhaps of some other really fundamental entities? The answer still lies beyond our present frontiers.

Questions

(30-2) 1. A pattern of detectors scattered over several acres all simultaneously register a sudden burst of electrons of such intensity that it is estimated about 3×10^{10} electrons and positrons were in the shower. Assuming that this entire shower was caused by one cosmic-ray particle, what is the least possible kinetic energy it could have had?

2. A cosmic-ray particle from space with a kinetic energy of 10^{15} eV strikes a nucleus in the upper atmosphere. Assuming that all its energy ultimately goes into electron production, how many electron-positron pairs can be produced?

(30-3) 3. When a proton and an antiproton annihilate, what is the wavelength of the photons produced?

4. In bombarding a target with a 6.2-GeV particle to make a proton–antiproton pair, how much energy is left over after the collision, representing KE of particles, photons, etc.?

(30-5) 5. Consider a neutral pion traveling at nearly the speed of light. How far could it be expected to move before decaying into a pair of photons?

6. Suppose a muon is created in a cloud chamber, and travels at nearly the speed of light. Is it likely that its decay into an electron (and a trackless neutrino) could be seen in the same chamber?

31

Biophysics

31-1 The Effects of Size

We have seen that the deciding factor for determining critical size in nuclear bombs and reactors is the surface/volume ratio. Living organisms are likewise energy-converting machines, and for them too S/V is important, as was hinted in Question 7 of Chapter 29. Whether mouse or man, an organism produces heat from the work of muscles and from the myriad chemical reactions that constantly take place. This heat production occurs throughout the body and is thus roughly proportional to the volume—or to the mass—since the density of organic tissues is never far from 1 g/cm^3. The internally generated heat can only be dissipated through the body surface (including the lung surface). For objects of about the same shape, volume is proportional to d^3—where d is any linear dimension—and surface area is proportional to d^2. The S/V ratio is thus proportional to d^2/d^3, or $1/d$, as we have seen in Chapter 29.

This idea can be readily applied to members of the animal kingdom, all of whom maintain about the same temperature and have about the same density. Hummingbirds, which have a large surface-to-volume ratio, have to metabolize at a terrific rate (a rate, incidentally, that is just about the same as the power production per unit weight in a helicopter). On the other hand, large animals can be very economical in their internal heating system. If an elephant metabolized at the same rate as a hummingbird, it would soon be a roasted elephant, since its body temperature would rise to about that of a kitchen oven. Figure 31-1 graphs the metabolic rate for several different animals. .

To maintain a normal body temperature, the tiny shrew with its large S/V ratio, must eat several times its weight each day; the elephant, enjoying the economy of his large size, is content with a few hundred pounds of vegetation, a small fraction of his weight of several tons. Hummingbirds, even without the energy expenditure of flying, would be unable to survive without feeding from dark to dawn, if subject to their normal daytime heat losses. They therefore go into a state of hibernation

FIG. 31-1 The relationship between mass and rate of heat production in animals ranging in size from hummingbird to elephant. The graph is extended far to the right to include the sun.

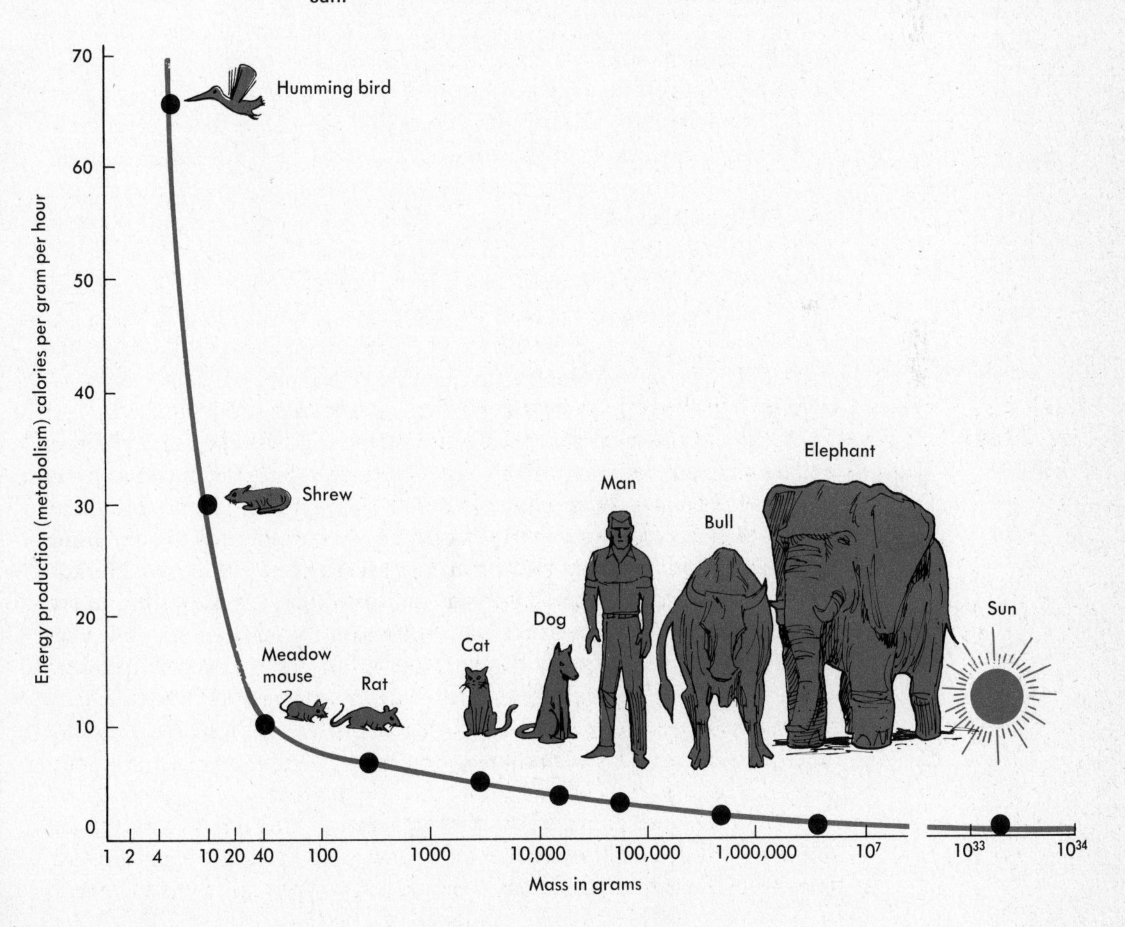

each night, in which their body temperatures drop to only slightly above that of the air, and their rate of metabolic activity is correspondingly reduced.

Figure 31-1 also includes the rate of energy production in the sun, which amounts to only 1.6×10^{-4} cal/(g-hr), averaged over its entire interior.

There are many other aspects of *scaling*—the comparison of similar objects differing in size. Swift, being more interested in social values, quite properly ignored the effects of scaling in his stories of *Gulliver's Travels*. The giant Brobdingnagians, built on the scale of a foot for each of Gulliver's inches, would stand 72 feet tall—and yet were of perfectly human proportions.

Let us first consider the size—the thickness, that is—of our own bones as they have been determined after millions of years of evolution. As we have seen in early chapters, the stresses on bones and joints are enormous. It might seem an advantage for the bones to be thicker, heavier, and stronger than they are. Imagine an ancient ancestor who had such thicker bones, through a chance mutation or an unusual gene combination. The added weight and inertia would make him a slower runner and a less nimble climber—and the first to be caught and eaten by a pursuing carnivore. Conversely, bones thinner than the optimum, although they should contribute to greater nimbleness and speed, are also more liable to breakage—which leads to the same dismal end. Optimum design, whether of man or machine, is always a delicate balance of advantages and disadvantages.

The Brobdingnagian, being 12 times Gulliver's height, would weigh $(12)^3 = 1728$ times as much. His bones, being 12 times thicker, would have a cross-sectional area $(12)^2 = 144$ times as large. The stresses in his bones would thus be $1728/144 = 12$ times as great as Gulliver's. The giant might be barely able to stand but any attempt to walk or run or bend would bring him crashing down in a tangle of shattered bones.

In view of these differences resulting from differences in size, should we be surprised that a cat and a cougar—about 4 times the size of a cat in linear dimensions—should both be able to jump to about the same height? Let us compare the work done in contracting the jumping muscles of both animals. The cougar's muscles have a cross-sectional area $(4)^2 = 16$ times that of the cat, and can thus exert a force 16 times as great. The shortening of the muscle is proportional to its length and is 4 times as much for the cougar. The work done equals Fd, and is therefore $16 \times 4 = 64$ times as much for the cougar as for the cat. And since (if the two animals are similar in shape) the cougar weighs $(4)^3 = 64$ times as much as the cat, we have, from

$$Fd = mgh$$

that h must be about the same for both.

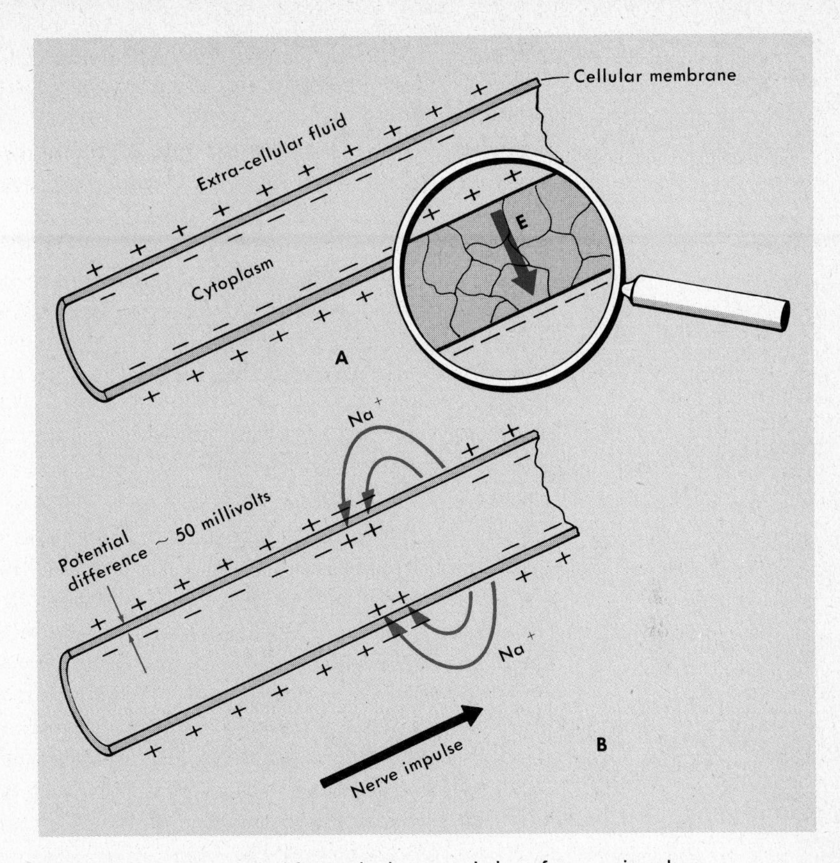

FIG. 31-2 Electrochemical factors in the transmission of a nerve impulse.

31-2 Conduction of Nerve Impulses

The transmission of an impulse along a nerve is an electrical phenomenon, but very different from the conduction of a wire, and much more complex. Nerve cells are long cylinders (Fig. 31-2) enclosed by an insulating membrane. Within the cylinder, the cytoplasm is an electrolyte—that is, a solution containing ions—the extracellular fluid surrounding the cylinder is also an electrolyte. Ions of sodium (Na^+) and potassium (K^+) seem to play the principal roles in conduction. The cytoplasm contains more K^+ than the extracellular fluid; contrary to what we might expect, most of the Na^+ ions remain on the outside in the extracellular fluid. This is apparently due to the action of an electrochemical *sodium pump*, which ejects Na^+ ions from the cytoplasm. As seen in the magnifying glass in Fig. 31-2A, the ions must be moved against the electric field in the membrane to do this. Just what the sodium pump is, and how it accomplishes this, is still shrouded in mystery and largely unknown.

Figure 31-2B shows a pulse traveling down a nerve cell from left to right. The pulse can be initiated by an electrical, chemical, thermal, or mechanical stimulus; normally it will be electrical, through a complex connector called a *synapse* at the end of the cell, which connects this nerve to others. The stimulus causes the membrane to allow a sudden surge of Na^+ ions to pass through it, thereby making the *inside* of the membrane temporarily positive. The sodium pump immediately goes to work again to eject the excess Na^+, while more Na^+ passes through the membrane on the "downstream" side. In this way, a pulse travels down the elongated cell at a speed of about 30 m/s. If this is a motor nerve, the impulse passes through another synapse into a contractile muscle cell. Here, a similar pulse travels down the muscle cell, causing it to contract as it passes. It is indeed fortunate that all this takes place automatically. It would be very difficult to play, say, Chopin's "Minute Waltz" if our conscious minds had to oversee all this complex detail for every finger movement.

31-3 Electrocardiographs and Electroencephalographs

The heart is a marvelous mechanism whose four chambers must pump in accurately synchronized sequence for the 2 or 3 billion cycles of operation in the average human lifetime. The muscle contractions of the heart are controlled by nerve impulses like those described in the last section. These impulses, consisting of changes in electric potential traveling along the nerves, cause small electric currents to circulate throughout the neighboring tissues. Passing through the electrical resistance of the skin, these currents result in microvolt *IR* drops that can be measured and recorded.

The electrocardiograph (ECG) is merely a set of sensitive galvanometers that measure the cyclically varying potential differences between electrodes placed against the skin. After electronic amplification the galvanometer responses are generally used to actuate pens moving across a moving strip of paper to record these changing potentials. (See Fig. Q16 in Chapter 8.)

Although their function is not to control muscles, similar rhythmic electric potentials originate in the outer cortex of the brain. Here again, these potential differences and their variations are roughly duplicated in a changing pattern of potential differences on the scalp. From a group of electrodes placed on the scalp the electroencephalograph (EEG) can record variations in potential that aid in the diagnosis of brain injuries, diseases, or malfunctions.

31-4 Cardiac Pacemakers

The heart, like any other complex machine involving interrelated chemical, electrical, and mechanical functions, can go wrong in many different ways as the result of disease, injury, or deterioration. Some malfunctions involve the electrical triggering mechanism that controls the 70 to 80 cycles/min beat and coordinates the contraction of various parts of the heart muscle.

In normal heart function the controlling pulse of electric potential originates in a cluster of specialized cells called the *sinoatrial node*. With appropriate time delays built in for the various parts of the heart, this sinoatrial (SA) node keeps the heart beating steadily at 70 to 80 beats per min—unless, of course, exercise, fright, or other stimuli put appropriate chemical messengers into the bloodstream that call for a faster rate. This is such an important mechanism that it is provided with two backup systems. If the SA node fails in its function, a secondary potential generator, the *atrioventricular* (*AV*) *node*, normally triggered by the SA pulses, takes over independently at a rate of about 50 pulses per minute, with reduced coordination between the contractions of the four chambers. Failure of the AV node passes the timing duties on to scattered centers in the heart muscle; these have but little coordination and result in a ragged beat of about 30 per minute—barely enough to keep the organism precariously alive.

Electrical devices (*pacemakers*) have been developed to apply regularly timed potential pulses to the heart to replace a faltering or intermittent function of the SA node. The controlled periodic pulse originates in the charging of a capacitor (Fig. 31-3). The cycle must start somewhere, so imagine the battery has just been connected in the circuit. The device D does not now conduct a current, and charge from the battery flows through the high resistance R into the plates of the capacitor C. As charge accumulates in the capacitor, the potential across it increases steadily ($V = Q/C$), and so does the potential across D. When V reaches a certain value, D becomes conducting, and the capacitor charge rapidly leaks off through it, dissipating its energy in I^2R heat loss in the resistance S. D again becomes nonconducting and the cycle repeats, with a regular period determined by the emf of the battery, the resistance R, and the capacitance C. The potential applied by electrodes at the sinoatrial node is the capacitor potential, and varies as described, building up to a maximum, and then falling quickly to near zero as D becomes conducting. This serves very satisfactorily to trigger the proper functioning of the heart.

FIG. 31-3 Schematic basic circuit of a cardiac pacemaker.

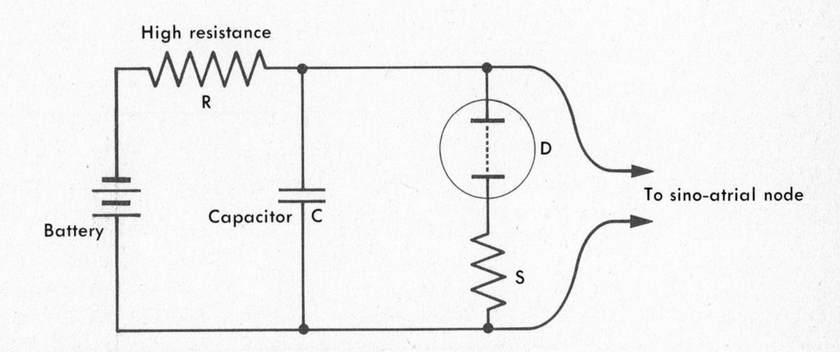

The circuitry is much more complex than the bare essentials shown in Fig. 31-3. The SA node often continues intermittent functioning; it would be most confusing and unsatisfactory if the heart were to receive triggering stimuli from both the SA node and the pacemaker. For this reason, the pacemaker period is set longer than the natural period of the SA node, so that if the node *does* fire, it will do so before the pacemaker does. If so, an auxiliary circuit picks up the potential pulse of the node, and prevents the pacemaker from producing its own.

31-5 Electric Shock

Ohm's law applies to organic tissues as well as it does to wires and resistors. If you touch the positive and negative terminals of a battery with your two hands, a current will flow through your arms and upper torso, including the heart. The amount of current, $I = V/R$, will of course depend on the voltage and the resistance of the body between the points of contact. Any attempt to calculate the current passing through the body is very uncertain due to the uncertainty of the body resistance R. We can consider R to be the total of three resistances in series: the internal resistance of the muscles and other tissues, plus the resistances of the two patches of skin where the contacts are made. The internal resistance from hand to hand to or from hand to foot we can take to be 500 Ω or so.

Skin resistance, however, is very uncertain and depends on the area of contact, as well as the dryness or dampness of the skin. For a light finger touch with dry skin, the resistance may be a few hundred thousand ohms; a hand grasping a wire, $5 \times 10^4\ \Omega$ if dry, or $10^4\ \Omega$ if damp; sweaty hands holding a pair of pliers or cutters, $10^3\ \Omega$; a foot immersed in water, 100 Ω. An open cut or broken blister will reduce the skin resistance to nearly zero.

Obviously, for any given voltage, the current may vary over a wide range, with widely varying effects. Table 31-1 is a crude and approximate summary of effects that are difficult to measure or even estimate for the human animal.

The phenomenon of *fibrillation* of the heart is worth some additional comment. If the shock current, usually fluctuating in both location and intensity, is intense enough, it initiates thousands of false impulses along

TABLE 31-1

Current	Effect
0.001 A = 1 mA	Barely felt
0.01 A = 10 mA	Painful shock
0.02 A = 20 mA	Muscle contraction; breathing difficult
0.05 A = 50 mA	Breathing stopped
0.1 A = 100 mA	Cardiac fibrillation—*FATAL*
0.2 to a few A	Cardiac paralysis—usually *not* fatal
A few amperes	Tissue burning

the nerves controlling the heartbeat. Any signals from the SA or AV nodes are lost in the resulting chaos, so that the many parts of the heart continue to contract and relax with irregular periods totally without coordination. No blood is pumped, and death follows within a few minutes. But if the shock current is larger, it puts the entire heart in a state of steady contraction. When this paralyzing current is interrupted before damage is done, the controlling nodes can pick up their rhythmic signals, and regular beating is often resumed, even without efforts at resuscitation.

Fibrillation may result from peculiar heart malfunction, without the stimulus of electric shock. The *defibrillator*, so beloved of TV writers for the many medical shows, sends a measured surge of current for a predetermined time, between two electrodes placed on either side of the chest. This, like a severe accidental shock, contracts the entire heart to wash out the chaotic signals, and allows the SA node to resume its rhythmic direction of the heart's activity.

31-6 Radiocative Isotopes as "Tracers"

Although large amounts of radiation are invariably harmful to any living organism, smaller amounts, if carefully controlled and administered, will do no perceptible damage. This fact has made it possible for biophysicists and biochemists to use certain radioactive isotopes deliberately in many kinds of biological research.

The radioactive isotope $_{53}I^{131}$ (half-life, 8 days), for example, is absorbed more strongly by the thyroid gland than by other tissues. Because I^{131} decays by β emission, shortly after a patient has received a dose of it, sensitive detectors of the emitted electrons can accurately show the size and location of the hidden gland. Continuing observations of the radiation can supply important information about the thyroid's activity and functioning.

This same isotope of iodine (as well as a few others) has also proved useful in the diagnosis of otherwise almost inaccessible brain tumours. It is incorporated into the complex molecules of certain organic dyes that are absorbed much more strongly by the tumor tissues than by normal tissue. Observation of the emitted radiation can then give the radiologist a good idea of the size and location of the tumor.

Besides these and many other biological applications, tracers find many industrial uses. As an example, a small amount of a radioactive isotope of iron may be incorporated in the steel of a bearing to be tested. After a period of running, the lubricating oil can be checked for radioactivity caused by a microscopic wearing away of the steel bearing. Tests of this sort are very delicate and accurate.

31-7 Medical Uses of Radiations

The medical profession has always been quick to seize on discoveries in the physical sciences and turn them to the personal advantage of mankind. Within months of the discovery of X rays, physicians were utilizing them in examination and diagnosis. The use of X-ray photographs to show internal bone structure is now so commonplace that they scarcely require mention. Bones absorb the X rays much more than do the soft skin and muscle tissue, and thus throw a shadow of their outline on the film. Different kinds of soft tissues absorb X rays differently, and X-ray pictures showing some soft-tissue detail can be made, if a less penetrating longer-wavelength radiation is used.

The ability of a material to absorb X rays depends very strongly on the atomic number of the material, the absorption being greater for higher atomic numbers. For example, muscle tissue is composed mostly of oxygen, carbon, and hydrogen, with smaller amounts of nitrogen and other elements. All these elements have atomic numbers that are low, and X rays and γ rays penetrate muscle and other soft tissues with comparatively small absorption. Bones, however, although they contain appreciable amounts of water, are composed largely of calcium phosphate, $Ca_3(PO_4)_2$. Both Ca and P have higher atomic numbers (20 and 15), and for this reason bone absorbs X rays much more strongly than soft tissue does.

In order to see, by X rays, the outline or the functioning of the stomach or intestines, the doctor customarily induces his patient to drink a concoction containing barium sulfate ($BaSO_4$), which is relatively opaque to X rays because of the high atomic number of barium.

By causing several beams of X rays to converge on a malignant growth, it is possible to kill undesirable tissues without causing fatal damage to surrounding normal material. Similar irradiation of tumors may be secured by introducing thin hollow gold needles filled with radon gas. The radon quickly decays through several steps to ${}_{83}Bi^{214}$, which emits high-energy (1.8 MeV) γ rays (called "hard" radiation), the effect of which in ionizing tissue is similar to that of X rays.

Increasing use is being made of γ-emitting isotopes, particularly Co^{60}. This cobalt isotope is formed from common Co^{59}, placed in a nuclear reactor where it is converted by neutron bombardment:

$$Co^{59} + n \rightarrow Co^{60}.$$

Cobalt-60 decays into $Ni^{60} + \beta^- + 2\gamma$. The β electrons are easily filtered out, and the γ photons each have an energy of more than 1 MeV, corresponding to hard X rays from a very high-voltage tube. The cobalt is kept in a heavy lead container; a hole opened in its wall allows a beam to escape. Its uses are similar to those of X rays.

31-8 Biological Effects of Radiation

The inevitably growing number of nuclear reactor installations for producing electric power has focused public attention on the possible effects of radiation. There can be no argument that γ, β, or α radiation can harm any living organism. All radiation produces a certain amount of ionization and the breaking of bonds in complex large molecules. This is of course harmful to some extent, and we are rightly concerned with how extensive this damage might be expected to be.

In considering radiation damage to a living organism, we should distinguish between two types of damage:

1. *Pathological damage* to the organism exposed to the radiation, which might lead to irreparable damage or death of the organism.

2. *Genetic damage* to the reproductive organs that might not seriously affect the organism itself, but could do harm to successive generations.

For any measure of the effect of radiations, there must be a basis for measuring the radiation itself. The original unit for this measure is the *roentgen*, named for the discoverer of X-rays. One roentgen is the amount of X-ray or γ radiation which will ionize dry air to the extent that 1 esu of both electrons and positive ions are produced per cubic centimeter. Another unit of somewhat broader application is the *rad*; it is applied to the effects of both electromagnetic radiation and particle radiation. A biological tissue (or any other material) receiving a dose of 1 rad has absorbed 100 ergs of energy per gram from the radiation.

Unfortunately, the damage to human tissue cannot be satisfactorily measured merely by the energy it has absorbed. Gamma- or X-rays are very penetrating. Their ionizing effects are likely to be widely separated in different cells, so that the body has a good chance to repair the damage. Electrons are also penetrating, and the effect of β-radiation is similar to that of X-rays. The more massive protons are more effective ionizers than electrons, so their ionizing effects are more closely bunched together, and this more localized damage to cells is not so readily repaired. Neutrons do little direct ionizing themselves, but a neutron may be absorbed into an atomic nucleus, converting it into a radioactive species that will cause ionization as it decays. If not absorbed in this way, its own spontaneous decay $n \rightarrow p + e$ releases about 0.8 MeV as kinetic energy. Alpha particles are the most damaging of all if they reach susceptible tissues. The relative effectiveness of different radiations is measured by their *relative biological effectiveness* (RBE), which can be *very* roughly tabulated:

X- and γ-rays	RBE = 1
electrons	1
protons	5
neutrons	5
α-particles	20.

These RBE figures cannot be taken as being any more than approximate estimates, as a great deal depends on how and where the body receives the radiation. The most dangerous source of β-radiation, for example, is probably strontium-90. This is one of the most plentiful of the fragments produced in the fission of uranium, thorium or plutonium, and emits 0.5 MeV electrons. It is in the same bivalent group II as calcium, and can take the place of calcium in food and milk derived from contaminated soil, and through them be absorbed into the body. Here it readily substitutes for the calcium in bone, and with its 28-year half-life, can damage the bone-marrow that produces red blood-cells. Fortunately, now that France and China are the only nations still conducting atmospheric testing of nuclear bombs, the danger of Sr^{90} in radioactive fallout is steadily lessening.

Alpha particles earn their high RBE only if they reach susceptible tissue. Their penetration is very short-range and nearly all are stopped by the outer layers of the skin, which rapidly flakes off normally. Although external α-radiation thus does little damage, some α-emitters can be very dangerous if absorbed internally in food, drink or air. For this reason Pu^{239}, with its increasing prevalence as a nuclear fuel, must be closely safeguarded. It emits 5 MeV α-particles, and its selective absorption and retention in bone marrow, where red blood corpuscles are manufactured, makes it one of the most deadly of poisons.

To get a more realistic measure of radiation dosage than the roentgen or the rad, researchers have taken into account the fact that it takes, on the average, less energy to ionize biological tissues than molecules of air; and also the relative biological effectiveness of different types of radiation. From this comes the unit called the *rem* (roentgen equivalent, man), and the millirem (1 mrem $= 10^{-3}$ rem).

Pathological damage to individuals requires large radiation doses that could be expected only in nuclear warfare, or in the immediate vicinity of a catastrophic accident in nuclear industry or research. A single dose to the whole body of 800 rem, more or less, is lethal; one of 100 rem may produce leukemia or cancer, with eventual death. It should be noted that the same amount of radiation spread over a lifetime would have effects much less drastic.

Genetic effects are probably of more importance. Radiation damage to sperm or ova, and to the cells that produce them, causes hereditary mutations, or abnormalities in the progeny. Mutations, except for perhaps one in many millions, are admittedly harmful, and, since they are passed down from generation to generation by the hereditary mechanism, they will lead sooner or later to the death of one of the descendants. Thus, while in the process of Darwinian evolution, which was based on the struggle for existence and survival of the fittest, the few beneficial mutations led to a slow improvement of the species, in a balanced human society where the life of each individual is carefully preserved, mutations are overwhelmingly likely to be harmful. Many mutations are of course recessive—that is, they have no visible effect until perhaps many generations later, when a defective chromosome in the ovum is paired with a similar defect in a sperm cell.

Genetic effects are difficult to measure, even approximately. Nearly the only data available are from the survivors of the Hiroshima and Nagasaki bombings, and from small experimental animals. Man's generations are so long that centuries must pass before much dependable data are available from the human bombing survivors; extrapolation from rats to man is subject to unknown effects of scaling, biological differences, and other unknown effects that contribute to its unreliability.

31-9 Radiation Exposure

Radiation damage is not a new thing that our modern technologies have only recently added to the list of hazards of living. We, like our prenuclear ancestors, are constantly bombarded from many natural sources. These include cosmic radiation, 40 mrem/yr at sea level plus 1 mrem/yr for each 100 ft of elevation; radiation from the uranium, thorium, and K^{40}* in the rock and soil beneath us, 15 mrem/yr; from the water we drink, the food we eat, and the air we breathe, 30 mrem/yr; and from the stone and plaster and wood of the houses we live in, about 50 mrem/yr. To this you can add another 1 mrem for each 1500 miles of jet plane flight; 100 mrem for a chest X ray; 20 mrem for a dental X ray; and up to 2000 mrem for a gastrointestinal X ray. The United States average is about 200 mrem/yr.

As a matter to be put in proper perspective, a person living a mile from a nuclear plant, for 24 hr/day, would receive an added dosage of 0.5 mrem/yr. At a distance of 5 miles or more the effect is not measurable.

---

* About 1.2×10^{-4} of all potassium consists of the radioactive isotope K^{40}, a β emitter with a half-life of 1.4×10^{9} y.

32

Geophysics

32-1 The Structure of the Atmosphere

We live immersed in an ocean of air on which our lives depend. This dependence goes beyond the mere existence of the atmosphere—without its ceaseless physical and chemical activities, the world would be largely, if not completely, uninhabitable.

Figure 32-1 gives a general indication of the temperature and density of the atmosphere as they vary with altitude in the earth's temperate regions. The density curve (red line, scale at the top) seems quite reasonable. We should expect the pressure and accordingly the density to decrease steadily with altitude, as it does.

The temperature curve, however (black line, bottom scale), shows some curious changes. In the troposphere—the lowest 12 km or so—there is a steady fall in temperature with increasing altitude. This, too, seems reasonable. Imagine a cubic meter, say, of air at sea level. If we raise this to the lower pressure of a higher altitude, it will expand; in expanding it must do work against the surrounding atmosphere, and hence will have less energy per molecule and a correspondingly lower

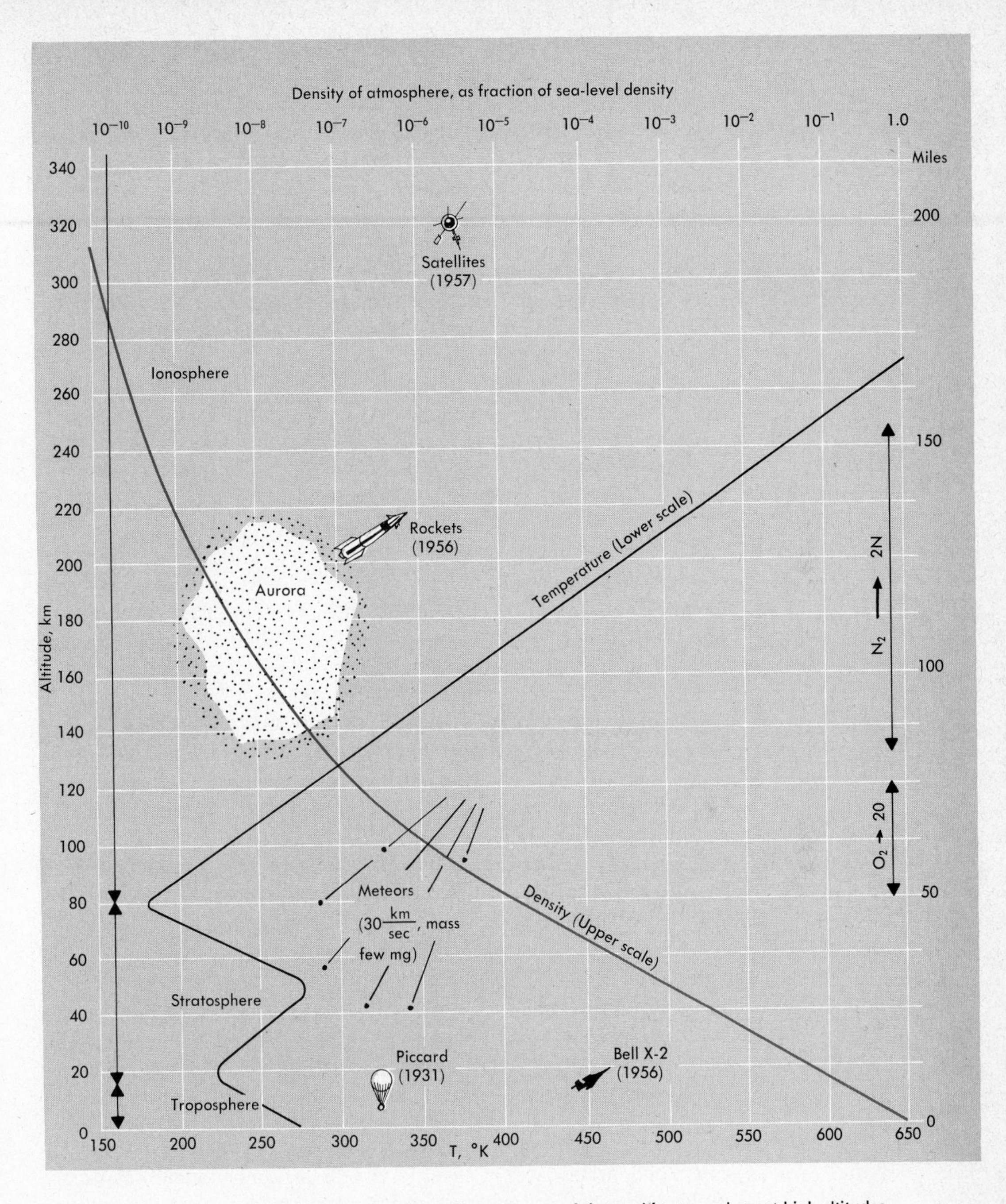

FIG. 32-1 Density and temperature of the earth's atmosphere at high altitudes.

temperature. If the actual temperature accurately followed the curve we could calculate for this expansion, the troposphere would be perfectly stable, with no part of it having any tendency either to rise or to sink in its surroundings. Fortunately, this is generally not the case.

The maximum intensity of the sun's radiation is in the visible part of the spectrum. Air is nearly transparent to this radiation, so that little of its energy is absorbed and little heating of the air results from it. The ground, however, absorbs much of the solar radiant energy and is warmed by it; the ground in turn warms the air near the surface. This warmed air expands, becomes less dense, and rises, to be replaced by cooler air from above it. In this way, there is a convective circulation normally taking place throughout the troposphere.

What the meteorologist calls a *temperature inversion* occurs when the ground is cooler than the air above it, so that this convective circulation does not take place. This is often the situation when the ground is snow-covered. Without convection currents to lift the air and carry it away, automobile exhaust fumes, chimney smoke, and other atmospheric pollutants linger near the surface in a stagnant layer.

Above the troposphere is the *stratosphere*, extending from 12 to 50 km, in which the temperature actually rises with increasing altitude. Since the warmer air is already on top, there is little or no convection to cause any vertical motion in this layer. Apparently, the situation is different here, with a source of heat at the *top* of the region. This source is a concentration of ozone (O_3), composed of three oxygen atoms rather than the common diatomic oxygen molecule (O_2). Ozone strongly absorbs ultraviolet radiation of $\lambda = 2000$ to 3000 Å. In this high-altitude atmosphere, the O_2 molecule is split into two atoms by absorbing radiation of $\lambda = 2000$ Å or less; an atom can then collide with an unbroken molecule to form ozone: $O + O_2 \rightarrow O_3$.

The energy absorbed by this ozone layer also heats the *mesosphere* from the bottom, so it behaves like the troposphere. The temperature falls, as we should expect, with increasing altitude, and there is also some convective circulation up to a height of about 80 km.

At this altitude the *ionosphere* begins, and the temperature again rises. The temperature graph cannot be taken literally here—about all that can be said is that the temperature rises with increasing altitude. Ordinary thermometers are useless in air of such low density. Near the ground our instruments and bodies receive about 10^{25} molecular impacts per square centimeter each second, and a thermometer reaches the temperature of its surroundings in a few minutes. At a height of 220 km, air density is only a billionth as great as at sea level, there are only a billionth as many impacts, and heat transference to or from a satellite or a thermometer from this source is completely negligible. The temperature of any material object is determined only by a balance between the radiation it receives from the sun, and its own radiation into space. Heat comes, as at lower atmospheric levels, from absorption of UV by $O_2 \rightarrow 2O$. But there is little ozone in these upper levels. With the atoms or molecules 1000 times farther apart then they are at sea level, there are very few collisions to allow for their recombination.

As we have seen, the upper atmosphere's absorption of short-wavelength radiation plays an important part in determining the temperature of the atmosphere and in protecting us from the harmful effects of the short and energetic ultraviolet. The infrared is also of great importance to the heat balance at the earth's surface but it works in the opposite direction. If the earth is to maintain the same average temperature century after century, it must radiate into space as much energy as it receives from the sun. The earth, however, radiates at a temperature of about 300°K rather than the sun's 6000°K. Let us refer back to Wien's law, which says that λ_{max} is inversely proportional to the absolute temperature of the radiating body. The most intense wavelength from the sun is about 5500 Å, so we have

$$\frac{\lambda_{max}(\text{earth})}{5500} = \frac{6000}{300}; \qquad \lambda_{max}(\text{earth}) = 110{,}000\ \text{Å}.$$

Such long infrared waves are strongly absorbed by both CO_2, composing about 3×10^{-4} of the air in the troposphere, and water (H_2O), also confined almost entirely to the troposphere, varying from nearly zero to about 4 percent. Thus much of earth's radiation is absorbed by the atmsophere, which then reradiates it, part out into the cold of space, and part back to the earth. The atmosphere, particularly if humid or cloudy, acts in this way as an insulating and radiating blanket to reduce the net heat loss from the surface at night. This is the *greenhouse effect*, so called because the glass of a greenhouse, like the atmosphere, is transparent to the short waves from sunlight, and opaque to the long waves reradiated from the warm interior. The heat is trapped inside, and the greenhouse may be even more than comfortably warm on a bright winter day, without any artificial heating. Most of us are more familiar with the same effect in a closed car in winter sunshine. We notice, too, that nights stay warm when humidity is high; in desert air, there is a sharp and extreme drop in temperature when the sun sets, due to the transparency of the dry air to the earth's infrared radiation.

32-2 Circulation in the Troposphere

All our winds, all our storms, and the atmospheric transport of heat energy and moisture, take place in the troposphere. The thickness of the troposphere varies with latitude and with seasonal changes, but we can take eight miles as a fair average. This is $\frac{1}{1000}$ of the earth's diameter. We can better visualize its scale if we imagine it to be on a desk-model globe of the earth, 10 inches in diameter. On this, we could represent the troposphere by a single coat of varnish, $\frac{1}{100}$ inch thick.

Without air circulation in this thin layer, most of the earth's surface

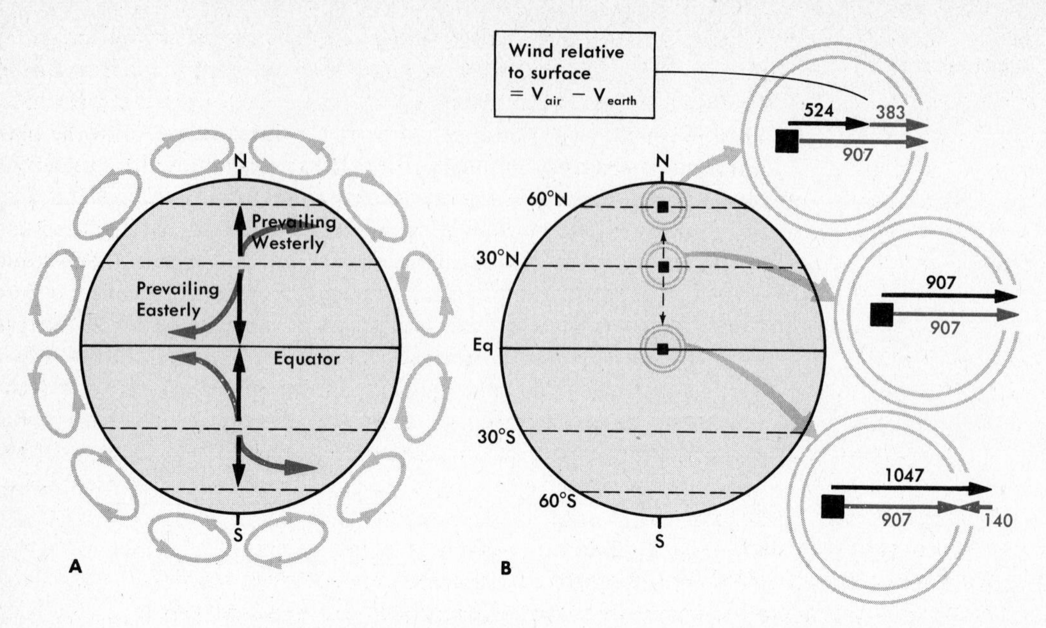

FIG. 32-2 Origin of the fictitious Coriolis force, causing deflection of winds blowing over the surface of the rotating earth.

would be either uninhabitably hot or uninhabitably cold. Figure 32-2A sketches (on a very exaggerated vertical scale) the simplified main features of the north–south air circulation pattern.

The main driving force for this circulation is of course the heating of the air by the warm land and water in the equatorial region. This air continually rises and spreads out to the north and south in the upper troposphere. As this warmer air cools, it sinks downward again in the vicinity of 30° latitude. The zone of sinking is very vaguely located, and is subject to considerable fluctuation and seasonal variation. The second circulation loop at higher latitude is even more irregular and uncertain, and the polar circulation, as shown in the drawing, often does not exist in any definite form.

Nevertheless, *if the earth did not rotate*, we should find winds blowing directly toward the equator in the lower latitudes, and winds blowing toward the poles in the higher latitudes of the temperate zone. These winds are shown as black arrows in Fig. 32-2A. The earth does rotate, however, and the effect of the rotation is shown by the red arrows in the drawing. If, at any latitude, we face in the direction toward which the wind is blowing, it is deflected to the right in the northern hemisphere, and to the left in the southern hemisphere. Thus, from the equator to latitude 30°±, the wind blows from the east; from 30°± to 60°±, it is from the west.

To see why this should be, let us consider a sample of air rotating with the earth at about 30° latitude—i.e., it is stationary on the earth's surface. If we take the radius of the earth to be 4000 mi, the radius of the circle in which a spot at 30° latitude moves is 4000 cos 30°; at 1 rev /24hr, its speed is $2\pi \times 4000 \cos 30°/24 = 907$ mi/hr. In Fig. 32-2B the black vector indicates the velocity of the earth; the red vector, the velocity of the air at this point. Now, let us completely ignore all friction and move our air sample northward to 60° latitude. The air still retains its easterly momentum, and is therefore still moving toward the east at 907 mi/hr. The earth beneath it, however, now has an easterly speed of only $2\pi \times 4000 \cos 60°/24 = 524$ mi/hr. From the point of view of the local inhabitants, the air is a west wind, moving toward the east at $907 - 524 = 383$ mi/hr. This deflection to the right, converting a south wind into a west wind, is shown by the curved red arrow in Fig. 32-2A. We can pursue the same argument for the other zones, as shown in the drawing. Friction does of course play an important part in air movement, but the direction of deflection remains qualitatively the same.

We have looked at this from the point of view of a stationary outsider, watching the earth and its winds move beneath us. If we want to refer these deflections to our more normal position, with an accelerated frame of reference attached to the rotating earth, we can explain these deflections by inventing another fictitious force (as we did earlier for centrifugal force). This is the *Coriolis force*, named for a nineteenth-century French scientist, which depends on the angular velocity of the rotating reference, and on the velocity of the moving body relative to it. We will not attempt to deal with it quantitatively.

32-3 Cyclones and Anticyclones

Unfortunately for the meteorologist, the flow of air over the earth's surface is not as smooth and regular as our oversimplified model might imply. There are many irregularities caused by the difference between the temperatures of water and land, large recently heated or cooled areas on the continents, mountain ranges, and so on. The flow of air might better be compared with the flow of water in a rocky stream, full of unpredictable swirling eddies, some large and some small.

A common type of atmospheric eddy is seen on weather maps, and is responsible for most of the day-to-day changes in our weather. These eddies are the circulation patterns around the "lows" and "highs" that (in the temperate zones) progress across the maps from west to east, with the prevailing winds. Figure 32-3 shows the circulation of air around these low-pressure and high-pressure areas as they occur in the northern hemisphere. Surrounding air moves toward the low pressure, and as it moves, it is deflected to the right by the Coriolis force. The result is a general counterclockwise rotation several hundred miles in diameter, called a *cyclone* by meteorologists. Around a region of high pressure, the air movement is away from it, and the Coriolis deflection causes a clockwise rotation called an *anticyclone*. In the southern hemisphere where

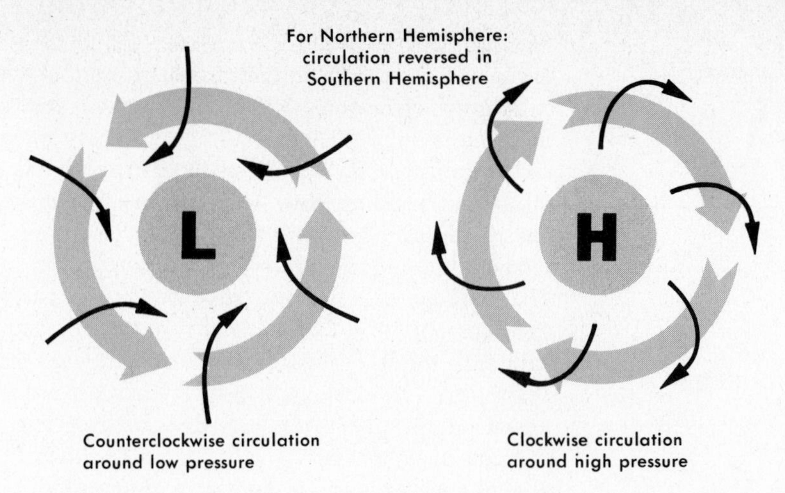

FIG. 32-3 Coriolis deflection as the cause of the circulation of air around centers of low and high pressure.

moving air is deflected to the left, these directions of rotation are of course reversed.

32-4 Weather Energies

"Weather" is so common that we seldom have any appreciation of the enormous energies involved. Air seems so light and insubstantial that the word "airy" is a synonym for these properties. At sea level, 1 cm^3 of air has a mass of only 1.2 mg, which is little enough. On a larger scale $1\ m^3 = 10^6\ cm^3$, to give it a mass of 1.2 kg. For a cubic kilometer, which contains $10^9\ m^3$, the mass is 1.2 million metric tons. A cubic mile, which we can perhaps visualize more easily, weighs 5.5 million of our ordinary 2000-lb tons. These are masses that have enormous kinetic energies when moving as wind.

We are accustomed to being awestruck at the unbelievable energy released in the explosion of a nuclear bomb. Let us use this as an energy unit—a 20-kiloton bomb releases the energy of 20,000 tons of TNT, or about 10^{14} joules. It has been estimated that at any moment, the kinetic energy of the many winds blowing all over the earth is equivalent to that of about 7 million of these nuclear bombs.

On a somewhat more modest scale, consider a gentle summer local rainstorm that drops 1 inch of rain on a small area only 10 miles square. This amount of water can be easily figured:

$$100\ \text{mi}^2 \times 1\ \text{inch} \times \left(\frac{1.6\ \text{km}}{1\ \text{mi}}\right)^2 \times \left(\frac{10^5\ \text{cm}}{1\ \text{km}}\right)^2 \times \frac{2.54\ \text{cm}}{1\ \text{inch}} = 6.5 \times 10^{12}\ \text{cm}^3.$$

This water has a mass of 6.5×10^{12} g or 6.5×10^{9} kg. In condensing from water vapor into raindrops, the latent heat of vaporization is released into the atmosphere. This amounts to

$$6.5 \times 10^{12}\ \text{g} \times \frac{540\ \text{cal}}{\text{g}} \times \frac{4.18\ \text{J}}{\text{cal}} = 1.5 \times 10^{16}\ \text{J}.$$

We should also calculate the potential energy of the water droplets, formed at an average altitude that we can assume to be 1000 m. This is all converted into heat, through air friction as the raindrops fall. For this,

$$mgh = 6.5 \times 10^{9}\ \text{kg} \times \frac{9.8\ \text{N}}{\text{kg}} \times 1000\ \text{m} = 6.4 \times 10^{13}\ \text{J}.$$

This is only a few thousandths of the heat of vaporization, so we can justifiably ignore it. In terms of our nuclear bomb units, this modest summer shower releases atmospheric energy equivalent to

$$\frac{1.5 \times 10^{16}}{10^{14}} = 150\ \text{nuclear bombs.}$$

It is less damaging only because the energy is released over several hours, rather than a few microseconds.

The devastation of hurricanes is well-known. They are similar to the ordinarily gentle cyclones circulating about a low-pressure region, but are much more energetic and generally concentrated in a smaller area. The source of their energy is the release of heat of vaporization. They always form around a region of low pressure located over the ocean. The inflowing air is laden with water vapor from its travel over the warm waters; its condensation as it rises and cools brings torrential rains. The air surrounding the central low has the normal counterclockwise rotation and corresponding angular momentum. Like the spinning girl in Fig. 5-9, conservation of angular momentum increases the angular, as well as the translational, velocity of the moving air as it approaches the center of the hurricane. Winds of 100 to 150 mi/hr, accompanied by deluges of rain, can cause enormous damage if these whirling storms pass over inhabited land. The rate of energy release of a well-developed hurricane may be equivalent to the explosion of 10 nuclear bombs per minute.

32-5 The Deeper, the Hotter

As deep mines and oil wells penetrate for as much as a few miles into the earth, the temperature of the rock steadily rises. In different locations, this temperature increase ranges from less than 10°C/km to more than

50°C/km, for an average value near 30°C/km. This rate of increase cannot continue very deeply. If it did, we would arrive at a temperature of about 200,000°C at the earth's center, and the outward conduction of heat would raise the surface temperature to a point at which life would not be possible. The general concensus is that the earth's central temperature is only a few thousand degrees.

The source of this internal heat almost certainly comes largely from the decay of radioactive elements concentrated in the relatively light rocks of the earth's outer crust. A ton of ordinary granite contains about 9 g of uranium and 20 g of thorium, whose decay over the ages provides a large and steady source of heat. The deeper, denser rock under the crust must contain much less radioactivity than this; otherwise the surface temperature would be a great deal higher than it is.

In many places, often in the neighborhood of present or recent volcanic activity, the subsurface heat is abnormally high and close to the surface. In Iceland, Italy, New Zealand, and California, wells drilled into these hot rocks have yielded steam or hot water that has been used for heating and power generation. More of these geothermal heat sources will no doubt be tapped in the future, although locations that provide water or steam hot enough and pure enough to be used in efficient power generation are probably very rare.

32-6 The Earth's Interior

Deep oil wells penetrate only a few of the 4000 miles between us and the center of the earth. To get more information about the earth's interior, we must rely on a less direct approach.

Destructive as they are, earthquakes are of great help to scientists in their study of the interior of our globe, since earthquake waves originating in some point of the earth's crust propagate to other points on the surface of the globe through its deep interior. As we have seen earlier, there are two kinds of waves that propagate through a continuous medium:

1. *P waves* (pressure or push waves) are longitudinal waves that can propagate equally well through a solid and through a fluid medium.
2. *S waves* (shear or side waves) are transverse waves that can propagate through solids but not through liquids.

When an earthquake wave from a distant disturbance arrives at the surface of the earth, the motion of the ground in the case of P waves will be in the direction of propagation, while in the case of S waves the motion will be perpendicular to it. This permits us to distinguish between P and S waves by registering the movements of the ground by means of a very sensitive instrument known as a *seismograph*.

Seismographs are based on the principle of inertia, according to which each body at rest tends to preserve its state of rest. A simple seismograph,

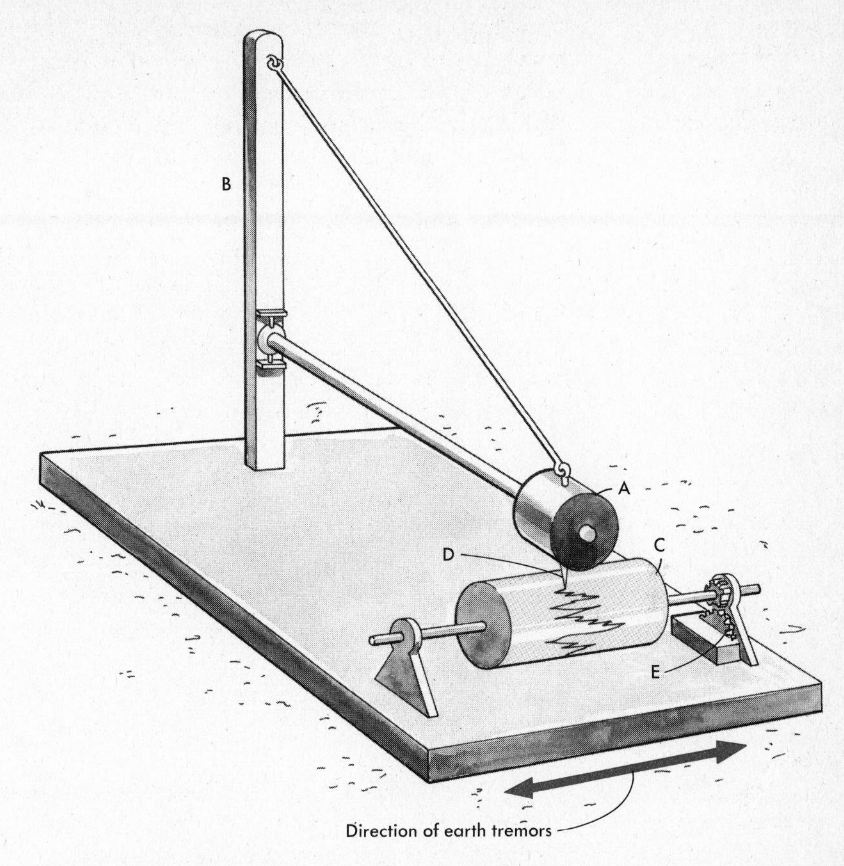

FIG. 32-4 Schematic diagram of a simple seismograph. A massive weight A is suspended from a vertical support B. C is a rotating cylinder driven by a clock mechanism E; D is a pen that marks the rotating cylinder as earth vibrations move it back and forth beneath the nearly stationary mass A.

known as the *horizontal pendulum*, is shown in Fig. 32-4 and consists essentially of a heavy weight *A* that can move with very little friction around the vertical support *B*. If the ground on which this instrument is installed is jerked by an earthquake wave in a direction perpendicular to the vertical plane passing through the support and the weight, the weight remains immovable because of its large inertia, and the displacement of the stand relative to the resting weight is registered on the rotating drum *C*. (A system of levers to magnify the relative motion between mass *A* and drum *D* is not included on the drawing). Two such instruments installed at right angles to each other give complete information about the horizontal displacements produced by the earthquake wave, and there are also seismographs that register the vertical displacements. The seismograms produced by these instruments permit us to carry out a complete analysis of the arriving disturbance.

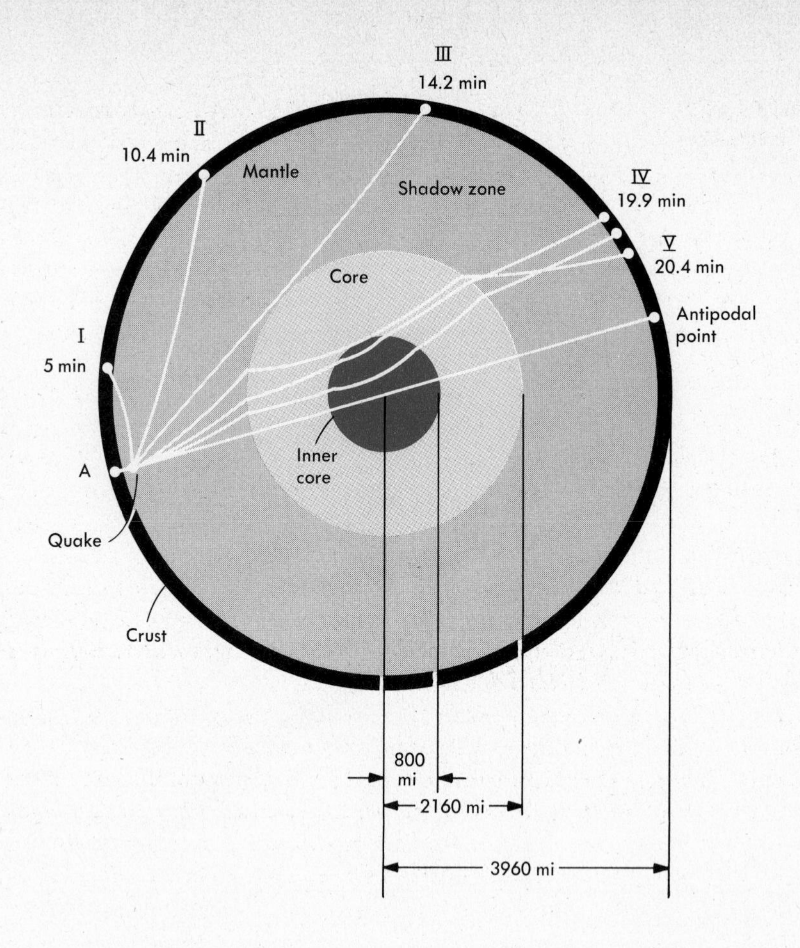

FIG. 32-5 The lines of propagation of an earthquake disturbance originating at a point near the top of the mantle.

Let us consider now what would be observed by seismic stations scattered all over the world when a sufficiently strong earthquake originates in some point of the earth's crust. Figure 32-5 shows the propagation lines of the disturbance originating beneath point *A*. At stations I, II, and III, located within 100° from the center of the disturbance, both *P* and *S* waves are observed, which proves that the material of the earth possesses the property of an elastic solid (capable of transmitting a shear) up to very great depths. Although these rocks are at temperatures well above what their melting points are on the surface, great pressure gives them a rigidity greater than steel in the transmission of seismic vibrations.

Beyond station III, about 100° from the earthquake origin, there is a ring-shaped zone extending to station IV, about 145° from the earthquake, in which neither *S* nor *P* waves can be felt. In the circular cap some 35° in radius directly opposite the earthquake, whose circumference passes through station IV, only the longitudinal *P* waves are received.

To explain this, seismologists have concluded that the central portion of the earth is occupied by a liquid core that cannot transmit the transverse *S* waves. With this assumption, the existence of the shadow zones can be easily understood by inspecting the rays of earthquake waves shown in Fig. 32-5. Due to a rapid increase of pressure and density with depth, the velocity of seismic waves (both *P* and *S*) also increases and causes the propagation lines to bend slightly toward the surface. Thus the waves arriving at station III, after passing very close to the surface of the fluid core, are the last ones that still propagate all the way through the plastic material of the rocky mantle. The rays that propagate deeper inward will hit the surface of the core, and while *S* waves will not be able to propagate beyond this point at all, *P* waves will be refracted as shown. After the second refraction at the exit point from the liquid core, the *P* waves will finally arrive at the surface much closer to the antipodal point and thus produce a ring-shaped shadow zone.

Detailed analysis of the transmission of earthquake waves and calculation of their velocities at different depths has led to the conclusion that the earth has an onion-like structure consisting of a number of concentric shells of differing properties.

The outermost of these is the *crust*, whose foundation is a layer of the dense, dark-colored rock *basalt*. Floating on the basalt layer, the continents are made of the lighter granite and sedimentary rocks. The thickness of the crust varies from 3 to 5 miles beneath the ocean basins, to 20 to 30 miles beneath the continents. Below the crust, the next 1800 miles is the *mantle*; within this are the liquid *core* and the solid *inner core*.

32-7 Floating Continents

The material of which the mantle is made is thought to be similar to that of the basalt underlying the crust. The great heat and pressure deep in the earth, however, have perhaps changed its crystal structure to give it the property of *plasticity*. This property can be demonstrated by some pieces of pitch, a naturally occurring asphalt that is as brittle as glass. A piece of pitch struck with a hammer shatters into a handful of jagged pieces. Toss these pieces into a beaker and in a few days or weeks—depending on the temperature and the grade of pitch—it will have flowed into a single mass that fills the beaker in a smooth, level layer. Rigid and solid to a quickly applied force, it nevertheless flows like a very viscous fluid under the long, steady pull of gravity. On a larger and longer scale, the behavior of the mantle material is similar to that of pitch or sealing wax. To the rapid earthquake waves it is rigid and elastic. Yet, in response to steady forces over thousands of years, it flows like a viscous liquid.

The title of this section, "Floating Continents," is apparently one that is literally true. The light rock of the continents floats on the denser plastic mantle like a slab of wood on water.

Archimedes' principle works just as well on a large scale as it does in a laboratory beaker, and can be applied to the structure of the earth

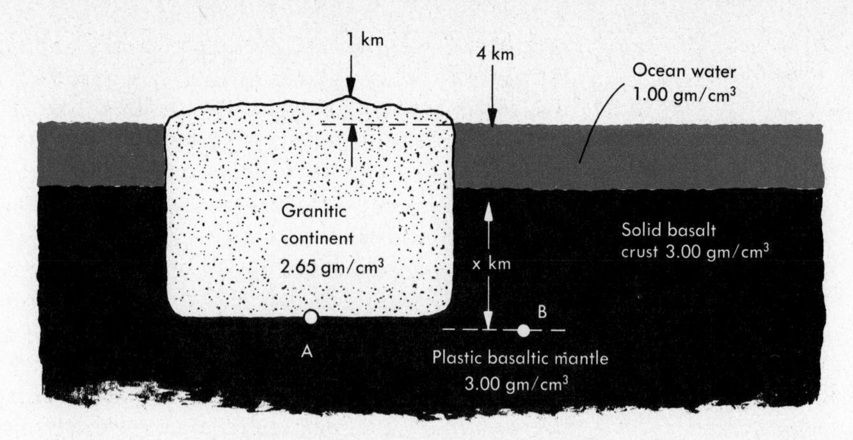

FIG. 32-6 Conventionalized diagram of a continent supported by its buoyancy.

itself. Figure 32-6 is a highly schematic diagram of a continent floating on a sea of denser plastic mantle material. The average altitude of a continent above sea level is about 1 km, as shown, and the average depth of the oceans is about 4 km. Point A, beneath the continent, and point B, beneath the ocean, are drawn to be at the same level, and the pressure at these two points must be the same. If it were not, the plastic mantle material would flow from the higher-pressure point to the lower until they were equalized.

So, let us calculate (from $p = hd$) the pressures at A and B and set them equal:

$$p_A = (x + 4 + 1) \times 2.65 = 2.65x + 13.25$$

$$p_B = x \times 3.00 + 4 \times 1.00 = 3.00x + 4.00.$$

So

$$2.65x + 13.25 = 3.00x + 4.00$$

$$0.35x = 9.25$$

$$x = 26 \text{ km.}$$

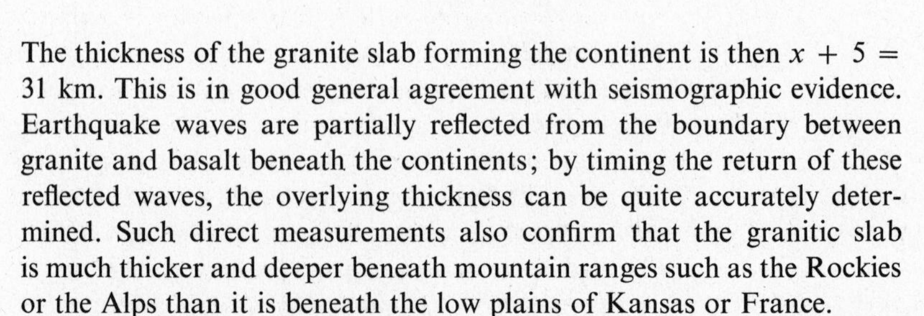

The thickness of the granite slab forming the continent is then $x + 5 = 31$ km. This is in good general agreement with seismographic evidence. Earthquake waves are partially reflected from the boundary between granite and basalt beneath the continents; by timing the return of these reflected waves, the overlying thickness can be quite accurately determined. Such direct measurements also confirm that the granitic slab is much thicker and deeper beneath mountain ranges such as the Rockies or the Alps than it is beneath the low plains of Kansas or France.

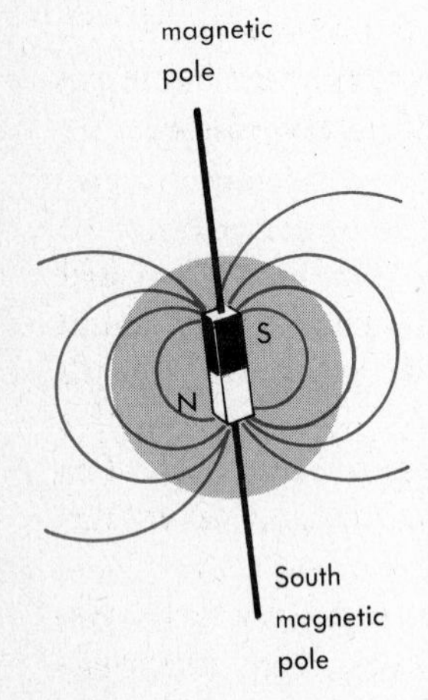

FIG. 32-7 The main features of the earth's magnetic field are similar to those that would be caused by a giant bar magnet (or an equivalent circulating current) at the earth's center.

The adjustment of the earth's crust under the shifts of mass on its surface has played a very important role in the evolution of the face of our planet. For example, considerable isostatic adjustment took place during the glacial periods, when thick sheets of ice covered much of North America and Europe. The weight of the ice caused the northern regions of these continents to sink deeper into the plastic layer of basalt underneath. At present, when most of the ice has retreated, the depressed parts of the continents are slowly rising toward their pre-Ice Age level, and we can notice a slow regression of the seas along many shorelines of the northern countries.

32-8 Terrestrial Magnetism

The earth's magnetic field, in both magnitude and direction, can be easily and accurately measured not only at the surface, but above it, and below it in deep mines. The field is essentially what we might expect to be produced by a giant bar magnet aligned as shown in Fig. 32-7. The axis of the magnet and its field lie about 11° from the geographical axis of rotation—close enough to make the uncorrected magnetic compass a useful, if approximate, direction indicator except at high latitudes.

For many reasons the idea of a huge permanent magnet deep within the earth is of course completely unacceptable. There can be no doubt that the magnetic field is produced by some sort of circulating electric current. Although we cannot completely rule out the effect of slow convective circulation in the plastic mantle, it seems more likely that a circulation of this sort would occur in the fluid core, which is no doubt well-ionized at its high temperatures. Any explanatory theory must take into account the higher thermal speed of the electrons, the effects of Coriolis forces on the electrons and on the more sluggish positive ions, and the details of heating to produce any convective currents in the first place. It is a problem that has not yet been satisfactorily solved. There are other complications, too, that will be apparent in the next section.

32-9 Fossil Magnetism

The earth's magnetic poles wander a bit, and the strength of the field varies somewhat from decade to decade, but there has been no substantial change since the beginning of dependable records in the seventeenth century. We are accordingly accustomed to thinking of the earth's magnetic field as something relatively constant and permanent. The investigation of magnetic mineral crystals in old rocks, however, has rudely changed this comfortable picture.

We can in general consider rocks to have been formed in two ways. To represent one class (the sedimentary rocks), let us consider the formation of a sandstone in a shallow sea. Rivers and streams constantly erode the land, and carry out to sea fine sand grains and other minerals, including a common oxide of iron called *magnetite*. As the name implies, these elongated crystals are quite magnetic, and align themselves parallel to a magnetic field (Fig. 32-8A). As they slowly settle to the bottom, they line up in the direction of the earth's magnetic field. As the sediments

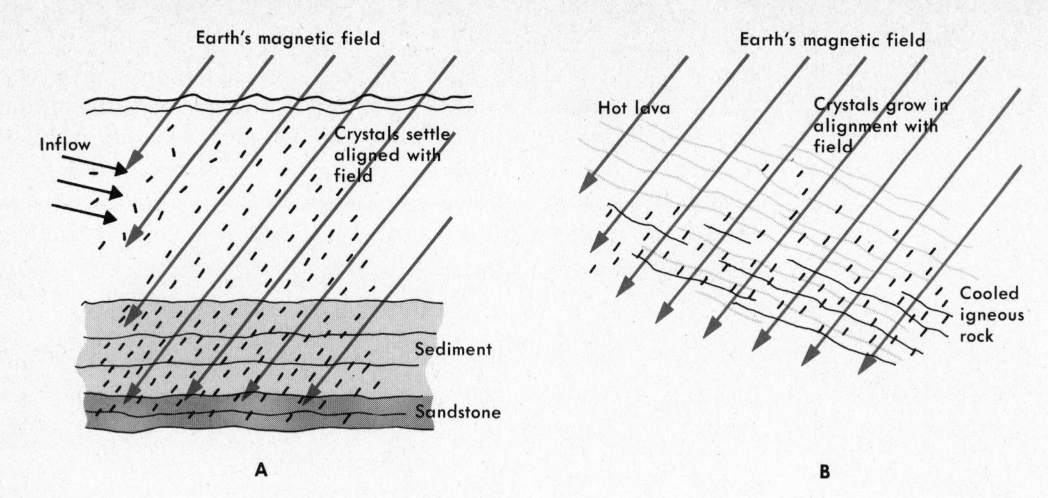

FIG. 32-8 The direction of the earth's magnetic field, preserved as "fossil magnetism" in the rocks formed at the time.

slowly build up and become consolidated into sandstones, these magnetic indicators are locked into the direction of the earth's magnetic field as it was when they were deposited.

Rocks that have solidified from a molten state are *igneous* rocks. As an example, take a dark iron-bearing lava poured out over the land in a volcanic eruption. Under proper conditions, similar magnetic minerals will crystallize out as the lava cools and hardens. These tiny magnetic crystals grow along the direction of the earth's magnetic field (Fig. 32-8B) as it was when the rock cooled.

Geophysicists carefully collect specimens of such rocks, noting of course their exact orientation while still in place. In the laboratory, the direction in which the rocks are magnetized then indicates the direction of the earth's magnetic field when they were formed. Data from a large number of such examples of fossil magnetism taken from many locations have demonstrated a peculiar phenomenon. At somewhat irregular intervals, ranging from about 10^5 to 10^6 years, the direction of the earth's magnetic field is reversed! We must therefore visualize the still unknown currents in the earth's core as periodically dying out—from the effect of friction, perhaps,—and then starting up again in the opposite direction.

33

Astrophysics

33-1 Inside the Sun

Direct observations of the sun are limited to the surface of the photosphere, and no human eye can penetrate into its deep interior. Yet we know more about the central regions of the sun than about the core of the earth, in spite of the fact that the earth is right under our feet and the sun is a hundred million miles away! The reason for this strange fact is that in both cases we have to make conclusions on the basis of theoretical considerations that require knowledge of the properties of matter under the physical conditions existing in the deep interiors of these two celestial bodies. In the case of the earth, we have to deal with the physical properties of a molten substance subjected to extremely high pressures inaccessible in our laboratories. Being unable to study these properties experimentally, we also lack any reliable theoretical predictions concerning them, since, as mentioned earlier in the book, the theory of the liquid state of matter is extremely difficult. It would seem that in the case of the sun's interior, where the temperature and pressure exceed by a large margin those encountered in the interior of the earth, the situation would be even worse. But this is not so. It turns out, in fact, that just because of these tremendously high temperatures that exist in the solar interior, the properties of matter become very simple and easily predictable on the basis of our present knowledge concerning atomic structure. The reason for this is that, *under the conditions existing in the solar interior, all molecules and atoms that form ordinary matter are almost completely broken up into their constituent parts.*

Consider a glass of water that is brought gradually to a higher and higher temperature. At room temperature, we have a complicated structure in which water molecules composed of hydrogen and oxygen atoms are held together by intermolecular cohesive forces. With the increase of the temperature, the molecules will be torn apart and we obtain water vapor, in which individual molecules fly freely through space and very rarely collide with one another. The physical properties of water vapor, being subject to classical gas laws, are much simpler to describe than those of liquid water. However, the properties of individual molecules (such as an absorption spectrum) are still almost as complicated as those in the liquid state. When the temperature rises to still higher values, violent thermal collisions between the water vapor molecules will break them up (*thermal dissociation*) into individual atoms of hydrogen and oxygen. In this state, the optical properties of the gas become more predictable, since they pertain now to individual atoms of hydrogen and oxygen, which are simpler than the composite molecules of water. Now, if the temperature goes still higher, the increasing violence of thermal collisions will begin to strip the individual atoms of their electronic shells (*thermal ionization*). At the temperature of 6000°K that prevails on the surface of the sun, most hydrogen atoms are broken up into protons and electrons, while the atoms of oxygen are stripped of two or three electrons from their outer shells. At the still higher temperatures that are encountered in the solar interior, oxygen atoms, as well as the atoms of all other heavier elements, will be almost completely stripped of all their electron shells, and there will be a *mixture of bare nuclei and free electrons involved in a violent thermal motion.*

This situation simplifies the picture quite a bit. First of all, we can expect that no matter how high the density, solar matter can be considered as an *ideal gas*. Indeed, as we have seen, an ideal gas is characterized by the smallness of the individual particles as compared with their relative distances. When a material consists of atoms and molecules that are about 10^{-8} cm in diameter, its particles must be packed tightly together when its density is about that of water. But once the atoms are broken into nuclei and free electrons, which have respective diameters of only 10^{-12} to 10^{-13} cm, the situation becomes entirely different, and the properties of gas will be retained up to much higher densities.

Knowing the physical conditions (temperature, density, etc.) on the surface of the sun and the laws governing the gaseous material in its interior, we can calculate, step by step, the change of these conditions as we proceed below the solar surface and toward its center. Such calculations were first carried out by the famous British astronomer Arthur Eddington, and gave us the first insight into the inside of our sun as well as of the other stars. The central temperature of the sun proves to be in the neighborhood of 20,000,000°K!

33-2 Energy Production in the Sun

In the middle of the last century, shortly after the discovery of the law of equivalence of mechanical energy and heat, a new theory of solar energy production was proposed simultaneously by H. von Helmholtz in Germany (1821–1894) and Lord Kelvin in England (1824–1907). They indicated that, being a giant sphere of hot gas that loses its energy through radiation from the surface, the sun is bound to contract, though very slowly. During this contraction, the forces of gravity acting between different parts of the solar body do mechanical work that is then turned into heat, according to the equivalence principle of mechanical and thermal energy. It has been calculated that, being much more powerful than chemical reactions (such as burning), this process could maintain the sun at its present luminosity for a couple of hundred million years. At the time of Kelvin and Helmholtz, a few hundred million years seemed to be a sufficiently long period to explain the events of historical geology, and so their point of view was accepted without objection. We know now, however, that life on the earth must have existed for at least half a billion years, and the solid crust and the oceans must be much older still. Thus the gravitational contraction of the sun certainly could not have provided enough energy to maintain the surface of the earth and its oceans comfortably warm for the origin and evolution of living organisms.

The discovery of radioactivity and the recognition of the fact that the energy stored within atomic nuclei exceeds, by a factor of a million, the energy liberated in ordinary chemical transformations, threw an entirely new light on the problem of solar-energy sources. If the sun could have existed for several thousands of years fed by an ordinary chemical reaction (burning), nuclear energy sources are surely rich enough to supply an equal amount of energy for billions of years. The trouble is, however, that in natural radioactive decay, the liberation of nuclear energy is extremely slow. In order to explain the observed mean rate of energy production in the sun (2 erg/g/s), we would have to assume that the sun is composed almost entirely of uranium, thorium, and their decay products. Thus we are forced to the conclusion that the liberation of nuclear energy inside the sun is not an ordinary radioactive decay, but rather some kind of induced nuclear transformation caused by the specific physical conditions in the solar interior. It is natural to expect that the factor responsible for the induced nuclear transformations is the tremendously high temperature existing in the solar interior. Indeed, at a temperature of 2×10^7°K, the kinetic energy of thermal motion is 4.2×10^{-9} erg per particle, which, expressed in electric units, amounts to 3 KeV. This is considerably smaller than the energies ordinarily used in the experiments on nuclear bombardment (1 MeV and up), but we must take into account that whereas artificially accelerated nuclear projectiles rapidly lose their initial energy, and have only a small chance to hit the target nucleus before coming out of the game, thermal motion continues indefinitely and the particles involved in it collide with each other for hours, centuries, and billions of years. Calculations carried out in this direction in 1929 by the Austrian physicist F. G. Houtermans and the British astronomer R. Atkinson led to the conclusion that, at the temperatures existing in the solar interior, thermonuclear reactions between hydrogen nuclei (protons) and the nuclei of other light elements can be expected to liberate sufficient amounts of nuclear energy to explain the observed radiation of the sun.

33-3 Solar Fusion Reactions

Although the work of Houtermans and Atkinson proved, beyond any doubt, that the energy production inside the sun is due to thermonuclear reactions between hydrogen and some light elements, the exact nature of these reactions remained obscure for another decade because of the lack of experimental knowledge concerning the result of nuclear bombardment by fast protons. However, with the pioneering work of Cockcroft and Walton on artificially accelerated proton beams and subsequent work in this direction, enough material was collected in this field to permit the solution of the solar-energy problem. One possible solution was proposed in 1937 by H. Bethe in the United States and C. von Weizsacker in Germany (independently of one another) and is known as the *carbon cycle*, while the other possibility was conceived by an American physicist, Charles Critchfield, and is known as the *H-H reaction*. The net total result of both reactions is the transformation of hydrogen into helium, but it is achieved in a different manner in each reaction.

In the carbon cycle, the carbon atom can be considered as a "nuclear catalyst" that helps unite 4 independent protons into a single α particle by capturing them one by one and holding them together until the union is achieved. After 4 protons are caught and the newly formed α particle is released, we get back the original carbon atom that can go again through the next cycle. The series of reactions constituting the carbon cycle is shown in Fig. 33-1. Proceeding clockwise from noon, we have a collision between a C^{12} and a proton, to form N^{13} plus a gamma photon. N^{13} is unstable and decays to C^{13} and a positron. (Not shown on the diagram is the fate of the positron, which of course is the annihilation $e^+ + e^- \rightarrow 2\gamma$). The C^{13} and a colliding proton now produce the stable $N^{14} + \gamma$. Another unstable isotope, O^{15}, plus another γ is produced by the reaction of N^{14} with another proton. O^{15} decays to N^{15}, with the

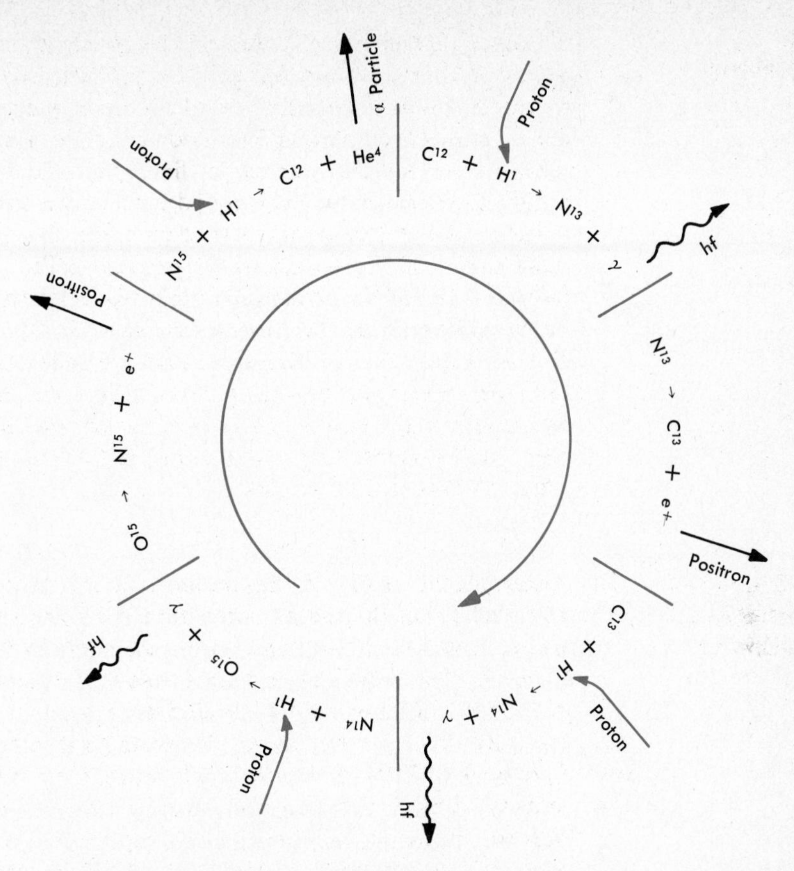

FIG. 33-1 Diagram of the nuclear reactions in the carbon cycle, responsible for most of the energy production of massive stars.

emission of another positron that is immediately annihilated. In the last step of the cycle, N^{15} and a proton combine to create an α particle plus a new C^{12}, ready to start the cycle again.

It must not be thought that each collision that takes place causes the reaction indicated. In fact, if we relied entirely on the principles of classical physics, none of the four fusion reactions of the carbon cycle could ever occur. The proton energies at 20×10^6 degrees are simply not enough to allow them to approach close enough to feel the pull of the nuclear binding forces of the strongly positive carbon and nitrogen nuclei. Quantum mechanics, however, is a matter of probabilities, rather than any definite yes or no. In 1928, George Gamow, and, independently, E. U. Condon and R. Gurney, developed the theory of what is called the *tunnel effect*. This gives the nuclear physicist a way to calculate the small but very definite probability that the de Broglie waves of a particle can penetrate or "tunnel" a barrier such as that of the electrostatic repulsion between nucleus and proton. This theory, with data obtained from laboratory measurements obtained by bombarding targets with high-energy protons, enables the probability for each type of collision reaction to be accurately calculated. The probabilities of the carbon-cycle reactions are very small. A carbon nucleus undergoes billions of impacts per second from the surrounding protons; but the "successful" impacts occur so seldom that it takes a C^{12} nucleus 6×10^6 years to complete the cycle of Fig. 33-1.

Quantitative calculations showed that the carbon cycle could account for only a small part of the sun's energy production. The other series of fusion reactions, the proton–proton series, proved to be much more appropriate for the temperatures at the sun's center:

$$H^1 + H^1 \to D^2 + e^+ \qquad\qquad H^1 + H^1 \to D^2 + e^+$$

$$D^2 + H^1 \to He^3 + \gamma \qquad\qquad D^2 + H^1 \to He^3 + \gamma$$

$$He^3 + He^3 \to He^4 + 2H^1 + \gamma.$$

The overall reaction is the same as that of the carbon cycle: $4H^1 \to He^4 + 2e^+ + 3\gamma$. But because of the smaller nuclear charges it is more effective at 20×10^6 degrees. Calculations show that its rate of energy production added to that of the carbon cycle equals the energy production of the sun. In the many stars hotter than the sun, however, the carbon cycle is the principal energy source.

Although astrophysicists are convinced that these two processes are responsible for the energy produced in the great majority of stars, an uncomfortable element of doubt or question has recently entered the picture. In our diagrams of the two energy-producing cycles, we have omitted the neutrino that must be emitted with each of the two β^+. So a flood of neutrinos must pour out from the center of the sun, pass through the sun almost without obstruction, and then fall on the earth. In order to measure this expected flux of solar neutrinos, equipment has been set up, similar in principle to that used by Reines and Cowan in their experimental detection of this elusive particle. To everyone's surprise, the number of solar neutrinos measured in this way has been only a fraction of the number to be expected. At this writing, the cause of the discrepancy is not understood.

33-4 Sunspots and Magnetism

When Galileo revolutionized astronomy in 1610 by turning his new telescope skyward, among the first things he observed were strange dark spots on the surface of the sun. We know now that these spots come and go in cycles. At the beginning of a cycle a few occasional small spots can be seen at a solar latitude of about 30°, apparently moving from west to east as the sun rotates at a speed of something like once a month. As the cycle progresses, more and more and larger spots, often in groups, appear at lower latitudes, until the cycle gradually fades away after about 11 yr with a few small spots near the equator. The spots are dark only in contrast with the brighter surface that surrounds them. With a temperature of perhaps 4500°K, Stefan–Boltzmann's law tells us the spots are only about a third as luminous as their 6000°K surroundings.

In Chapter 24, Section 7, we have mentioned the dark absorption (Fraunhofer) lines that can tell the spectroscopist what elements are present in the solar chromosphere. But this is not the full extent of their information. We have seen, also, that different values of the magnetic quantum number m have no effect on the energy of an electron—unless the electron is in a magnetic field. In a magnetic field the ordinarily single line is split into a number of lines whose separation depends on the strength of the field. By measuring this line splitting, the spectroscopist can determine not only the strength but the general direction of the magnetic field surrounding the absorbing atoms.

When sunspots appear in pairs or groups (Fig. 33-2), the leading major spot is opposite in magnetic polarity to the trailing spot. If, for example, a pair is observed in the northern hemisphere of the sun, with the N pole leading and the S pole trailing, all the pairs or groups in the northern hemisphere will have this same N–S sequence throughout the 11 years of the cycle. In the southern hemisphere the polarity is reversed to S–N. And peculiarly, during the next 11 years these polarities are reversed. It will be S–N in the northern hemisphere, and N–S south of the solar equator. The actual length of the sunspot cycle is thus about 22 years, rather than 11.

It seems inescapable that in the hot, dense gas beneath the sun's surface there are circulating currents—perhaps similar to our earthly cyclones, hurricanes, and tornadoes—that break through to the surface and appear as lower-temperature spots. There are a number of theories as to the nature and cause of these hypothetical currents, but none satisfactory enough to be generally accepted by solar astronomers. Whatever the cause, there is a cycle on the sun in which the magnetic fields of the spots build up and then die down and start up again in the opposite direction. It is tempting to speculate that there must be some basic similarity to the unknown circulation in the earth's fluid core, which causes the buildup, dying out, and reversal of the earth's magnetic field.

33-5 The Evolution of Stars

Since the energy radiated by the sun is due to the continuous transformation of hydrogen into helium in its interior, the sun evidently cannot shine for an eternity, and is bound to run out of fuel sometime in the future. It is estimated that during the 5 billion years of its existence, the core of our sun has used about half of its original supply of hydrogen, so that it still has enough nuclear fuel for another 5 billion years. What will happen 5 billion years from now, when our sun comes close to the

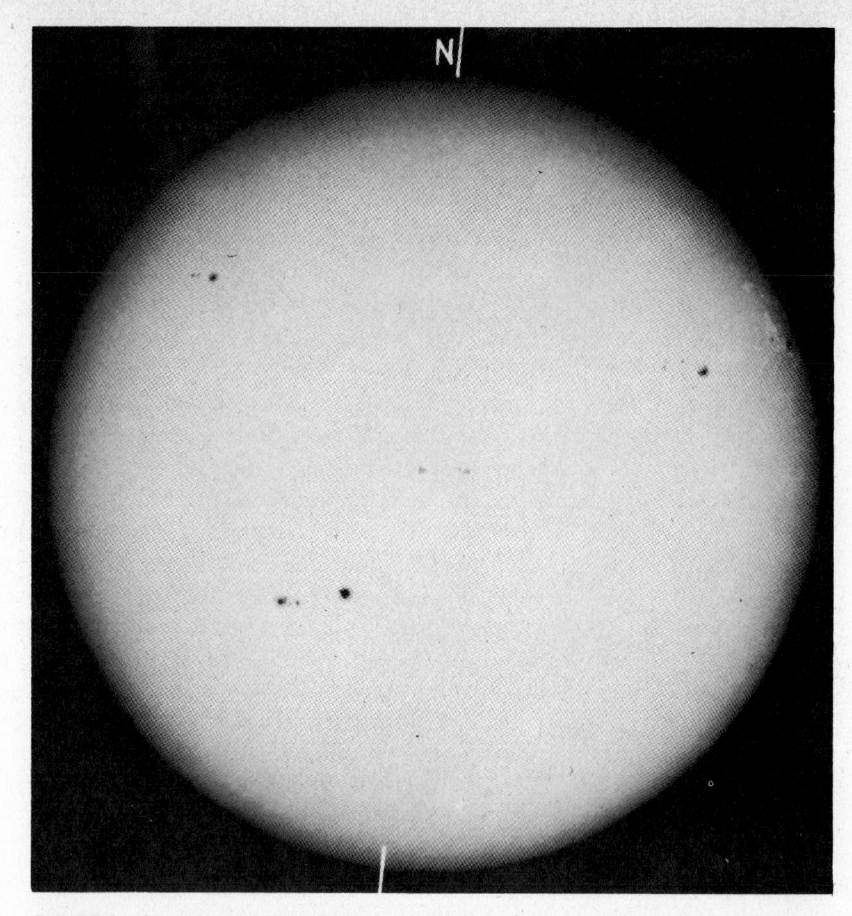

FIG. 33-2 Three groups of sunspots. In the northern hemisphere the large leading spots are of south magnetic polarity, and the trailing spots are north magnetic poles. The large leading spot in the southern hemisphere is a north magnetic pole, and the trailing spots are south poles. These polarities will be reversed in the next 11-year half-cycle. (*Photograph by Aerospace Corporation San Fernando Solar Observatory, courtesy of U.S. National Oceanic and Atmospheric Administration, Environmental Research Laboratories.*)

end of its resources? To answer this question, we have to remember that thermonuclear reactions proceed almost exclusively near the center of the sun, where the temperature is the highest. Thus the shortage of nuclear fuel will be felt first in the central regions of the sun, where all the originally available hydrogen will have been transformed into helium. We can easily visualize that this will result in a rearrangement of things in the solar interior in such a way that the high-temperature region will move to the interface between the "burned-out core" and the outer layers that still contain enough hydrogen to maintain a nuclear fire. The internal structure of the sun, therefore, will be transformed from a so-called *point-source model* (energy source in the center) to a *shell-source model*, in which the energy is liberated in a thin spherical shell that separates the burned-out core from the rest of the solar body (Fig. 33-3).

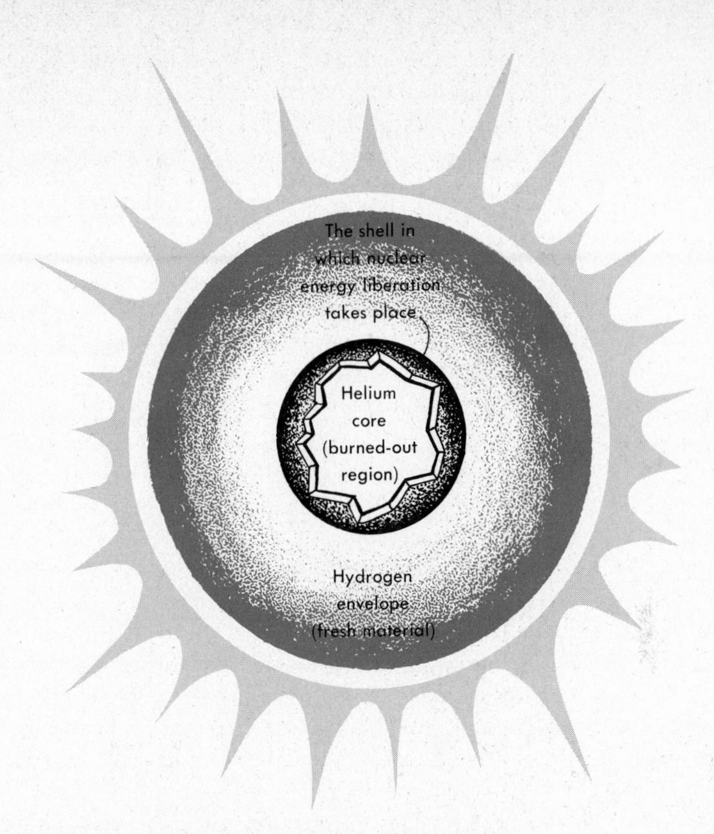

FIG. 33-3 The shell-source model for giant red stars.

As more and more hydrogen is consumed, the "shell" will move outward from the center, as does a ring of fire that has been started by a carelessly dropped match in a dry grass field.

It was suggested by Charles Critchfield and George Gamow, and later confirmed by the more detailed calculations of M. Schwarzschild and his associates, that the formation of such a shell source inside the sun (or any other star) must result in a steady growth of the star's size and in a gradual increase of its luminosity. In fact, within a few hundred million years after the shell source is formed, the diameter of the sun is expected to become as large as the orbit of Venus, and its luminosity will increase by a factor of between 10 and 20, making the oceans on the earth boil violently. After this last effort, the sun will begin to shrink and fade out again, until it becomes quite faint and insignificant. But there is no reason for immediate panic—we still have 5 billion years to go!

This will no doubt be the end of all life on the earth, but it is not the end for the sun and the other stars that follow a similar course in their evolution. When the shell-burning has exhausted the star's hydrogen by converting it all to helium, this method of energy production must come to an end. The energetic collisions that cause internal pressure die down, the pressure of the outward-streaming radiation diminishes, and the outer regions of the swollen giant star now have nothing to support them. At this time, we must go back to call on the nineteenth century ideas of Kelvin and Helmholtz.

The gravitational pull of its own mass causes the star to collapse inward; its gravitational PE is converted to compress and heat the central region to hundreds of million degrees. At this temperature, the fusion of alpha particles can occur. For more massive stars, the central temperature is higher, and fusion reactions can produce even heavier elements, up to the neighborhood of iron. Since such elements have the maximum binding energy (Fig. 28-2), fusion reactions can produce none heavier.

This last is a period of instability for evolving stars. They vary in brightness, often at irregular intervals; stars somewhat more massive than the sun may suddenly increase their brightness by 100 or 100,000 or more, explode some of their outer material into space, and then fade back. This may cause a star too dim to be seen to become visible—for this reason it is called a *nova*—"new star." Even more violent explosions in a more massive star may cause it to become a *supernova*, in which a substantial part of its mass is blown away, and its brightness temporarily increases by a factor of many millions. It is thought that elements of higher atomic number than iron are formed in these gigantic explosions.

As far as we know, after this period of instability, there are three possible fates for a star, depending on its mass. With the exhaustion of its resources for nuclear fusion, it may again undergo a gravitational collapse that reduces a star the size of the sun to the size of the earth. Its central density may rise to a million or more times the density of water. Several hundred of these *white dwarfs* have been identified. Because of their small size and surface area, they are not very luminous, and most of them are no doubt invisible even to the most powerful telescopes. They are probably very plentiful, but most of them we will never see.

We visualize the interior of a white dwarf to be composed of the protons and neutrons of nuclei, with electrons equal in number to the protons, all compressed close together. For some stars, this is not enough. So much gravitational energy is released in their collapse that the protons and electrons are forced together to become neutrons. We thus get a *neutron star* composed entirely (except for a thin outer layer) of neutrons, with a density of as much as 10^{14} g/cm^3—the density of the nuclear fluid of the droplet model. This reduces a star the size of the sun to a sphere of 5- to 10-km radius. Neutron stars were predicted by theorists without much hope that anything so small could ever be observed. Later, however, pulsating radio signals were observed by radio astronomers, coming from highly localized spots in the sky. These signals had very regular periods ranging from a small fraction of a second to a few seconds, and their source was then an unanswered question. Now everyone (or *almost* everyone) is agreed that they originate from rapidly rotating neutron stars. Even for such extreme specimens as neutron stars, conservation of angular momentum must hold. If a star like the sun. rotating sedately at 1 rev/month, were to be shrunk to a radius of 10 km, it would have to spin very rapidly indeed to conserve its original angular momentum; with this small size, its intense gravitational field would more than counteract any tendency of centrifugal force to spin off material. It is thought that irregularities in its magnetic field accelerate the ionized gases around the neutron star as it spins, causing them to emit electromagnetic waves whose intensity rises and falls with the same period as the rotation of the star. A substantial number of these *pulsars* are optically visible.

An ever more extreme form of matter is thought to be possible. The density of the so-called "nuclear fluid" is not necessarily the ultimate density. Recent experimental evidence has suggested localized scattering centers within nucleons—quarks, perhaps?—which leads to the picture that nuclear particles, like the larger atoms, may also be composed mainly of empty space. With this picture in mind, it is easy to imagine that the release of even more gravitational energy could compress a star to a density orders of magnitude greater than that of a neutron star. As a star is compressed, its surface gravitational field increases. Beyond a certain point, even photons cannot escape, even less any sort of real particle. A star that reaches this stage is completely cut off from any observation. It cannot emit light or any other radiation by which it could be observed. Its gravitational field, however, would persist. Such a star is called a *black hole*, and at least one has been tentatively identified. It is a member of a double-star pair, each revolving about their common center of gravity. The black hole is of course invisible, but its presence is deduced from the gravitational effects it produces on its visible companion.

Reference Tables

Standard Decimal Prefixes to Units

Prefix	Symbol	Meaning	Prefix	Symbol	Meaning
tera	T	$\times 10^{12}$	deci	d	$\times 10^{-1}$
giga	G	$\times 10^{9}$	centi	c	$\times 10^{-2}$
mega	M	$\times 10^{6}$	milli	m	$\times 10^{-3}$
kilo	k	$\times 10^{3}$	micro	μ	$\times 10^{-6}$
hecto	h	$\times 10^{2}$	nano	n	$\times 10^{-9}$
deka	da	$\times 10$	pico	p	$\times 10^{-12}$
			femto	f	$\times 10^{-15}$
			atto	a	$\times 10^{-18}$

Important Physical Constants

c (speed of light) $= 2.9979 \times 10^{8}$ m/s
e (electron charge) $= 1.602 \times 10^{-19}$ C $= 4.80 \times 10^{-10}$ esu
h (Planck's constant) $= 6.625 \times 10^{-34}$ J-s; $h/2\pi = 1.054 \times 10^{-34}$ J-s
G (gravitational constant) $= 6.670 \times 10^{-11}$ N-m^2/kg^2
N_0 (Avogadro's number) $= 6.022 \times 10^{23}$ atoms/g-atomic weight

Data on Particles

Particle	*Mass* (kg)	*Mass* (MeV)	*Charge* (C)
electron	9.11×10^{-31}	0.511	-1.602×10^{-19}
proton	1.673×10^{-27}	938.4	$+1.602 \times 10^{-19}$
neutron	1.675×10^{-27}	939.6	0
alpha	6.642×10^{-27}	3725	$+3.204 \times 10^{-19}$

Conversion Factors

Length

1 inch = 2.54 cm exactly
1 foot = 0.3048 m
1 mile = 1.609 km
1 nautical mile = 1.151 mi = 1852 m
1 light-year = 9.46×10^{12} km = 5.88×10^{12} mi
1 Angstrom unit (Å) = 10^{-10} m = 0.1 nm = 3.937×10^{-9} inch

Area

1 in^2 = 6.452 cm^2 = 6.452×10^{-4} m^2
1 ft^2 = 929 cm^2 = 0.0929 m^2
1 mi^2 = 2.590×10^6 m^2 = 259 hectares
1 acre = 0.405 hectare

Volume

1 in^3 = 16.39 cm^3
1 ft^3 = 0.02832 m^3
1 quart = 946 cm^3 = 0.946 liter

Weight

1 lb = 4.448 N
1 lb = weight of 453.6 g or 0.4536 kg

Electric and Magnetic

1 esu = 3.336×10^{-10} C
1 C = 2.998×10^9 esu
1 gauss = 10^{-4} weber/m^2

Nuclear Mass-Energy

1 amu	= 931 MeV	= 1.492×10^{-10} J	= 1.660×10^{-27} kg
1.074×10^{-3} amu	= **1 MeV**	= 1.602×10^{-13} J	= 1.783×10^{-30} kg
6.70×10^9 amu	= 6.24×10^{12} MeV	= **1 J**	= 1.113×10^{-17} kg
6.02×10^{26} amu	= 5.61×10^{29} MeV	= 8.99×10^{16} J	= **1 kg**

Miscellaneous Relationships

60 mi/hr = 88 ft/s = 96.5 km/hr
1 radian = 57.296° = 57° 17′ 45″
1 horsepower = 746 watts
1 lb/in^2 = 6.895×10^4 dyne/cm^2 = 6.895×10^3 N/m^2
1 standard atmosphere = 1.013×10^6 dynes/cm^2 = 14.7 lb/in^2
1 standard gravity = 9.8062 m/s^2 = 32.173 ft/s^2
1 kilowatt-hour = 3.60×10^6 J
1 calorie = 4.18 J

SINES, COSINES, AND TANGENTS

Angle	Sine	Cosine	Tangent	Angle	Sine	Cosine	Tangent
0°	0.000	1.000	0.000				
1°	.017	1.000	.017	46°	0.719	0.695	1.036
2°	.035	0.999	.035	47°	.731	.682	1.072
3°	.052	.999	.052	48°	.743	.669	1.111
4°	.070	.998	.070	49°	.755	.656	1.150
5°	.087	.996	.087	50°	.766	.643	1.192
6°	.105	.995	.105	51°	.777	.629	1.235
7°	.122	.993	.123	52°	.788	.616	1.280
8°	.139	.990	.141	53°	.799	.602	1.327
9°	.156	.988	.158	54°	.809	.588	1.376
10°	.174	.985	.176	55°	.819	.574	1.428
11°	.191	.982	.194	56°	.829	.559	1.483
12°	.208	.978	.213	57°	.839	.545	1.540
13°	.225	.974	.231	58°	.848	.530	1.600
14°	.242	.970	.249	59°	.857	.515	1.664
15°	.259	.966	.268	60°	.866	.500	1.732
16°	.276	.961	.287	61°	.875	.485	1.804
17°	.292	.956	.306	62°	.883	.469	1.881
18°	.309	.951	.325	63°	.891	.454	1.963
19°	.326	.946	.344	64°	.899	.438	2.050
20°	.342	.940	.364	65°	.906	.423	2.145
21°	.358	.934	.384	66°	.914	.407	2.246
22°	.375	.927	.404	67°	.921	.391	2.356
23°	.391	.921	.424	68°	.927	.375	2.475
24°	.407	.914	.445	69°	.934	.358	2.605
25°	.423	.906	.466	70°	.940	.342	2.747
26°	.438	.899	.488	71°	.946	.326	2.904
27°	.454	.891	.510	72°	.951	.309	3.078
28°	.469	.883	.532	73°	.956	.292	3.271
29°	.485	.875	.554	74°	.961	.276	3.487
30°	.500	.866	.577	75°	.966	.259	3.732
31°	.515	.857	.601	76°	.970	.242	4.011
32°	.530	.848	.625	77°	.974	.225	4.331
33°	.545	.839	.649	78°	.978	.208	4.705
34°	.559	.829	.675	79°	.982	.191	5.145
35°	.574	.819	.700	80°	.985	.174	5.671
36°	.588	.809	.727	81°	.988	.156	6.314
37°	.602	.799	.754	82°	.990	.139	7.115
38°	.616	.788	.781	83°	.993	.122	8.144
39°	.629	.777	.810	84°	.995	.105	9.514
40°	.643	.766	.839	85°	.996	.087	11.43
41°	.656	.755	.869	86°	.998	.070	14.30
42°	.669	.743	.900	87°	.999	.052	19.08
43°	.682	.731	.933	88°	.999	.035	28.64
44°	.695	.719	.966	89°	1.000	.017	57.29
45°	.707	.707	1.000	90°	1.000	.000	

Answers to Odd-Numbered Questions

Chapter 1

1. (a) 5.63×10^2; (b) 1.2×10^{-3}; (c) 4.39×10^7; (d) 7.19×10^{-7}.
3. (a) 10^{-4}; (b) 10^4; (c) 2×10^{-6}; (d) $0.0043 \times 10^4 = 43$.
5. (a) about 10^4; (b) about 10^7; (c) the diameter of the earth.
7. (a) 4×10^2; (b) 3.95×10^9
9. 5.99×10^{26}.
11. (a) 2.16×10^8; (b) 1.728×10^{-15}; (c) 2.25×10^{10}.
13. (a) 10^4; (b) 3×10^4; (c) 3.16×10^4; (d) 3.16×10^4; (e) 8×10^4; (f) 6.80×10^6.
15. (a) 10^3; (b) 3×10^3; (c) 4.64×10^3; (d) 4.64×10^3; (e) 4×10^3; (f) 3.16×10^3.
17. (a) 10^6; (b) 10^6; (c) 10^{-9}.
19. 0.318 cm.
21. 1.276×10^9 cm $= 1.276 \times 10^7$ m $= 1.276 \times 10^4$ km.
23. 6.97×10^5.
25. 13.97 ft.
27. 0.984 ft.

Chapter 2

1. 47 lb.
3. (a) yes; (b) 183 lb.
5. Yes, $\Sigma\tau = 0$
7. 111.1 lb and 88.9 lb.
9. 4.41 ft in front of rear axle.
11. 1.67 lb.
13. (a) $W = 200$ lb; (b) $T = 700$ lb, $G = 500$ lb.
17. (a) 15.0 units; (b) 26.0 units.
19. (b) 1.643 S, 18.52 E.
21. 18.60, 5.1° S of E.
23. 11.5° S of E, 7490 lb.
25. (a) 500 lb; (b) 400 lb.
27. (a) 248 lb; (b) $P_h = 215$ lb, $P_v = 76$ lb.
29. (a) 423 lb; (b) $S_h = 407$; $S_v = 79$; $S = 415$ lb.
31. 28.9 lb downward and to the right, 20.7° from vertical; applied 7.02 ft from left end.
33. 2450 lb.
37. 7.43 ft.
39. 97.1 lb.
41. 62.4 lb/ft^3.
43. 7.80 g/cm^3.
45. (a) 0.81 g/cm^3.
47. (a) 64.3 lb/ft^3; (b) 643 lb/ft^2 = 4.47 lb/in^2; (c) 2.89×10^5 lb.
49. 379 g/cm^2.
51. 333 lb.
53. (a) 250 g; (b) 250 g; (c) 250 g; (d) 250 cm^3; (e) 250 cm^3; (f) 3.00 g/cm^3
55. (a) 200 cm^3; (b) 7.60 g/cm^3
57. (a) 100 cm^3; (b) 3.00 g/cm^3; (c) 0.80 g/cm^3.
59. 0.48 g/cm^3.
61. level unchanged.
63. 2981.0 g.

Chapter 3

1. (a) 0.625 mi/hr/s or 0.917 ft/s^2; (b) 1027 ft.
3. -8×10^{12} cm/s^2.
5. (a) 4×10^{-6} s; (b) 24 cm.
7. (a) 4.38×10^{11} cm/s^2; (b) 5.50×10^6 cm/s.
9. (a) 6.4×10^{15} cm/s^2; (b) 1.60×10^8 cm/s.
13. 75 N; 7500 N.
15. 0.200 slug.
17. (a) 0.375 m/s^2; (b) 0.15 N.
19. 0.934 N.
21. (a) 2.94×10^5 N; (b) 4.41×10^4 N; (c) 0.0832 m/s^2; (d) 136 m.
23. 0.50 s.
25. 144 ft or 44.1 m.
27. (a) 6 m/s^2; (b) 60 N.
29. (a) 1.68 ft/s^2; (b) 2.18 s.
31. 18.95 lb.
33. 1.90 m/s^2.
35. 2.21 m.

37. (a) 4.75 N; (b) 5.22 N.
39. 390 N.
41. (b) 6.0 ft/s^2.
43. 13.5 ft/s^2 or 4.14 m/s^2 or 414 cm/s^2.
45. 9.17 ft/s^2 or 2.81 m/s^2.
47. 30.8 lb.
51. 12.8 ft/s.
53. 7.49 s.
55. 5 N.
57. (a) 20.4 m; (b) 141 m.
59. (a) 22.5 ft; (b) 53.7 ft/s.
61. (a) 37.2 s; (b) 9660 m.

Chapter 4

1. 1.32×10^6 ft-lb.
3. (a) 2.45×10^6 J; (b) 2.45×10^6 J.
5. (a) 4500 ft-lb; (b) $-13{,}500$ ft-lb; (c) 5500 ft-lb; (d) $-12{,}500$ ft-lb.
7. 3.92×10^5 ergs.
9. 6.4×10^{-15} erg, or 6.4×10^{-22} J.
11. 121 ft.
13. 1.89×10^5 m/s.
15. 3.2×10^7 ergs = 3.2 J.
17. 96 ft-lb.
19. 284 lb.
21. 4.00 ft/s.
23. 8600 dynes/cm^2.
25. 283 cm^3/s.
27. 1021 W.
29. 162 horsepower.
31. 1524 W; 2178 W.
33. 703 lb.
35. 5.40×10^6 J.
37. 4.44×10^3 m/s.
39. 1.00 s.
41. 10,300 ft/s or 7020 mi/hr.
43. 13 cm/s.
45. $v_p = 1.6 \times 10^5$ cm/s east, $v_\alpha = 6 \times 10^4$ cm/s east.
47. (a) 0.559 ft/s; (b) 0.375 ft/s.
49. 0.098 kg/s.
51. 7.55×10^6 dynes or 75.5 N.
53. 3.57 lb.

Chapter 5

1. 1.83 radians.
3. (a) 7.17×10^{-4} rad/hr; (b) 6.67×10^4 mi/hr.
5. (a) 3142 rad/s; (b) 1.57×10^4 cm/s = 157 m/s.
7. 13.09 rad/s.^{2}
9. (a) 75 rev/s^2; (b) 150 rev.
11. 2.5 slug-ft^2.
13. 10.7 kg-m^2.

15. 0.56 lb-ft.
17. 2×10^6 g-cm^2.
19. 106.7 lb.
21. 918 lb.
23. (a) 2.00×10^{20} N; (b) 2.04×10^{16} metric tons.
25. (a) 171.5 cm/s; (b) 14,700 m ergs; (c) 58,800 m ergs; (d) 73,500 m ergs; (e) 0; (f) 73,500 m ergs; (g) 383 cm/s.
27. 4.22 rad/s or 40.3 rev/min.
29. 1.48 J.
31. $\sqrt{(10/7)\,gh}$.
33. 2/3 translational, 1/3 rotational.
35. 4/3 slug-ft^2.
37. KE (arms in) = 9/4 times KE (arms out).
39. 1.33×10^4 min = 222 hr.
41. to the right.
43. 1789 rev/min.

Chapter 6

1. 21.3 N.
3. 245 cm/s^2 or 2.45 m/s^2 or 8 ft/s^2.
5. N-m^2/kg^2 or m^3/(kg-s^2).
7. 108 lb.
9. 86 pj.
11. (a) 7000 km = 7.00×10^6 m; (b) 7.55×10^3 m/s = 7.55 km/s.
13. (a) 4.17×10^{10} J; (b) one-third as much.
15. (a) 1.25×10^8 J; (b) -1.25×10^8 J.
17. (a) 2.85×10^{12} J; (b) 2.85×10^{12} J; (c) -5.70×10^{12} J; (d) -2.85×10^{12} J.
19. (a) 20.9 mi/s; (b) 3.48 mi/s.
21. (a) $v_e^2 = 2v_0^2$; (b) $\sqrt{2} = 1.414$; (c) 7.92 km/s.
23. 20.3 rev/min.
25. 850 m/s.
27. 9.65 days.
29. (a) 1.00 day; (b) 4.23×10^4 km = 2.63×10^4 mi; (c) 22,300 mi = 3.59×10^4 km.

Chapter 7

1. (a) 25.5 kg/cm^2; (b) 2.50×10^6 N/m^2.
3. 9.80×10^8 dynes/cm^2.
5. 2×10^{-5}.
7. 2.08×10^{12} dynes/cm^2.
9. 4.40×10^6 dynes = weight of 4.49 kg.
11. (a) 10/7; (b) 20/21 (from metric moduli).
13. 1.60×10^{-15} s.
15. (a) 0.2 cyc/s; (b) 5 s; (c) 0.2 rev/s; (d) 1.26 rad/s.
19. (a) 2.79 ft/s; (b) 11.7 ft/s^2.
21. 108 cyc/min.
23. 59.2 cyc/min.
25. 4.98 cyc/s = 299 cyc/min.
27. 55.9 cm.

29. 0.817 hr = 49 min.
31. 1.22×10^{-2} rad = 0.70°.
33. No. Natural frequency is 240.1 vibr/min.

Chapter 8

1. slug/ft.
3. 1.12×10^4 cm/s.
7. 0.52 + 1.81 = 2.33 s.
9. At t = 1.04 s and 4.66 s.
11. (a) upper side; (b) upper side.
13. 4.74×10^{14} cycles/s.
15. 0.6 ft or 7.2 in.
17. 1000 ft/s.
19. λ = 24 ft, v = 408 ft/s, f = 17.0 cycles/s.
21. (a) 12 cm; (b) 2.88×10^4 cm/s.
23. (a) 2.50 cm; (b) in phase, to make antinode; (c) $\lambda/2$ = 1.25 cm; (d) 7 antinodes; (e) 6 nodes.
25. No; no node is equidistant from the two generating points.

Chapter 9

1. 1115 ft/s.
3. 825 ft.
5. 2200 ft.
7. 640 rev/min.
9. 600 vibr/s.
11. 18.8 in.
13. (a) 412 vibr/s; (b) 1236 vibr/s; (c) 2060 vibr/s.
15. (a) 880 vibr/s; (b) 1760 vibr/s.
17. 0.625 ft or 18.75 cm. (v = 1100 ft/s or 330 m/s).
19. 1600.
21. 1750 m.
23. λ = 3.30 mm; length of train = 33 cm.
25. sin α = 0.30; 2α = 35°.
27. 8660 ft.
29. 110 ft/s.
31. 1326 vibr/s.
33. (a) 10^{-12} watts/cm^2; (b) 10^4 s = 2.78 hr.
35. 0.90 or 90 percent.
37. 46 dB.

Chapter 10

1. 235.4°F.
3. 60°C and 140°F.
5. 110°F.
7. (a) 623°K; (b) 321.9°K; (c) 23°K; (d) 183°K.
9. 1013 millibars.
11. 4470 in^3.
13. (a) 21.6 lb/in^2; (b) 8.6 lb/in^2.
15. 72.1 lb/in^2.

17. 50.8°C.
19. 250.074 ft.
21. (a) Cooled; (b) −56.4°C.
23. 200.24 cm^3.
25. 18.9 cm^3.
27. 1.13×10^8 calories.
29. 403 g.
31. 86.7°C.
33. 83.0°C.
35. (a) 227 kg; (b) 21 kg.
37. 56.3°C.
39. 201 s.
41. 0.22 s. (This is obviously a case where wrong assumptions lead to a wrong answer).

Chapter 11

3. 0.50°C.
5. 7.18×10^4 cal.
7. (a) 4.92×10^3 cal; (b) 1.12°C.
9. 65.2°C.
11. 501 m/s. (Remember heat of fusion of lead, given in Sec. 10-8.)
13. 4.33×10^4 s, or 12 hr 2 min.
15. 0.35.
17. Eff. goes from 0.559 to 0.597; the 0.038 increase is 0.068 of the original 0.559.
19. (a) 0.699; (b) 0.489; (c) 10.22×10^7 J/s; (d) 2.45°C.
21. +0.58 cal/degree.

Chapter 12

1. 7.4 diameters.
3. 0.029 cm/s.
5. (a) 1.21; (b) 1.1
7. 377 m/s.
9. $v_K = 0.67\ v_O$.
11. (a) $v_B = 1.23\ v_A$; (b) 2N/3.
13. 3.08×10^4 cm/s.
15. (a) more; (b) more; (c) increased.

Chapter 13

1. (a) 577 s; (b) 667 s.
3. (a) 0.982; (b) 1.004; (c) 1.0018; (d) 0.994; (e) 0.99968.
5. (a) $3 \times 10^{-7} + 6 \times 10^{-15}$ s; (b) $3 \times 10^{-7} + 12 \times 10^{-15}$ s; (c) 6×10^{-15} s.
7. 9.17 km.
9. 8.9×10^{-8} km = 0.089 mm.
11. 2.24×10^8 m/s.
13. 3.47×10^{-11}.
15. 3.83×10^{-27} kg.
17. $0.974\ c = 2.92 \times 10^8$ m/s.
19. $0.652c = 1.96 \times 10^8$ m/s.

21. 0.464×10^{-9} kg = 0.464 μg.
23. 3.6×10^{26} W.
25. (a) 8.87×10^{40} J; (b) 9.85×10^{23} kg; (c) 9.85×10^{23} kg.
27. 218 ft.
29. 2081.

Chapter 15

1. 0.75 dyne repulsion.
3. 4.62×10^{-9} C.
5. 2.77×10^{-9} m.
7. (b) Both $+10^{-9}$ or -10^{-9} C (or 3 esu); (c) $+4 \times 10^{-9}$ and -2×10^{-9} C (or +12 and −6 esu); or -4×10^{-9} and $+2 \times 10^{-9}$ C (or −12 and +6 esu).
15. 7.74×10^{3} N/C, east.
17. (a) 1.60×10^{8} N/C, toward the electron; (b) 5.12×10^{-11} N, toward the electron.
19. 1.5 cm, or 1.5×10^{-2} m from the smaller charge, measured toward the larger charge.
21. 2.64×10^{-2} N.
23. − 387 J/C, or − 387 V.
25. 1.38×10^{-19} J.
27. (a) No; (b) 6×10^{-3} m from 7×10^{-10} C charge, toward other charge; and 7.5×10^{-3} m from 7×10^{-10} C charge, away from other charge.
29. (a) −3920 V; (b) +1110 V; (c) 5030 V; (d) 1.006×10^{-4} J.
31. C/6.
33. (a) 10^{-9} C; (b) 10^{4} V/m or 10^{4} N/C.
35. (a) 10^{-9} C; (b) 50 V; (c) 10^{4} V/m or 10^{4} N/C.
37. Weaker.
39. 2×10^{-9} C.
41. 6.64×10^{-10} F = 664 pF.

Chapter 16

1. 2500 Ω.
3. 0.064 A.
5. 20 Ω.
7. 0.4 A.
9. (a) 2.1 Ω; (b) 0.71 A; (c) 0.071 V; (d) 1.43 V; (e) 0.36 V.
11. 1.45 Ω.
13. 2/3 Ω.
15. $I_4 = 2.00$ A; $I_6 = 1.33$ A; $I_{12} = 0.67$ A; $I_B = 4.00$ A.
17. (a) $I_{0.4} = I_{2.5} = I_B = 2.00$ A; $I_3 = 1.40$ A; $I_7 = 0.60$ A; (b) 9.20 V; (c) 5.00 V; (d) 4.20 V.
19. (a) 25 W; (b) 40 s.
21. 707 V.
23. (a) 12 Ω; (b) 1200 W.
25. (a) 6.40 Ω, in series with magnet; (b) 40 W = 0.57 of total.
27. The 500-W heater. It draws 222 W, while the larger heater draws only 111 W.
29. 7260 s.

Chapter 17

1. 212 A/m.
3. 750 A/m.
5. 0.24 A.
7. 2.67×10^{-4} weber/m^2.
9. Toward the east.
11. 4.80×10^{-18} N, away from the wire.
13. (a) 1.60×10^{-16} N; (b) 1.76×10^{14} m/s^2.
15. (a) Both move under effect of a force perpendicular to velocity; (b) Grav. force set $= mv^2/r$; (c) 5.68 mm.
17. (a) 1.92×10^{-17} J; (b) 6.49×10^6 m/s; (c) 6.16×10^{-4} weber/m^2.
19. 5.30×10^{-3} weber/m^2.
21. 5.85×10^{-3} weber/m^2.
23. $H = 5 \times 10^3$ A/m; $B = 6.29 \times 10^{-3}$ weber/m^2.
25. 1.18 weber/m^2.
27. 0.6 N.
29. 1.6 weber/m^2.
31. 1.60×10^{-17} A.
33. (a) 124,700 Ω; (b) in series with galv. coil.
35. 0.120 Ω, in parallel with galv. coil.
37. (a) Each force $= 3.60 \times 10^{-4}$ N; (b) directions opposite.
39. 2940 A.
41. 2.50×10^{-3} V.
43. (a) 1.60×10^{-3} V; (b) no.
45. (a) 7.07×10^{-3} V; (b) 1.77×10^{-4} A.
47. 7.16 weber/m^2 = 71,600 gauss.
49. 12 V; 0.75 A.
51. \$133 − \$21 = \$112.

Chapter 18

1. Ag^+ and NO_3^-; $\pm 1e$ charge $= 1.60 \times 10^{-19}$ C.
3. (a) Pb, 82, 207.19; Zn, 30, 65.37; (b) 3.17; (c) 3.17 times as large; (d) same.
5. 3.49×10^{22} (Pb), 4.60×10^{22} (Zn).
7. (a) 1.00797 g; (b) 107.870 g; (c) 8.998 g.
9. (a) 29,520 C; 33.00 g.
11. 0.648 lb.
13. 2.00×10^{-15} N.
15. 4.80×10^{-15} N.
17. 3240 V.
19. (a) 1.408×10^{-14} N; (b) 1.408×10^{-14} N.
21. (a) 3.64×10^{-12} J or 3.64×10^{-5} erg; (b) Same as (a).
23. About 3000 inches, or 250 ft.

Chapter 19

1. 1.9×10^{19} km.
3. 1500 rev/min.
5. 3 feet.
9. (a) 6 cm, R, I; (b) 6.86 cm, R, I; (c) 9.60 cm, R, I; (d) 12 cm, R, I; (e) ∞, no image; (f) −12 cm, V, E; (g) −3 cm, V, E.
11. (a) 6 in; (b) virtual; (c) erect; (d) 7.2 cm.
13. 10.3 cm.

15. $31°40'$ or $31.66°$.
17. $42°16'$ or $42.26°$.
19. (a) $28°36'$ or $28.61°$; (b) $50°$.
21. $70°11'$ or $70.19°$.
23. $34°21'$ or $34.36°$.
25. $28°42'$ or $28.72°$.
27. (a) $39°52'$ or $39.87°$; (b) $60°40'$ or $60.67°$.
29. $21°28'$ or $21.47°$.
31. (a) 6 cm, R, I; (b) 6.86 cm, R, I; (c) 9.60 cm, R, I; (d) 12 cm, R, I; (e) ∞, no image; (f) -12 cm, V, E; (g) -3 cm, V, E.
33. 34.7 mm.
35. 8.47 in.
37. 18 in. beyond diverging lens.
39. 15 in.
41. 40 cm from object, toward the screen; or 20 cm from first lens, towards the screen.

Chapter 20

1. (a) 1.714×10^8 m/s; (b) 337 nm.
3. 4.42×10^{-2} cm or 0.442 mm.
5. 9.33×10^{-5} cm or 933 nm or 9330 Å. It is not visible light.
7. 5.06 cm.
9. (a) $5.06/2 = 2.53$ cm; (b) $5.06 \times 2/3 = 3.37$ cm.
11. 2.552×10^{-5} in $= 6482$ Å.
13. (a) 1.26×10^{-5} cm; (b) Any odd number $\times$ 1.26×10^{-5} cm.
15. No reflection; black.
21. About 4×10^{-7} m, or 4000 Å.

Chapter 21

1. Noseward.
3. -60 cm.
5. 16.7 cm.
7. (a) 5.0; (b) -0.50; (c) 7.9; (d) -40.
9. (a) 18.9 mm; (b) 6.7 mm.
11. Very nearly 8 times as much.
13. Very nearly 1/100 s.
15. 1/100 s at $f/5.6$.
17. (a) 8.78×10^{-4} cm; (b) 8.78×10^{-4} cm; (c) $r = 1.22\lambda \times f$-number.
19. 7.32×10^{-6} rad $= 1.51$ seconds.
21. 1.26 in.
23. (a) 5 cm; (b) 2 in.
25. 62.5.
27. 20 mm.
29. (a) 4.47×10^{-6} rad $= 0.92$ seconds; (b) 55 seconds $= 0.92$ minute; (c) Not quite; (d) 15.4 mm, to give magnification of 78X.
31. 45X.
33. (a) Yes, 3.24×10^{-5} cm $< 1\mu$; (b) 225X; (c) 4×10^{-6} rad $= 1.38 \times 10^{-2}$ minute; (d) Yes.
35. (a) $11°27'$ or $11.44°$; (b) $23°23'$ or $23.38°$; (c) $36°31'$ or $36.53°$.
37. (a) Third order; (b) yes.

Chapter 22

1. (a) 1.009×10^{-4} cm; (b) 2.91×10^{-5} cm.
3. 2420°K.
5. (a) $5.61 \times 10^8 W/m^2$; (b) $3.86 \times 10^6 W/m^2$.
7. 1013 W.
9. (a) visible; (b) UV; (c) UV.
11. (a) 9.88×10^{-21} erg $= 9.88 \times 10^{-28}$ J; (b) 9.95×10^{-13} erg $= 9.95 \times 10^{-20}$ J; (c) 3.32×10^{-6} erg $= 3.32 \times 10^{-13}$ J.
13. 6.63×10^{-2} cm.
15. 1.001×10^{-11} erg $= 1.001 \times 10^{-18}$ J.
17. (a) 7.49×10^{-12} erg $= 7.49 \times 10^{-19}$ J; (b) 1.099×10^{-11} erg $= 1.099 \times 10^{-18}$ J; (c) 3.50×10^{-12} erg $= 3.50 \times 10^{-19}$ J.
19. 6.69×10^{14} Hz.
21. 2.06 Å.

Chapter 23

1. 7.55×10^{14} Hz.
5. 8.5×10^{-10} m, or 8.5 Å.
7. 1.80×10^{-11} m, or 0.18 Å.
9. -8.72×10^{-20} J.
11. (a) $+8.72 \times 10^{-20}$ J; (b) -17.44×10^{-20} J.
13. PE $= -1.938 \times 10^{-18}$, KE $= +9.69 \times 10^{-19}$, E $= -9.69 \times 10^{-19}$ J.
15. (a) 1.876×10^{-6} m; (b) No—it is IR; (c) Paschen.
17. Each line has $f = 9 \times$ frequency of corresponding hydrogen line.

Chapter 24

1. (a) $l = 0, 1, 2, 3$; (b) $l = 2$.
5. (a) 1.7×10^{-23} cm^3; (b) 3.0×10^{-23} cm^3.
7. $n = 1, l = 0, m = 0, s = -1/2$; $n = 1, l = 0, m = 0, s = 1/2$; $n = 2, l = 0, m = -1, s = -1/2$.
9. $l = 0, 1, 2$.
11. $1s^2, 2s^2, 2p^6, 3s^2, 3p^3$.
13. (a) Z = 5 through 10; (b) Z = 11 and 12.
15. (a) $n = 5, l = 2$; (b) $n = 4, l = 1$; (c) $n = 4, l = 3$.
17. 1.55×10^5 V.

Chapter 25

1. 6.06×10^{-10} m $= 6.06$ Å.
3. 7.28×10^6 m/s.
5. 9.99×10^{-10} m $= 9.99$ Å.
7. (a) 2.41×10^{-17} J; (b) 151 V.
9. (a) $v_p = 2v_\alpha$; (b) $p_\alpha = 2p_p$; (c) $\lambda_p = 2\lambda_\alpha$.
11. 3.31×10^{-12} m $= 0.0331$ Å.
13. 2.00×10^{-11} m $= 0.200$ Å.
15. $\pm 3.52 \times 10^{-30}$ m.
17. $\pm 1.26 \times 10^{-10}$ m $= 1.26$ Å.
19. (a) 0.5×10^{-6} m; (b) 0.126 m/s; (c) No.

Chapter 26

1. Ca: 20p, 20n. Al: 13p, 14n. Ra: 88p, 138n. Se: 34p, 46n.
3. 300 eV or 4.8×10^{-17} J.
5. 65.4

7. ${}_{7}N^{14}$, ${}_{7}N^{15}$, ${}_{17}Cl^{35}$, ${}_{17}Cl^{37}$. (These are now often written in a different format: ${}^{14}_{7}N$, ${}^{15}_{7}N$, ${}^{35}_{17}Cl$, ${}^{37}_{17}Cl$).
9. ${}_{92}U^{232}$, ${}_{89}Ac^{222}$, ${}_{86}Rn^{217}$.
11. ${}_{83}Bi^{211}$, ${}_{84}Po^{210}$, ${}_{95}Am^{243}$.
13. ${}_{84}Po^{216} \rightarrow {}_{2}He^{4} + {}_{82}Pb^{212}$.
15. ${}_{17}Cl^{36} \rightarrow {}_{-1}\beta^{0} + {}_{18}Ar^{36}$.
17. ${}_{17}Cl^{36} + {}_{-1}\beta^{0} \rightarrow {}_{16}S^{36}$.
19. ${}_{83}Bi^{209}$.
21. (a) 6.89×10^{-8} erg; (b) 2.89×10^{-9} cm $= 0.289$ Å.
23. 1.50×10^{5} eV $= 2.40 \times 10^{-7}$ erg $= 2.40 \times 10^{-14}$ J.
25. $KE_{\alpha} = 2KE_{Ne+}$.
27. 12.5 mg.
29. A little more than 140 years. (141.6 years).
31. (a) 2μg; (b) no; (c) a *very* small amount, but theoretically, yes.
33. About 17,100 years ago.
35. (a) 1.79 earth years; (b) 2103 A.D.

Chapter 27

1. ${}_{5}B^{10} + {}_{2}He^{4} \rightarrow {}_{7}N^{13} + {}_{0}n^{1}$.
3. (a) ${}_{17}Cl^{35} + {}_{1}H^{1} \rightarrow {}_{2}He^{4} + {}_{16}S^{32}$; (b) ${}_{3}Li^{6} + {}_{0}n^{1} \rightarrow {}_{2}He^{4} + {}_{1}H^{3}$.
5. (a) ${}_{14}Si^{28}$; (b) ${}_{0}n^{1}$; (c) ${}_{0}n^{1}$; (d) ${}_{14}Si^{30}$; (e) ${}_{2}He^{4}$.
7. 29.6 cm.
9. (a) 8.76×10^{5} m/s; (b) 6.42×10^{-16} J $= 6.42 \times 10^{-9}$ erg; (c) 4.01×10^{3} eV.
11. 1.50×10^{5} eV $= 0.150$ MeV $= 2.40 \times 10^{-7}$ erg $= 2.40 \times 10^{-14}$ J.
13. 40 revolutions.
15. 5.46×10^{-8} s.
17. 9.16×10^{6} cyc/s $= 9.16$ megahertz or 9.16 MHz.
19. 2.77×10^{7} m/s.
21. 0.96 m or 96 cm.
23. Mass and period both increased by 51 percent.
25. (a) 0.511 MeV; (b) 4.70×10^{4}.

Chapter 28

1. (a) 1.05; (b) 1.30; (c) 1.43.
3. (a) 0.04216 amu $=$ 39.25 MeV; (b) 39.25 MeV.
5. 5.61 MeV/nucleon.
7. 0.00938 amu/nucleon $=$ 8.73 MeV/nucleon.
9. 7.87 MeV/nucleon.
11. (a) 0.00130 amu $=$ 1.21 MeV; (b) bombarding particle has more energy than products.
13. (a) 0.00300 amu $=$ 2.79 MeV; (b) products have more energy than bombarding particle.
15. (a) 0.00741 amu $=$ 6.90 MeV; (b) bombarding particle has more energy than products.

Chapter 29

1. About 190 MeV.
3. (a) ${}_{41}Nb^{105}$ and ${}_{51}Sb^{128}$; (b) 3e + 3n; (c) 1e + 4n.
7. Mouse; it has a larger S/V.
9. (a) 0.89; (b) reduced to less than 1.3 percent after 2 collisions.

11. e/m of $U^{238}F_6^+ = 0.9915 \times e/m$ of $U^{235}F_6^+$.
13. (a) 10^{21} ergs; (b) 6.2×10^{26} Mev; (c) 3.1×10^{24} atoms; (d) 1.2 kg; (e) 1.1 g.
15. (a) 2.15×10^{13} cal; (b) 1.46×10^{13} cal. In the English-speaking countries such ratings are nearly always in British Thermal Units: 1 BTU = 252 calories. In these units, (a) 8.53×10^{10} BTU; (b) 5.79×10^{10} BTU.

Chapter 30

1. 4.9×10^{-3} J $= 3.1 \times 10^{16}$ eV.
3. 1.32×10^{-15} m $= 1.32 \times 10^{-5}$ Å.
5. 7.8 m.

Index

C

D

E

F

G

H

I

J

K

L

M

T

U

V

W

X

Y

Z